Y0-BCR-552

HANDBOOK OF LARGE TURBO-GENERATOR OPERATION AND MAINTENANCE

Books in the IEEE Press Series on Power Engineering

Principles of Electric Machines with Power Electronic Applications, Second Edition
M.E. El-Hawary

Pulse Width Modulation for Power Converters: Principles and Practice
D. Grahame Holmes and Thomas Lipo

Analysis of Electric Machinery and Drive Systems, Second Edition
Paul C. Krause, Oleg Wasynczuk, and Scott D. Sudhoff

Risk Assessment for Power Systems: Models, Methods, and Applications
Wenyuan Li

Optimization Principles: Practical Applications to the Operations of Markets of the Electric Power Industry
Narayan S. Rau

Electric Economics: Regulation and Deregulation
Geoffrey Rothwell and Tomas Gomez

Electric Power Systems: Analysis and Control
Fabio Saccomanno

Electrical Insulation for Rotating Machines: Design, Evaluation, Aging, Testing, and Repair
Greg Stone, Edward A. Boulter, Ian Culbert, and Hussein Dhirani

Signal Processing of Power Quality Disturbances
Math H. J. Bollen and Irene Y. H. Gu

Instantaneous Power Theory and Applications to Power Conditioning
Hirofumi Akagi, Edson H. Watanabe and Mauricio Aredes

Maintaining Mission Critical Systems in a 24/7 Environment
Peter M. Curtis

Elements of Tidal-Electric Engineering
Robert H. Clark

Handbook of Large Turbo-Generator Operation and Maintenance, Second Edition
Geoff Klempner and Isidor Kerszenbaum

HANDBOOK OF LARGE TURBO-GENERATOR OPERATION AND MAINTENANCE

Geoff Klempner
Isidor Kerszenbaum

Mohamed E. El-Hawary, *Series Editor*

IEEE Press

A JOHN WILEY & SONS, INC., PUBLICATION

IEEE Press
445 Hoes Lane
Piscataway, NJ 08855

Published by John Wiley & Sons, Inc., Hoboken, New Jersey
Published simultaneously in Canada.

Library of Congress Cataloging-in-Publication Data is available.

ISBN 978-0470-16767-0

Printed in the United States of America.

10 9 8 7 6 5 4 3 2 1

To our families:
Susan Klempner,
Jackie, Livi, and Yigal Kerszenbaum

CONTENTS

PREFACE

It is not uncommon for a large utility to have units of disparate size, origin, and vintage in its fleet of generators. Among its dozens of generators, there might be some from the 1950s or 1960s and some with their original asphalt or thermoplastic windings. These, and later units, may be running with and without magnetic retaining rings. Some might have thermoelastic windings of all sorts, with or without asbestos; they might be hydrogen-cooled or air-cooled, have split-stator windings, be self-excited or different types of externally excited, steam-driven or combustion-driven, and the list goes on and on. Now, take that diversity and include units operating in 50 and 60 Hz grids, built by Western, Asian, and Eastern European manufacturers to different standards. This is what you may find in some of the new independent, deregulated power producers that, in addition to building new plants, have purchased entire fleets of older units in several countries around the globe.

The reasons why one may find so many "old" units still in operation are not difficult to discern. First of all, a typical generator is made with an intent to last no less than 30 years or so. Second, replacing an operating unit is very capital intensive and, thus, done only when a catastrophic failure has occurred or some other major failure of the machine that renders continuous operation not economically viable. Third, although expected to last 30 years, large turbogenerators are known to have their lives extended far beyond that, if well maintained and operated. Sometimes that also requires replacing a major component, such as the armature winding and/or a rotor winding (or the entire rotor!). Significant changes in design tend to occur every few years, for different components. For instance, a history of the insulation systems encountered in generators shows that every few years there is some big change resulting in increased ratings. These changes typically derive from the adoption of a new materials such as the change from magnetic to nonmagnetic material for retaining rings. Not all changes are always positive. Some new designs end up being reversed or revised after experience unmasks significant defects in them.

There are countless scraps of information about the operation, maintenance, and troubleshooting of large turbogenerators in many publications. All vendors at one stage or another have produced and published interesting literature about the operation of their generators. In particular, the technical information letters put out by some manufacturers (called different names by different vendors) offer a wealth of detailed

O&M topics. Institutions such as EPRI in the United States, CIGRE, IEC, ANSI, IEEE, and other national standards cover various aspects of the operation and maintenance of generators in general, but offer no specifics that may help troubleshoot a particular unit. It is difficult to obtain from those sources a condensed and operational set of insights useful to the solution of a given problem with a specific machine. It is no wonder then that with so many dissimilar units in operation and such a variegated experience, we are often forced to call the "experts," who tend to be folks almost as old as the oldest units in operation. These are individuals who have crawled, inspected, tested, and maintained many diverse generators over the years. In doing so, they have retained knowledge about the different design, material, and manufacturing characteristics, typical problems, and most effective solutions. This type of expertise cannot be learned in a classroom.

Unfortunately, not every company retains an individual with the breadth and depth of expertise required for troubleshooting all its units. In fact, with the advent of deregulation, many small nonutility (third-party) power producers operate small fleets of generators without the benefit of in-house expertise. In lieu of that, they depend heavily on OEMs and independent consultants. Large utilities in many places have also seen their expertise dissipate, not to a small extent because of a refocus of management priorities. All these developments are occurring at the same time that these units are called to operate in a more onerous environment. Economic dispatch in a deregulated or semideregulated world results in an increased use of double-shifting and load-cycling.

Some effort has been made over the years to capture the experts' knowledge and make it readily available to any operator. This effort took the shape of *expert systems.* However, adaptation of these computer programs to the many different types of generators and associated equipment in existence has proved to be the Achilles heel of this technology.

This book is designed to partially fill the gap by offering a comprehensive view of the many issues related to the operation, inspection, maintenance, and troubleshooting of large turbine generators. The contents of this second edition have been significantly enhanced and many new additional topics included. All of the information in the book is the result of many years of combined hands-on experience of the authors. It was written with the machine's operator and inspector in mind, as well as providing a guide to uprating and life enhancement of large generators. Although not designed to provide a step-by-step guide for the troubleshooting of large generators, it serves as a valuable source of information that may prove to be useful during troubleshooting activities. The topics covered are also cross-referenced to other sources. Many such references are included to facilitate those readers interested in enlarging their knowledge of a specific issue under discussion. For the most part, theoretical equations have been left out, as there are several exceptionally good books on the theory of operation of synchronous machines. Those readers who so desire can readily access those books. Several references are cited. This book, however, is about the practical aspects that characterize the design, operation, and maintenance of large turbine-driven generators, and a significant number of practical calculations used commonly in maintenance and testing situations have been added.

Chapter 1 ("Principles of Synchronous Machines") provides a basis of theory for electricity and electromagnetism upon which the machines covered in this book are based. As well, the fundamentals of synchronous machine construction and operation are also discussed. This is for the benefit of generator operators who have a mechanics background and are inclined to attain a modicum of proficiency in understanding the basic principles of operation of the generator. It also comes in handy for those professors who would like to adopt this book as a reference for a course on large rotating electric machinery.

Chapters 2 and 3 ("Generator Design and Construction" and "Generator Auxiliary Systems") contain a very detailed and informative description of all the components found in a typical generator and its associated auxiliary systems. Described therein are the functions that the components perform, as well as all relevant design and operational constrains. Some additional insight into design methods and calculations are also provided.

Chapter 4 ("Operation and Control") introduces the layperson to the many operational variables that describe a generator. Most generator–grid interaction issues and their affect on the machine components and operation are covered in great detail.

Chapter 5 ("Monitoring and Diagnostics") and Chapter 6 ("Generator Protection") serve to introduce all aspects related to the on-line and off-line monitoring and protection of a large turbogenerator. Although not intended to serve as a guideline for designing and setting up the protection systems of a generator, they provide a wealth of background information and pointers to additional literature.

Chapters 7 ("Inspection Practices and Methodology"), leads off the second part of the book with a look at preparing for a hands-on inspection of large generators. The chapter discusses the issues of concern for both safety of personnel and the equipment as well as the types of tools and approaches used in inspecting large generators. This chapter also contains a collection of most inspection forms typically used for inspecting turbogenerators. These forms are very useful and can be readily adapted to any machine and plant.

Chapter 8 ("Stator Inspection"), Chapter 9 ("Rotor Inspection"), and Chapter 10 ("Auxiliaries Inspection") constitute the core of this book. They describe all components presented in Chapters 2 and 3, but within the context of their behavior under real operational constraints, modes of failure, and typical troubleshooting activities. These chapters provide detailed information on what to look for, and how to recognize problems in the machine during inspection. Chapters 8 and 9 also contain some basic formulas and procedures for some of the various activities that occur during inspection, maintenance, and testing of large generators.

Chapter 11 ("Generator Maintenance Testing") contains a comprehensive summary of the many techniques used to test the many components and systems comprising a generator. The purpose of the descriptions is not to serve as a guide to performing the tests-there are well-established guides and standards for that—rather, they are intended to illustrate the palette of possible tests to choose from. Provided as well is a succinct explanation of the character of each test and explanations of how they are carried out.

Chapter 12 ("Maintenance Philosophies") is included to provide some perspective to the reader on the many choices and approaches that can be taken in generator and auxiliary systems maintenance. Often, there are difficult decisions on how far to take maintenance. In some cases, only basic maintenance may be required, and on other occasions it may be appropriate to carry out extensive rehabilitation of existing equipment or even replacement of components. This chapter discusses some of the issues that need to be considered when deciding on what, how much, and where to do it. Along with the regular maintenance aspects, other important issues like uprating and long-term storage are also addressed.

We hope that this book will be not only useful to the operator in the power plant but also to the design engineer and the systems operations engineer. We have provided a wealth of information obtained in the field about the behavior of such machines, including typical problems and conditions of operation. The book should also be useful to the student of electrical rotating machines as a complementary reference to the books on machine theory.

Although we have tried our best to cover each topic as comprehensively as possible, the book should not be seen as a guide to troubleshooting. In each case in which a real problem is approached, a whole number of very specific issues only relevant to that very unique machine come into play. These can never be anticipated or known and thus described in a book. Thus, we recommend the use of this book as a general reference source, but that the reader should always obtain adequate on-the-spot expertise when approaching a particular problem.

We remain intent on updating the contents of this book from time to time, from our own experience as well as from that of others. Therefore, we would welcome from the readers their comments, which they can submit to the publisher, for incorporation in future editions.

GEOFF KLEMPNER
ISIDOR KERSZENBAUM

Toronto, Ontario, Canada
Irvine, California
August 2008

ACKNOWLEDGMENTS

The contents of this book are impossible to learn in a class. They are the result of personal experience accumulated over years of working with large turbine-driven generators. Most of all, they are the result of the invaluable long-term contribution of coworkers and associates. Each author was motivated by an important individual at an early stage of his career, and by many outstanding individuals in the profession over subsequent years. Attempting to mention all these people would lead to the unintended omission of some.

The authors are most indebted to the IEEE Press for reviewing the second edition proposal and supporting its publication. They also wish to express their sincere gratitude to the technical reviewers, Robert Hindmarsh and Nils Nilsson, for painstakingly reviewing the final manuscript and making numerous useful remarks. The authors also would like to thank the members of the editorial departments of the IEEE Press and John Wiley & Sons, the reviewers, and all others involved in the publication of this book for their support in making its publication possible.

Finally, but certainly most intensely, the authors wish to thank their immediate families for their continuous support and encouragement.

G.K.
I.K.

I

THEORY, CONSTRUCTION, AND OPERATION

1

PRINCIPLES OF OPERATION OF SYNCHRONOUS MACHINES

The synchronous electrical generator (also called *alternator*) belongs to the family of electric rotating machines. Other members of the family are the direct-current (dc) motor or generator, the induction motor or generator, and a number of derivatives of all these three. What is common to all the members of this family is that the basic physical process involved in their operation is the conversion of electromagnetic energy to mechanical energy, and vice versa. Therefore, to comprehend the physical principles governing the operation of electric rotating machines, one has to understand some rudiments of electrical and mechanical engineering.

Chapter 1 is written for those who are involved in operating, maintaining, and trouble-shooting electrical generators, and who want to acquire a better understanding of the principles governing the machines' design and operation, but who do not have an electrical engineering background. The chapter starts by introducing the rudiments of electricity and magnetism, quickly building up to a description of the basic laws of physics governing the operation of the synchronous electric machine, which is the type of machine to which all turbogenerators belong.

1.1 INTRODUCTION TO BASIC NOTIONS ON ELECTRIC POWER

1.1.1 Magnetism and Electromagnetism

Certain materials found in nature exhibit a tendency to attract or repel each other. These materials, called *magnets,* are also called *ferromagnetic* because they include the element iron as one of their constituent elements.

Magnets always have two poles: one called *north,* the other called *south.* Two north poles always repel each other, as do two south poles. However, north and south poles always attract each other. A *magnetic field* is defined as a physical field established between two poles. Its intensity and direction determine the forces of attraction or repulsion existing between the two magnets.

Figures 1.1 and 1.2 are typical representations of two interacting magnetic poles and the magnetic field established between them.

Magnets are found in nature in all sorts of shapes and chemical constitution. Magnets used in industry are artificially made. Magnets that sustain their magnetism for long periods of time are denominated "permanent magnets." The magnetic field produced by the north and the south pole of a permanent magnet is directional from north to south (see Fig. 1.3). These are widely used in several types of electric rotating ma-

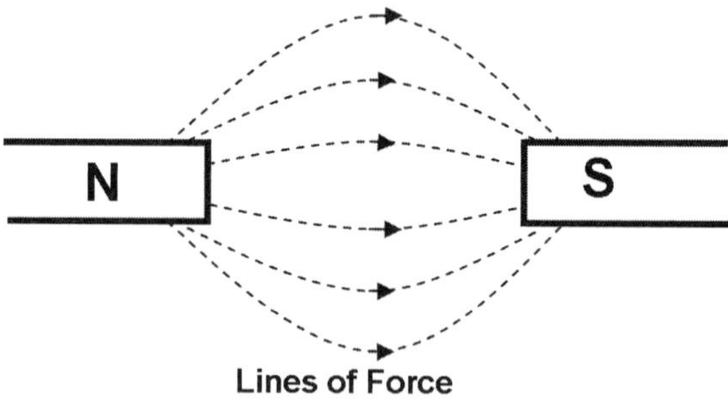

Fig. 1.1 Schematic representation of two magnetic poles of opposite polarity, with the magnetic field between them shown as "lines of force."

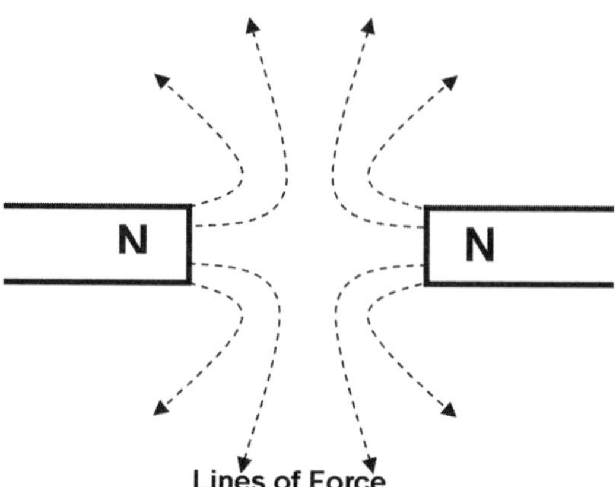

Fig. 1.2 Schematic representation of two north poles and the magnetic field between them. South poles will create similar field patterns, but the lines of force will point toward the poles.

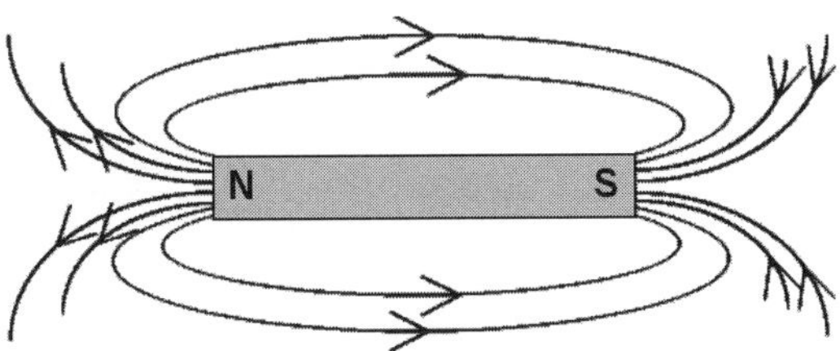

Fig. 1.3 Schematic representation of a "permanent magnet" showing the north and south poles and the magnetic field between them. Note that the magnetic field flows from north to south (outside the magnet).

chines, including synchronous machines. However, due to mechanical as well as operational reasons, permanent magnets in synchronous machines are restricted to those with ratings much lower than large turbine-driven generators, which is the subject of this book. Turbine-driven generators (for short: turbogenerators) take advantage of the fact that magnetic fields can be created by the flow of electric currents in conductors. See Fig. 1.4.

A very useful phenomenon is that forming the conductor into the shape of a coil can augment the intensity of the magnetic field created by the flow of current through the conductor. In this manner, as more turns are added to the coil, the same current produces larger and larger magnetic fields. For practical reasons, all magnetic fields created by current in a machine are generated in coils. See Fig. 1.5.

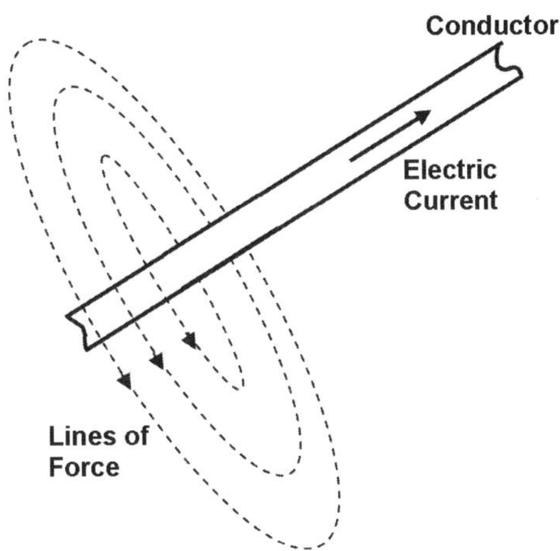

Fig. 1.4 Schematic representation of a magnetic field created by the flow of current in a conductor. The direction of the *lines of force* is given by the "law of the screwdriver": mentally follow the movement of a screw as it is screwed in the same direction as that of the current; the lines of force will then follow the circular direction of the head of the screw. The magnetic lines of force are perpendicular to the direction of current.

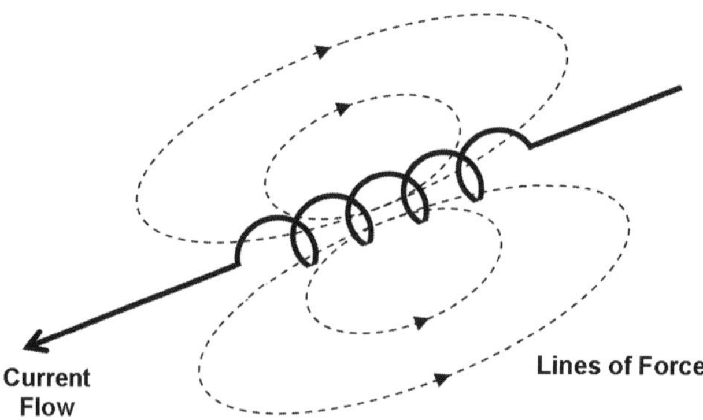

Fig. 1.5 Schematic representation of a magnetic field produced by the flow of electric current in a coil-shaped conductor.

The use of coils to amplify the magnetic field intensity requires a significant amount of control so that the resulting flux is produced in an effective way. When the coil is operating in air, the magnetic field direction, shape, and intensity depends on the number of turns in the coil, the size of the coil, and the direction of electric current flow in the coil winding. The flux produced is basically divided into two types. One is the *effective flux* that links the entire coil and does the useful work, and the other is the *leakage flux* which is a more localized effect and does no useful work. In fact, the leakage flux creates additional losses that make the coil less efficient, electromagnetically speaking (see Fig. 1.6). The principles illustrated here become very important later on as we discuss the magnetic field in the generator and stray losses.

To use as much of the flux produced in a coil as effectively as possible, highly *magnetically permeable* materials—basically, *ferromagnetic* materials—are used to capture and direct the flux so that the amount of leakage flux is minimized. This allows the coil to do more useful work and keeps losses to a minimum. Iron in various derivatives is by far the most widely used material because it has all the magnetic characteristics required, is structurally suitable, and cost-effective. When an "iron" core is used within the coil, and current is flowing, the magnetic field produced is shaped effectively and the iron core essentially becomes a north–south magnet in the process (see Fig. 1.7). This is why rotor forgings and stator cores of generators are made of iron. The iron allows the principles discussed above to become a reality and is one of the reasons generators can be built to at least 98.5% efficiency.

1.1.2 Electricity

Electricity is the flow of positive or negative charges. Electricity can flow in electrically conducting elements (called *conductors*), or it can flow as clouds of ions in space or within gases. As will be shown in later chapters, both types of electrical conduction are found in turbogenerators. See Fig. 1.8.

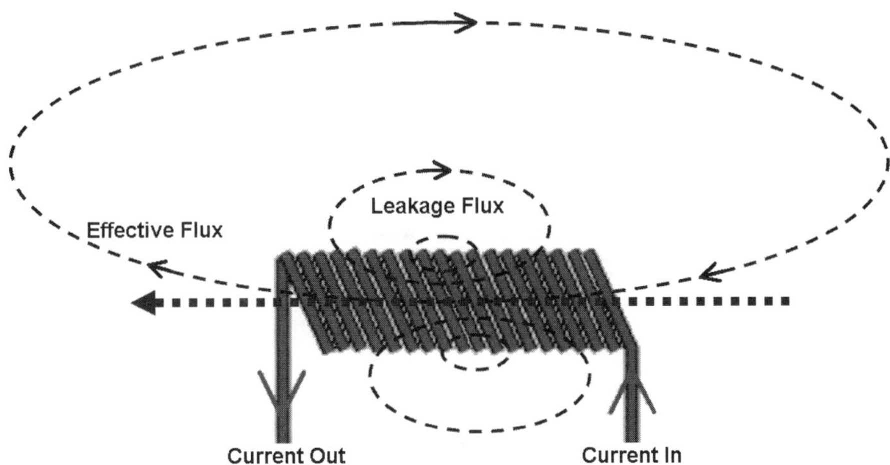

Fig. 1.6 Schematic representation of a magnetic field produced by the flow of electric current in a coil-shaped conductor operating in air, showing the effective and leakage flux components of the magnetic field produced.

1.2 ELECTRICAL–MECHANICAL EQUIVALENCE

There is an interesting equivalence between the various parameters describing electrical and mechanical forms of energy. People with either electrical or mechanical backgrounds find this equivalence useful to the understanding of the physical process in either form of energy. Figure 1.9 describes the various forms of electrical–mechanical equivalence.

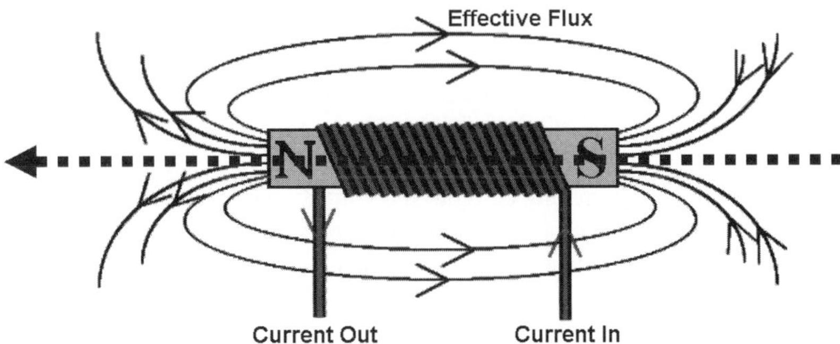

Fig. 1.7 Schematic representation of a magnetic field produced by the flow of electric current in a coil-shaped conductor with an "iron" core. The majority of the field produced is effective flux and the leakage field is reduced to a minimum.

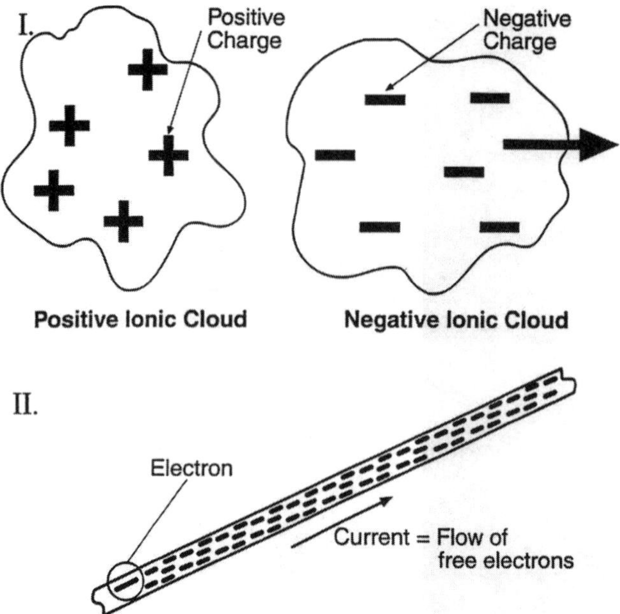

Fig. 1.8 Electricity. (I) Ionic clouds of positive and negative currents. The positive clouds are normally atoms that lost one or more electrons; the negative clouds are normally free electrons. This effect can be found inside the generator as *partial discharge* in the stator winding. (II) The flow of electrons inside a conductor material, for example, the copper windings of the rotor and stator.

1.3 ALTERNATING CURRENT (ac)

As will be shown later, alternators operate with both alternating-current (ac) and direct-current (dc) electric power. The dc can be considered a particular case of the general ac, with frequency equal to zero.

The frequency of an alternating circuit is measured by the number of times the currents and/or voltages change direction (polarity) in a unit of time. The hertz is the universally accepted unit of frequency, and measures cycles per second. One Hz equals one cycle per second. Alternating currents and voltages encountered in the world of industrial electric power are for all practical purposes of constant frequency. This is important because *periodic* systems, namely systems that have constant frequency, allow the currents and voltages to be represented by *phasors*.

A phasor is a rotating vector. The benefit of using phasors in electrical engineering analysis is that it greatly simplifies the calculations required to solve circuit problems.

Figure 1.10 depicts a phasor of magnitude **E**, and its corresponding sinusoidal trace representing the instantaneous value of the quantity *e*. The magnitude **E** represents the maximum value of *e*.

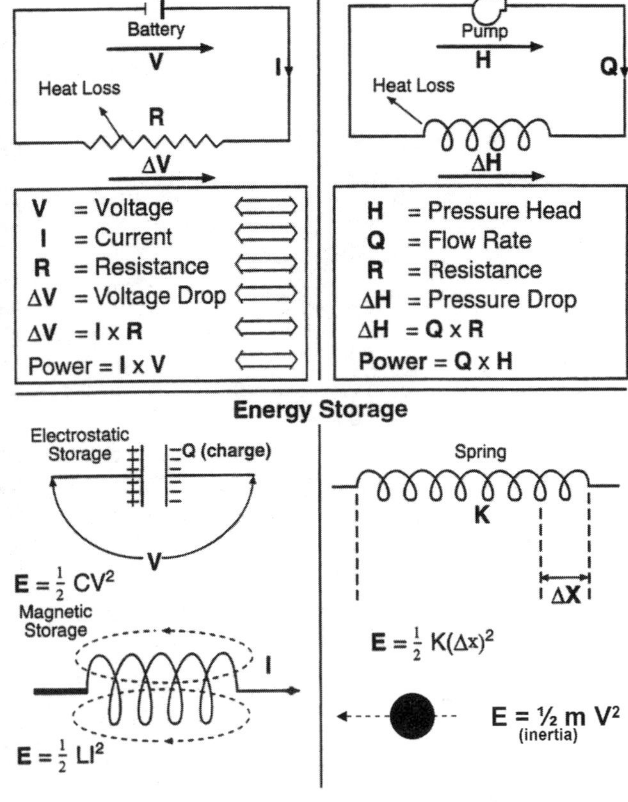

Fig. 1.9 Electrical–mechanical equivalence.

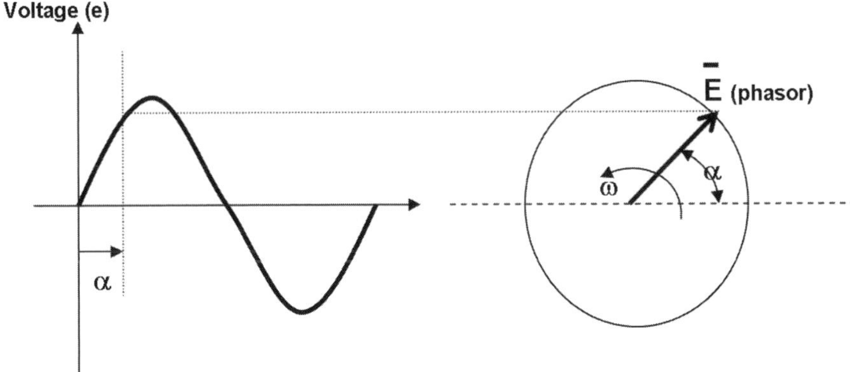

Fig. 1.10 A phasor **E** that can represent the voltage impressed on a circuit. The phasor is made of a vector with magnitude proportional to the magnitude of **E**, rotating at a constant rotational speed ω. The convention is that phasors rotate counterclockwise. The vertical projection of the phasor results in a sinusoidal representing the instantaneous voltage *e* existing at any time. In the graph, $\alpha = \omega \times t$, where *t* is the time elapsed from its zero crossing.

When a sinusoidal voltage is applied to a closed circuit, a current will flow in it. After a while, the current will have a sinusoidal shape (this is called the *steady-state* current component) and the same frequency as the voltage. An interesting phenomenon in periodic circuits is that the resulting angle between the applied voltage and the current depends on certain characteristics of the circuit. These characteristics can be classified as being *resistive, capacitive,* and *inductive.* The angle between the voltage and the current in the circuit is called the *power angle.* The cosine of the same angle is called the *power factor* of the circuit or, for short, the PF.

Note: As will be shown latter, in synchronous machines the term power angle is used to identify a different concept. To avoid confusion, in this book the angle between the current and the voltage in the circuit will therefore be identified by its "power factor."

In the case of a circuit having only resistances, the voltages and currents are *in phase,* meaning that the angle between them equals zero. Figure 1.11 shows the various parameters encountered in a resistive circuit. It is important to note that resis-

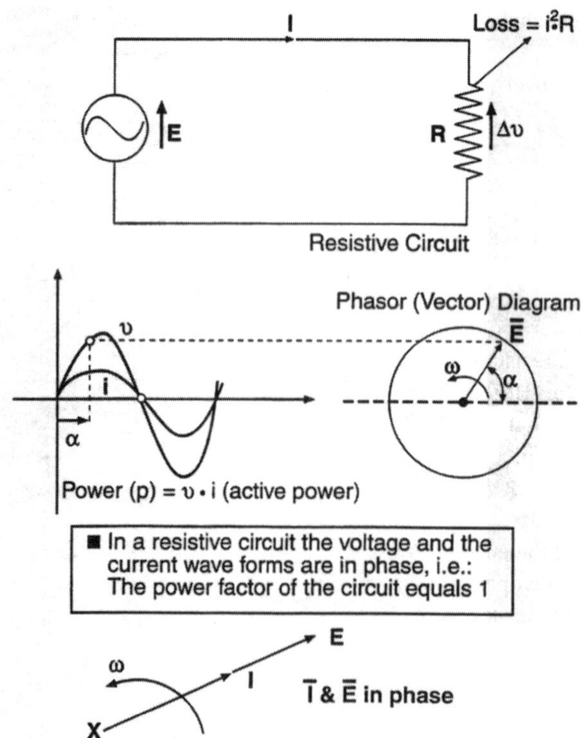

Fig. 1.11 Alternating circuits (resistive). Schematic representation of a sinusoidal voltage of magnitude **E** applied on a circuit with a resistive load *R*. The schematics show the resultant current *i* in phase with the voltage *v*. It also shows the phasor representation of the voltage and current.

tances have the property of generating heat when a current flows through them. The heat generated equals the square of the current times the value of the resistance. When the current is measured in amperes and the resistance in ohms, the resulting power dissipated as heat is given in watts. In electrical machines, this heat represents a loss of energy. It will be shown later that one of the fundamental requirements in designing an electric machine is the efficient removal of these resistive losses, with the purpose of limiting the undesirable temperature rise of the internal components of the machine.

In resistive circuits, the instantaneous power delivered by the source to the load equals the product of the instantaneous values of the voltage and the current. When the same sinusoidal voltage is applied across the terminals of a circuit with capacitive or inductive characteristics, the steady-state current will exhibit an angular (or time) displacement vis-à-vis the driving voltage. The magnitude of the angle (or power factor) depends on how capacitive or inductive the load is. In a purely capacitive circuit, the current will lead the voltage by 90°, whereas in a purely inductive one, the current will lag the voltage by 90° (see Fig. 1.12).

A circuit that has capacitive or inductive characteristics is referred to as being a *reactive* circuit. In such a circuit, the following parameters are defined:

S: The *apparent power* — $\mathbf{S} = \mathbf{E} \times \mathbf{I}$, given in units of volt-amperes or VA.

P: The *active power* — $\mathbf{P} = \mathbf{E} \times \mathbf{I} \times \cos\varphi$, where φ is the angle between the voltage and the current. **P** is given in units of watts.

Q: The *reactive power* — $\mathbf{Q} = \mathbf{E} \times \mathbf{I} \times \sin\varphi$, given in units of volt-amperes-reactive or VAR.

The active power **P** of a circuit indicates a real energy flow. This is power that may be dissipated on a resistance as heat, or may be transformed into mechanical energy, as will be shown later. However, the use of the word "power" in the definition of **S** and **Q** has been an unfortunate choice that has resulted in confounding most individuals without an electrical engineering background for many years. The fact is that *apparent power* and *reactive power* do not represent any measure of real energy. They do represent the reactive characteristic of a given load or circuit, and the resulting angle (power factor) between the current and voltage. This angle between voltage and current significantly affects the operation of an electric machine, as will be discussed later.

For the time being, let us define another element of ac circuit analysis: the *power triangle*. From the relationships shown above among **S**, **P**, **Q**, **E**, **I**, and φ, it can be readily shown that **S**, **P**, and **Q** form a triangle. By convention, **Q** is shown as positive (above the horizontal), when the circuit is *inductive,* and vice versa when *capacitive* (see Fig. 1.13).

To demonstrate the use of the power triangle within the context of large generators and their interaction with the power system, we need to consider a one-line schematic that includes the generator, transmission system, and the connected load at the end (see Fig. 1.14).

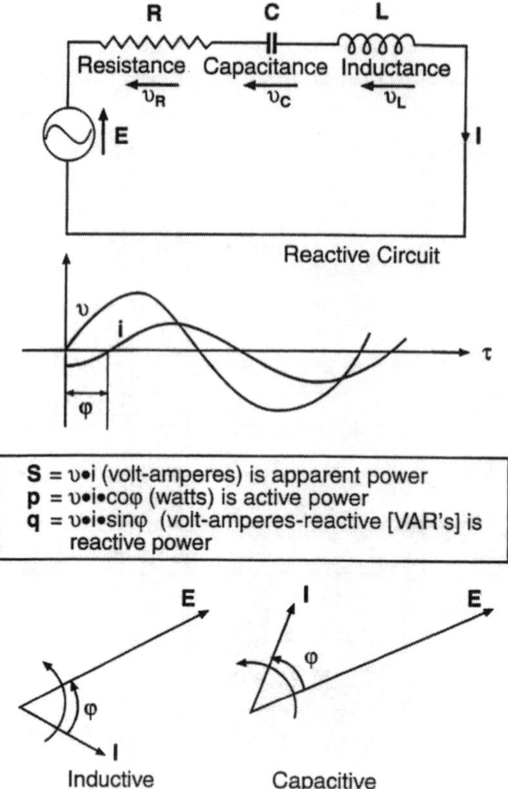

Fig. 1.12 Alternating circuits (resistive–inductive–capacitive). Here, the sinusoidal voltage **E** is applied to a circuit comprised of resistive, capacitive, and inductive elements. The resulting angle between the current and the voltage depends on the value of the resistance, capacitance, and inductance of the load.

If we now consider an actual load for the simple system of Fig. 1.14, we can calculate the current drawn by the load and the voltage required from the generator source to compensate for all the line losses and voltage drop across the line. Two cases are provided to illustrate the effect of a purely resistive load versus a load with a reactive component included (see Figs. 1.15 and 1.16). Working out the required voltage from the generator for the two different loads by the power triangle method shows how reactive loads greatly affect the power system operation and the generation requirements. Reactive power compensation is a large part of generator operation and affects generator design in a significant way, as will be discussed later on in Chapter 2. There is a delicate balance between generation and load that is clearly shown by the two cases presented and the comparison of operational results (see Fig. 1.17).

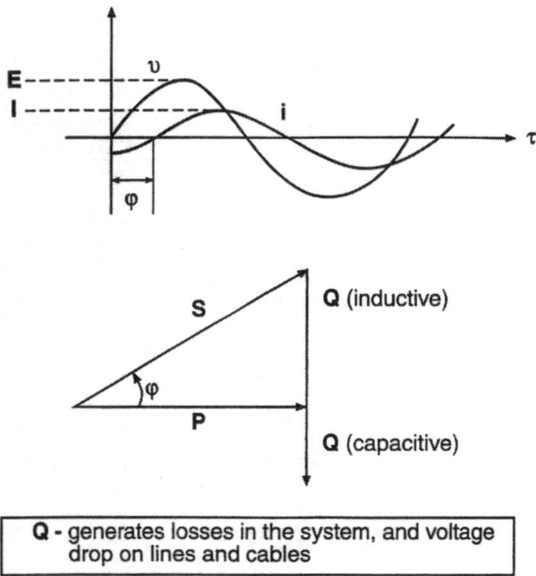

Fig. 1.13 Definition of the "power triangle" in a reactive circuit.

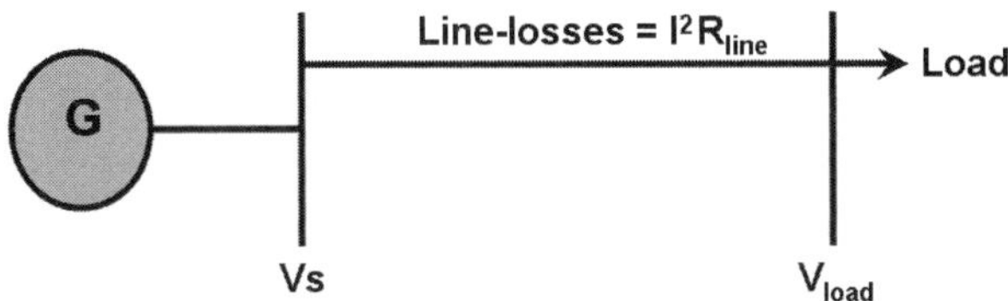

Simple power system showing a generator, bus, line and load.

V_{load} = 1,000 volts

Line resistance (R) = 10 Ω (resistive)

Line reactance (X) = 10 Ω (inductive)

Line impedance (Z) = $\sqrt{(10^2 + 10^2)}$ = 14.14 Ω

Fig. 1.14 Schematic of a simple system in one-line form. The voltage required at the load (so that it will operate correctly) is given as 1000 volts. The transmission line resistance and reactance are provided and the line impedance calculated as shown, using the power triangle approach.

Case 1 – Load = 100 kW (Unity Power Factor {i.e. Cos φ =1} **)**

$P = \sqrt{3} * I * V * \cos\varphi$

$\Rightarrow$ I = 100,000 watts /1,000 volts /1.73 /1 = **57.8 amperes**

Losses in the line ($I^2 R_{line}$) = 57.8^2 amperes * 10Ω = **33.4 kW**

Voltage drop along the line = $I*Z = \sqrt{(I * R)^2 + (I * X)^2}$ = **817 volts**

$\Rightarrow$The required delivery voltage at the source (Vs) is:

$Vs = \sqrt{(V_{load} + IR)^2 + (IX)^2}$

$Vs = \sqrt{(1{,}000 + 578)^2 + (578)^2}$ = **1,680 volts**

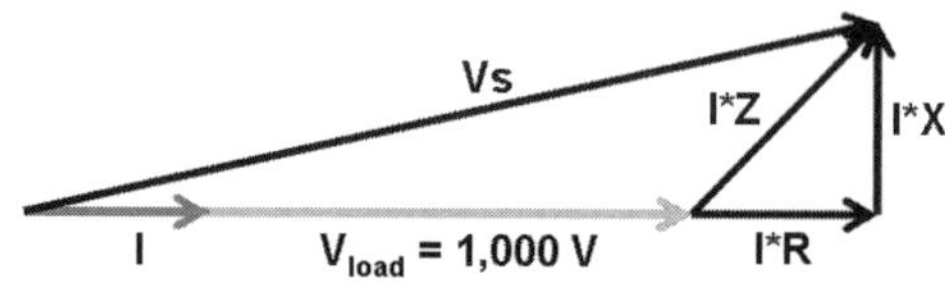

Fig. 1.15 Case 1. The load is purely resistive in this example, and the system is operating at the “unity” power factor.

Case 2 – Load = 100 kW & 50 kVAR-inductive (lagging power factor)

$S = \sqrt{3} * I * V = \sqrt{100{,}000^2 + 50{,}000^2}$ = 111,803 VA (Volt-Amperes)

$\Rightarrow$ I = 111,803 VA /1,000 volts /1.73 = **64.55 amperes**

Losses in the line ($I^2 R_{line}$) = 64.55^2 amperes * 10Ω = **41.6 kW**

Voltage drop along the line = $I*Z = \sqrt{(I * R)^2 + (I * X)^2}$ = **913 volts**

$\Rightarrow$The required delivery voltage at the source (Vs) is:

$Vs = \sqrt{[V_{load} + IR(\sin\varphi+\cos\varphi)]^2 + [IX(\cos\varphi-\sin\varphi)]^2}$

$Vs = \sqrt{[1{,}000+645.5(\sin\varphi+\cos\varphi)]^2 + [645.5(\cos\varphi-\sin\varphi)]^2}$ = **1,892 volts**

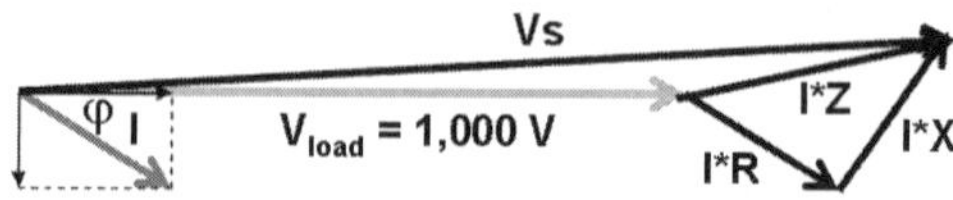

Fig. 1.16 Case 2. The load is resistive and inductive in this example, and the system is operating in the “lagging” power factor range.

Load	100 kW	100 kW & 50 kVAR
Power consumed by the load	100 kW	100 kW
Current	57.8 A	64.6 A
Line losses	33.4 kW	41.6 kW
Voltage drop along the line	817 V	913 V
Required delivery voltage at generating end	1,680 V	1,892 V

Fig. 1.17 A comparison of Case 1 and Case 2 shows that although the "real" power consumed is the same, the addition of the reactive component in Case 2 has caused an increase in current drawn from the generator, an increase in line losses, a higher volt drop across the line, and, therefore, a higher voltage required from the generator source.

The above examples show that there is a considerable demand placed on the generator to operate the various loads on a system. In reality, the generator terminal voltage $\mathbf{V}_s$ is constant, plus or minus 5% by design. As the load increases or decreases, the current from the generator changes significantly and the voltage drop on the system $\mathbf{V}_{load}$ requires compensation (Fig. 1.18). Therefore, the second major function of the generator, after production of "real" power, is to produce "reactive" power to help control the grid's voltage and frequency, which will also be discussed later in Chapter 4.

1.4 THREE-PHASE CIRCUITS

The two-wire ac circuits discussed above (called *single-phase* circuits or systems), are commonly used in residential, commercial, and low-voltage–low-power industrial applications. However, all electric power systems to which industrial generators are connected are *three-phase* systems. Therefore, any discussion in this book about the "power system" will refer to a three-phase system. Moreover, in industrial applications the voltage supplies are, for all practical reasons, balanced, meaning that all three-phase voltages are equal in magnitude and apart by 120 electrical degrees. In those rare events in which the voltages are unbalanced, the implications for the operation of the generator will be discussed in other chapters of this book.

Three-phase electric systems may have a fourth wire, called "neutral." The "neutral" wire of a three-phase system will conduct electricity if the source and/or the load are unbalanced. In three-phase systems, two sets of voltages and currents can be identified. These are the *phase* and *line* voltages and currents.

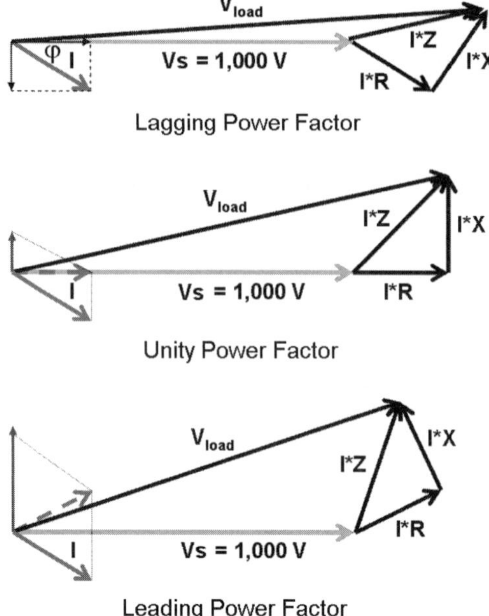

Fig. 1.18 The effect on the voltage drop as the circuit goes from lagging through unity to leading power factor operation is shown.

Figure 1.19 shows the main elements of a three-phase circuit. Three-phase circuits can have their sources and/or loads connected in *wye* (*star*) or in *delta* forms. (See Fig. 1.20 for a wye-connected source feeding a delta-connected load.)

Almost without exception, turbine-driven generators have their windings connected in wye (star) form. Therefore, in this book the source (or generator) will be shown wye-connected. There are a number of important reasons why turbogenerators are wye-connected. They have to do with considerations about its effective protection as well as design (insulation, grounding, etc.). These will be discussed in the chapters covering stator construction and operations.

On the other hand, loads can be found connected in wye, delta, or a combination of the two. This book is not about circuit solutions; therefore, the type of load connection will not be brought up herein.

1.5 BASIC PRINCIPLES OF MACHINE OPERATION

In Section 1.1, basic principles were presented showing how a current flowing in a conductor produces a magnetic field. In this section, three important laws of electromagnetism will be presented. These laws, together with the law of energy conservation, constitute the basic theoretical bricks on which the operation of any electrical machine is based.

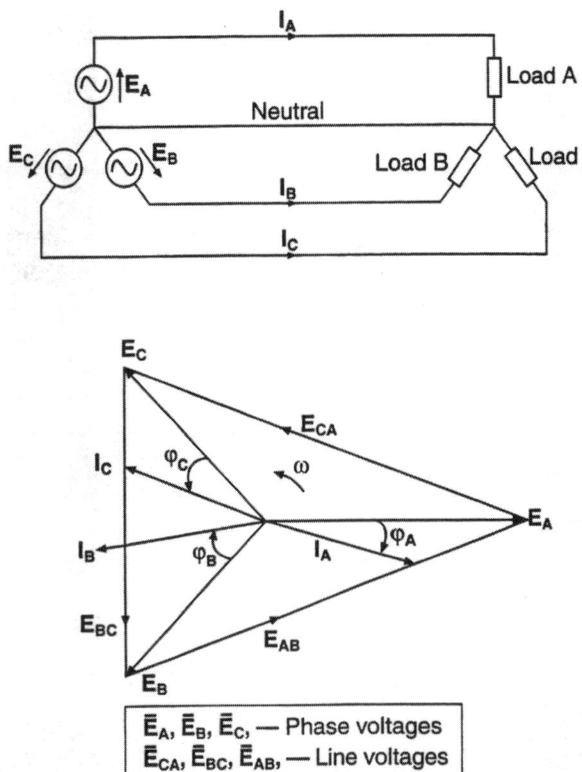

Fig. 1.19 Three-phase systems. Schematic depiction of a three-phase circuit and the vector (phasor) diagram representing the currents, voltages, and angles between them.

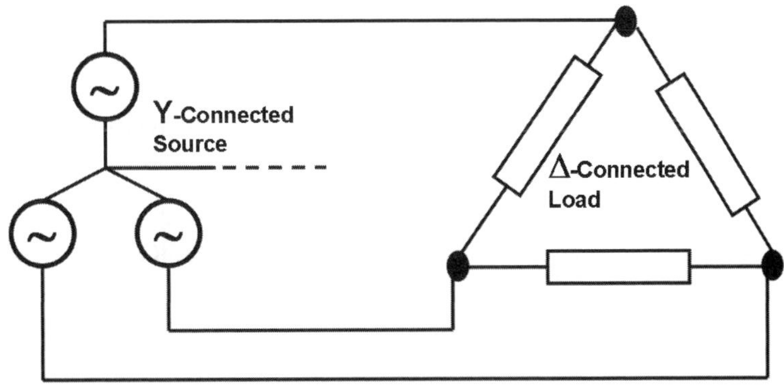

Fig. 1.20 A "wye-connected" source feeding a "delta-connected" load.

1.5.1 Faraday's Law of Electromagnetic Induction

This basic law, due to the genius of the great English chemist and physicist Michael Faraday (1791–1867), presents itself in two different forms:

1. A moving conductor cutting the lines of force (flux) of a constant magnetic field has a voltage induced in it.
2. A changing magnetic flux inside a loop made from a conductor material will induce a voltage in the loop.

In both instances, the *rate of change* is the critical determinant of the resulting differential of potential. Figure 1.21 illustrates both cases of electromagnetic induction, and also provides the basic relationship between the changing flux and the voltage induced in the loop, for the first case, and the relationship between the induced voltage in a wire moving across a constant field, for the second case. The figure also shows one of the simple rules that can be used to determine the direction of the induced voltage in the moving conductor.

1.5.2 Ampere–Biot–Savart's Law of Electromagnetic Induced Forces

This basic law is attributed to the French physicists Andre Marie Ampere (1775–1836), Jean Baptiste Biot (1774–1862), and Victor Savart (1803–1862). In its simplest form, this law can be seen as the "reverse" of Faraday's law. Whereas Faraday's law predicts a voltage induced in a conductor moving across a magnetic field, the Ampere–Biot–Savart law establishes that a force is generated on a current-carrying conductor located in a magnetic field.

Figure 1.22 presents the basic elements of Ampere–Biot–Savart's law as applicable to electric machines. The figure also shows the existing numerical relationships, and a simple hand-rule to determine the direction of the resultant force.

1.5.3 Lenz's Law of Action and Reaction

Both Faraday's law and Ampere–Biot–Savart's law neatly come together in Lenz's law, written in 1835 by the Estonian-born physicist Heinrich Lenz (1804–1865). Lenz's law states that electromagnetic-induced currents and forces will try to cancel the originating cause.

For example, if a conductor is forced to move, cutting lines of magnetic force, a voltage is induced in it (Faraday's law). Now, if the conductors' ends are closed together so that a current can flow, this induced current will produce (according to Ampere–Biot–Savart's law) a force acting upon the conductor. What Lenz's law states is that this force will act to oppose the movement of the conductor in its original direction.

Here, in a nutshell, is the explanation for the *generating* and *motoring* modes of operation of an electric rotating machine! This law explains why, when a generator is

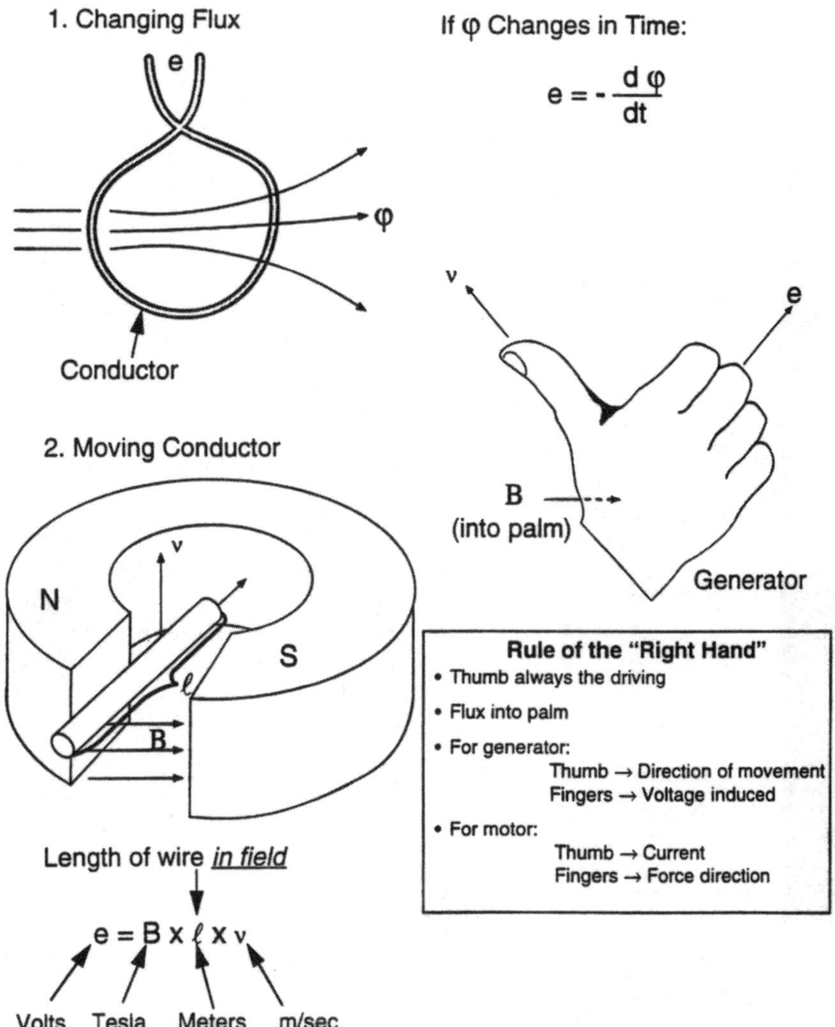

Fig. 1.21 Both forms of Faraday's basic law of electromagnetic induction. A simple rule (the "right-hand" rule) is used to determine the direction of the induced voltage in a conductor moving across a magnetic field at a given velocity.

loaded (more current flows in its windings, cutting the magnetic field in the gap between rotor and stator), more force is required from the driving turbine to counteract the induced larger forces and keep supplying the larger load. Similarly, Lenz's law explains the increase in the supply current of a motor as its load increases.

Figure 1.23 neatly captures the main elements of Lenz's law as it applies to electric rotating machines.

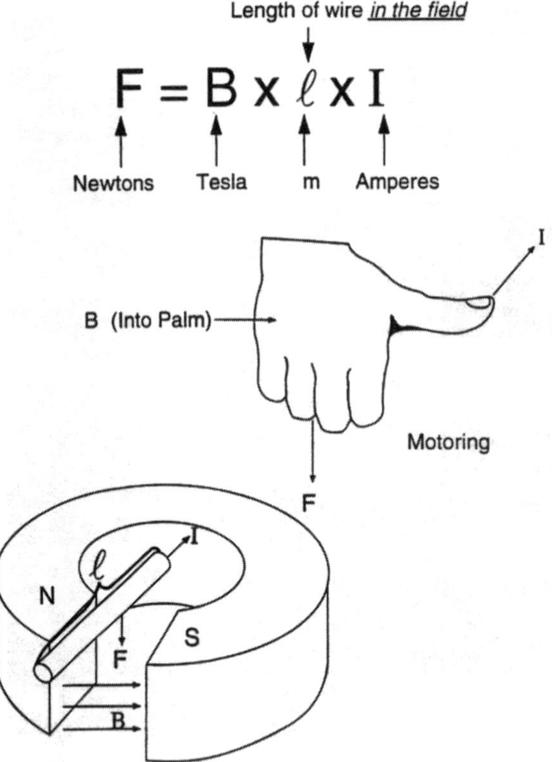

Fig. 1.22 The Ampere–Biot–Savart law of electromagnetic-induced forces as it applies to electric rotating machines. Basic numerical relationships and a simple rule are used to determine the direction of the induced force.

1.5.4 Electromechanical Energy Conversion

The fourth and final physical law that captures, together with the previous three, all the physical processes occurring inside an electric machine, is the "principle of energy conversion." Within the domain of the electromechanical world of an electric rotating machine, this principle states that:

> All the electrical and mechanical energy flowing into the machine, less all the electrical and mechanical energy flowing out of the machine and stored in the machine, equals the energy dissipated from the machine as heat.

It is important to recognize that although mechanical and electrical energy can go into or out of the machine, the heat generated within the machine always has a negative sign, that is, heat generated in the machine is always released during the conversion process. A *plus* sign indicates energy going in; a *minus* indicates energy going out. In the case of the stored energy (electrical and mechanical), a plus sign indicates

Example

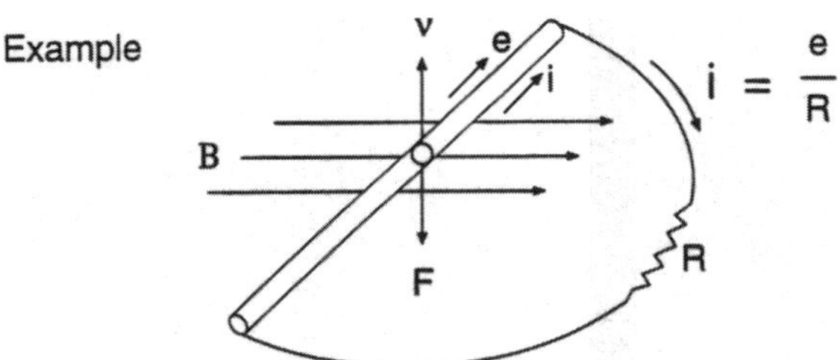

1. The upwards moving conductor in a magnetic field induces a voltage (Faraday)
2. Closing the circuit generates a current
3. The current creates a force opposing the movement (Ampere and Lenz)

Hint: Use the rule of the palm to show the direction of "F"

This phenomenon explains the torque applied by the generator on the turbine, when the unit is loaded

Induced currents and forces will try to cancel the originating cause

Fig. 1.23 Lenz's law as it applies to electric rotating machines. Basic numerical relationships and a simple rule are used to determine the direction of the induced forces and currents.

an increase of stored energy, whereas a minus sign indicates a reduction in stored energy.

The balance between the various forms of energy in the machine will determine its efficiency and cooling requirements, both critical performance and construction parameters in a large generator. Figure 1.24 depicts the principle of energy conversion as applicable to electric rotating machines.

1.6 THE SYNCHRONOUS MACHINE

At this point, the rudiments of electromagnetism have been presented, together with the four basic laws of physics describing the inherent physical processes coexisting in any electrical machine. Therefore, it is the right time to introduce the basic configuration of the synchronous machine, which, as mentioned before, is the type of electric machine that all large turbine-driven generators belong to.

1.6.1 Background

The commercial birth of the alternator (synchronous generator) can be dated back to August 24, 1891. On that day, the first large-scale demonstration of transmission of ac

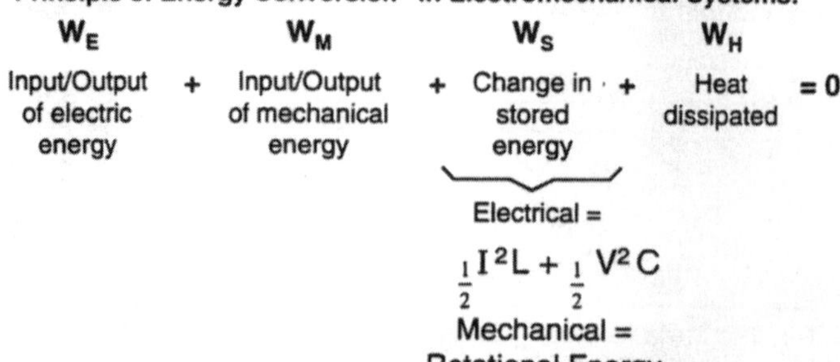

"Principle of Energy Conversion" In Electromechanical Systems:
W_E W_M W_S W_H
Input/Output of electric energy + Input/Output of mechanical energy + Change in stored energy + Heat dissipated = 0
Electrical =
$\frac{1}{2} I^2 L + \frac{1}{2} V^2 C$
Mechanical =
Rotational Energy

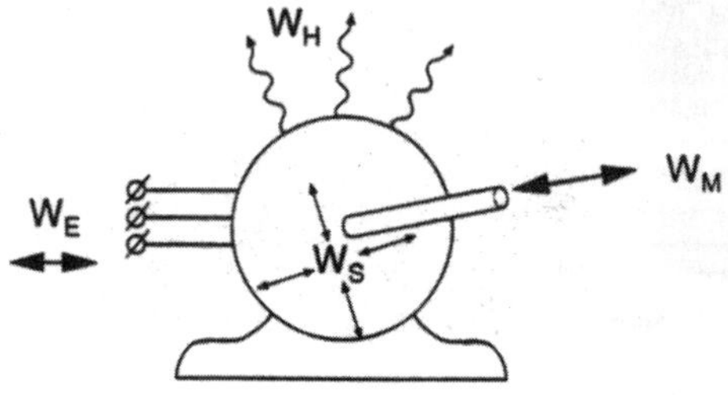

W_H is always negative (i.e. heat is always released during the conversion process)

W_E, W_M and W_S can have "+" or "-" signs

W_E and W_M with a "plus" means input to the machine
"minus" means output from the machine

W_S with a "plus" means increase of stored energy
"minus" means decrease of stored energy

Fig. 1.24 Principle of energy conversion as applicable to electric rotating machines.

power was carried out. The transmission extended from Lauffen, Germany, to Frankfurt, about 110 miles away. The demonstration was carried out during an international electrical exhibition in Frankfurt. This demonstration was so convincing about the feasibility of transmitting ac power over long distances that the city of Frankfurt adopted it for their first power plant, commissioned in 1894. This happened about 113 years before the writing of this book (see Fig. 1.25).

The Lauffen–Frankfurt demonstration, and the consequent decision by the city of Frankfurt to use ac power delivery, were instrumental in the adoption by New York's Niagara Falls power plant of the same technology. The Niagara Falls power plant became operational in 1895. For all practical purposes, the great dc versus ac duel was over. Southern California Edison's history book reports that its Mill Creek hydro plant is the oldest active polyphase (three-phase) plant in the United States. Located in San Bernardino County, California, its first units went into operation on September 7,

Fig. 1.25 The hydroelectric generator from Lauffen, now in the Deutches Museum, Munich. (Reprinted with permission from *The Evolution of the Synchronous Machine* by Gerhard Neidhofer, 1992, Asea Brown Boveri.)

1893, placing it almost two years ahead of the Niagara Falls project. One of those earlier units is still preserved and displayed at the plant.

It is interesting to note that although tremendous development in machine ratings, insulation components, and design procedures has occurred for over one hundred years, the basic constituents of the machine have remained practically unchanged.

The concept that a synchronous generator can be used as a motor followed suit. Although Tesla's induction motor replaced the synchronous motor as the choice for the vast majority of electric motor applications, synchronous generators remained the universal machines of choice for the generation of electric power. The world today is divided between countries generating their power at 50 Hz and others (e.g., the United States) at 60 Hz. Additional frequencies (e.g., 25 Hz) can still be found in some locations, but they constitute the rare exception.

Synchronous generators have continuously grown in size over the years (see Fig. 1.26). The justification is based on simple economies of scale: the output rating of the machine per unit of weight increases as the size of the unit increases. Thus, it is not uncommon to see machines with ratings reaching up to 2000 MVA in the most recent times, with the largest normally used in nuclear power stations. Interestingly enough, the present ongoing shift from large steam turbines as prime movers to more efficient gas turbines is resulting in a reverse of the trend toward larger and larger generators, at least for the time being. Transmission system stability considerations also place an upper limit on the rating of a single generator.

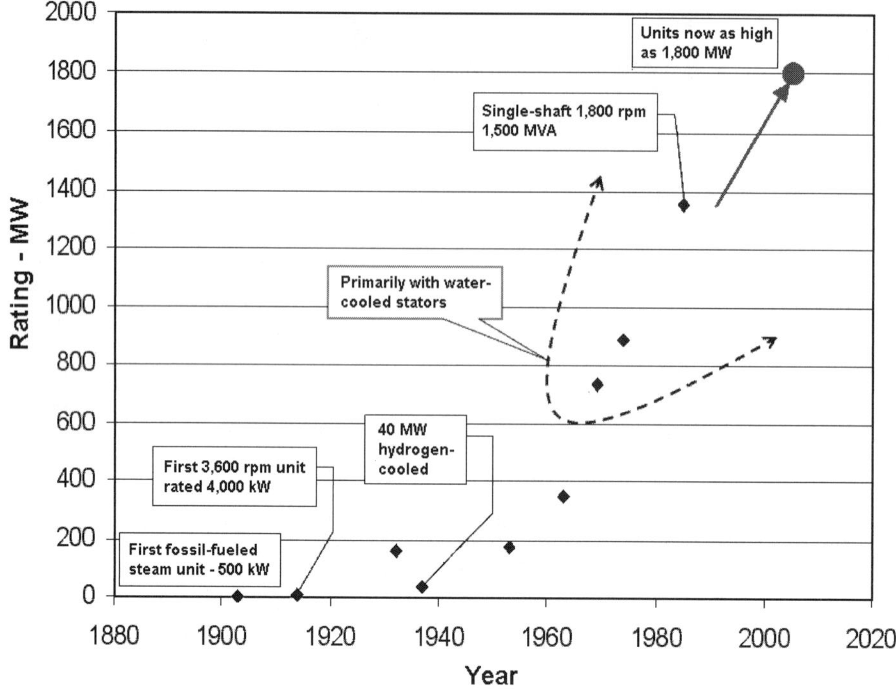

Fig. 1.26 "Growth" graph, depicting the overall increase in size over the last century, of turbine-driven generators.

1.6.2 Principles of Construction

Chapter 2 includes a description of the design criteria leading to the construction of a modern turbogenerator, as well as a detailed description of all components most commonly found in such a machine. This section is limited to the presentation of the basic components comprising a synchronous machine, with the purpose of describing its basic operating theory.

Synchronous machines come in all sizes and shapes, from the miniature permanent magnet synchronous motor in wall clocks, to the largest steam-turbine-driven generators of up to about 2000 MVA. Synchronous machines are one of two types: stationary field or rotating dc magnetic field.

The stationary field synchronous machine has salient poles mounted on the stator—the stationary member. The poles are magnetized either by permanent magnets or by a dc current. The armature, normally containing a three-phase winding, is mounted on the shaft. The armature winding is fed through three slip rings (collectors) and a set of brushes sliding on them. This arrangement can be found in machines up to about 5 kVA in rating. For larger machines—all those covered in this book—the typical arrangement used is the rotating magnetic field.

The rotating magnetic field (also known as revolving field) synchronous machine has the field winding wound on the rotating member (the *rotor*) and the *armature* wound on the stationary member (the *stator*). A dc current, creating a magnetic field that must be rotated at synchronous speed, energizes the rotating field winding. The rotating field winding can be energized through a set of slip rings and brushes (external excitation) or from a diode bridge mounted on the rotor (self-excited). The rectifier bridge is fed from a shaft-mounted alternator, which is itself excited by the pilot exciter. In externally fed fields, the source can be a shaft-driven dc generator, a separately excited dc generator, or a solid-state rectifier. Several variations on these arrangements exist.

The stator core is made of insulated steel laminations. The thickness of the laminations and the type of steel are chosen to minimize eddy current and hysteresis losses, while maintaining required effective core length and minimizing costs. The core is mounted directly onto the frame or (in large two-pole machines) through spring bars. The core is slotted (the slots are normally open), and the coils making up the winding are placed in the slots. There are several types of armature windings, such as concentric windings of several types, cranked coils, split windings of various types, wave windings, and lap windings of various types. Modern large machines typically are wound with double-layer lap windings (more about these winding types in Chapter 2).

The rotor field is either of salient-pole (Fig. 1.27) or nonsalient-pole construction, also known as round rotor or cylindrical rotor (Fig. 1.28). Nonsalient-pole (cylindrical) rotors are utilized in two- or four-pole machines, and, very seldom, in six-pole machines. These are typically driven by steam or combustion turbines. The vast majority of salient-pole machines have six or more poles. They include all synchronous hydrogenerators, almost every synchronous condenser, and the overwhelming majority of synchronous motors.

Nonsalient-pole rotors are typically machined out of a solid steel forging. The winding is placed in slots machined out of the rotor body and retained against the large centrifugal forces by metallic wedges, normally made of aluminum or steel. The *retaining rings* restrain the end part of the windings (end windings). In the case of large machines, the retaining rings are made out of steel.

Large salient-pole rotors are made of laminated poles retaining the winding under the pole head. The poles are keyed onto the shaft (spider-and-wheel structure). Salient-pole machines have an additional winding in the rotating member. This winding, made of copper bars short-circuited at both ends, is embedded in the head of the pole, close to the face of the pole. The purpose of this winding is to start the motor or condenser under its own power as an induction motor, and take it unloaded to almost synchronous speed, when the rotor is "pulled in" by the synchronous torque. The winding also serves to damp the oscillations of the rotor around the synchronous speed and is, therefore, named the *damping winding* (also known as *amortisseur*).

This book focuses on large turbine-driven generators. These are always two- or four-pole machines having cylindrical rotors. The discussion of salient-pole machines can be found in other books. (See the Additional Reading section at the end of this chapter.)

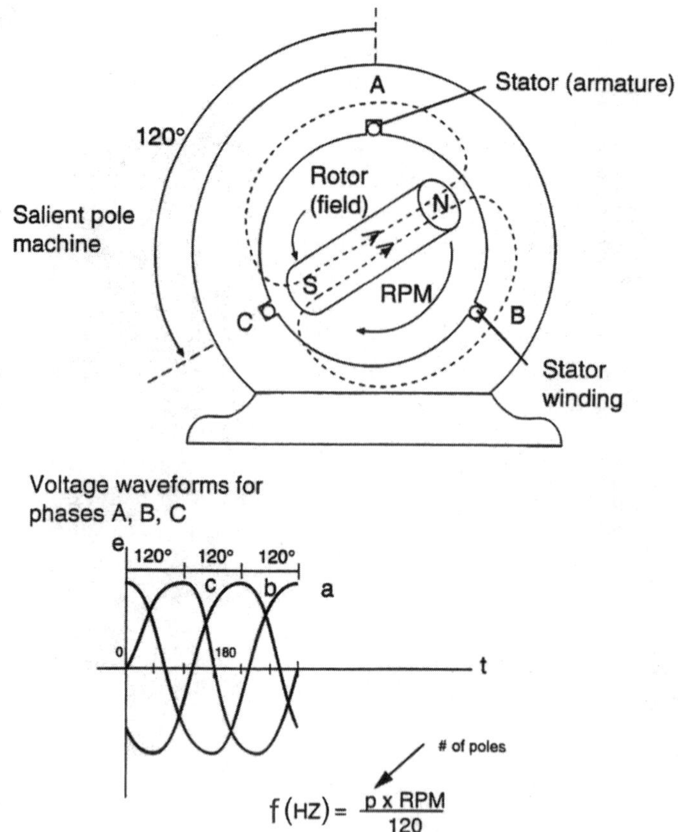

Fig. 1.27 Synchronous machine construction. Schematic cross section of a salient-pole synchronous machine. In a large generator, the rotor is magnetized by a coil wrapped around it. The figure shows a two-pole rotor. Salient-pole rotors normally have many more than two poles. When designed as a generator, large salient-pole machines are driven water turbines. The bottom part of the figure shows the three-phase voltages obtained at the terminals of the generator, and the equation relates the speed of the machine, the number of poles, and the frequency of the resulting voltage.

1.6.3 Rotor Windings

In turbogenerators, the winding producing the magnetic field is made of a number of coils—single-circuit, energized with dc power fed via the shaft from the collector rings riding on the shaft, and positioned outside the main generator bearings. In *self-excited* generators, the shaft-mounted exciter and rectifier (diodes) generate the required field current. The shaft-mounted exciter is itself excited from a stationary winding. The fact that, unlike the stator, the rotor field is fed from a relatively low-power, low-voltage circuit has been the main reason why these machines have the field mounted on the ro-

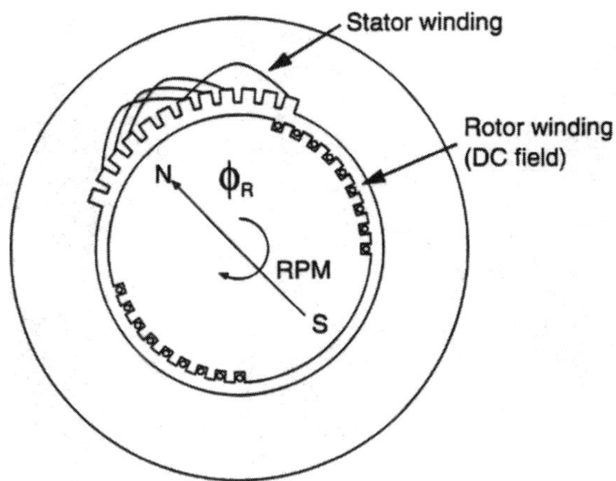

Production of MMF and Flux

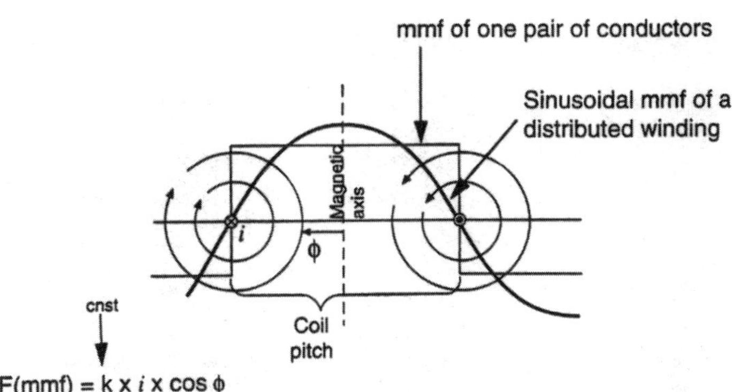

Fig. 1.28 Schematic cross section of a synchronous machine with a cylindrical round rotor. This is the typical design for all large turbogenerators. Here, both the stator and rotor windings are installed in slots, distributed around the periphery of the machine. The lower part shows the resulting waveforms of a pair of conductors and that of a distributed winding. The formula gives the magneto-motive force (mmf) created by the windings.

tating member and not the other way around. Moving high currents and high power through the collector rings and brushes (with a rotating armature) would represent a serious technical challenge, making the machine that much more complex and expensive.

Older generators have field supplies of 125 volts dc. Newer ones have supplies of 250 volts and higher. Excitation voltages of 500 volts or higher are common in newer machines. A much more elaborate discussion of rotor winding design and construction can be found in Chapter 2.

1.6.4 Stator Windings

The magnitude of the voltage induced in the stator winding is, as shown above, a function of the magnetic field intensity, the rotating speed of the rotor, and the number of turns in the stator winding. An actual description of individual coil design and construction, as well as how the completed winding is distributed around the stator, is meticulously described in Chapter 2. In this section, a very elementary description of the winding arrangement is presented to facilitate the understanding of the basic operation of the machine.

As stated above, coils are distributed in the stator in a number of forms, each with its own advantages and disadvantages. The basic goal is to obtain three balanced and sinusoidal voltages having very little harmonic content (harmonic voltages and currents are detrimental to the machine and other equipment in a number of ways). To achieve a desired voltage and MVA rating, the designer may vary the number of slots and the manner in which individual coils are connected, producing different winding patterns. The most common winding arrangement is the lap winding, shown in Figure 1.29 for a four-pole machine and in Figure 1.30 for a two-pole machine.

A connection scheme that allows great freedom of choice in designing the windings to accommodate a given terminal voltage is one that allows connecting sections of the

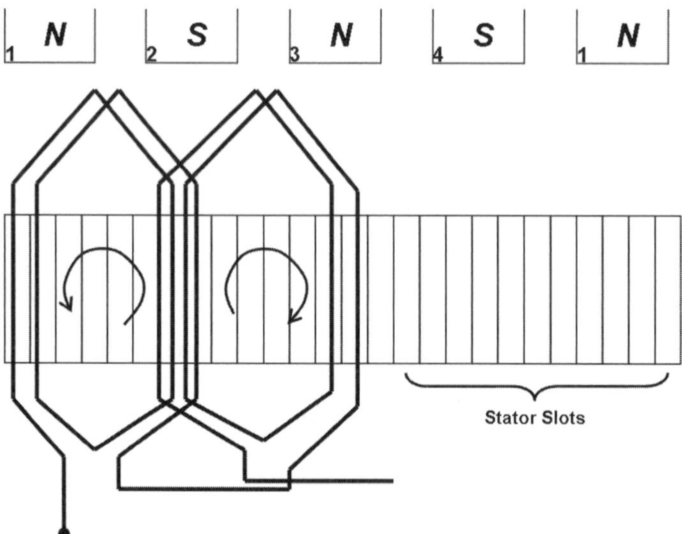

Fig. 1.29 "Developed" view of a four-pole stator, showing the slots, the poles, and a section of the winding. The section shown is of one of the three phases. It can be readily seen that the winding runs clockwise under a north pole, and counterclockwise under a south pole. This pattern repeats itself until the winding covers the four poles. A similar pattern is followed by the other two phases, but located at 120 electrical degrees apart.

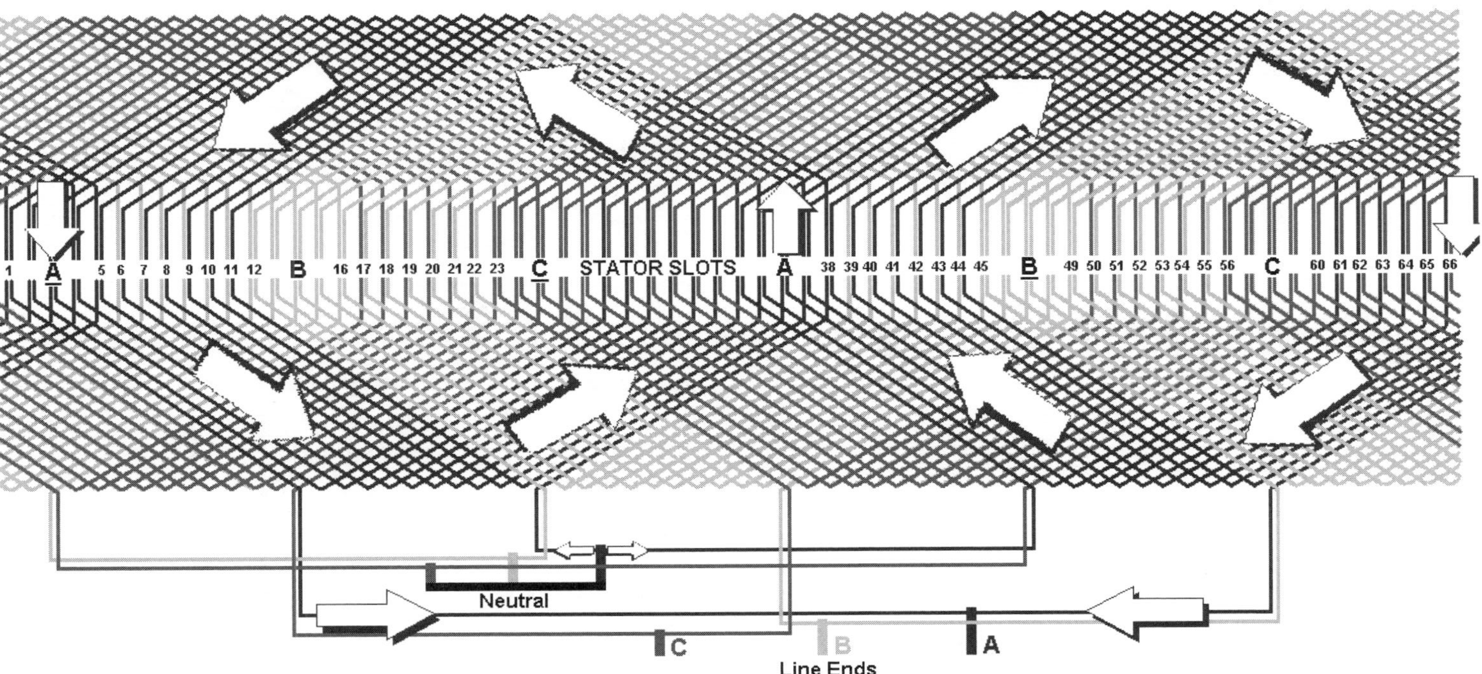

Fig. 1.30 "Developed" view of a two-pole stator, showing the slots, the poles, and the three winding phases. It can be readily seen that the winding runs clockwise under a north pole and counterclockwise under a south pole. A similar pattern is followed by the other two phases, but located 120 electrical degrees apart. (See color insert.)

winding in parallel, series, and/or a combination of the two. Figure 1.31 shows two typical winding arrangements for a two-pole generator.

1.7 BASIC OPERATION OF THE SYNCHRONOUS MACHINE

For a more in-depth discussion of the operation and control of large turbogenerators, the reader is referred to Chapter 4. In this chapter, the most elementary principles of operation of synchronous machines will be presented. As mentioned above, all large turbogenerators are three-phase machines. Thus, the best place to start describing the operation of a three-phase synchronous machine is a description of its magnetic field.

Earlier, we described how a current flowing through a conductor produces a magnetic field associated with that current. It was also shown that by coiling the conductor, a larger field is obtained without increasing the current's magnitude. Recall that if the three phases of the winding are distributed at 120 electrical degrees apart, three balanced voltages are generated, creating a three-phase system.

Now, a new element can be brought into the picture. By a simple mathematical analysis, it can be shown that if three balanced currents (of equal magnitudes and 120 electri-

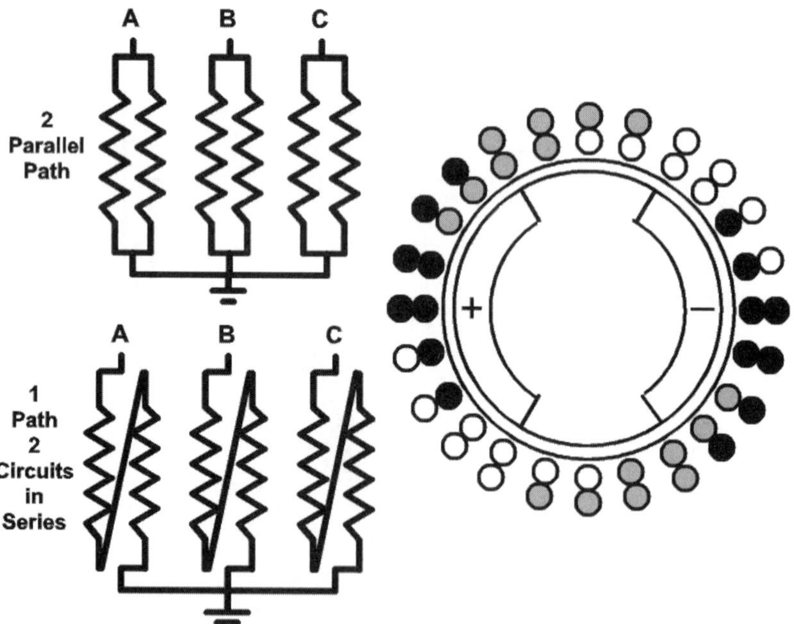

Fig. 1.31 Schematic view of a two-pole generator with two possible winding configurations: (1) a two parallel circuits winding and (2) A two series connected circuits per phase. On the right, the three phases are indicated by different tones. Note that some slots only have coils belonging to the same phase, whereas in others, coils belonging to two phases share the slot.

cal degrees apart) flow in a balanced three-phase winding, a magnetic field of constant magnitude is produced in the airgap of the machine. This magnetic field revolves around the machine at a frequency equal to the frequency of the currents flowing through the winding (see Fig. 1.32). The importance of a three-phase system creating a constant field cannot be stressed enough. The constant magnitude flux allows hundred of megawatts of power to be transformed inside an electric machine from electrical to mechanical power, and vice versa, without major mechanical limitations. It is important to remember that a constant-magnitude flux produces a constant-magnitude torque. Now, try to imagine the same type of power being transformed under a pulsating flux (and, therefore, pulsating torque), which is tremendously difficult to achieve.

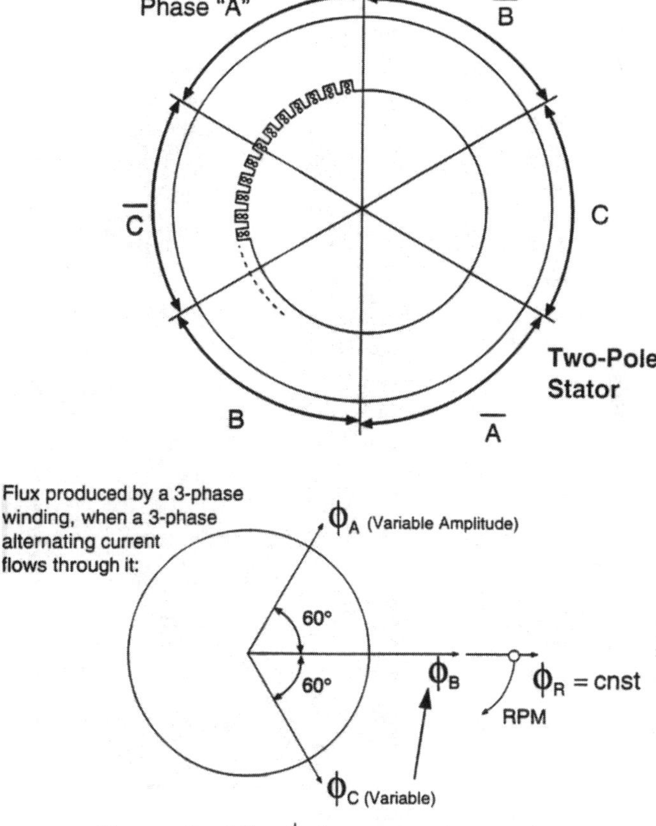

Fig. 1.32 Production of stator rotating field. A constant magnitude and constant rotational speed magnetic flux is created when three-phase balanced currents flow through a three-phase symmetrical winding. The sketch is for a two-pole winding; however, similar result applies for any number of pairs of poles.

It is convenient to introduce the fundamental principles describing the operation of a synchronous machine in terms of an *ideal* cylindrical-rotor machine connected to an *infinite* bus. The infinite bus represents a bus bar of constant voltage, which can deliver or absorb active and reactive power without any limitations. The ideal machine has zero resistance and leakage reactance, infinite permeability, and no saturation, as well as zero reluctance torque.

The production of torque in the synchronous machine results from the natural tendency of two magnetic fields to align themselves. The magnetic field produced by the stationary armature is denoted as $\boldsymbol{\phi}_s$. The magnetic field produced by the rotating field is $\boldsymbol{\phi}_f$. The resultant magnetic field is

$$\boldsymbol{\phi}_r = \boldsymbol{\phi}_s + \boldsymbol{\phi}_f$$

The flux $\boldsymbol{\phi}_r$ is established in the airgap (or gasgap) of the machine. (Bold symbols indicate vector quantities.)

When the torque applied to the shaft equals zero, the magnetic fields of the rotor and the stator become perfectly aligned. The instant torque is introduced to the shaft, either in a generating mode or in a motoring mode, a small angle is created between the stator and rotor fields. This angle (() is called the torque angle of the machine.

1.7.1 No-Load Operation

When the ideal machine is connected to an infinite bus, a three-phase balanced voltage $\mathbf{V}_1$ is applied to the stator winding (within the context of this work, three-phase systems and machines are assumed). As described above, it can be shown that a three-phase balanced voltage applied to a three-phase winding evenly distributed around the core of an armature will produce a rotating (revolving) magneto-motive force (mmf) of constant magnitude $\mathbf{F}_s$. This mmf, acting upon the reluctance encountered along its path, results in the magnetic flux $\boldsymbol{\phi}_s$ previously introduced. The speed at which this field revolves around the center of the machine is related to the supply frequency and the number of poles by the following expression:

$$n_s = 120\left(\frac{f}{p}\right)$$

where

f = electrical frequency in Hz

p = number of poles of the machine

n_s = speed of the revolving field in revolutions per minute (rpm)

If no current is supplied to the dc field winding, no torque is generated, and the resultant flux $\boldsymbol{\phi}_r$, which in this case equals the stator flux $\boldsymbol{\phi}_s$, magnetizes the core to the extent that the applied voltage $\mathbf{V}_1$ is exactly opposed by a counterelectromotive force (cemf) $\mathbf{E}_1$. If the rotor's excitation is slightly increased and no torque is applied to the shaft, the rotor provides some of the excitation required to produce $\mathbf{E}_1$, causing an equivalent reduction of $\boldsymbol{\phi}_s$. This situation represents the underexcited condition shown in the no-load condition (a) in Figure 1.33. When operating under this condition, the machine is said to behave as a *lagging condenser,* meaning that it absorbs reactive power from the network. If the field excitation is increased over the value required to produce $\mathbf{E}_1$, the stator currents generate a flux that counteracts the field-generated flux. Under this condition, the machine is said to be *overexcited,* shown as the *no-load* condition (b) in Figure 1.33. The machine is behaving as a leading condenser; that is, it is delivering reactive power to the network.

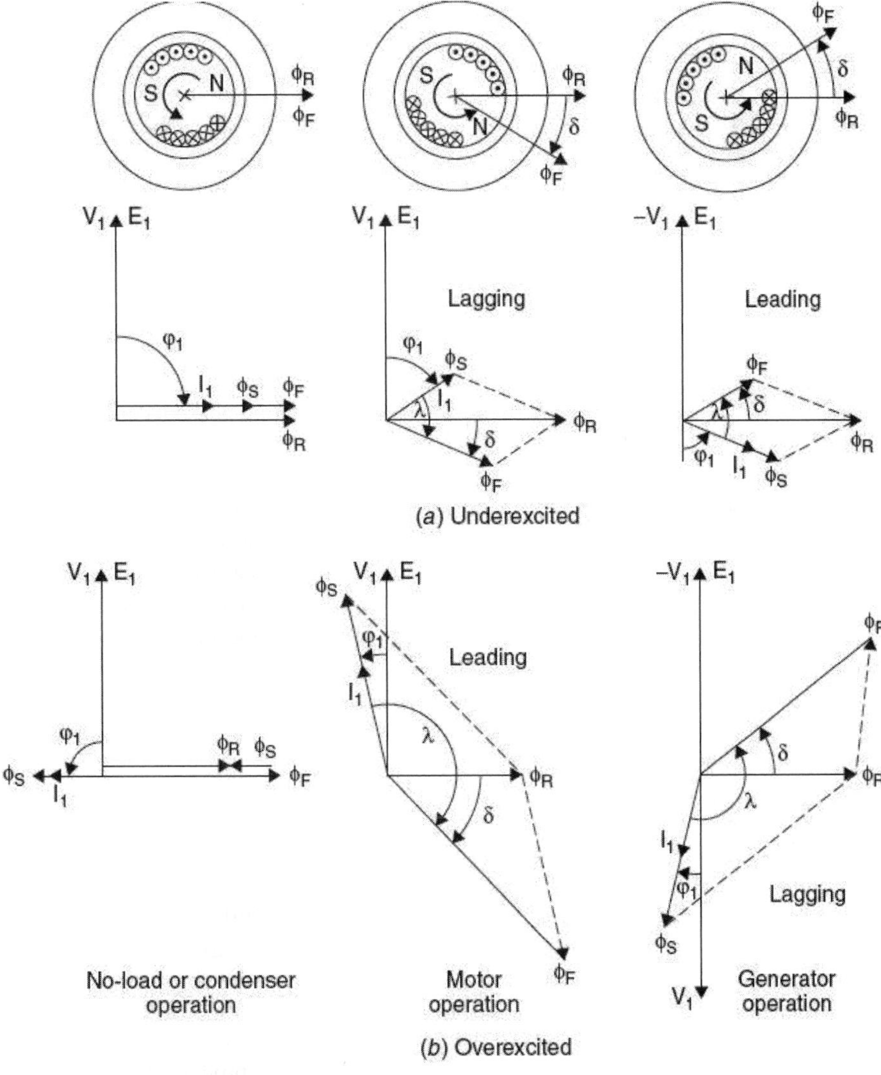

Fig. 1.33 Phasor diagrams for a synchronous cylindrical-rotor ideal machine.

Under no-load conditions, both the torque angle (λ) and the load angle (δ) are zero. The *load angle* is defined as the angle between the rotor's mmf ($\mathbf{F}_f$) or flux ($\boldsymbol{\phi}_f$) and the resultant mmf ($\mathbf{F}_r$) or flux ($\boldsymbol{\phi}_r$). The load angle is the most commonly used because it establishes the torque limits the machine can attain in a simple manner (discussed later). One must be aware that in many texts the name *torque angle* is used to indicate the load angle. The name torque angle is also sometimes given to indicate the angle between the terminal voltage ($\mathbf{V}_1$) and the excitation voltage ($\mathbf{E}_1$). This happens because the leakage reactance is generally very much smaller than the magnetizing reactance, and therefore the load angle δ and the angle between $\mathbf{V}_1$ and $\mathbf{E}_1$ are very similar. In this book, the more common name *power angle* is used for the angle between $\mathbf{V}_1$ and $\mathbf{E}_1$. In Figure 1.33, the power angle is always shown as zero because the leakage impedance has been neglected in the ideal machine.

It is important at this stage to introduce the distinction between electrical and mechanical angles. In studying the performance of the synchronous machine, all the electromagnetic calculations are carried out based on electric quantities; that is, all angles are electrical angles. To convert the electrical angles used in the calculations to the physical mechanical angles, we use the following relationship:

$$\text{Mechanical angle} = \left(\frac{2}{p}\right) \text{Electrical angle}$$

1.7.2 Motor Operation

The subject of this book is turbo-generators. These units seldom operate as motors. (One such example is when the main generator is used for a short period of time as a motor fed from a variable-speed converter. The purpose of this operation is for starting its own prime-mover combustion turbine.) However, this section presents an introductory discussion of the synchronous machine and, thus, the motor mode of operation is also covered.

If a breaking torque is applied to the shaft, the rotor starts falling behind the revolving-armature-induced magneto-motive force (mmf) $\mathbf{F}_s$. In order to maintain the required magnetizing mmf $\mathbf{F}_r$, the armature current changes. If the machine is in the underexcited mode, *motor operation* in Figure 1.33*a* represents the new phasor diagram.

On the other hand, if the machine is overexcited, the new phasor diagram is represented by *motor operation* in Figure 1.33*b*. The active power consumed from the network under these conditions is given by

$$\text{Active power} = V_1 \times I_1 \times \cos \varphi_1 \qquad \text{(per phase)}$$

If the breaking torque is increased, a limit is reached in which the rotor cannot keep up with the revolving field. The machine then stalls. This is known as "falling out of step," "pulling out of step," or "slipping poles." The maximum torque limit is reached when the angle δ equals $\pi/2$ electrical. The convention is to define δ as negative for motor operation and positive for generator operation. The torque is also a function of the magnitude of $\boldsymbol{\phi}_r$ and $\boldsymbol{\phi}_f$. When overexcited, the value of $\boldsymbol{\phi}_f$ is larger than in the underexcited condition. Therefore, synchronous motors are capable of greater mechanical output when overexcited. Likewise, underexcited operation is more prone to result in an "out-of-step" situation.

1.7.3 Generator Operation

Let us assume that the machine is running at no load and a positive torque is applied to the shaft; that is, the rotor flux angle is advanced ahead of the stator flux angle. As in the case of motor operation, the stator currents will change to create the new conditions of equilibrium shown in Figure 1.33, under *generator operation*. If the machine is initially underexcited, condition (a) in Figure 1.33 results. On the other hand, if the machine is overexcited, condition (b) in Figure 1.33 results.

It is important to note that when "seen" from the terminals, with the machine operating in the underexcited mode, the power factor angle (φ_1) is leading (i.e., $\mathbf{I}_1$ leads $\mathbf{V}_1$). This means that the machine is absorbing reactive power from the system. The opposite occurs when the machine is in the overexcited mode. As for the motor operation, an overexcited condition in the generating mode also allows for greater power deliveries.

As generators are normally called upon to provide VARs together with watts, they are almost always operated in the overexcited condition.

1.7.4 Equivalent Circuit

When dealing with three-phase balanced circuits, electrical engineers use the *one-line* or *single-line* representation. This simplification is allowed because in three-phase balanced circuits, all currents and voltages, as well as circuit elements are symmetrical. Thus, by "showing" only one phase, it is possible to represent the three-phase system, as long as care is taken in using the proper factors. For instance, the three-phase balanced system of Figure 1.19 or Figure 1.20 can be represented as shown in Figure 1.34. Hereinafter, when describing a three-phase generator by an electrical diagram, the one-line method will be used.

The most convenient way to determine the performance characteristics of synchronous machines is by means of equivalent circuits. These equivalent circuits can become very elaborate when saturation, armature reaction, harmonic reactance, and other nonlinear effects are introduced. However, the simplified circuit in Figure 1.35 is conducive to obtaining the basic performance characteristics of the machine under steady-state conditions.

In Figure 1.35, the reactance X_a represents the magnetizing or demagnetizing effect of the stator windings on the rotor. It is also called the *magnetizing reactance*. R_a rep-

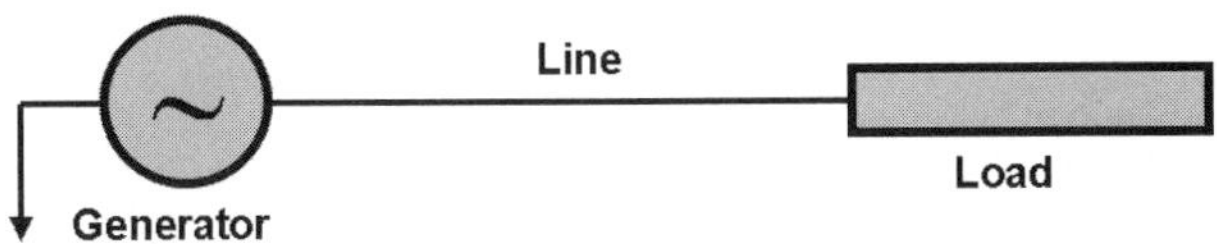

Fig. 1.34 One-line representation of the circuit shown in Figures 1.19 and 1.20.

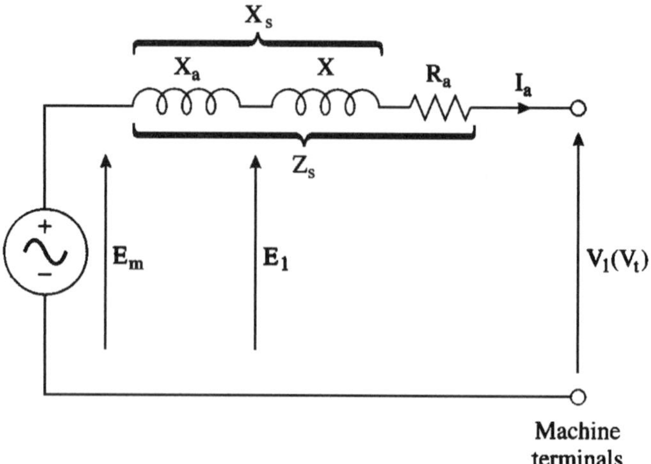

Fig. 1.35 Steady-state equivalent circuit of a synchronous machine. X = leakage reactance, X_a = armature reaction reactance, $X_s = X_a + X$ = synchronous reactance, R_a = armature resistance, Z_s = synchronous impedance, $V_1(V_t)$ = terminal voltages, and E_m = magnetizing voltage.

resents the effective resistance of the stator. The reactance X represents the stator leakage reactance. The sum of X_a and X is used to represent the total reactance of the machine, and is called the *synchronous reactance* (X_s). Z_s is the *synchronous impedance* of the machine. It is important to remember that the equivalent circuit described in Figure 1.35 represents the machine only under steady-state conditions.

The simple equivalent circuit of Figure 1.36 suffices to determine the steady-state performance parameters of the synchronous machine connected to a power grid. These parameters include voltages, currents, power factor, and load angle. The *regulation* of the machine can be easily found from the equivalent circuit for different load conditions by using the regulation formula:

$$\Re(\%) = 100(V_{\text{no-load}} - V_{\text{load}})/V_{\text{load}}$$

For a detailed review of the performance characteristics of the synchronous machine, in particular the turbo-generator, the reader is referred to Chapter 4.

Note: Regulation in a generator indicates how the terminal voltage of the machine varies with changes in load. When the generator is connected to an infinite bus (i.e., a bus that does not allow the terminal voltage to change), a change in load will affect the machine's output in a number of ways. (See Chapter 4 for a discussion of this topic.)

1.7.5 Machine Losses

In Section 1.5.4, the balance of energy in an electric machine was discussed. As part of the discussion, reference was made to the fact that current flows through the machine's

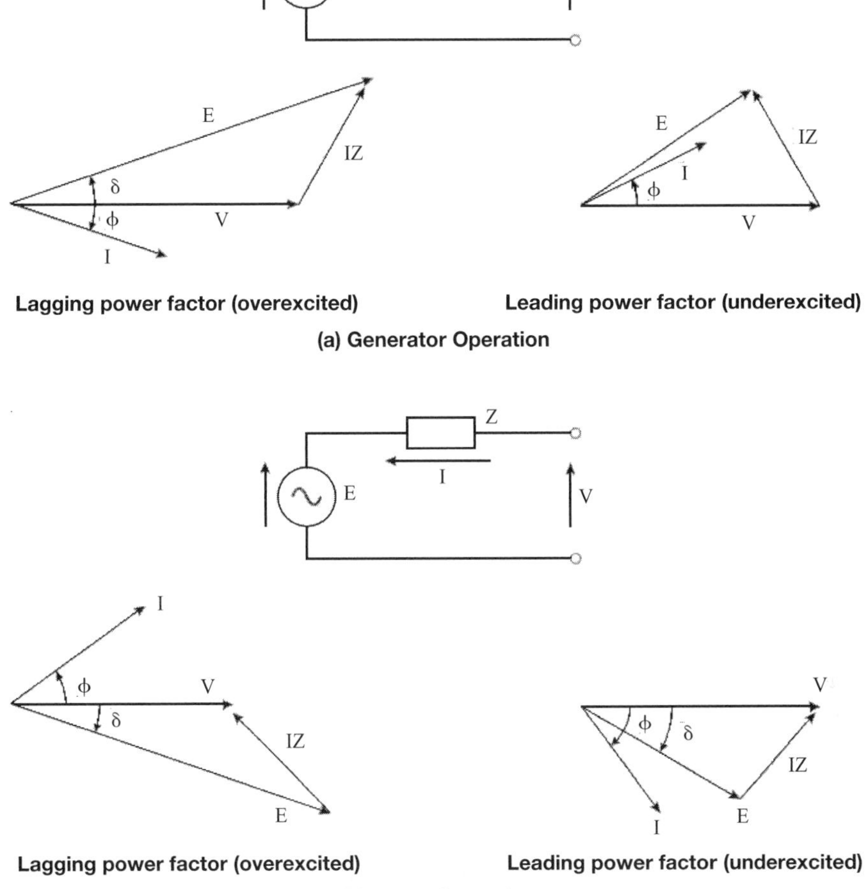

Fig. 1.36 Steady-state equivalent circuit and vector diagram.

conductors generating heat (a loss). However, there are a number of other sources within a working alternator that produce heat and, thus, losses. The following is a list of some of those sources of losses:

Winding Losses (Copper Losses)

- I^2R stator loss
- I^2R rotor loss
- Eddy and circulating current loss in windings (parasitic currents induced in the windings)

Iron Losses

- Mainly stator losses due to hysteresis loss and eddy current loss in stator laminations

Parasitic Eddy Losses

- Induced currents in all metallic components (bolts, frame, etc.)

Mechanical Losses

- Friction and windage loss
- Losses in fans and rotor and stator cooling vents
- Losses in bearings

Exogenous Losses

- Losses in auxiliary equipment
  - Excitation
  - Lubrication oil pumps
  - H_2 seal oil pumps
  - H_2 and water cooling pumps
- Iso-phase or lead losses

In the following chapters, these losses and their origin, control, and consequences to the machine's design and operation will be covered in detail.

ADDITIONAL READING

A wealth of literature exists for the reader interested in a more in-depth understanding of synchronous machine theory. The following is only but a very short list of readily available classic textbooks describing the operation and design of synchronous machines in a manner accessible to the uninitiated.

1. D. Zorbas, *Electric Machines—Principles, Applications, and Control Schematics.* West, 1989.
2. M. G. Say, *Alternating Current Machines.* Pitman Publishing, 1978.
3. T. Wildi, *Electrical Machines, Drives and Power Systems.* Prentice-Hall.
4. V. del Toro, *Electric Machines and Power Systems.* Prentice-Hall, 1985.
5. M. Liwschitz-Garik and C. C. Whipple, *Electric Machinery,* Vols. 1–2. Van Nostrand.
6. A. E. Fitzgerald, C. Kingsley and S. Umans, *Electric Machinery.* McGraw-Hill, 1990.
7. For a text describing the practical issues related to operation and maintenance of both turbogenerators and hydrogenerators, see Isidor Kerszenbaum, *Inspection of Large Synchronous Machines,* IEEE Press, 1996.

2

GENERATOR DESIGN AND CONSTRUCTION

The focus of this chapter is on the construction of the generator and its major individual components. In addition, issues that significantly influence the design of the various generator components are discussed. The class of generators under consideration are steam and gas turbine-driven generators, commonly called turbo-generators. These machines are generally used in nuclear and fossil-fueled power plants, cogeneration plants, and combustion turbine units. They range from relatively small machines of a few megawatts (MW) to very large generators with ratings up to 1800 MW. The generators particular to this category are of the two- and four-pole design employing round rotors, with rotational operating speeds of 3600 or 1800 rpm in North America, parts of Japan, and Asia, and 3000 or 1500 rpm in Europe, Africa, Australia, Asia, and South America.

The basic function of the generator is to convert mechanical power, delivered from the shaft of the turbine, into electrical power. Therefore, a generator is actually a rotating mechanical energy converter. The mechanical energy from the turbine is converted by means of a rotating magnetic field produced by direct current in the copper winding of the rotor or field, which generates three-phase alternating currents and voltages in the copper winding of the stator (armature). The stator winding is connected to terminals, which are in turn connected to the power system for delivery of the output power to the system.

As the system load demands more active power from the generator, more steam (or fuel in a combustion turbine) needs to be admitted to the turbine to increase power output. Hence, more energy is transmitted to the generator from the turbine, in the form of a torque. This torque is mechanical in nature, but electromagnetically coupled to the power system through the generator. The higher the power output, the higher the torque between turbine and generator.

The power output of the generator generally follows the load demand from the system. Therefore, the voltages and currents in the generator are continually changing based on the load demand. The generator design must be able to cope with large and fast load changes, which show up inside the machine as mechanical forces and temperatures change. The design must, therefore, incorporate electrical-current-carrying materials (i.e., copper and/or aluminum), magnetic-flux-carrying materials (i.e., highly permeable steels), insulating materials (organic and/or inorganic), structural members, and cooling media (i.e., gases and liquids), all working together under the operating conditions of a turbo-generator.

Since the turbo-generator is a synchronous machine, it operates at one very specific speed to produce a constant system frequency of 60 or 50 Hz, depending on the frequency of the grid to which it is connected. As a synchronous machine, a turbo-generator employs a steady magnetic flux passing radially across an airgap that exists between the rotor and the stator. (The term "airgap" is commonly used for air- and gas-cooled machines.) For the machines in this discussion, this means a magnetic flux distribution of two or four poles on the rotor. This flux pattern rotates with the rotor, as it spins at its synchronous speed. The rotating magnetic field moves past a three-phase, symmetrically distributed winding installed in the stator core, generating an alternating voltage in the stator winding. The voltage waveform created in each of the three phases of the stator winding is sinusoidal. The output of the stator winding is the three-phase power, delivered to the power system at the voltage generated in the stator winding.

In addition to the normal flux distribution in the main body of the generator, there are stray fluxes at the extreme ends of the generator that create fringing flux patterns and induce stray losses in the generator. The stray fluxes must be accounted for in the overall design.

Generators are made up of two basic members, the stator and the rotor, but the stator and rotor are themselves constructed from numerous parts. Rotors are the high-speed rotating member of the two, and they undergo severe dynamic mechanical loading as well as the electromagnetic and thermal loads. The most critical components in the generator are the retaining rings mounted on the rotor. These components are very carefully designed for high-stress operation. The stator is stationary, as the term suggests, but it also sees significant dynamic forces in terms of vibration and torsional loads, as well as the electromagnetic, thermal, and high-voltage loading. The most critical component of the stator is arguably the stator winding because it is a very high-cost item and it must be designed to handle all of the harsh effects described above. Most stator problems occur with the winding.

From the previous discussion, it becomes obvious that there are many issues to consider in generator design and each of these influences the performance of the overall machine. Design issues of high-voltage insulation, electrical currents (ac and dc), magnetic flux, heat production and cooling, mechanical forces, and vibrations all must be accounted for and made to work together for proper operation of a large generator.

(For a more in-depth discussion of the theory behind the operation of turbo-generators, refer to Chapters 1 and 4).

2.1 STATOR CORE

The stator core is made up of thin [0.014 inch (0.355 mm) or 0.019 inch (0.483 mm)] sheets of electrical-grade steel, with 3% to 4% silicon or grain oriented (see Fig. 2.1 and Fig. 2.2). There are numerous terms for these sheets, such as coreplate, punchings, laminations, or laminates. They are segmented, meaning that generally from 10 to 24 laminates (depending on whether the machine is two- or four-pole) are laid side by side to form a full 360 degree ring layer. Each of these layers is staggered relative to the locations of adjacent layers above and below, by the butted radial edges of the adjacent laminates in each ring layer (see Fig. 2.3). This staggering has a significant effect in increasing the mechanical integrity of the stator core as an assembled unit. This feature also has a beneficial effect in reducing shaft voltages on the rotor due to magnetic circuit dissymmetry.

Each lamination is insulated on both sides with an organic or inorganic compound of very thin dimension, generally about 0.001 of an inch. Organic compounds include varnish and inorganic coatings can be a layer of oxidation. The purpose of the inter-

Fig. 2.1 Individual coreplate segment or laminate (lamination) for a 4-pole stator core.

Fig. 2.2 Individual coreplate segment or laminate (lamination) for a two-pole stator core. This laminate is from the first packet of the stator core and shows clearly the slit-tooth arrangement for reduction of eddy currents and core-end heating effects. Also, this laminate is from an axially hydrogen-cooled core and so has numerous ventilation circles and elliptical holes punched in it. These line up to form full-length core cooling vents.

laminar insulation is to confine any induced eddy currents to a path along the same lamination where it is induced, without bridging into neighboring laminations. This has the effect of increasing the resistance to the eddy currents and reducing their magnitude, with an overall reduction of eddy-current loss and associated temperature rise. There are abnormal operating events such as overfluxing of the core (discussed elsewhere in this book) that can elevate the core temperature enough to compromise the interlaminar insulation. This can result in complete failure of the core.

The core is built up from thin laminates (laminations) to limit eddy-current losses in the core iron due to the alternating flux induced in the core during operation. Eddy-current activity in the core ends is further increased, due to stray and end-leakage flux from axial impingement on the core teeth in the end region, which will be discussed further on in this chapter. To reduce the eddy-current effect due to axial flux impingement in the core ends and its subsequent increase in core-end heating, the core teeth are slit, up to several inches axially inward from each end of the core (see Fig. 2.3).

The stator core of a turbo-generator contains tens of thousands of core laminations in the smaller machines and hundreds of thousands of laminations in the largest machines, which must be held tightly together, especially when the generator is laid sideways when installed and operating. To keep the laminated ring segments in line with each successive layer, they are fitted onto keybars in a stator frame structure (see Figs. 2.3 and 2.4). Then, to consolidate the core as a solid and stiff mass, they are clamped

Fig. 2.3 Lamination stacking in the stator, also showing the keybars at the extreme back of the core and vacant "through bolt" holes in the mid-yoke area of the core back.

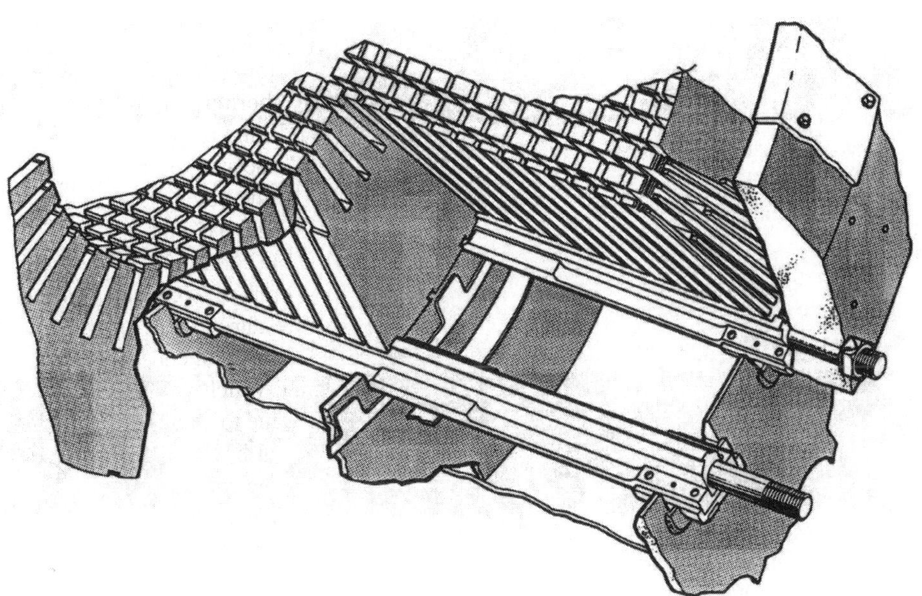

Fig. 2.4 Core-to-keybar mounting arrangement in the stator frame, with pressure plate at the end for clamping the core tight. (Courtesy of General Electric.)

axially. There are two basic methods used to achieve the axial clamping force. The most common is to use the keybar structure on which the core is mounted by dovetail slots in the core laminates at the core back, in conjunction with a large full-ring pressing plate at each end of the stator. The other is to additionally use "through bolts" installed through holes in the core yoke area (See Fig. 2.3) that extend the full axial length of the stator core and through the "pressing plate." In both of these methods, pressure is distributed over the end face or surface at both ends of the stator, using the large pressing plates.

In the area of the core teeth, however, the stator winding extends out of the core ends and, therefore, the core cannot be pressed directly by the pressing plate in that area of the core. For this reason, substantially strong fingers are installed on top of the core teeth at the ends, between the end-core laminate and the pressing plate to extend the pressing plate pressure to the "fingers" and the core teeth. This way, the required pressure is extended to the inner edge or bore portion of the stator core, as well as the yoke area. The intent is to provide equal pressure over the whole end surface and transmit it over the full axial length of the core to hold it tight so it will remain solid over its design life. In addition to the high-pressure loading in the factory of up to 250 psi, cores are often bonded between laminates as well, to assure the mechanical integrity of the core.

The support of the core must also accommodate the machine torque inherent with generator operation. Steady load appears as a steady torque to the core, but transients introduce peak torques for which the machine design must accommodate. These torques are transmitted to the stator frame via the keybars installed at the back of the core where the core dovetail slots are located. The keybars are in turn mounted on the stator frame to support the entire stator structure.

Because of the high levels of force and vibration experienced during normal and abnormal operation, mechanical damping arrangements or special spring mountings are often employed to minimize adverse affects to the core and the frame itself or to other parts of the stator.

In a two-pole generator, the stator core experiences a rotating inward magnetic pull at the two diametrically opposite locations of maximum airgap flux density, thus deflecting the core a few thousandths of an inch out of round to form an oval shape that is undetectable visually (see Fig. 2.5). However, the result is a primarily four-node vibration mode at a frequency that is twice the power system frequency (commonly called the "twice per rev" frequency). The resulting vibrations can be significant and, therefore, the generator must be isolated in such a way that the vibration is damped and not transmitted to the foundation. This is accomplished by mechanical dampers or spring-mounting the core. The detailed designs of the various manufacturers vary substantially, but they all must provide vibration isolation while supporting the weight of the core and handling both the steady and the peak transient torques developed in the stator core.

Cooling of the core is accomplished in large generators with the use of hydrogen gas. Radial "ducts" are provided in the core for this purpose. The losses generated in the core are dissipated to the cooling gas at the surface of the radial ducts. The width of the ducts and the thickness of the core packages are chosen as required by the ventila-

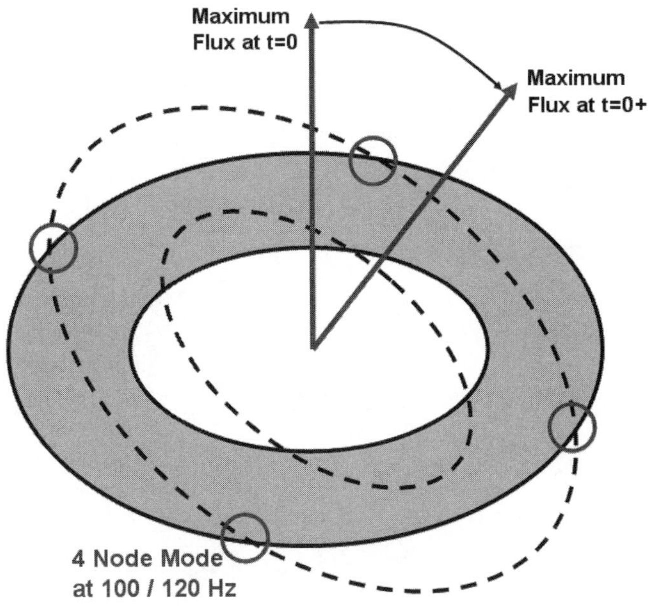

Fig. 2.5 Schematic representation of a two-pole generator core deformation as the flux rotates at synchronous speed. The actual deformation is invisible to the naked eye, and it is not more than a fraction of a millimeter. However, the impact on the vibrations of the frame is such that two-pole machines require the core to be mounted on springs or spring-like systems to protect the frame from the vibrations.

tion needs of the machine and the temperature permitted in the core. To allow for the radial ventilation ducts, "space blocks" are secured to a thicker core laminate to form a radial gas path from the stator bore to the back of core area, and vice versa (see Fig. 2.6).

Slots are provided in the stator core for installation of the stator winding. Core material is removed to accommodate the winding and, hence, the tooth flux densities are very high in this area. This affects the generated losses and heating in the core and the winding that the machine cooling must dissipate.

The portion of the core below the bottom of the winding slots is called the *yoke* or *core-back* area. The level of magnetic saturation and mechanical vibration governs the radial depth of the yoke in a two-pole generator. Generators with a relatively small back-of-core depth (four-pole) are usually lower fluxed machines. Machines with a large back-of-core depth (two-pole) generators are generally high-flux machines. Attention must be paid to both types in terms of the oval deformation effect on the stator core as well as the magnetic saturation and vibration.

With directly cooled stator winding machines, the armature reaction and stray flux are high and require flux shielding at the stator core ends to minimize the losses in the

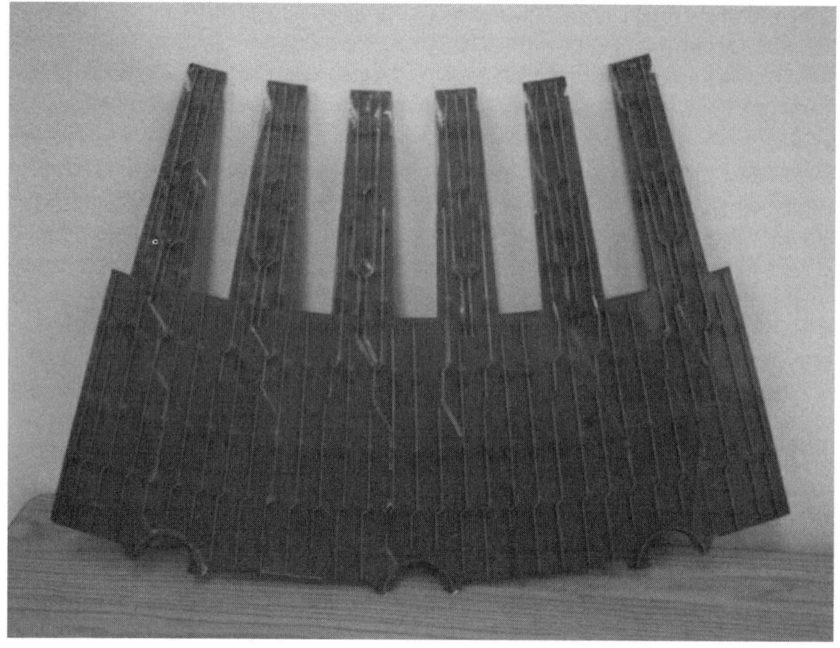

Fig. 2.6 Thick coreplate segment with space blocks.

core ends and the subsequent higher temperatures. Flux shielding is accomplished either by a copper, current-carrying shield or by a laminated magnetic steel flux-carrying shunt. This is discussed in more detail in Section 2.5.

2.2 STATOR FRAME

The basic purpose of the stator frame is to provide support for the stator core and to act as a pressure vessel for the hydrogen cooling gas in hydrogen-cooled generators. It also is segregated internally to create a ventilation circuit within the generator (see Fig. 2.7). In hydrogen-cooled units, the stator frame also supports the hydrogen coolers (heat exchangers) used to remove the heat absorbed by the hydrogen. In some larger generators, hydrogen gas is pressurized up to as high as 90 psi in more recent times.

The stator frame includes an outer shell, commonly called a *wrapper plate,* to which circumferential ribs and the keybars are attached. On the outside of the wrapper, there is a welded structure of footings attached, to secure the generator to a foundation. The footings and frame mounts are required to carry the weight of the generator and the rotational torque loads from the ovalizing effect of the magnetic fields in the generator. The frame structure must also be capable of withstanding abnormal events from the power system and generator faults, which cause high transient stresses in the frame.

Fig. 2.7 Stator core and frame cross section. (Courtesy of Siemens-Westinghouse.)

Since the frame provides the basic support for the stator core, it must be able to move with the core expansion and contraction from heating and the magnetic pulls associated with the rotating flux patterns in the core. To accommodate all this, the core-to-frame mechanical coupling is usually done with some flexibility installed. In turbogenerators, the frame design is usually accommodated with some isolation assembly or spring mounting of the core. This helps to dampen the inherent core and frame vibrations and keep them isolated from the foundation.

Frame stiffness and natural frequencies of vibration are important parameters due to the once-per-revolution (60 or 50 Hz) and twice-per-revolution (120 or 100 Hz) characteristics of the generators in conjunction with the stimulus from the power system frequency. Therefore, great care is taken to ensure that the natural frequencies of the core and the frame are not near 60 (50) or 120 (100) Hz.

To provide stiffness for the outer shell of the frame or casing, there are circumferential ribs welded to the wrapper at spaced axial intervals over the length of the stator. These are designed to give the stator frame the strength it needs for its intended purpose of supporting the core and acting as a pressure vessel for the hydrogen cooling gas (in H_2-cooled machines). The entire frame structure is dimensioned to ensure the correct strength and to avoid the natural frequencies of the once- and twice-per-revolution characteristics of the generator. The type of steel used in the frame is generally highly weldable material with good strength and low-temperature ductility (i.e., mild steel) to contain the internal hydrogen gas pressure.

The frame will have some inherent weak points that must be accounted for in the design. These result from the cutouts within the frame structure that connect the various portions of the ventilation path through the frame structure. Some of these include the hydrogen cooler ports, terminal-box structure, and instrumentation feed-throughs. The ventilation path must be provided, though, to direct cooling gas from the exit of the hydrogen coolers to the various parts of the stator core, the rotor, and the terminals, and then back through the hydrogen coolers to begin the circuit again. Of course, the sizing of the cutouts and cooling passages is determined by the amount of cooling required in each part of the generator. Similar arrangements exist in air-cooled machines.

Stator frames are also designed for lifting and handling. Once a machine is built, it must be delivered to a site and to do this requires transportation by any number of means such as a large truck for smaller machines and by rail and ship for large stators. The method of lifting is generally by craning. To do this, trunnion plates are bolted onto the side of the stator frame, which is generally a large plate itself and, thus, the crane cables can be attached there for lifting. The weight of some of the largest stators can reach up to 500 tons. It is, in fact, the transportation mode that governs the maximum size that a generator can be built to. There is no point in building a machine so big that it cannot be transported to the generation site. Therefore, such things as overall weight of the stator and the transport system must be accounted for, as well as the overall size dimensions. Some of the things to consider are clearance to railroad bridges, tunnels, station platforms, and other obstructions along the route.

There is also another issue with large generator design that seems to be minimally considered during the design phase, undoubtedly due to size and cost considerations. This is the issue of accessibility to the core back and other generally inaccessible areas of the stator. Regardless of the discussion above on size and transportation, the inescapable reality is that all large generators require maintenance and need to be accessible for inspection and to carry out any repairs or modifications that may be required in future. More often than one would like, problems such as core and frame vibrations occur, resulting in the need to inspect for damage and make repairs or modifications, and the core and keybar interface area is largely inaccessible. Attention to some "designed in" accessibility should be considered to accommodate future maintenance and inspections, although it is recognized that such accessibility would affect machine size and cost.

2.3 FLUX AND ARMATURE REACTION

The rated apparent power of a generator is proportional to the flux and the armature reaction, in the well-known relationship

$$\mathrm{MVA} = K M_{\mathrm{a}} \Phi P f$$

where

MVA = rated apparent power
K = a proportionality constant
M_{a} = armature reaction
Φ = magnetic flux per pole at rated voltage
P = number of poles

This is really the same as the product of the stator current and the stator terminal voltage. The stator or armature current is proportional to the armature reaction. The stator voltage is proportional to the flux. The field winding ampere-turns or field current at rated load is directly related to the level of armature reaction. Calculation of the flux per pole is described in Section 2.4 and the calculation of armature reaction is as follows:

$$\text{Armature reaction, } \mathbf{M_a} = \frac{N_{ph}}{\sqrt{2}P} \times \frac{N_{st}/N_{ph}/k}{K_p K_d} \times \mathbf{I}_a, \text{ ampere-turns}$$

where
N_{ph} = number of phases
P = number of poles
N_{st} = number of turns (number of slots)
k = number of parallel paths
K_p = winding pitch factor
K_d = winding distribution factor
$\mathbf{I}_a$ = stator current in amperes

One other basic relationship that governs the rating of a generator is the output coefficient. Simply put, the output of the generator increases with the square of the diameter of the rotor or stator bore, and with the length of the machine, based on the following relationship:

$$\text{Output coefficient} = \frac{\text{MVA}}{D_b^2 LS} \qquad \text{MVA min/m}^3$$

where
MVA = rated apparent power
D_b = diameter of the stator bore
L = length of the active iron in the stator
S = speed of the rotor in rpm

Specific generator ratings are accommodated in machine design by trading off the levels of magnetic flux against the level of armature reaction. The actual component dimensions as described above also play a role in optimizing designs of large generators. Therefore, a specific rating can be achieved by a relatively high value of flux and a low level of armature reaction, and vice versa, or some combination in between. Increasing the generator output at a specific combination of flux and armature reaction can also be done by making the machine longer. Using all these factors, one can design a machine to fit any output rating desired. However, when one parameter changes, it affects all the other parameters, some marginally but some others significantly.

Two additional formulas that help to describe the output of the generator in relative terms are *specific magnetic loading* and *specific electric loading.* These two formulas in their more basic form can be multiplied together to produce the *output coefficient* above:

$$\text{Specific magnetic loading} = \frac{P\Phi}{D_b L} \qquad \text{webers/m}^2$$

$$\text{Specific electric loading} = \frac{I_{st} N_c}{D_b k} \qquad \text{A/m}$$

where
P = number of poles
Φ = flux per pole in webers
D_b = stator bore diameter in meters
L = core iron effective length in meters
I_{st} = stator, single-phase current in amperes
N_c = total number of stator conductors
k = number of parallel paths in stator circuit

Using the above formulas, one can compare basic machine design outputs to determine which is more highly loaded in specific terms. For instance, if a machine is prone to high core-end heating, the specific electric loading of the generator is likely to be high relative to other machines, indicating that high stray losses are present. High stray losses can directly affect core-end flux penetration and, subsequently, the level of core-end heating.

Machines with a high level of flux require a relatively large volume of iron to carry the flux and a relatively small amount of copper to carry the stator and field currents. Such machines tend to be larger and more costly to build. Machines with a low level of flux require a relatively small volume of iron to carry the flux but a relatively large volume of copper in their windings. Such machines are termed "copper rich," and they increase the problem of heat removal from the windings. These machines tend to be smaller and less costly to build.

The per-unit transient and subtransient reactances, which play a significant role in the electrical performance of the generator connected to the power system, tend to be low with high-flux levels. The higher-flux generator will, therefore, tend to have a somewhat better inherent transient stability. It will also tend to have higher per-unit transient currents during severe disturbances and, therefore, higher winding forces and torques, than a lower-flux machine. To limit fault currents in the generator and, hence, the forces and torques, minimum values of subtransient reactance are usually specified. The subtransient reactance is a function of the stator leakage reactance and the effects of the rotor amortisseur or damper winding. In large turbo-generators with solid-pole rotors, the winding effects are negligible and the amortisseur contribution very small. To achieve a required higher reactance, one can build a stator with more slots and make the slots deeper and narrower to set the stator bars further away from the air gap (see Fig. 2.8). This has the effect of increasing the short-circuit ratio and improving stability, but requires a larger and more costly machine.

Mechanically, the larger higher-flux generator will have a larger, likely longer, rotor. Balancing can be more difficult with these. The moment of inertia of the larger rotor will help limit overspeed on loss of load.

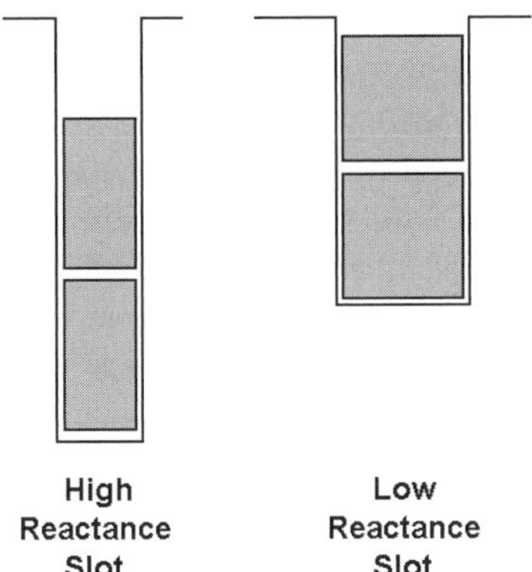

Fig. 2.8 Comparison of high- and low-reactance slot configurations.

2.4 ELECTROMAGNETICS

The electromagnetic circuits of both a two-pole and a four-pole turbo-generator are shown schematically in Figure 2.9*a* and *b*. The cross-sectional views presented in these figures show an airgap separating the slotted outer surfaces of both the rotor and the stator. The major elements of the magnetic circuit, as shown, are the solid steel rotor (including the rotor winding teeth/slots and poles and the main body below the slots and poles), the airgap (which constitutes the principal reluctance in the circuit), and the laminated steel stator core (including the stator teeth/slots and stator yoke below the slots).

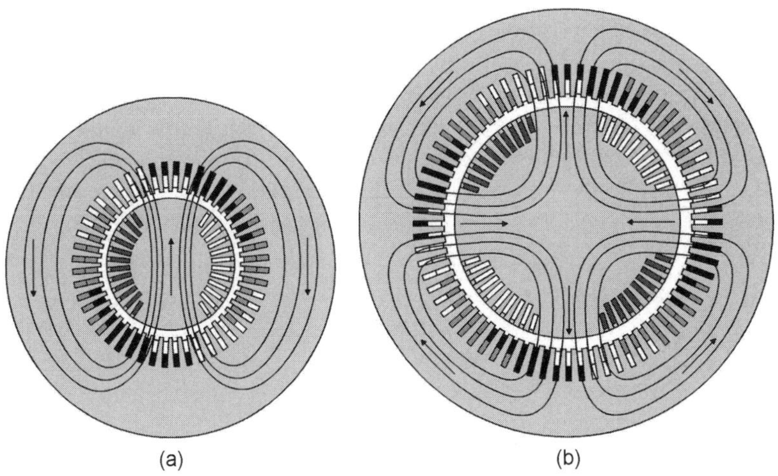

Fig. 2.9 (*a*) Two-pole generator flux pattern. (*b*) Four-pole generator flux pattern.

The airgap is the annular region between the rotor body and the stator core and probably has the largest influence on the electromagnetic design of the generator. Although the airgap is large to accommodate insertion of the rotor and its larger-diameter retaining rings, it is small in relative terms to the rest of the magnetic circuit of the generator. It has a major influence with regard to the reluctance of the total magnetic circuit and, hence, the overall stability of the generator. The airgap greatly affects the steady-state stability of the generator when connected to the power system by simple variation of the length of the space between the stator and rotor outer surfaces. The length of this airgap is used to determine the *short-circuit ratio* (SCR), which is calculated as described elsewhere in this book.

In practical terms, this means that the longer the airgap, the higher is the magnetic circuit reluctance, and, therefore, the higher the short-circuit ratio. Furthermore, the generator will tend to be more stable, producing higher ampere-turns (A-T) to achieve the required level of magnetic flux across the airgap. In real terms, this means more field current is required.

A reasonable rule of thumb for the ampere-turns of the generator as a whole is that the airgap generally accounts for about 90% of the total ampere-turns produced by the rotor. The remainder of the iron in the total magnetic circuit uses the other 10% and yet accounts for the majority of the electromagnetic flux path. This is because of the high permeability of the iron and high reluctance of air or hydrogen in the airgap. Therefore, a larger generator is required for higher apparent power output if the SCR ratio is to remain constant. This is because a larger rotor is required to handle the extra field current for the higher output and the airgap would be required to be about the same size to maintain the constant SCR. The airgap always needs to be large enough to permit insertion of the rotor through the stator bore with sufficient clearance for safe handling, taking into account the larger diameter of the retaining rings. This may limit the minimum possible SCR in some generators.

Electromagnetic finite element analysis is the preferred method to determine the actual magnetic field and its distribution in the machine. An example of a two-pole generator analysis on an *open circuit* is shown in Figures 2.10*a* and *b*, and at *full load* in Figures 2.11 and 2.12.

Although detailed generator design work usually requires finite element analysis for accuracy and refinement, for some calculations such as the required excitation level for a *flux test* on the generator stator core, a hand calculation of the total magnetic flux per pole in the generator is all that is needed, and it is determined as follows:

$$\text{Machine flux, } \Phi = \frac{V_{LL}k}{\sqrt{3}\left(\dfrac{2\pi}{\sqrt{2}}\right)fk_w N_p} = \frac{V_{LL}k}{7.7fk_w N_{ph}} \quad \text{webers}$$

where

V_{LL} = line-to-line stator terminal voltage in volts

k = number of stator winding parallel paths per phase

f = frequency

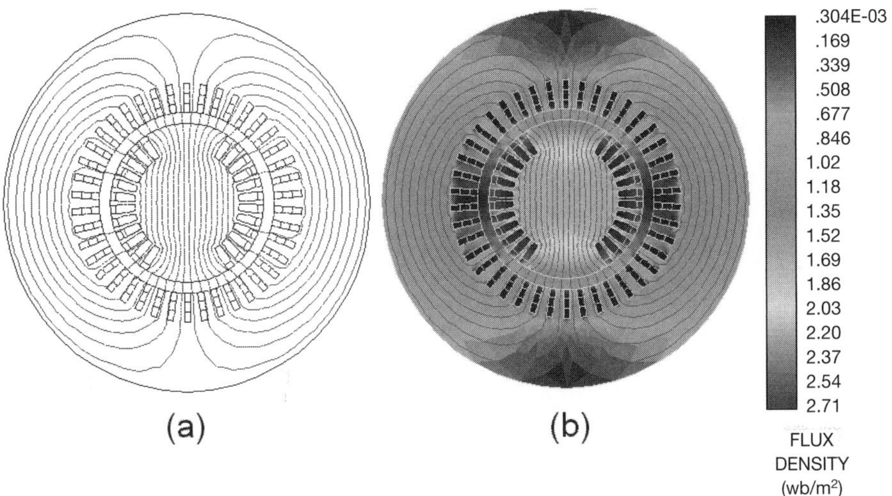

Fig. 2.10 (*a*) Two-pole generator, open-circuit flux distribution. (*b*) Two-pole generator, open-circuit flux density distribution. (See color insert.)

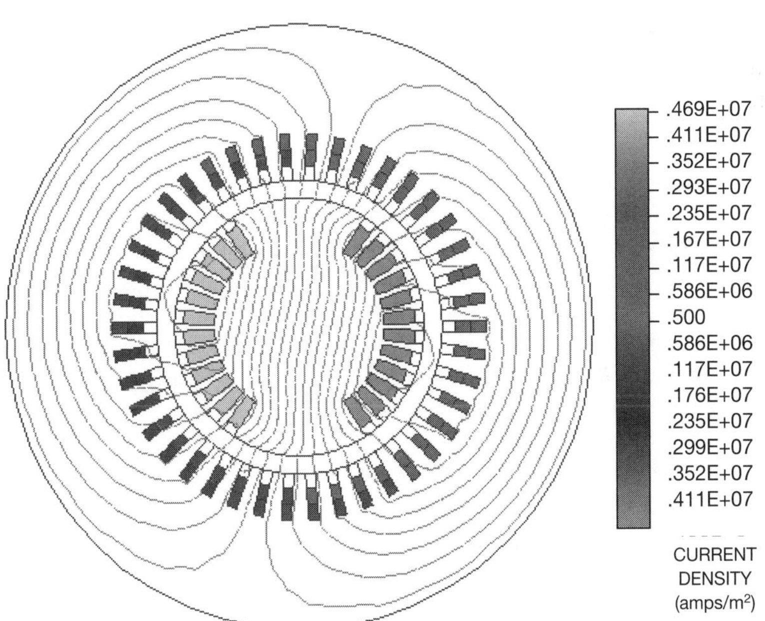

Fig. 2.11 Two-pole generator, full load current density and flux distribution. (See color insert.)

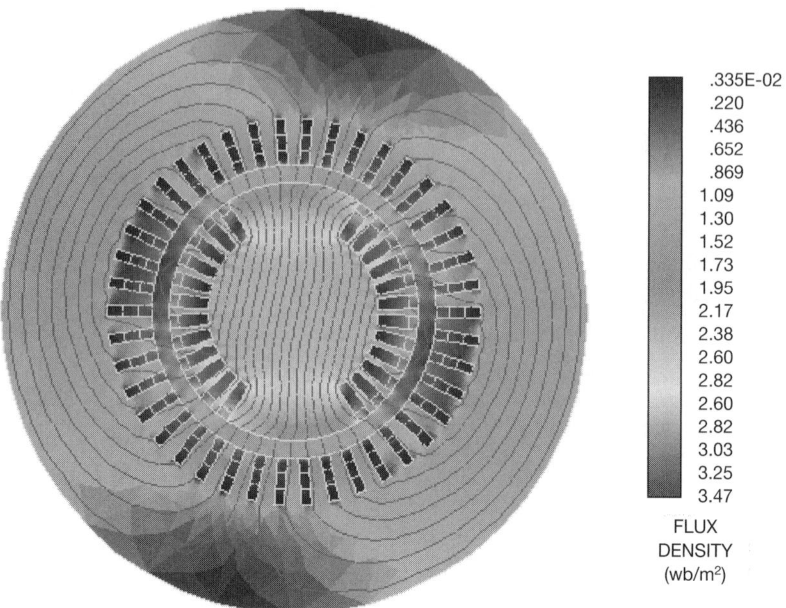

Fig. 2.12 Two-pole generator, full load flux and flux density distribution. (See color insert.)

k_w = stator winding factor (includes pitch and distribution)
N_{ph} = number of stator winding turns per phase

Note: The *winding factor* of the machine is largely concerned with reducing harmonic effects and wave shaping. It is comprised of the *pitch* and *distribution factors* (see Fig. 2.13). The pitch factor is determined from a winding diagram and depends on the number of slots separating the distance (the *coil span*) between connection from top and bottom bars in series, that is, a top bar in slot 1 connected to a bottom bar in slot 20 for a 48 slot, two pole, two parallel path machine would have a pitch of 19/24. The distribution factor is concerned with the arc distance over which the top and bottom bars for one pole of the winding are distributed, that is, the number of stator bars on one side of a parallel path in the winding as distributed over the total number of slots in the machine, or the eight top bars and the eight bottom bars per parallel side in the 48 slot machine example, would be distributed over 12 slots. The distribution factor is the ratio of the vector sum divided by the arithmetic sum of the stator coil emf's for this distribution.

To work out the winding factor (k_w) from the pitch (k_p) and distribution (k_d) factors, the following formula is used:

$$k_w = k_d \times k_p = \frac{\sin(\beta/2)}{\eta \sin(\gamma/2)} \times \sin(\rho\pi/2)$$

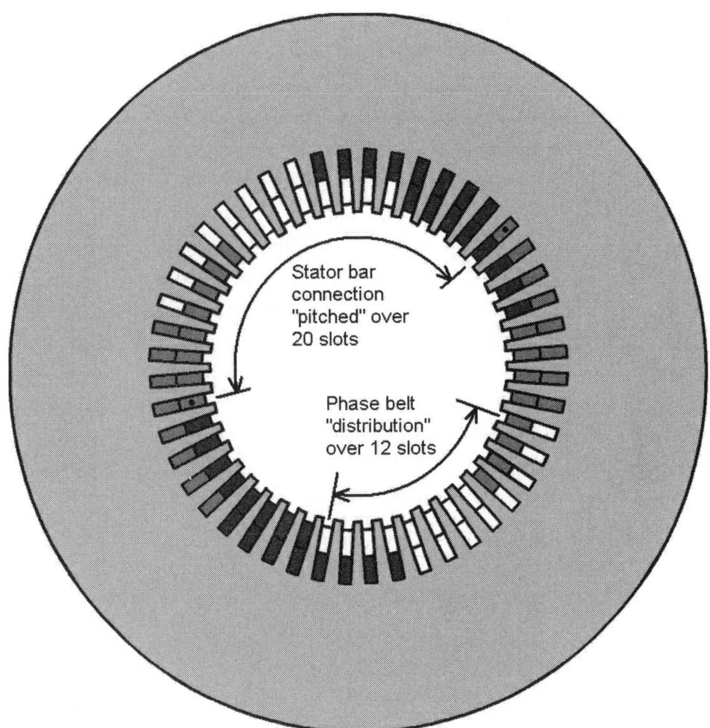

Fig. 2.13 Two-pole generator winding configuration showing pitch and distribution factors.

where

$\beta = \pi$/number of phases $= \pi/3$

η = number of slots/number of poles/number of phases

$\gamma = \pi$/(number of slots/number of poles)

ρ = stator winding pitch (from winding diagram)

The above formulas provide the basic level of machine flux required to achieve rated line-to-line terminal voltage in a generator, given a specific winding configuration. This formula will be elaborated on in Chapter 11 and an example provided for determining excitation levels in a *flux test.*

In the open-circuit example of Figure 2.10, the flux pattern is completely symmetrical about the pole axis of the rotor. Although the flux path includes the stator, the stator winding is on open circuit, and no current is flowing. Therefore there is no back emf (electromotive force) from the stator winding, and no electromagnetic torque coupling between the stator and rotor windings, and hence no load angle.

In the case where the generator is connected to the system, there is current in the stator winding (see Figs. 2.11 and 2.12), and significant torque is developed. As the turbine drives the rotor (in the counterclockwise direction in the example shown), the electromagnetic coupling between the stator and rotor windings tries to pull the rotor back in line with the axis of the stator poles. This difference in position of the stator and rotor pole axes creates a load angle that can be varied by changing the power output from the turbine and the field current for magnetic coupling between the stator and rotor. Increased field current pulls the rotor back toward the direct axis in the clockwise direction.

Generators are made with different power factor ratings. The most common are 0.90 and 0.85 lagging. Two machines of the same MVA rating will have different capability design parameters for the two different power factors. The 0.85 power factor machine will require more field current to achieve the same power at the 0.85 power factor. Hence, the machine is somewhat larger to accommodate a rotor that can handle more field current and is more costly to build. It is easy to see that design optimization to make the best utilization of the magnetic materials is a design priority.

The flux density becomes the driving factor for the amount of stator core material that is required. As can be seen from Figure 2.10*b* and Figure 2.12, the flux densities are different between open circuit and full load, but only marginally higher on load. However, there is considerable redistribution of the flux when the machine is on load, due to the stator currents. On open circuit, the stator core does not approach the electromagnetic loss limits of the iron, which are typically in the 2 tesla range in the stator teeth and under 1.5 tesla in the stator core back. Higher densities will be found in the rotor, but they are dc and magnetostatic in nature, and so do not cause high losses. That is to say, they are unidirectional as far as the rotor is concerned and so there are no hysteresis or eddy current losses in the rotor body due to the main flux. It is the alternating effect in the stator that designers are concerned with in this instance. Heating of the rotor components is a concern, but more for the effects of parasitic rotor surface and *negative-sequence* heating, as well as losses generated in the nose of the retaining ring due to the stator slot ripple effect (i.e., the variation of the main field due to the slotting of the stator). These will be discussed later in the book.

2.5 END-REGION EFFECTS AND FLUX SHIELDING

In addition to the electromagnetics of the main flux distribution across the airgap and in the main body of the stator and rotor, there are end-region effects of the flux produced. The end-region effects arise from the end windings of the stator and rotor, and the core-end fringe effects, as shown in an example of a magnetostatic finite element analysis of a generator end-region in Figure 2.14. The stray fluxes from these effects enter the stator core in the teeth and just behind the bottom of the slot, in an axial direction, and induce eddy currents in the core teeth the back of the core (see Fig. 2.15). In addition, the axial flux quickly turns in the direction of the main radial flux and adds to it. Hence, from this additional flux there is additional heating in the core ends (due to the sum of the main and stray fluxes and eddy-current heating) and magnetic saturation of the core end. Both effects vary with the power factor and ro-

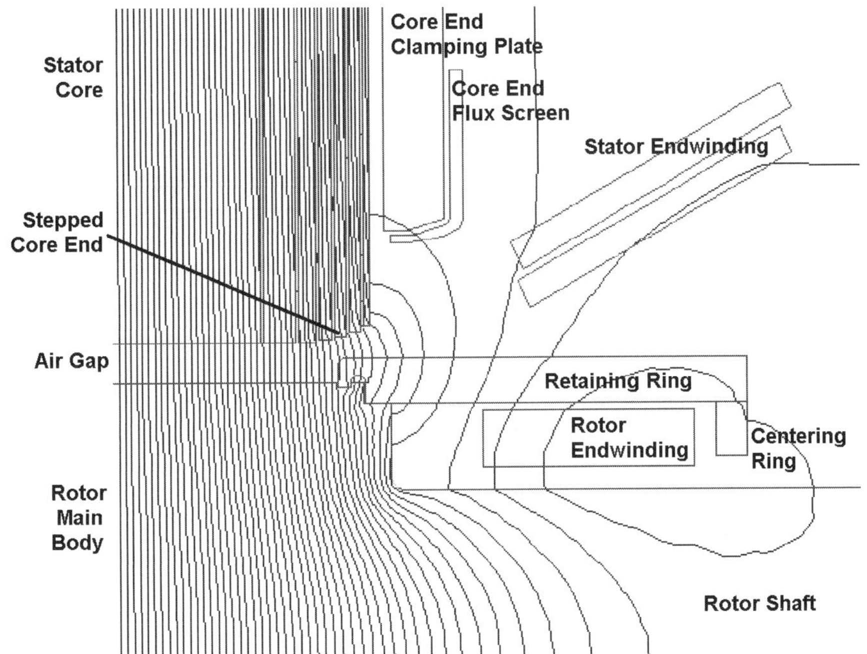

Fig. 2.14 Side view cross section showing the stray end effects of the magnetic field.

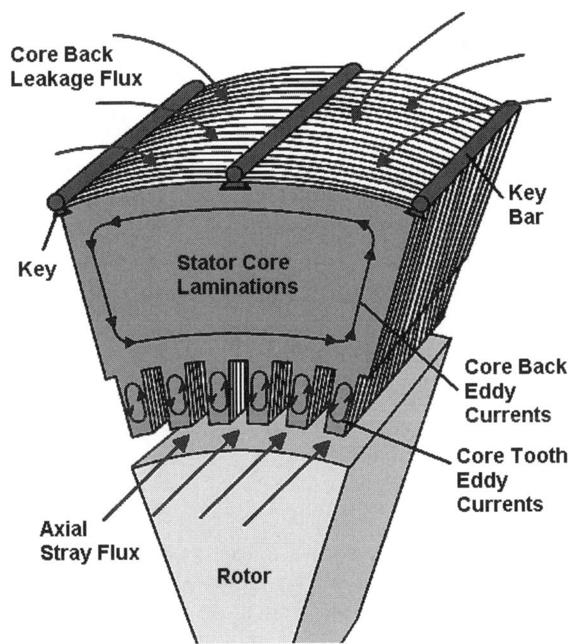

Fig. 2.15 View in 3D of the stray end effects of the magnetic field on the stator core end.

tor angle of the generator because of the change in interaction between the stator and rotor magnetic fields. In the lagging power factor range, these two fluxes tend to oppose and reduce the heating effect. In the leading power factor range, the fluxes tend to sum up (vector summation) and create higher losses in the core end (see Figs. 2.16 and 2.17).

To compensate for the increased heating effect, special slotting, stepping of core packets, and ventilation methods are employed on the core ends. In addition, special flux shielding or shunting is employed to stop excessive flux from entering the core in the axial direction. This is also shown in Figure 2.14.

Another design feature shown in Figure 2.14 is the thinner end packets of the stator core. These packets are usually much thinner than the main core body packets and there is more ventilation in this area. The thinner packets and increased ventilation add

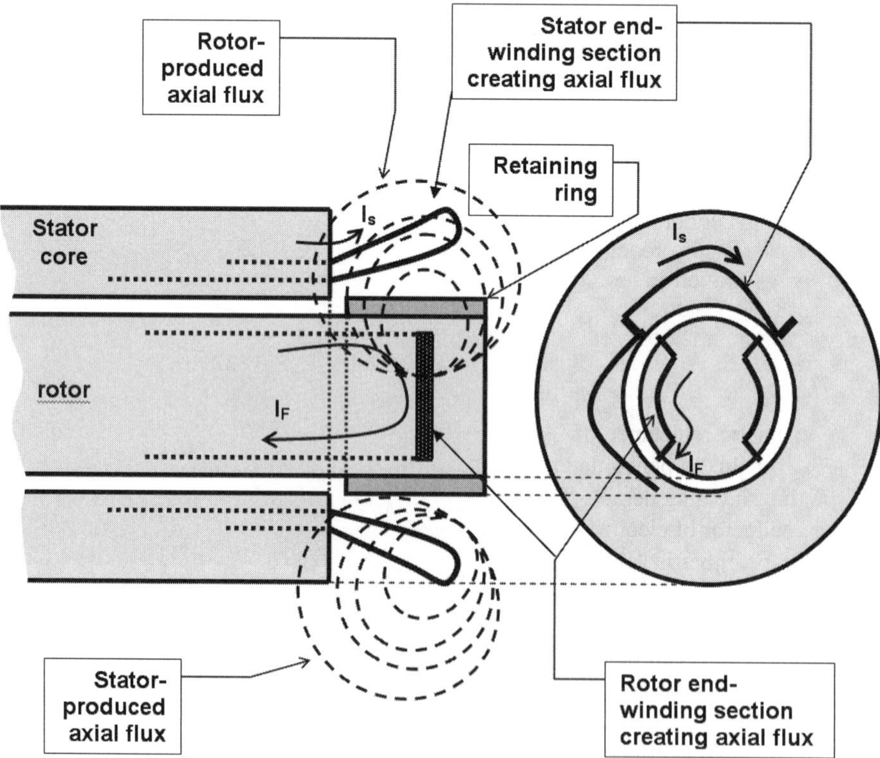

Fig 2.16 The figure schematically shows the end portion of both stator and rotor windings. In both cases, the end-winding have a component perpendicular to the axis of the generator. This component gives origin to an axial flux, part of which impinges on the stator core-ends. For simplicity-sake, the rotor-produced axial flux is shown on the top half of figure, and the stator-produce flux is shown on the bottom half of the figure. Obviously both fluxes act together in all regions of the end-core.

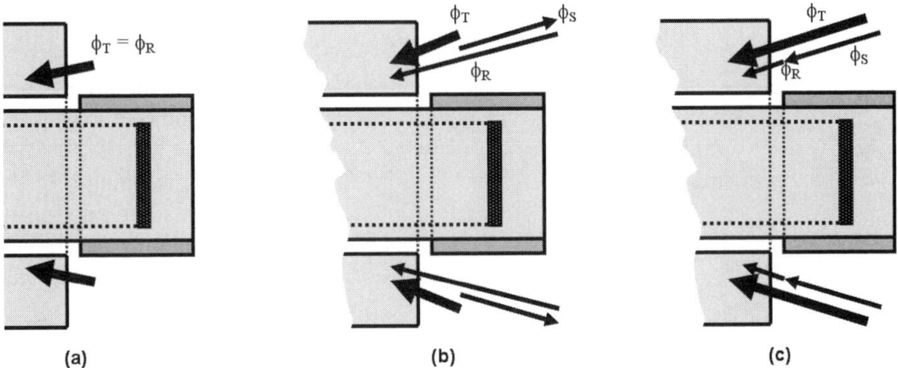

Fig. 2.17 The rotor- and stator-produced axial fluxes shown in Figure 2.16 are represented in this figure as vectors. The representation includes three cases. Case (a): No-load operation. In this condition, the only contribution to end-core flux is from the rotor (ϕ_R), which at rated voltage equals (ϕ_T), the total end-core axial flux (represented by the thick arrow). Case (b): Lagging operation. In this condition the rotor is overexcited, that is, most of the airgap flux (and end-core axial flux) is produced by the *field* winding. The stator-produced flux tends to oppose the rotor's flux (this phenomenon is called the armature reaction). The resulting total axial flux at the end-core region is somewhat smaller than in the no-load case. Case (c): Leading operation. In this condition the rotor is underexcited, that is, most of the airgap flux (and end-core axial flux) is due to the armature reaction. The resulting total axial flux at the end-core region is larger than in the other modes of operation. As the generator is made more and more underexcited, the end-core axial flux resultant increases to the extent that harmful temperatures of the core ends are reached.

to the magnetic reluctance of the core end in the axial direction and help reduce the tendency for flux to penetrate the core end axially.

The flux shielding depicted in Figure 2.14 is done with the use of copper, since it is a good conductor of electrical current but a poor conductor of magnetic flux. The flux shield can be either isolated from ground or grounded to the core-end clamping plate. Eddy currents are induced in the core-end flux shield or screen and cause the copper shield to heat up (See Fig. 2.18). The flux shield requires cooling because of the losses generated and this can be either gas or water cooling. The copper flux shield reduces the axial flux into the core end and the losses in the core that would otherwise result. Thus, high losses are generally associated with the flux shield and show themselves in a significant heating effect in that component. This is accounted for in the core-end design and the cooling circuit.

One additional design feature employed on most generators, in the core end, to reduce the higher losses in the stator teeth, is to split the teeth into smaller sections by cutting slits in the radial direction, away from the slot bottom (See Fig. 2.19). This reduces the eddy current effect in the teeth and, hence, the losses and heating effect. The slitting is usually done up to an axial distance into the core of about six inches (15 cm),

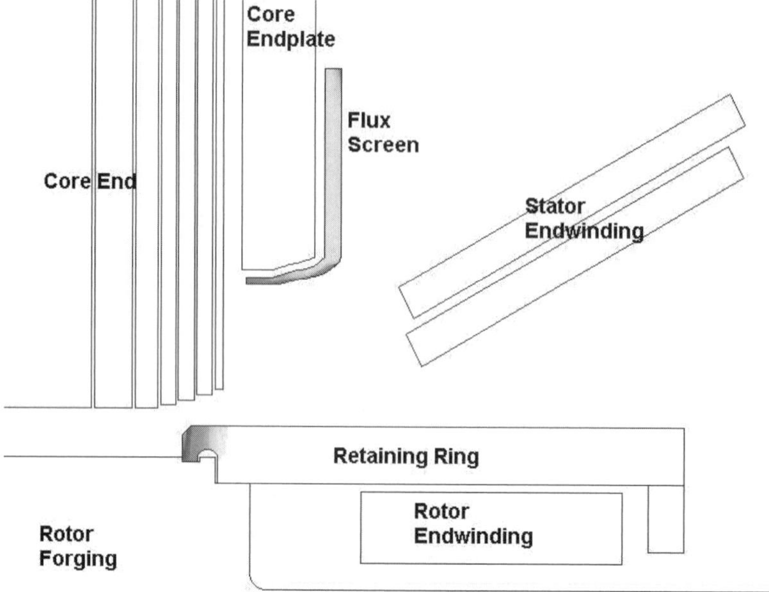

Fig. 2.18 Eddy-current heating in the flux screen and retaining ring nose are shown by the shaded areas from an eddy-current finite element analysis where there are increased losses. The loss in the retaining ring nose depends on the stator slot harmonics and, therefore, is a function of the distance between the nose of the retaining ring and the stator core, and the width of the stator slots and core teeth.

Fig. 2.19 Slitting of the stator teeth in the stepped end region of the core.

more or less, depending on the overall end-region design. The number of slits per tooth may also vary from one to three, generally.

The other method of shielding the axial flux from the core end is to use a flux shunt (See Fig. 2.20). This is done with a highly magnetically permeable and laminated arrangement of additional coreplate that diverts end-region flux from the core end, to avoid excessive heating in the stator-core iron. The high losses produced in the flux shunt are accounted for in the core-end design and cooling circuit of the machine.

Another significant end-region effect that can occur in a machine is *back-of-core burning.* This occurs when there is arcing at the core back between the core and keybar. To stop arcing from occurring, a *shorting-strap* is welded from keybar to keybar around the circumference of the stator and at both ends (see Fig. 2.21). Essentially, these straps short any currents flowing in the keybars to the adjacent keybar, as in a squirrel-cage rotor type of arrangement. This keeps the currents from flowing through the core when passing from keybar to keybar. The currents in the keybars are due to the main flux leaking out of the back of core and linking the keybars (see Fig. 2.15). It is more pronounced in the core-end region because of the additional flux present in that region of the machine. The actual core back burning occurs when there is current interruption between the core and the keybars due to vibration effects. Core and frame vibration, especially when the two components are vibrating out of phase, causes a make-and-break contact at the core ends where the keybar currents try to transfer to the

Fig. 2.20 Laminated flux shunt on the stator core end.

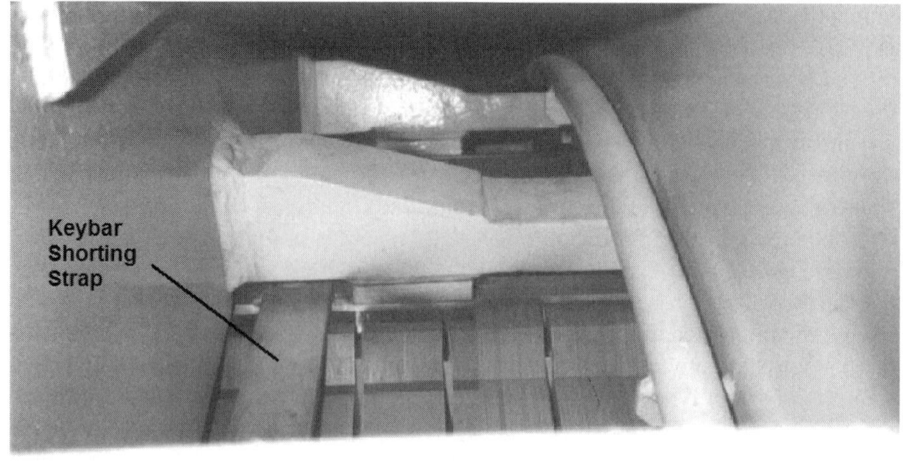

Fig. 2.21 Keybar shorting strap.

next keybar, circumferentially, and arcing and burning results. If positive contact were maintained, no burning would occur.

One of the other reasons to ensure that the core end is well flux shielded is so that it does not become oversaturated in magnetic flux. Oversaturation has the effect of increasing back-of-core burning, especially in the leading power-factor range.

2.6 STATOR CORE AND FRAME FORCES

As discussed above, the principle function of the stator core is to carry electromagnetic flux. The core must handle magnetic field flux densities in the stator teeth and in the core-back or yoke area. The magnetic field is revolving, so it creates an alternating voltage and current effect in the generator components, which is a source of high losses and heating. This alternating effect also causes vibration of the core at the rotational frequency and with harmonics, due to the nature of the flux patterns. In a two-pole generator, the driving frequency is 60 (50) Hz and there is a 120 (100) Hz (twice per revolution) component due to the four-node pattern of the flux and the rotational speed of 3600 (3000) rpm. In a four-pole machine, the driving frequencies and harmonics produce the same result as in the two-pole machine, due to the rotational speed being half that of the two-pole machine [i.e., a rotational speed of 1800 (1500) rpm] and an eight-node pattern is produced.

The four- and eight-node patterns can be seen in Figure 2.9a and b, where there are two and four areas of high-flux density in the core-back area and two and four areas of minimum density, at any given point in time, as the flux patterns rotate at the rated speed. This causes the core to be distorted by the electromagnetic pull, in and out, in the radial direction. The result is vibration of the core and, subsequently, the frame. Because of the difference in the nodal patterns between the two- and four-pole machines, the vibration levels of the four-pole machine is substantially less than the two-pole. The vibration level of a four-pole machine can be as low as 10% that of a two-pole machine.

Because of the inherent vibration and the large forces involved, the core must be held solidly together so that there are no natural frequencies near the once- and twice-per-revolution forcing frequencies. Designers take great care to ensure that the natural frequencies of the core are not near 60 (50) or 120 (100) Hz. It is desirable to keep the natural frequencies at least (20% away from the once- and twice-per-revolution frequencies.

There also is a large rotational torque created by the electromagnetic coupling of the rotor and stator across the airgap. This is in the direction of rotor rotation. The torque due to the magnetic field in the stator core iron is transmitted to the core frame via the keybar structure at the core back. Therefore, the stator frame and foundation must be capable of withstanding this torque, as well as large changes in torque when there are transient upsets in the system or the machine. Instantaneous changes can cause impacting between the core and frame components, and severe damage can result if the structure is not designed to handle these massive forces.

The natural vibration inherent in the core must also be accounted for in the core-to-frame coupling. Heating and cooling effects in the core and frame materials will also affect this coupling and vibration, due to differences and rates of thermal expansion and contraction in the core and frame components. Impacting and looseness between the core and the frame will allow arcing to occur between the keybars and core. Make-and-break contact at the core-to-keybar interface can result in significant burning at the core back and melting of the core iron, in addition to back-of-core leakage flux and end-iron effects. Again, shorting straps are employed at the back of core to short the current flowing in the keybars. Therefore, if the core-to-keybar contact is not sufficient, arcing and burning may occur.

2.7 STATOR WINDINGS

The stator winding is made up of insulated copper conductor bars that are distributed around the inside diameter of the stator core, commonly called the stator bore. The winding is installed in equally spaced slots in the core to ensure symmetrical flux linkage with the field produced by the rotor. Each slot contains two conductor bars, one on top of the other (see Fig. 2.22 and Fig. 2.23). These are generally referred to as top and bottom bars. Top bars are the ones nearest the slot opening (just under the wedge) and the bottom bars are the ones at the slot bottom. The core area between slots is generally called a core tooth.

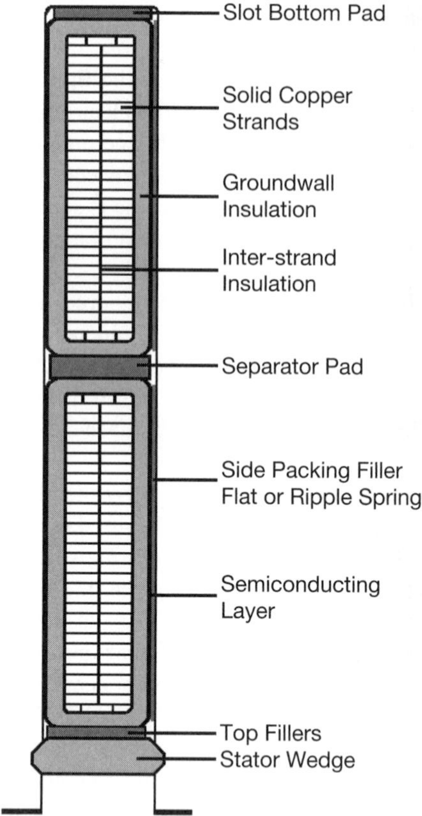

Fig. 2.22 Stator slot cross section with "indirectly cooled" stator bars and wedging system installed.

The stator winding is then divided into three phases, which are almost always wye connected. Wye connection is done to allow a neural grounding point and for relay protection of the winding. The three phases are connected to create symmetry between them in the 360 degree arc of the stator bore. The distribution of the winding is done in such a way as to produce a 120 degree difference in voltage peaks from one phase to the other, hence the term "three-phase voltage." Each of the three phases may have one or more parallel circuits within the phase. The parallels can be connected in series or parallel, or a combination of both if it is a four-pole generator. This will be discussed in the next section. The parallels in all of the phases are essentially equal, on average, in their performance in the machine. Therefore, they each "see" equal voltage and current, and magnitudes and phase angles, when averaged over one alternating cycle.

The stator bars in any particular phase group are arranged such that there are parallel paths that overlap between top and bottom bars (see Fig. 2.24). The overlap is stag-

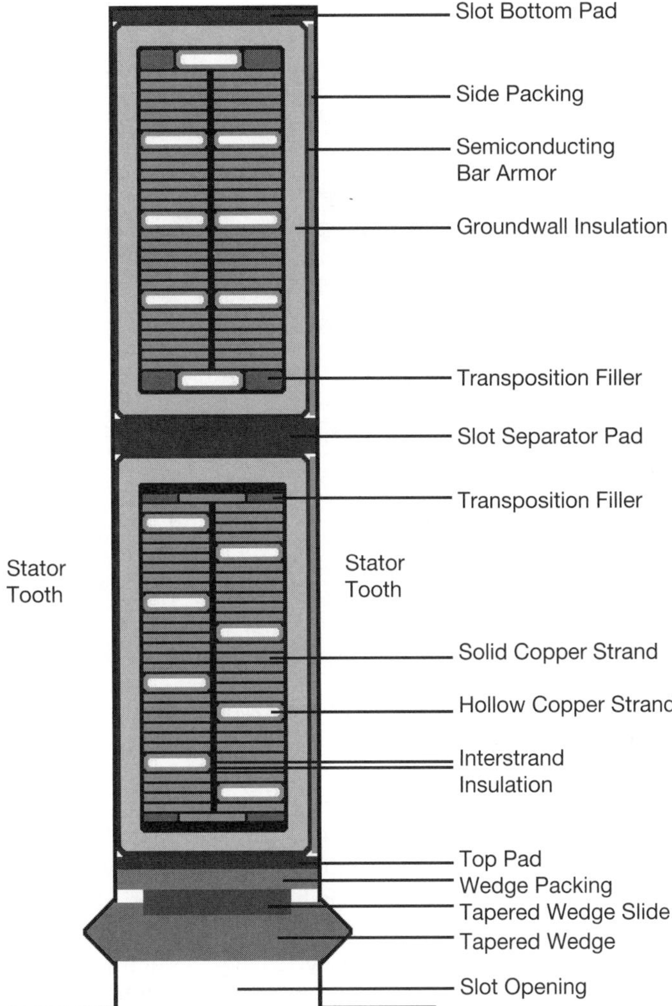

Fig. 2.23 Stator slot cross section with "direct water cooled" stator bars and wedging system installed.

gered between the top and bottom bars. The top bars on one side of the stator bore are connected to the bottom bars on the other side of the bore in one direction, whereas the bottom bars are connected in the other direction on the opposite side of the stator. This connection with the bars on the other side of the stator creates a "reach" or "pitch" of a certain number of slots. The pitch is, therefore, the number of slots that the stator bars have to reach in the stator bore arc, separating the two bars to be connected. This is always less than 180 degrees and it is done to assist in reducing the harmonics induced in the stator winding.

Fig. 2.24 Stator end winding showing a two-pole winding overlap in the ends and the phase connectors at the excitation end of the stator.

Once connected, the stator bars belonging to the same phase form a single coil or turn. The total width of the overlapping parallels is called the "breadth." The combination of pitch and breadth create a "winding or distribution factor." The distribution factor is used to minimize the harmonic content of the generated voltage. In the case of a two-parallel-path winding, these may be connected in series or parallel outside the stator bore, at the termination end of the generator (see Fig. 2.24). The connection type will depend on a number of other design issues regarding current-carrying ability of the copper in the winding. This will be discussed in the next section.

In a two-parallel-path, three-phase-winding, alternating voltage is created by the action of the rotor field as it moves past these windings. Since there is a plus and minus, or north and south, to the rotating magnetic field, opposite-polarity currents flow on each side of the stator bore in the distributed winding. The currents normally flowing in large turbo-generators can be on the order of thousands of amperes. Due to the very high currents, the conductor bars in a turbo-generator have a large cross-sectional area. In addition, they usually have one single turn per bar, as opposed to motors or small generators that have multiple-turn bars or coils. These stator or conductor bars are also very rigid and do not bend unless significant force is exerted on them.

The high current capacities of copper in the stator bars generate significant heat. The losses due to the flowing currents are called I^2R losses in the winding. Controlling the losses in the stator winding requires careful design consideration because of the variance in magnetic field from the stator bore toward the slot bottom. The magnetic field tends to be more intense toward the top of the slot and, therefore, the top bars generally produce more heat than the bottom bars. Within the bars themselves, there are also eddy currents flowing in each bar caused by the localized-leakage magnetic field.

To reduce the effect of the eddy currents within each individual stator bar, the conductors are made up of numerous copper "strands" (see Fig. 2.25*a*, *b*, and *c*). This is similar to the reasoning behind the stator core being made up of laminations rather than a solid mass of steel. However, although the strands are insulated from one another in the bar, they are eventually connected at each end of the stator bar. Therefore, additional circulating current could flow from the top to the bottom strands in a single bar. This is due to the difference in the magnetic field from the top to bottom of the slot. To reduce the effect of the circulating currents, the strands are "Roebel transposed" in each bar (see Figs. 2.26 and 2.27). Roebel transposition of the copper strands refers to the repositioning of each strand in the stator bar stack such that it occupies each position in the stack at least once over the full length of the stator bar.

Roebel transpositions are mainly 360 and 540 degree. A 360 degree transposition means that each strand occupies each position once over the length of the bar, and a 540 degree transposition means that each strand occupies each position one-and-a half times. The 360 transposition is generally done in the slot only and the 540 transposition is done out to the very ends of the stator bars, and in the curved end-winding portion as well.

There is another problem with circulating currents that occurs in double-stack stator bars, that is, designs where there are two separate Roebel-transposed stacks side by

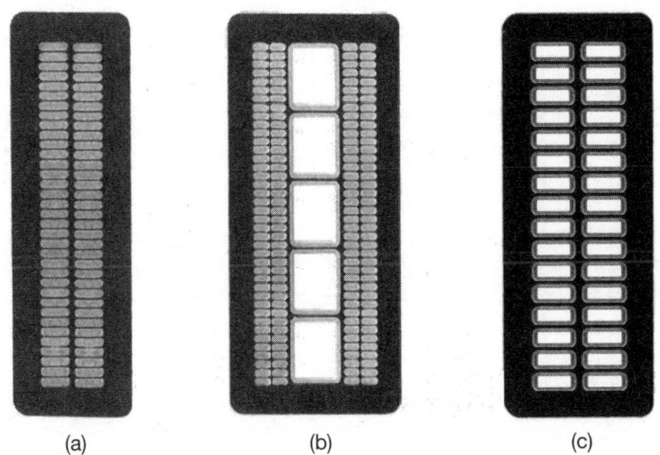

Fig. 2.25 Stator conductor bar cross sections. (*a*) Indirectly cooled stator conductor bar; (*b*) directly gas-cooled stator conductor bar; (*c*) directly water-cooled stator conductor bar. (Courtesy of General Electric.)

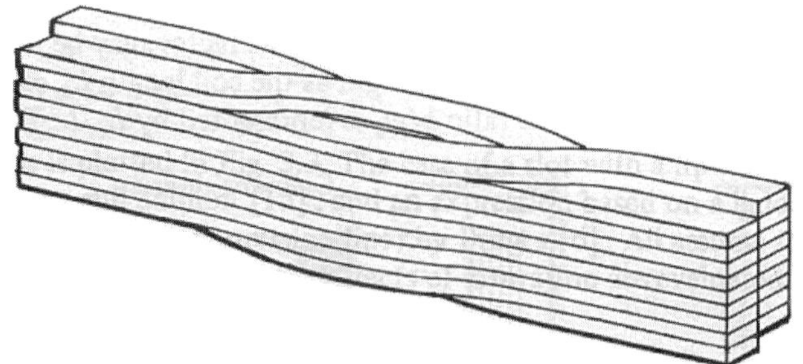

Fig. 2.26 Roebel transposition, 3D view. (Courtesy of General Electric.)

side, thus giving four strand widths in the bar. Although it appears that the stacks are so close together that there would be no difference in magnetic field from one side of the bar to the other, this is not true. In fact, there is a significant difference, because the magnetic field does cut between the stacks such that a certain amount of circulating current occurs in a double-stack stator bar. The amount of circulating current from one stack to the other in a single bar can cause temperature differentials from on stack to the other up to 10°C on average. Figure 2.28 shows a normal type of temperature profile for a double-stacked stator bar with separate Roebel transpositions for each stack. The temperature difference from one side to the other has the overall effect of reducing

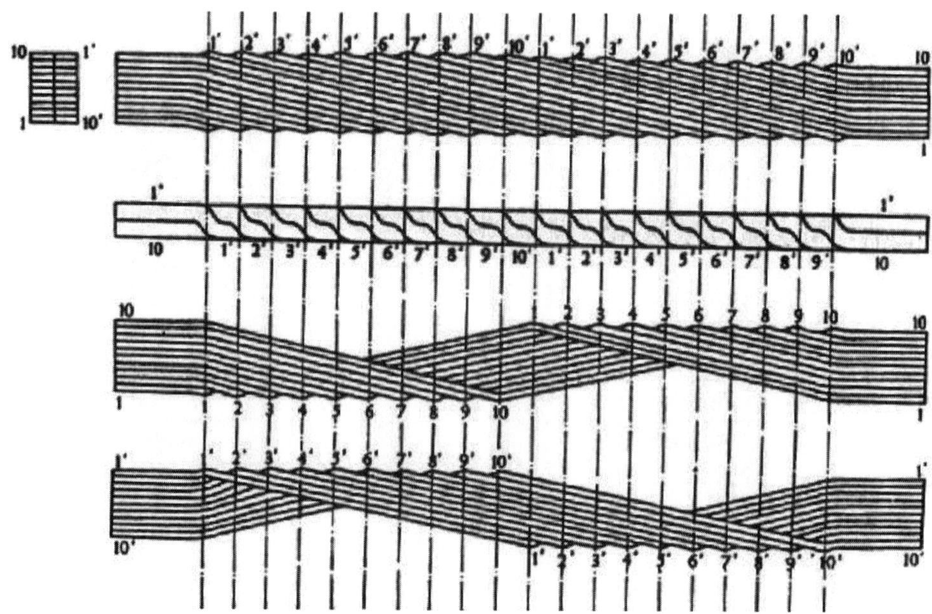

Fig. 2.27 Roebel transposition map. (Courtesy of Alstom Power Inc.)

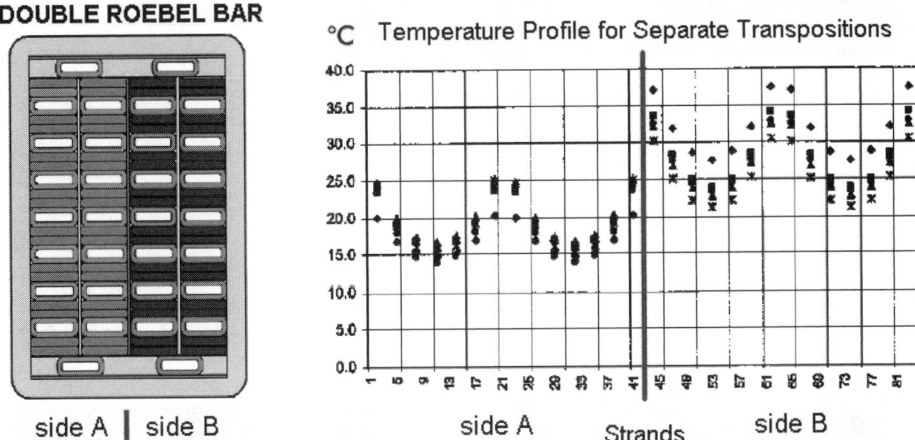

Fig. 2.28 Temperature profile of a double-stack stator bar with separate Roebel transposed stacks. The average temperature difference between side A and side B of the bar example shown is about 10°C. (Courtesy of Alstom Power Inc.)

the available stator current output because the maximum hot-spot temperature is raised by about 10°C during operation. Obviously, eliminating this temperature difference would allow higher output from the same slot dimensions of a bar if the temperature hot spots could be reduced.

A method has been developed that does allow for a more even temperature distribution across the strand stack and it is termed cross roebel (see Fig. 2.29). The method simply has all strands transposed such that they occupy both sides of the bar stack and not just one side. In this method, the two strand stacks are balanced electromagnetically from side to side as well, and this eliminates the stack-to-stack circulating currents (see Fig. 2.30).

There are many ways of designing stator conductor bars, depending on the size and cooling method required for the machine. Cooling is particularly critical in designing machines for higher outputs. In this regard, *direct cooling* is the most desirable type of cooling because it increases the generator stator's current-carrying capability consider-

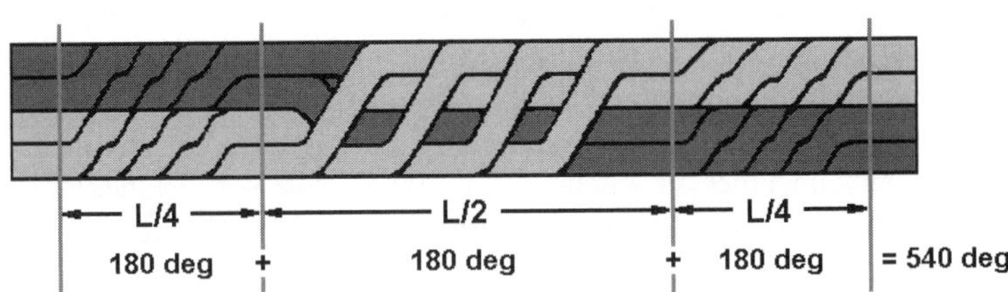

Fig. 2.29 Cross Roebel transposition of a double-stack stator bar. (Courtesy of Alstom Power Inc.)

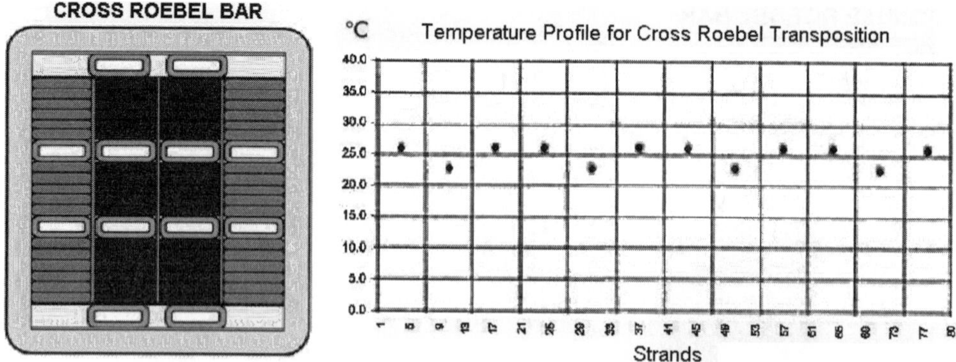

Fig. 2.30 Cross Roebel transposition-temperature profile of a double-stack stator bar. All cooling strands are temperature equalized due to the elimination of stack-to-stack circulating currents. (Courtesy of Alstom Power Inc.)

ably. The advantage of this is to reduce flux levels and, hence, the physical size and weight of the generator. The basic limit for conventionally cooled generators (i.e., indirect cooling with gas) is generally in the 400 MVA range. However, there are now newer high thermal conductivity ground-wall insulation systems that are reported to allow up to 600 MVA [1]. With direct conductor cooling it is now possible to build generators up to 1900 MW [2].

In indirectly cooled machines, the strands within the conductor bars are all solid and the heat generated in the conductors is removed by conduction through the ground-wall insulation to the stator core. The size of the generator is significantly limited by the temperature conduction through the ground-wall insulation to the stator core.

In directly gas-cooled bars, the gas passes from end to end in rectangular ducts with low conductivity and nonmagnetic metal walls. These cooling ducts are insulated from the copper conductor strands since they are assembled within the conductor bar stack alongside the actual electrical conductors. This type of direct cooling is more effective than indirect cooling, which is dependent on heat transfer through the ground-wall insulation, and allows a large-output machine to be built.

In direct water-cooled bars, the copper strands are made hollow, to carry liquid coolant. The stands are generally rectangular in shape to allow stacking and they are each individually insulated from one another and Roebel transposed. There are numerous different stator bar stand arrangements, including all hollow strands and a mix of solid and hollow (see Fig. 2.31*a*, *b*, and *c*). In the mixed arrangement, the hollow strands are evenly interspersed among solid strands. The solid stands are generally thinner so that they may have lower eddy-current loss. It is also possible to make them thinner because there is no concern of crushing the coolant path and restricting coolant flow. The strands can be arranged in various combinations to produce more efficient winding designs and, hence, even larger-output generator designs than with direct gas-cooled windings.

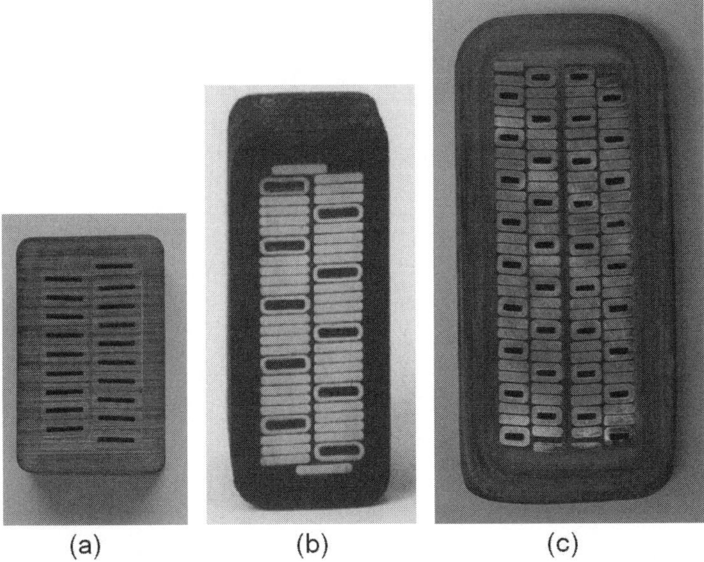

Fig. 2.31 Various stator bar designs for directly cooled generators. (*a*) Single stack—all hollow strands; (*b*) single stack—mixed strands; (*c*) double stack—mixed strands.

In directly cooled stators, it is possible to increase the current density in the copper winding of the stator to achieve higher ratings. Trade-offs are also made between slot sizes and winding configurations to find the optimum terminal voltage level versus the current flowing in the stator winding, all in consideration with keeping magnetic flux densities in the stator iron at manageable levels.

Because the stator current densities in directly cooled windings are so much higher than in indirectly cooled windings, designers must also consider the effect of transients and temperature rise. Considerations of reactance and stability also come into play and, therefore, so do short-circuit ratio and excitation performance.

Some modern generator designs mix solid copper stands for conduction of the electrical current and hollow stainless steel strands for carrying the coolant (see Fig. 2.32). This design has been in service for the last 30 years and has been successful. The use of stainless steel strands for cooling has eliminated certain industry problems of copper erosion and corrosion in the stator bars. The mixed steel and copper stator bars also tend to be more rigid than fully copper bars and allow higher wedging pressures in the slot.

In direct water-cooled machines, the cooling method dictates the need for an external system to remove the heat picked up by the stator cooling water after it passes through the stator winding. Therefore, an external system is attached to the generator that employs heat exchangers to accomplish this function. To circulate the water, pumps and a piping system are provided. In addition a *filtering system* is provided to remove any large particles suspended in the SCW (stator cooling water) that can cause blockage within the stator windings inside the generator. Since the water is in contact

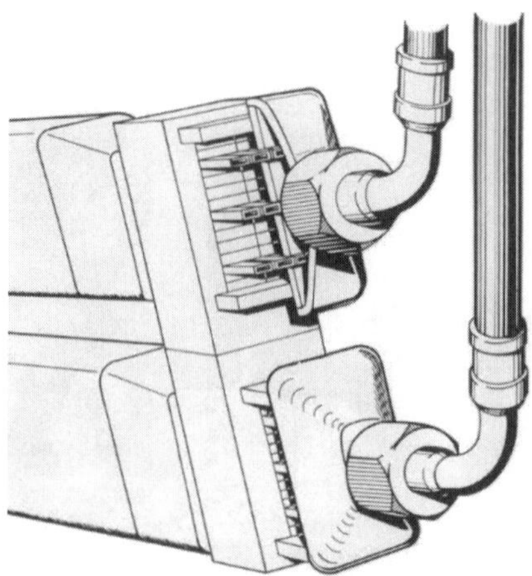

Fig. 2.32 Stainless steel cooling strand type of bar. (Courtesy of Alstom Power Inc.)

with current-carrying copper conductors, which are also operating at voltage levels from ground potential up to 27 kV, the water must be kept absolutely as pure as possible to avoid flashovers by conduction through the water. To maintain pure water, a deionizing system is provided.

The basic functions of electrical insulation in the stator winding are to maintain ground insulation between the conductors and the stator core and other grounded objects, and to maintain insulation between turns of multiturn coils and between the strands within a turn.

The ground-wall insulation must be designed to withstand line-to-line ac voltages over the entire life of the generator. In addition, it must be capable of withstanding overvoltages from system faults. The turn insulation must withstand normal coil voltage over its lifetime, with substantial short-time overvoltages in the event of a steep-front voltage surge. Strand insulation is exposed to only a few volts with brief overvoltages during occasional high-current transients.

A high-resistance coating or "semiconducting" system is applied to the ground insulation in the slot to control the voltage distribution over the length of the slot (see Fig. 2.22 and Fig. 2.23). In addition, a special "grading" system is applied to the bars over a short distance of several inches at the bar exit from the slot, at the end of the core, to produce a gradual voltage drop in the end winding and prevent destructive electrical discharges in the transition area from the ground potential at the core to high voltage levels out in the end winding. To ensure good contact between the stator bar and the core in the slots, a side-packing filler is also generally inserted along side both top and bottom stator bars (see Fig. 2.22 and Fig. 2.23). The side filler is impregnated with semiconducting material to assist with the electrical contact to the stator core. The base material is usually made up of strong resin-filled woven glass material. It may be a flat piece (See Fig. 2.33), but a *ripple-spring filler* is now commonly used to ensure continual pressure and contact (see Fig. 2.34).

Fig. 2.33 Flat side-packing piece with semiconducting impregnation.

Due to the current flowing in the stator bars, there is a reaction force in each slot, which varies according to the level of current and direction of flow at any instant. This creates forces between bars that are both repulsive and attractive at any give time in the alternating cycle. Therefore, the slot section of a stator conductor bar "sees" significant and constant vibration forces at the twice-per-revolution frequency. This is due to the "cross-slot" flux produced by the normal load current. The stator bars tend to vibrate in the slot, a phenomenon called "bar bouncing." Therefore, the stator bars must be tightly wedged in the slot to minimize the relative motion and avoid fretting damage from contact against themselves and the stator core and bar packing systems. Stator windings have been known to fail quickly once they becomes loose in the slot.

2.8 STATOR WINDING WEDGES

There are many different wedging systems employed by different manufacturers, but all have the common purpose of keeping the stator bars tight in the slot (see Figs. 2.35–38).

Fig. 2.34 Side-packing ripple spring with semiconducting impregnation.

Fig. 2.35 Single-piece flat stator slot wedges.

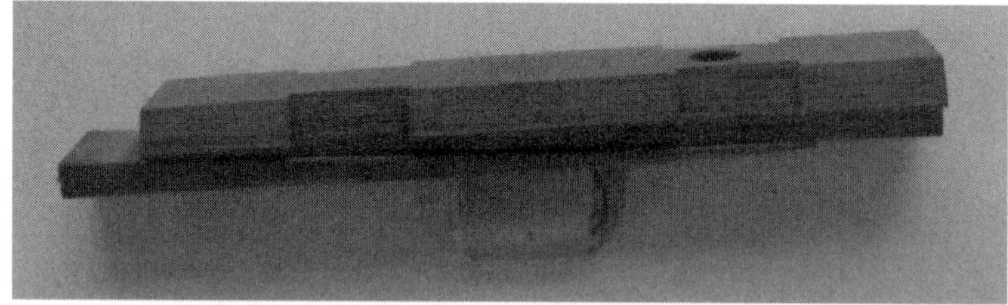

Fig. 2.36 Double-piece tapered stator slot wedge system.

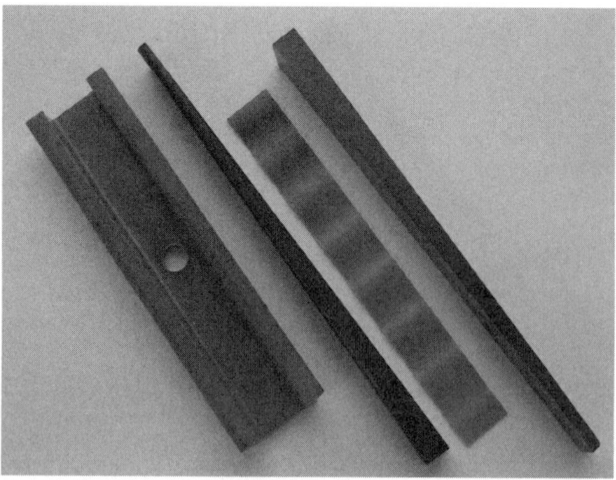

Fig. 2.37 Four-part stator-slot wedge system with channel wedge two tapers and radial ripple spring.

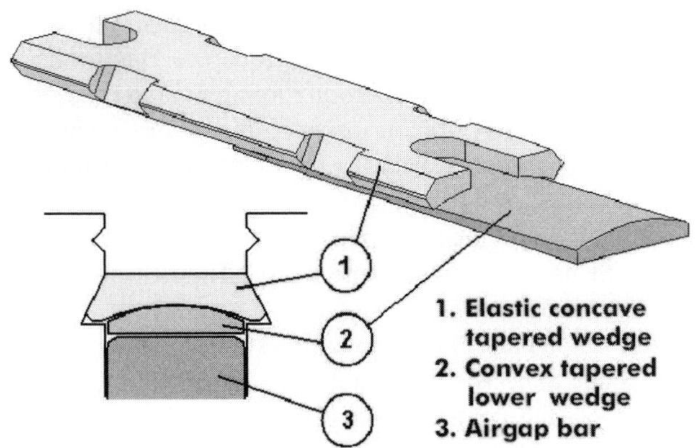

Fig. 2.38 Concave/convex stator-slot wedging system. (Courtesy of Alstom Power Inc.)

Stator-bar bouncing is one of the reasons that ensuring that stator wedges are tight in the slots is so important. The resulting vibrations of the bars in the slot due to looseness can quickly wear the ground-wall insulation on the bar right through to the copper and cause a stator ground-fault failure.

Maximum instantaneous *bar bounce force* per unit length of stator bar in the slot occurs when top and bottom bars in the same slot are in phase and carrying maximum stator current:

$$\text{Total bar bounce force, } F_{\text{total}} = \downarrow F_{\text{top bar}} + \downarrow F_{\text{bottom bar}}$$

where

$$\downarrow F_{\text{top bar}} = \frac{3\mu_0 I^2}{w_s k^2}, \qquad \text{N/m}$$

$$\downarrow F_{\text{bottom bar}} = \frac{\mu_0 I^2}{w_s k^2}, \qquad \text{N/m}$$

therefore,

$$F_{\text{total}} = \frac{3\mu_0 I^2}{w_s k^2} + \frac{\mu_0 I^2}{w_s k^2} = \frac{4\mu_0 I^2}{w_s k^2}, \qquad \text{N/m}$$

where

μ_0 = $4\pi \times 10^{-7}$ H/m
I = stator phase current in amperes
w_s = stator slot width in meters
k = number of parallel stator circuits per phase

This force is toward the bottom of the stator slot and is also sinusoidal in nature, due to the fact that it is proportional to the current squared. This further means that the force is associated with a twice-per-revolution forcing function and produces vibration at 120 (100) Hz, similar to the vibration forces on the stator core and frame. Since the magnetic field in the slot is highest near the top of the slot and diminishes toward slot bottom, it can be shown that the resulting difference between the forces on the top and bottom bars is substantial. In fact, the top bar forces are three times that of the bottom bar forces when both bars are in the same phase. The net effect for maximum bar bounce forces is described above.

Wedging of the stator bars, however, is not strictly concerned with just the bar bouncing effects. Since there is considerable heat generated in a stator bar, there are also thermal expansion and contraction and insulation shrinkage issues to consider. Thermal expansion and contraction can easily loosen bars in the slot if they are not wedged properly, and the heat impact on the insulation systems can also be a factor if the insulation is not preshrunk prior to wedging.

2.9 END-WINDING SUPPORT SYSTEMS

In addition to the slot, significant forces are present in the end regions of the stator winding as well. The end-winding geometry is also complex and requires a support structure that is flexible in certain modes and stiff in others, all at the same time, to restrain the end winding under all modes of normal and abnormal operation. In addition, the strong electric fields in the end region require that nonconducting supports be used. Most support systems use blocks, tension devices, and rings, which together with the bars themselves form a substantially rigid structure. Support in the radial direction is generally made to be very stiff, to keep vibration levels minimized. In the axial direction, it is required that the end-winding structure be allowed to move axially to accommodate the axial thermal expansion of the slot section of the winding.

Sudden phase-to-phase short circuits are the most significant transient behaviors in which excessive forces are developed in the stator winding. These must be accounted for in the design of the winding and in its support structures in the slot and the end windings. Spacers, blocks, and wedges associated with the stator end winding should be made of material that will not buckle, shrink, absorb moisture, or otherwise allow the windings to become loose and unsupported. All parts of the stator end winding and associated connections and support structures should be designed so that they will be capable of withstanding full line-to-line and three-phase short circuits at the generator terminals.

Vibration forces in the end winding of a large turbine generator under normal load are also high, and must be kept under control to ensure that there is no wear incurred on the end-winding as a consequence of rubbing or impacting. Thermal cycling and shrinkage effects can also promote advanced loosening and high vibration. The maximum vibration level of the end windings and associated support structures, once the machine is installed and operating, should be less than 50 μm peak to peak. This is unfiltered, with no natural resonances in the frequency ranges of 50–75 Hz and 100–140 Hz for a 60 Hz system (40–65 Hz and 80–120 Hz for a 50 Hz system).

Figure 2.39 is a schematic representation of a direct water-cooled stator's end winding and support system. Figure 2.40 shows the end-winding support system of an indirectly cooled two-pole stator and Figure 2.41 shows the end-winding support system of a direct water-cooled four-pole stator.

2.10 STATOR WINDING CONFIGURATIONS

Stator windings are designed with a trade-off between operating voltage and current-carrying ability. This goes back to the basic MVA relationship, which is a combination of the stator terminal voltage and the stator winding current. For the same level of MVA, as the terminal voltage of the winding is increased, the stator current required is reduced. The opposite is also true. As the terminal voltage is reduced, the stator current would have to be increased to keep the MVA rating constant.

The above relationship has significant consequences for generator design, but it is also quite useful in allowing optimization of any particular generator design. For instance, a two-pole generator may have each of the two parallel circuits of each phase of the stator winding connected as two parallel paths (see Fig. 2.42*a*) or as two paths connected in series (see Fig. 2.42*b*). If the connection is parallel, the terminal voltage tends to be lower and the stator current higher. For the same MVA rating, if the connection of the stator winding is in series, the terminal voltage will be higher and the current lower. The physical consequence of this is that the higher voltage machine requires a thicker ground-wall insulation to withstand the higher voltage. For the paral-

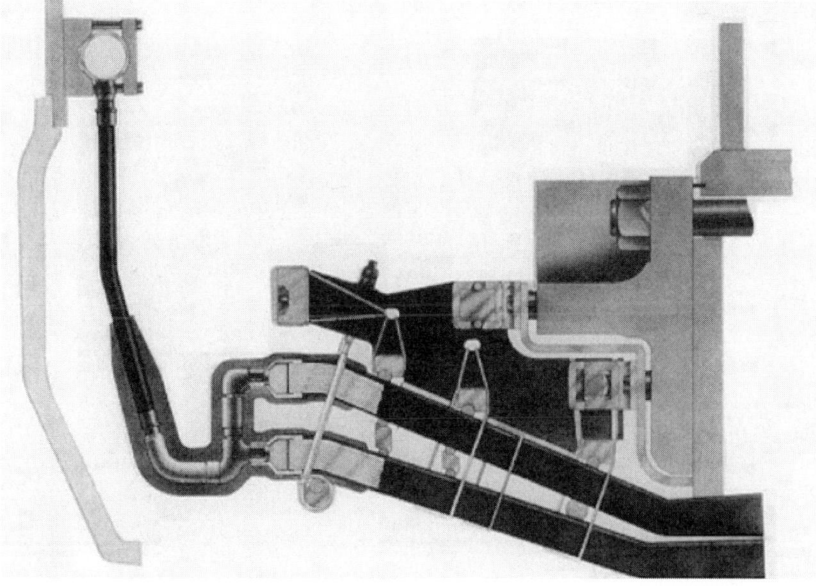

Fig. 2.39 Stator end-winding support system. (Courtesy of General Electric.)

Fig. 2.40 End-winding support system of an indirectly hydrogen-cooled two-pole turbo-generator.

Fig. 2.41 End-winding support system of a direct water-cooled four-pole turbogenerator.

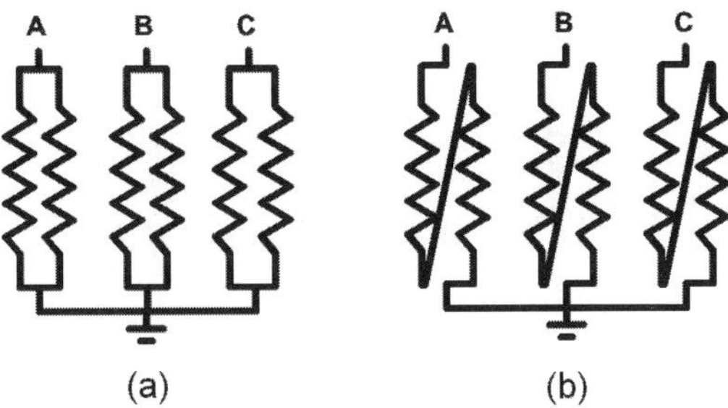

Fig. 2.42 Types of parallel-series winding combinations for a two-pole generator.

lel-connected winding, there would need to be a large amount of copper and increased cooling to accommodate the higher stator current.

In four-pole machines there are four parallel paths in the stator winding. The two most common winding connection configurations are all four parallels connected in parallel (see Fig. 2.43), and two parallel paths comprised of two of the parallels connected in series (see Fig. 2.44). There are some four-pole machines, however, that have special jumper connections to arrange the four poles in a three parallel path arrangement (see Fig. 2.45). All of these issues and configurations described above al-

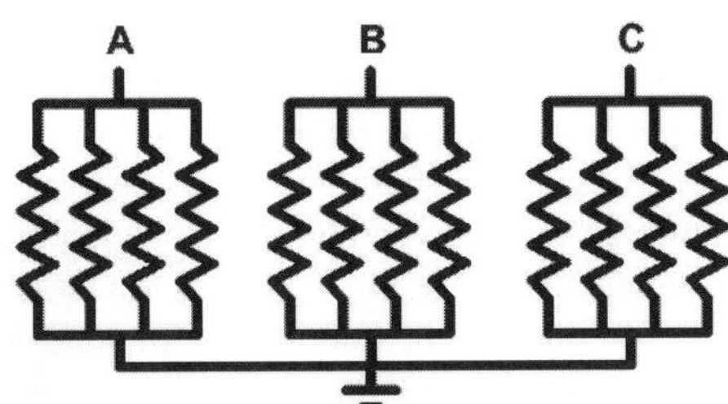

Fig. 2.43 Four parallel paths.

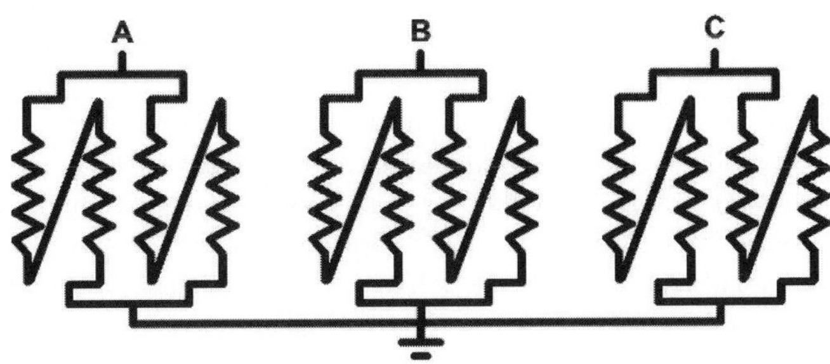

Fig. 2.44 Two parallel paths with two series.

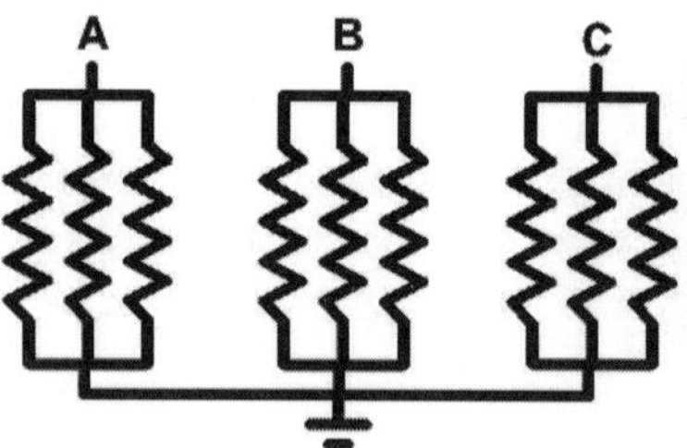

Fig. 2.45 Three parallel paths.

low flexibility in design to achieve a machine with a smaller overall size, lower cost, and lowest losses for best efficiency.

2.11 STATOR TERMINAL CONNECTIONS

All generators require a means to deliver the power produced inside the machine, out to the main transformer, via an isolated phase bus (IPB) system consisting of three individually enclosed, high-voltage and high-current-carrying leads. Small combustion-turbine generators, mainly used for "peaking" purposes, often have cables between the generator terminals and the main step-up transformer.

Since there are three phases in the generator, three-phase lead connections are required, commonly called stator terminal connections. These are used to make the connection from the stator winding inside the generator, out through the generator frame and casing, to the isolated phase bus system. Each stator terminal carries the same current as the sum of the currents of all the parallels in a single phase. Therefore, the terminals are subject to high losses and heating, and in large units must be force-cooled as well. This is usually done either by the internal cooling gas in the generator casing, or by water-cooling as part of the stator winding cooling water system. Since the terminals are also at the rated voltage of the generator, they are insulated conductors, and, generally, the same type of materials used for the stator windings are used for the terminals as well (see Fig. 2.46).

In addition to the high-voltage terminals, there are also three neutral terminals or bushings that make up the common connection point at the zero voltage or neutral ends of the stator winding phases. Although these are essentially at zero or ground potential, they do carry the full stator current that the high-voltage bushings carry and so must be given the same cooling as the high-voltage terminals. They are also insulated from ground, except at the actual connection or "star" point, to ensure no circulating currents or faults occur anywhere else in the winding system.

Furthermore, in all enclosed generators having an internal atmosphere of hydrogen, great care must be taken to ensure a gas-tight seal where the high-voltage and neutral bushings exit the stator casing. Hydrogen is a very dangerous gas and can self-ignite when it leaks from a pressure vessel. The flame is invisible by its nature. Also, any

Fig. 2.46 Generator terminal bushings. Current transformers (CTs) can also be seen around each bushing (three CTs per phase).

leaking hydrogen can collect in the enclosure below the terminals and create an explosive mixture with the oxygen in the air in the compartment. Hydrogen and its containment will be discussed further on in this chapter.

2.12 ROTOR FORGING

The rotor forging is generally a one-piece solid steel forging (see Figs. 2.47 and 2.48), but there are rotors built in sections and locked together in a *spigot* type arrangement. Two-piece rotors are no longer common due to the advanced technology in steelmaking and the ability to make even the largest rotor forging from a single-piece forging.

The type of material used in rotor forgings is highly permeable magnetic steel to carry the flux produced by the rotor winding. And because the rotor is a dynamic component operating at high speed, the materials are highly stressed and must have considerable strength to carry the copper winding and operate under high mechanical and thermal loading. Very high stresses occur in the rotor-slot tooth roots, shrink-fit area, and, generally, where there are machined radii. The types of stresses that a rotor forging is subjected to are high-strain, low-cycle mechanisms during start-up and shutdown; torsional stresses in operation and during faults; and high-cycle fatigue due to

Fig. 2.47 Two-pole forging.

Fig. 2.48 Four-pole forging.

rotation and self-weight bending. Safety factors for stress are usually on the order of 150% at 20% overspeed. This is done to ensure that cracks do not initiate in any part of the rotor under any of the types of modes of operation that the rotor might encounter.

One of the difficulties in designing turbo-generator rotors is that they are very long, thin high-speed components. Two-pole rotors have greater length-to-diameter ratio than four-pole rotors, but, regardless, they both operate above the second critical speed. Balancing then becomes a critical issue for these types of rotors, in addition to the high stresses. The difficulty in achieving good balance arises from all the other components installed in the rotor forging that undergo thermal loading and that must be allowed to expand and move while in operation. Also, there are insulations systems that must do their job under high-speed operation.

The forging acts as a main structure of the rotor, but has many different components installed to accomplish its overall function. Within the forging itself, there are the main body in which the rotor winding slots, *pole-face crosscuts,* and axial slots are machined; the shaft that is the main support for the rotor on the bearings; the turbine coupling; the hydrogen sealing surfaces; the main lead slots in the shaft; and the collector-ring assembly portion. Additional components mounted in or on the rotor consist of the copper (and very rarely aluminum) winding and insulation system, winding slot wedges, end-winding retaining rings, balancing or centering rings, end-winding blocking, fans or blowers, main leads and terminal studs, collector rings, and a collector-ring cooling fan.

The rotor main body of the forging consists of a pole area and slots machined axially to carry the copper winding and associated insulation. Teeth are created between the machined slots, and these are highly stressed due to the loading of the copper when the rotor is at speed. Grooves or dovetails are machined axially near the top of the winding slots, to accommodate wedges that hold the copper winding and insulations in the slot. Therefore, the wedges are under high load as well when the rotor is at speed.

It is the main body of the rotor forging that carries the flux, in both the body of the forging in the pole and under the winding area of the rotor. Because the flux is dc, only the magnetostatic saturation characteristics of the forging steel are relevant with respect to the main body of the rotor. The design of the rotor must ensure that there is enough magnetic material to carry the design level of flux for the generator and yet have enough material to accommodate all the stresses and loads discussed above. There are other ac flux effects that come into play, but these are associated with leakage fluxes and interaction with the stator. These account for surface heating effects in the rotor forging and other installed components. Such ac effects arise from cross-slot leakage flux, negative-sequence operation, motoring, slip, and so forth.

With regard to the cross-slot leakage flux, one of the issues designers must account for when selecting the size of the generator airgap is the width of the stator slots. During operation, there is slot-to-slot leakage flux in the stator and it is important that this leakage flux not link with the surface of the rotor and cause additional and excessive pole-face losses. The way to avoid this is to ensure that the slot width of the stator is never larger than the airgap. Having the airgap shorter in length than the slot width will allow the slot to slot leakage flux to cross the airgap and link the next slot through the highly permeable steel in the rotor surface, causing additional losses there. Therefore, the ratio of stator slot width to airgap length should be less than 1. This is one of the reasons many motor stators have nearly closed slots at the opening, so that the inherently small airgap of these machines can be achieved without concern for pole-face losses. Figure 2.49 shows the cross-slot leak flux effect in the airgap of a large generator as compared to a motor. Figure 2.50 shows the cross-slot leak flux effect in the airgap of a large generator when the airgap is smaller than the stator slot width.

The shaft portion of the rotor forging supports the entire weight of the forging and all installed components of the rotor. The diameter of the shaft is chosen so that it will support the rotor adequately under all modes of operation and maintain good vibration characteristics of the rotor and the entire turbo-generator line when coupled to the turbine. As mentioned previously, turbo-generator rotors generally operate above the second critical speed and the shaft diameter largely governs the natural frequencies of the rotor.

Along with the natural frequencies of the rotor, diameter, plus the rotor mass and speed, also affect the *rotor inertia*. Rotor inertia is important during system events with regard to system stability and is used in determining some of the effects on the system interaction with the machine when transient events occur. It is generally determined by the following relationship:

$$\text{Rotor intertia, } H = \frac{0.231 \times 10^{-6}\, WR^2N^2}{\text{MVA}}, \qquad \text{kg-m}^2$$

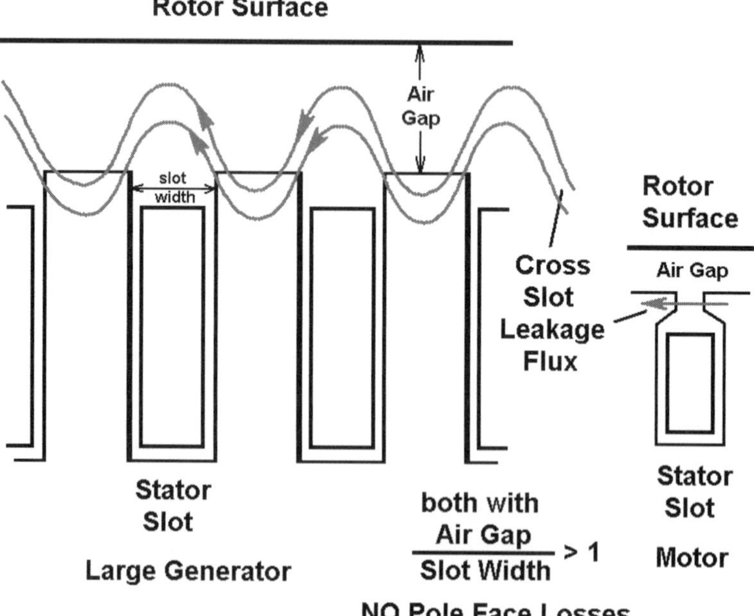

Fig. 2.49 Cross-slot leak flux effect in a large generator airgap as compared to a motor.

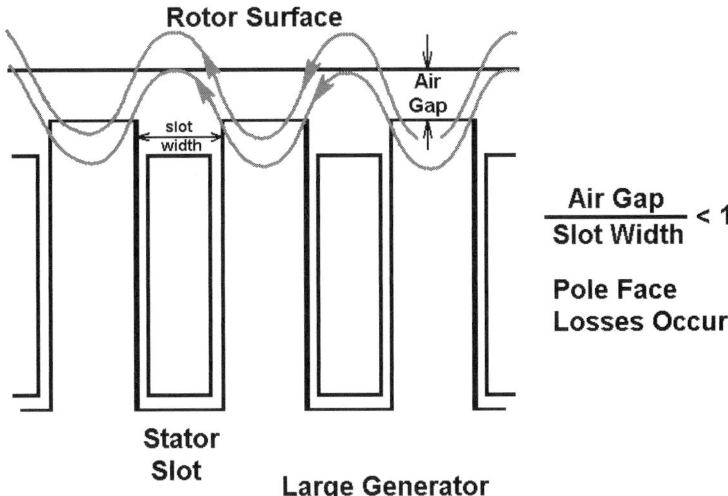

Fig. 2.50 Cross-slot leak flux effect in a large generator airgap when the airgap is smaller than the stator slot width. Pole-face losses will occur, causing rotor surface heating from induced parasitic currents.

where

W = rotor weight in kg
R = radius of gyration in meters
N = rotor speed in rpm
MVA = generator apparent power

The shaft sits in *friction* (*sleeve*) *bearings,* which must also be sized to adequately support the rotor. The interaction of the rotor in the bearings is critical to proper operation of the rotor. Many issues must be considered, such as vertical and horizontal stiffness of the bearing supports, oil film thickness, shaft diameter, bearing length, torsional stresses, and alignment. All these things greatly affect what the critical speeds will be and the rotor balance.

Within the rotor shaft, a borehole is often found machined at the center of the rotor forging through its full axial length. This was done mostly in the past because of impurities and porosity in the forgings, which tended to concentrate in the center. The borehole serves two purposes. The first is to remove the material defects and the second is to provide an access for performing boresonic (ultrasonic) inspection of the rotor bore. In modern forgings, the material manufacturing processes are so improved that a borehole is not generally required for the full length of the rotor. Forgings without boreholes also have more magnetic material available to reduce the overall diameter necessary to achieve a specific rating, but this is generally marginal in effect. All machines have certain length of the shaft bored to accommodate the excitation *bore bars* (more about this later in this chapter).

One other important issue in forging design is shaft torsional oscillations. Specialized torsional monitoring equipment has even been developed to monitor various generator parameters to allow estimation of the effect of system-stress events on shaft life. Likewise, torsional oscillations are monitored with regard to unstable operation. Such things as power system subsynchronous resonance, sudden short circuits, and load rejections can cause transient torques in the rotor and significantly affect forging life. These can stimulate torsional natural frequencies and cause the rotor to go unstable. Such events as subsynchronous resonances can even cause the stress in the shaft to go beyond the torsional endurance limit of the forging material. Failures due to this mechanism have occurred in past. All these issues must be considered in designing the rotor forging.

The torque on the rotor as seen at the rotor surface in the airgap is as follows:

$$\text{Specific airgap torque (at rotor surface)}, T = BA\frac{\pi}{2}D_r L, \qquad \text{N-m}$$

where

$$\text{Flux density}, B = \frac{P\Phi}{\pi D_r L}, \qquad \text{webers/m}^2$$

$$\text{Specific electric loading}, A = \frac{N_{ph} N_c I_a}{\pi D_r}, \qquad \text{A/m}$$

and

P = number of poles
Φ = flux per pole in webers
D_r = rotor diameter in meters
L = core iron effective length in meters
N_{ph} = number of phases
N_c = number stator conductors per phase
I_a = stator phase current in amperes

The above formula is an elegant result derived from basic physical principles. However, the typical user of a turbo-generator can make use of the following very simple expression for the shaft torque:

$$\text{Torque} = \frac{\text{MW output}}{\text{Speed} \times \text{Efficiency}}$$

The efficiency of a typical turbo-generator is above 98%; thus, an approximate and conservative simplification of the equation yields:

$$\text{Torque} \cong \frac{\text{MW output}}{\text{Speed}}$$

Using SI units:

$$\text{Torque (Nm)} = \frac{\text{KW output} \times 60{,}000}{2\pi\,\text{RPM}}$$

This torque on the rotor is significant and creates torsional stresses in the forging all along its length. The torsional stresses are difficult to analyze due to the complex geometry of the various sections of the rotor forging. Torsional stresses are basically shear stresses in the rotor shaft due to the twist in the forging that is created by the action of the rotor's magnetic coupling to the stator magnetic field, as opposed to the opposite force imposed by the steam flow to the turbine. Increasing the steam flow to the turbine causes the rotor load angle to increase and, hence, MW load to increase, and produces increased mechanical torque in the shaft. The magnetic coupling between the rotor and stator is what inhibits the rotor from running away and keeps the turbine generator system in synchronous equilibrium. Increasing or decreasing the rotor magnetic field causes the load angle to increase or decrease, but does not actually change the torque applied, only the angle of the torque, or, in electrical terms, the power factor and reactive power output of the machine.

In terms of the actual stresses created in the rotor forging, they are greatest in areas of smallest shaft diameter and where there are critical radii in the shaft. Also, there are rotor shafts with solid forgings and, in the past, forgings with a hollow bore along the entire length. Generally, the only real way to work out rotor stresses is with detailed finite element analysis, however, if one is most concerned with the simple stresses in the cylindrical portion of the shaft, the following relationships may be used to obtain a rough estimate of the shaft's torsional stress for solid and hollow bore forgings (see Fig. 2.51).

$$\text{Solid forging torsional stress, } \tau_{max} = \frac{2T}{\pi r_o^3}, \qquad \text{N/m}^2$$

$$\text{Hollow bore forging torsional stress, } \tau_{max} = \frac{2T}{\pi r_o^3\left(1 - \dfrac{d_i^4}{d_o^4}\right)}, \qquad \text{N/m}^2$$

where

T = specific airgap torque in N-m

r_i = hollow bore inner radius in meters

r_o = shaft outer radius in meters

d_i = hollow bore inner diameter in meters

d_o = shaft diameter in meters

The nature of rotors also dictates that there are pole areas and winding areas in the main body. Referring back to Figure 2.47 for the two-pole rotor, it can be seen that the rotor would be naturally weaker in the axis that cuts the center of both winding areas. The pole axis would, therefore, be much stiffer. Equalization of the stiffness between

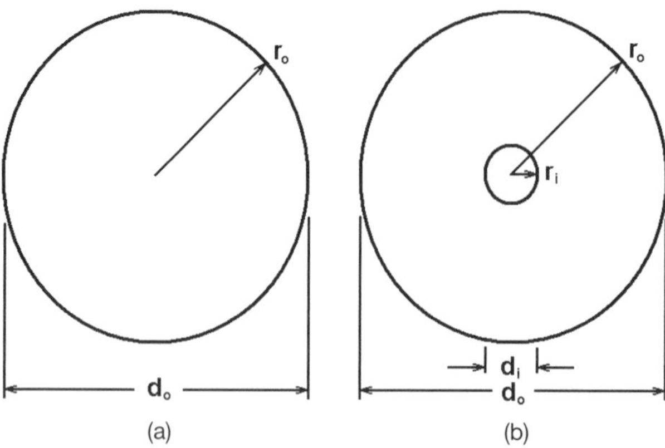

Fig. 2.51 (a) Solid shaft geometry. (b) Hollow bore shaft geometry.

these two axes is accomplished by either crosscuts in the transverse direction—across the pole face—or additional axial slots, or a combination of both (see Fig. 2.52). The various methods of stiffness equalization are employed to keep unequal flexure and vibration to minimal levels.

In the four-pole rotor designs, symmetry is less of an issue (see Fig. 2.48), and in many cases no equalization is needed. However, some four-pole rotors do employ axi-

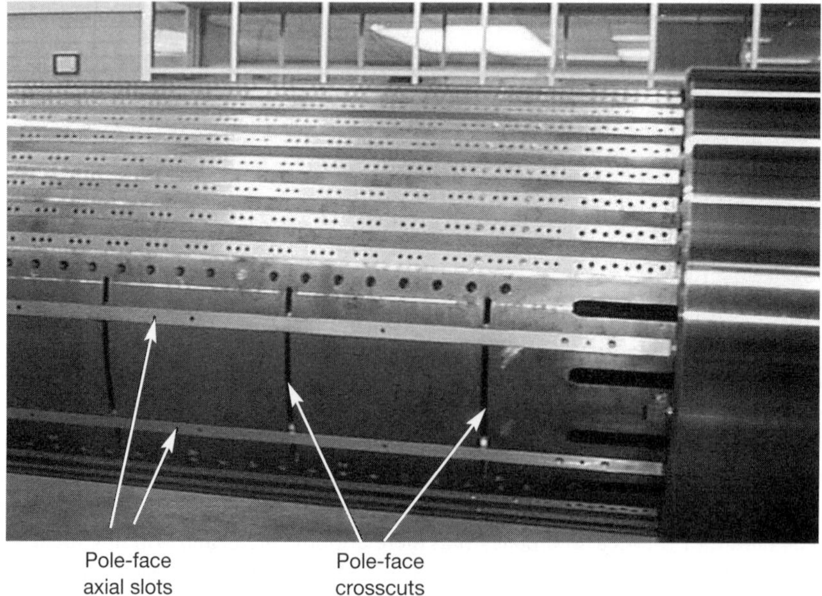

Fig. 2.52 Rotor stiffness equalization by crosscuts and axial slots in the pole face.

al wedges, but these are more for carrying negative-sequence currents and dampening torsional oscillations. These issues will be discussed later in the book.

2.13 ROTOR WINDING

The rotor winding is installed in the slots machined in the forging main body and is distributed symmetrically around the rotor between the poles (See Fig. 2.53). The winding itself is made up of many turns of copper to form the entire series-connected winding. All of the turns associated with a single slot are generally called a coil. The coils are wound into the winding slots in the forging, concentrically in corresponding positions on opposite sides of a pole. The series connection essentially creates a single multiturn coil overall, which develops the total ampere-turns of the rotor (which is the total current flowing in the rotor winding times the total number of turns).

There are numerous copper-winding designs employed in generator rotors, but all rotor windings function basically in the same way. They are configured differently for different methods of heat removal during operation. In addition, almost all large turbo-generators have copper windings directly cooled by air or hydrogen cooling gas.

Cooling passages are provided within the conductors themselves to eliminate the temperature drop across the ground insulation and preserve the life of the insulation material. Some of the design variations are rather significant, as shown in the rotor winding slot cross-sectional sketches in Figure 2.54.

In an axially cooled winding, the gas passes through axial passages in the conductors, being fed from both ends, and is exhausted to the airgap at the axial center of the rotor. In other designs, radial passages in the stack of conductors are fed from subslots machined along the length of the rotor at the bottom of each slot. In the air-gap pickup method, the cooling gas is picked up from the airgap and cooling is accomplished over a relatively short length of the rotor; the gas is then discharged back to the airgap. The flow of the cooling gas in this design is diagonal in nature rather

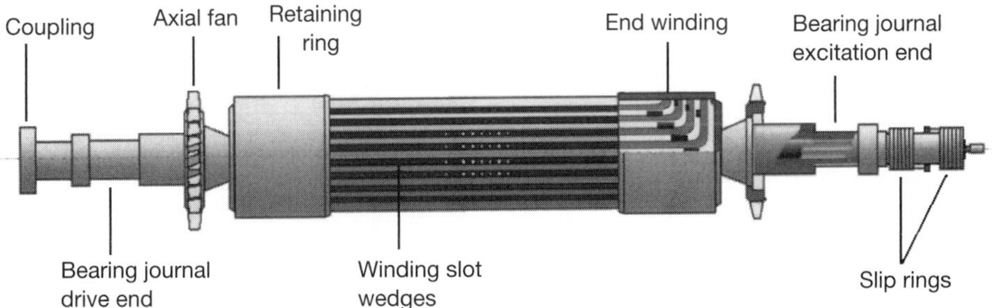

Fig. 2.53 Rotor winding arrangement showing the major components riding on the forging of a standard turbo-generator rotor. This example is of an externally fed field winding (via the two slip rings shown on the right end of the rotor, mounted on the shaft. (Courtesy of Alstom Power Inc.)

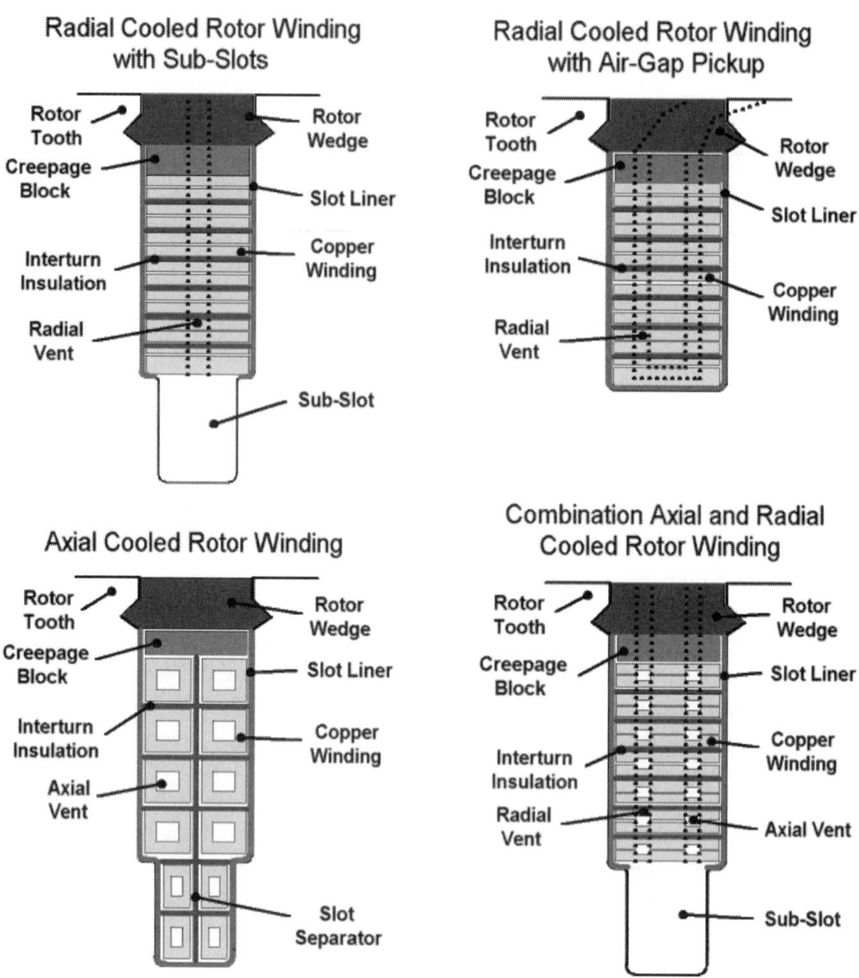

Fig. 2.54 Several rotor winding slot cross-sectional arrangements.

than radial or axial as in most designs. The cooling of the end regions of the winding varies from design to design, as much as that of the slot section. In smaller turbine generators the indirect cooling method is used (similar to indirectly cooled stator windings), by which the heat is removed by conduction through the ground insulation to the rotor body.

The other design feature that is common to most windings is that they are assembled as individual straps of copper insulated from one another, and installed one strap at a time. A recent variation on this has been to manufacture and assemble a set of straps as one consolidated half-coil and then insert the half-coil into the rotor forging slot as one piece (see Figs. 2.55 and 2.56). The individual turns and interturn insulation are bonded to one another, and the ends are then brazed to the matching half-coil of the

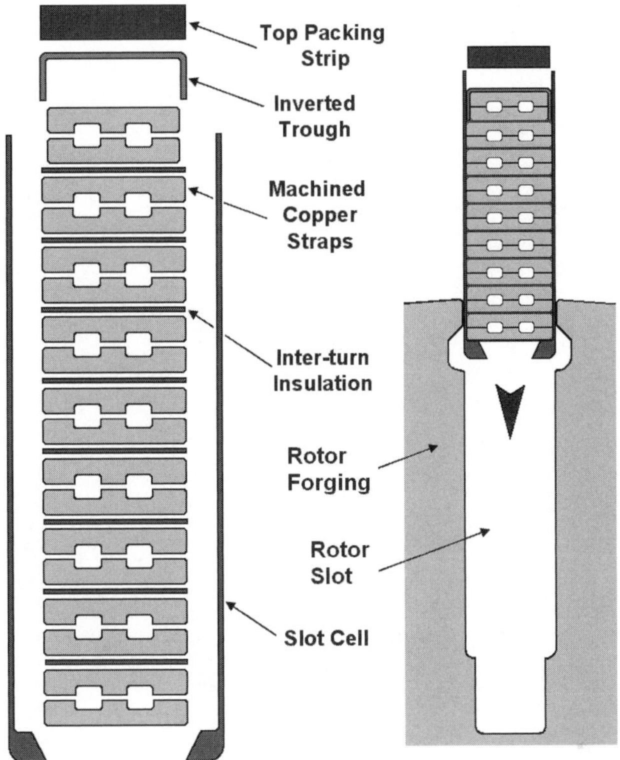

Fig. 2.55 Rotor winding slot cross-sectional arrangement of a consolidated half-coil. (Courtesy of Alstom Power Inc.)

opposite pole, one by one (see Fig. 2.57). This design and assembly method is done in this way to avoid movement of the individual turns against one another and, hopefully, eliminate the possibility of shorted turns and minimize the possibility of rotor winding thermal vibration issues.

Once the winding is installed in the forging slots, it is held in place by metallic wedges, in a similar manner as the stator windings. The difference is that the rotor winding loading on the wedges is far greater due to centrifugal forces at speed. The wedges, therefore, are subjected to a tremendous static load from these forces and bending stresses because of the rotation effects. The wedges in the rotor are not generally a tight fit in order to accommodate the axial thermal expansion of the rotor winding during operation.

There are many available designs and configurations for the end-winding construction and ventilation methods. As in the rotor slots, the copper turns in the end winding must be isolated from one another so that they do not touch and create shorts between turns. Therefore, packing and blocking are used to keep the coils separated and in their relative position as the rotor winding expands and contracts from thermal effects dur-

Fig. 2.56 Fully bonded, consolidated half-coil ready for installation in the rotor forging slot.

ing operation (see Figs. 2.58 and 2.59). To restrain the end-winding portion of the rotor winding during high-speed operation, retaining rings are employed to keep the copper coils in place.

As in the stator, insulation is required to isolate the rotor winding from the rotor forging and the retaining rings, which are essentially at ground potential. In addition, the turns within each winding coil must be separated since they are wound in series. The insulation system must be designed to carry out its insulating function and survive the harsh mechanical duty imposed by the rotation forces in operation.

Rotation imposes a huge centrifugal load on the insulation system. These mechanical effects are further exacerbated by the temperature changes in the winding, which occur when the generator is excited and loaded. Cycling of the load causes temperature cycling, during which the conductors expand and contract. This can promote artificial aging and wear out the insulation system. The degree to which the conductors are locked in place by the centrifugal force affects the actual motions that the insulation must be designed to accommodate.

Field voltage is dc and can reach as high as 700 V on the larger machines. However, the insulation must be capable of handling the field-forcing duty, which is general-

Fig. 2.57 Opposite pole, consolidated half-coils being brazed together in the end winding overhang.

Fig. 2.58 Two-pole rotor winding.

Fig. 2.59 Four-pole rotor winding.

ly twice the rated field voltage. Normally, the turn-to-turn voltage is only a few volts, with only brief occurrences of higher voltages. It is thus generally sufficient to provide mechanical separation between turns.

The capability of the field winding is expressed in ampere-turns (A-T). The total cross-sectional area available for copper in each slot is subdivided into turns, all of which are connected in series to form a coil. The number of field turns per pole multiplied by the current in the winding equals the total A-T. The current density in the copper determines the total loss to be dissipated and, hence, the temperature of the winding. For constant current density, as the number of turns increases, the copper area per turn (and, hence, the current per turn) decreases proportionally, but the total A-Ts per pole remains the same. Since field voltage is proportional to the number of turns, it does not affect the A-Ts.

The major design criterion for the A-T capability of the field winding is the temperature of the conductors. Increasing this capability may be done by using improved insulation materials, which are capable of higher temperatures, or by improving the cooling system, or by increasing the total area available for copper in the rotor cross section.

2.14 ROTOR WINDING SLOT WEDGES

The wedges that hold the rotor winding in the slots are sometimes also complex in design, but always highly stressed (see Figs. 2.60, 2.61, and 2.62). The wedges must hold the copper winding and its insulation systems in place at high rotational speeds, and allow cooling gas to pass through them. The wedges generally have cooling vents machined into them, which reduces their effective strength. High cooling-gas temperatures can also affect wedge strength if the temperatures begin to affect the creep life of the material.

The wedges are generally made of lightweight materials, such as aluminum or brass, in the winding slots. This area does not generally carry a useful magnetic flux, so the wedges do not need to be made of magnetic material. In some designs, however, the first winding slot wedge next to the pole may be made of a magnetic steel material

Fig. 2.60 Short rotor wedge.

Fig. 2.61 Airgap pickup rotor wedge.

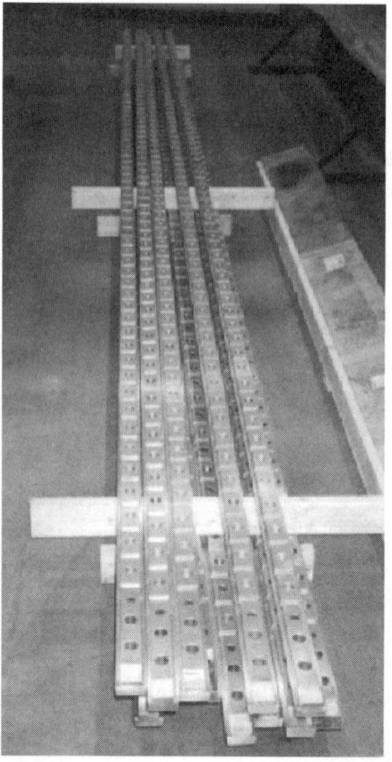

Fig. 2.62 Long rotor wedge (aluminum).

to improve flux distribution and lower the flux density in the pole area. This helps reduce the excitation requirements for the generator.

The wedges do not usually sit tight in the slots. They have a loose fit, relatively speaking, to allow the copper winding underneath to expand axially during operation. Expansion of the copper winding under load can create an enormous axial shear force in the winding slots because of the direction of copper growth. The overall design of the rotor and the wedges must take this movement into consideration.

2.15 AMORTISSEUR WINDING

Most modern rotors employ a *damper* (also called *amortisseur* or *damping*) *winding* to dampen torsional oscillations and provide a path for induced currents to flow. The amortisseur winding is essentially a separate winding installed under the rotor wedges and retaining rings that is connected in a way similar to the squirrel-cage of an induction motor. It produces an opposing torque when currents flow in it, and this helps dampen torsional oscillations and add to the stability of the rotor during system stress events. In some instances, where full-length aluminum wedges are used in the rotor, these may serve additionally as part of the damper winding. Also, some designs use the retaining rings as the shorting connection at the end of the rotor, instead of a dedicated component. Figure 2.63 shows a particular type of amortisseur. Photographs of other types of amortisseurs can be seen in Chapter 9.

In addition to the above, the damper winding can help prevent *negative-sequence* and *motoring currents* from flowing in the rotor forging and causing overheating damage. The negative-sequence rating or current-carrying ability of any rotor design is largely dependent on the arrangement and effectiveness of the amortisseur winding. (More about this in Chapters 6 and 9.)

2.16 RETAINING RINGS

Retaining rings are generally the most highly stressed component in the generator. They are required to hold the end-winding copper of the rotor winding against centrifugal loading during operation. There is one ring at each end of the rotor to do this, and the rings are shrunk-fit onto the body of the rotor forging. There are many types of ring designs and fit types as well. Some rings have a *barrel* type fit and others a *castellated* fit (See Figs. 2.64 and 2.65). All require some form of locking arrangement to inhibit rotational and, subsequently, axial movement as the rotor operates at speed (see

Fig. 2.63 Amortisseur winding.

Fig. 2.64 Barrel-fit type retaining ring (balance ring not installed).

Fig. 2.65 Castellated-fit type retaining ring with balance ring installed.

Fig. 2.66). The axial growth of the copper winding creates an additional force on the retaining rings, which tends to push them out, away from the end of the rotor main body. In addition, all these fits have some form of "shrink-fit" application to hold the ring onto the forging in the radial direction, even at overspeed conditions. Overspeed must be considered because the retaining rings become highly stressed even at rated speeds. They are generally designed to retain their shrink-fit on the rotor body up to 120% of their rated speed.

The rings are under tremendous stress from the loading of the copper and their own weight. As a rule of thumb, the loading of the copper generally accounts for one-quarter to one-third of the stress in retaining rings and the other two-thirds to three-quarters are the rings' own "hoop stress." The rings also are under considerable stress at rest because of the shrink-fit and the nature of the ring shape at rest. In operation, the distribution of the copper loading on the underside of the rings is not completely even, and this, in conjunction with variations in shrink-fit stiffness from pole to winding face, can cause an ovalizing effect on the rings from standstill to speed. To compensate for the ovalizing effect at the non-shrink-fit end of the rings, a *balance* or *centering* ring is shrunk-fit inside the retaining-ring (see Fig. 2.65). It is used to produce stiffness in the radial direction at that end of the ring, since it does not have the forging shrink area to keep it concentric. From these issues, it can be seen that bending stresses come into play from standstill to operation at speed, as the rotor undergoes deformation in this range (see Fig. 2.67). As a result, retaining rings are subjected to the high-strain, low-cycle effects of start/stops, as well as high-cycle stress modes in operation. Because the retaining rings are so highly stressed, they are also designed with a safety factor of 150% up to 20% overspeed.

Fig. 2.66 Castellated-fit type retaining ring with locking tab shown. The locking tab stops the ring from rotating during operation.

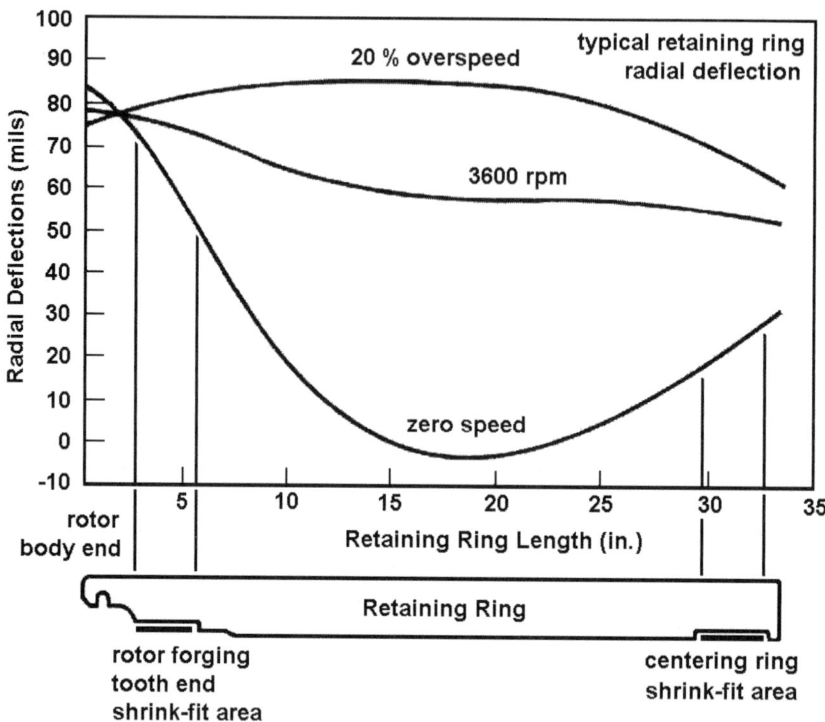

Fig. 2.67 Retaining-ring deflection as a function of speed. (From EPRI Special Report EL/EM-5117-SR, "Guidelines for Evaluation of Generator Retaining-rings," April 1987. Courtesy of EPRI.)

At this point, it is worthwhile to discuss how an estimate of retaining-ring hoop stress is determined, due to the critical nature of retaining rings. The starting point is the fact that any ring needs to be capable of 20% overspeed and still remain within a 150% safety factor. Therefore, the copper loading of the end winding on the retaining ring and the ring's own hoop stress must be within the capability of the ring material (i.e., 18 Mn–18 Cr for most modern units) considering the overspeed and safety factors included.

18 Mn–18 Cr material is on average capable of up to 1200 MPa in terms of stress. This means that at rated speed the ring has to be designed to be loaded up to no more than about 533 to 600 Mpa, considering both hoop stress due to its own weight and the copper loading, to give it the 150% safety factor margin at 20% overspeed. The basic estimation method for this follows a fairly simple formula as follows:

$$\text{Total stress on retaining ring} = \text{hoop stress of ring} + \text{weight of copper load, MPa}$$

$$\text{Retaining ring hoop stress} = \text{18–18 ring material density } (7800\ \text{kg/m}^3) \times \omega_r^2 \times R_r^2$$

$$\text{Copper loading on the ring} = \text{copper density } (8400\ \text{kg/m}^3) \times \omega_r^2 \times R_c^2$$

where
$\omega_r = 2\pi f$
f = frequency of rotation
R_r = retaining-ring mean radius in meters
R_c = copper end winding mean radius in meters

As an example, consider the retaining ring and copper end winding configuration of the two-pole rotor in Figure 2.68. This rotor has a mean ring radius of 0.55 meters and a mean copper radius of 0.4 meters. If the frequency of rotation is 60 Hz, then the ring hoop stress works out to $7800(2\pi\ 60)^2 \times 0.55^2 = 335\ \text{MN/m}^2$ or Mpa, and the copper loading would be $8400(2\pi\ 60)^2 \times 0.4^2 = 191$ MPa. Summing the two produces a total stress on the ring of approximately 526 MPa at rated speed. For the 20% overspeed condition, the above values for the ring and the copper need to be recalculated at 72 Hz. This produces a total stress on the ring of 758 MPa. The overspeed stress value times 1.5 (150%) works out to approximately 1137 MPa. Since the 18 Mn–18 Cr material is generally good to 1200 Mpa, the design of this ring meets the 150% safety factor at the 20% overspeed condition.

The series of photos in Figures 2.70 to 2.73 show the additional collateral damage from a burst retaining ring. In Figure 2.73, the total destruction of the machine can be seen.

The ring material is critical as well because of the high stresses. Both magnetic and nonmagnetic steel materials have been used, but nonmagnetic materials are the most common for large generators. Retaining rings made of magnetic steel essentially introduce magnetic material into the airgap at the stator core end and reduce the reluctance of the airgap in that area. The result can be an oversaturation of magnetic flux in the core end, causing core end heating problems, if the end region of the generator is not designed to account for the use of magnetic rings. This effect is more pronounced as

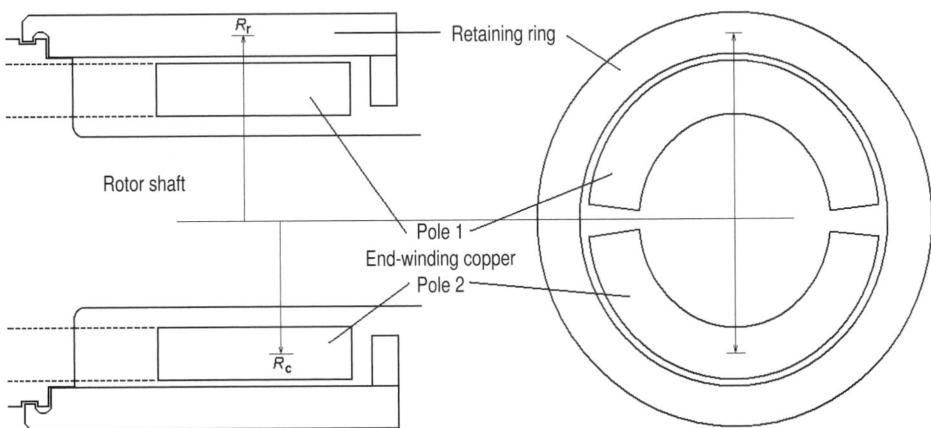

Fig. 2.68 Retaining-ring dimensions for estimation of hoop stress and copper end winding loading.

Fig. 2.69 This photo shows a burst retaining ring that had a tapered nose at the shrink-fit region. It was simply not robust enough at the shrink-fit and shows the importance of the hoop stress calculation and getting the stresses correct.

Fig. 2.70 This photo shows the overall effect from a burst retaining ring. The ring is nowhere to be seen and was completely destroyed.

Fig. 2.71 This photo shows how destructive the force was as the copper winding actually ripped the rotor wedges out of the slots and tore the rotor teeth apart as well.

Fig. 2.72 Here the torsional effects of the failure can be seen by the shearing of the rotor coupling bolts.

Fig. 2.73 The stator was completely destroyed as the rotor winding came out of the slots and contacted the iron core. Rotation was halted in about one revolution, abruptly!

the power factor approaches leading because of the interaction of the end region stray fluxes.

Nonmagnetic materials have been the main choice for retaining rings because of their electromagnetic high reluctance. Historically, nonmagnetic materials were not always as strong as magnetic materials, hence, this is one of the reasons magnetic rings have been used in past. It goes back to the relationship for increasing the diameter of the rotor to achieve a higher output machine, but the trade-off is the strength limit of the material. For a constant operating speed, the larger the ring diameter, the higher the stress in the ring. Obviously, there is a limit to the operable ring diameter, based on the capability of the ring material and the need for overspeeds and safety factors.

Today, there are newer and better nonmagnetic retaining-ring materials that are as strong as magnetic materials, so they are the materials of choice. The most common material used is 18% Mn–18% Cr (also called 18 Mn–18 Cr or simply 18–18). This material has the additional benefit of being highly resistant to aqueous stress corrosion pitting and cracking, which most other ring materials are not. Prior to the 18 Mn–18 Cr rings, the most common nonmagnetic material was 18 Mn–4 Cr or 18 Mn–5 Cr. There are problems with these materials when moisture contamination is present. There have been some rings that had cracks initiate and eventually fail, causing a catastrophic failure of the entire generator. The 18 Mn–18 Cr material is highly resistant to this problem but is not immune if halides or copper ions are present in any moisture attack on the rings. There have been a few reported cases of pitting and cracks initiating in these rings [4,5].

One other issue about retaining rings is that the rings are larger in diameter than the rotor's main body diameter. This is because after the rings are shrunk-fit onto the rotor body, the rotor must still be capable of performing its function as a rotating magnet. The airgap must, therefore, be large enough to accommodate the retaining rings when the rotor is installed. As discussed earlier, this has a significant effect on the short-circuit ratio and other electrical parameters of the generator. Figure 2.74 shows a retaining-ring with a lifting arrangement.

2.17 BORE COPPER AND TERMINAL CONNECTORS

Current is supplied to the rotor winding by means of twin copper conductors running from radial terminal connectors next to or under the excitation end's retaining ring, through the shaft center bore, and out to the radial connectors associated with the slip/collector rings (see Fig. 2.75). The two copper conductors are isolated from each other and from the rotor forging, since they are at dc potential, in the normal operating range up to 700 volts dc and twice to three times that under the field-forcing operation of the exciter. There are two copper conductors, so current is fed in one and out the other. One conductor is at *plus* polarity and the other at *negative* polarity. Because the bore copper is generally not force cooled due to its enclosed location, the conductors are substantially sized to minimize the current density.

The connections to the bore copper are made to both the slip rings and to the internal rotor winding by means of radial terminals connectors or "studs" (see Fig.

Fig. 2.74 A retaining ring with a lifting attachment.

2.75). In some designs, these may be force cooled, and not in others. Connections are made on opposite sides of the rotor to maintain balance in the overall rotor when at speed.

Due to the internal hydrogen atmosphere of the larger machines, a sealing arrangement is also required on the inner set of terminal studs to ensure that the hydrogen gas does not leak past the studs and into the rotor bore. Should the hydrogen gas leak at this location, its most likely exit is at the excitation end of the shaft and, possibly, into the slip/collector ring enclosure. Sparking at the slip-rings can ignite the hydrogen if an explosive mixture occurs, so great care is taken to ensure a good seal of the outer terminals. In some designs, a further seal is established at the inner set of terminal connectors and at the excitation end of the bore copper in the bore. This is for double protection but is not a common design.

2.18 SLIP/COLLECTOR RINGS AND BRUSH GEAR

A dc current is supplied to the rotor winding to create the rotating magnetic field. This can be done by a brushless excitation system or by a set of positive and negative slip or collector rings.

Brushless excitation will be discussed under excitation systems in Chapter 3. In brief, the shaft-mounted exciter produces a polyphase ac output from its winding as it rotates. The output is rectified by rotor-mounted rectifiers, and current is delivered to the rotor winding directly without requiring slip/collector rings.

For the slip/collector ring type of current delivery system, the rings are shrink-fit mounted on the rotor shaft, atop an insulating sleeve, which is generally made of epoxy glass or mica-based system. Each ring is opposite in polarity to the other, since one conducts current into the rotor winding and the other out. The current transfer to

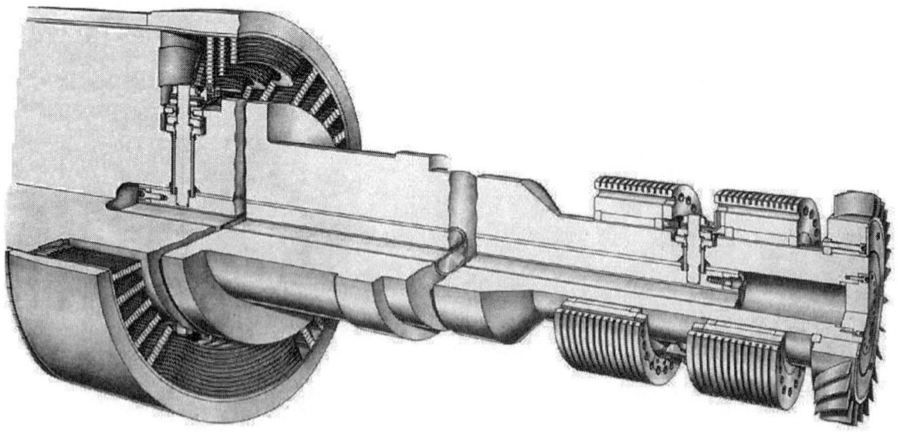

Fig. 2.75 Bore copper and terminal stud connector arrangement. (Courtesy of Alstom Power Inc.)

the rings takes place at high speed, so a sliding contact surface is needed on the rings. Conduction to the rings takes place by graphite-loaded brushes that slide along the rotating surface of the rings as the rotor spins (see Fig. 2.76). Good contact is difficult to maintain if the surface of the rings is not adequate or the brushes are improperly prepared for operation. The brushes are spring-loaded to maintain a consistent pressure against the ring surface during operation.

The friction between the ring and brush surfaces and the I^2R across the brush-collector contact resistance generates significant heat. To keep the rings and brushes cool during operation, there are helical grooves cut into the ring surface to vary the contact surface in operation and allow cooling air to flow. However, the main reason for the helical grooves is to reduce air pressure under the brushes. This allows the brushes to remain in better contact with the collector ring. In some designs, there are additional vents in the rings to ensure cooling airflow. In addition, the slip/collector ring area is generally enclosed and force cooled. The cooling air is filtered due to the carbon byproduct produced.

2.19 ROTOR SHRINK COUPLING

The generator rotor must be coupled to the turbine to be able to rotate and to transmit the rotational torque developed while on load. The coupling is either part of the machined forging or a separate shrunk-on arrangement (see Fig. 2.77). Bolt holes are drilled out for mating of the turbine to the generator rotor.

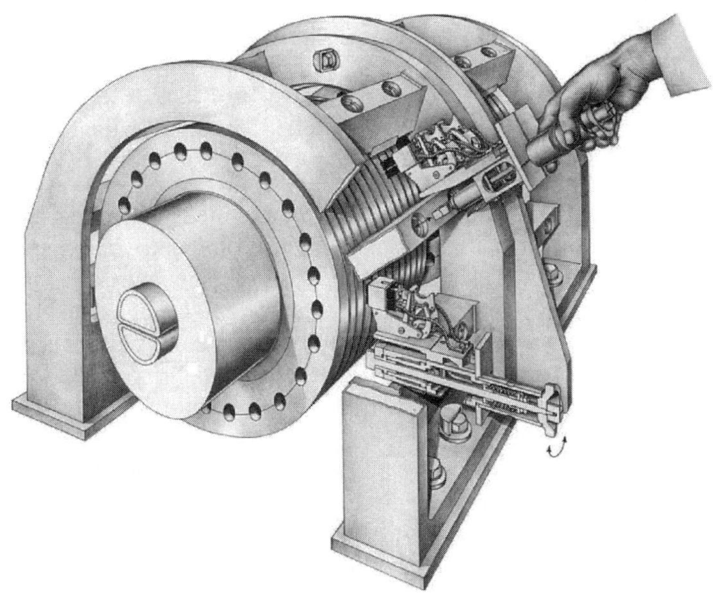

Fig. 2.76 Brush gear. (Courtesy of Alstom Power Inc.)

Fig. 2.77 Shrink coupling showing tooth wheel where the turning-gear apparatus engages the rotor.

In many cases, the turning gear for low-speed rolling (*turning gear* operation) during cooling periods is machined into the shrink coupling as a set of toothed gears (see Fig. 2.77).

2.20 ROTOR TURNING GEAR

All generator rotors are generally equipped with a turning gear arrangement of some kind. The turning gear is required to slowly roll the rotor when the generator is taken off line and begins to cool down. The forging is extremely long, thin, and heavy. A bow may develop in the rotor shaft while the rotor is still cooling down if it is not rotated. The turning gear reduces this possibility.

Two-pole rotors may range in weight from a few tons on very small machines, up to 80 or 90 tons on the larger generators. Four-pole rotors of the largest size can be as much as 200 tons in weight.

The turning gear (Fig. 2.78) is also used to roll the rotor before start-up on steam so that there is no sudden inertial torque from standstill. The weight of large rotors can also cause rotors to bow if suddenly turned. In addition, a rotor may shift in its bearings if not lubricated properly and cause damage to the bearing surface. To prevent this, a "jacking oil" system is also used to lift the rotor before the turning gear is started for run-up.

Fig. 2.78 Turning gear.

One of the other uses of turning gear is for cross-compound generators, where there is a high- and a low-speed line in the generating unit. In such machines, it is common to use magnetic coupling between the high-speed and low-speed generators, to pull the low-speed rotor up to speed as the high-speed rotor is turned on steam. A turning gear is required to start both rotors rolling and keep torsional inertia minimized, as well as to keep the rotors in synchronism during the starting phase. Mismatches in speed will cause one rotor to act as an asynchronous motor and damage the forging and wedges by heavy currents flowing in them.

2.21 BEARINGS

All turbo-generators require bearings to rotate freely with minimal friction and vibration. The main rotor body must be supported by a bearing at each end of the generator for this purpose. In some cases where the rotor shaft is very long at the excitation end of the machine to accommodate the slip/collector rings, a "steady" bearing is installed outboard of the slip/collector rings. This ensures that the excitation end of the rotor shaft does not create a wobble that transmits through the shaft and stimulates excessive vibration in the overall generator rotor or the turbo-generator line.

There are generally two common types of bearings employed in large generators: "journal" and "tilting pad" bearings (Figs. 2.79 and 2.80). Journal bearings are the

Fig. 2.79 Journal bearing.

Fig. 2.80 Tilting-pad bearing. (Courtesy of Alstom Power Inc.)

most common. Both require lubricating and jacking oil systems, which will be discussed in Chapter 10.

When installing the bearings, they must be aligned in terms of height and angle to ensure that the rotor sits in the bearing correctly. Such things as shaft catenary must be considered and preloading or shimming of the bearings to account for the difference when the rotor is at standstill and at speed. Getting any of these things wrong in the assembly can cause the rotor to vibrate excessively and damage either the rotor shaft or the bearing itself. Generally, a "wipe" of the bearing running surface or Babbitt metal results.

2.22 AIR AND HYDROGEN COOLING

Many of the internal generator components do not have the capability in their design to have direct liquid cooling and, yet, they incur substantial losses during operation. In addition, there is the problem of rotation of the rotor and the windage and friction that goes with it. Therefore, large generator designs need a cooling medium that has good heat transfer properties and low windage and friction characteristics.

Turbo-generators employ either air or hydrogen as the internal cooling medium of the generator. Air is used in the smaller machines (nowadays up to about 300 MVA and growing), but hydrogen is the most effective gas for ventilating a rotating machine and is used in the larger machines to achieve higher ratings. Generally, hydrogen is used in all large turbine generators and most of the medium-size machines, but it has been also used in some smaller generators.

When hydrogen is used as a coolant in generators, it is supplied at a purity of approximately 98% or better. It is usually maintained from a continuous supply of commercial-grade hydrogen of high purity. It is necessary to maintain a large supply for filling the generator after overhauls and to replace gas lost during operation. Hydrogen consumption occurs in the generator by absorption into the seal oil and through small leaks in the hydrogen coolers, stator winding, rotor terminal stud seals, or out of the casing. A pressure regulator holds the hydrogen pressure at the rated level specified by the generator design.

Hydrogen's density at 98% purity is on the order of one-tenth that of air at a comparable pressure. This reduces the fan and windage losses to an extremely low value. Because of this, it is possible to increase hydrogen pressure in machines to as high as 75 psi relative to atmospheric pressure. Because of low windage and friction, the higher pressure does not compromise efficiency.

The main benefit of increasing the hydrogen pressure is that it greatly increases the heat removal capability of the hydrogen. Hydrogen's properties are such that its heat transfer coefficient is 50% more effective than that of air at the same pressure. Therefore, hydrogen is much more effective in removing heat from a surface.

The heat capacity per unit volume (the product of specific heat at constant pressure and density) of hydrogen is approximately equal to that of air at the same pressure. Therefore, the temperature rise of hydrogen would be approximately the same as that

TABLE 2.1 Relative cooling properties of the various cooling mediums used in turbo-generators

Coolant	Specific heat	Specific density	Flow volume	Cooling capacity
Air	1.00	1.00	1.00	1.00
H_2 @ 30 psi	14.36	0.21	1.00	3.00
H_2 @ 45 psi	14.36	0.26	1.00	4.00
H_2 @ 60 psi	14.36	0.35	1.00	5.00
Oil	2.09	848.00	0.012	21.00
Water	4.16	1000.00	0.012	50.00

of air if the same volume flow rate of the two gasses were used to remove the same amount of heat. The temperature rise is substantially reduced because the fan and windage loss is reduced in hydrogen. Table 2.1 compares the cooling properties of a hydrogen, oil, and water versus air.

The hydrogen is circulated throughout the generator by shaft-mounted fans or blowers (see Figs. 2.81 and 2.82). The hot rotor gas is discharged to the airgap, after having absorbed the heat from the field winding losses. The hydrogen is also circulated through the core and stator terminals and then back to the coolers for cooling and recirculation. To remove or introduce hydrogen in the generator, an external system is connected that in most cases employs CO_2 for hydrogen purging on removal and air purging when admitting hydrogen into the machine. This ensures that an explosive mixture of hydrogen and oxygen cannot happen, as would be the case if purging were done with air.

Instrumentation is also generally provided for monitoring of hydrogen gas purity, dewpoint, and temperatures. Air-cooled turbine generators are commonly open ventilated, taking air from outside the machine and discharging the warm air back to the outside in another location.

2.23 ROTOR FANS

The cooling medium inside the generator is required to circulate through the various components of the machine to pick up the generated heat. Cooling gas circulation is accomplished by the use of rotor fans or blowers. There are a number of variations in fan design that are used. The two main types are simple one-stage axial fans (see Fig. 2.81) and radial flow fans (see Fig. 2.82). There is generally one installed at each end of the rotor, although there are many single-fan designs as well. In addition, multi-staged blowers are widely used on generator rotors for cooling-gas circulation in the generator (see Fig. 2.83 and Fig. 2.84).

Rotor fans are highly stressed components that can affect rotor balance as well. Great care is taken in the design of these components to ensure good fatigue life and symmetry of design so that balance of the rotor is not adversely affected in operation.

Fig. 2.81 Axial rotor fan.

2.24 HYDROGEN CONTAINMENT

Since hydrogen is highly combustible in the right air/hydrogen mix, it must be maintained at purity well above 75% or below 4% air. Purity of 98% and above is commonly maintained. It is important to keep the hydrogen content of the air outside the generator at a low level to ensure safe operation of the power plant. To minimize hydrogen

Fig. 2.82 Radial (centrifugal) rotor fan.

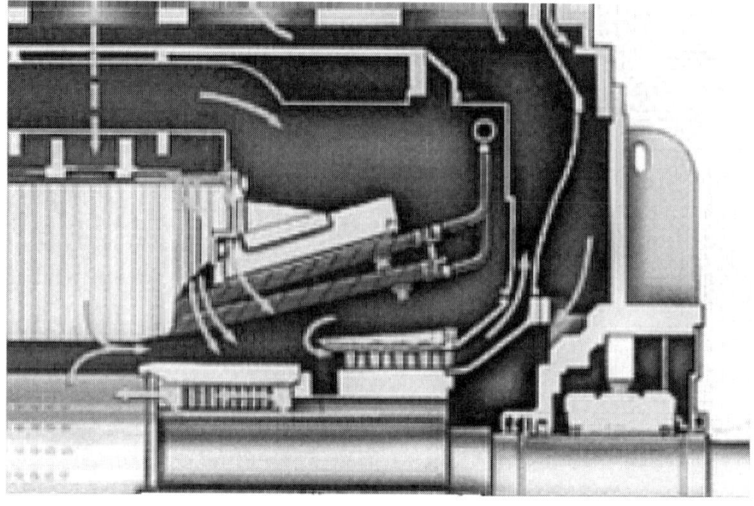

Fig. 2.83 Axial multistaged blower configuration with rotor installed.

escaping from the generator, great care is required in design, installation, and operation. The stator frame provides the primary containment of the hydrogen.

There are two main sealing locations associated with the rotor to keep the hydrogen within the machine from escaping. The first location is the rotor bore, where seals are installed in the shaft to seal the hole in the shaft where the dc field bore copper is installed. This is between the slip rings and the field winding. The seals are generally

Fig. 2.84 Rotor with five-stage axial blower.

made of rubbery material, which under pressure (sometimes from a nut) expands, filling the area between the bolt and shaft, connecting the conductor in the hollow of the shaft to the slip rings.

Some rotors have only one set of seals close to the collectors, whereas other rotors have a second set of seals where the leads exit the shaft, under the retaining rings.

The integrity of the rotor terminal seals is normally checked during major overhauls. Depending on the design of the rotor, some can be pressure tested through a nipple permanently installed on the shaft, or by placing a can over the shaft extension and collector assembly, tightly sealed against the rotor forging, and pressurizing the can.

The second set of seals is to prevent hydrogen from escaping along the shaft where the rotor extends out from the stator bore. A close-clearance oil seal is provided between the stator end doors and the rotor shaft for this purpose. The design must accommodate axial motion of the shaft up to 2 inches (5 cm) or more due to variations in steam turbine temperatures. There are two general types of seals: journal and thrust collar (see Figs. 2.85–88). Both types of seals are used by the different manufacturers to provide a shaft seal, but the journal-type seals are by far the most widely used. Oil is used to prevent the leakage of hydrogen along the shaft at each end of the generator, via the hydrogen sealing arrangements.

Seal oil is supplied at a pressure slightly higher than the hydrogen pressure, and in sufficient quantity to remove the viscous losses in the seals while maintaining proper temperature of the close-fitting parts. Hydrogen that is absorbed by the oil is removed by a "detraining" process, which some manufacturers do in a vacuum.

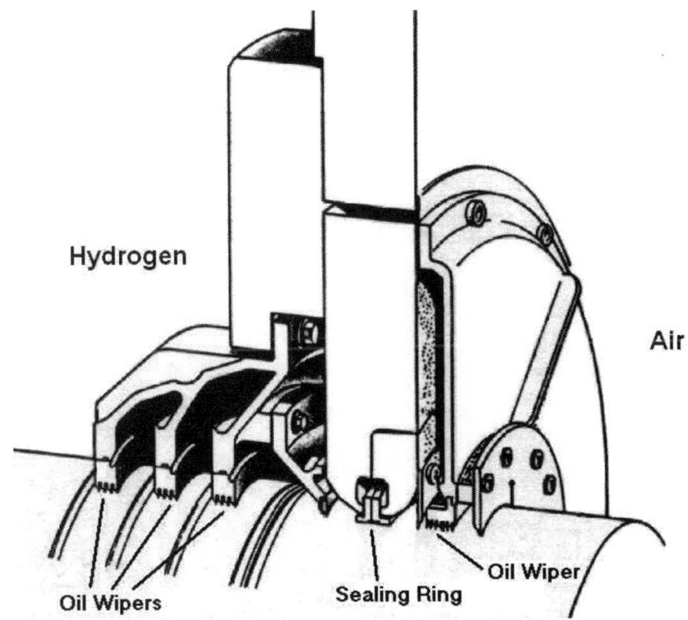

Fig. 2.85 Journal-type hydrogen seal. (Courtesy of Alstom Power Inc.)

Fig. 2.86 Journal-type hydrogen seals.

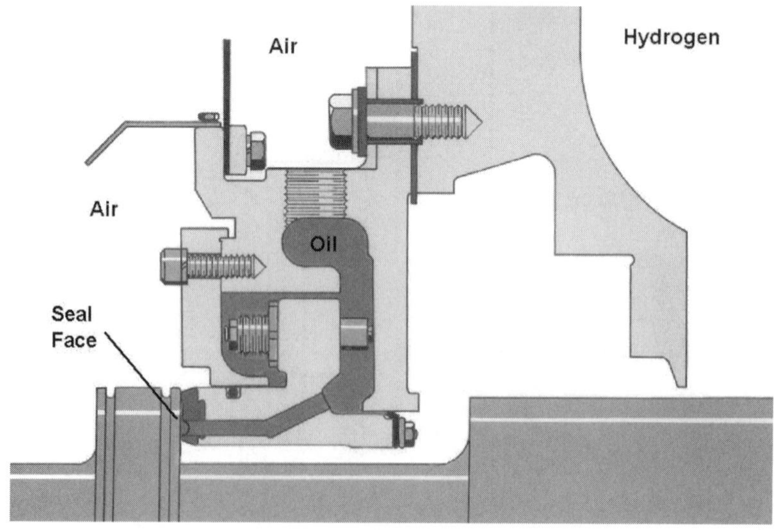

Fig. 2.87 Axial thrust collar. (Courtesy of Siemens-Westinghouse.)

Fig. 2.88 Axial thrust collar seal rings.

The seals are bracket mounted on the stator casing and designed to keep the hydrogen from escaping through the clearance between the moving shaft and the stationary frame, at both ends of the machine. Installing the seals inboard from the bearings achieves this purpose. Sometimes they are mounted on the same bracket as the bearings and sometimes separate, depending on the bearing design and configuration. When *pedestal bearings* are used, the seals are mounted on the brackets (end shields).

Journal-type hydrogen seals consist of a sealing ring supplied by oil under pressure. The sealing ring, with a clearance of about 2.5 to 5 mils (63.5 to 127 μm) per side, can move with the shaft as the shaft moves within the bearing clearances. However, the sealing ring cannot rotate with the shaft. The small clearance between seal ring and shaft is filled with oil at high pressure (several pounds/square inch above hydrogen pressure). The oil keeps the gas from escaping from the machine.

A requisite of the hydrogen seal assembly is that it does not allow humidity and other impurities to contaminate the hydrogen in the machine. There are several types of designs, typically classified as single- and double-flow oil systems. The sealing systems are critical to the operation of the hydrogen-cooled generator. The most obvious issue is safety. The hydrogen must be contained inside the generator and in concentrations that are not in the explosive range. Second, machines operating under hydrogen pressure are severely reduced in output capability if the hydrogen pressure cannot be maintained at its nominal value.

Fig. 2.89 Vertical-type hydrogen cooler.

2.25 HYDROGEN COOLERS

As the hydrogen cooling gas picks up heat from the various generator components within the machine, its temperature rises significantly. This can be as much as 46°C [3] and, therefore, the hydrogen must be cooled down prior to being recirculated through the machine for continuous cooling. Hydrogen coolers or heat exchangers are employed for this purpose (see Fig. 2.89).

Hydrogen coolers are basically heat exchangers mounted inside the generator in the enclosed atmosphere. Cooling tubes with fins are used to enlarge the surface area for cooling, as the hydrogen gas passes over the outside of the finned tubes. Raw water (filtered and treated) from a local river or lake is pumped through the tubes to take the heat away from the hydrogen gas and vent it outside the generator. The tubes must be extremely leak-tight to ensure that hydrogen gas does not enter into the tubes, since the gas is at a higher pressure than the raw water.

REFERENCES

1. M. Tari, K. Yoshida, and S. Sekito (Toshiba Corporation), and R. Brutsch, J. Allison, and A. Lutz (Von Rolttsola Corporation), "HTC Insulation Technology Drives Rapid Progress of Indirect-Cooled Turbo Generator Unit Capacity," presented at IEEE PES 2001 Summer Power Meeting, Vancouver, BC, Canada.

2. K. Sedlazeck, W. Adelmann, H. Bailly, I. Gahbler, H. Harders, U. Kainka, W. Weiss, B. Scholz, S. Henschel (Germany), R. Chianese, P. Hugh Sam, R. Ward, L. Montgomery (United States, Siemens Power Generation), U. Schuberth, H. Spies (Germany, Framatome Advanced Nuclear Power), "Influence of Customer's Specifications Upon Design Features of the EPR Turbogenerator," presented at CIGRE '2002, Paris, France.
3. IEEE/ANSI C50.13-1989, "Requirements for Cylindrical-Rotor Synchronous Generators."
4. H. Feichtinger, HFC-Consulting, Zürich, Switzerland; G. Stein and I. Hucklenbroich, VSG Energie- und Schmiedetechnik GmbH, Essen, Germany, "Case History of a 18Mn-18Cr Retaining-ring Affected by Stress Corrosion Cracks," presented at EPRI Generator Retaining-ring Workshop, December 8–9, 1997, Miami, FL.
5. A. G. Seidel, Houston Lighting & Power Company, "Surface Indications On 18Mn-18Cr Retaining-ring," presented at EPRI Generator Retaining-ring Workshop, December 8–9, 1997, Miami, FL.
6. ERPI Special Report EL/EM-5117-SR, "Guide for Evaluation of Generator Retaining Rings," April 1987.
7. J. M. Fogarty, GE Power Systems, "Connections between Generator Specification and Fundamental Design Principles," presented at IEMDC 2001, Cambridge, MA.
8. M. G. Say, *Alternating Current Machines*. Pitman Publishing, 1978.
9. A. E. Fitzgerald and C. Kingsley, *Electric Machinery*. McGraw-Hill, 1971.

3

GENERATOR AUXILIARY SYSTEMS

All large generators require auxiliary systems to handle such things as lubricating oil for the rotor bearings, hydrogen cooling apparatus, hydrogen sealing oil, demineralized water for stator winding cooling, and excitation systems for field-current application. Not all generators require all these systems and the requirements depend on the size and nature of the machine. For instance, air-cooled turbo-generators do not require hydrogen for cooling and, therefore, no sealing oil as well. On the other hand, large generators with high outputs, generally above 400 MVA, have water-cooled stator windings, hydrogen for cooling the stator core and rotor, seal oil to contain the hydrogen cooling gas under high pressure, lubricating oil for the bearings, and, of course, an excitation system for field current.

This chapter discusses the general nature of the five major auxiliary systems that may be in use in a particular generator:

1. Lubricating-oil system
2. Hydrogen cooling system
3. Seal-oil system
4. Stator cooling water system
5. Excitation system

Each system has numerous variations to accommodate the hundreds of different generator configurations that may be found in operation. But regardless of the generator design and which variation of a system is in use, they all individually have the same basic function as described in the first paragraph.

3.1 LUBE-OIL SYSTEM

The lubricating-oil (lube-oil) system provides oil for all of the turbine and generator bearings and also is the source of seal oil for the seal-oil system. The lube-oil system is generally grouped in with the turbine components and is not usually looked after by the generator-side personnel during maintenance. It is mentioned primarily for completeness.

The main components of the lube-oil system consist generally of the main lube-oil tank, pumps, heat exchangers, filters and strainers, centrifuge or purifier, vapor extractor, and various check valves and instrumentation. The main oil tank serves both the turbine and generator bearings and is often also the source of the sealing oil for the hydrogen seals. It is usually located under the turbines and holds thousands of gallons of oil.

There are a number of philosophies for designing the main lube-oil pump system. Some machines have a main shaft-driven pump to supply oil to the bearings under normal operation. Others machines employ an ac motor-driven pump to deliver the lubricating oil to the generator and turbine bearings. In most instances a dc motor-driven pump is used for emergency shutdown.

Heat exchangers are provided for heat removal from the lube oil. Raw water from a local lake or river is circulated on one side of the cooler to remove the heat from the lube oil circulating on the other side of the heat exchanger.

Full-flow filters and/or strainers are employed for removal of debris from the lube oil. Strainers are generally sized to remove larger debris and filters remove debris in the range of a few microns and larger. They can be mechanical or organic-type filters and strainers. Debris removal is important to reduce the possibility of scoring the bearing babbitt material or plugging of the oil lines.

A centrifuge or purifier is used to remove moisture from the oil. Moisture is also a contaminant to oil and can cause it to lose its lubricating properties.

3.2 HYDROGEN COOLING SYSTEM

Hydrogen is used for cooling in most large turbo-generators rather than air for several reasons:

- Inherently better heat transfer characteristics (approximately 14 times)
- Better heat transfer with higher hydrogen pressure
- Less windage and friction losses than air

- Suppression of partial discharges with increased hydrogen pressure
- Significant increases of the breakdown voltage of generator components

Although hydrogen is a very useful medium for cooling the generator internal components, it is very dangerous if not handled correctly. A dedicated system to handle the supply and control of the hydrogen atmosphere inside the generator is required. Since hydrogen is used at generator casing pressures up to 90 psig, the generator is also considered a pressure vessel. This requires various sealing arrangements to keep the hydrogen inside the machine. These sealing arrangements include the hydrogen seals from the stationary stator casing to the rotating shaft of the rotor, the terminal studs in the rotor forging, pipe flanges, instrumentation ports, and so on.

Supply of the hydrogen to the generator is generally provided by an on-site hydrogen manufacturing plant, or purchased in a pressure container and replenished periodically. Delivery to the generator is accomplished by a system of pipes, valves, and pressure regulators. Control is achieved by pressure regulators and gas purity and dew-point monitors. Dedicated hydrogen drying equipment is also sometimes used when seal-oil vacuum treatment is not provided on the seal-oil system. (This will be discussed in Section 3.3.)

In addition to the hydrogen, a separate supply system is required for CO_2 to purge the generator of hydrogen during filling and degassing. CO_2 is used because it is inert and will not react with the hydrogen. If the hydrogen in the generator were to be purged with air, this would encroach upon both the upper and lower explosive limits due to the combustible nature of a hydrogen/oxygen mixture. Hydrogen at high purity (above 90%) will not support combustion, and at this level there is no danger of explosion since the explosive range of a hydrogen/oxygen mixture is 4 to 75% hydrogen in air. To prevent the possibility of an explosive mixture when filling the generator with hydrogen for operation, air is first purged from the generator by CO_2, and the CO_2 is then purged by hydrogen. When degassing the generator for shutdown, hydrogen is first displaced by CO_2 and then the CO_2 is purged by air. This way, no explosive mixture of hydrogen and oxygen can occur. In some rare cases, other inert gases have been used, such as argon.

During operation, a gas pressure regulator automatically maintains the generator casing hydrogen pressure at a preset (rated) value. If hydrogen leaks occur, the pressure regulator admits additional hydrogen from the supply system until the predetermined pressure is restored. There is always a certain amount of expected leakage into the seal oil, through minute leaks, permeation through the stator winding hoses, and so forth, but most generators should be capable of continuous operation below 500 cubic feet per day loss. If the loss increases to 1500 cubic feet per day, the source of the leak should be investigated immediately and corrected.

A hydrogen gas analyzer is usually present to monitor the hydrogen purity, which should be maintained above 97%. Dew-point monitoring is sometimes provided to control the level of moisture inside the generator. The dew point is generally maintained below –10°C and should not be allowed to rise above 0°C at generator casing pressure. The best type of dew-point monitoring is a system that also works when the generator is offline. The main concern is actually not when the generator is running. During shutdown, when the internal component temperatures inside the machine cool down, there is increased possibility of moisture in the hydrogen condensing onto stationary components such as the 18 Mn–5 Cr retaining rings. Moisture will attack both the insulation systems and certain metal components inside the generator.

Inside the generator, the hydrogen picks up heat from the various components as it flows over and through such components as the stator core vents and rotor winding. Then it is routed to pass through heat exchangers inside the generator, where the hydrogen leaving the cooler outlet side has been reduced in temperature to complete another cycle of heat pickup as it goes through the same generator components again.

The hydrogen is maintained at the correct purity, dew point, and pressure by an external system that handles all of the above-discussed functions. The system is often a package unit provided with the generator and then connected to the generator by pipes and control wiring. An example of such a package system is shown in Figure 3.1. Some units are also supplied with dedicated hydrogen dryers as a separate unit (see Fig. 3.2 and Fig. 3.3).

In some plants, one may find an elaborate hydrogen cooling system panel where all hydrogen controls are located (see Fig. 3.4). All the pressure and temperature controls, plus the purging controls and the purity indicator, would be located in this panel. When

Fig. 3.1 Hydrogen cooling system—packaged unit. (Courtesy of Alstom Power Inc.)

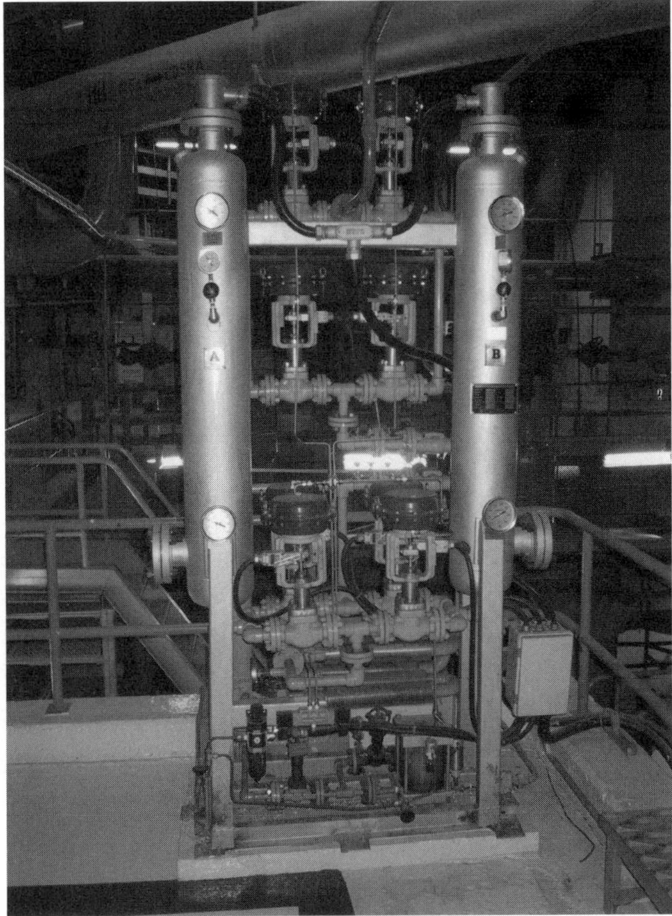

Fig. 3.2 Hydrogen dryer packaged unit with twin drying towers. One side is usually in service while the other side regenerates its desiccant.

working in the hydrogen control panel, sparkless tools made of bronze are required so that any hydrogen that may have leaked and collected inside the panel cannot be ignited.

3.3 SEAL-OIL SYSTEM

As previously mentioned, most large generators use hydrogen under high pressure for cooling the various internal components. To keep the hydrogen inside the generator, various places in the generator must be sealed to prevent hydrogen leakage to atmosphere. One of the most difficult seals to make is the juncture between the stator and the

Fig. 3.3 Another example of a hydrogen dryer packaged unit with twin drying towers. One side is usually in service while the other side regenerates its desiccant.

rotating shaft of the rotor. This is done by a set of hydrogen seals at both ends of the machine. Also, as previously described, the seals may be of the journal (ring) type or the thrust-collar type, one thing both arrangements have in common is the requirement of high-pressure oil in the seal to do the actual sealing. The equipment that provides the oil to do this is called the seal-oil system.

In general, the most common type of seal is the journal type. This arrangement uses pressurized oil fed between two floating segmented rings, usually made of bronze or babbitted steel. At the ring outlet, against the shaft, oil flows in both directions from the seals and along the rotating shaft. For the thrust-collar type, the oil is fed into a babbitted running face via oil delivery ports and makes the seal against the rotating thrust collar. Again, the oil flows in two directions, to the air side and the hydrogen side of the seals.

The seal-oil itself is actually a portion of the lube oil, diverted from the lube-oil system. It is then fed to a separate system of its own, with pumps, motors, hydrogen-de-

Fig. 3.4 Hydrogen control panel.

training or vacuum-degassing equipment, and controls to regulate the pressure and flow.

The seal-oil pressure at the hydrogen seals is maintained generally about 15 psi above the hydrogen pressure to stop hydrogen from leaking past the seals. The differential pressure is maintained by a controller to ensure continuous and positive sealing at all times when there is hydrogen in the generator.

One of the critical components of the seal-oil system is the hydrogen degasifying plant. The most common method of removing entrained hydrogen and other gases is to vacuum treat the seal oil before supplying it to the seals. This is generally done in the main seal-oil supply tank (see Fig. 3.7). As the oil is pulled into the storage tank under vacuum, through a spray nozzle, the seal oil is broken up into a fine spray. This allows the removal of dissolved gases. In addition, there is often a recirculating pump to recirculate oil back to the tank through a series of spray nozzles for continuous gas removal.

After passing through the generator shaft seals, the oil goes through the detraining sections before it returns to the bearing-oil drain. As a safety feature, there is often a second ac motor-driven pump or a dc motor-driven emergency seal-oil pump provided. The backup motor will start automatically on loss of oil pressure from the main seal-oil pump. This is to ensure that the generator can be shut down safely without risk to personnel or the equipment.

Figures 3.5 and 3.6 show two types of seal-oil systems. Figure 3.8 shows a typical seal-oil skid supplied with large generators.

3.4 STATOR COOLING WATER SYSTEM

The stator cooling water system (SCW) is used to provide a source of demineralized water to the generator stator winding for direct cooling of the stator winding and associated components. The SCW is generally used in machines rated at or above 300 MVA. Most SCW systems are provided as package units, mounted on a single platform, which includes all of the SCW system components. All components of the system are generally made from stainless steel or copper materials. See Figure 3.9 for a typical stator cooling water system mounted on a skid.

3.4.1 System Components

Pumps. Generally, ac motor-driven pumps are used to deliver the cooling water to the windings. In some instances, a dc motor-driven pump is used for emergency shutdown.

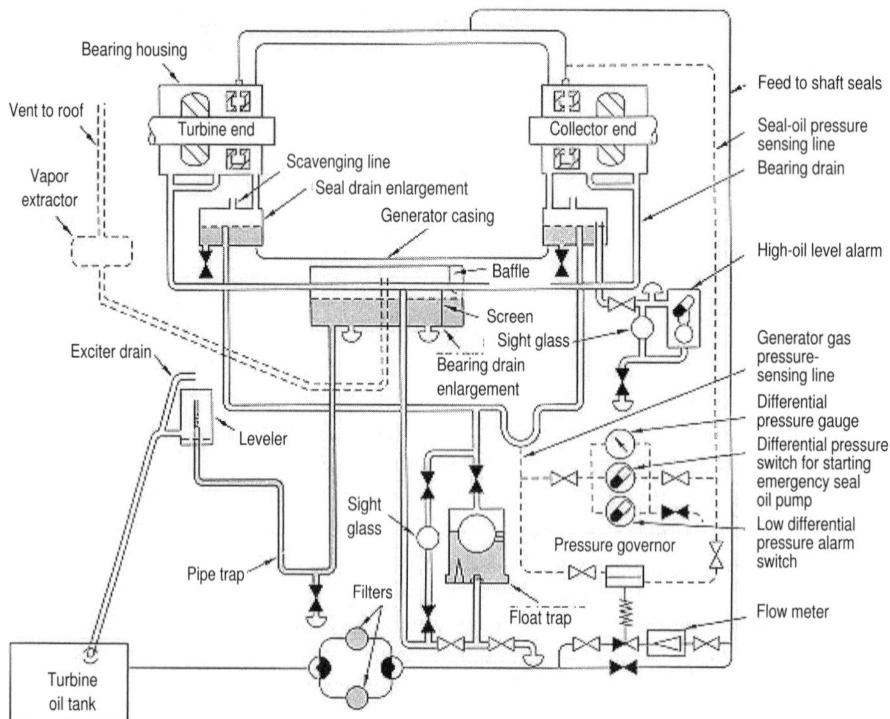

Fig. 3.5 Seal-oil system, scavenging type. (Courtesy of General Electric.)

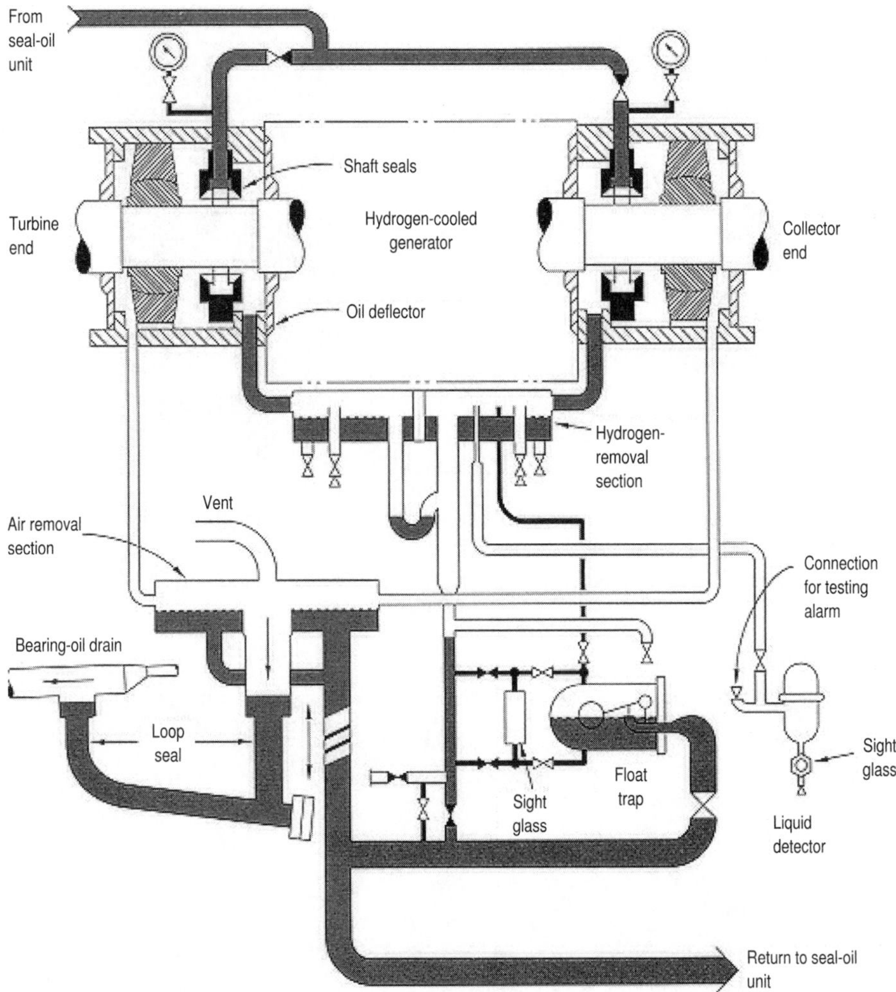

Fig. 3.6 Seal-oil system, vacuum type. (Courtesy of General Electric.)

Heat Exchangers. Heat exchangers are provided for heat removal from the SCW. Raw water from a local lake or river is circulated on one side of the cooler to remove the heat from the demineralized SCW circulating on the other side of the heat exchanger.

Filters and/or Strainers. Full-flow filters and/or strainers are employed for removal of debris from the SCW. Strainers are generally sized to remove debris in the 20 to 50 μ range and larger, and filters for debris in the range of 3 μ and larger. They can be mechanical or organic-type filters and strainers. Debris removal is important to reduce the possibility of plugging in the stator conductor bar strands.

Fig. 3.7 Hydrogen removal tank and associated piping under the generator for the system of Figure 3.6.

Fig. 3.8 Seal oil system, packaged unit. (Courtesy of Alstom Power Inc.)

Fig. 3.9 Stator cooling water system, packaged unit. (Courtesy of Alstom Power Inc.)

Deionizing Subsystem. A deionizing subsystem is required to maintain low conductivity in the SCW, generally on the order of 0.1 μS/cm. High conductivity can cause a flashover to ground in the stator winding, particularly at the Teflon hoses where an internal tracking path to ground exists. The system generally maintains a continuous bleed-off of 5% from the main SCW flow to keep the conductivity in the operable range.

Stator Cooling Water System Storage or Makeup Tank. In the event the SCW is lost, or the SCW system must be refilled after shutdown and draining, the system requires replenishing. Therefore, a storage tank to hold sufficient makeup water is required. Some systems are open to the atmosphere, whereas others maintain a hydrogen blanket on top of the water to keep the level of oxygen at a minimum. The storage tank arrangement is manufacturer specific, depending on the desired water chemistry.

Gas Collection and Venting Arrangement. Since no SCW system is leakproof, there is some ingress of hydrogen and natural collection of other gases such as oxygen in the SCW system. A means for venting off these gases is required. Generally, the excess gases are vented to the atmosphere. In some systems, the venting process is monitored and/or quantified and in other systems there is none. This is manufacturer specific (see Fig. 3.10).

Fig. 3.10 Stator cooling water system—H_2 gas release and alarm tank.

3.4.2 Stator Cooling Water Chemistry

Conductivity. High conductivity of the demineralized cooling water can cause electrical flashover to ground by tracking. Acceptable levels are generally maintained near 0.1 μS/cm.

Copper/Iron Content. The content of copper and iron in the SCW is normally less than 20 ppb. High concentrations of either could cause conductivity problems.

Hydrogen Content. When no leaks are present in the system, hydrogen content is at a minimum. High hydrogen in the SCW can cause gas locking if the leak rate is too high. Excessive venting of hydrogen is an indication of a high leak rate. High concentrations of hydrogen may also cause conductivity problems.

Oxygen Content. The dissolved oxygen content of the SCW is controlled to prevent corrosion of the hollow copper strands. Corrosion products can build up and block the cooling water flow. Oxygen at 200 to 300 μg/l produces the highest corrosion rate. The content of oxygen in the SCW is normally maintained at less than 50 ppb in hydrogen-saturated and low-oxygen-type systems, and without limit for open-vented or high-oxygen-type systems (see Fig. 3.11).

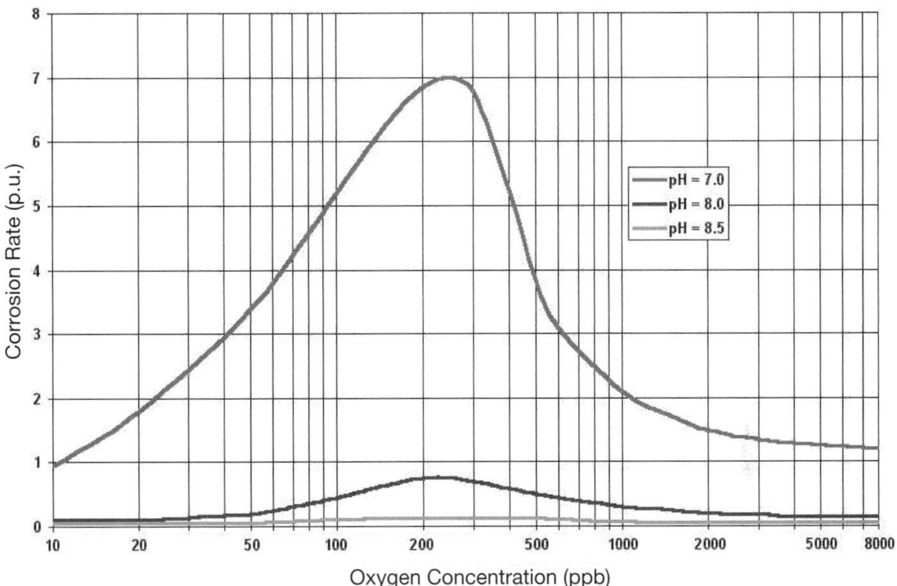

Fig. 3.11 Stator cooling water copper corrosion rate plotted against oxygen content and pH.

High-Oxygen System. High oxygen refers to air-saturated water with dissolved oxygen present in the SCW in the range of greater than 2000 μg/l (ppb) at STP. The high-oxygen system is based on the supposition that the surface of pure copper forms a corrosion-resistant and adherent cupric oxide layer (CuO) that becomes stable in the high-oxygen environment. The oxide layer is generally resistant to corrosion as long as the average water velocity is less than 15 ft/s. Figure 3.12 shows a typical high-oxygen system.

Low-Oxygen System. Low oxygen refers to SCW with a dissolved oxygen content less than 50 μg/l (ppb). The low-oxygen system is based on the supposition that pure copper does not react with pure water in the absence of dissolved oxygen. The upper limit is set by the corrosion rate that the water cleansing system can handle. The lower limit is set to the level where copper will not deposit on any insulating surface in the water circuit such as a hose. This is to avoid electrical tracking paths to ground. It has better heat transfer properties at the copper/water interface and a lower copper ion release rate. Figure 3.13 depicts a typical low-oxygen system.

pH Value. The pH value is manufacturer specific. Generally, there are two modes of operation: neutral and alkaline.

Neutral pH refers to a pH value of around 7. A neutral pH with low oxygen content less than 50 μg/l works best. Oxygen at 200 to 300 μg/l produces the highest corrosion rate, but high oxygen over 2000 μg/l will also work.

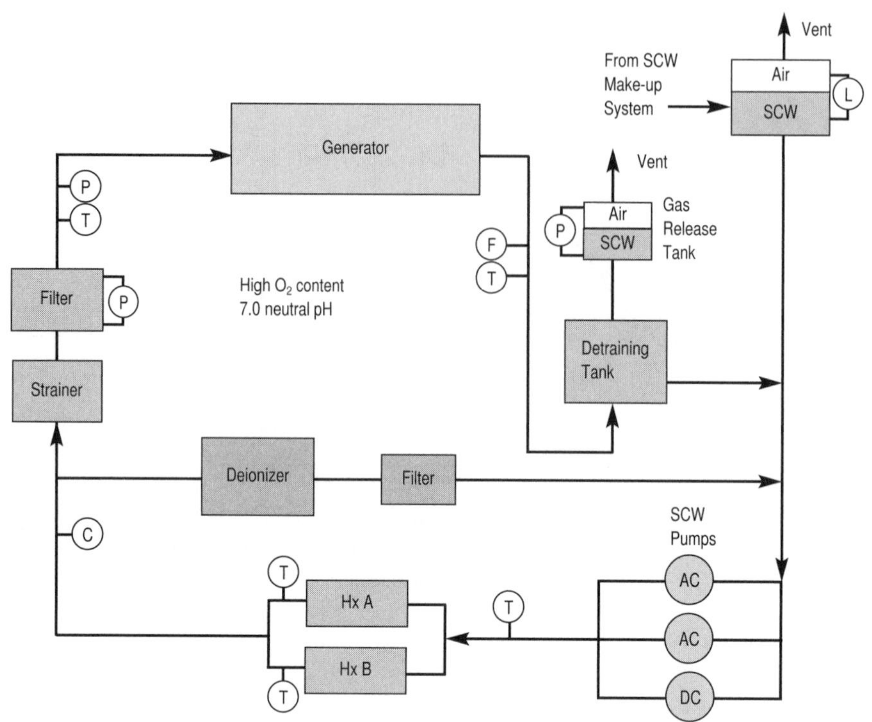

Fig. 3.12 Stator cooling water system, high-oxygen type.

Alkaline pH refers to a high pH value of around 8.5. This works for both low- and high-oxygen cooling water philosophies. This method requires an alkalizing subsystem to keep the pH at the proper level. Figure 3.14 shows such a system for pH control.

Deionizer Materials/Resins. Many deionizing systems use the mixed-bed type, employing both a strongly acidic cation resin and a strongly basic anion resin. The operation of the deionizing subsystem requires a small percentage of full coolant flow to pass through the deionizer on a continual basis. The percentage of full flow varies from system to system and can be found in the range of 5 to 20%.

3.4.3 Stator Cooling Water System Conditions

Stator Cooling Water Inlet Temperature. Generally, the SCW inlet temperature is maintained below 50°C, but normally operates in the 35 to 40°C range. It is desirable to keep the stator core, winding, and hydrogen gas temperatures in the same range to minimize thermal differentials.

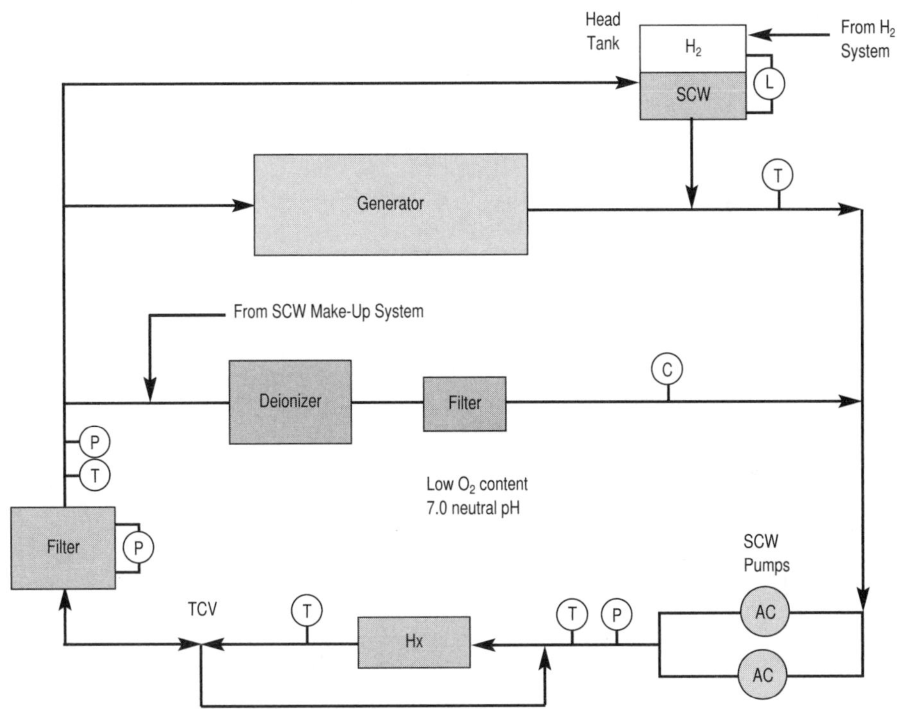

Fig. 3.13 Stator cooling water system, low-oxygen type.

Stator Cooling Water Outlet Temperature. The SCW outlet temperature is usually maintained below 75°C. The limit is about 90°C. Beyond that, overheating of the stator bars is likely to occur and, possibly, boiling of the SCW.

Pressure. The SCW inlet pressure to the generator is generally kept at a design level to ensure cooling water flow to the stator winding, but 5 psi below the hydrogen gas pressure to minimize to possibility of water leakage into the generator.

Stator Winding Differential Pressure. There will be a normal differential pressure that exists across the stator winding, based on design factors. A higher than normal differential cooling water pressure across the stator winding can indicate that there is a cooling water flow problem. This may be due to plugging, hydrogen gas locking, and so on.

Flow. Continuous cooling water flow is essential to carry generated heat away. Flow velocities are design specific and are based on such things as heat-carrying capacity of the water, cross-sectional flow area in each bar, and corrosion effects on the copper.

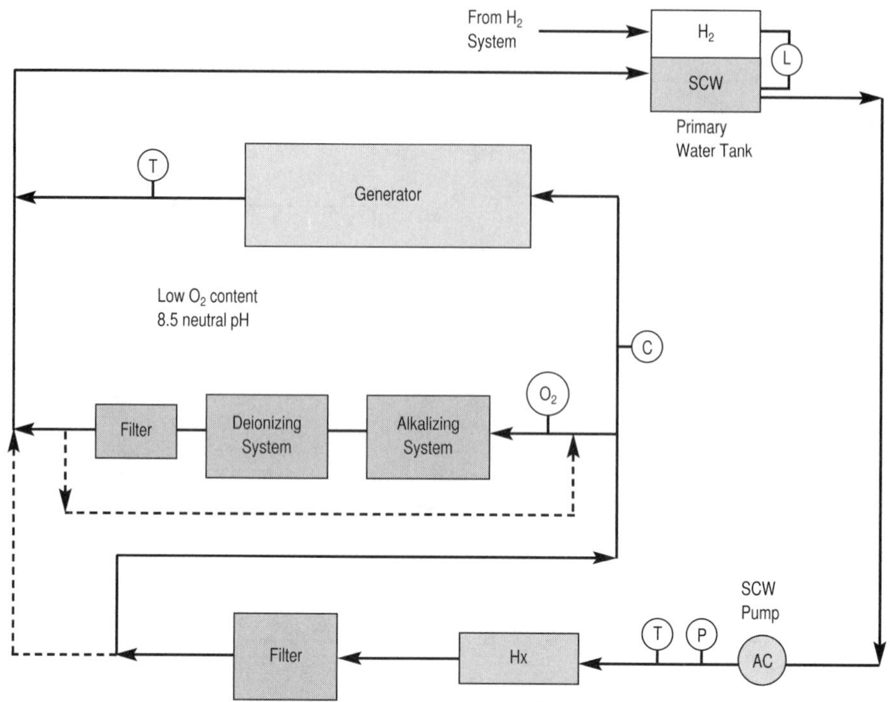

Fig. 3.14 Stator cooling water system, alkaline injection type.

Stator Cooling Water System Instrumentation. In all SCW systems, the various parameters involved are monitored at different points in the system. The general parameters monitored are: SCW inlet to generator temperature, SCW outlet from generator temperature, SCW inlet pressure to generator, SCW outlet pressure from generator, differential pressure across filters and strainers, system SCW flow, SCW conductivity, and storage and makeup tank levels. In addition, the raw service water for the SCW coolers is also generally monitored for inlet and outlet temperature of the coolers.

3.5 EXCITER SYSTEMS

For the generator to function as a generator, magnetizing current (or "excitation") must be supplied to the generator rotor winding. The excitation system provides this function. The system is designed to control the applied voltage and, thus, the field current to the rotor, which in turn gives control of the generator output or terminal voltage. Subsequently, this is what provides reactive power and power factor control between the generator and the system.

Voltage requirements range from very low dc levels up to 700 volts dc for the larger generators. The dc field currents may approach 8000 amps dc on the larger turbogenerators.

Excitation response time must be fast so that the automatic voltage regulator can control the generator during system disturbances or transients in which rapid changes of excitation are necessary. *Field forcing* is the term generally used for this mode of operation and requires the exciter to be capable of field forcing voltages from two to three times the rated dc field voltage. Therefore, the rotor winding must be insulated for these voltage levels.

3.5.1 Types of Excitation Systems

The three basic excitation system types are as follows:

1. Rotating
2. Static
3. Brushless

Rotating. Within the family of rotating exciters, there are numerous types that can be found operating on all types and sizes of generators. The basic kinds of rotating exciters are motor or shaft driven and separately self-excited or bus-fed systems. The subject of excitation systems is a book in itself, and it is not our intention to focus on exciters in this book. They are discussed in brief as an auxiliary system to the generator.

The shaft-driven rotating excitation system has been the most widely used excitation source in the past. It is used for small and large turbine generators. The basic configuration is a stationary armature and a rotating field. The ac output of the alternator is rectified by stationary diodes physically located off the generator and fed to the main rotor slip rings as dc current. Regulation of the current output is achieved by phase control of thyristors in the alternator field power circuit (see Fig. 3.15). Figure 3.16 shows the typical basic circuit for a shaft-driven excitation system.

Static. An external source of power (often the generator isolated phase bus) is used to supply ac power to an *excitation transformer*. The transformer output is fed to a three-phase controlled rectifier bridge for conversion to dc (see Fig. 3.17). The required generator field voltage is obtained by properly controlling the thyristor firing as in the rotating exciter. The standard control generally consists of an ac (auto) control mode for regulating generator terminal voltage, and a dc (manual) control mode for regulating exciter field voltage (see Fig. 3.18).

Brushless. The brushless excitation system is most widely used for gas turbine units with air-cooled generators, but also can be found in steam generators of every rating. It consists of a high-frequency ac generator with a rotating fused diode assembly, a static *voltage regulator* for excitation control, and a transformer power

Fig. 3.15 A six-pole rotating exciter with a reluctance rotor of the unit's high-frequency generator attached on the left is shown. The exciter slip rings are located on the right side, but covered with a protection barrier during overhaul. The slip rings carry the dc current out from the exciter to a busbar that feeds the main generator field through its own set of slip rings.

supply. The exciter is generally attached as an extension of the generator shaft. Rectifier components are mounted on the diode wheel assembly or, sometimes, housed inside the shaft. A three-phase transformer, backed up by the station battery for initial field flashing support and momentary fault current support, powers the regulator. The standard control generally consists of an ac (auto) and dc regulator (see Figs. 3.19 and 3.20).

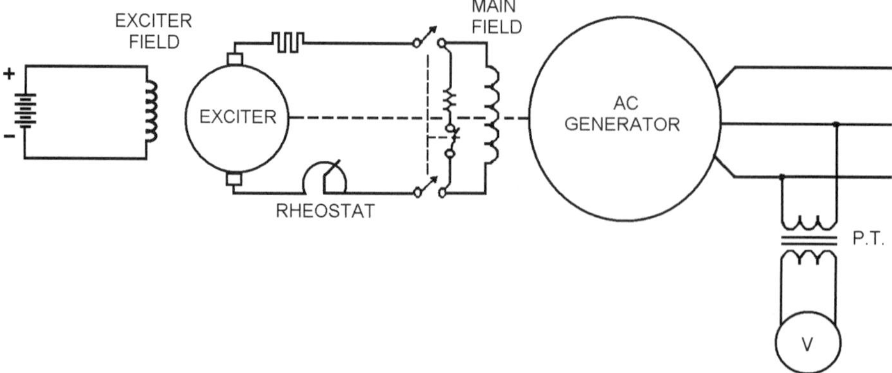

Fig. 3.16 Schematic of a shaft-driven, separately excited excitation system.

Fig. 3.17 Static exciter.

3.5.2 Excitation System Performance Characteristics

The principal function of the excitation system is to furnish dc power (direct current and voltage) to the generator field, creating the magnetic field. The excitation system also provides control and protective equipment that regulates the generator electrical output.

Excitation voltage is a key factor in controlling generator output. One desirable characteristic of an excitation system is its ability to produce high levels of excitation

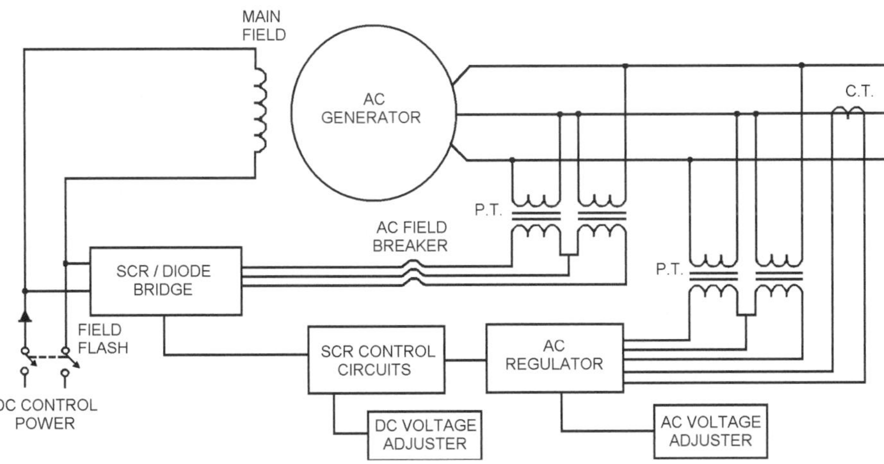

Fig. 3.18 Static exciter schematic.

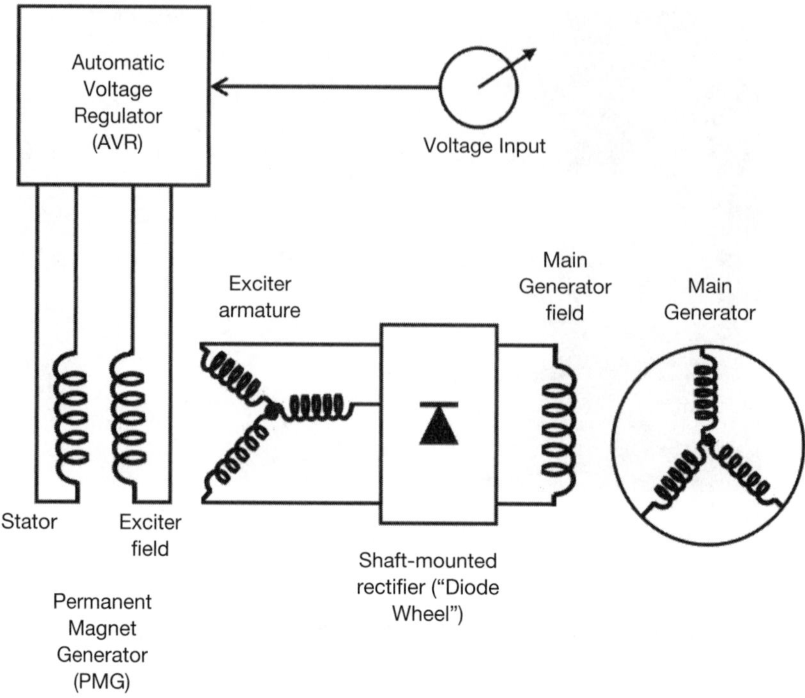

Fig. 3.19 Brushless exciter schematic.

voltage (the ceiling) very rapidly following a change in terminal voltage. IEEE defines a high initial response (HIR) excitation system as one that reaches 95% of the specified ceiling voltage in 0.1 second or less. For units tied into a power system grid, such quick action to restore power system conditions reduces the tendency for loss of synchronization.

The other important performance feature of an excitation system is the level or amount of ceiling voltage it can achieve. The *response ratio* is the term for quantifying the forcing or ceiling voltage available from the exciter. The response ratio is the average rate of rise in exciter voltage for the first one-half second after change initiation, divided by the rated generator field voltage. Thus, it is expressed in terms of per unit (pu) of rated field voltage. A standard level of exciter response ratio is 0.5 pu. This level has been found to be adequate for the large majority of industrial and utility applications. Power system studies have shown that some applications benefit from higher response ratios or more powerful exciters.

In general, it can be observed that conventional rotating exciters, such as the classical rotating and the brushless-type exciters have slower response times due to the time constants of the rotating magnetic components. In fast-acting static exciters, maximum exciter output is available almost instantaneously by signaling the controlling thyristors to provide full forcing.

Fig. 3.20 Brushless exciter rotor shown with diode wheel. (Courtesy of Electric Machinery.)

3.5.3 Voltage Regulators

The generator's voltage regulator is another important part of the excitation system and is often provided as a dual-voltage regulator. It is generally comprised of automatic (ac) and manual (dc) regulators. The ac regulator regulates the generator ac terminal voltage and the dc regulator regulates the exciter dc field voltage. The dc regulator controls the voltage regulator's output voltage in order to regulate the generator field current. The ac regulator controls the voltage regulator's output voltage in order to regulate the ac voltage at the generator terminals. The selection of ac or dc regulator control is by the operator or by protective relaying.

The regulator, which forms part of the excitation control system, typically provides additional control and protective functions for the main generator. Many of these functions are listed below. Some of these features mentioned below are not applicable to a brushless excitation system.

- Underexcitation limiter. Minimizes the generator end-iron heating and prevents excessive underexcitation operation and possible dynamic instability.
- Generator field maximum excitation limiter. Protects the generator field from overheating due to prolonged excessive field current.
- Underfrequency voltage limiter.
- Overvolts per hertz equipment. Protects the generator and connected magnetic apparatus from damage due to excessive magnetic flux at any frequency.

- Field ground detection.
- Field temperature monitoring.
- Shaft voltage suppression. Reduces the effects of capacitive-coupled voltages in the generator shaft from damaging bearings and other components in the turbine and generator.
- Field current and voltage transducers. Permits isolated voltage monitoring of generator field current and voltage.
- Reactive current compensation. Permits two or more parallel generators to operate together and share reactive current.
- Active and reactive current compensation. Permits holding constant voltage at some point in the power system remote from the generator.
- Voltage matching equipment. Permits safe synchronizing of the generator with a power system.
- Automation features. Permit computer control of starting, loading, and shutting down a generator.
- Power system stabilizer. Dampens power system oscillations by permitting terminal voltage to change in phase with changes in generator speed.

4

OPERATION AND CONTROL

Chapters 2 and 3 describe the very complex construction aspects of a turbo-generator and the peripheral equipment required for its operation. One can thus imagine that this is a very costly system, in addition to being a very critical component in a power plant. Given the importance of this machine, the need to operate it reliably for many years, and the initial large capital expenditure, is goes without saying that a large effort must go into preparing a comprehensive and detailed purchasing specification. Assuming this has been done, and a well-designed and well-manufactured generator has been delivered, the long-term availability and reliability of the unit will depend greatly on how the machine is operated and maintained. A well-designed and well-manufactured unit can become compromised by serious operational challenges, some of which, as will be discussed below, are outside the control of the plant operators. However, even in this case several ameliorating actions can be taken by a proactive station management. Other challenges are directly related to how the operators run the unit. All these operational aspects are discussed in this chapter; the very important maintenance issues are covered in Part II of this book.

4.1 BASIC OPERATING PARAMETERS

All generators are designed such that they have a *rating*. The rating of the machine is a series of parameters that describe the generator in engineering terms. These parameters tell about the available power output of the generator and its capability with regard to electrical, thermal, and mechanical limits. With enough experience, the trained person can also often infer other information about such things as the generator size and basic construction features.

Like any industrial apparatus, large generators are specified, designed, and constructed to meet a number of requirements. These requirements are predicated on customer needs, as well as mandatory industry standards and "best industry practice" guidelines. The requirements are given in the form of performance parameters and dimensional standards.

The performance parameters of a large generator are defined in a number of standards. In the United States, the leading standards defining generator performance variables are IEEE Std. 67 and ANSI/IEEE C50.13 and C50.12; see [1,2,16,17,18]. In other countries, these standards may also apply, in addition to IEC, CIGRE, and local codes, like VDE in Germany and others. In the following subsections, a definition and, when required, an explanation of all performance parameters is given.

4.1.1 Machine Rating

A generator is usually described by giving it a rating. This rating is given at the generator's capability point of maximum continuous power output. The terms generally used to provide the rating are as follows:

Apparent power	MVA	Megavolt-amperes
Real power	MW	Megawatts
Reactive power	MVAR	Megavolt-ampere reactive
Power factor	pf	A dimensionless quantity
Stator terminal voltage	V_t	Alternating voltage
Stator current	I_a	Alternating current
Field voltage	V_f	Direct voltage
Field current	I_f	Direct current
Frequency	Hz	Hertz
Speed	rpm	Revolutions per minute
Overspeed capability	rpm	Revolutions per minute
Hydrogen pressure	psi	Pounds per square inch
Hydrogen temperature	°C	Degrees centigrade
Stator winding insulation class		
Stator winding temperature rise		
Rotor winding insulation class		
Rotor winding temperature rise		
Short-circuit ratio		

Each of these parameters signifies a finite design quantity that describes a certain capability or limitation of the generator. In some cases, they also provide operating limits that, if exceeded, will cause excess stress in the generator (mechanical, thermal, or electrical) on one or more of its components.

All large generators are designed with these parameters in mind and they are all reflected in the design standards for generators [2]. There are specific ranges for the above-mentioned parameters, and these are outlined in the design standards and discussed in documents regarding good operating practice for large generators [1].

The ratings of large generators have increased dramatically over the years as designers have learned to incorporate newer and better materials in their designs and to optimize the use of the materials. The rate of increase of generator ratings over the years has been logarithmic (see Fig. 4.1).

Gas-turbine generators are presently being built with ratings up to approximately 400 MVA. Steam-turbine generators are presently being built with ratings up to approximately 1600 MVA, but there are designs up over 2000 MVA. An example of a nameplate that may be found on a large generator is shown in Figure 4.2.

4.1.2 Apparent Power

Apparent power refers to the rating of a turbine generator. In large generators, it is almost always given in units of megavolt-amperes (MVA), although it may be also be stated in kVA. Although machines are commonly talked about in terms of *real power* [almost always given in megawatts (MW), though it may also be stated in kW], it is the apparent power that best describes the rating. This is because the product of the voltage and the current (MVA) largely determines the physical size of a machine.

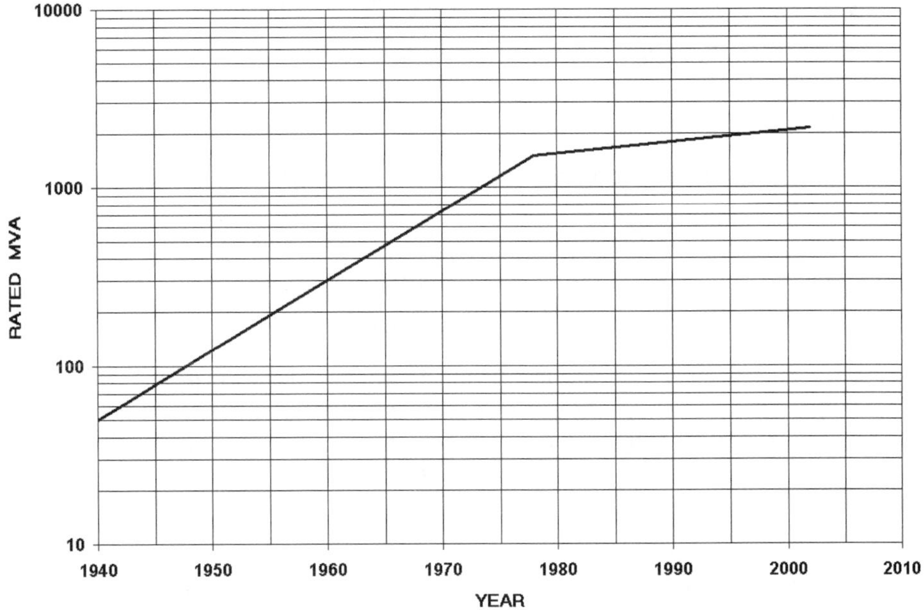

Fig. 4.1 Trend in MVA rating of large turbo-generators.

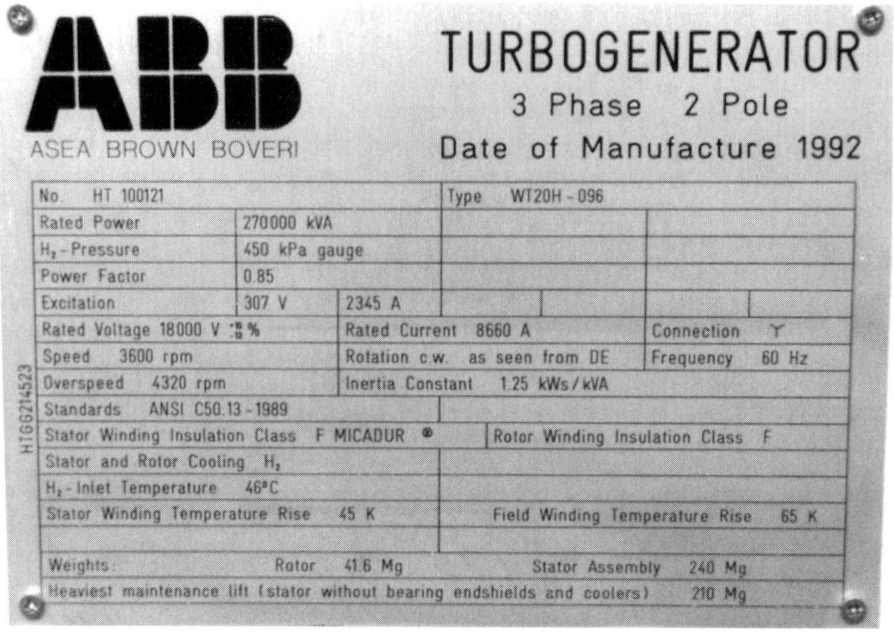

Fig. 4.2 Typical nameplate for a large turbo-generator.

In a three-phase power system, the MVA is given by the following expression:

$$\text{MVA} = \sqrt{3}\,(\text{Generator's line current in kA}) \times (\text{Line voltage in kV})$$

Alternatively,

$$\text{MVA} = 3\,(\text{Generator's line current in kA}) \times (\text{Phase voltage in kV})$$

Also,

$$\text{MVA} = \frac{\text{MW}}{\text{Power factor}}$$

Using the expressions above, one can find the maximum current that can be supplied by a generator at a given terminal voltage. This is important for sizing the conductors or busses that must carry the generator's energy into the system, as well as for setting protection relays. For a theoretical explanation about the origins of apparent power, see Section 1.3 in Chapter 1.

4.1.3 Power Factor

It was shown in Section 1.3 that the power factor is a measure of the angle between the current and the voltage in a particular branch or a circuit. In mathematical terms, the power factor is the cosine of that angle. Within the context of a generator connected to a system, the power factor describes the existing angle between the voltage at the terminals of the generator (**V**$_t$), and the current flowing through those terminals (**I**$_1$).

In the workings of generators, by definition, the angle between the current and the voltage is deemed positive when the current lags the voltage, and vice versa, it is defined as negative when the current leads the voltage. Therefore, the power factor is used to describe the generator as operating in the "lagging" or "leading" power factor range. A positive power factor indicates that the unit is operating in the lagging region—it is generating VARs. A negative power factor indicates the unit is operating in the leading region—it is absorbing VARs from the system. Additional terms for describing if the unit is producing or consuming VARs are *overexcited* or *inductive* for lagging power factor operation, and underexcited or capacitive for leading power factor operation. Unity power factor refers to a power factor of 1. It is common for generator operators to say the unit is "boosting" or "bucking" VARs. Boosting in this context is synonymous with overexcited or inductive, and bucking means underexcited or capacitive.

These different terms for defining the same mode of operation can be confusing to the uninitiated. A simple way out is just to remember that if the generator is overexcited (i.e., if field current is increased), it will export more VARs into the system. On the other hand, if it is underexcited (i.e., if the field current is reduced), the generator will absorb VARs from the system in order to maintain the required airgap flux density.

The *rated power factor* is the operating point that maximizes both watts and VARs delivered, and it is a design variable. Increasing excitation from that point onward requires the unit to significantly reduce the active output (watts), in order to remain within the allowable operating region (more about that later). For most turbo-generators the rated power factor is in the range of 0.85 to 0.90 lagging (overexcited).

The power factor (actually reflecting the flow of reactive power) has a big influence on the power system in that it can change the system's voltage. The change in voltage in turn affects the ability of the system to carry the required levels of power and, consequently, its stability. To illustrate this important concept, a very elementary example is offered. Figure 4.3 depicts a generator supplying a single radial circuit, with a load at the end of it.

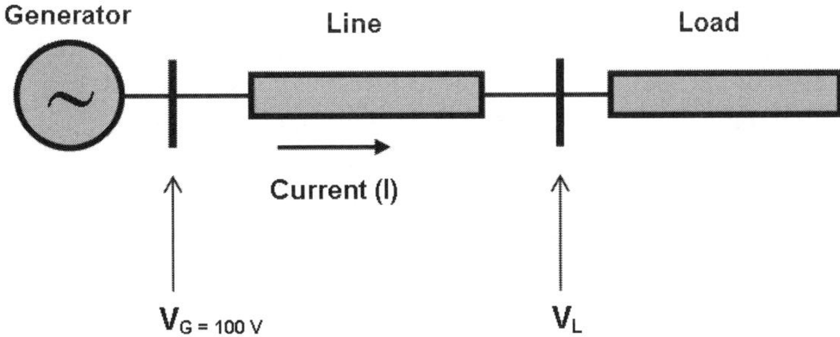

Fig. 4.3 Schematic representation of a generator feeding a load through a line.

Let us assume two cases: case 1 has a line impedance of $1 + j5$ ohms and a load of 5 ohms, and case 2 has the same line impedance but the load has, in addition to the 5 ohms resistance, a 5 ohms reactance (a reactance is denoted by preceding its value with the letter j). Assume that the generator's voltage is maintained at 100 volts in both cases.

Case 1. The impedance of the line is $1 + j5\ \Omega$ and the load is equal to $5\ \Omega$. The magnitude of the current delivered by the generator is then

$$\mathbf{I} = \frac{100}{\sqrt{(6^2 + 5^2)}} = 12.8 \text{ amperes}$$

The magnitude of the voltage at the load terminals is

$$\mathbf{V} = 12.8 \text{ A} \times 5\ \Omega = 64 \text{ volts}$$

And the power delivered to the load is

$$\mathbf{P} = 12.8^2 \text{ A} \times 5\ \Omega = 819 \text{ watts}$$

Case 2. In this case, the line impedance has not changed, but the load now has an additional inductive reactance of 5 ohms. The magnitude of the current delivered by the generator is now

$$\mathbf{V} = \frac{100}{\sqrt{(6^2 + 10^2)}} = 8.57 \text{ amperes}$$

The magnitude of the voltage at the load terminals is

$$\mathbf{V} = 8.57 \text{ A} \times \sqrt{(5^2 + 5^2)}\ \Omega = 60.6 \text{ volts}$$

And the power delivered to the load is

$$\mathbf{P} = 8.57^2 \times 5\ \Omega = 367 \text{ watts}$$

As a result of the addition of a load reactance, the power factor of the load has been reduced from unity to PF = 0.71, the voltage at the load terminals dropped 5%, and the real power delivered to the load is reduced by more than 50%. This simple exercise illustrates the significant impact on a system of an addition of inductive reactance (i.e., in reducing the power factor). Increasing the excitation of the generator in the simple case of this example would increase the generator terminal voltage, driving the load voltage higher and somewhat compensating for the voltage drop introduced by the reduced power factor.

4.1.4 Real Power

The *rated power* (in MW) of the generator is the product of the *rated apparent power* (in MVA) and the *rated power factor*. The turbine determines the rated power of the turbo-generator, as a whole unit.

The rated power of the generator is often specified and designed to be somewhat higher than that of the turbine to take advantage of additional output that may become available in the future from the turbine, boiler, or reactor. This parameter is measured and monitored to keep track of the load point of the machine and allow the operator to control the operation of the generator.

The MW overload of the generator is a serious concern. MW overload means that the stator current's limit has probably been exceeded, and this will affect the condition of the stator winding. The stator terminal voltage may also have been exceeded during overload, depending on the main transformer tap settings, but it is more commonly associated with stator current's overload. Excessive terminal voltage will affect core heating.

Transient MW events from the system or internally in the machine will also show up as transients in the stator current and/or terminal voltage.

4.1.5 Terminal Voltage

The *rated* or *nominal voltage* of a three-phase generator is defined as the line-to-line terminal voltage at which the generator is designed to operate continuously. The rated voltage of large generators is normally in the range of 13,800 to 27,000 volts. Generators designed to IEEE standards and equivalent standards are able to operate at 5% above or below rated voltage at rated MVA, continuously.

When special requirements of a power system dictate the need for a wider terminal voltage range, the manufacturer has to account for this in the generator design and produce a larger and more expensive machine. The cases where this type of variation is required depends on the location and requirements for interaction between the generator and the power system.

Monitoring of the generator terminal voltage is also critical and is done on a per-phase basis. It is required to ensure that there is voltage balance at all times, to avoid negative sequence type heating effects, and it is most critical during synchronizing of the generator to the system. The terminal voltage of the generator must be matched in magnitude, phase, and frequency to that of the system voltage before closing the main generator breakers. This is to ensure smooth closure of the breakers and connection to the system, and to deter faulty synchronization.

4.1.6 Stator Current

Stator current capability in large generators depends largely on the type of machine in question. In the simplest machines (i.e., the indirectly air-cooled generator) the capability of the stator winding defines the rated stator current.

The capability of an indirectly hydrogen-cooled generator winding is significantly sensitive to the hydrogen pressure within the machine. Reduced capabilities are commonly stated for below rated pressures, down to 15 psig (103 kPa) and at slightly above atmospheric pressure. Modern generators may be found operating with hydrogen pressures up to 90 psig (622 kPa). A direct hydrogen-cooled stator winding is directly dependent on hydrogen pressure. Capabilities are commonly stated in increments of 15 psi (103 kPa) below rated hydrogen pressure.

The capability of a water-cooled stator winding is not normally sensitive to hydrogen pressure. However, hydrogen pressure does affect the cooling and, therefore, the temperature of many parts of the generator in which the losses are proportional to the stator current (leads, core, etc.). For this reason, generator capability is usually expressed in increments of 15 psi (103 kPa) below rated hydrogen pressure. In Figure 4.4, the dependency of the generator's rating on the pressure of the cooling hydrogen can be seen.

4.1.7 Field Voltage

Increasing the field voltage increases the field current in proportion to the rotor winding resistance. The field voltage is monitored but not usually used for alarms or trips. It is used to calculate rotor winding resistance and, subsequently, the rotor winding average and hot-spot temperature. Automatic voltage regulator problems can cause the field voltage to become too high, and this in turn causes the excitation to increase beyond design limits.

4.1.8 Field Current

The capability of the rotor winding is generally determined by the field current at the rated apparent power, the rated power factor, and the rated terminal voltage. All of the capability considerations described for indirectly and directly hydrogen-cooled stator windings apply to the rotor winding as well.

As was pointed out in Chapter 1, Section 1.7.3, and in Section 4.1.3 above, increasing the field current will

- Augment the MVARs exported to the system
- Increase armature (stator) current if the unit is already in the boost or overexcited region
- Tend to increase the differential of potential at the machine's terminals

4.1.9 Speed

Unlike an induction machine, the synchronous generator can only generate power at one speed, called the *synchronous speed* of that unit. That unique speed is related to the system's frequency and the number of poles of the machine, by the following equation:

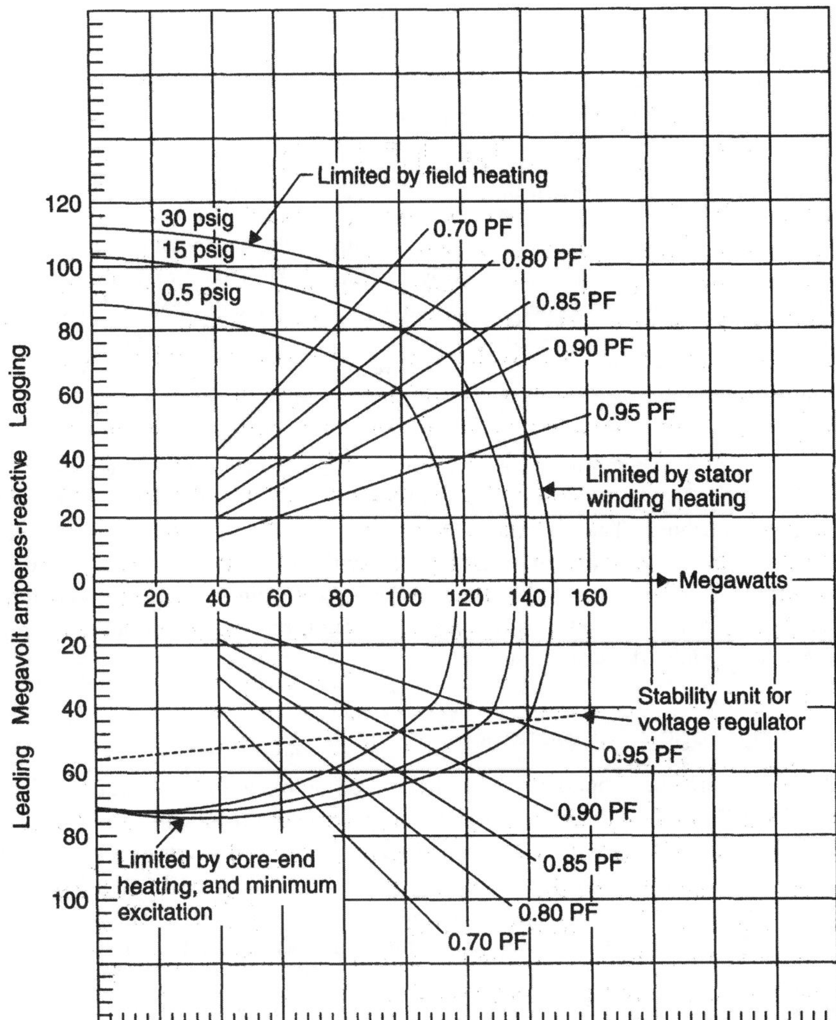

Fig. 4.4 Typical capability curves for a synchronous generator. (Copyright © 1987, Electric Power Research Institute. EPRI EL-5036, Power Plant Electrical Reference Series, Volumes 1-16, reprinted with permission.)

$$\text{Synchronous speed (rpm)} = 120 \times \frac{\text{System frequency (Hz)}}{\text{Number of poles}}$$

Practically all large turbo-generators are of the two- or four-pole design. Therefore, almost without exception, the following apply:

60 Hz system	3600 rpm for two-pole generators
	1800 rpm for four-pole generators
50 Hz system	3000 rpm for two-pole generators
	1500 rpm for four-pole generators

4.1.10 Hydrogen Pressure

The *rated hydrogen pressure* is the required pressure of the hydrogen in the generator when it is loaded to its nominal rating. It is commonly the maximum hydrogen pressure for which the generator is designed to operate. The range of rated hydrogen pressures for generators now being built is up to 90 psig (622 kPa). (The unit psig is pounds per square inch "gauge," relative to standard atmosphere.) For a discussion about the generator's capability dependence on the pressure of the hydrogen, see Section 4.1.6 above.

Regardless of the design hydrogen pressure for any given machine, the pressure is always maintained higher than the stator cooling water pressure in water-cooled stator winding type machines. The reason for this is to allow hydrogen to leak into the stator cooling water, where it can be more easily dealt with by hydrogen detraining and removal systems that are almost always found with such machines. Therefore, one of the sources of a drop in hydrogen pressure in the generator may be leakage into the stator cooling water system, if a leak exists.

4.1.11 Hydrogen Temperature

Similar to pressure, the temperature of the hydrogen cooling gas is also maintained at a specific level for proper cooling of the internal generator components. The hydrogen gas picks up heat in the generator components as it flows over the various parts of the machine internals and transfers that heat to raw water circulating though hydrogen coolers in the generator. Therefore, the gas entering the coolers is quite hotter than the gas leaving the coolers. These are generally referred to as "hot" and "cold" hydrogen gas temperatures.

Unlike the hydrogen pressure however, hydrogen temperature does not vary as widely and is governed by the generator design standards. Generally, the maximum allowable cold gas temperature is 46°C [2,8]. The hot gas temperature rise will vary depending on the generator cooling arrangement and the design of the hydrogen coolers. The cold gas operating set point is usually found between 30 to 40°C. A normal temperature difference between hot and cold hydrogen gas is around 15 to 25°C at the full load condition.

The hydrogen gas temperatures are usually maintained by an arrangement of coolers, four being the most common. A balance between these is then maintained by adjusting the flow of raw water through the coolers, and locking the inlet valves in those positions. The balance is generally kept as close as possible to under 2°C for the cold outlet gas temperature from the coolers. Further regulation in the generator hydrogen temperature is then done on a bulk cooling water basis by an overall temperature control valve for flow, recirculation, or both.

Temperature control of the hot and cold hydrogen gas is accomplished by installing thermocouples (TCs) or resistance temperature detectors (RTDs) in the gas path. These can then be monitored and set with alarm points to notify operators when limits are exceeded.

4.1.12 Short-Circuit Ratio

The short-circuit ratio (SCR) is defined as the ratio of the field current required to produce rated terminal voltage on the open circuit condition to the field current required to produce rated stator current on sustained three-phase short circuit, with the machine operating at rated speed. During operation, to maintain constant voltage for a given change in load, the change in excitation varies inversely as the SCR. This means that a generator with a lower SCR requires a greater change in excitation than a machine having a higher SCR, for the same load change.

The inherent stability of a generator in a power system is partly determined by its short-circuit ratio, which is a measure of the relative influence of the field winding versus the stator winding on the level of useful magnetic flux in the generator. The higher the short-circuit ratio, the less influence the changes in stator current have on the flux level and the more stable the machine tends to be. But the ratio will also be larger for the same apparent power rating and less efficient. However, machines with higher SCRs are not necessary the ones showing higher stability in a particular setting. There are other important factors such as the speed of response of the voltage regulator and excitation systems, match between the turbine and generator time constants, control functions, and the combined inertia of turbine and generator.

The short-circuit ratio for turbine generators built in recent years has been in the approximate range of 0.4 to 0.6.

4.1.13 Volts per Hertz and Overfluxing Events

The term "volts per hertz" has been borrowed from the operation of transformers. In transformers, the *fundamental voltage equation* is given by

$$V = 4.44\, f\, \mathbf{B}_{\mathrm{max}} \times \text{Area of core} \times \text{Number of turns}$$

where $\mathbf{B}_{\mathrm{max}}$ is the vector magnitude of the flux density in the core of the transformer.

By rearranging the variables, the following expression is obtained:

$$\text{Volts/Hertz (V/Hz)} = \frac{V}{f} = 4.44\ \mathbf{B}_{\mathrm{max}} \times \text{Area of core} \times \text{Number of turns}$$

Or, alternatively,

$$\mathbf{B}_{\mathrm{max}}[\text{tesla}] = \text{constant} \times \left(\frac{V}{f}\right)$$

Or, in another notation,

$$\mathbf{B}_{\max} \propto \frac{V}{\mathrm{Hz}}$$

The last equation indicates that the maximum flux density in the core of a transformer is proportional to the terminal voltage divided by the frequency of the supply voltage. This ratio is known as V/Hz.

A very similar set of equations can be written for the armature of an alternating-current machine. In this case, the constant includes winding parameters such as winding *pitch* and *distribution factors*. However, the end result is the same: in the armature of an electrical alternate current machine, the maximum core flux density is proportional to the terminal voltage divided by the supply frequency (or V/Hz).

In machines, as well as in transformers, the operating point of the voltage is such that for the given rated frequency the flux density is just below the knee of the saturation point.

Increasing the volts per turn in the machine (or transformer) raises the flux density above the knee of the saturation curve (see Fig. 4.5). Consequently, large magnetization currents are produced, as well as increases in the core loss due to the bigger hysteresis loop created (see Fig. 4.6). Additionally, large harmonic eddy currents are developed in the core and other metallic components. All these result in substantial increases in core and copper losses, and excessive temperature rises in both core and

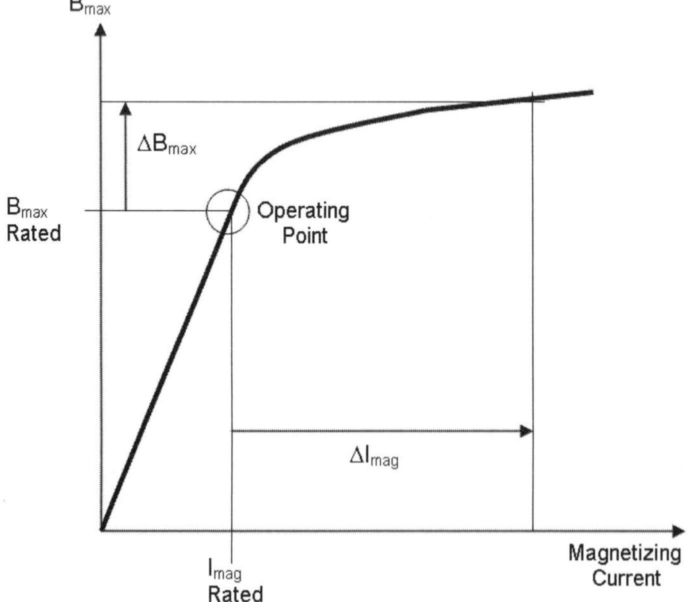

Fig. 4.5 Typical saturation curve for transformers and generators.

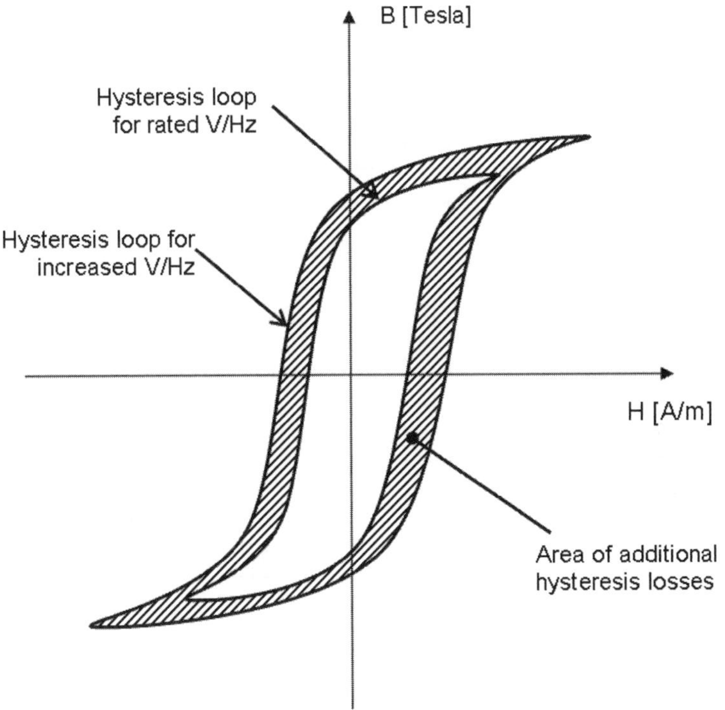

Fig. 4.6 Hysteresis losses under normal and abnormal conditions.

windings. If not controlled, this condition can lead to loss of the core interlaminar insulation, as well as loss of life of the winding insulation. In fact, if a unit becomes excessively overfluxed (i.e., the maximum V/Hz has been exceeded) for just a few seconds, complete failure of the core may result in short time or after some time of operation.

To obtain a better insight of a V/Hz incident, let us look at a simplified example. Consider a core made of 0.5 mm (19.7 mils) thick coreplate material with an operating flux density of 1.3 tesla in the core area (back of core) with a loss of 15 W/kg (see Figure 4.7), and a flux density of 2 tesla in the teeth area with a loss of 35 W/kg. If a sudden overfluxing event occurs with an intensity of 120% (this is well within the accumulated experience for this type of occurrence), the resultant flux densities and losses will be 1.68 Tesla and 25 W/kg in the core area, and 2.4 Tesla and 50 W/kg in the teeth.

Note. Almost all overfluxing events occur during run-up. Thus, the core losses will be lower than those shown in Figure 4.7, all else being equal, because core losses are driven by the frequency of the exciting current (as explained in previous sections). However, the main conclusions of the example remain valid.

The cooling system of a turbo-generator is carefully designed to remove the maximum expected losses under normal operating conditions. Any sudden loss increase of significant magnitude is absorbed by the surrounding components, with a resulting in-

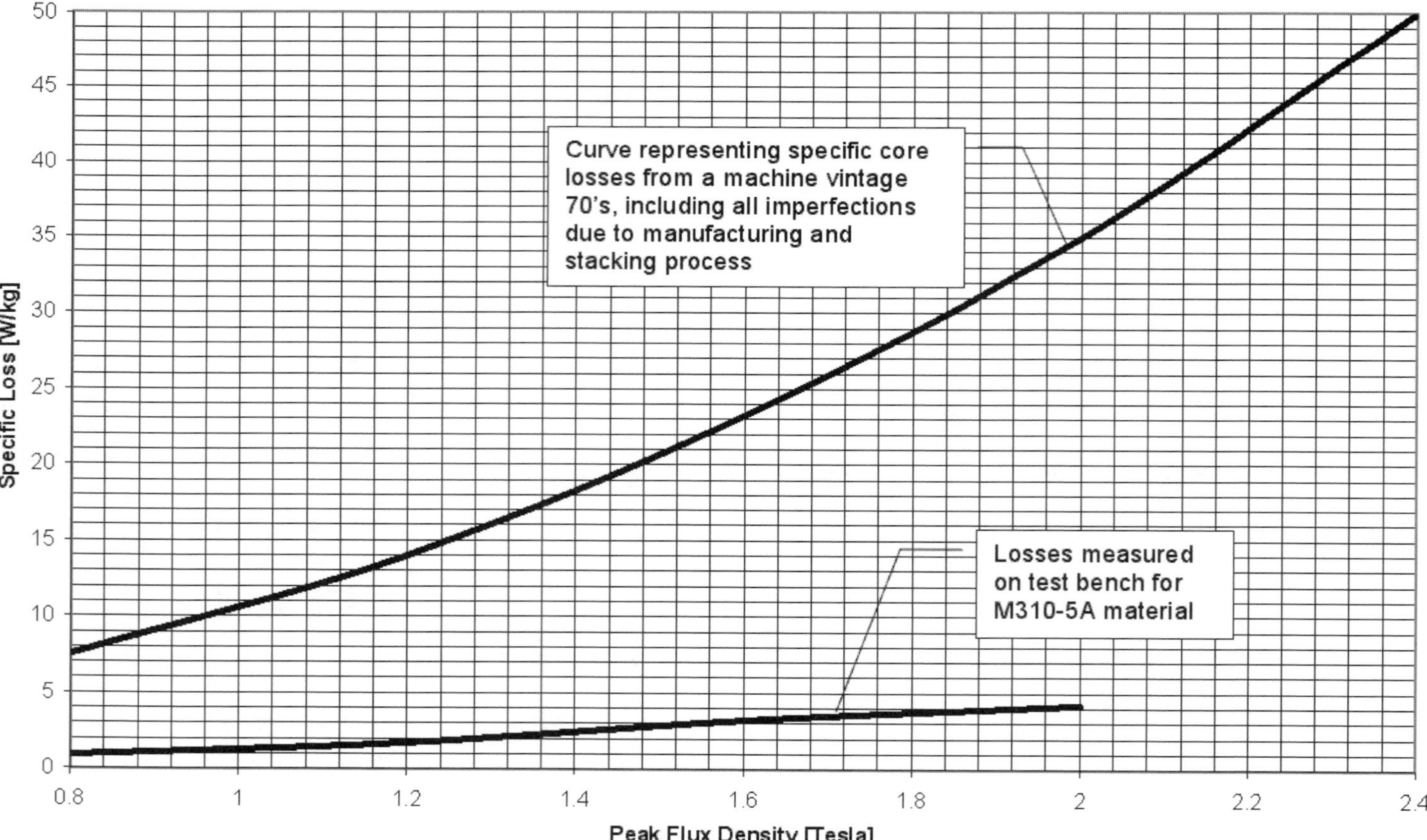

Fig. 4.7 Core loss versus peak flux density at 50Hz for 0.5mm thick lamination material, type M310-50A, on the test bench, and a typical assembled core, not necessarily the same steel but the same thickness, of a machine built in the 1970s.

crease of temperature. This imposes the type of time-based restrictions for operation under abnormal conditions that will lead to these sorts of temperature excursions. In most cases, with the aim of achieving a conservative estimate, adiabatic calculations are performed, that is, any additional loss is assumed to be captured by the pertinent components as retained heat. This calculation yields maximum temperatures for the abnormal event.

For the case at hand, the additional losses in the core region were equal to 10 W/kg, and in the teeth region they were equal to 15 W/kg. If an event with duration of 60 seconds is assumed, the total accumulated heat per kilogram of lamination material equals 600 joules in the core region and 900 joules in the teeth region. The specific heat of steel equals 452 joules/kg/°C; therefore, in 60 seconds this calculation would yield about a one degree increase in the temperature of the back of the core, and about a two degree rise in the teeth, not enough to be of any concern. Thus, the analysis made to this point explains long-term, low-level V/Hz effects (for instance a 10% overfluxing event of many minutes in duration). The resulting loss is a loss the machine has not been designed to cope with and, therefore, temperature rises ensue, with detrimental effects to the integrity of the lamination and winding insulation. However, so far the example does not explain the extensive damage to the core associate with high-intensity overfluxing events. In the case of high-intensity overfluxing events, accurate calculations must include precise consideration of large leakage fluxes created by the deep saturation of the core, and resulting circulating currents in the core-supporting bars and between bars and core at the end region of the core, as well as a very large increase of harmonic eddy currents in the core and core-supporting structure. The combination of all these effects can result in highly concentrated points of losses and temperature increase that may damage the insulation between the laminations in a number of spots. Once these damaged spots are created, they will give rise to larger than normal eddy currents. These augmented localized eddy currents and losses will continue to damage (and even melt) the metal during normal operation.

Figure 4.8 demonstrates in a simplified manner how breaches in the interlaminar insulation result in larger than normal eddy current and losses. Larger than normal means currents and losses substantially higher than those found when the machine operates within its designed parameters. Figure 4.9 carries the analysis in Figure 4.8 farther. It can be seen therein that with 100 or so divisions, the eddy-current losses are reduced to about 2% of the original losses. Farther divisions make this type of loss very small when compared to that of an equivalent single-body core. From the content of this paragraph, it becomes obvious how by damaging the insulation between the laminations in a few spots, the eddy-current losses can significantly increase, with consequential temperature rises and additional damage to the insulation, further increasing the amount of short-circuited core. This has the potential to becoming a runaway situation, leading to melted core material and a catastrophic failure. The issue of short-circuited laminations and consequential core damage due to increased eddy-current loss is broached in other places in this book when discussing foreign metallic objects left inadvertently on the core laminations in the bore, or metal parts (e.g., bolts and washers) becoming loose during operation and landing on the core in such a way that laminations are shorted.

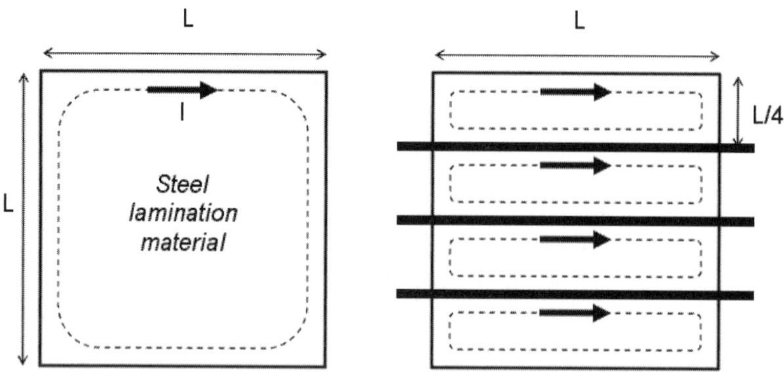

Due to *skin-effects* most current flows in the periphery of the conducting material. Therefore, the following applies:

- Induced volts ∝ area
- Loop resistance ∝ perimeter
- Current ∝ area / perimeter
- Losses = I^2R

For the full block:

- Induced volts ∝ L^2
- Loop resistance ∝ 4L
- Current ∝ L/4
- Losses ∝ $L^3/4$

For the same block divided into four insulated "laminations":

- Induced volts in one section ∝ $L^2/4$
- Loop resistance of one section ∝ 2.5L
- Current in one section ∝ L/10
- Losses in one section ∝ $L^3/40$
- Total loss for all four sections ∝ $L^3/10$

This means that by dividing the original metal sheet into four insulated laminations, the total eddy-current loss was reduced 2.5 times.

Fig. 4.8 Interlaminar insulation reduces eddy-current losses in the steel.

To obtain a better perspective of the dramatic impact of eddy-loss heating during a V/Hz event of large magnitude, let us look at how deep saturation of the core affects the flux distribution in the stator. Figures 4.10 and 4.11 depict how the leakage flux past the back of the core increases a great deal due to a large V/Hz event. From inspection of Figure 4.11 and Figure 4.9, it becomes obvious how the lamination losses are increased drastically by the laminations shorted by the supporting bars. For instance, in

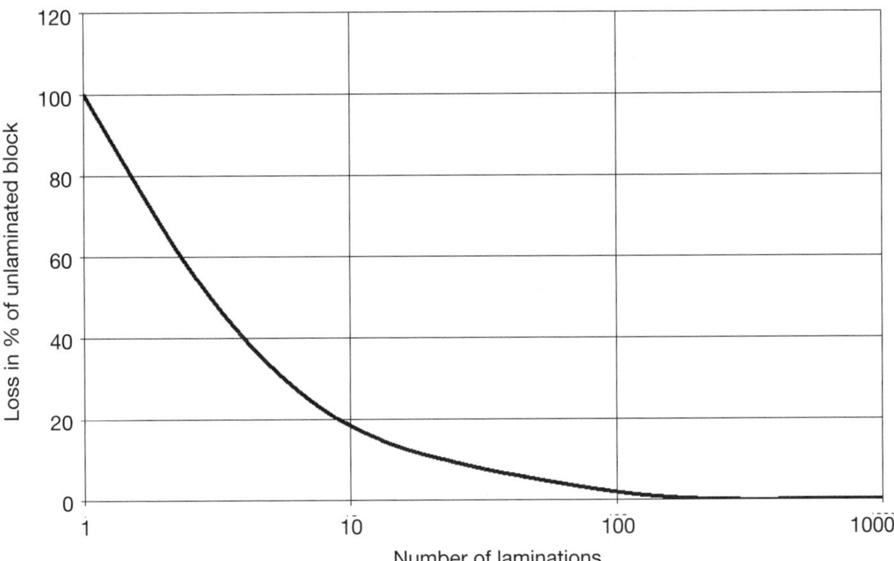

Fig. 4.9 Laminated-core eddy-current loss as percentage of full-block loss. This graph shows the cumulative effect of decreasing eddy-current losses by laminating the core steel.

the case presented in this section as an example, the additional eddy losses may be amplified by two or three orders of magnitude. So, instead of 10 W/kg in the core area and 15 W/kg in the teeth area, one may have thousands of W/kg of additional losses. For example, for a 7000 W/kg additional loss, the adiabatic temperature rise of the laminations over one minute is over 900°C. Reference [8] contains typical values of the temperature capability of lamination coatings, from which it can be seen that continuous capabilities are in the higher 100°C to lower 300°C range, and may reach into the 800s for certain coatings under ideal conditions of an inert gas atmosphere. These numbers will vary according to the type of coating (inorganic vs. organic) and manufacturer. However, it can be said that 900°C will damage most if not all insulation, creating the opportunity for short circuits to appear between laminations. These short circuits are the ones that facilitate additional eddy currents to flow after an overflux event, eventually causing more insulation damage until the core is destroyed. One can see from Figure 4.9 that if enough laminations participate in an intense overflux event, the temperatures attained by the core in a few seconds can easily reach over a few hundred degrees centigrade, impacting the long-term integrity of good coatings and partially destroying coatings that have already been compromised by age, previous V/Hz events, loose cores, and normal core vibration due to magnetostriction.

IEEE Standard 67 states that generators are normally designed to operate at rated outputs of up to 105% of rated voltage [1,16]. ANSI/IEEE C57 standards for transformers state the same percentage for rated loads and up to 110% of rated voltage at no load. In practice, the operator should make sure (by consulting vendor manuals and

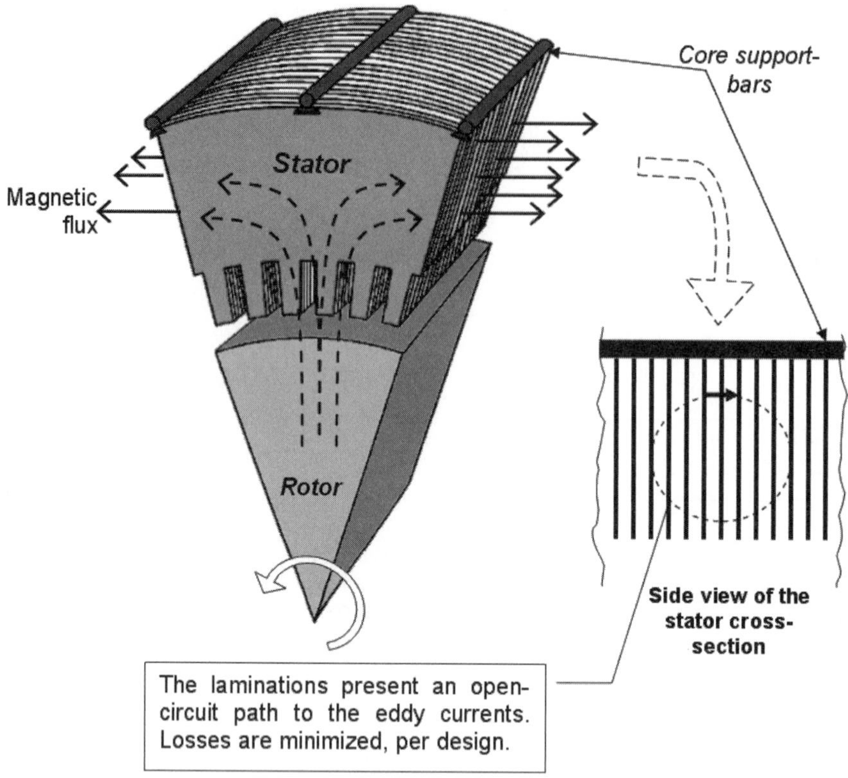

Fig. 4.10 Flux distribution in stator core during normal operation.

pertinent standards) that the machine remains below limits that may affect the integrity of both the generator and the unit transformer. For operation of synchronous machines at other than rated frequencies, refer to IEEE Std 67-2005 [16].

4.2 OPERATING MODES

4.2.1 Shutdown

Shutdown mode refers to the time when the generator is offline and not connected to the system. It also implies that the generator is at zero speed, with the main generator and field breakers open. Therefore, there is no energy flowing to or from the generator.

4.2.2 Turning Gear

Turning gear mode is the mode of operation when the generator's rotor is turned at low speed. Turning gear mode is generally used in two instances: (1) when the generator is to be put back online and (2) when the generator comes offline from operation.

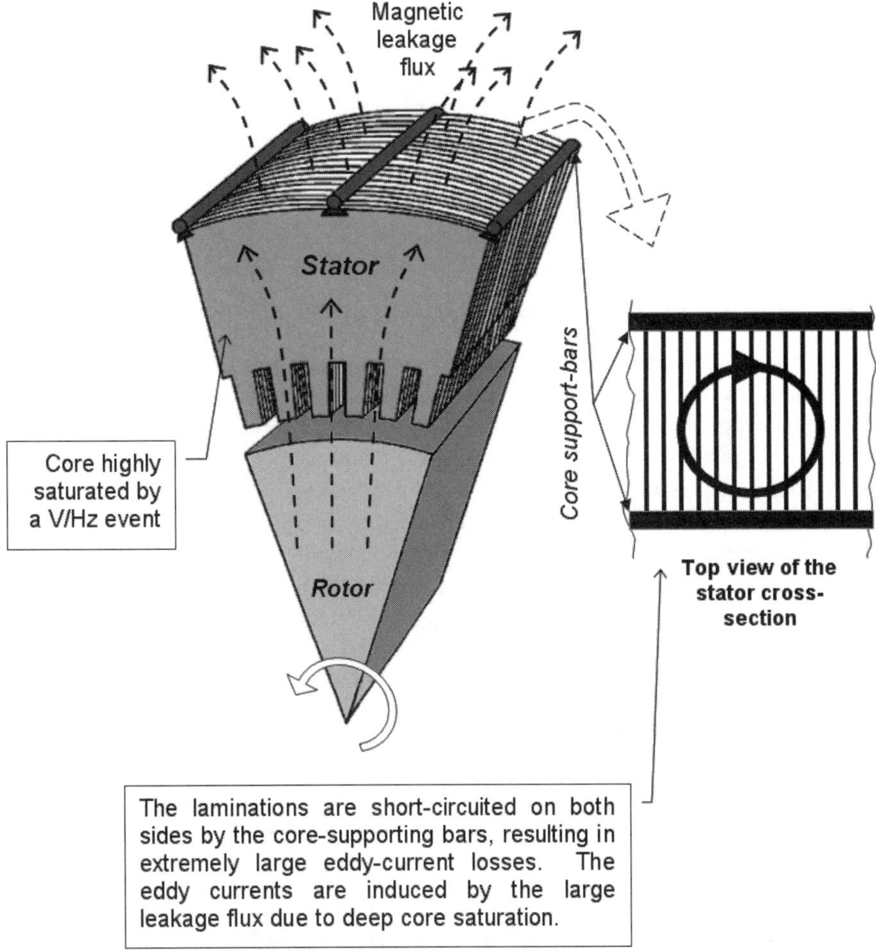
Magnetic
leakage
flux
Stator
Core highly
saturated by
a V/Hz event
Rotor
Core support-bars
Top view of the
stator cross-
section
The laminations are short-circuited on both sides by the core-supporting bars, resulting in extremely large eddy-current losses. The eddy currents are induced by the large leakage flux due to deep core saturation.

Fig. 4.11 Flux distribution in stator core during an overflux event.

In the first instance, the generator is usually started from rest with the turning gear, once high-pressure lifting oil has been applied to the bearings, and then brought up to turning gear speed. The generator is then rolled off the turning gear by firing the turbine, and brought up to rated speed. The turning gear assists in reducing the required starting torque on the rotor and allows a smooth start of the turbine (for those units without a starter motor).

When coming offline, the generator is in the "hot" condition and requires a cooling down period, on a slow roll, prior to being allowed to sit in one position for any length of time. This is done to eliminate the possibility of a permanent bow being established in the rotor forging. The turning gear is used to accomplish this slow roll during the cool-down period.

Depending on the manufacturer, the turning gear speed may typically be anywhere from 3 to 50 rpm. The typical design consists of an induction motor (e.g., sized 10 HP for a 155 MVA, 3600 rpm generator) linked by a gear-reduction system to the generator's shaft. Once the machine is accelerating by the force of the turbine, the gearbox delinks the motor from the generator. When performing a visual inspection of the inside of the machine, it is important to ascertain that the turning gear mechanism will not be energized inadvertently (more about this in the section about inspections). See Figure 4.12 for a typical turning gear arrangement.

4.2.3 Run-up and Run-down

Run-up refers to the period of speed increase from turning gear to rated speed by steam admission to the turbine, in the case of a steam-driven turbine, or by turbine firing, in the case of a combustion turbine. Alternatively, the unit may be accelerated by a "pony motor," or by a solid-state variable-speed drive temporarily driving the generator as a motor. Run-down refers to the period of speed decrease when the generator is taken offline and allowed to coast down from rated speed to the speed at which it is placed on the turning gear.

During both modes of operation, the generator goes through its critical speeds. There are generally two critical speeds in large generator rotors. These are the natural resonance frequencies of the generator rotor mounted on its bearings. The rotor can become damaged if allowed to spin at these speeds for any length of time. Therefore, to

Fig. 4.12 1250 MVA, four-pole, hydrogen-cooled generator. Shown is the shaft-mounted part of the generator's turning gear system. Not shown is the turning gear electric motor. The turning gear is being overhauled as part of the main outage undertaken by the unit.

avoid rotor damage, care is taken to run through these two frequency points fairly quickly.

4.2.4 Field Applied Offline (Open Circuit)

The condition of the generator when the field is applied but the machine is not connected to the system is referred to as the *open-circuit* condition. At open circuit, if the generator is spinning at its rated speed and the field's current magnitude equals the amperes field–no load (AFNL), the voltage at the generator's terminals will be the nominal voltage.

4.2.5 Synchronized and Loaded (Online)

Once the generator is at rated speed and rated terminal voltage, the sinusoidal waveform of the generator output must be matched to the system waveform by frequency, voltage level, and phase shift. The frequency and voltage level are achieved in the open-circuit condition when the generator is brought to rated speed and the field current is raised to the AFNL value (see previous section). The phase shift (or vector shift) is accomplished automatically by a "synchronoscope," which adjusts the generator output voltage to be in phase with that of the system, or manually by the operator. Once the generator is synchronized to the system, the main generator breaker is closed and the generator is connected to the system. At this point, loading the turbine will increase the generator's MW output. Power factor and reactive power output are adjusted by changes to the rotor field current. More about synchronizing the generator to the system can be found in Chapter 6. Table 4.1 contains a useful method to determine the generator operating mode, using indications of generator main (line) breaker status, field breaker status, rotor speed, and terminal voltage.

4.2.6 Start-up Operation

Following is a nonexhaustive list of activities that must be followed before attempting to start a generator:

- Make sure all protection is enabled and operational. In some protective schemes, a number of relays may have to have their trips curtailed during start-up. Make sure the OEM's operational instructions are followed to the letter.
- Do not attempt to reenergize the machine without an investigation after a protective relay has operated during a start-up.
- Follow OEM instructions regarding prewarming.
- Follow OEM instructions regarding application of the field current and turbine speed.
- Establish clear procedures when energizing cross-compound machines.
- Watch the maximum terminal voltage on open-breaker operation.
- Establish clear and safe synchronizing procedures and follow them carefully.

Table 4.1 Generator operating modes

	Generator breaker	Field breaker	Rotor speed (rpm)	Terminal voltage
Shutdown	Open	Open	0	0
Turning gear	Open	Open	0 < rpm < 50	0
Run-up/run-down	Open	Open or closed/open	50 < rpm < rated	0
Field-applied open circuit	Open	Closed	Rated or lower	Rated or lower
Synchronized and loaded	Closed	Closed	Rated	Rated ±5%

Note: Turning gear speed is manufacturer dependent. Conditions in the table are typical for any industrial generator.

Prewarming. Prewarming of the turbine-generator unit is designed to maintain mechanical stresses within the turbine and the generator within acceptable levels. It is necessary to prewarm generator rotors, especially those of an older vintage, so that the iron temperature is greater than the "FATT values," at which the material is much more brittle. Sudden loading of a cold unit will stress certain components much more than the application of a gradual load. Prewarming also has the effect of curtailing the thermal differentials within critical components of both turbine and generator. In some cases, when certain problems exist, it might be advisable to enhance the prewarming operation. The operator should closely follow the OEM's instructions regarding prewarming. For additional information, refer to IEEE Std. 502-1985, Section 9.5 on turbine rotor prewarming [3].

4.2.7 Online Operation

Following is a nonexhaustive list of activities that must be followed during the operation of a generator:

- The unit must remain within its capability curve at all times.
- Voltage regulators and power system stabilizers (when applicable) should be in operation at all times.
- All protection and monitoring devices must be in fully functional condition and always in operation.

Typical generator trips are:

- Stator phase-to-phase fault
- Stator ground fault
- Generator motoring
- Volts/Hz
- Loss of excitation
- Vibration (if unit is not closely monitored by personnel)

Other protective functions/systems might also trip the unit, according to specific unit requirements and design.

Note. See Section 4.10 for an example of generator operating instructions provided by a generating company.

4.2.8 Shutdown Operation

Following is a nonexhaustive list of activities that must be followed during the shutdown of a generator:

- The turbine should be tripped *before* the generator.
- Make sure the generator does not *motor* the turbine. (However, in some instances, especially in nuclear plants, the system is designed such that the generator can "synchronously" motor the turbine. The turbine still has vacuum applied and the reactor is brought back up to power within a given period; however, this can be many hours and it is a special requirement, not a standard case.)
- Attain electrical separation by following clearly established safe routines.
- Place the unit on the turning gear as deemed necessary, for cooling out of the turbines and to avoid bowing of the rotor shaft during cooling-down periods.

4.3 MACHINE CURVES

4.3.1 Open-Circuit Saturation Characteristic

The *open-circuit saturation curve* for the generator provides the characteristic of the open-circuit stator terminal voltage as a function of field current, with the generator operating at rated speed.

At low voltage and, hence, low levels of flux, the major reluctance (magnetic resistance) of the magnetic circuit is the airgap. In the linear portion of the open-circuit curve, terminal voltage and flux are proportional to the field current. This portion of the open-circuit saturation curve, which is linear, is called the "airgap line." At higher voltages, as the flux increases, the stator and rotor iron saturate, and additional field current is required to drive magnetic flux through the iron. This is due to the apparent higher reluctance of the magnetic circuit. Hence, the upper part of the curve bends away from the airgap line at an exponential or logarithmic rate, dependent on the saturation effect in the stator and rotor. Without the presence of iron in the circuit, the airgap line would continue on linearly, meaning that the terminal voltage and machine flux would increase in linear proportion to the increase in field current. (Figure 4.13 shows the open-circuit saturation curve.)

4.3.2 Short-Circuit Characteristic

The short-circuit characteristic curve is a plot of stator current (from zero up to rated stator current) as a function of field current, with the stator winding terminals short-

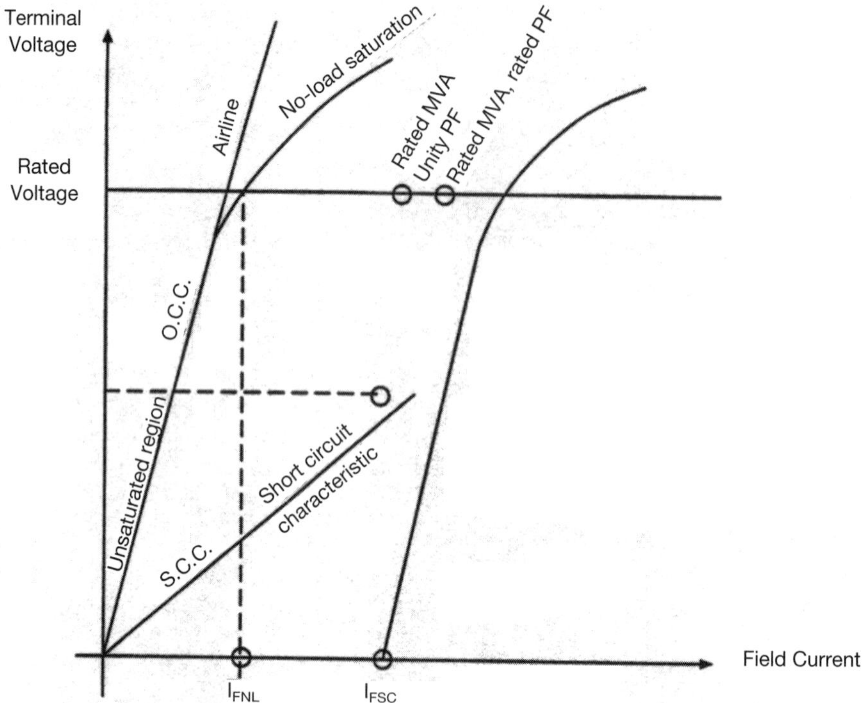

I_{FNL} = Field current required to produce open-circuit rated voltage
I_{FSC} = Field current that produces rated armature current with short-circuited terminals

The S.C.C. relates almost exclusively to the "armature reaction" of the machine

$$\text{Short-circuit ratio (SCR)} = \frac{I_{FNL}}{I_{FSC}}$$

In turbo-generators (most) SCR = 0.5 – 0.6

Fig. 4.13 Open-circuit characteristics (OCC). Also shown are short-circuit curves (SCCs) and other points of interest as well as a graphic definition of the short-circuit ratio (SCR). The SCC relates almost exclusively to the "armature reaction" of the machine.

circuited and the generator operating at rated speed. The short-circuit curve is usually plotted on the same graph along with the open-circuit curve. The short-circuit characteristic is for all practical purposes linear because in this short-circuit condition the flux levels in the generator are below the level of iron saturation.

The short-circuit curve is also called the "synchronous impedance curve" because it is the synchronous impedance of the generator that determines the level of the stator current for the machine. This can be readily seen by inspection of Figure 1.35 in Chapter 1. It can be seen in the figure that when $\mathbf{V_t} = 0$, the entire internal generated voltage ($\mathbf{E_m}$) is dissipated across the synchronous impedance ($\mathbf{Z_s}$). The synchronous impedance is highly dependent on the *armature reaction* of the machine ($\mathbf{X_a}$).

Both open- and short-circuit characteristics are shown in Figure 4.13. The figure also presents a number of typical acronyms that are commonly encountered when discussing machine characteristics.

4.3.3 Capability Curves

Capability curves are plots of apparent power capability (MVA), at rated voltage, using active power (MW) and reactive power (MVAR) as the two principal axes. Circumferences drawn with their centers at the origin represent curves of constant stator current. A capability curve (see Fig. 4.4) separates the region of allowed operation, inside the curve, from the region of forbidden operation, outside the curve.

On an *x-y* graph where the x axis represents MW and the *y* axis represents MVAR, a circumference represents a constant MVA. If, as in the case of a machines capability curve, the voltage is kept constant (at rated value), then a circumference also represents a constant-current trajectory. On the same graph, any line starting at the intersection of the axis represents a particular power factor (see Fig. 4.14). Different parts of the capability curve are limited by different machine components. There is a part limited by field winding capability, a part limited by stator winding capability (the circular part), and a part limited by core-end heating, as shown in Figure 4.4.

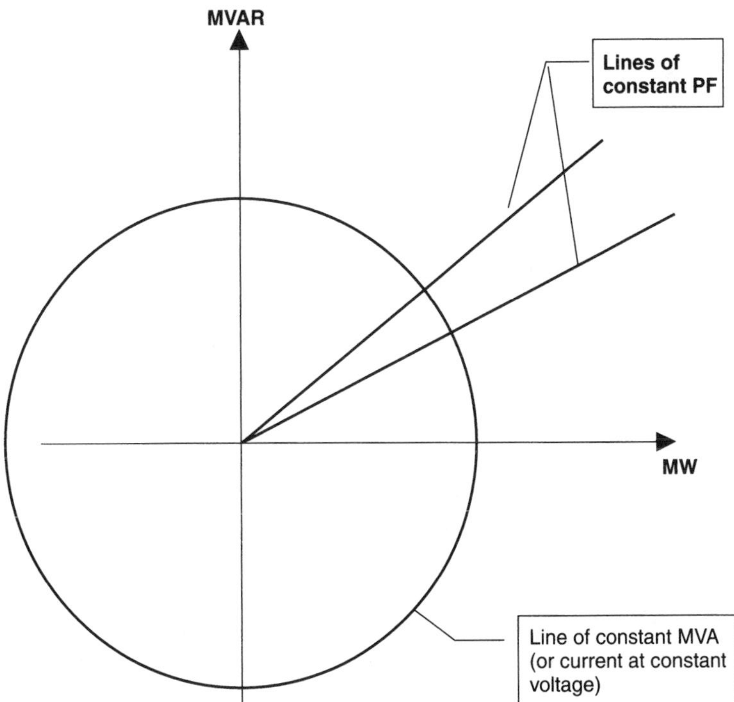

Fig. 4.14 Constant MVA, current, and power factor plotted on a MW-MVAR graph.

As the power factor is varied from fully overexcited, through unity, to fully underexcited, first the field current, then the stator current, and then the stator core ends are limiting. Accordingly, curves that define a turbo-generator's capability have three segments that pictorially describe the effect of the capability of the three machine components.

Furthermore, the capability curve represents the fact that the maximum temperature of the machine components during operation depends on the pressure of the hydrogen in hydrogen-cooled generators. This is shown as a set of curves, each for a given hydrogen pressure, up to the rated pressure. A similar set of curves can be drawn for various water-inlet temperatures in directly or indirectly water-cooled machines. This topic is discussed in Section 4.1.6.

Construction of Approximate Reactive Capability Curve. As stressed often in this book, operators should always maintain the generator within the capability curves of the machine. Given the importance of this, it is rare for the capabilities of a generator to be unavailable. However, for those rare occasions, one can construct an approximated capability curve for the lagging region by following the guide provided in ANSI/IEEE C50.30-1972 or later, IEEE Std. 67. For convenience sake, the method is shown in Figure 4.15.

Limits Imposed by the Turbine and the System. The turbine and the generator are designed to operate as a unit. As was stated earlier in this book, the generator rating is almost always designed to be somewhat larger than the turbine. This fact is shown on the operator's screen as a line inside the maximum MW output of the unit at unity PF (see Fig. 4.16). In addition, system stability issues may limit the number of MVARs the unit can import when operating in the leading PF region. This fact is shown as a line crossing the leading portion of the capability curve (see Fig. 4.16). Therefore, the "working" capability curve of the entire unit represents a combination of generator, turbine, and system constraints.

Pay attention to the fact the orientation of Figure 4.16 is different than that of Figure 4.4. In the United States and many other countries, it is common to show the MW axis on the horizontal. However, in Canada, the United Kingdom, Australia, and some other countries, it is common to present the capability curves with the MW on the vertical axis and the lagging MVAR on the right side of the horizontal axis. Figure 4.16 also presents all parameters as per unit (pu) of rated values.

Capability Curves Adjustments for Nonrated Terminal Voltage. As discussed in Section 4.20, most generators allow a ± 5% voltage deviation from nominal volts. Capability curves' behavior must be understood when attempting to operate at maximum ratings below rated voltage. For instance, if operating at –5% voltage, the armature current should not be increased beyond its rated value, but the leading section of the capability curve shrinks by about the same 5%. The lagging (field-dominated) section expands by a similar amount. The opposite is true when operating at +5% voltage. See Figure 4.17.

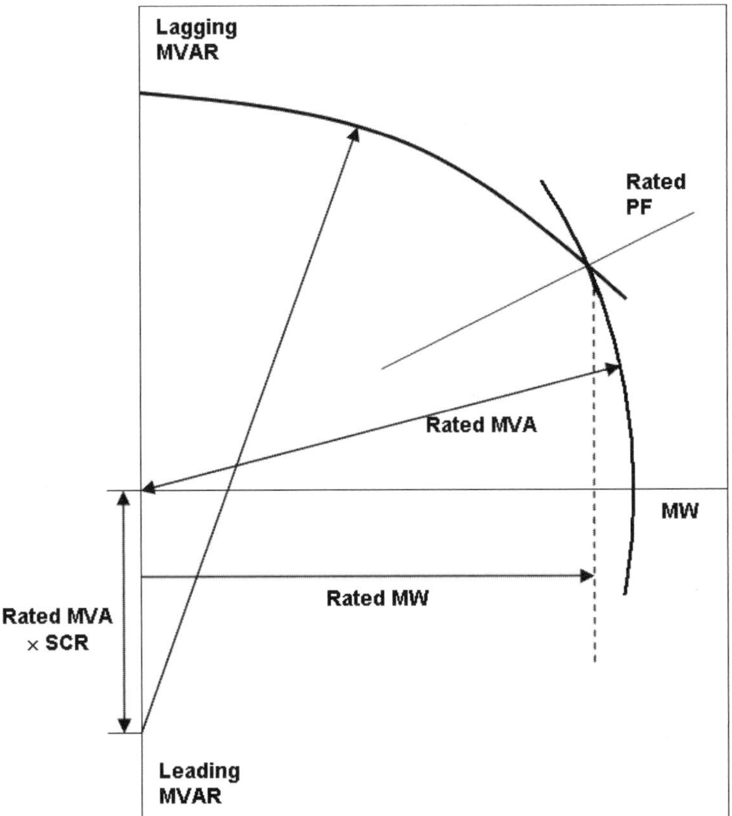

Fig. 4.15 Construction of approximate reactive capability curve, per IEEE Std. 67. The top curve is drawn after the intersection of the rated MVA circumference and the rated PF line. The leading (bottom) part of the curve is too dependent on the specific machine design to be drawn by any general algorithm.

4.3.4 V-Curves

V-curves provide the apparent power (MVA) as a function of field current, plotted for various constant power factors, holding speed and stator voltage at the rated values. Horizontal lines represent constant stator current. The rating of the generator is the intersection of the line for rated apparent power (1.0 per unit) and the curve for rated power factor (usually 0.85 or 0.9 lagging). All constant-power-factor curves converge at a common point at zero apparent power. This is at the field current for rated voltage, open circuit.

Vertical and horizontal lines can also be shown for the field and stator winding capabilities at varying hydrogen pressures. The reduction in capability caused by stator core-end heating at low levels of excitation, below 0.95 power factor, leading, can also

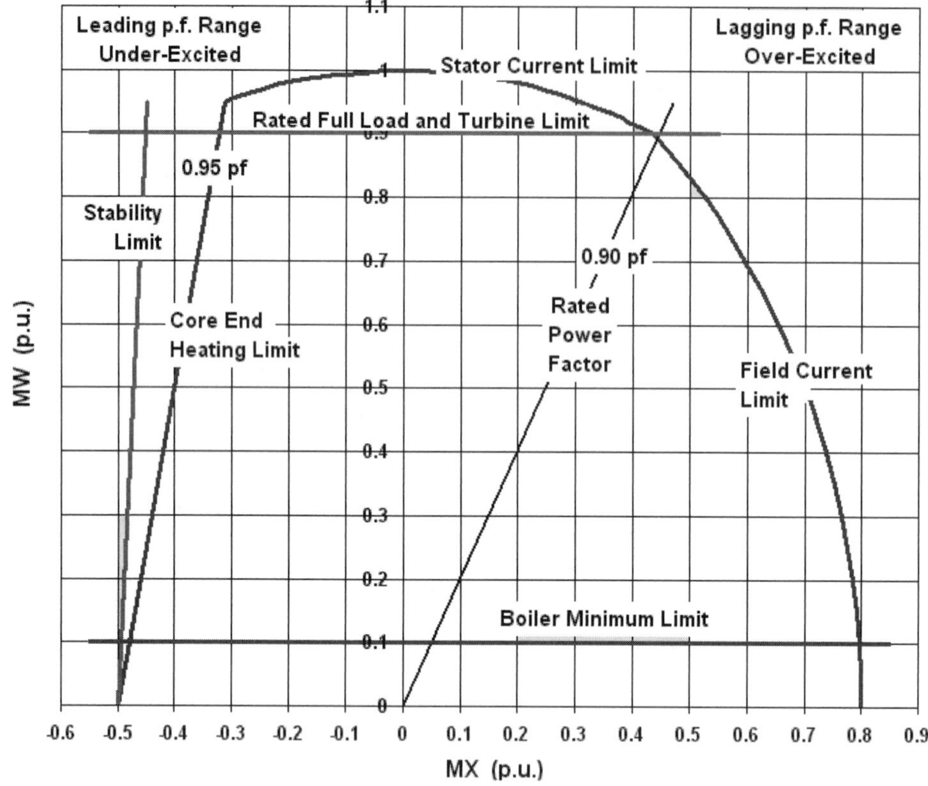

Fig. 4.16 Capability characteristics of a generator showing turbine and power system stability constraints. The curve is shown with the MW on the vertical axis. This is common in Canada, the United Kingdom, Australia, and a few other countries.

be included, as can be done for turbine- and system-imposed limits. See Figure 4.18 for a typical V-curve.

4.4 SPECIAL OPERATING CONDITIONS

4.4.1 Unexcited Operation ("Loss-of-Field" Condition)

Operation without field current is potentially dangerous and can occur under a number of circumstances. The following are the most common two:

1. *Loss of field during operation.* If for some reason the field current goes to zero while the generator is connected to the system, the machine starts acting as an induction generator. The rotor operates at a speed slightly higher than synchronous speed and slip-frequency currents are developed. These penetrate deep into

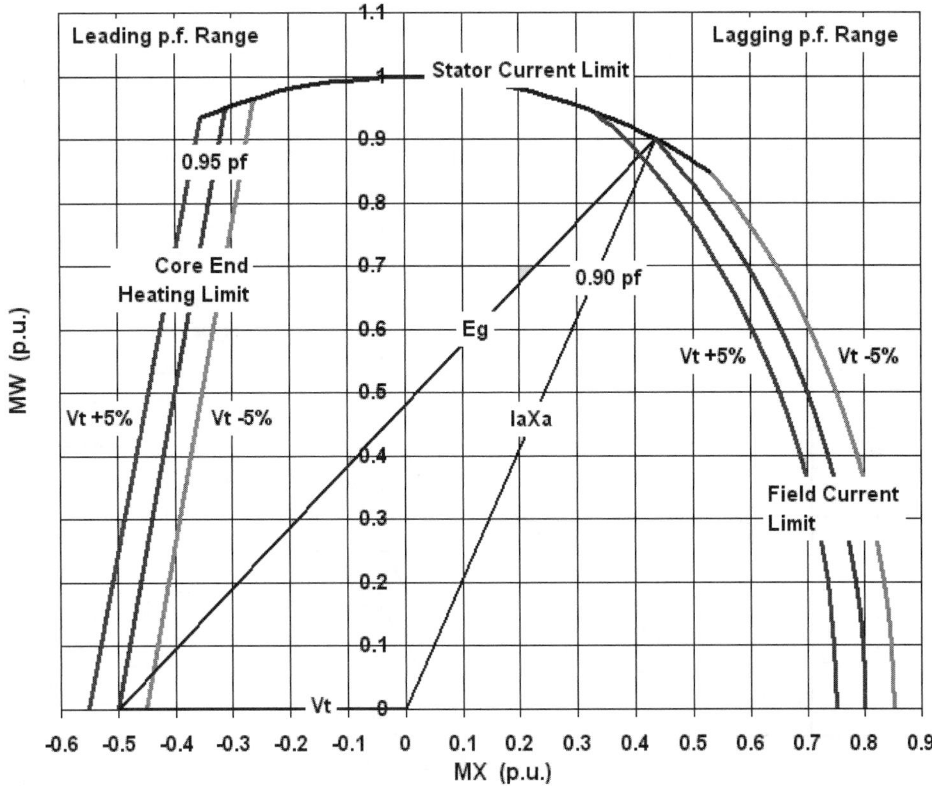

Fig. 4.17 Capability characteristics of a generator showing adjustments for operation over and under rated terminal volts (± 5% of rated).

the rotor body because they are of low frequency (this does not represent the skin effect discussed in case 2, below). Severe arcing between rotor components and heavy heating may result. The ends of the stator core also experience heating due to stray fluxes in the end region, more severely than for operation at underexcited power factor. Protection is commonly provided to prevent or minimize the duration of this mode of operation, by the so-called loss-of-field relay.

2. *Inadvertent energization.* If a generator is at rest and the main generator three-phase circuit breaker is accidentally closed, connecting it to the power system, the magnetic flux rotating in the airgap (or gas gap) of the machine at synchronous speed will induce large currents in the rotor. The stator windings will also be subjected to high electromagnetic forces at the time the main breaker is inadvertently closed. The rotor then will tend to start rotating as an induction motor. The very high currents induced in the rotor will tend to flow at its surface, in the forging, wedges, and retaining rings. As the rotor accelerates, the currents will penetrate deeper and deeper. The maximum damage occurs while the speed is

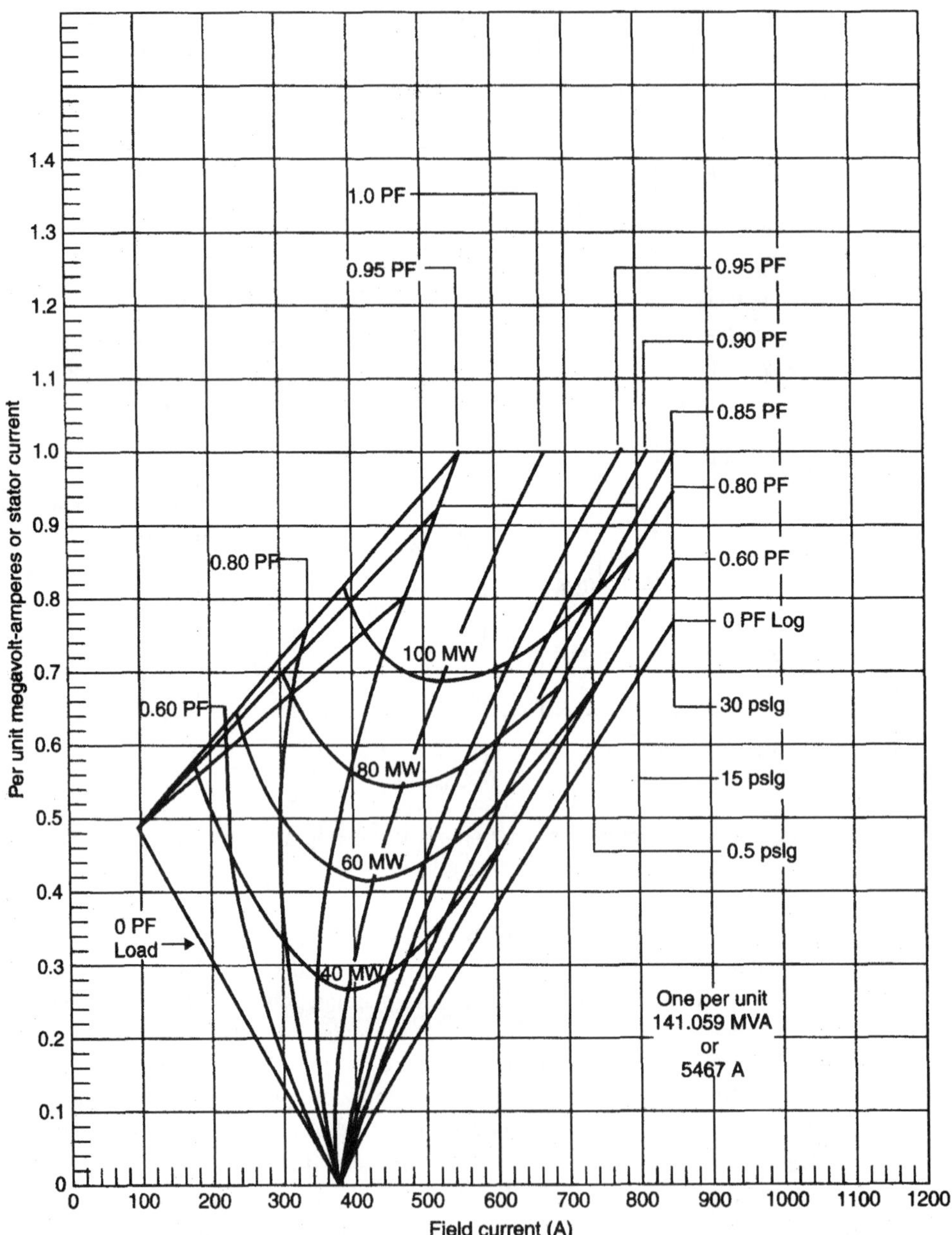

Fig. 4.18 Typical V-curves for generator operation. (Copyright © 1987, Electric Power Research Institute.)

low and the large currents concentrate in a thin cross section around the surface of the rotor (due to the skin effect). The temperatures generated by the large currents, flowing in a relatively small cross section of the rotor, create very large temperature differentials and large mechanical stresses within the rotor. Areas most prone to damage are at the ends of the *circumferential flex slots*. Other areas are the wedges and in the body-mounted retaining rings, the area where the rings touch the forging and the end wedges (see Figs. 4.19 and 4.20)

The initial stator current supplied from the power system will also be very high, but the most vulnerable part of the generator is the rotor. As the rotor speed rises, stresses increase at the same time that the temperatures of the stressed regions increase due to circulating rotor body currents. Generators have been destroyed from this event, as extreme temperatures reduce the component material strengths. The internal rotor components are so weakened that they cannot handle the applied loads any longer. The result can be that the rotor wedges or retaining rings fail. Therefore, protection is needed for the generator, even when it is out of service, to prevent or at least limit motoring from rest. Overheated ends of the circumferential flex slots can over time develop cracks in the forging, compromising its integrity.

Heating of the ends of the stator core is strongly affected by stray magnetic flux in the end region. This field is complex and is affected by the magnitudes and angular positions of the current in the stator and rotor windings.

In addition to the above consequences, the turbine is connected to the generator, and if there is no vacuum on the turbine at the time, this can result in overheating of the

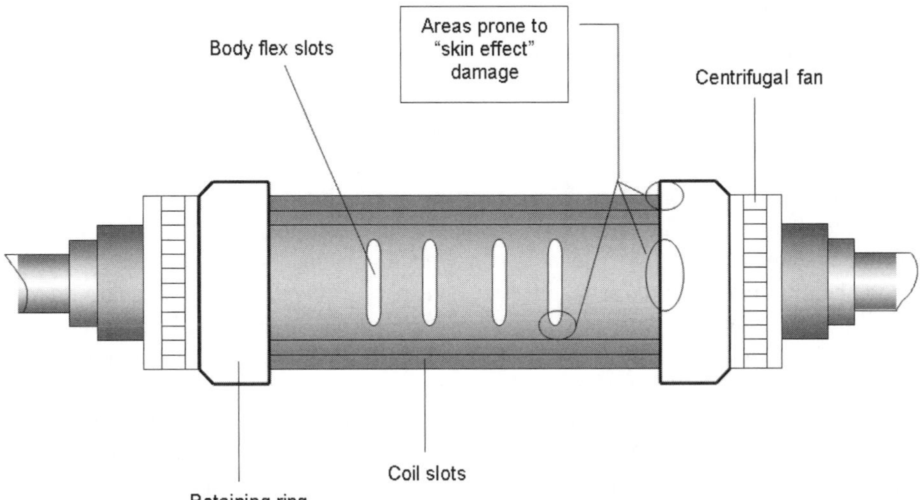

Fig. 4.19 Schematic representation of a turbo-generator's rotor and the areas most prone to be damaged by the "skin currents" generated during inadvertent energization events.

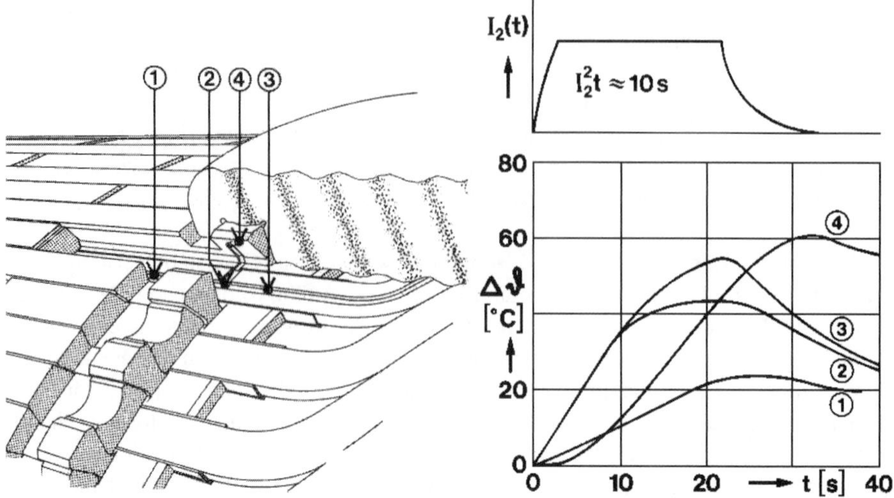

Fig. 4.20 Temperature rise measured at the end of the rotor body during short-term unbalanced load operation. (I_2 given as per unit). (Reproduced with permission from *Design and Performance of Large Steam Turbine Generators,* 1974, Asea Brown Boveri.)

turbine blades and casing. Another consequence could be wiped bearings and overheated rotor journals if there is no oil to the bearings at the time of the event.

4.4.2 Negative-Sequence Currents

A three-phase balanced supply voltage applied to a symmetrical three-phase winding generates a constant-magnitude flux in the airgap of the machine, which rotates at synchronous speed around the circumference of the machine (see Fig. 1.32). In addition, the slots and other asymmetries within the magnetic path of the flux create low-magnitude space harmonics (i.e., fluxes that rotate in both directions) of multiple frequencies of the fundamental supply frequency. In a synchronous machine under normal operation, the rotor rotates in the same direction and speed as the main (fundamental) flux.

When the supply voltage or currents are unbalanced, an additional flux of fundamental frequency appears in the airgap of the machine. However, this flux rotates in the opposite direction from the rotor. This flux induces in the rotor windings and body voltages and currents with twice the fundamental frequency. These are called *negative-sequence currents* (I_2). The negative-sequence terminology derives from the vector analysis method of symmetrical components. This method allows an unbalanced three-phase system to be represented by *positive, negative,* and *zero sequences.* The larger the imbalance, the higher is the negative-sequence component.

There are several abnormal operating conditions that give rise to large currents flowing in the forging of the rotor, rotor wedges, teeth, end rings, and field windings of synchronous machines. These conditions include unbalanced armature current (producing negative-sequence currents), inadvertent energization of a machine at rest, and asynchronous motoring or generation (operation with loss of field), producing alternate stray rotor currents. As was shown in the previous section, the resultant stray rotor currents tend to flow on the surface of the rotor, generating $(I_2)^2R$ losses with rapid overheating of critical rotor components. If not properly controlled, serious damage to the rotor will ensue. Of particular concern is damage to the end rings and wedges of round rotors (see Figs. 4.19 and 4.20).

All large synchronous machines have (or should have) installed protective relays that remove the machine from operation under excessive negative-sequence currents. To properly "set" the protective relays, the operator should obtain maximum allowable continuous negative-sequence I_2 values from the machine's manufacturer. The values shown in Table 4.2 are contained in IEEE C50.13-2005 as values for continuous I_2 current to be withstood by a generator without injury, while exceeding neither rated MVA nor 105% of rated voltage.

When unbalanced fault currents occur in the vicinity of a generator, the I_2 values of Table 4.2 will probably be exceeded. In order to set the protection relays to remove the machine from the network before damage is incurred, but avoiding unnecessary relay operation, manufacturers have developed the so-called $(I_2)^2t$ values. These values represent the maximum time in seconds a machine can be subjected to a negative-sequence current. In the $(I_2)^2t$ expression, the current is given as per unit of rated stator current. These values should be obtained from the manufacturer. Table 4.3 shows typical values given in the standard [17].

Figure 4.21 depicts in graphic form the last two rows of Table 4.3, representing the negative sequence capability of generators with direct-cooled stator windings.

4.4.3 Off-Frequency Currents

There are sources, in the generator and the power system, of currents at frequencies other than that of the power system. For example, current components at higher fre-

Table 4.2 Values of permissible I_2 current in a generator

Type of generator	Permissible I_2 as % of rated stator current
Salient pole	
Without connected amortisseur winding	5
With connected amortisseur winding	10
Cylindrical rotor	
Indirectly cooled (rotor)	10
Directly cooled (rotor)	—
to 350 MVA	8
351 to 1250 MVA	8 – (MVA – 350)/300
1251 to 1600 MVA	5

Table 4.3 Values of permissible $(I_2)^2t$ in a generator

Type of machine	Permissible $(I_2)^2t$
Salient-pole generator	40
Salient-pole condenser	30
Cylindrical rotor generator (the subject of this book)	
Indirectly cooled	30
Directly cooled up to 800 MVA	10
800 to 1600 MVA	10 – (0.00625)(MVA – 800)

quencies could be produced by transformer saturation and by incompletely filtered harmonic currents from rectifiers or inverters. Current components at frequencies lower than that of the power system have been produced by resonance between power-factor-compensating series capacitors (used to increase the power handling capability of long ac transmission lines) and the inductance of generators and transformers. This is commonly known as subsynchronous resonance.

Off-frequency currents interact with the useful flux in the generator to produce pulsating torques felt by the combined turbine-generator shaft system. If the frequency of one component of the pulsating torque is identical to the torsional natural frequency of any mode of vibration of the complex shaft system, destructive vibration could result. The degree of damage depends on the mode shape and the level of the current damping present.

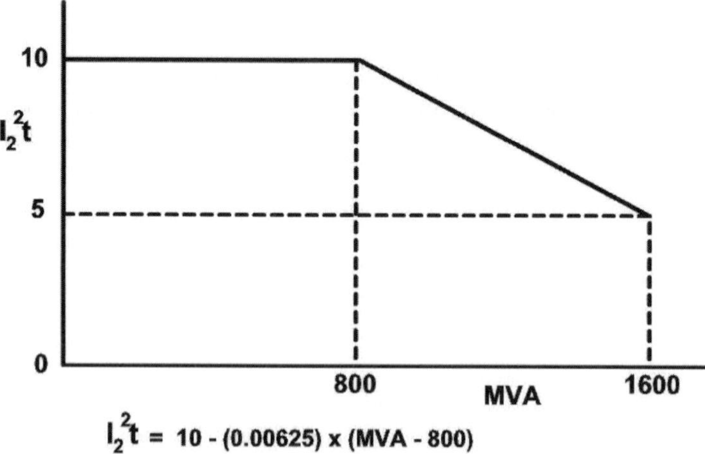

Fig. 4.21 Graphic representation of the negative-sequence capability of generators with directly cooled stator windings, according to IEEE C50.12-2005 and IEEE C50.13-2005.

At the present state of the art, it is not possible to calculate the higher natural frequencies accurately. Hence, designing to avoid higher stimulating frequencies may not be feasible. However, since the resonance peaks tend to be sharp (due to low damping), the likelihood of matching stimulus and response to a frequency is low, but the consequences of a match may be severe.

The frequency of the torque due to subsynchronous resonance is variable, depending on the level of series capacitance compensation being used at the time. It is necessary to avoid those frequencies that would stimulate a rotor torsional natural frequency, or to block the current when at a potentially damaging frequency from reaching the generator. Subsynchronous resonance in a power system was first studied following the destruction in California of two new shafts belonging to generators rated at several hundreds of megawatts. In that state, there are long transmission lines compensated with series capacitors. The use of power system stabilizers, together with the voltage regulator, can suppress these types of harmonics.

4.4.4 Load Cycling and Repetitive Starts

It is well known in the power industry that load cycling represents a long-term onerous mode of operation. Turbines and generators "like" to be in a steady-state condition, meaning where the temperatures in the machine are stable. Any situation in which load is changed significantly will result in relatively large changes in temperatures. It is the transition time between the steady states that embraces an amalgam of problems. For instance, when load is increased suddenly, the conductors will rise in temperature first, followed by the core and other components. As the temperature differentials increase momentarily, so do the mechanical stresses induced.

Another consequence of a change in temperature is related to the fact that the copper conductors expand and contract more than the iron core and frame. Sometimes, this results in a "ratcheting" effect, by which the conductor or, more often, its supporting system, is partially "stuck" and does not fully return to its original position. This problem shows up quite frequently in rotors, with resulting deformation of the field winding. Figure 4.22 clearly shows the results of such a ratcheting effect in a 90 MVA, hydrogen-cooled, 3600 rpm unit.

Figure 4.22 by no means depicts the only problem resulting from excessive load cycling. Other problems are loosening of stator wedges, looseness of the stator core, weakening of the stator end-winding support system, cracking of conductors, weakening of frame support systems, leakage of hydrogen through gasket degradation, accelerated deterioration of rotor and stator insulation, and so forth.

By far the most onerous load cycling is the complete start-and-stop operation. Numerous units nowadays start and stop every day. This type of operation stresses all those elements enumerated above to the extreme. Retaining rings, in particular spindle-mounted rings, are significantly stressed during start–stop operation due to the flexing they undergo when going from rest to full speed, and vice versa. Recognition of this accelerated deterioration of machines operated with many starts and/or load cycling demands that the inspection intervals be significantly shorter than for units

Fig. 4.22 Rotor field end winding of a 90 MVA, 3600 rpm, hydrogen-cooled generator. The top turn of one coil has become distorted and elongated during operation all the way across the gap separating it from the neighboring coil, resulting in a severe shorted-turns condition. This is the result of ratcheting during cycling and insufficient blocking.

operated under base-load conditions. The rotor teeth at the slot exits also require detailed inspection, by nondestructive-examination-type methods, for cracking due to considerable load-cycling stresses from changes in speed and in conjunction with the expansion and contraction of the retaining rings from zero to full speed.

4.4.5 Overloading

In Section 4.3.3, the need to remain within the capability curves of the machine at all times was stressed. Nonetheless, if a severe overload situation is reached, the need to schedule an inspection of the windings of the machine as soon as possible ought to be considered. Bear in mind that the heating developed in a conductor is proportional to the square of the current. Thus, a 10% overload condition will increase the heat generated in that conductor by about 20%. The temperature will change also in a similar fashion. However, the expected life of insulation is approximately halved for every 8 to 10°C increase in temperature (the Arrhenius Law, after Svante August Arrhenius, 1859–1927). Thus, long-term operation at moderate overloads or short-time severe overloads can markedly reduce the expected life of a machine's insulation systems.

4.4.6 Extended Turning-Gear Operation

In Section 4.2.2, the benefit of turning gear operation was stated. However, turning-gear operation has inherent disadvantages. In particular, long periods of turning on gear may induce production of copper dusting in the rotor field conductors. The rotor field coils are rather heavy, and when radial clearances are present in the slot, slow rotation of the rotor results in the coils "falling" and "rising" in the slots (Fig. 4.23). This movement and the banging between turns (specifically in those designs with double copper conductors) will eventually lead to copper dusting. Copper dust has the potential to create shorted turns in the rotor field (Fig. 4.24).

How fast the rotor rotates, its dimensions, and the type of conductor are all factors in determining if and how much copper dusting will result from extended turning-gear operations. It is, therefore, important that the operator learn how the machine will be affected from this mode of operation.

A simple calculation can deliver the minimum turning-gear speed that would eliminate most of the movement that results in the formation of copper dust. It is possible to change the turning-gear speed. Therefore, a unit that spends significant time in turning-gear mode (due to operational or/and dispatch-mode characteristics), and that is prone to the formation of copper dust (double copper bar per turn and not consolidated), could benefit from having its turning-gear speed set above the minimum required

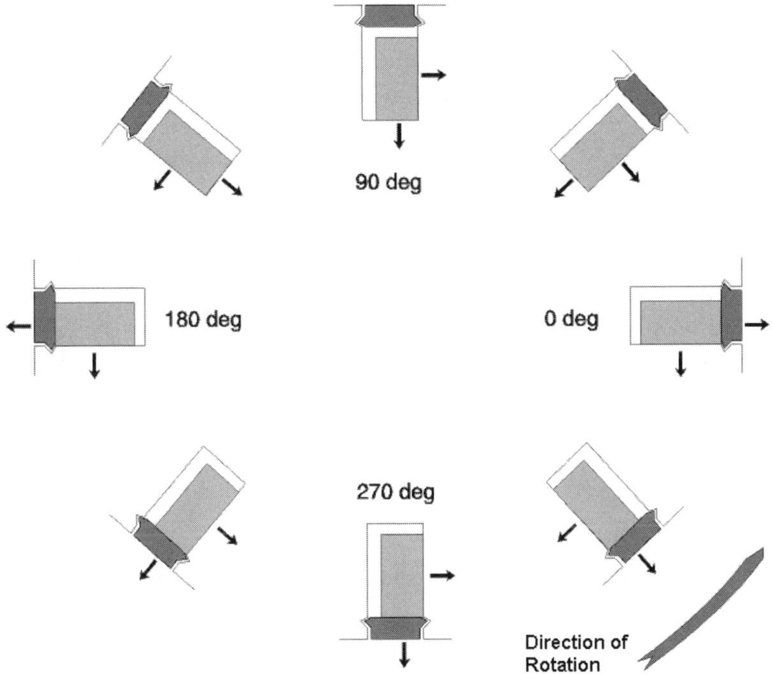

Fig. 4.23 Schematic representation of the movement of the coil in the slot of a rotor when rotated at low speed (turning-gear operation). The continuous pounding of the heavy copper bar against the slot-sides, generates, over time, copper dust in those coils where copper bars are in touch with each other.

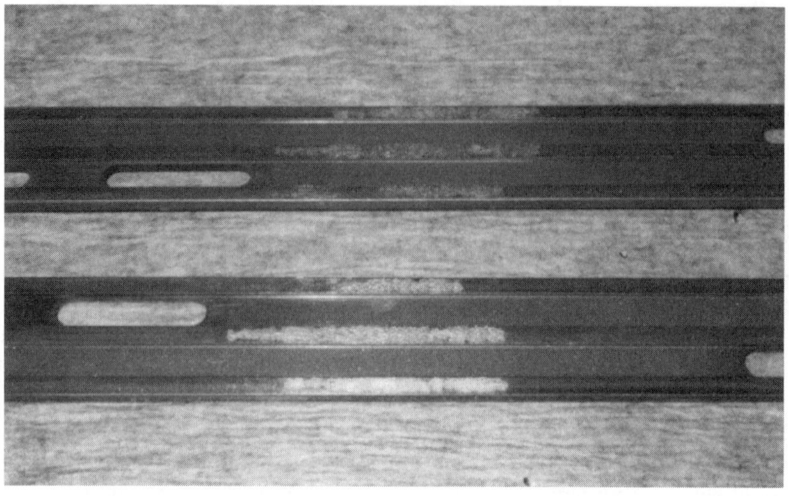

Fig. 4.24 Severe erosion on two halves of a conductor due to continuous pounding and the generation of copper dust.

to eliminate the bouncing of its copper bars in the rotor winding. The calculation described below is derived from a real case example.

Example. Calculation of minimum turning-gear speed to eliminate bar bouncing in a rotor for a 1300 MVA, four-pole rotor with a radius to the center of the bottom bar equal to 0.635 meters follows. The reader should first refer to Figure 4.25.

The maximum bounce occurs when the slot is coincident with the vertical axis of the rotor. At that point, the centrifugal force (F_c) acting on a conductor bar counteracts the weight (W). For the bars not to bounce, the centrifugal forces on each bar must be equal or higher than the weight of that bar. It is very simple to show that the lowest centrifugal force acts on the bottom bar (because its radius is the smaller). So the calculations must be carried out for the bottom bar. The equation that gives the minimum rotor rpm (revolutions per minute) that keeps all bars pressed toward the outer edge of the slot is

$$\text{Minimum rpm} = (896/R_B)^{1/2}$$

where R_B is the radius to the center of the bottom bar and is given in meters.

Going back to our example,

$$\text{Minimum rpm} = (896/0.635)^{1/2} = 37.6 \text{ rpm}$$

This means that if the turning-gear speed is no less than 37.6 rpm, the formation of copper dust is almost completely eliminated.

Two-pole rotors have inherently smaller radii. Therefore, the turning-gear speed will be higher. The same applies for rating of the generator. The lower the rating, the

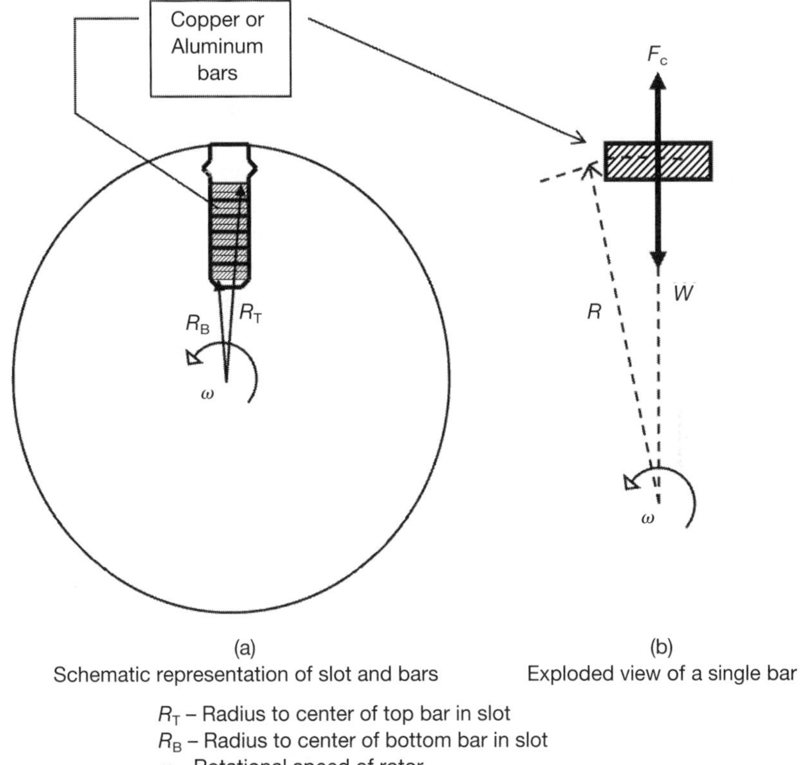

(a)
Schematic representation of slot and bars

(b)
Exploded view of a single bar

R_T – Radius to center of top bar in slot
R_B – Radius to center of bottom bar in slot
ω – Rotational speed of rotor
F_c – Centrifugal force acting on bar
W – Weight of bar

Fig. 4.25 Forces on the rotor winding inside the slot.

smaller the diameter, and the faster it must rotate in turning-gear mode to avoid copper dusting.

In most cases of two-pole rotors, increasing the speed is not practical because the required speed to stop copper conductor movement is too high for other design issues such as those associated with the turbine rotor blades. Therefore, the rotor turning gear speed is sometimes reduced to a few rpm to minimize the number of cycles that the conductors would be subject to.

4.4.7 Loss of Cooling

On occasion, a unit is inadvertently run without cooling medium (mainly water), for some period of time. It has happened more than once that someone walking on a turbine deck and seeing external paint blistering and flaking away from a generator's casing discovers that the unit was running without its water-cooling system in operation. This problem may result in serious overheating of the windings and, perhaps irreversible, damage. After such an event, the unit ought to be removed from service and opened for careful inspection of the windings.

What type of damage may occur under loss of cooling operation is largely predicated on the kind of insulation system. For example, thermoplastic systems (asphaltic) will deform under severe heating. Oozing of the asphalt may occur. These conditions can be very onerous. On the other hand, thermoelastic systems will be more resilient. Though a loss of expected life of the insulation might have occurred, the overall situation of the winding may be satisfactory for long-term operation.

There are a number of mechanical problems that may also result from high temperatures attained during loss-of-cooling operation, such as severe misalignments and damage to the end-winding support systems.

4.4.8 Overfluxing

Overfluxing occurs when a generator is operated beyond its maximum continuously allowed V/Hz. In Section 4.1.13, an elaborate description of overfluxing was included. There it was also indicated that overfluxing could destroy a large core in seconds.

Most common instances of severe overfluxing occur while the machine is being run up to speed prior to synchronization with the system. Under those conditions, any misoperation of the excitation system, voltage regulator, or the voltage and current sensing systems may cause the excitation to go to "ceiling" and the terminal voltage go much higher than nominal. If the speed (and, thus, frequency) is rated or lower, the V/Hz so attained may cause the machine to be well beyond its capability to withstand heating. Also, when the speed is below rated there is reduced cooling because the rotor-fan capability is reduced at lower speed and this will exacerbate the situation further. Core melting and/or core-insulation damage are probable results.

Once the machine is connected to the system, the probability of sudden damage due to overfluxing is very low. In any event, it is truly important that V/Hz protection is properly designed and set. Moreover it is important to design the voltage-sensing scheme for the excitation in such a way that loss of a single potential transformer winding will not result in a V/Hz event. ANSI/IEEE C37.106 has a good discussion on the subject and presents examples of how to design and set the protection schemes [4].

Figure 4.26 reproduces the V/Hz withstand curves of a number of manufacturers for purposes of illustration only. For your specific machine, consult the OEM when setting the protection.

Voltage Buildup during Acceleration. As stated above, the vast majority of serious overfluxing events occur during the activity of bringing the unit into synchronism with the grid. The following discussion describes the process.

With the generator terminal breaker open (open-circuit condition), the differential of potential across the terminals of the machine are given by the following expression (see Chapter 1):

$$\mathbf{E} = \mathbf{K} \times \mathbf{B} \times \omega$$

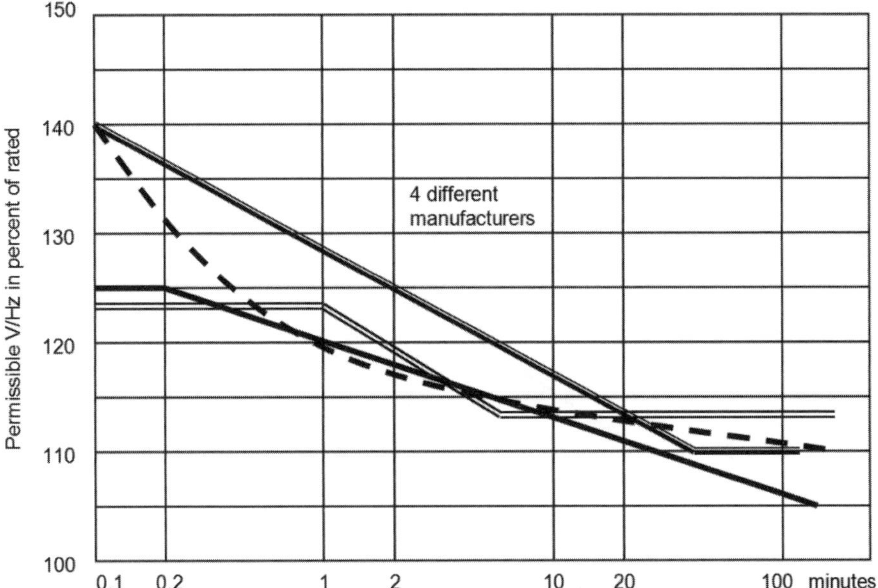

Fig. 4.26 Permissible V/Hz curves (or V/Hz withstand characteristics) of four manufacturers. The permissible area is *below* the curves.

where **E** is the voltage generated across the terminals of the machine, **K** is a constant that depends on the generator's construction details (such as number of slots, poles, diameter of gap, winding factors, and so forth), **B** is the flux density across the airgap (also called gas gap), and ω is the rotational speed of the machine. However, neglecting saturation effects, the flux density is directly proportional to the field current ($\mathbf{I_F}$), and ω can be expressed in revolutions per minute (rpm). Then the equation can be rewritten as follows:

$$\mathbf{E} = \mathbf{K} \times \mathbf{I_F} \times \text{rpm}$$

Thus, it can be said that the generated potential at the terminals of the generator is directly proportional to the field current and the speed or rotation:

$$\mathbf{E} \; \alpha \; \mathbf{I_F} \times \text{rpm}$$

Let us assume that a constant field current ($\mathbf{I_F}$) is applied on the machine as it accelerates from rest to full speed. Figure 4.27 depicts the voltage generated as the machines ramps up in speed.

Figure 4.28 illustrates how an infinite number of output voltage traces can be drawn, each one relating the output voltage, the speed, and the field current. The figure also shows the open-circuit field current ($\mathbf{I_{FOC}}$), which represents the value of field

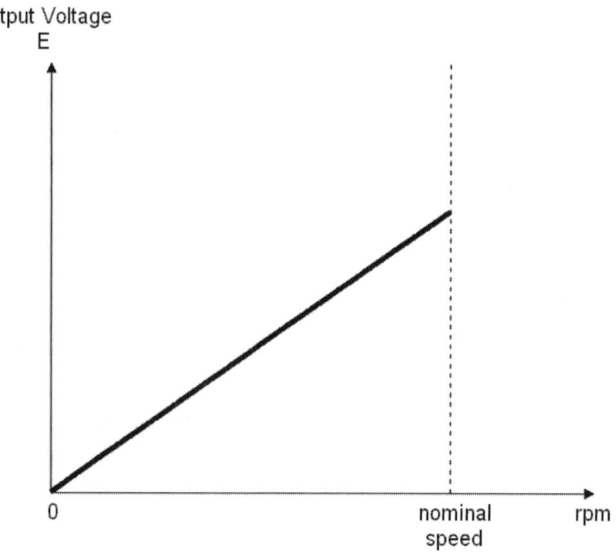

Fig. 4.27 Open-circuit terminal voltage with constant field current applied during acceleration (saturation effects neglected).

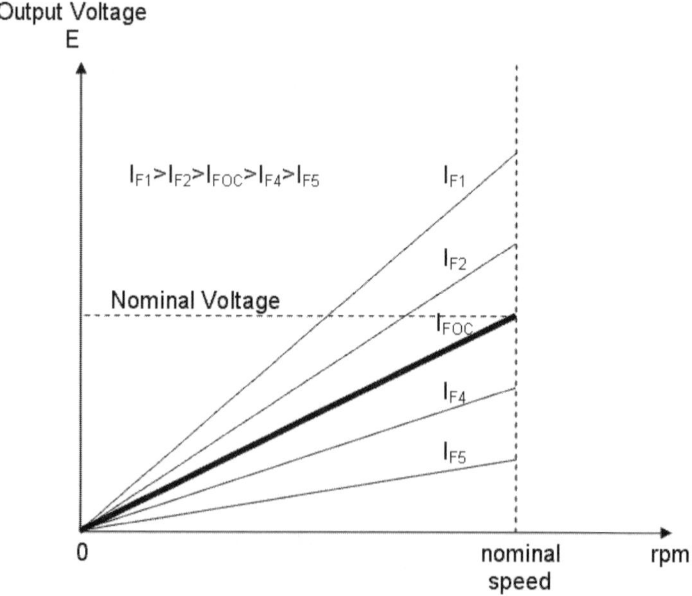

Fig. 4.28 Open-circuit terminal voltage for various levels of field current applied during acceleration.

current that, when applied to the generator with the terminals open, will result in nominal (rated) voltage at the terminals of the machine at nominal (rated) speed.

It is important to note that the voltage versus speed line that corresponds to $\mathbf{I}_{FOC}$ and matches the nominal output voltage of the generator at full speed also represents the nominal V/Hz (or V/rpm) line for all speeds between zero and rated speed. The maximum V/Hz line allowed during continuous operation is 5% higher than that, based on the fact that, per international standards, generators are designed for nominal voltage ± 5% for continuous operation. Figure 4.29 shows the nominal and maximum V/Hz bounds.

In practice, field current is applied not at rest but at full speed or somewhat below that. Earlier practice, before the advent of the *automatic voltage regulator* (AVR), or with the advent of the first primitive AVRs, the application of current to the field (by closing a *field breaker*), was carried out after the unit attained its rated speed. As technology moved forward, AVRs became able to build up voltage in a controlled manner, to the extent that field current could be and now is applied in many cases with the unit still ramping up speed. There are number of advantages to an earlier and well-controlled application of field current, as explained below. Figure 4.30 illustrates a typical trajectory for the voltage generated in the terminals of an open-circuit generator as it speeds up from rest to rated speed.

One must keep in mind that so far, saturation effects have been left out from the discussion. In reality, there is a limit to how much voltage can be built across the termi-

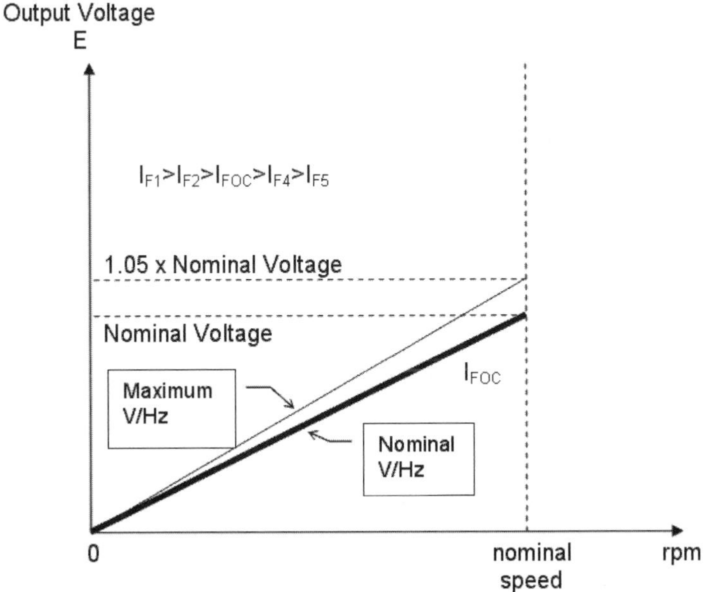

Fig. 4.29 Nominal and maximum V/Hz bounds.

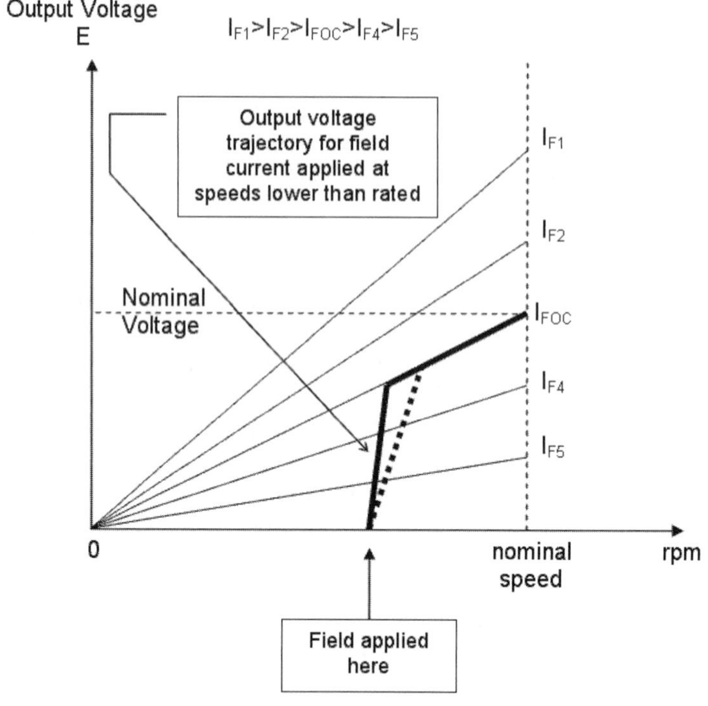

The straight line trajectories are obtained when both the machine's rate of speed increase and the AVR rate of voltage increase are constant.

The rising broken line to the right of the full line would obtain if the speed were to increase at a quicker rate. The opposite would happen for slower rates of speed increase.

Fig. 4.30 Open-circuit terminal voltage trajectory with field current applied before rated speed.

nals of the machine when maximum excitation current is applied. Figure 4.5 shows a typical behavior for the saturation of a core. Depending on the design of the machine, open-circuit voltage may, for all practical purposes, have a ceiling anywhere from 130% to 160% of nominal voltage.

Figure 4.31 tries to capture the three-dimensional characteristics of the equation shown above, tying the output voltage to the rotor speed and the field current.

Referring to Figure 4.30, let us now assume that during start-up the field current is overapplied, that is, the value of $\mathbf{I_{FOC}}$ is exceeded. This situation is depicted in Figure 4.32. Obviously, under these conditions the V/Hz safe curve is exceeded. In addition, the nominal voltage may or may not be exceeded.

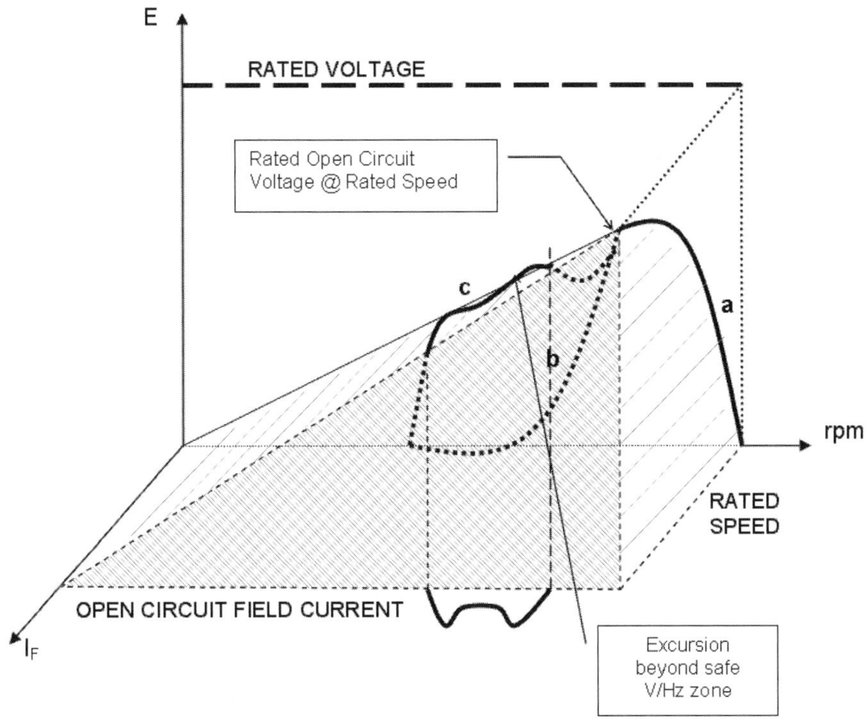

The Pentahedron formed by connecting the center of the *E-I_F-rpm* coordinates with the rated values of open circuit voltage and speed, and open circuit field current, define the area of safe V/Hz operation.

Curve "a" shows the voltage trajectory when the field is applied at rated speed.
Curve "b" shows the voltage trajectory when the field is applied at a lower-than-rated speed.
Curve "c" shows the voltage trajectory when the field is applied at a lower-than-rated speed, and the V/Hz safe region is breached.

Cooling limitations (at low speeds) are not considered in this analysis.

Fig. 4.31 Three-dimensional view of V/Hz safe operation during start-up.

From the discussion above, an interesting argument can take place about the differences in consequences to the integrity of the core by experiencing an overflux event of the same duration and intensity (i.e., same field current) happening at different values of rotor speed. Figure 4.33 shows two traces representing overflux events of equal magnitude and duration. It is clear from the figure that case 2, though having the same overflux intensity, has additional detrimental effects related to the higher voltage at-

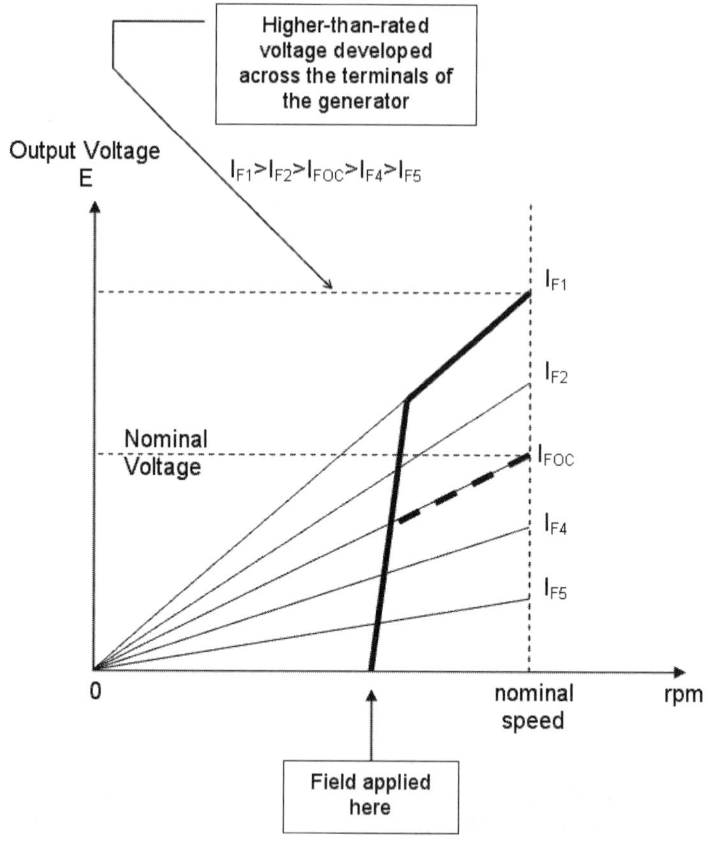

Figure 4.32 Overexcited open-circuit terminal voltage during start-up.

tained (a de facto high pot). An in-depth analysis of the attributes of each event would have to take into account secondary effects, such as the fact that at lower speeds (i.e., frequencies) the actual hysteresis and eddy losses are lower, but cooling is somewhat poorer. The existing literature does not provide a differentiation of V/Hz detrimental effects based on the machine's rpm at the time of the event. The common practice is to use the same curves (see Figure 4.26) for both full speed and reduced speed events. However, as can be seen in Figure 4.34, the speed at which the overfluxing event takes place has a nonnegligible effect that should be taken into consideration in future research. For example, in Figure 4.34 the ratio of the loss at 60 Hz (1.6 W/kg) is roughly 33% higher than the loss at 50Hz (1.2 W/kg). Therefore, in a 60 Hz generator, an over-

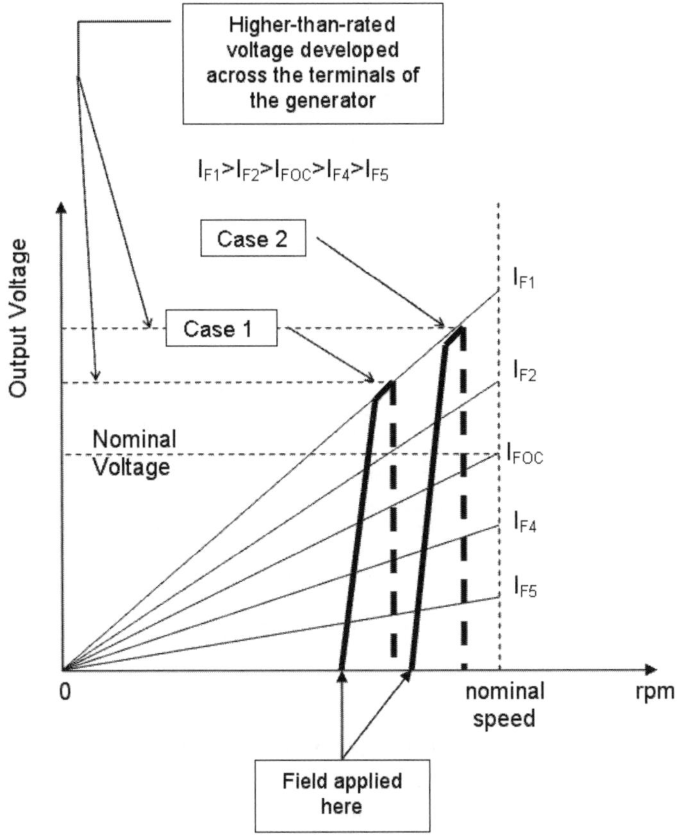

Fig. 4.33 Open-circuit overflux events at different speeds.

flux event at 80% of speed would have a significantly lower impact than an event at full speed (all else being equal). Nevertheless, as stated before, this fact is not captured in the literature. One must say, though, that by assuming the worst possible scenario, a healthy conservative approach can be adopted.

The reader who wants to extend his or her knowledge of lamination material and laminate insulation issues is referred to the book by Beckley [8].

Figure 4.35 was drawn to clarify the issue of overreach into regions of operation with potential damage to the machine. In the figure, the area of excessive V/Hz can be seen as the region above the open-circuit voltage given by $\mathbf{I_{FOC}}$. The area of excessive voltage is above 105% of nominal voltage. In the figure, it becomes clear how the two events (cases 1 and 2) described above affect the same machine to different extents.

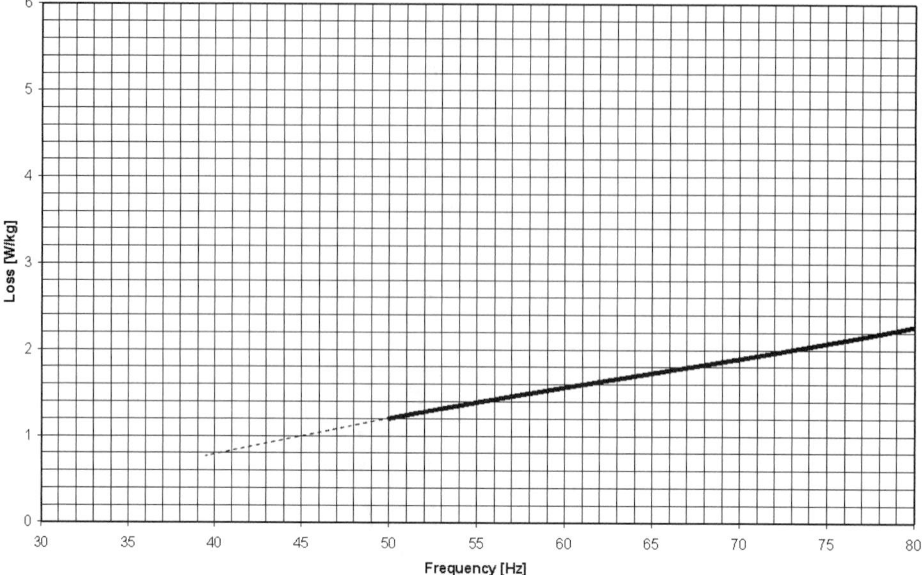

Fig. 4.34 Loss versus frequency at 1 tesla for M310-50A lamination material at 0.5 mm thickness.

Summarizing the aforementioned, it can be said that the vast majority of overfluxing events occur with the unit in start-up mode (not synchronized to the grid), and that similar overexcitation events may result in different stress on the core and winding insulation, depending on at what speed these events occur.

4.4.9 Overspeed

A typical industry rule is that rotors of turbo-generators are designed to withstand a 120% spin test. Any significant overspeed can damage the rotor components to the extent that a new rotor or major parts (e.g., retaining rings) are required. The protection against overspeed must be well designed and set because a severe overspeed condition can be unforgiving. See Figure 4.36 for the unpalatable results of such an event.

4.4.10 Loss of Lubrication Oil

The result of a loss of lubrication oil during operation can be catastrophic. Such events are not unheard of, but only few result in severe loss of equipment. For this reason, lube-oil systems offer redundancy. In general, the backup is provided by an additional ac or dc motor, or both, operating a backup pump. Oftentimes, shaft-mounted pumps provide critical lubrication for as long as the turbine rotates.

Failure of this system can be very costly in materiel and lost production. Whatever the system, it is very important that it is not forgotten to check the lube-oil system

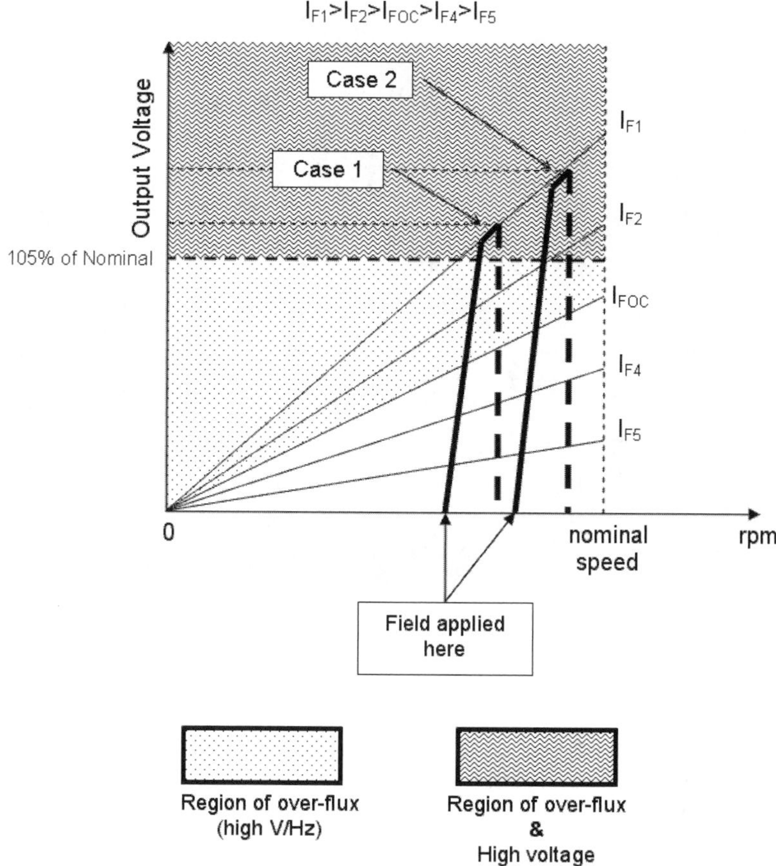

Fig. 4.35 Regions of excessive V/Hz and overvoltage on the open-circuit voltage versus speed curve.

when carrying out periodic inspections and operational testing of the unit's support systems.

4.4.11 Out-of-Step Synchronization and "Near" Short Circuits

Both out-of-step synchronization and short circuits occurring in or in the vicinity of the generator (in particular, between the generator's terminals and the main step-up transformer) can result in severe damage to the unit: sheared coupling and frame supporting bolts; damaged stator coils, end-windings and end-winding supports; rotor end-winding deformation; shaft cracked or sheared; bushings damaged; and so forth.

The severity of the damage depends, for example, on generator rating, angle between system and generator voltage vectors at the moment of synchronization, and type of short circuit (phase-to-phase or three-phase). The presence of isolated phase busses

Fig. 4.36 Catastrophic failure of a turbo-generator unit due to overspeed (loss of governor control). This is a view into the generator (whatever is left of it).

(IPBs) makes a three-phase bolted short circuit on the terminals of any large generator an event with negligible probability of occurrence. Nonetheless, faults within the machine itself and on the main step-up transformer, or very close to the high-voltage side of it, can seriously damage the unit (both generator and turbine). Therefore, before a new attempt is made to synchronize the unit after a major out-of-step mishap, or in the aftermath of a strong and near short circuit, the unit should be opened at both ends for visual inspection. Alternatively, when easy access can be achieved for visual inspection, such as removing coolers, opening of the end shields might be avoided.

Out-of-step synchronization protection is provided by a couple of protective devices. See Chapter 6 on generator protection. For additional discussion on sudden short circuits, see Section 4.5.7.

4.4.12 Ingression of Cooling Water and Lubricating Oil

On occasion, a water leak develops during operation. Also, a large quantity of oil might be found to have leaked into the generator during operation or start-up. Both of these situations present the possibility of short- and long-term deterioration and damage of many components, such as some types of retaining rings susceptible to water stress–corrosion cracking and winding and winding-support system deterioration from excessive oil presence. Both issues will be discussed amply in the chapters covering monitoring and diagnostics, as well as stator and rotor inspection. Let us say here that these events require attention, and the severity of the situation determines how soon.

4.4.13 Under- and Overfrequency Operation (U/F and O/F)

U/F and/or O/F operation indicate that the unit is operating slightly under or over the rated frequency of the system. Interestingly, the turbine is, in general, more sensitive to this condition than the generator.

In the case of the generator, as well as the transformers connected to the generator, the main concern is that while running under frequency and at the highest allowable terminal volts, the machine may move beyond the permissible V/Hz region. This condition should result in an alarm generated by the protective scheme, so that the operator has an opportunity to correct it (lowering terminal voltage or removing the unit from operation).

However, as stated above, it is the turbine element that is more sensitive to U/F and O/F operation. The main areas of concern are the turbine blades moving into one of their natural frequencies, resulting in accelerated metal fatigue. It was once common knowledge that steam turbines are more sensitive than combustion turbines. This is not the case nowadays. To keep up with the ongoing drive to improve combustion turbine efficiencies, modern-day design practices for these units result in reduced margins, making them less tolerant of O/F and U/F operation, oftentimes less so than steam turbines. The protection of the turbine is done directly by protective devices on the turbine's panel sensing speed and load conditions (which are always set to the vendor's specifications) and, indirectly, by O/F and U/F protection monitoring of output voltage.

To prevent damage to the turbine, it is imperative that this protection be set up and maintained properly. Manufacturers provide curves for permissible regions of operation and the maximum accumulated time for the lifetime of the unit, and for periods of operation in the restricted zones. The maximum accumulated time over the unit's lifetime is, in general, given in minutes (usually within a few seconds to a few tens of minutes). Figure 4.37 gives an illustration of such withstand curves. In all cases, operators must refer to the equipment's manufacturer for obtaining the correct information on their units.

4.5 BASIC OPERATION CONCEPTS

4.5.1 Steady-State Operation

A turbo-generator can be seen as a nonlinear combination of magnetically coupled windings, airgap reluctance, and the electrically conductive mass of the rotor body (which acts as a distributed winding). Its electrical characteristics when it is operating in a steady-state fashion are very different from those when conditions are changing. The turbo-generator has different characteristics under slow changes than it does under rapid changes. For many conditions, a turbo-generator can be represented as a reactance in series with a voltage source. That reactance takes on different values for different operating conditions. In addition to the familiar concept of a reactance as it functions in an electric circuit, there are magnetic considerations that are useful in describing the operation of a synchronous machine. An inductance (which is multiplied

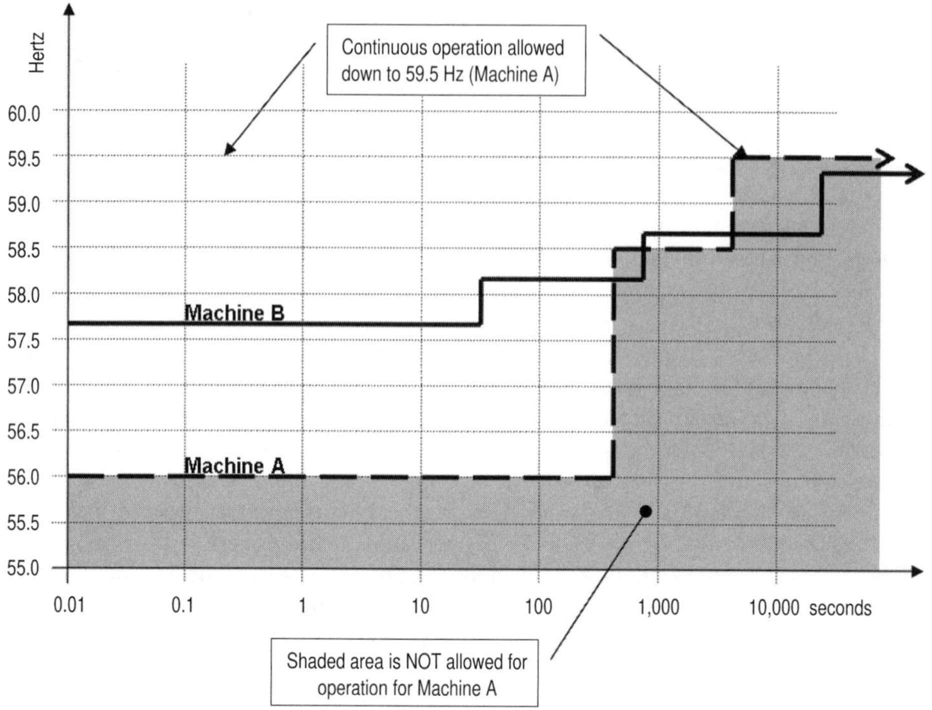

Fig. 4.37 Permissable underfrequency operation curves for machines A and B. For machine A, above the broken line operation is allowed below 59.5 Hz, subject to the maximum total accumulated time over the life of the machine; below the broken line (the shaded area) operation is not allowed. Similar reasoning applies to Machine B. The graphs can be extended upward for higher rated frequencies (60 Hz in this case). The graphs so obtained for O/F operation are similar to those for the U/F region.

by the angular frequency to obtain the reactance) can be defined as the flux linkages produced by one ampere of current. Thus, the reactance is a measure of the ease with which current produces flux in the machine.

When the generator is operating in a steady load-carrying condition, it appears to the power system as a voltage source connected to the generator terminals through the generator's synchronous impedance (Fig. 1.35, Chapter 1). The generator resistance is negligible, and it is common to consider only the generator's reactance, in this case the synchronous reactance X_s.

During steady-state operation, a component of flux (φ_A in Fig. 4.38 or ϕ_S in Fig. 1.33) is produced by the stator current, and passes through the same magnetic circuit as that for the flux produced by the rotor field winding (φ_{DC} in Fig. 4.38 or ϕ_F in Fig. 1.33). This is an effective flux path, and a relatively high value of reactance may be expected, in the range of 1.5 to 2.1 per unit. The per-unit synchronous reactance is approximately equal to the reciprocal of the short-circuit ratio.

- The flux produced by the armature distorts the main flux produced by the DC rotating field
- The amount of change/distortion depends on *Load and Power Factor*

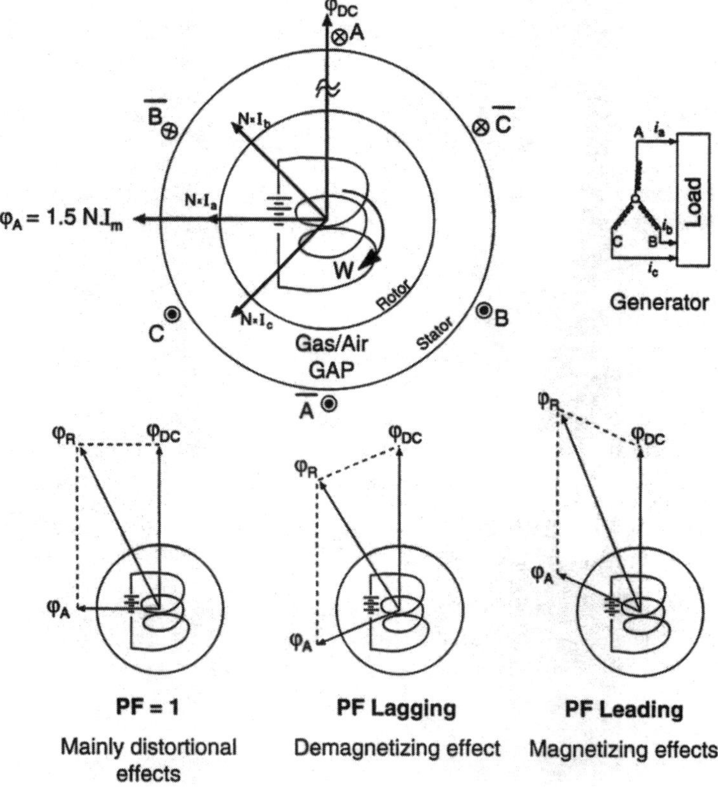

Fig. 4.38 Armature reaction. The top part of the figure shows how the resulting flux from the fluxes generated by a three-phase balanced winding (where three-phase balanced currents flow) is constant and of value equal to 1.5 times the maximum value of the flux produced by each phase. This resultant flux rotates at synchronous speed. The bottom part of the figure shows how the stator-produced flux affects the rotor-produced flux for unity, leading, and lagging power factors. This is the "armature reaction" effect. ϕ_R = Resulting flux in machine, ϕ_A = Aramature produced flux, ϕ_{DC} = DC field flux.

The stator-produced flux acts together with the rotor-produced flux to create the total "useful" (meaning linking both windings) flux, called the resultant flux (φ_R in Fig. 4.38 or ϕ_R in Fig. 1.33). The way the stator-produced flux affects the rotor-produced flux is called the "armature reaction" of the machine. This can be clearly seen in Figure 4.38, where the bottom of the figure presents how the armature reaction affects the rotor-produced flux for three power-factor conditions: unity, leading, and lagging.

The armature reaction of the generator affects the voltage regulation of the machine (i.e., how the terminal voltage changes as the load changes, all other things remaining the same; see Fig. 4.39). With lagging power factors, the armature reaction tends to accentuate the voltage drop in the machine, requiring additional dc current to be supplied by the exciter for compensation. How much armature reaction exists in a machine is the result of design compromises.

4.5.2 Equivalent Circuit and Vector Diagram

Section 1.7 in Chapter 1 introduced the reader to the most basic description of synchronous machine operation. In this section, the concept will be further developed, and the use of vector analysis will be illustrated with a few very basic examples.

Figure 4.40 presents the alternator's basic equivalent circuit that can be used by any individual to solve simple application problems. The *fundamental circuit equation* in Figure 4.40 relates machine variables to the connected system's current and voltage (at the generator's terminals). Figure 4.41 shows the vector representation of the fundamental circuit equation in the case of a synchronous machine acting as a generator. Figure 4.41 also shows the definition of regulation as it applies to an alternator.

4.5.3 Power Transfer Equation between Alternator and Connected System

The power transfer equation is one of the basic equations in electric power engineering. It states: "The power transmitted between two points in an ac circuit is equal to the

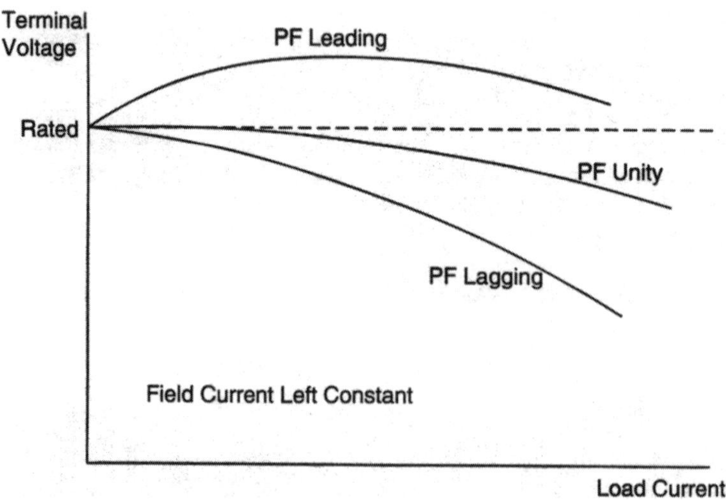

Fig. 4.39 How the armature reaction affects the output voltage of a generator for unity, leading, and lagging power factors.

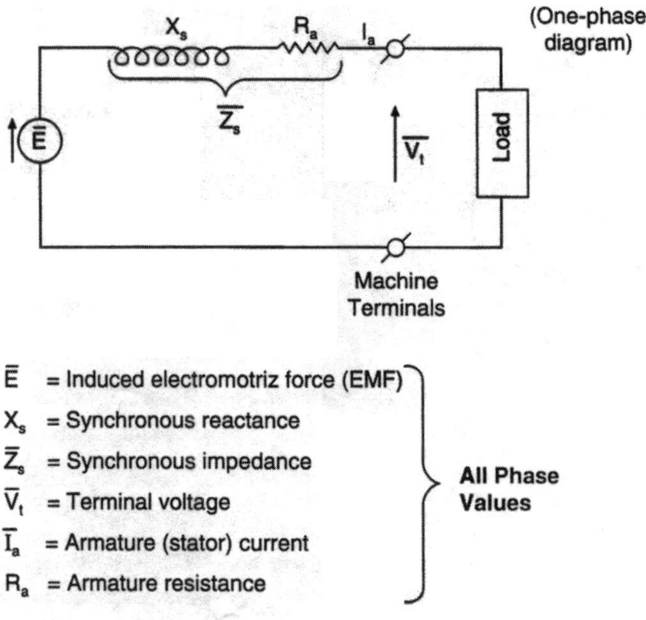

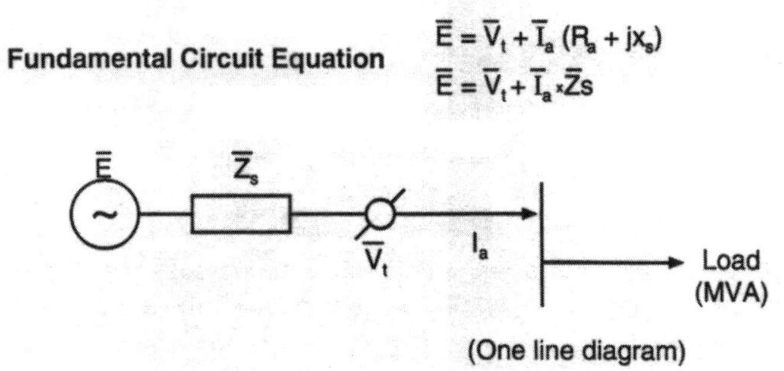

Fig. 4.40 Generator equivalent circuit. The equivalent circuit diagram of a synchronous machine developed in Figure 1.35 is reproduced here. Also shown is the one-line representation of the generator behind its synchronous impedance and the fundamental circuit equation.

product of the magnitude of the voltages at both ends, times the sine of the angle between the two voltages, divided by the reactance between the two points."

The maximum power that a circuit can deliver between two points is, thus, when the sine of the angle between the voltages equals 1, meaning the angle between the voltages equals 90°. Figure 4.42 illustrates the power transfer function as it applies between two electric machines, and between an alternator and the electric power system.

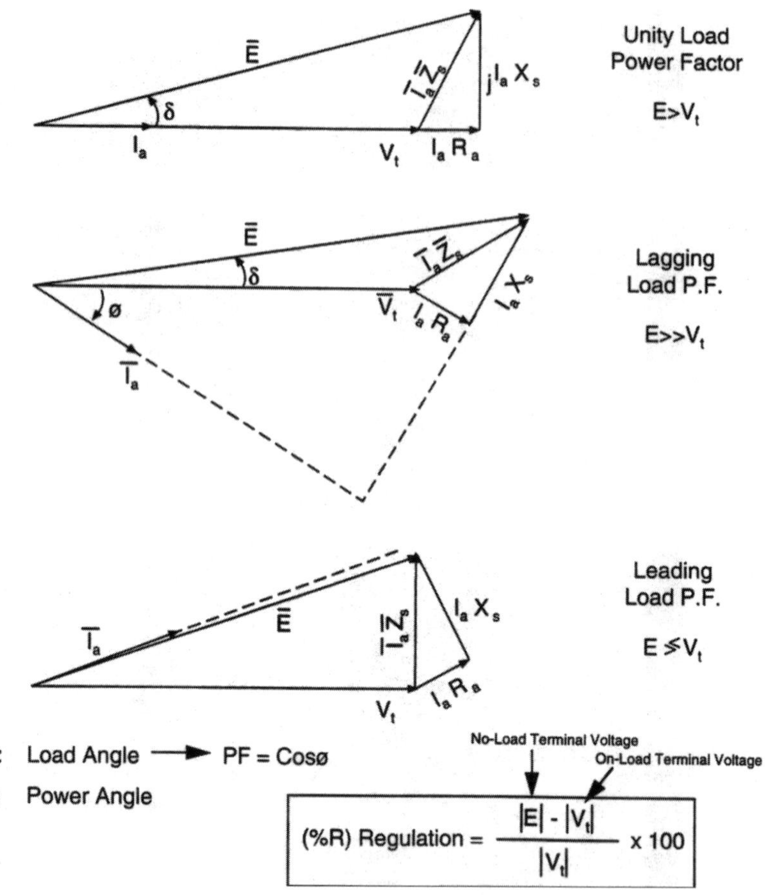

Fig. 4.41 Vector representation of the fundamental circuit equation in the case of a generator, for various power factor conditions. Also shown is the formula for the calculation of the regulation.

4.5.4 Working with the Fundamental Circuit Equation

The following two simple circuit problems with the generator connected to the system illustrate how the fundamental circuit equation, the power transfer equation, the active power equation, and a little basic trigonometry can be used to obtain solutions. Figure 4.43 captures those equations in a vector diagram.

Case 1: Change in Excitation. A generator is supplying power to the system. Now let us assume that the excitation is changed but the turbine's output is not changed. Additionally, the system may be assumed to be much larger than that of the generator ("infinite" system) so that the frequency of the system (hence, the genera-

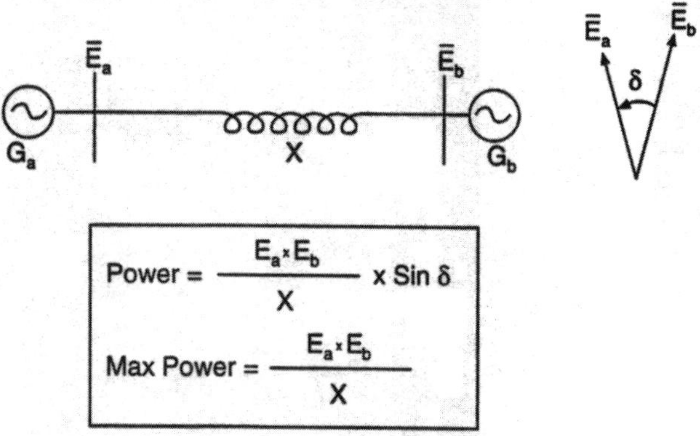

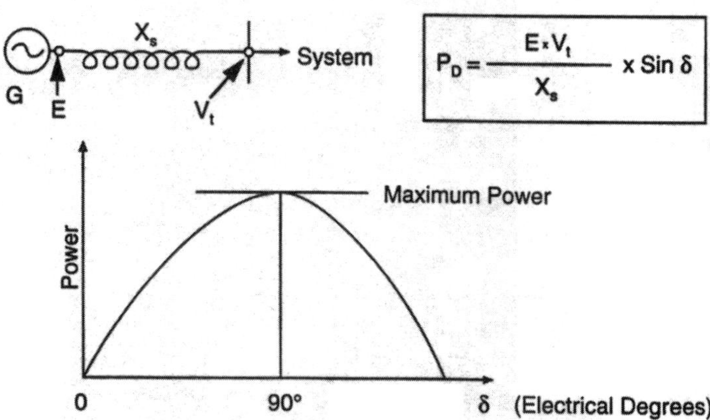

Fig. 4.42 Power transfer function applied to the power transferred between two electric machines, and between a generator and the power system.

tor's speed) and the voltage at the terminals do not change. Under these circumstances, it is desired to estimate how the power factor PF and the armature current $\mathbf{I}_a$ change.

The solution of this simple problem can be found by inspection of the vector diagram in Figure 4.44. The voltage induced in the machine (**E**) multiplied by the terminal voltage ($\mathbf{V}_t$) and by the sine of the angle between them (δ) represents the power transferred from the machine to the terminals (power transfer equation):

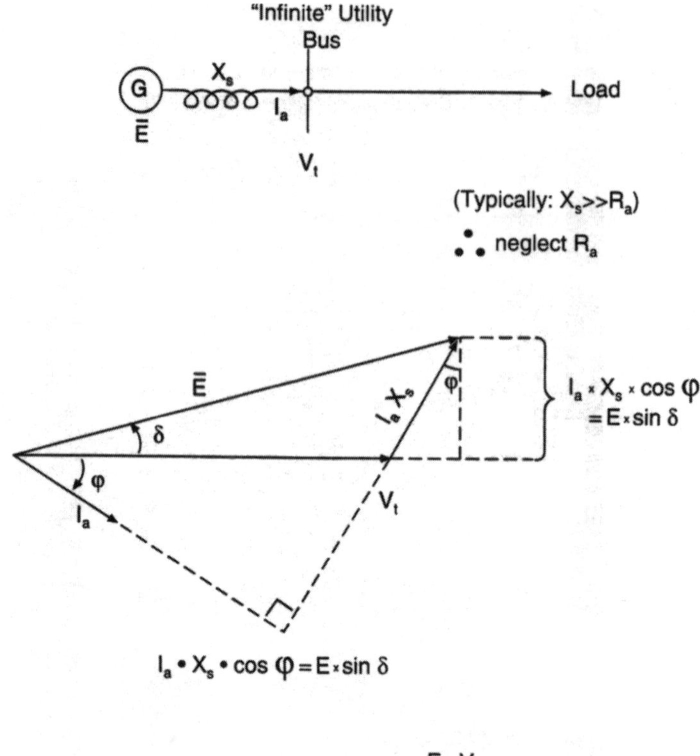

$$\text{Power Delivered} = V_t \times I_a \times \cos\varphi = \frac{E \times V_t}{X_s} \times \sin\delta$$

* In an "infinite" bus, V_t taken as constant
* E assumed linear with I_F for small changes of I_F

Fig. 4.43 Simple load change and excitation change calculation. The two basic equations can be combined to solve most steady-state problems of an electric machine connected to a power system (using the power transfer and the active power equations).

Neglecting generator losses,

$$\mathbf{E} \times \mathbf{V} \times \sin(\delta) = \text{Power delivered} = \text{Turbine's output (constant)}$$

However, since as was stated above the terminal voltage does not change, we have

$$\mathbf{E} \times \sin(\delta) = \text{constant}$$

But $\mathbf{E} \times \sin(\delta)$ is the vertical projection of $\mathbf{E}$. Clearly, changing the field current changes $\mathbf{E}$. So, if $\mathbf{E} \times \sin(\delta)$ must remain constant, then δ must change in such a way that the vertical projection is still the same.

Let's assume I_F grows from I_{F1} to I_{F2} and turbine power output is not changed: How will the P.F. and armature current change?

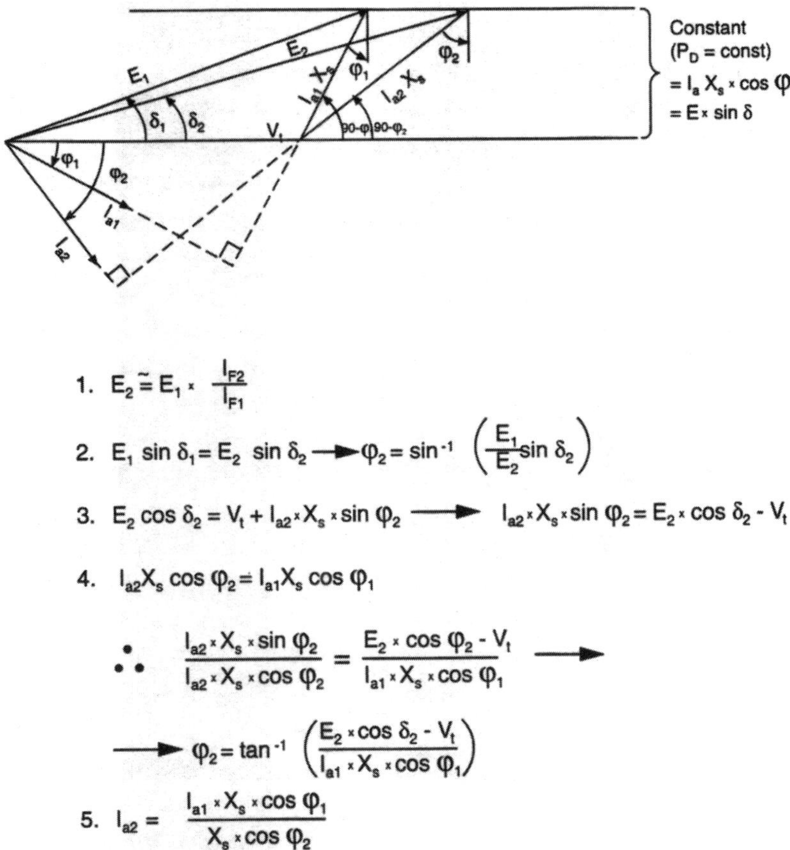

Constant
(P_D = const)
$= I_a X_s \times \cos\varphi$
$= E \times \sin\delta$
E_1
E_2
δ_1
δ_2
V_t
φ_1
φ_2
I_{a1}
I_{a2}
$I_{a1} X_s$
$I_{a2} X_s$
$90-\varphi_1$
$90-\varphi_2$
1. $E_2 \cong E_1 \times \frac{I_{F2}}{I_{F1}}$
2. $E_1 \sin\delta_1 = E_2 \sin\delta_2 \rightarrow \varphi_2 = \sin^{-1}\left(\frac{E_1}{E_2}\sin\delta_2\right)$
3. $E_2 \cos\delta_2 = V_t + I_{a2} \times X_s \times \sin\varphi_2 \rightarrow I_{a2} \times X_s \times \sin\varphi_2 = E_2 \times \cos\delta_2 - V_t$
4. $I_{a2} X_s \cos\varphi_2 = I_{a1} X_s \cos\varphi_1$
$\therefore \frac{I_{a2} \times X_s \times \sin\varphi_2}{I_{a2} \times X_s \times \cos\varphi_2} = \frac{E_2 \times \cos\varphi_2 - V_t}{I_{a1} \times X_s \times \cos\varphi_1} \rightarrow$
$\rightarrow \varphi_2 = \tan^{-1}\left(\frac{E_2 \times \cos\delta_2 - V_t}{I_{a1} \times X_s \times \cos\varphi_1}\right)$
5. $I_{a2} = \frac{I_{a1} \times X_s \times \cos\varphi_1}{X_s \times \cos\varphi_2}$

Fig. 4.44 Change of excitation. The solution of a simple problem of a generator connected to an "infinite" system, where only the excitation is increased from $\mathbf{I_{F1}}$ to $\mathbf{I_{F2}}$. The changes in current and power factor are deduced.

Finally, we know that the power delivered equals the product of the terminal voltage, times the current, times the power factor, cos(φ). By combining both equations and introducing a little trigonometry, the solution to the problem can be found.

Figures 4.45a and 4.45b present a simple numerical example for case 1. (Recommended Exercise: Repeat this simple example for your generator, using MVA, volts, frequency, and field current as they apply to any given load point. After calculating the new PF and armature current, use the vendor's V-curves of your machine to calculate the new PF and current, and compare these with the calculated values.)

Example

A13.8 kV generator, rated 500 MVA, is delivering 250 MW @ 0.8 PF lagging if the excitation is increased by 10% what are the changes in PF and load amps? Assume infinite bus;steam inlet unchanged, and Xs = 125%

Solution

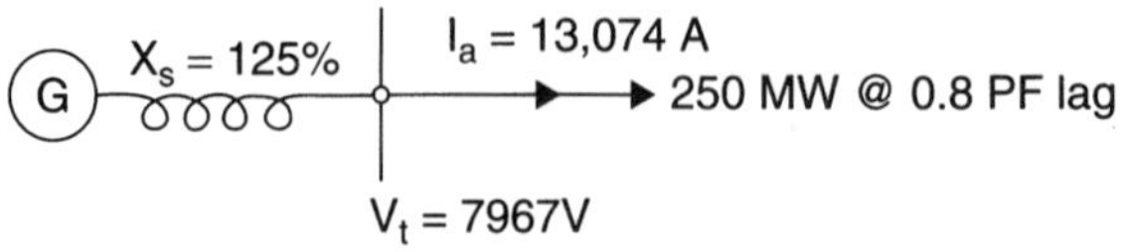

$$\text{(Phase value) } V_t = \frac{13{,}800}{\sqrt{3}} = 7967 \text{ volts}$$

$$I_a = \frac{P}{\sqrt{3} \times V_{L\text{-}L} \times PF} = \frac{250 \times 10^6}{\sqrt{3} \times 13{,}800 \times 0.8} = 13{,}074 \text{ A}$$

$$X_{BASE} = \frac{KV^2{}_{RATED}}{MVA_{RATED}} = \frac{13.8^2}{500} = 0.38\ \Omega$$

$$X_s = 1.25 \times X_{BASE} = 1.25 \times 0.38 = 0.48\ \Omega$$

$$\varphi_1 = \cos^{-1} 0.8 = 37°$$

$$PD = 3\,\frac{7967 \times E_1}{0.48} \times \sin\delta_1$$

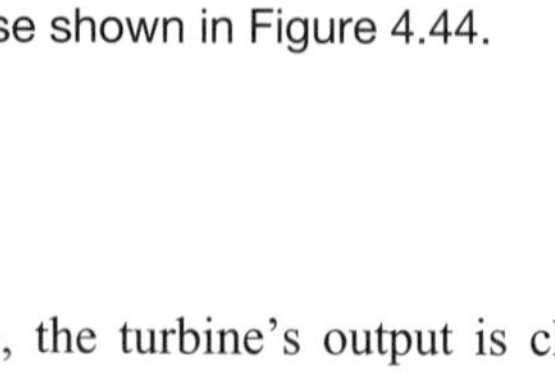

$$\longrightarrow E_1 \sin\delta_1 = \frac{250 \times 10^6 \times 0.48}{3 \times 7967} = 5020$$

Fig. 4.45a Numerical example for the case shown in Figure 4.44.

Case 2: Change in Power. In this instance, the turbine's output is changed while feeding an "infinite" system. Thus, the terminal voltage and frequency are kept constant by the system. The excitation field is also kept constant.

In this case, the fact the excitation is kept constant means that **E** is constant. Figure 4.46 shows how from the power transfer equation applied to this case it is obvious that δ must change with the power **P**. This fact, and a little geometry, lead to a simple solution of the problem. Figure 4.47 provides a simple numerical example of finding the change in current and power factor of a generator feeding an "infinite" power system when the excitation is kept constant and the turbine's output is increased.

$$E_1 \cos \delta_1 = V_t + I_{a1} \times X_s \times \sin \varphi_1$$

$$\frac{E_1 \times \sin \delta_1}{E_1 \times \cos \delta_1} = \tan \delta_1 = \frac{5{,}020}{7{,}967 + 13{,}074 \times 0.48 \times \sin 37°} = 0.43$$

$$\therefore \tan^{-1}(0.43) = \delta_1 = 23°$$

$$E_1 = \frac{5{,}020}{\sin 23°} = 12{,}771 \text{ V}$$

$$\delta_2 = \sin^{-1}\left(\frac{E_1}{E_2} \sin \delta_1\right) = \sin^{-1}\left(\frac{E_1}{1.1\, E_1} \times \sin 23° = 21°\right)$$

$$\varphi_1 = \tan^{-1}\left(\frac{E_2 \times \cos \delta_2 - V_t}{I_{a1} \times X_s \times \cos\varphi_1}\right) = \tan^{-1}\left(\frac{1.1 \times 12{,}771 \times \cos 21° - 7{,}967}{13{,}074 \times 0.48 \times \cos 37°}\right) \cong 46°$$

$$I_{a2} = \frac{I_{a1} \times X_s \times \cos\varphi_1}{X_s \times \cos\varphi_2} = \frac{13{,}074 \times 0.48 \times 0.8}{0.48 \times \cos 46°} = 15{,}056 \text{ Amps}$$

Conclusions

By increasing field current by 10% ⟶

* Power factor moved from 0.8 to 0.7 (LAG)
* Armature current increased from 13,074A to 15,056A (15% increase)

Fig. 4.45b Continuation of numerical example for the case shown in Figure 4.44.

4.5.5 Parallel Operation of Generators

Most large generators are connected to a common switchyard bus via the main step-up transformer, whereas smaller machines, mainly below 100 MVA, may be found connected directly to a common bus, and from there to a step-up transformer.

When two or more generators have their terminals connected to the same bus, a number of issues may arise. The first is the existence of circulating currents. As in the case of transformers connected in parallel, generators in parallel are affected by circulating currents if voltages and impedance do not match. In the case of generators, there is an additional degree of freedom not found in transformers: the angle of the voltage between both machines. Any mismatch will introduce significant circu-

Let's assume the turbine's output is increased from PD1 to PD2
How will PF and armature current change?

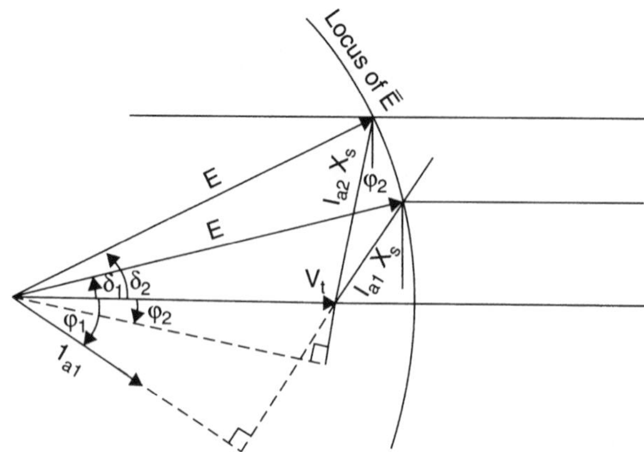

1. $P_D = \underbrace{\frac{V_t \times E}{X_s}}_{\text{cnst}} \times \sin\delta \longrightarrow \sin\delta \propto P_D$

$$\therefore \sin\delta_2 = \sin\delta_1 \times \frac{P_{D2}}{P_{D1}}$$

2. $P_D = \sqrt{3} \cdot V_t \, I_a \cos\varphi \longrightarrow I_{a2} \times \cos\varphi_2 = \frac{P_{D2}}{P_{D1}} I_{a1} \cdot \cos\varphi_1$

3. From the figure: $E \cdot \cos\delta_2 = V_t + 1_{a2} \times X_s \times \sin\varphi_2 \longrightarrow I_{a2} \times \sin\varphi_2 = \frac{E\cos\delta_2 - V_t}{X_s}$

$$2 \oplus 3 \longrightarrow \varphi_2 = \tan^{-1}\left(\frac{P_{D1}\,(E \times \cos\delta_2 - V_t)}{P_{D2} \times X_s \times I_{a1} \times \cos\varphi_1}\right)$$

4. $I_{a2} = \frac{P_{D2}}{\sqrt{3}\, V_t \cos\varphi_2}$

Fig. 4.46 Change in power. The solution of a simple problem of a generator connected to an "infinite" system, for which only the power is changed and excitation is kept constant. The changes in current and power factor are deduced.

lating currents, resulting in an exchange of VARs between the units. This results in unwanted losses and curtailment of available output from at least one of the units. Thus, it is important that the operators control the units' parameters in such a way that circulating currents are kept to a minimum. Figure 4.48 shows how the circulating current is calculated.

Interestingly, circulating currents between two or more generators tend to reduce

Example

A 13.8 kV generator rated 500 MVA, is delivering 250 MW @ 0.8 PF lagging
If the output power is increased by 10%, what are the changes in PF and load amps?
Assume infinite bus; excitation unchanged and Xs = 125%

Solution

From Previous example we know: V_t/ph = 7967V; I_{a1} = 13,074A
E = 12,771V
$X_s = 0.48\ \Omega$

$$\varphi_2 = \tan^{-1}\left(\frac{P_{D1}\,(E\cos\delta_2 - V_t)}{P_{D2}\times X_s\times I_{a1}\times\cos\varphi_1}\right)$$

$$= \tan^{-1}\left(\frac{250\,(12{,}771\times 0.9 - 7967)}{1.1\times 250\times 0.48\times 13{,}074\times 0.8}\right)$$

$$= \tan^{-1} 0.64 = 32.5°$$

From $E\sin\varphi_1 = I_a X_s \cos\varphi_1$

$\longrightarrow \delta_1 = 23°$

From $\sin\delta_2 = \sin\delta_1 \times \dfrac{P_{D2}}{P_{D1}}$

$\longrightarrow \delta_2 = \sin^{-1}(\sin 23° \times 1.1) = 25.4°$

$\cos\delta_2 = 0.9$

$$\cos\varphi_2 = \cos 32.5° = .0.84$$

$$I_{a2} = \frac{P_{D2}}{\sqrt{3}\,V_t\cos\varphi_2} = \frac{250\times 10^6\times 1.1}{\sqrt{3}\times 13{,}800\times 0.84} = 13{,}696\text{ A}$$

- PF increased from 0.8 to 0.84
- Armature current increased from 13,074 to 13,696A (5%)

Fig. 4.47 Numerical example for the case shown in Figure 4.46.

the angle of the terminal voltages of the units. The explanation is beyond the scope of this book, but can be found in Chapter 10 of reference [5]. However, if there is a tendency to increase the angle—for instance, one turbine delivering more power than the other—then a "hunting" situation might be established between the units. These types of situations can be controlled by a fine-tuned AVR and operator input.

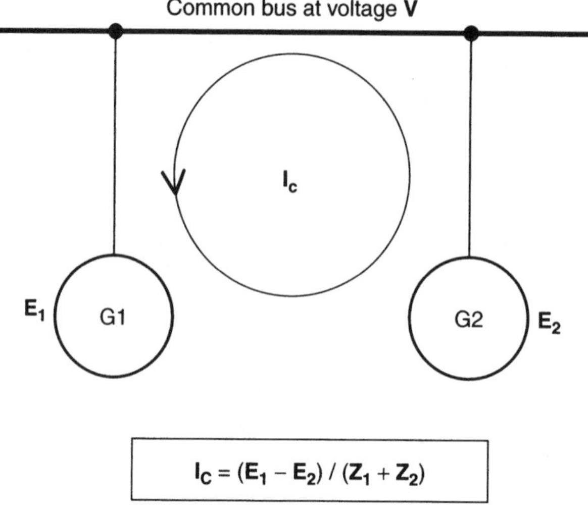

- In the figure, all bold letters represent vector variables.
- $\mathbf{E_1}$ and $\mathbf{E_2}$ represent the back-emf in each generator, i.e., the voltage generated in the armature, before the drop across the leakage reactance.
- $\mathbf{Z_1}$ and $\mathbf{Z_2}$ represent the synchronous impedance.
- $\mathbf{I_c}$ is the circulating current.

Conditions for Synchronization are:

1. Same phase sequence
2. Same voltage
3. Remaining within:
   - Maximum frequency slip
   - Maximum phase angle

Fig. 4.48 Calculation of circulating current between two generators connected directly to the same bus. Also shown are the conditions required for safe synchronization between a generator and the system.

4.5.6 Stability

One of the most fundamental concerns when operating industrial generators (and synchronous machines in general) is that they may become "unstable" and, eventually, "out-of-step" (also known as "slipping a pole or poles"). As explained in Chapter 1, the operation of a synchronous machine is predicated on the rotor and stator fluxes aligning themselves and rotating together at synchronous speed. When the machine is loaded, a *torque angle* appears between both fluxes. Similarly, a power angle appears between the voltage induced in the machine (**E**) and the terminal voltage ($\mathbf{V_t}$). Recall from Section 4.5.3 that the power transfer equation determines the power flow in the machine, which is given by

$$\mathbf{P} = \mathbf{E} \times \mathbf{V_t} \times \sin(\delta)$$

Thus, the maximum power the machine can deliver is,

$$\mathbf{P}_{\text{max}} = \mathbf{E} \times \mathbf{V_t}$$

This maximum power will occur when the internal generated voltage and the terminal voltage are 90° apart. However, if additional load is applied to the unit, resulting in the voltages being pushed apart beyond 90°, the capability of delivering the required power (and torque) will not be satisfied, and the rotor will come out of synchronism. This phenomenon, called *out-of-step* or *slipping poles,* is extremely onerous. Generators can suffer extreme damage under this condition. Therefore, it is the practice to operate a generator with its internal angle not reaching beyond 60 electrical degrees. Figure 4.49 presents a simplified mechanical equivalent of slipping poles.

The maximum transfer of power limit applies to any branch or element of the circuit in which a reactance separates two voltages. For a broader perspective of this issue, let us examine it first from a system's perspective.

Figure 4.50 depicts a simple transmission system comprising two lines connecting two busses. Both lines are transferring power $\mathbf{P}_0$ from bus A to bus B. The top of Figure 4.50 shows that under that condition, the steady-state point $\mathbf{P}_0$ is well within the maximum power transfer capability of the two lines, meaning the lines can absorb a relatively large increase in transmitted power from A to B, without any stability concern. In mathematical terms, this is indicated by the angle $\delta_0 \ll 90°$.

Now let us assume that line 2 breaker opens following a fault on it (see bottom of Fig. 4.50). The moment line 2 opens, the maximum capability to transfer power from A to B is given by the lower curve representing the capability of line 1. However, the power being transferred is still $\mathbf{P}_0$. The new equilibrium point (indicated by δ_1) comes very close to the maximum capability of the system. Thus, a relatively small increase in load will throw the system into disarray. The system is now denoted as being *unstable* or *marginally stable.*

A similar treatment can be applied to the generator delivering power to a system. Figure 4.51 shows a generator feeding a power system. At normal operation, the maximum capability of the system to transfer power is denoted by the higher curve in Figure 4.51*a*. Shown there is the operating internal angle δ_0, which is significantly lower than 90°. As the system experiences a fault on one of its lines, the load $\mathbf{P}_2$ is removed and the generator feeds only the remaining $\mathbf{P}_1$. Now, the turbine does not (cannot) change its output instantaneously (the turbine keeps "pushing" watts into the system), so δ advances toward 90° as the system tries to find a new equilibrium. The excess power between what the turbine delivers and the output of the generator goes

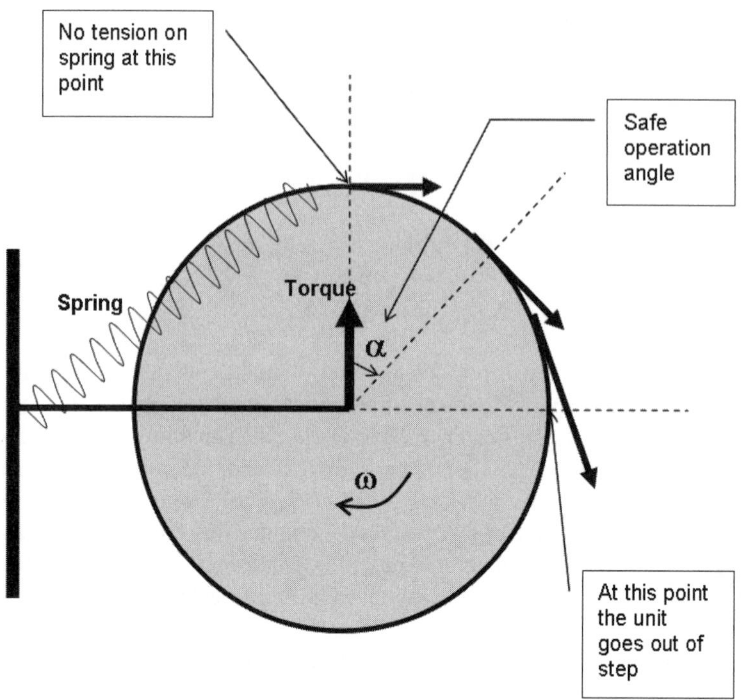

Fig. 4.49 Out-of-step mechanical conceptualization.

into accelerating the unit's rotors and is converted into spinning energy. Depending on the power transfer capability of the remaining system and the ratio between $\mathbf{P}_1$ and $\mathbf{P}_2$, the generator may or may not remain stable. If it does not, it will slip a pole (see Fig. 4.49) or, if the protection is adequate, it will be removed from operation. In some cases, the system may recover fully or partially (shown by the middle curve in Fig. 4.51*a*). In that case, there is a greater chance that the generator will stay connected and stable. Mathematically, calculating the areas between the intersection of the power transfer curves and the output power can provide an estimation of the stability (Fig. 4.51*b* and *c*). These areas represent the additional spinning energy that has gone into the rotors during acceleration. This energy must return to the system once the generator is again stable (i.e., its speed is the system's synchronous speed). For a more in-depth study of stability issues, the reader is referred to a number of texts (e.g., [6]).

Transients and Subtransients. In the context of power system applications, a transient state occurs while a system is undergoing major changes. This may be due to, for instance, faults, switching on or off large loads, or loosing large chunks of generation. At the same time, internally, the generator is also undergoing significant changes. Under such unsteady conditions, the changing flux produced by the chang-

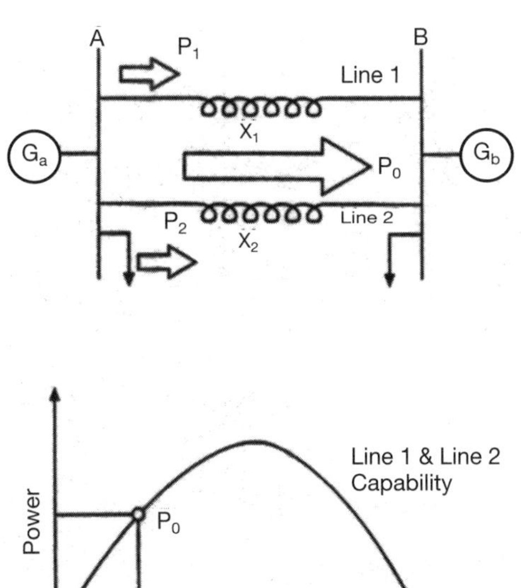

Condition Before Fault

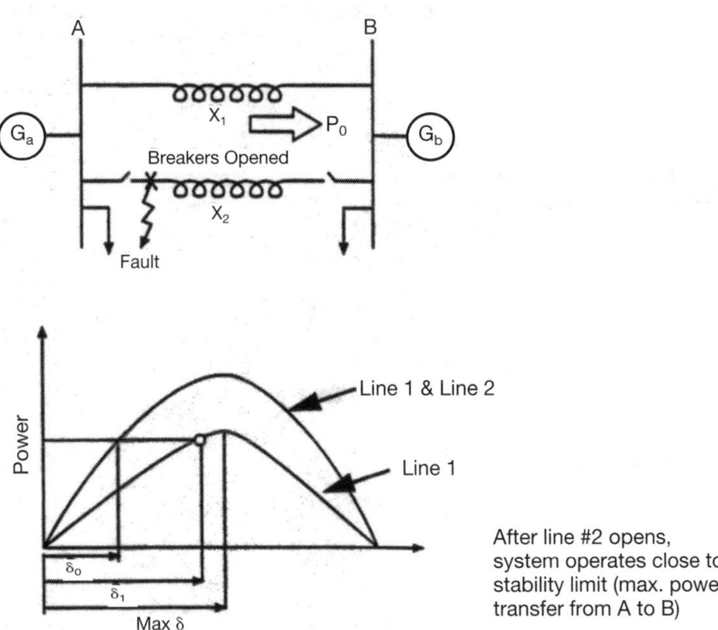

Fig. 4.50 Power system stability case with two lines and two busses, before and after the fault.

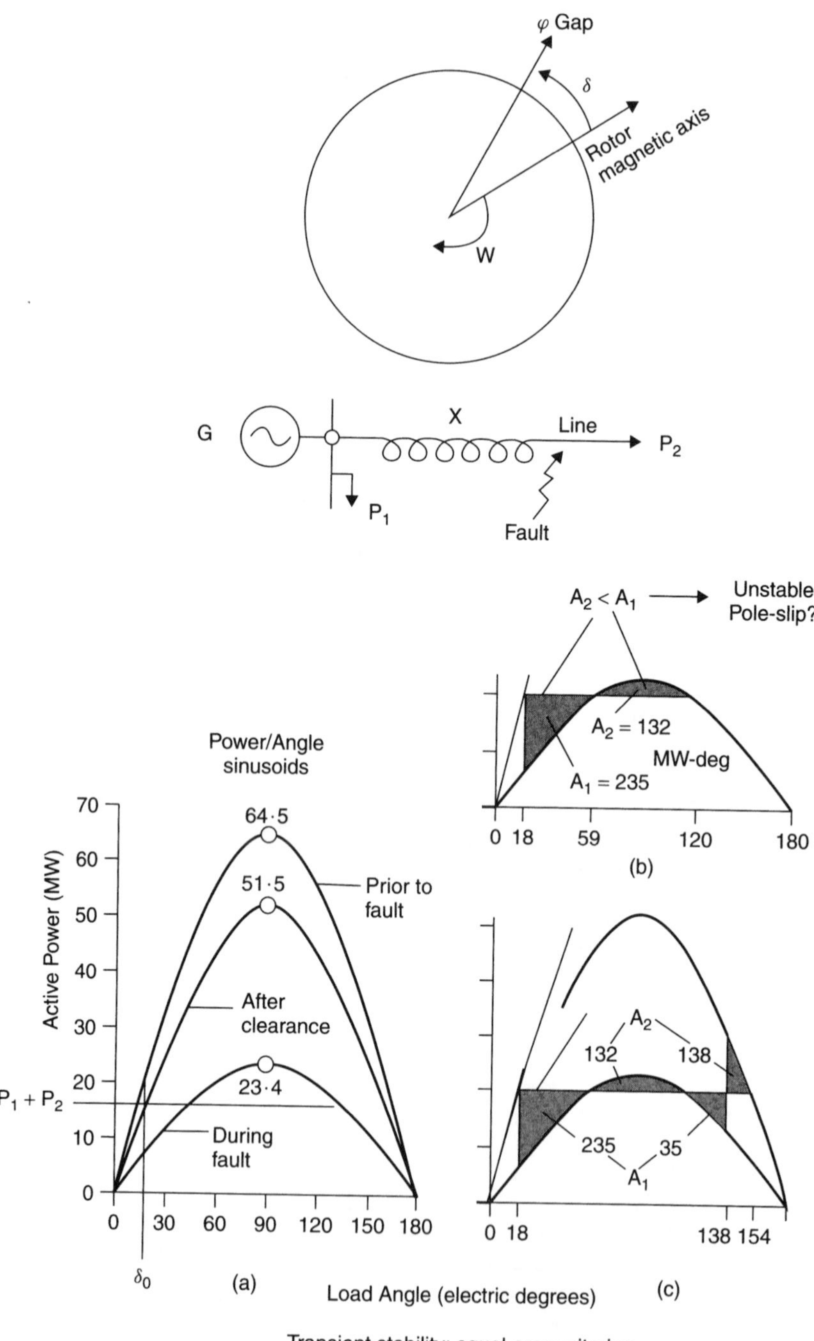

Fig. 4.51 Simple case of generator stability, from the generator perspective.

ing stator current in the direct axis (parallel to the pole faces of the rotor) induces a voltage in the field winding, resulting in a field current that opposes the change in flux and, hence, the change in stator current. This makes it more difficult for the stator-produced magnetic flux to pass through the rotor poles than in the steady-state condition. Under the transient condition, only the leakage flux paths of the stator and field windings are available, meaning fewer flux linkages per stator ampere. The result is that the generator looks like a reactance in the range of 0.15 to 0.35 per unit, which is much smaller than the synchronous reactance. This is called transient reactance, often denoted by X'_d.

The transient reactance is important to understanding transient stability, which, as stated above, is the ability of the power system to recover from a short circuit that has been interrupted, perhaps by circuit-breaker action. "Subtransient" is used to describe a rapidly changing condition that may last one to four cycles (0.016–0.07 s in a 60 Hz system). In this case, the magneto-motive force (mmf) of the stator winding changes so rapidly that it causes currents to arise in the rotor body as well as in the field winding, all of these opposing the change in stator current. This restricts the stator-produced flux to the stator leakage paths and to the very surface of the rotor. Therefore, the generator appears as a smaller reactance, in the range of 0.10 to 0.25 per unit. This is called the subtransient reactance, often denoted by X''_d. The subtransient reactance is commonly used to calculate the maximum current following a nearby sudden short circuit.

Stator and Rotor Transient Capabilities. During sudden increases of rotor and/or stator currents, the respective windings are subjected to additional losses. Given the short duration of these occurrences, the maximum temperature attained in the conductors is calculated by assuming no additional heat transfer from the conductor to the surrounding medium. Figure 4.52 shows the typical transient capabilities of rotors and stators of turbo-generators, as given by the standards. From the plot of Figure 4.52, we obtain the data listed in Table 4.4.

4.5.7 Sudden Short Circuits

If a short circuit occurs suddenly in the power system near a turbine generator, a high-current transient ensues, which is of interest for several reasons. In the design of the turbine-driven generator, winding forces and torques experienced by the stator and torques on the rotor system must be adequately accommodated. Also, external buses and circuit breakers that must carry and interrupt the current must be adequately specified.

For a sudden short circuit at the stator terminals, the exciter is assumed to be a source of constant voltage; it is not controlled by the voltage regulator. In addition, the generator appears to react in a linear fashion in terms of electrical and magnetic circuits.

Each winding in the generator traps the flux, linking it at the instant of short circuit. The relationship is such that the flux linking such a winding does not change instantaneously. A large direct current suddenly appears in each phase of the stator winding in

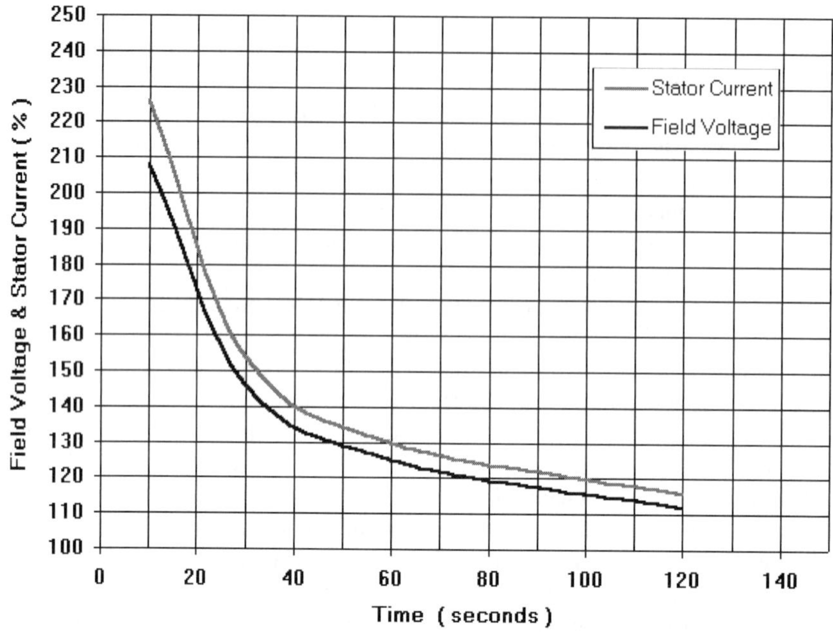

Fig. 4.52 Stator and rotor transient capability.

proportion to the flux linking it at the instant of short circuit, in order to sustain that flux. Since there is no source of direct current in the stator winding, it decays exponentially to zero in accordance to the stator time constant T_a (in 0.14-0.5 s). Large direct currents also arise in the field winding and in the rotor iron circuit to sustain the flux trapped in them at the time of the short circuit. The field current decays exponentially according to the transient time constant T'_d (in 0.4-1.6 s) to the steady value supplied by the exciter. The rotor iron current decays in accordance with the subtransient time constant, T''_d (0.01-0.02 s) to zero, since there is no source for direct current in the rotor iron circuit.

Table 4.4 Stator and rotor transient capability, permissible transient operation

		Stator		
Time (s)	10	30	60	120
Stator current (%)	226	154	130	116
		Rotor		
Time (s)	10	30	60	120
Field voltage (%)	208	146	125	112

Source: IEEE C50.13

Therefore, both a decaying trapped flux in the stator and a decaying trapped flux rotating with the rotor are present. Because of relative motion, the stator flux produces a decaying alternating current of power-system frequency in all elements of the rotor, and the rotor flux produces a decaying alternating current of the same frequency in the stator winding.

At the instant of short circuit, the value of the dc component of current in each phase is equal and opposite to the instantaneous value of the ac component. Thus, there is no sudden change in current.

4.6 SYSTEM CONSIDERATIONS

Numerous industry standards have been developed, both nationally and internationally, that specify the required performance of a turbo-generator. These standards define limiting temperatures at rating, required characteristics, and steady and transient conditions that must be successfully tolerated. Such standards are found in IEEE, ANSI, IEC, BS, VDE, and other industry publications.

With regard to the turbo-generator, its primary requirement is to provide electric power continuously or for peak-load periods as needed, and to do so reliably and economically. A generator is also normally required to provide voltage support to the system by supplying the needed reactive power. The rated power factor assures that the generator will have adequate ability to carry out this function.

The rating normally defines the continuous duty required of the generator. A temperature class is assigned to the generator, which defines the thermal capability of the electrical insulation systems of the stator and field windings. Turbine generators are generally class B, F, or H, which implies a hot-spot capability of 130, 155, or 180°C, respectively, as prescribed in industry standards (which also specify limiting observable temperature rise over cold coolant temperatures for each class).

In sum, the general temperature rises are as follows, but the precise values should be taken from reference [2].

	Class B	Class F
Indirect air-cooled stator windings	85	110
Direct air-cooled stator windings	80	100
Indirect H_2-cooled stator windings	70–85	90–105
Direct H_2-cooled stator windings	70	90
Direct water-cooled stator windings	50	50
Indirect air-cooled rotor windings	85	105
Direct air-cooled rotor windings	60–80	75–95
Indirect H_2-cooled rotor windings	85	105
Direct H_2-cooled rotor windings	60–80	75–95
Direct water-cooled rotor windings	50	50

Note: The temperature rises given above are all relative to the input temperature of the cooling medium, which is 40°C.

The wave shape of the stator voltage must be very nearly sinusoidal to avoid certain environmental concerns such as telephone interference. To this end, it is common to specify a limiting *telephone influence factor* (TIF), which is calculated from the harmonic content of the voltage by using a weighting-factor curve that reflects the frequency response characteristics of telephone systems. A limiting deviation factor may also be specified. This is a measure of the maximum deviation that the stator voltage has relative to a sine wave.

A *voltage response ratio* is specified for the excitation system to be compatible with the stability needs of the power system. A turbine generator must also be able to operate successfully in a real power system where the ideal is not always achievable. Therefore, other conditions that may be experienced by a turbine generator must be accounted for in the specification of its required capability, as discussed below.

4.6.1 Voltage and Frequency Variation

A generator must be able to deliver rated apparent power at terminal voltages deviating from the rated value by up to ± 5%, according to current international standards. These standards also specify the frequency range over which rated output can be delivered, which the manufacturer must provide.

4.6.2 Negative-Sequence Current

The three phases of a power system are not perfectly balanced in voltage and impedance. Accordingly, a small amount of steady negative-sequence current is produced. The standards specify a maximum steady negative-sequence current that must be tolerated. This value, which varies among the various types and sizes of turbine generators, is based on an economic evaluation of system needs and generator rotor heating characteristics (see Section 4.4.2).

A disturbance may occur on one phase of the power system, which is then isolated by circuit breakers. The event may subject turbine generators in the vicinity to a large negative-sequence current for a brief period. Recognizing the economics of providing tolerance for the rotor heating that would result from such an event, the industry standards require that a turbine generator be capable of withstanding a prescribed value of I_2^2t, where I_2^2 is the square of the per-unit value of the negative-sequence component of current, integrated over the period of exposure t in seconds.

Calculation of Negative-Sequence Currents. As shown in Chapter 1, any three-phase symmetrical system can be shown as a group of phasors (vectors rotating at constant speed) representing currents or voltages. For example, Figure 4.53*a* represents such a balanced system of phasors. On the other hand, once a fault in the grid occurs, or the grid is unbalanced for whatever reason, current phasors will show as unbalanced in magnitude and in phase angles. Figure 4.53*b* shows such a situation (the phasors in the figure could represent either currents or voltages).

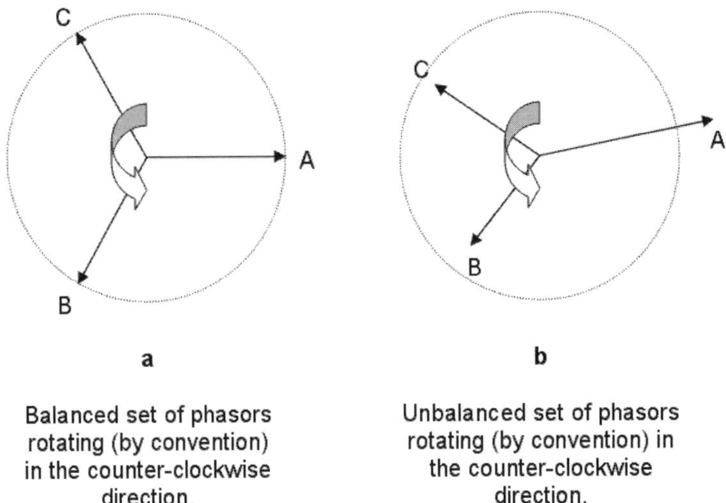
C
A
B
a
Balanced set of phasors rotating (by convention) in the counter-clockwise direction.
C
A
B
b
Unbalanced set of phasors rotating (by convention) in the counter-clockwise direction.

Fig. 4.53 Sets of balanced and unbalanced three-phase phasors.

To avoid the loss of simplicity introduced by the one-phase equivalent circuit approach in solving three-phase circuits, *symmetrical-sequence components* were added (first developed by Charles LeGeyt Fortescue, 1876–1936) to the bag of tools electrical engineers use in solving these types of problems. What the theory of symmetrical-sequence components states is that any three-phase unbalanced system of vectors (representing currents or voltages or anything else) can be replaced by three sets of three balanced phasors each: one rotating in the same direction as the original vectors, called the *positive-sequence set;* one rotating in the opposite direction, called the *negative-sequence set;* and one pulsating at the same frequency as the original set, called the *zero-sequence set.* Figure 4.54 shows the original unbalanced set of vectors, together with the set of symmetrical component.

Figure 4.54 shows nine symmetrical-component phasors. However, the same as symmetrical three-phase circuits, unsymmetrical circuits can also be solved using only one phase with the aid of the symmetrical-sequence components. Therefore, when calculating the negative-sequence components, only one phase is required. For simplicity, the phasor representing phase A is always chosen. Graphical methods exist for finding the symmetrical-sequence components, and these can be found in many books. Here, the mathematical method is introduced by way of a simple example.

EXAMPLE FOR CALCULATION OF SYMMETRICAL-SEQUENCE COMPONENTS. Let us assume that the following unbalanced system represents currents in a given point in a circuit:

$$\mathbf{A} = 10\text{ kA } \angle 0°$$

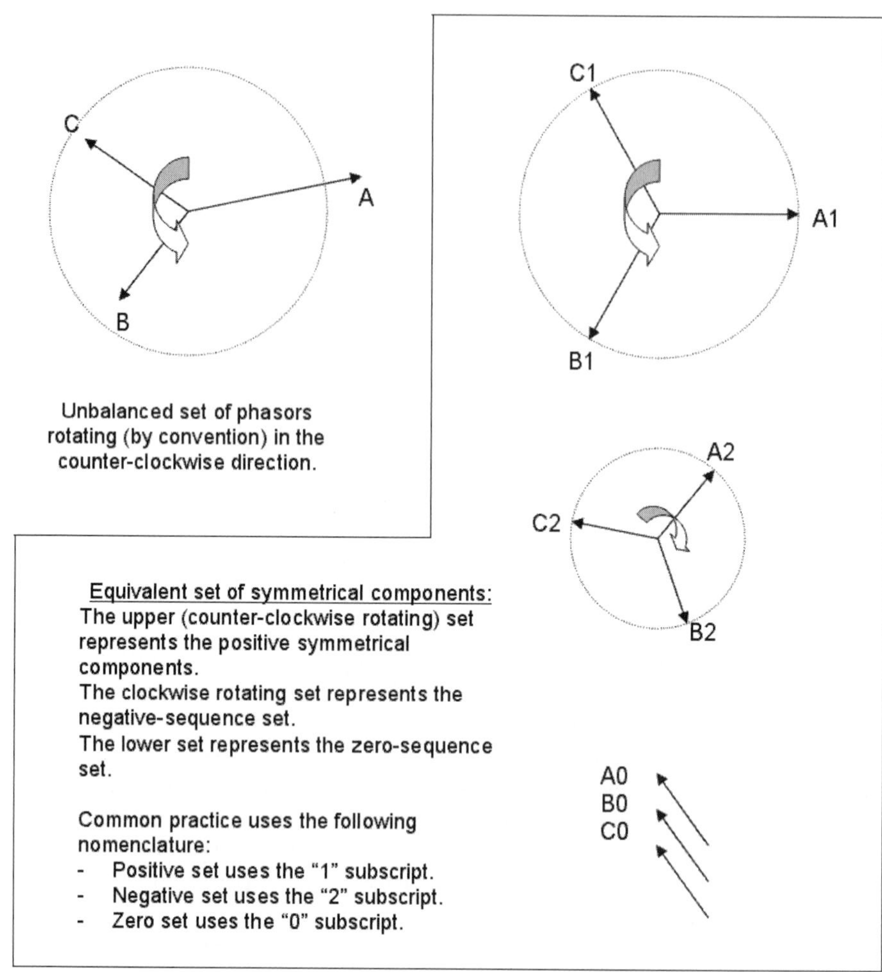

Fig. 4.54 An unbalanced set of three-phase phasors and its symmetrical component equivalent.

that is, Phase A has a magnitude of 10 kA and lies at a zero-degree angle from the horizontal. "A" is then the reference phasor:

$$\mathbf{B} = 12\text{ kA } \angle 250^\circ$$

$$\mathbf{C} = 7\text{ kA } \angle 110^\circ$$

To find $\mathbf{A_1}$, $\mathbf{A_2}$, and $\mathbf{A_0}$ (i.e., the positive, negative, and zero sequence phase A phasors), the following equations are used:

$$\mathbf{A_1} = (\tfrac{1}{3})\,(\mathbf{A} + a \times \mathbf{B} + a^2 \times \mathbf{C}) \tag{1}$$

$$\mathbf{A_2} = (\tfrac{1}{3})\,(\mathbf{A} + a^2 \times \mathbf{B} + a \times \mathbf{C}) \tag{2}$$

$$\mathbf{A_0} = (\tfrac{1}{3})\,(\mathbf{A} + \mathbf{B} + \mathbf{C}) \tag{3}$$

Take note that $\mathbf{A_1}$, $\mathbf{A_2}$, $\mathbf{A_0}$, **A**, **B**, and **C** in the above equations represent vector quantities, that is, they have a magnitude and an angle. The operator "*a*" in the equations has the following meaning:

- *a* rotates the vector 120° in the counterclockwise direction (i.e., +120°)
- a^2 rotates the vector 240° in the counterclockwise direction (i.e., +240°)

SOLUTION. Let us write all the variables required for the solution as per the equations above, both in polar and rectangular form. The transformation from polar to rectangular form of a vector is:

$$\mathbf{A} = A\ \angle\text{angle} = A\ [\cos(\text{angle}) + j\ \sin(\text{angle})]$$

In the present example (vector-quantities are shown in bold):

$$\mathbf{A} = 10\ \angle 0° = 10\ [\cos(0°) + j\ \sin(0°)] = 10 + j0$$

$$\mathbf{B} = 12\ \angle 250° = 12\ [\cos(250°) + j\ \sin(250°)] = -4.10 - j11.28$$

$$\mathbf{C} = 7\ \angle 110° = 7\ [\cos(110°) + j\ \sin(110°)] = -2.39 + j6.58$$

$$a \times \mathbf{B} = 12\ \angle(250° + 120°) = 12\ [\cos(370° + j\ \sin(370°)] = 11.82 + j2.08$$

$$a^2 \times \mathbf{B} = 12\ \angle(250° + 240°) = 12\ [\cos(490°) + j\ \sin(490°)] = -7.71 + j9.19$$

$$a \times \mathbf{C} = 7\ \angle(110° + 120°) = 7\ [\cos(230°) + j\ \sin(230°)] = -4.50 - j5.36$$

$$a^2 \times \mathbf{C} = 7\ \angle(110° + 240°) = 7\ [\cos(350°) + j\ \sin(350°)] = 6.89 - j1.22$$

Now, let us plug these values in equations (1), (2), and (3) above:

$$\mathbf{A}_1 = (\tfrac{1}{3})\ (10 + j0 + 11.82 + j2.08 + 6.89 - j1.22) = \tfrac{1}{3}\ (28.71 + j0.86)$$

$$\mathbf{A}_2 = (\tfrac{1}{3})\ (10 + j0 - 7.71 + j9.19 - 4.50 - j5.36) = \tfrac{1}{3}\ (-2.21 + j3.83)$$

$$\mathbf{A}_0 = (\tfrac{1}{3})\ (10 + j0 - 4.10 - j11.28 - 2.39 + j6.58) = \tfrac{1}{3}\ (3.51 - j4.7)$$

To convert back from rectangular to polar coordinates, the following formula is employed. If **A** is a vector such as

$$\mathbf{A} = a + jb$$

then

$$|\mathbf{A}| = \sqrt{(a^2 + b^2)}$$

$$\angle\mathbf{A} = \arctan\,(b/a)$$

In our example:

$$\mathbf{A_1} = (\tfrac{1}{3})\sqrt{(28.71^2 + 0.86^2)}\ \angle\arctan\,(0.86/28.71) = 9.57\ \angle 1.72°$$

$$\mathbf{A_2} = (\tfrac{1}{3})\sqrt{(2.21^2 + 3.83^2)}\ \angle\arctan\,(3.83/{-2.21}) = 1.47\ \angle 120°$$

$$\mathbf{A_0} = (\tfrac{1}{3})\sqrt{(3.51^2 + 4.7^2)}\ \angle\arctan\,(3.51/{-4.7}) = 1.96\ \angle{-53.24}°$$

CHECKING THE RESULT. Another key equation with symmetrical-sequence components is the following:

$$\mathbf{A} = \mathbf{A_1} + \mathbf{A_2} + \mathbf{A_0} \tag{4}$$

Equation (4) states that the sum of the three symmetrical-sequence components equals the original reference phasor. In the present example, the sum of the symmetrical-sequence components found by calculation must equal vector **A** (= 10 ∠0().

Then,

$$\mathbf{A_1} + \mathbf{A_2} + \mathbf{A_0} = (\tfrac{1}{3})\,[(28.71 + j0.86) + (-2.21 + j3.83) + (3.51 - j4.7)] = 10 + j0 = \mathbf{A}$$

This proves our calculations are correct. Figure 4.55 shows the original phasor A and its associated symmetrical-sequence components.

The phasor diagram above schematically shows the original phasor A and it associated symmetrical sequence components.

Fig. 4.55 Symmetrical-sequence components.

The main objective of the calculation was to find out the negative-sequence component. The negative-sequence component is normally defined per unit or as a percentage of the average of all three original vectors. In our case,

$$|\mathbf{A_2}| = 1.47/[(10 + 12 + 7)/3] = 0.152 \text{ pu } (= 15.2\%)$$

This type of calculation would be required while analyzing how a generator is affected by the negative-sequence component during a steady-state grid unbalanced condition. In fact, things are a bit simpler, because during grid unbalanced conditions, the angles between the phasors can be always taken to be equal to 120°. Also, during unbalanced grid conditions without a fault, zero-sequence components are almost nonexistent, making the search for the negative-sequence component much easier, as shown in the following paragraph. Under these steady-state conditions (grid unbalance), the value found for the negative-sequence component is used together with Table 4.2 to evaluate whether the machine remains within its safe operating region, based on the published permissive negative-sequence current values.

The solution of the previous example included a *zero-sequence* component. It is interesting to note that this component is mostly nonexistent when calculating short-circuit currents in any generator with high-impedance grounding (or in the grid, when a fault is not present). This applies to all large turbo-generators discussed in this book. There is a simpler method to find out the negative-sequence current component when there are no zero-sequence currents flowing. This method can be found in reference [1]. The method, in the form of a graph, is reproduced in Figure 4.56 for convenience of the reader.

EXAMPLE OF CALCULATING NEGATIVE-SEQUENCE COMPONENT FOR THE CASE OF A SHORT CIRCUIT ON THE TERMINALS OF THE GENERATOR. Let us consider the case of a large turbo-generator connected to a bank of three single-phase step-up transformers (SUTs). The generator is a four-pole, 60 Hz, direct-water-cooled, 25 kV, 1300 MVA unit. When the unit is delivering rated power, a major fault occurs on phase A of the step-up transformers. The fault causes a full short circuit across the low-voltage winding. Figure 4.57 shows the currents flowing prior to the fault and during the fault. It is required to calculate the negative-sequence current component and to evaluate if the unit remained within the negative-sequence current tolerance capability curves published in the standards. For this transient event, Table 4.3 must be used.

SOLUTION. Inspection of Table 4.3 reveals that for a 1300 MVA machine, the maximum permissive $(I_2)^2t$ is equal to $10 - (5/800) \times (1300 - 800) = 6.88$.

The most practical approach would be to read the values of the generator terminal currents during the fault, directly off the unit's digital fault recorder (DFR). One might expect all units this size to have installed DFRs. Then, those current values would be converted to per unit, and using the graph in Figure 4.56, the negative-sequence component of the fault current could be found. Taking also the duration of the fault from the DFR, the $(I_2)^2t$ is then calculated and compared with the value of 6.88. Depending

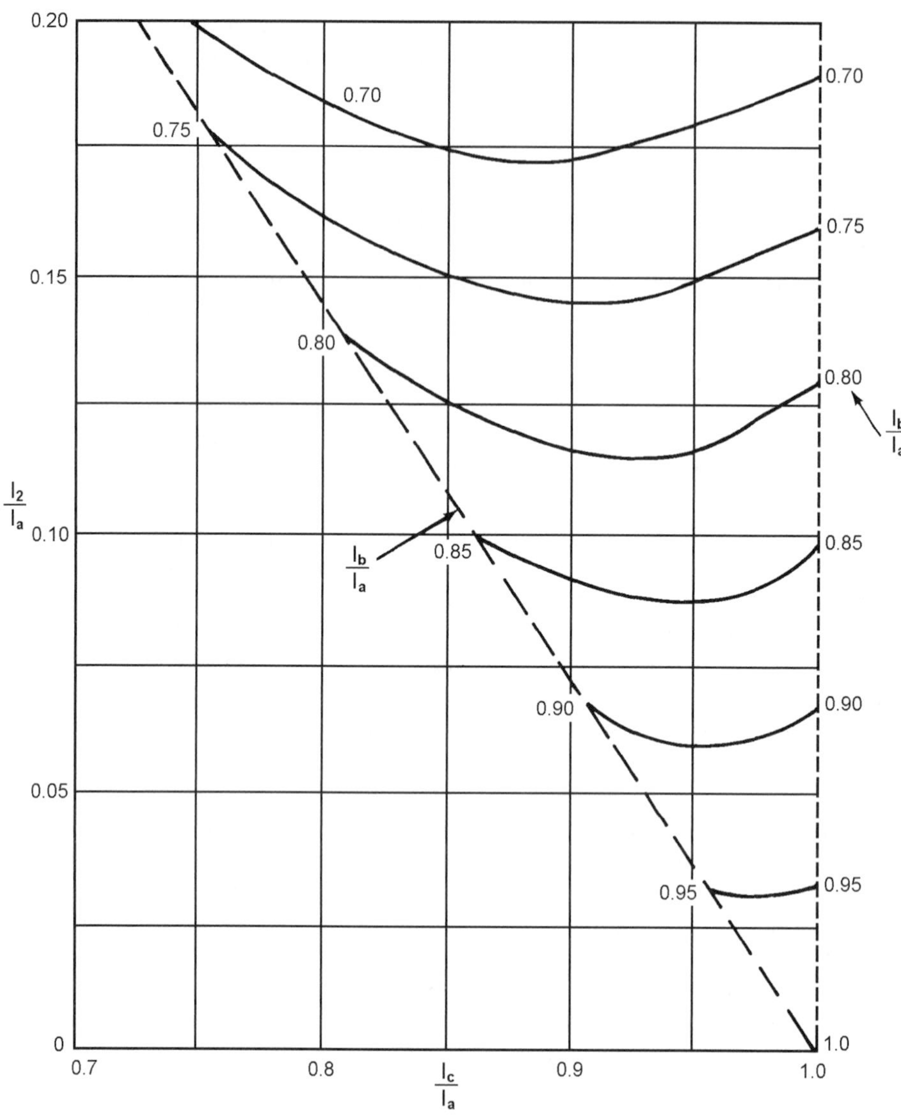

I_2 = Negative-sequence current
I_a = Largest of three-phase currents
I_b = Smallest of three-phase currents
I_c = Third phase current of interim value

(All currents in per unit or in amperes)

Fig. 4.56 Negative-sequence current calculation (only good for the case of negligible zero-sequence component). (From IEEE Standard 67-2005.)

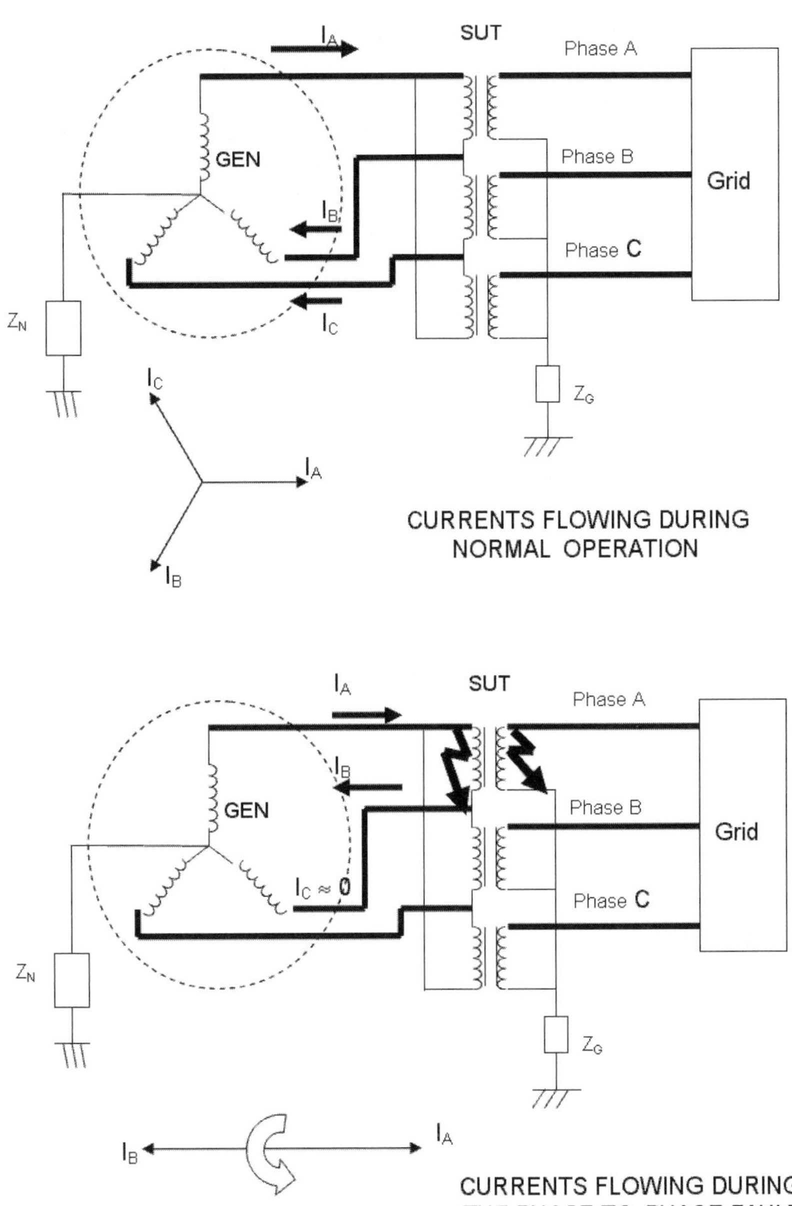

Fig. 4.57 Generator subjected to a phase-to-phase short circuit on its terminals.

on what the result of the comparison is, plans for close-up inspection of the rotor components may or may not be warranted. For the purpose of this example, let us assume that data from the DFR is not available. In this case, one must calculate the fault currents. Hopefully, by looking at what relays cleared the fault and their settings, one can estimate the duration of the fault. Let us assume that the fault cleared by the differential relays lasted six cycles (= 0.1 seconds). The following expression found in any good book on power systems analysis (for instance, reference [9]), gives the phase-to-phase subtransient short-circuit current at the terminals of a generator:

$$I_k'' = 1.08\ V_{LL}/(X_d'' + X_2) \cong 1.08\ V_{LL}/(2\ X_d'')$$

In the present example, $X_d'' = 0.22$ pu. Therefore, $I_k'' = 1.08 \times 1/(0.44) = 2.45$ pu.

At this stage, a few adjustments ought to be made. For instance, it is known that at the start of the fault a dc component will be present. Depending on a number of factors, this dc offset may result in the current being at the onset of the fault, close to twice its ac value. Although the dc offset decays rapidly, it can be assumed to be present until the fault is cleared. A conservative approach may entail taking the dc offset current and adding it up, or some of it, to the value I_k'', to obtain an equivalent short-circuit current. For instance, the value of 1.8 is taken by some authors (reference [9]). In this case, the short-circuit current will be $I_k'' = 1.8 \times 2.45$ pu $= 4.42$ pu.

To calculate the negative-sequence current component, equation (2) presented earlier in this section, is used:

$$I_2 = I_{k2}'' = (\tfrac{1}{3})\ (4.42\ \angle 0° + a^2 \times 4.42\ \angle 180°)\ \text{pu} = (\tfrac{1}{3})\ (4.42)\ (1\ \angle 0° + 1\ \angle 420°)$$
$$= 2.55\ \angle 30°\ \text{pu}$$

that is, the magnitude of the negative-sequence current is equal to 2.55 pu.

The rated current of the generator = 1300 MVA/(1.732 × 25 kV) = 30 kA. Therefore, the kA value of the negative sequence current is

$$I_2 = 30 \times 2.55 = 76.5\ \text{kA}$$

Then,

$$(I_2)^2 t = (2.55)^2\ \text{pu} \times 0.1\ \text{sec} = 0.65 \lll 6.88$$

which is well within the *permissive* region of operation. However, this does not take into account the fact that the magnetic field (and, thus, the output current) of the generator does not go down to zero immediately upon opening of the main breaker. The field current is discharged via a shunt resistor or the exciter circuit (in solid-state excitations). The actual decay may take several seconds. Let us say we expect the field to dissipate within 1 second. In that case, the total $(I_2)^2 t$ can be approximately calculated in the following way:

$$(I_2)^2 t\ (\text{including field discharge time}) \cong 0.65 + (2.55/2)^2 \times 0.5 = 1.46$$

In the calculation, half of the value of the original negative-sequence component is used, as an average for the decaying current, and half the time of dissipation of the field. The result is still within the permissive calculated value of 6.88.

The effect that both the duration of the fault and the field dissipation have on the calculations can be seen in Figure 4.58. The figure shows that without considering the field decay, there is about a 0.85 second margin, that is, the fault, at that level of negative-sequence current, could last roughly and extra 0.85 seconds without crossing the permissive limit. With the field decay included in the analysis, the margin shrinks to about 0.37 seconds.

The authors want to make sure the reader understands that this is a simplified example. For a real-case incident, all factors related to the fault should be investigated, and the generator's vendor should be consulted. Different people have differing approaches for calculating these quantities, and the responsible engineers in charge of analyzing the event should take a more comprehensive look at the situation and available information.

It is important to note that the fact the calculations may show the unit has not crossed into the permissive $(I_2)^2t$ region per se does not indicate that the rotor did not develop some damage. Bear in mind that the design formulas in place with each manufacturer to assure compliance with this (and other) criterion are not infallible, and certainly more so with older machines.

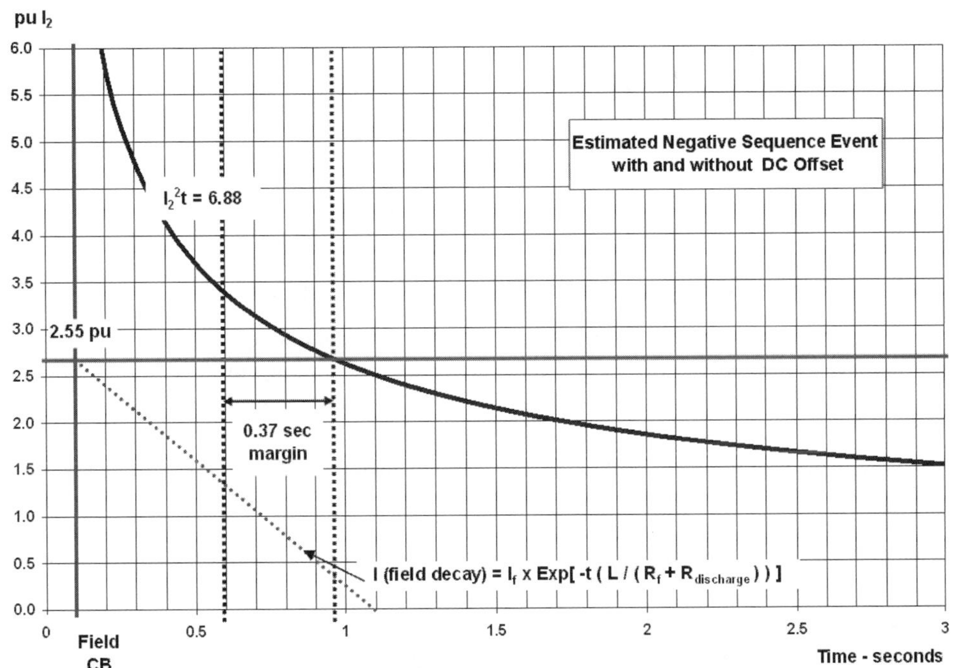

Fig. 4.58 Example calculated during a fault versus permissive $(I_2)^2t$ of 6.88 (1300 MVA machine rating).

4.6.3 Overcurrent

The stator and field windings may withstand periods of overcurrent. For example, if the system voltage drops for a brief period, the excitation system may be called upon to apply ceiling voltage to the field winding. The field current will rise according to its time constant from the initial value to a higher-than-rated value. The higher-than-normal current in both windings would result in a brief excursion to higher-than-normal temperatures. Accordingly, industry standards require that a generator be capable of operating at specified levels of overcurrent in the stator and field windings for a prescribed period of time (see Section 4.5.6).

4.6.4 Current Transients

Current transients may occur in a power system, for example, due to a sudden short circuit or due to switching when the voltages of the circuits to be connected are unequal in magnitude or phase angle. The high currents produce high electromagnetic forces in the stator winding in the end regions and in the slots. They also result in transient torques felt by the rotor and the stator. To ensure that the turbo-generator has the necessary ruggedness, U.S. industry standards require that it be capable of withstanding, without mechanical injury, a three-phase terminal, sudden short circuit while at load and at 105% voltage.

4.6.5 Overspeed

A turbo-generator is required to be able to withstand a brief overspeed test to 120% of rated speed to ensure that the rotor system is mechanically sound.

4.7 GRID-INDUCED TORSIONAL VIBRATIONS

4.7.1 Basic Principles of Shaft Torsional Vibration

One system- (grid-) related problem that often arises but is seldom recognized as such is damage to the generator from grid-induced torsional vibrations. In 1971, a unit in the Mohave Generating Station incurred severe shaft damage (twice!) due to grid-induced torsional vibration during commissioning. This event is often noted as the first major recorded grid-induced torsional incident that resulted in serious generator damage (Figure 4.59). The said incident, eventually classified as a *subsynchronous resonance* (SSR) case of grid-induced vibration also became the catalyst for a significant effort in elucidating these types of problems. A cursory search of the technical literature yields a very large number of academic papers addressing the issue of turbo-generator shaft torsional torques. All the same, the industry in large tends to experience periods of great interest (normally just after a well-known and important event), and periods of lassitude. A significant event that occurred in 2004 with two large nuclear-driven four-pole machines (Figure 4.60) has reawakened the interest of the industry in this topic. A number of key references are included at the end of this chapter. Most of those have a

Fig. 4.59 Shaft failure in a two-pole machine from a fossil-fuel plant (Mohave Genating Station).

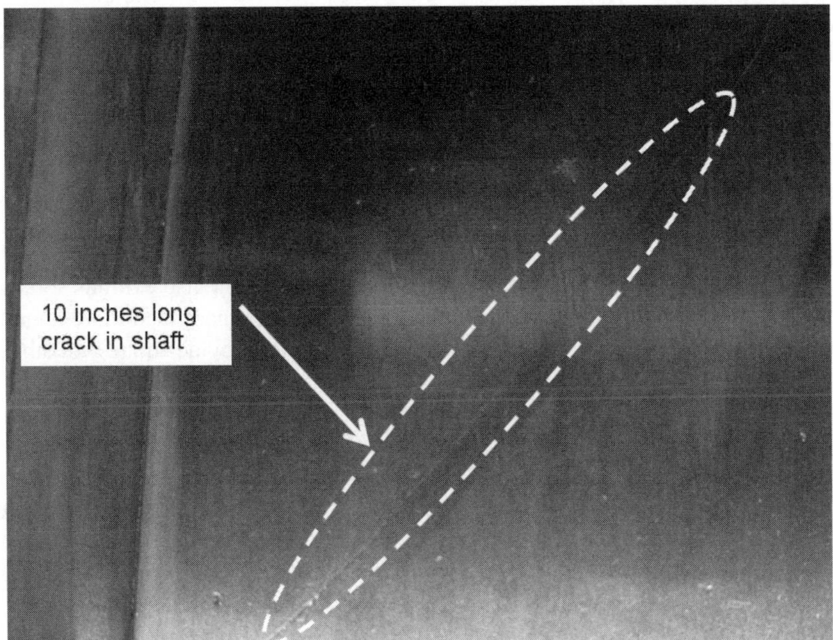

Fig. 4.60 Shaft cracks in a four-pole machine from a nuclear plant.

number of additional references for the reader interested in an in-depth understanding of the underlying physical foundations of this subject.

There are a number of reasons why grid-induced torsional vibration is so elusive. First of all, most units in operation do not have torsional vibration monitoring that is responsive to this type of vibration. Second, grid-induced vibration, although always present to a small degree, is seldom of such a magnitude that it poses a risk to the integrity of the unit. Third, when grid-induced torsional events are of a magnitude that constitutes a risk to integrity of the rotating plant, they tend to be of very small duration and are not captured by recorders, such as DFRs (digital fault recorders). In fact, unless the unit trips during the grid event, most operators would not notice that the unit underwent a severe torsional stress event. Fourth, torsional vibration, unlike lateral vibration, is, typically, not felt on the turbine deck, thus, intense torsional events may go unnoticed. Interestingly, most occurrences of damage from torsional sources happen in the turbines, specifically the low-pressure turbines with those long blades. Therefore, almost all recorded incidents of damage are in steam-fired units. Combustion gas turbine generators (CTGs) have brushless excitation and a single-shaft turbine-compressor, thus there is little room for the type of torsional oscillation that occurs in the long steam turbine trains that include up to a dozen bearings and several couplings. Generator damage tends to be less catastrophic than in LP turbines, with the exceptions of such failures as noted above, and a few others.

The complexity of the theory of torsional vibrations is beyond the scope of this practical book. However, an attempt is being made herein to present the most rudimentary principles of torsional vibration. The Reference section at the end of this chapter includes a number of entries where the reader can find additional references.

4.7.2 Spring Model of a Turbo-Generator Shaft Train

In its most simplified view, the main components of a turbo-generator train can be represented for the purpose of analyzing torsional oscillations as a number of masses connected by springs. The masses represent the main mass of each component, and the springs represent the much thinner shafts connecting those main masses. Figure 4.61 depicts this simplified view. In reality, each major component also exhibits some torsional-spring characteristics. Thus, an exact model will be that much more complex. For simple, low-frequency analysis, the simplified model provides quite workable results.

The meanings of resonance, natural frequency, and mode are critical to understanding the behavior of the physical model shown in Figure 4.61. *Resonance* is the periodic movement that a system of mass and springs will attain due to a short external stimulus. *Natural frequency,* in its most basic definition, is the frequency of vibration that a system will attain if all variable external stimuli are removed. This is also called the natural frequency of the system. At its natural frequency, continued vibration requires a minimum of energy. Figure 4.62 depicts the main characteristic parameters in the simplest linear and torsional models.

The systems in Figure 4.62 are so-called "one degree of freedom" systems. More complicated systems, such as the one for the turbo-generator train of Figure 4.61,

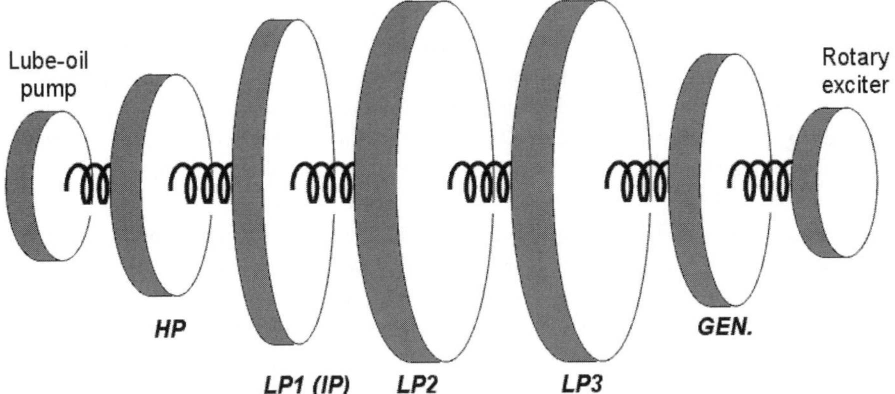

Fig. 4.61 Simplified torsional model of a turbo-generator train.

have a number of natural frequencies. This multiplicity of natural frequencies is key to understanding the issue of grid-induced torsional vibrations in a turbo-generator set.

The meaning of *mode* comes as an extension of the aforementioned. Once a resonating system reaches its steady-state condition (i.e., it is resonating purely in it natural frequency), the shape of the deflection pattern is called the *mode shape.* A com-

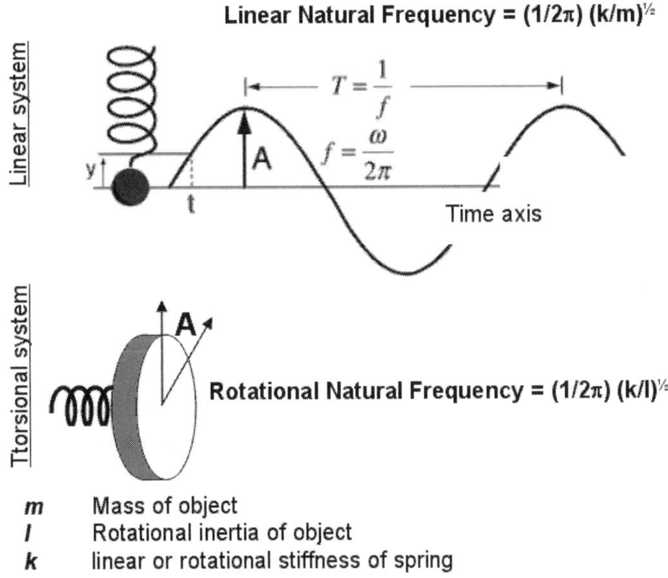

Fig. 4.62 Natural frequencies of both a simple linear and torsional mass–spring systems.

plex system that has a number of natural frequencies will exhibit different deflection patterns (mode shapes) for each of its resonant frequencies. Each is given a *mode number;* the lowest mode is always for the lowest natural frequency. A *node* is also defined as a characteristic point in the deflection for each mode. Nodes are also given numbers. Figure 4.63 provides an example of a torsional system of three masses and two springs. Modes and nodes are noted in the figure. Each node is located at the center of the lumped masses. Mode-1 represents a lateral displacement of the entire system at frequency equal to zero. Mode-2 shows the system displacement equal in both directions, with zero amplitude at the center (Node-2). This would be the first *critical frequency,* that is, coincident with the first natural frequency of the system (Mode-1), with frequency equal to 1, and not considered a resonance mode. Mode-3, coincident with the second critical frequency, shows a pattern of deflection in which both ends bend in one direction and the center in another. There are two locations where the deflection equals zero.

The locations where the displacement is zero are interesting because any external stimuli applied to the system at those points will not have any effect on the system's behavior.

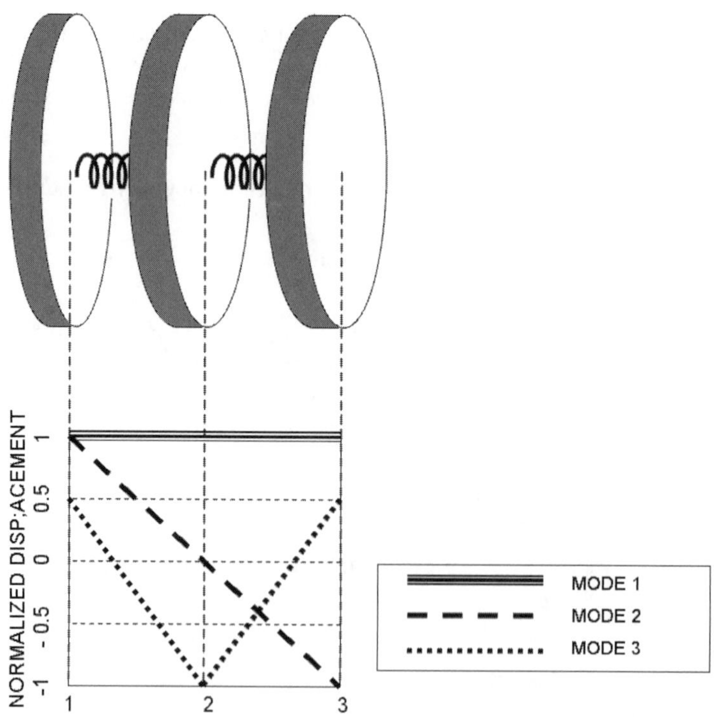

Fig. 4.63 Simple torsional system of three masses and two springs showing all the different modes and nodes.

4.7.3 Determination of Shaft Torque and Shaft Torsional Stress

The calculation of these stresses is rather complicated and beyond the practical approach of this book. The expertise required almost never resides within a single plant. The OEM and specialist in the field are the most qualified to carry out calculations on torques and stresses on the moving plant. Although subsynchronous analysis can be done successfully with lumped masses and spring models, an analysis for higher harmonics (supersynchronous) requires complex models of the entire train, and, often, the approach that yields the better model is the one in which the parameters of the model are taken from actual tests.

Information about the torsional model of a unit can be obtained by creating short circuits at the vicinity of the generator and reading the torsional response in a number of locations of interest along the train. This type of test is like a "bump" test, eliciting responses in a wide range of frequencies. Another test is carried out by injecting different frequencies through the excitation system. In such a way, responses to specific frequency stimulus can be obtained. For additional information, refer to the references at the end of this chapter.

4.7.4 Material Changes Due to Torsional Vibrations

Metallic shafts may fracture due to a sudden application of torque beyond their maximum capacity to carry torque, or may fail due to torsional fatigue due to numerous twisting cycles. There are a number of mechanisms driving this type of failure, such as high-cycle fatigue and low-cycle fatigue, and specific crack initiators such as shaft fillet areas, or fretting fatigue areas such as shaft keyways. Also crucial is the issue of *critical* crack and crack growth. The references herein have a wealth of information on this subject. A most important fact is that torsional fatigue is lifetime limited. The graph in Figure 4.64 provides a view of loss of life of a shaft due to cyclic torsional oscillation. The graph shows, for each strain level applied to a shaft, the statistical expected number of cycles before crack initiation. This graph will be different for different metal alloys and geometries.

4.7.5 Types of Grid-Induced Events

There are a number of grid events that may result in torsional oscillations of the generators. Depending on a number of factors, such as the intensity of the event as measured at the generator terminals and the type of generator prime driver characteristics, the grid-induced events may or may not produce deleterious results (i.e., damage to some of the unit's components). Some "events" are really not events, but conditions that may be permanent fixtures on parts of the power system. The following subsections list some of the grid events and conditions that may induce torsional vibrations in generators.

Short Circuits. The strength of the torsional effect in a given generator emanating from a specific grid short circuit depends on a number of factors, such as electrical

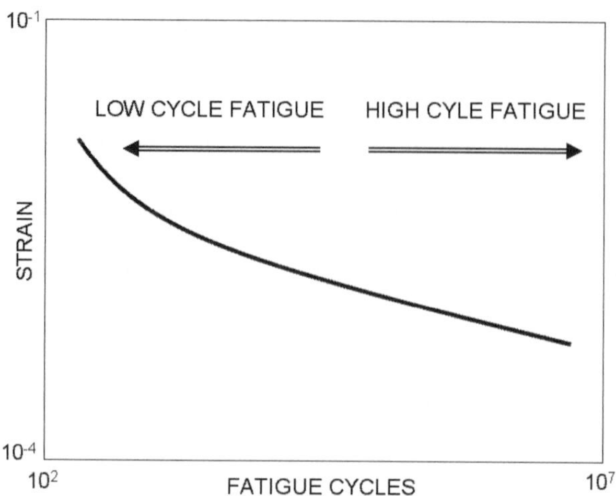

Fig. 4.64 Typical life endurance of a shaft under periodic torsional strain.

distance to the plant, voltage level of the affected system, speed of fault clearing, whether or not there is automatic reclosing, type of short circuit (one phase to ground, phase to phase, or three phase), and whether or not there are series and/or shunt capacitors on that system. Auto reclosing has long been identified as having a major potential magnifying effect on the oscillations of a generator. This is because the torsional damping (free decay of oscillations following the removal of the stimuli) is rather small in large turbo-generators. In fact, it is measured in seconds. Automatic reclosing is normally applied within half a second or so after the breakers open on a fault, thus, depending on when in the oscillation waveform the voltage is reapplied, the amplitude of these oscillations may be increased, with the potential of taking the generator out of step. For this reason, auto reclosing has been eliminated from some grids in a number of countries. Grid short circuits can induce very large power and frequency swings in generators. Although some units may ride through grid events without any damage, others, older or weaker, may not. Figure 4.65 shows the large voltage, frequency, and power swings from such an event.

Negative-Sequence Occurrences. These normally arise during a short circuit or another grid imbalance. Large short circuits at the terminal of the generator (or on one of the transformers connected to the generator) will give rise to large negative-sequence currents to a degree that may require removal of the rotor for inspection (this is discussed in other sections of this book). However, most negative-sequence events will be within the capability of the generators. Figure 4.66 shows a negative-sequence event due to a large generator remaining connected to the grid with only two poles of the main breaker closed. The data for the graph in the figure was taken in the switchyard of a power plant many kilometers away.

Figure 4.67 depicts a real case of a long-duration negative-sequence event of low magnitude (due to a generator in another plant motoring on two phases) that caused the

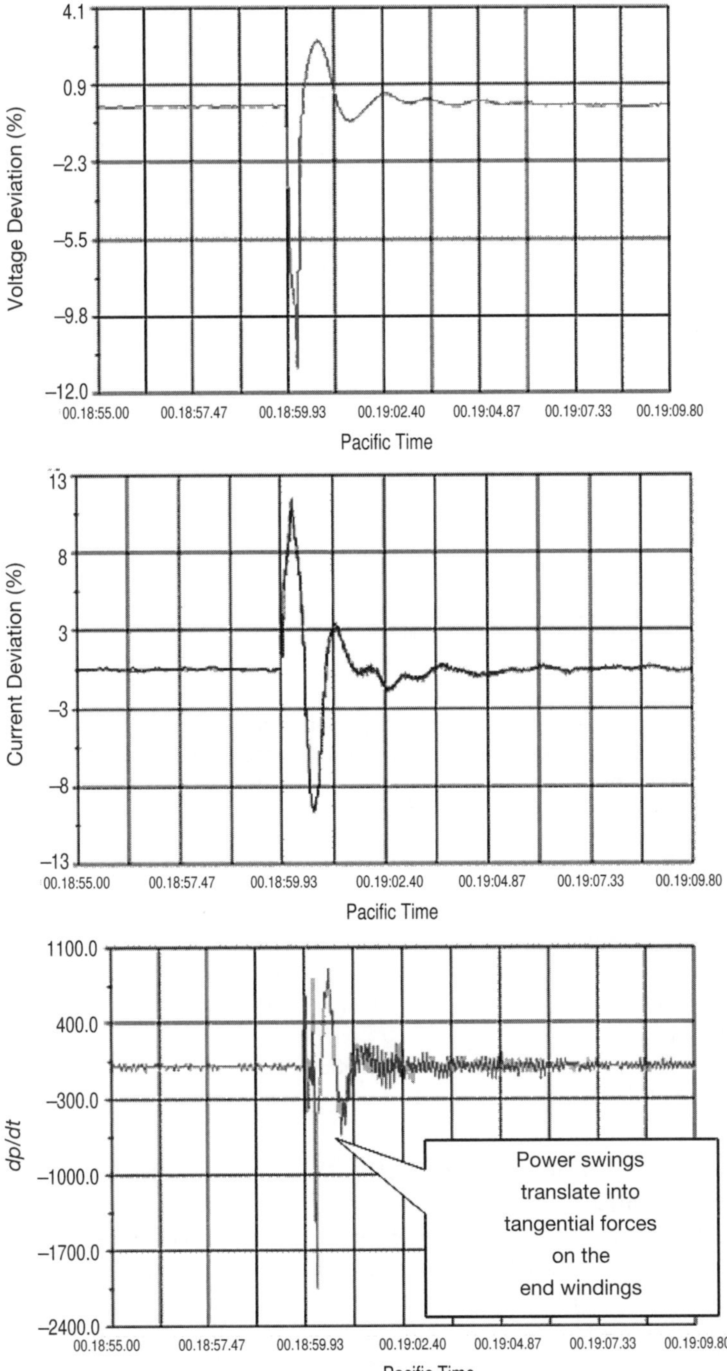

Fig. 4.65 Large oscillations due to a grid event, as measured in the switchyard of a nuclear power plant. Switchyard voltage is 230 kV and the power swings indicated on the graph are actual power changes measured at the output of one of the station generators.

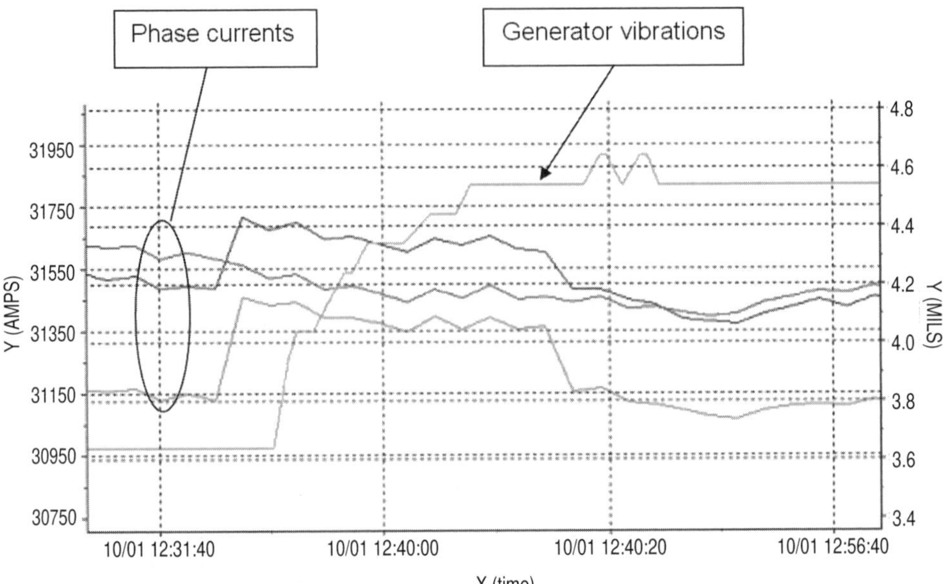

Fig. 4.66 Negative-sequence event due to generator motoring on two phases. Although the negative-sequence currents were within the values given in the IEEE standards, the generator developed rotor-shorted turns during the 13-minute-long event, due to a preexistent weakened field winding.

development of shorted turns in the field of a generator. This event was repeated a few months later, with similar results, that is, development of additional shorted turns in the field winding. This rotor winding was weakened by a previous loss-of-lube-oil event. This example clearly shows that torsional vibrations do impose strains on turbines and generators.

Negative-sequence events and their potential detrimental effects are discussed in other sections of this book.

Subsynchronous Resonance (SSR). As noted above, the Mohave generators were the first to suffer severe damage due to this type of grid condition. SSR is the result of electrical resonance in long transmission lines equipped with series capacitors. A lot of research was done after the problem was first identified, and now it is clearly understood. Solutions normally take the shape of power system stabilizers (PSS), devices installed together with the AVR. The function of the PSS is to counteract the SSR condition by acting upon the excitation of the generator. The PSS is set for a particular grid configuration, and it is important to check if the setting remains correct once the grid conditions change over time, or due to temporary transmission-line realignment. A wrongly set PSS may create a situation worse than that without the device in operation.

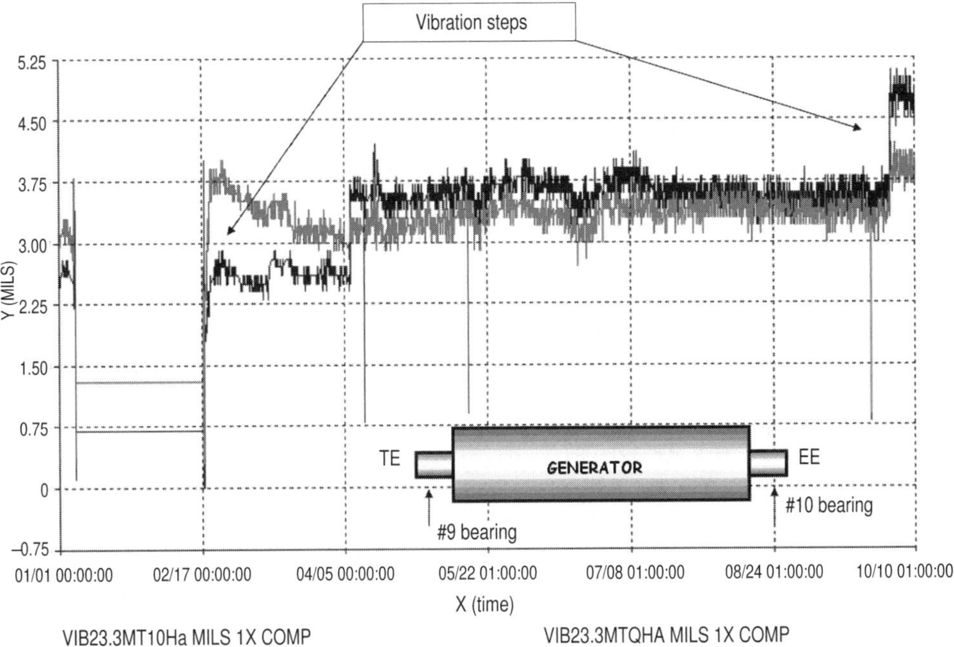

Fig. 4.67 Shorted-turns development in a 1300 MVA, four-pole unit, during a long, moderate negative-sequence event.

Line Switching Operations Line switching that interrupts or engages large blocks of load have the ability to create temporary system swings that may induce torsional oscillations of significant magnitude in the machine. For the most part, line switching does not constitute torsional stimuli of great importance. However, line switching that results in changes in load of a generator of more than 50% is considered to have the potential of resulting in damage to the unit.

Solid-State Loads The modern world, and its vastly growing number of electronic devices, in particular, power electronics, results in more harmonic load being carried by electrical power systems. A grid with a significant harmonic content has the potential to induce torsional oscillations. Nevertheless, unless the power electronic devices (such as HVDC converting stations) are installed in the main transmission system in the electrical vicinity of the generator, it is not considered a probable cause of concern.

Steel Mills Large steel mills in the proximity of generating units have been found in some cases to induce torsional vibration in generators. There are a number of solutions that have been developed. Each case needs to be addressed according to its own situation.

In addition to outside-the-station origins of torsional stimuli, there are a number of in-the-station events or condition, covered in the following subsections, that may result in generators having undesirable torsional oscillations.

Faults on Equipment Connected to the Generator Transformers, lines, busses, potential transformers, current transformers, arresters all prone to failure, creating large short-circuit currents that, due to their proximity to the generating units, can create large torsional transients. Any such fault should elicit intense scrutiny of the condition of the generator before the unit is returned to operation.

Hunting between Two or More Units Under certain conditions, "hunting" (low-frequency interactive oscillation) between units in the same station or in the near vicinity can occur. Hunting is mainly a result of inadequate excitation control. Techniques for avoiding this type of oscillation are well known. Hunting is mainly "seen" as a swing of VARs between two or more units. Sometimes, some MW swing is also observed. The period of this oscillation is in the several seconds range and, thus, does not pose any type of torsional fatigue stress on the units participating in the hunting. However, it does present some problems to the electrical system and, therefore, is an undesirable phenomenon. Figure 4.68 shows a simple scheme that would eliminate hunting between two units at the same plant.

Out-of-Phase Synchronization This is one of the most onerous events a unit can experience. Out-of-phase synchronization is a step load applied to the generator when the main breaker is closed. From the point of view of torsional vibration, it is like applying a "bump" stimulus to the generator. This is a wide-band stimulus that will elicit response in a number of resonance frequencies. Some may reach damaging amplitude if the step load is large enough. Induced, electromagnetic and mechanical transients can last up one second or so. A severe out of phase synchronization incident can render a unit inoperable for a long time. Thus, proper protection against this type of occurrence is essential. This topic is covered in Chapter 6.

Out-of-Step Event Also called "slipping poles," this is another very serious event that will cause oscillations among the unit's components. Protection against this type of operating condition is also covered in Chapter 6.

Load Rejection When a large amount of load is lost (for instance, if the plant becomes an island due to some grid event), a negative-load step is applied to the generator. This event will also induce torsional oscillations but, for the most part, these events can be handled successfully by the generators.

4.7.6 Monitoring of Torsional Vibration Events

Figure 4.69 presents view of typical grid conditions over a period of about two years. From inspection of the figure, the reader can see that "abnormal" events containing significant swings in power (dp/dt) and frequency (df/dt) are quite common. Each one of

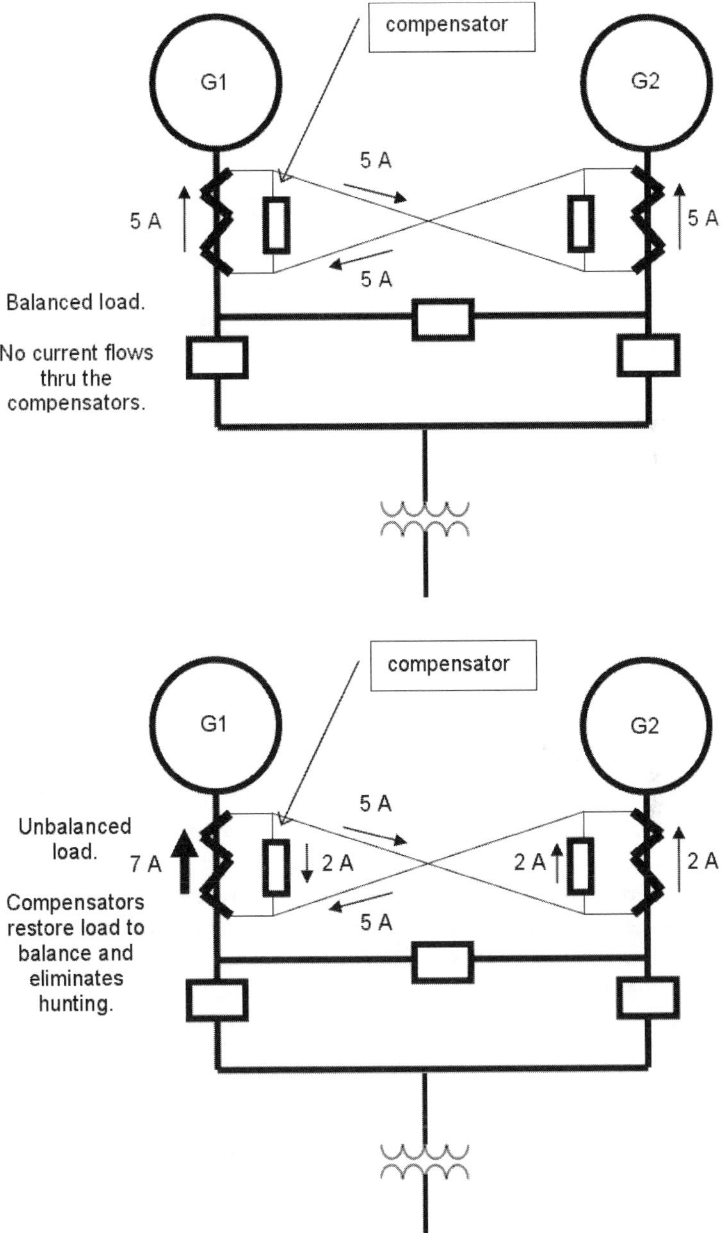

Fig. 4.68 Antihunting excitation-control scheme.

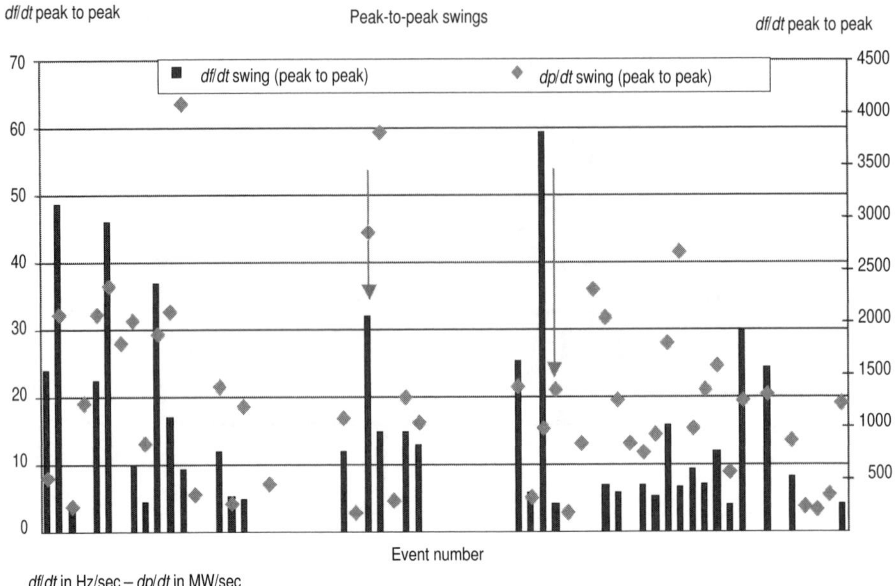

Fig. 4.69 System abnormal events as noted in the 230 kV switchyard of a large power plant.

these events introduces torsional vibrations in the generators that are connected to the grid where those disturbances occur; however, for the great majority of cases, they "ride" them without any lasting effect. There is a debate going on in some quarters about the effect that deregulation has on the power quality of the grid, that is, is it getting worse or not? This is certainly a subject beyond the scope of this book, but certainly of serious consequence for the long-term reliability of generators connected to these grids.

Actual monitoring of torsional vibration in generators is mostly carried out on those units (in particular turbines) that for one reason or another are considered susceptible to damage due to this phenomenon. Torsional vibrations are measured by a number of methods, one of these using strain gauges strategically placed along the shaft. For machines monitored in such a way, sophisticated torsional models are almost always prepared by the vendor and/or consultants. With these models, and the monitored readings, loss-of-life calculations can be made for various areas along the unit.

4.7.7 Industry Experience and Alleviation Techniques

As stated above, one can find significant literature on the subject of vibration, although not too much awareness at the plant level, unless specific problems related to torsional vibration have been experienced. There is an important aspect that has not received enough attention: do commonly experienced grid disturbances affect in a negative way the long-term reliability of the unit? For instance, do torsional vibrations that are not of a magnitude large enough to damage the unit's components during a single event reduce shaft life due to accumulated torsional fatigue? Is the loss of life significant enough that it may reduce the expected life of the unit (including extension well beyond the 30 years or so that these units are expected to serve after purchase)? Some plants in which conditions exist for significant torsional oscillations have on-line monitoring devices and models of the power train that allow calculating loss of life due to high-cycle or low-cycle metal fatigue, but most do not. Lately, EPRI researched this issue, and a number of technical reports were published (see references at the end of this chapter). Interested readers can explore the subject further by looking at the references and talking to people in other plants and leading consultants.

4.8 EXCITATION AND VOLTAGE REGULATION

4.8.1 The Exciter

The exciter supplies direct current to the field winding of the generator, at whatever voltage is required to overcome the resistance of the winding. The rating of the exciter is specified as its output power, current, and voltage corresponding to the rating of the generator, taking into account the temperature limits of the generator's field winding. The exciter rating generally has some margin over this requirement, as defined when the generator is designed.

The most common type of exciter used in early years was the commutator-type dc generator. This is very rarely used for new generators today. Any of the following systems usually supply the newer turbine generators:

- A shaft-driven alternator with solid-state diode rectifiers
- A solid-state, thyristor-based rectifier supplied by a transformer, deriving its power from the power system or from the generator's output
- A shaft-driven alternator with its output winding on the rotor, its output rectified by rotating solid-state rectifiers (commonly called a "brushless exciter")

The normal function of the exciter is to provide the proper level of direct current to the generator field winding, as required for the apparent power being supplied to the system, the terminal voltage, and the power factor of the generator load. In addition, the exciter must also be able to produce a ceiling voltage (which is the maximum exciter voltage) and to operate at that condition for a specified brief period, as required by the *voltage response ratio,* which is specified in excitation system's specification. The voltage response ratio is a measure of the change of exciter output voltage in 0.5 s when a change in this voltage is suddenly demanded.

When the exciter is a rotating machine driven by the generator shaft, it becomes part of the turbo-generator shaft system. It must be designed to accommodate axial motions due to thermal expansion of the turbine and generator rotors, and vertical motions of the generator shaft due to bearing-oil film and thermal expansion of the generator bearing support.

4.8.2 Excitation Control

Steady State. With the turbine's governor fixed and, therefore, the active power output of the generator fixed, and with the configuration of the power system fixed, an increase in exciter output, that is, in generator field current, causes the stator voltage to try to rise. This changes the power factor and causes the reactive power delivered by the generator to increase. While the turbine governor responds to provide the power needed by the system, the exciter enables the generator to provide the needed reactive power and, thus, to help provide the needed voltage support in the system.

The control system includes a voltage regulator that causes the generator's field current to be at whatever level is required to maintain the stator terminal voltage at a selected value. The control system also can be instructed to hold the generator field current at a desired value when voltage regulator is not needed. This is done by "manual control."

A lower limit is provided so that the field current is not reduced to the point where stator core-end heating becomes excessive (underexcited condition) or stability margins are compromised. An upper limit is provided so that the capability of the exciter and that of the generator field winding are not exceeded.

Volts-per-hertz protection is commonly provided to prevent the level of the magnetic flux in the generator and in the unit step-up transformer from exceeding safe levels. A volts-per-hertz control is occasionally specified to adjust generator excitation so as to avoid overfluxing.

Transient. The ability of the excitation system to change the generator field voltage rapidly may be important to system stability. Stability may be difficult to achieve when the system supplied has relatively high reactance; for example, when a long transmission line separates a generator from its load. In such a situation, providing an excitation system with a high voltage response ratio may help in the system's design. It can help reduce major expenses in additional transmission-line construction.

A relatively new concept made possible in part by the use of thyristor power rectifiers is the *high-initial-response* excitation system. In such a system, the output voltage of the exciter changes almost instantly on command, enhancing system stability.

Another concept in excitation control function is the *power system stabilizer* (PSS). It operates to enhance stability in situations where one power system may swing at low frequency relative to another (i.e., subsynchronous resonance conditions).

4.9 PERFORMANCE CURVES

4.9.1 Loss Curves

Figure 4.70 shows the typical calculated loss curves for a 215 MW, hydrogen-cooled, 3600 rpm generator.

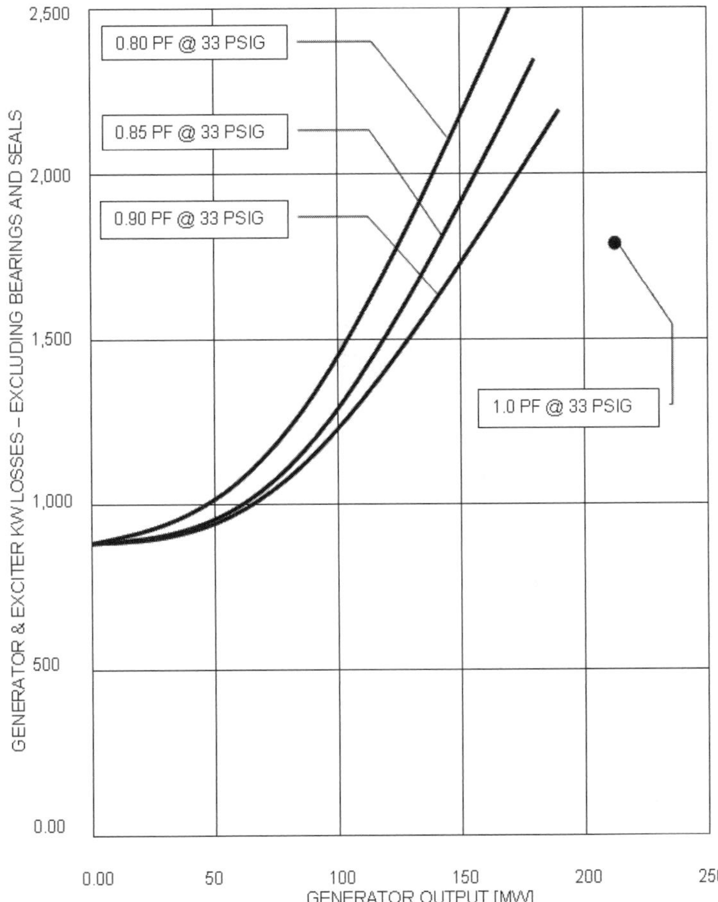

Fig. 4.70 Calculated loss curves for a 215 MW, 18 kV, 0.85 PF, 3600 rpm, hydrogen-cooled generator. Bearing and seal losses are excluded from the graph and are given as 286 kW.

4.9.2 Efficiency Curves

Figure 4.71 shows typical calculated efficiency curves for a 215 MW, hydrogen-cooled, 3600 rpm generator.

4.10 SAMPLE OF GENERATOR OPERATING INSTRUCTIONS

Like for any other critical apparatus, generating stations develop *operating instructions* for their generators. Operating instructions must be tailored to the specific conditions of a given plant, equipment, and system requirements, but they ought to capture the best industry practices, as developed over many years of experience.

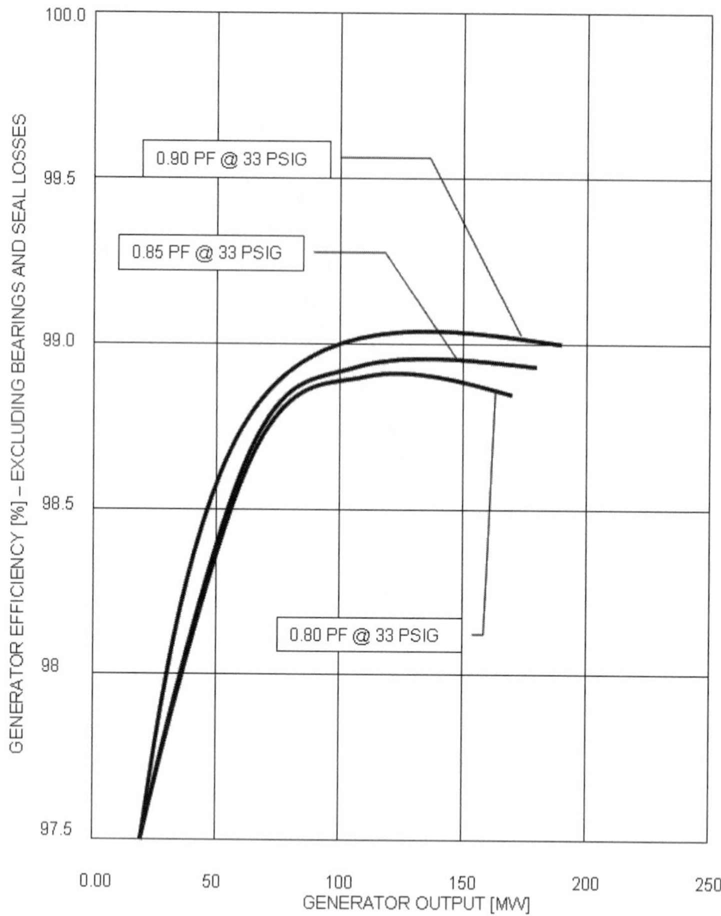

Fig. 4.71 Calculated efficiency curves for a 215 MW, 18 kV, 0.85 PF, 3600 rpm, hydrogen-cooled generator. Bearing and seal losses excluded from the graph, and are given as 286 kW.

Following is an example of a company's operating instructions. The reader should only look at these as an example of how such operating instructions can be outlined. He or she must remember that each case may be different; therefore, each operator must develop operating instructions that are specifically tailored to his/her plant and equipment.

MAIN GENERATOR OPERATING INSTRUCTIONS

I. Purpose

The purpose of this order is to establish guidelines for the operation of main generators and the associated equipment during normal and emergency conditions.

II. General

To ensure the continuing operational integrity of generators, generator temperatures, vibration and the various generator support systems must be closely monitored. Any abnormalities must be reported and investigated in a timely manner.

III. Start-up Operation

At no time should excitation interlocks or relay protection be disabled or made nonautomatic for the purpose of establishing a generator field. Also, a generator field should not be reestablished after operation of a generator protective relay until a thorough investigation has been completed.

On generators requiring field prewarming, the manufacturer's instructions and established local procedures should be followed relative to maximum allowable field current.

A generator field should not be applied or maintained at turbine speeds above or below that recommended by manufacturer's instructions. On cross-compound units for which a field is applied while on the turning gear, extreme caution must be exercised. Should either or both shafts come to a stop, the field should immediately be removed to prevent overheating damage to the collector rings.

After the field breaker is closed, the generator field indications should be closely monitored. If a rapid abnormal increase occurs in field current and/or terminal voltage, immediately open the field breaker and inspect the related equipment for proper working conditions before reclosing the breaker.

During off-line conditions, at no time is field current to be greater than 105% of that normally required to obtain rated generator terminal voltage at rated speed and no load.

When synchronizing the generator to the system, the synchroscope should be rotating less than one revolution every 30 seconds and should be 5 degrees or less from being in-phase when the switchyard circuit breaker is closed. Synchro-acceptors will typically not allow closing of the switchyard circuit breakers when the synchroscope is rotating faster than one revolution every 15 seconds. It is also critical that incoming and running voltages be matched as closely as possible when synchronizing the generator to the system; the allowable variation is plus or minus 10%.

IV. Shutdown Operation

The unit should be removed from the line via the turbine emergency trip push button. Closure of the turbine steam valves will initiate antimotoring relay operation and opening of the generator circuit breakers, field breakers, and auxiliary generator or transformer breakers.

If the limit-switch circuitry fails to function properly, the unit may fail to clear itself electrically after closure of the steam valves. A generator will then drive the turbine in a motoring mode of operation. This condition will overheat the LP turbine blades (typical withstand time of 10 minutes) unless operator action is taken as follows to isolate the unit electrically:

A. Verify that steam valves are closed (to prevent turbine overspeed when generator switchyard circuit breakers are opened).
B. If unit auxiliaries are not on the reserve transformer, open the auxiliary generator or transformer circuit breakers (automatic transfer to reserve auxiliary transformer should occur).
C. Switch voltage regulators to manual operation and reduce field to no load.
D. Open generator switchyard circuit breakers.
E. Open generator field breakers.

The generator field should not be reestablished for the turbine coast-down until verification that the automatic voltage regulator is tripped and the field rheostat is at the no-load position. The exciter field breaker should be tripped open at a preset generator speed per the manufacturer's instructions.

During off-line conditions, at no time is field current to be greater than 105% of that normally required to obtain rated generator terminal voltage at rated speed and no load.

V. System Separation

If during system trouble the unit is separated from the system, close attention to the field current/generator terminal voltage must be maintained, particularly when rated speed is not being continuously maintained. Reductions in speed can result in overexcitation (volts/Hz) as the voltage regulator attempts to maintain rated voltage.

VI. On-line Operation

Generators should be operated within their capability curves, which limit loading and field current (as related to VAR output) at various levels of hydrogen pressure. Operation beyond these limits will result in generator overheating.

During normal operation, the voltage regulator and, where applicable, power system stabilizer (PSS) should continuously be in service. To prevent a significant weakening of the generator field in the event that the voltage regulator is lost, the field rheostat should be set at approximately 80% of generator field nameplate value ("nominal full load").

If conditions require the unit be on manual voltage control, the unit will be operated off AGC and on the turbine solid block or load limit. Unless directed otherwise by the ECC, the terminal voltage should always be maintained high enough to boost VARs, especially if the voltage regulator is out of service.

Loss-of-excitation protective relays are designed to separate a unit from the system in the event the generator field is lost. This serves to protect the generator from overheating. If the field is lost and the unit is not automatically tripped by relay protection, immediately trip the unit manually.

Normal "180 Hz third harmonic" residual voltage should be indicated on the generator stator ground voltmeter. The amount of this residual voltage will vary with load and should be read by pushing the 10% push button. The absence of residual voltage indicates the one of the following may exist:

A. The generator neutral potential circuit is not complete.
B. There is a ground near the neutral end of the generator stator windings.

To ensure that this condition is detected as soon as possible, periodic monitoring of the residual voltage by operators at consistent intervals must be accomplished. Any abnormalities should be reported and investigated promptly.

VII. Field Grounds

An internal generator field ground could be caused by high-temperature electrical arcing, insulation damage, or shorted turns and other severe damage in the generator field. Upon detection of a field ground, field temperature and generator bearing vibration levels must be closely monitored for changes. If a 25% change in the field temperature is observed (increase or decrease), or if there is a notable increase in generator bearing vibration levels, the unit should be removed from service immediately.

Each station should have detailed and up-to-date operating instructions that address locating and isolating field grounds for each individual unit type. Timely and thorough investigation should be performed to identify and isolate the source of the field ground. Considering the numerous peripheral components associated with the excitation system, it is very possible that the ground is outside the generator and can be corrected without removing the unit from service. The following should be accomplished:

A. Operators transfer from automatic to manual voltage control.
B. Operators pull fuses to isolate noncritical generator-field-related circuits.
C. Test technicians verify proper operation of the generator field ground relay.
D. Electricians inspect brush rigging and all associated apparatus.
E Test technicians lift wires as possible, to further isolate generator field related dc circuits.

If the ground cannot be isolated, the unit should be removed from service for further investigation.

VIII. Power System Stabilizers (PSS)

This equipment is designed for aiding system stability by introducing a supplementary control signal into a continuous-acting voltage regulator. This has the effect of improved damping of power system swings. The power system stabilizer signal is a reflection of system speed deviation, as detected by a turbine shaft-driven tachometer or a frequency transducer at the generator voltage regulator potential transformer's (POT) secondaries.

To comply with system requirements and improve electrical system stability, the power system stabilizers should be maintained in service as much as possible. Some machines have overly sensitive power system stabilizers (PSS output is more active than desired) below approximately 25% load. On these units, the PSS should be placed in service at approximately 25% load and removed from service below 25% load.

IX. Voltage Regulators

Voltage regulator instability is evident when the regulator output meter and the unit varmeter swing between buck and boost. The swings may increase in magnitude. If this occurs, the voltage regulator must be removed from service. Disturbances elsewhere in the system are typically indicated by large initial voltage regulator output and unit varmeter swings, which then dampen out within a few seconds.

When transferring between manual voltage control and voltage regulator control, first verify that the voltage regulator output is at zero "differential" or "nulled." Improperly removing the voltage regulator from service (i.e., field rheostat improperly set) could result in reduction of the generator field, bucking VARs, and a trip of the unit from loss of excitation relays.

When mobile spare exciters are in use for unit excitation, they are not equipped with voltage regulators. Generator voltage control is strictly manual by use of the mobile exciter field rheostat or voltage adjuster.

If the exciter field should become demagnetized or reversed on amplidyne-type voltage regulators, the amplidyne may be used to restore magnetism with correct polarity. The amplidyne should first be checked for proper polarity by tuning the amplidyne control switch to the test position and operating the voltage adjuster to obtain an amplidyne voltage of at least 100 volts boost. With the field breaker closed and the exciter at running speed, the amplidyne control switch should then be turned to the "on" position until the exciter voltmeter begins to read upscale, indicating that magnetism with the correct polarity has been restored. Nonamplidyne equipped stations should have detailed operating instructions on how to restore loss of exciter magnetism for their particular excitation systems.

X. Moisture Intrusion

Even small amounts of moisture inside generators can result in reduced dielectric capability, stress corrosion pitting of retaining rings, or lead carbonate production and plating of the machine surfaces. The following recommendation should be adhered to, to maintain the generators in a dry condition:

A. The backup seal-oil supply from the bearing-oil system is to be used only in emergency situations.
B. Hydrogen dryers must be maintained and serviced.
C. Desiccant is to be monitored and replaced or regenerated as needed.
D. Hydrogen cooler and inner-cooled-coil leaks are to be reported and repaired in a timely manner.
E. Moisture detectors should be routinely checked for proper calibration/operation.

Generators equipped with nonmagnetic 18 Mn–5 Cr retaining-rings should be operated as dry as possible to reduce the possibility of stress corrosion pitting damage to the rings. Those units should be taken offline at the first indications of moisture intrusion (dew point higher than 0°C) and any required repairs completed before returning the unit to service.

XI. Recommended Routines

To maintain the operating integrity of main generators, the following should be performed during routine inspection rounds:

A. Check hydrogen purity levels normal and adjust as needed.
B. Check that seal-oil system is operating properly and maintaining proper differential pressure.
C. Check that hydrogen dryers are in service and check desiccant and regenerate as needed.
D. Check liquid level detectors for accumulations of water or oil. Report and monitor any abnormalities.
E. Check stator, field, and gas-path temperatures. Report and monitor any abnormalities.
F. Check generator residual ground voltage. Report and monitor any abnormalities.
G. Check collector ring areas for broken or arcing brushes.

XII. Protective Relaying

Generator Differential Protection. Generator differential relays compare the secondary currents from current transformers (CTs) installed on the neutral end of the generator windings to current transformers installed on the output side of the generator or the output side of the generator circuit breaker. If an internal phase-to-phase or three-phase fault occurs between the neutral and output CTs, the current flows will not balance and the differential relay will instantaneously actuate to trip the unit offline. *The unit should not be reenergized following a generator differential trip until the cause of the relay operation can be determined and resolved by engineers or technicians.*

Generator Stator Ground. Generator stator ground schemes protect the generator, isolated-phase buses, generator bus circuit breaker (if included), arrestors and surge capacitors (if included), and the primary windings of potential, auxiliary, and main step-up transformers from breakdowns in the insulation system to ground. Typically, the stator ground relays are set to operate in 1.0 second for a 100% ground-fault condition. Typically, the relay senses voltage on the secondary side of the generator stator-grounding transformer, which is connected between the generator neutral and the station ground grid. Under normal operation, the relay will see a few volts of the third harmonic (180 Hz for 60 Hz machines) due to the nonsymmetry of the stator core iron and zero volts at normal system frequencies (50 or 60 Hz). The stator ground relays are usually desensitized at 180 Hz and are normally set to operate for a 5% ground at system frequencies.

The unit should not be reenergized following a generator stator ground trip until the cause of the relay operation can be determined and resolved by station engineers or technicians.

Generator Bus Ground Detectors. Units equipped with generator circuit breakers require a second ground detector scheme to protect the generator circuit breaker, isophase buses on the transformer side of the circuit breaker, and the primary windings of potential, auxiliary, and main step-up transformers from ground-fault conditions when the generator circuit breaker is open. This scheme normally uses a wye/broken-delta connection with a voltage relay installed on the secondary side to sense ground conditions. Under normal conditions when the unit is running, this scheme will detect a few volts of the third harmonic and zero volts at running frequency. However, the scheme can become unstable under certain conditions (neutral instability or three-phase ferro-resonance) causing a blown fuse(s) in the ground-detecting transformers. A blown fuse may cause the ground detector relay to actuate. Also, coordination between the stator and bus ground detector schemes is difficult and, depending on the design, the station may not be able to quickly ascertain which side of the generator circuit breaker has the ground condition.

Preferred designs alarm only for bus ground-detector schemes or alarm when the generator circuit breaker is closed, and trip with a short time delay when the generator circuit breaker is open. This prevents the unit from tripping for blown fuse conditions and allows operators to quickly determine which side of the generator circuit breaker has the ground condition. With these designs, the generator stator ground relay will trip the unit for a ground on either side of the generator circuit breaker when the unit is online. If the ground is on the generator side of the circuitbreaker, the ground will clear after tripping. If not, the bus ground detector will sound an alarm or trip the unit main step-up transformer after generator tripping. For the alarm-only schemes, the bus section should be immediately deenergized by the operators through the appropriate switching to mitigate the possibility of the low-level ground fault current developing into a damaging high-current phase-to-phase or three-phase short circuit.

The unit and/or the transformer should not be reenergized following a generator bus ground trip until the cause of the relay operation can be determined and resolved by engineers or technicians.

Loss of Excitation. Synchronous generators are not designed to be operated without dc excitation. Unlike induction machines, the rotating fields are not capable of continuously handling the circulating currents that can flow in the rotor forging, wedges, amortisseur windings, and retaining rings during underexcited or loss-of-field operation. Consequently, loss-of-excitation relays are normally included in the generator protection package to protect the rotor from damage during underexcited operation. Impedance-type relays are usually applied to automatically trip the unit with a short time delay whenever the var flow into the machine is excessive. Limits in the automatic voltage regulator should be set to prevent loss-of-excitation relaying whenever the voltage regulator is in the automatic mode of operation.

The machine should not be resynchronized to the system following a loss-of-excitation trip until an investigation has been completed to determine the cause of the relay operation. Given the complexity of modern excitation systems, unexplained events are not that uncommon. Engineers or technicians should inspect the physical excitation system, verify calibration of the loss-of-excitation relays, check the dc resistance of the field windings, and review any available data acquisition monitoring that would verify the operating condition. The unit can then be started for test and proper operation of the excitation system can be ascertained by operations before synchronizing the unit to the system.

Overexcitation. Overexcitation (volts/Hz) relays are applied to protect the generator from excessive field current and overfluxing of the generator stator core iron. Typically, generators are designed to handle a full load field with no load on the machine for 12 seconds before the stator iron laminations become overheated and damaged. The relays are often set to trip the unit in 45 seconds at 110% volts/Hz and 2.0 seconds at 118% volts/Hz. The term volts/Hz is used to cover operation below normal system frequencies (50 or 60 Hz) at which generators and transformer can no longer withstand rated voltages. At normal system frequencies, the relays will operate in 45 seconds at 110% voltage and in 2.0 seconds at 118% voltage. Generators are continuously rated for operation at 105% voltage and transformers for continuous operation at 110% voltage. Consequently, the generators are the weak link, and safe operation for generators will, in most cases, automatically protect unit transformers that are connected to generator buses.

The unit should not be resynchronized to the system following a volts/Hz trip until an investigation has been completed to determine the cause of the relay operation. However, given the complexity of modern excitation systems, unexplained events are not that uncommon. Engineers or technicians should inspect the physical excitation system, verify calibration of the volts/Hz relays, and review any available data acquisition monitoring that would verify the operating condition. The unit can then be started for test and proper operation of the excitation system can be ascertained by operations before synchronizing the unit to the system.

Reverse Power. Reverse power or antimotoring relays are often applied for both control purposes and for protective relaying. In the control mode, they are typically used to automatically remove units from service during planned shutdowns and to ensure that prime movers have no output before isolating units electrically to prevent overspeed conditions. In the protection mode, they are used to protect turbine blades from windage overheating and sometimes to protect combustion turbine units from flameout conditions. The reverse power or antimotoring protective relays should have enough time delay before tripping to allow for synchronizing excursions (typically around 6 seconds). Motoring is not damaging to generators as long as proper excitation is maintained. In steam turbines, the LP turbine blades will overheat from windage. Typically, steam turbine blades can withstand motoring conditions for 10 minutes before damage.

Following a reverse power or antimotoring protection trip, the operators should determine if the trip was caused by control instability by reviewing the recorder or DCS trending of unit megawatt output. If control instability is evident, engineers or technicians should investigate and resolve the problem before resynchronizing the unit. If control instability is not evident, engineers/technicians should check the calibration of the protective relays before returning the unit to service.

Negative-Sequence Protection. Unbalanced phase current flow in generator stators causes double-frequency reverse rotation currents to circulate in the rotor body that can damage the rotor forging, wedges, amortisseur windings, and retaining rings. Cylindrical rotor generators designed according to ANSI standards are capable of continuously carrying 10% negative phase sequence current. This roughly corresponds to an operating condition in which two phases are carrying rated current and the third phase has 70% of rated current. Depending on the design of the rotor (indirectly or directly cooled) generators with two phases at rated current and no current in the third phase can carry this imbalance for 90 to 270 seconds before damage occurs to the rotor components. Accordingly negative-phase sequence relays are necessary to protect generator rotors from damage during all possible operating conditions, including phase-to-phase and phase-to-ground faults on the transmission system.

Some negative-sequence overcurrent relays provide an alarm function with a pick-up value set somewhere below the trip point. This alerts the unit operator to a negative-sequence condition prior to a trip. If the negative-sequence alarm is initiated, the operator should take the following action:

1. Notify the transmission dispatcher of the negative-sequence condition and find out if there are any electrical problems on the transmission system. When a negative-phase sequence alarm is activated, operators should also check the phase currents for balance. In addition to off-site causes for imbalance, open conductors, disconnect, or breaker poles at the site can cause the imbalance condition. If no abnormalities exist, notify the dispatcher that load will be reduced on the generator until the alarm clears.
2. The generation should be taken off AGC, and load should be reduced until the alarm clears.
3. Engineers or technicians should verify calibration of the negative-phase sequence relay and review any data-acquisition-monitoring devices (protective relay digital storage or DCS trends) to verify that the unit operated with a significant current imbalance.
4. If the alarm is coincident with any electrical switching in the switchyard or within the plant, the device should be opened, and if the alarm clears, the apparatus in question should be investigated for proper operation.

Following a negative-phase sequence trip, the unit should not be returned to service until an investigation is completed. Engineers or technicians should prove the calibration of the negative-phase sequence relay, and review any data-acquisition-monitoring devices (protective relay digital storage or DCS trends) to verify that the unit operated with a significant current imbalance. It may have been caused by an electrical system fault that did not clear promptly because of faulty circuit breakers or protective relays.

Under/Overfrequency Protection. Almost always, the under/over (O/U) frequency protection of the unit is there to protect the turbine before it protects the generator the turbine tends to be more sensitive to off-frequency operation that the generator. Therefore the U/O frequency protective devices are in general set to protect the turbine.

Operators should be aware of the frequency limitation for their particular turbines, and not operate them outside of the manufacturer's recommended limits under any circumstances.

In the event of a unit trip by the operation of a U/O frequency relay, the unit should not be returned to service until the system frequency stabilizes within acceptable limits and the protection is found to be properly calibrated.

XIII. Routine Operator Inspections

To maintain the operating integrity of generators, the following checks should be routinely performed by operations when making their daily inspection rounds:

- Check hydrogen purity levels (where applicable) and adjust gas flow to the purity monitor as required.
- Check that the seal-oil system is operating properly. Verify that the proper pressure differential between the seal-oil and hydrogen gas systems is maintained (where applicable).
- Check that the hydrogen dryers are in service (where applicable), operating properly, and the desiccant is in good condition.
- Check the liquid detectors for accumulation of water or oil.
- Verify proper water flow to hydrogen or stator coolers (where applicable).
- On generators equipped with water-cooled stator coils, verify proper flow, conductivity, and differential pressure between the water and the hydrogen gas systems.
- Check the stator, gas, and field temperatures.
- Check the bearing vibration levels.
- Check the generator stator ground scheme for proper residual or third harmonic voltages.
- Check the brush rigging (where applicable) for broken, vibrating, and arcing brushes.
- When the rotor is stopped or on the turning gear, the brush rigging area should be checked periodically for hydrogen leakage (where applicable). Hydrogen gas can leak through the bore conductors and accumulate in the brush rigging areas when the unit is off line.
- Check the shaft grounding brush(es) or braid(s) to verify physical integrity. In those units where the grounding brushes/braids are not visually accessible, refer to the maintenance guideline (EMG-1) for periodic maintenance.
- Verify that the field-winding ground-fault detection system is operational.

- Anomalies found during the routine inspections should be monitored and work orders prepared to resolve problems noted.

REFERENCES

1. IEEE Std 67-1990, "IEEE Guide for Operation and Maintenance of Turbine Generators."
2. IEEE/ANSI C50.13-1989, "Requirements for Cylindrical-Rotor Synchronous Generators."
3. IEEE Std 502-1985, "IEEE Guide for Protection, Interlocking, and Control of Fossil-Fueled Unit-Connected Steam Stations."
4. ANSI/IEEE C37.106-1987, or later, "Protection for Power Generating Plants."
5. M. G. Say, *Alternating Current Machines.* Pitman Publishing, 1978.
6. P. Anderson and A. A. Fouad, *Power System Control and Stability.* IEEE Press, 2003.
7. IEEE Std 1-2000, "IEEE Recommended Practice—General Principles for Temperature Limits in Rating of Electrical Equipment and for the Evaluation of Insulation."
8. Philip Beckley, "*Electrical Steels for Rotating Machines,*" IEE Power and Energy Series 37, Published by the Institute of Electrical Engineers, London, England, 2002.
9. Richard Roeper, *Short-Circuit Currents in Three-Phase Networks,* Siemens–Pitman, 1972 (Pitman ISBN 0-273-31884-5, Siemens ISBN 3-8009-1095-0).
10. EPRI Technical Report 1013460: "Torsional Interaction between Electrical Network Phenomena and Turbine-Generator Shafts," November 2006.
11. EPRI Techncial Report 1011679: "Steam Turbine-Generator Torsional Vibration Interaction with the Electrical Network," November 2005.
12. Duncan N. Walker, *Torsional Vibration of Turbomachinery*, McGraw-Hill, 2003.
13. S. F. Dorsey and G. P. Smedley, "The Influence of the Fillet Radius on the Fatigue Strength of Large Steel Shafts," presented at IME-ASME Internal Conference on Fatigue of Metals, 1956.
14. J.S. Joyce, T. Kulig, and D. Lambrecht, "Torsional Fatigue of Turbine-Generator Shafts Caused by Different Electrical System Faults and Switching Operations," presented at IEEE 1978 Winter Power Meeting.
15. L. Kilgore, E. R. Taylor, Jr., D. G. Ramey, R. G. Farmer, and A. L. Schwalb, "Solutions to the Problems of Subsynchronous Resonance in Power Systems with Series Capacitors," in *Proceedings of the American Power Conference,* Vol. 35, pp. 1120–1128, 1973.
16. IEEE Std 67-2005, "IEEE Guide for Operation and Maintenance of Turbine Generators."
17. IEEE C50.13-2005, "IEEE Standard for Cylindrical-Rotor 50 Hz and 60 Hz Synchronous Generators Rated 10 MVA and Above."
18. IEEE C50.12-2005, "IEEE Standard for Salient-Pole 50 Hz and 60 Hz Synchronous Generators and Generator/Motors for Hydraulic Turbine Applications Rated 5 MVA and Above."

5

MONITORING AND DIAGNOSTICS

Generator operation should be kept within design limits for optimum performance and to maintain reliability and longevity of the equipment. Monitoring of the online performance of the generator is done by using installed sensors and instrumentation; therefore, good sensor information is critical in making a correct diagnosis.

Good sensor information includes ensuring that the right sensors are installed in the right places, and that they are in good working order. In addition, the information from the individual sensors is often used in conjunction with other monitoring information, to make a more detailed and useful diagnosis. For instance, if all stator winding temperatures are hotter than normal and the stator current is above maximum allowable, then one would conclude that there is an overload situation that may be correctable by nothing more than reducing load. However, if the same sensors are in alarm when the machine is perhaps only at three-quarters load, then one would conclude that some other problem is present, such as low cooling water flow or plugging. At this point, it is the additional sensor information that one would look at in trying to diagnose the problem. But the initial diagnosis was actually done by a combination of information from two sensors: the stator winding temperatures and the stator current.

It is easy to see how monitoring can help avoid major failures before they happen, by early warning of problems. It is also easy to see that the more extensive the monitoring is, the more that can be determined during operation. This allows more flexibility in operation by knowing more about the performance of the machine. It may even be possible to extend generator life by adjusting the operation to avoid known operating regimes or ranges that cause some generator parameters to exceed their limits.

It is important that, for whatever sensors are installed in any particular generator, the most efficient use is made of the information from each sensor. There are numerous types generator-monitoring systems and/or approaches to generator monitoring, but they are not always equal in their effectiveness. Some machines are minimally equipped with installed sensors, whereas others have as much installed instrumentation as possible, including specialized monitoring devices. Generally, the highly equipped machines are large nuclear units for which outage time is extremely costly.

This chapter covers some of the different approaches to generator monitoring and the level of monitoring that can be accommodated. Included is a description of the various monitoring devices presently available, along with how they are generally used to determine abnormal operation, and how they could be further used in various types of monitoring systems.

5.1 GENERATOR MONITORING PHILOSOPHIES

Generator online monitoring and diagnostics covers a wide range of approaches, from minimal monitoring with few sensors and simple alarms up to elaborate expert systems with extensive diagnostic capability.

The level of system sophistication for the most basic monitoring allows the operator simply to keep track of generator operation by periodically checking various operating parameters on the gauges and indicators provided. Some sensors that are connected to an alarm system can also give warning when a static high limit is reached.

As hardware and software capability in data loggers and computers has progressed, the ability to provide better monitoring has increased dramatically in recent years. Computers can now handle many more sensors at one time and scan them for information far more frequently. This is because of faster CPUs, increased memory capability, and more sophisticated data acquisition hardware.

In addition to the hardware, software capability and flexibility has also grown exponentially. Software improvements have provided the industry with the ability to take simple sensor inputs and produce much more detailed information about machine performance from them. We can now combine sensor inputs to provide "intelligent" indications of problems that would otherwise not be foreseeable. This is done by computer modeling that predicts how various generator components will react during load changes and operating events. The use of such techniques allows closer tracking of sensors and the ability to diagnose problems at a much earlier stage in their progression.

The information gathered from sensor readings by the monitoring systems can now more readily be stored as an archived history of the performance of a machine. This can be used for long-term trending and maintenance management of the equipment.

In addition, the use of graphical user interfaces has allowed more meaningful presentation of the data collected, so that operators can interpret the information faster and more accurately. Readings are presented in both numeric and graphical form.

Short-term trends are used to compare various operating parameters as operators attempt to diagnose problems based on such things as temperature rise with increase in load.

Because of the significant advances in computers, generator monitoring has become very sophisticated. Along with this sophistication comes a high price tag to install complex so-called *expert systems*. In some cases, high-level sophistication is not necessary. A utility must assess its needs based on the equipment under consideration. A large 1200 MVA nuclear unit may warrant the installation of an expert system, whereas a 100 MVA air-cooled machine may only need a more basic level of monitoring to suit the particular needs and philosophy of the user.

When deciding on the correct approach for generator monitoring of any machine, it is always a good idea to first understand what the needs are. Then, one must understand the monitoring options that may be employed and know the cost to provide them. The next few sections of this chapter provide descriptions of some of these types of monitoring approaches that may be used on large generators in operation today.

5.2 SIMPLE MONITORING WITH STATIC HIGH-LEVEL ALARM LIMITS

Simple monitoring implies that the generator itself has very few sensors installed, and that only the most necessary and basic operating parameters are selected for permanent monitoring. Alarms are generally set at static high limits.

Generally, all generators have their main electrical parameters connected to a unit computer so that the operators are aware of the load point of the machine and where they are operating in relation to the limits of the generator (e.g., DCS, SCADA). The main electrical parameters include megawatts, reactive power, stator phase currents, terminal voltage, frequency, field current, and field voltage. All of these have operating limits that, if exceeded, can cause damage to one or more of the generator components.

In addition to the electrical parameters, there are operating values that must be monitored to ensure that the generator operating limits are followed. Some of the more critical parameters include shaft speed, hydrogen gas pressure and temperature, lube- and seal-oil temperatures and pressures, stator cooling water temperature, pressure and conductivity, bearing vibration, and raw service water temperature. These critical parameters tell the operator something about the condition of the generator or its components. In addition, all have specific operating limits that, when they are exceeded, have certain consequences.

Not all parameters have the same level of priority. For example, exceeding shaft speed limits by 10% while operating at steady load will have far greater consequences in terms of machine damage than exceeding a component temperature or bearing vibration limit by the same percentage.

In addition not all parameters have to be monitored to guarantee proper operation of the generator. For example, all of the stator cooling water outlet temperatures being normal during the on-load condition tells an operator that there must be water flow in the system. Therefore, it is not absolutely necessary to monitor the stator cooling water flow itself. Monitoring the stator cooling water outlet temperature is sufficient to safely operate the generator and detect when problems are occurring.

There is also a wide range in the number of sensors installed by the various manufacturers in their generators over the years. There are some machines with no core or stator winding thermocouples (TCs) or resistance temperature detectors (RTDs) installed, and some that have as many as two dozen core TCs and an RTD in every stator winding slot and on every stator winding outlet hose. The variation is extensive and is generally dependent on the manufacturer, but the operator may request additional monitoring sensors to be installed during manufacture if they are willing to pay more to get them. Most manufacturers will oblige.

5.3 DYNAMIC MONITORING WITH LOAD-VARYING ALARM LIMITS

Regardless of what type and number of sensors are installed and connected to the monitoring system, it is always a good idea to get the most out of what is monitored. A more effective way to get the best use of installed sensors is by dynamic monitoring. Dynamic monitoring simply means having alarm limits that change in relation to the generator load point, rather than waiting for a high-level alarm limit to be reached. Relying on static high limits can sometimes mean that a problem has progressed too far to allow any meaningful corrective action by the time the alarm limit is exceeded and the operator is notified by the alarm.

The premise behind dynamic monitoring is to mathematically predict what a particular sensor or group of sensors should be reading at any operating point and compare it with the actual sensor reading. The difference between the two can be closely monitored, and if the deviation is more than a previously determined limit, it can be alarmed or brought to the attention of the operator. The obvious advantage is that much earlier warning can be obtained. Also, it makes it possible to look for long-term problem trending, as well as more immediate failure modes.

To do this, a mathematical model of the generator parameter being monitored must be available. Then it has to be customized to the particular machine being monitored. Subsequently, this mathematical model becomes an artificial sensor or an indicator of a problem. The following is a brief description of how this type of sensor is constructed.

One of the best examples of an artificial indicator built from sensor readings into a mathematical model is that of water-cooled stator winding hose outlet temperature measurement. To build this indicator in its simplest form, one must look at what affects the stator winding temperature during all modes of operation, but specifically when the generator is connected to the system and loaded.

In the case of the stator winding hose outlet sensor, we are not concerned with stator winding temperature when the generator is offline since no current flows in the winding. Fault current could flow when the generator is on open circuit and a failure of the ground wall insulation occurs. However, this is a case in which generator ground fault relay protection comes into play, and stator winding temperature monitoring is a secondary issue. The main concern is the temperature of the stator winding when the machine is online and stator current is flowing in the winding.

To begin, the stator winding hose outlet temperature will be at least that of the water inlet temperature. Therefore, the first component of a stator bar hose outlet temperature, T_{out}, will be the temperature of the cooling water in, T_{in}. (For direct hydrogen-cooled stator windings, the coolant inlet temperature is derived from the cold hydrogen gas temperature.)

Stator bar temperature will increase as electrical current flows in the copper of the winding. The relationship of temperature to electrical current is well known as $T \propto I^2$. Therefore, if the generator is at full load while the stator current is theoretically at its maximum (I_{ref}), then the temperature of the stator bar hose outlets will be some temperature above the cooling water inlet temperature. The difference between the cooling water inlet and outlet temperatures will be the temperature rise, dT_{ref}, at this reference load, due to the heat input from the stator bar I^2R losses. The temperature difference between T_{out} and T_{in} will obviously change as the generator loading (operating stator current, I_s) is increased and decreased. Applying the relationship $T \propto I^2$, we can use I_s and I_{ref} in the form $(I_s/I_{ref})^2$ to account for generator load changes. Therefore, the basic formula to calculate stator winding hose outlet temperatures can be written as

$$T_{out} = T_{in} + dT_{ref}\left(\frac{I_s}{I_{ref}}\right)^2$$

In the relationship above, we can see that the portion of the function $(I_s/I_{ref})^2$ is equal to one, as it should be, when fingerprinting of the stator winding temperatures is done at the reference load. As I_s becomes lower, at lower loads, the temperature calculated for T_{out} will decrease proportionally [1]. Using the formula, the difference between the measured reading and the calculated value can be closely monitored. An alarm value (e.g., 5°C) can then be added to the calculated value to produce the dynamic alarm limit as follows:

$$T_{alarm} = T_{out} + 5°\text{C}$$

If the deviation is more than the calculated alarm limit, $T_{alarm,}$, it is then brought to the attention of the operator (see Figs. 5.1 and 5.2).

It should be noted that the preceding algorithms are in their simplest form. Other factors must be included for complete accuracy. The manufacturer generally knows these factors. For example, the stator bar expected temperature calculation can also be enhanced, to include a factor allowing for variable coolant flow in water-cooled stator windings. Likewise, for direct hydrogen-cooled stator windings, this model can be enhanced to incorporate changes in hydrogen gas pressure. When implementing these types of models, the utility should consult the manufacturer before implementation.

Now consider the direct hydrogen-cooled stator winding. If we do not include the gas pressure factor, the model allows diagnosis of overheating due to drops in casing

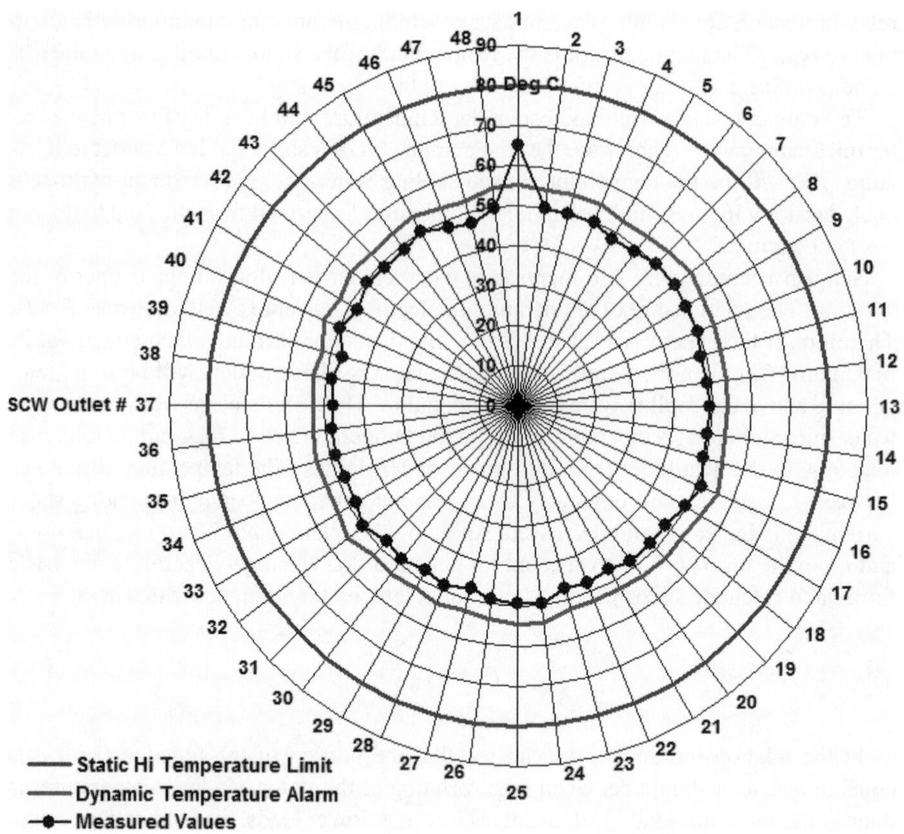

Fig. 5.1 Polar graph of instantaneous temperature magnitude for all hose outlets monitored.

hydrogen pressure. If we include the factor, then the model will always account for casing pressure changes and never be able to predict temperature rises on pressure variation, because the model is now further dynamic on pressure as well as coolant inlet temperature and stator current. Therefore, to determine if one of the model variables is affecting the outlet temperature, one either needs to depend on separate indicators to advise that a limit has been exceeded or use combinations of the basic model with each of the variables sequentially removed.

The distinct advantage of using this type of indicator in conjunction with direct sensor readings is the capability to predict expected values over the entire load and power factor range of the generator, and compare them to the actual readings. This allows a much improved and closer degree of monitoring on specific generator components, rather than simply relying on a maximum limit before an alarm is incurred. Using this dynamic monitoring method, one can look for deviations of only a few degrees above normal (for temperature relationships), at any load, and be provided with much faster

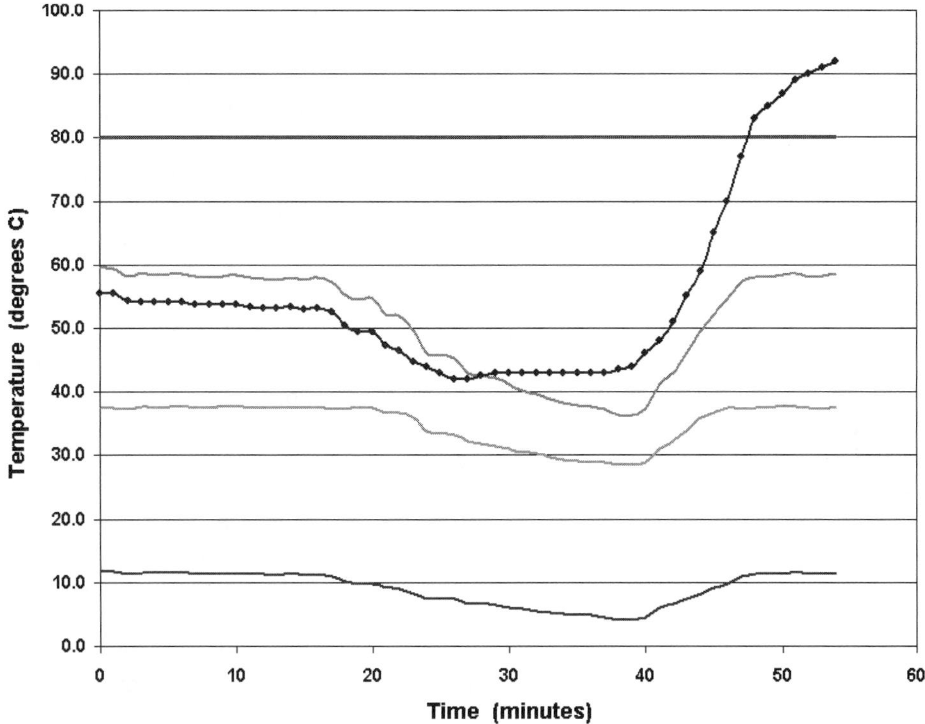

Fig. 5.2 One hose-outlet sensor indicating plugging as temperature in relation to time. Legend: —— = high limit, —— = dynamic alarm level, -•-•- = measured SCW outlet temperature, —— = SCW inlet temperature, —— = average stator current, Kamps.

warning of impending problems in the generator, long before measured parameters get anywhere near their absolute limits.

5.4 ARTIFICIAL INTELLIGENCE DIAGNOSTIC SYSTEMS

Monitoring systems with diagnostic capability using artificial intelligence software are more commonly referred to as *expert systems.* The purpose of an expert system for monitoring and diagnostics of large steam turbine-driven generators is to collect, analyze, and interpret generator and auxiliary system sensor information, and provide early diagnosis of developing problems in the generator and its associated systems. The expert system should provide an easily understandable description of the suspected problem and recommendations to correct it, or bring the unit to a safe operating condition in a timely and appropriate manner.

The main advantage of an expert system is its ability to look at all available sensor data in real time, correlate it as an "expert" would, and continuously update the diagnosis based on changing sensor readings. This allows operators to react quickly at the onset of the majority of generator problems that may be experienced during operation and avert major failures. In addition, an expert system gives station maintenance engineers a tool to closely monitor and log the performance of the generator, and make better maintenance decisions from the data collected during operation.

There are a variety of types of expert systems in use today on large rotating machines. Their deterministic capabilities rely on such methods as rule-based systems, pattern recognition, neural networks, and Bayesian belief networks. Regardless of the type of expert system implemented, all have a number of common elements. These include a *knowledge base* containing the equipment facts, component relationships, and mathematical models; data acquisition hardware and software for sensor inputs; AI (artificial intelligence) software, more commonly referred to as an inference engine, to perform the reasoning function between the knowledge base and sensor inputs; a graphical user interface to allow the operator to interact with the system; installation software to allow changes and updates to be made to the system by the experts; and in some cases a simulator for offline testing and training.

Within the various elements of the expert system, some interesting and novel techniques have been developed in recent years to provide accurate analysis of impending problems. These include mathematical modeling techniques used for logical and probabilistic determination of large generator problems. Also, methods of combining sensor inputs to create mathematical indicators of problems and techniques for dynamic tracking of problem indicators over the full power factor and load range of the generator have been developed.

An expert system generally consists of a computer for monitoring and processing of data, external data acquisition hardware for collection of the generator and auxiliary systems sensor inputs, and the software that forms the basis of the expert system. The computer is usually a fairly powerful type of machine such as a workstation. This is mainly because real-time expert systems have large memory requirements and need fast CPU speeds. However, based on the present trends in computer development, it is now possible to put even the larger expert monitoring systems on a PC.

The data acquisition system is used to collect raw sensor data from the generator and auxiliary systems. It can consist of a stand-alone data logger or the equivalent, or the existing unit computer data acquisition system.

The number of sensors monitored will vary, depending on the particular generator and how extensively it is instrumented. The expected readings of the monitored sensors are determined during installation and configuration of the expert system, by "heat run" tests on the generator at various loads and power factors to fingerprint the expected machine's behavior. The fingerprint data is used to produce scaling factors for specific formulas developed to track sensor inputs over the entire load and power factor range (i.e., dynamic tracking), and to set maximum sensor limits according to insulation class, machine rating, and other machine-specific parameters. Such formulas include stator winding temperature as a function of coolant inlet temperature and stator phase current as previously described, and stator core temperatures as a function of MW, MVAR, and cold hydrogen gas temperature and pressure.

Examples of sensors are those providing direct temperature readings from thermocouples and RTDs, pressure readings, voltage measurements, current measurements, and equipment status (breaker open/closed, pump on/off, tank level high/low, etc.). These are generally the instruments that are hard-wired directly to the data acquisition system. The readings are used in their raw form both in terms of the measurement value and units.

Within the software will be found the AI software, the knowledge base, third-party software for such things as the graphical user interface, installation software, and simulation software.

Within the knowledge base are the general and specific generator information on problems and indicators of problems, which the AI software must process. The knowledge base is generally a refined database that attempts to incorporate industrywide commonality in generator operation and troubleshooting. In addition, the specific information on the particular generator and its auxiliary systems being monitored are structurally mapped into the database.

The knowledge base consists of the possible generator and auxiliary system problems. Attached to the problem network or table are the indicators that consist of as many sensors and problem indicators as available from the installed generator instrumentation. The problem indicators are, in effect, the sensor inputs or combinations of sensor inputs that convey the information to the problem set that some operating parameter or limit has been exceeded, and that a real problem is occurring. The AI software determines this, as it processes the information from deviations between actual sensor readings and expected readings.

The expert system software uses the sensor inputs to look for deviations in readings that indicate generator problems and reports the relevant information to the monitor for operator interaction. Dynamic monitoring lends itself to the expert system application extremely well and takes even greater advantage of this technique. The example given in the section on dynamic monitoring has shown how the individual hose outlet temperature from one stator bar is calculated, so that it may be compared to the actual sensor reading. However, each stator bar is also associated with a particular stator winding parallel and phase. This is very important to consider since some stator winding problems are not simply related to only one stator bar, but to a particular parallel, phase, or the whole winding. An expert system can handle this complexity extremely well as follows.

On a two-pole, two-parallel path stator winding, for example, one of the phase parallels could be affected if a parallel connector on the collector end-ring bus were open circuited. This would cause all the current on the associated phase to divert from the open parallel to the other parallel. In effect, there would be no current flow in the one parallel, and the other parallel would be attempting to carry full-phase current and subsequently overheat. We would, therefore, expect to see all the bars associated with the overheated parallel in high-temperature alarm, and the other bars to go down in temperature since no stator current is flowing.

Let us say, for the same winding configuration, that one phase loses all cooling water flow due to plugging of the coolant path, which is sometimes made possible by the configuration of the water delivery system to the winding. We would then expect to see all the bars associated with this phase in high-temperature alarm since the phase is still carrying current but is not being cooled.

Finally, consider the whole winding in the case in which the cooling water flow is greatly reduced but is still flowing. The temperature monitoring indicators for the stator bars will see normal stator current and inlet water temperature, but reduced flow. Therefore, all the stator bar hose outlet temperatures should be in alarm since all will be reading higher than the calculated expected value for each bar.

The point of the last three examples is that we do not wish to have every stator bar hose outlet temperature in alarm when the root problem is not related to the bars themselves. Therefore, a further method is required to establish that the problem is not with the stator bars but is rather rooted in the connectors or the cooling-water delivery system.

To do this, we can use the stator winding diagram to form a sensor network and map out which bars belong to each of the three phases, and, subsequently, which of these are in each parallel [1]. Using the stator bar temperature models, the expert system can then reason that the problem is related to, for example, the red phase only because the winding mapping tells it that only the bars in the red phase are overheating. Therefore, the expert system would report simply that the red phase is overheating, rather than all the stator bars in that phase. One can then use graphical abilities of the computer to track the temperature of the affected phase in comparison to the other phases and to load changes.

The degree of overheating can be determined by further using the individual bar temperature models to give the average temperature of the bars in the affected phase, or the hottest bar in the affected phase, and so on. Handling of the temperature reporting is discretionary, and is simply a matter of choice in this case. Hence, there is additional flexibility built in with this type of approach.

5.5 MONITORED PARAMETERS

All generator-monitoring systems require sensors connected to the generator and its auxiliary systems to provide a diagnosis of problems that may be occurring. These sensors are various types of instrumentation that are installed to measure some parameter that is important to the safe operation of the machine. A sensor can come in many forms such as a thermocouple, pressure gauge, or level switch. In this section, we discuss the basic parameters monitored in generators and their auxiliary systems, and the effect of operation outside the limits of the parameter. A brief description of the types of sensors or instruments used for each parameter is included, as well as what each detects and how they are used.

In addition to the basic individual parameter or sensor information, some sensors can be related to another sensor to form an indication of a problem or existing generator condition. Here we describe how certain of these relevant sensors can be used in conjunction with others, to provide some additional diagnosis of the generator condition.

In discussions of generators, in general, the machine is broken down by components and subsystems that have common elements. The sensor information in this chapter will be presented organized by subsystems of the generator.

5.5.1 Generator Electrical Parameters

Generator Output Power. Generator output power is the real MW power output from the generator. It is a function of the stator terminal voltage, current, and the generator power factor as follows:

$$\text{MW} = \text{MVA} \times \text{PF} = \sqrt{3} \times V_t \times I_a \times \text{PF}$$

where V_t is the terminal line voltage and I_a is the terminal current. The power outputs, both active power (MW) and reactive power (MVAR), are monitored by voltage and current signals taken from the generator potential and current transformers. The signals are processed to provide the MW and MVAR information and displayed and recorded in the main control room, to keep track of the load point of the machine and allow operator control of the generator.

MW overload on the generator is the main concern, and often this means that the stator current limit has been exceeded. High stator current will affect the condition of the stator winding. Excessive stator terminal voltage can also result in an overload condition, but this is generally alarmed and relay protected to ensure that an overvoltage situation does not occur. Excessive terminal voltage will affect core heating.

Generator Reactive Power. Generator reactive power is the volt-amperes-reactive (MVAR) output from the generator. As with the MW of the generator, the MVARs are also a function of the stator terminal voltage and current. However, the MVARs are principally determined by the field current input to the rotor. Therefore, when MW is held constant, varying the field current will change the MVAR. The relationship is as follows:

$$\text{MVAR} = \sqrt{(\text{MVA}^2 - \text{MW}^2)}$$

The MVAR loading of the machine must also be monitored, since it also has operating restrictions.

Exceeding the maximum MVAR loading means that the field current limit on the rotor has probably been exceeded in the lagging power factor range. This will cause rotor-winding overheating. In addition, the stator terminal voltage can also be exceeded during excessive VAR loading and cause stator core overfluxing and, hence, core heating.

Exceeding the minimum MVAR loading means that the field current on the rotor has been reduced to a very low level, such that the generator is operating in the leading power factor range. When the MVARs are reduced beyond design limits, the possible problems that can occur are stator core-end overheating, exceeding the minimum terminal voltage limit, and loss of stability from slipped poles.

Three Stator Phase Currents. The three stator phase currents are monitored via the generator current transformers. Different classes of current transformers may be employed for metering and relaying requirements. They are located close to the generator winding around the terminal bushings. The analog output from the current transformers, approximately 5 A at rated generator output current, may be displayed and/or recorded on the main control panel.

The three phase currents flowing in the stator winding produce I^2R losses, which directly affect the temperature of the winding. Excessive current will cause excessive temperature, proportional to the square of the current. In addition, vibration and bar-bouncing forces are induced in the stator windings in proportion to the square of the current flowing.

Temperature and vibration affect the electrical and mechanical integrity of both the interstrand and ground-wall insulations, and the stator bar surface coatings. The mechanical integrity of the copper strands is also affected by temperature, in terms of absolute temperature and thermal cycling, and by the mechanical effects of bar bouncing.

The stator currents are monitored and used to provide indication of stator current overload and phase current unbalance, and they can be used to calculate the negative-sequence currents flowing in the rotor.

Stator Terminal Voltage. The stator voltage is monitored at the generator potential transformers. Different classes of potential transformers may be employed for metering and relaying requirements. The stator voltage is generally displayed and or recorded in the main control room.

Terminal voltage is a function of magnetic flux, rotor speed, and the stator-winding configuration. Excessively high voltage on the stator winding can break down the ground-wall insulation and deteriorate the stator bar surface coatings due to electrical breakdown phenomena. In addition, if the voltage becomes too high, the stator core may become overheated by an "overfluxing" condition. The induced eddy currents in the individual laminations would become excessive under such conditions, overheat the interlaminar insulation, and eventually melt the stator iron if the voltage is over the limit and held for too long a time. (For a discussion of this topic, refer to the discussion in Chapter 4, Section 4.4.8). There is an exponential relationship in which the higher the overvoltage is, the less time the stator core insulation can withstand it. The capability of the interlaminar insulation varies from one machine to another, but nowadays most are generally built to class F.

The stator terminal voltage is monitored to look for abnormal terminal voltage, either too high or too low, and to monitor the degree of phase voltage unbalance. Monitoring of the generator terminal voltage is critical during synchronizing of the generator to the system. The terminal voltage of the generator must be matched in magnitude, phase, and frequency to that of the system voltage before closing the main generator breakers. This is to ensure smooth closure of the breakers and connection to the system, with no malsynchronization occurring. Additionally, overfluxing occurrences tend to happen during start-up operations, before connection to the grid.

Field Current. Field current can be measured by a dc shunt installed in series in the excitation supply circuit or by dc CTs on the power circuit at the excitation system. The measurement is brought back to the control room for monitoring and is also used with the filed voltage to calculate the rotor winding average temperature.

The current flowing in the rotor winding is direct current and produces I^2R losses, which directly affect the temperature of the copper winding. The temperature of the rotor winding cannot be measured directly but, rather, by resistance using the measured field current and the measured field voltage to calculate the resistance of the rotor winding.

Excessive field current will cause field winding overheating directly and high terminal voltage. A field current that is too low can cause core-end heating and/or decrease of the terminal voltage to below its minimum allowable value. In addition, if the field current is reduced too far, magnetic coupling between the stator and rotor fields will be weakened such that loss of synchronism occurs from slipped poles. This is more commonly referred to as instability.

Field Voltage. Field voltage is measured at the excitation system supply to the rotor, and is generally taken as the voltage across the slip rings. The measurement is brought back to the control room for monitoring and is also used with the field current to calculate the rotor winding average temperature.

Increasing the field voltage increases the field current proportional to the rotor winding resistance. It is monitored but not usually used for alarms or trips. It is used to calculate rotor winding resistance and, subsequently, the rotor winding average and hot-spot temperature. AVR problems can cause the field voltage to become too high and, in turn, cause the excitation to increase beyond design limits.

Field voltage may be increased in times of system events to boost generator voltage by "field forcing." This term is used to describe increasing the field voltage to approximately double its normal value, on a short-term basis. The capability of any rotor during field forcing depends on the winding and cooling design. The relationship is generally an exponential one over time.

Frequency. The frequency of the generator output voltage is monitored at the generator potential transformers. The frequency is usually displayed on the main control panel. Frequency is measured in cycles per second or hertz (Hz). It refers to the electrical frequency of the generator. It is monitored for abnormal deviation from the system frequency, which is 60 and 50 Hz, depending on location in the world.

Frequency is more a consideration for turbine blade, hydrogen seal, and bearing operation. There are many subsynchronous vibration modes associated with low-frequency operation that could cause these components to fail, and must be avoided. In terms of generator operation, over- and underfrequency are the main concerns.

Overfrequency is most often the result of an instantaneous load reduction when the generator is synchronized to the system, or from excessive turbine torque when the generator is in the open-circuit condition before synchronization with the grid.

During on-load operation, fast load reductions may cause the current in the stator winding to decrease rapidly and the terminal voltage to likewise rise rapidly due to a high level of field excitation still applied. In this case, automatic action is taken to decrease the steam input to the turbine, to match the load requirements, and to quickly reduce field current to keep the terminal voltage within limits (i.e., the automatic voltage regulator).

Underfrequency is generally caused by a system event rather than the generator itself. The effect on the generator, however, is almost always an attempt by the system to extract excessive current from the stator and to drag the rotor speed down. This also has the effect of depressing the stator terminal voltage. To offset this, the excitation system for the generator will generally go into "field forcing" to try to maintain rated terminal voltage. Therefore, there is a possibility of sustaining overheating in both the stator and rotor windings during this type of event. Protection against overloading in these components is usually provided.

Volts per Hertz. Volts per hertz (V/Hz) is the ratio of terminal voltage to generator electrical frequency. It is actually used as a generator supplementary start protection during the open-circuit condition rather than for on-load operation. It is primarily in place to protect the generator from overfluxing during open-circuit operation. During the open-circuit condition, core overfluxing and, hence, overheating, is possible if an overvoltage or underfrequency condition occurs.

In the online state, however, should the terminal voltage rise beyond the design capability of the stator core, there is a danger of overfluxing the core in that condition as well. (This topic is extensively covered in other sections of this book; see the Index.)

Negative Sequence. Negative sequence generally refers to negative-sequence currents in the generator rotor induced by imbalance in the three stator-phase currents. These currents flow on the surface area of the rotor in the forging and wedges and are skin-effect currents that occur at twice rated frequency component. The unbalanced condition may be due to a machine or system problem but is more often system related rather than due to a problem in the generator itself.

There are two components of negative sequence to consider. The first is the continuous I_2 component, which refers to the amount of phase unbalance the generator can tolerate for an infinite operating period. The second is the transient component, called I_2^2t, which refers to the degree of short-term phase unbalance that the generator can withstand.

For large steam turbine generators, a typical continuous I_2 value of 8 would be normal. This means that the generator could carry a continuous phase imbalance in the stator winding of 8% or 0.08 pu of the rated stator current without damaging any of the generator components, specifically the rotor. A typical transient value for I_2 (the I_2^2t component) would be 10. This means that the generator could withstand 100% or 1 pu phase imbalance for 10 seconds (i.e., 1 pu phase current times 10 seconds = an I_2^2t value of 10) [2]. The overall relationship for negative-sequence capability of a generator is expressed exponentially as shown in Figure 5.3.

There is always a small natural degree of imbalance in these three-phase currents, but they are not harmful below the continuous I_2 value. When the degree of imbalance becomes significant, it appears as 120 (100) Hz currents flowing in the surface of the rotor body and wedges, which can overheat the rotor forging and/or wedges. The symptoms may arise as high vibration due to thermal imbalance in the rotor, and eventual component failure should overheating occur. Relay protection is generally provided to detect the level of negative-sequence currents and initiate a generator trip.

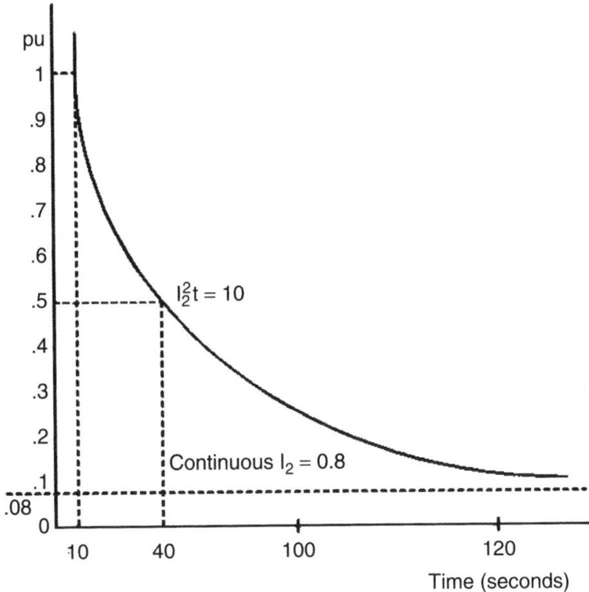

Fig. 5.3 Negative-sequence inverse time relationship.

5.5.2 Stator Core and Frame

Core Temperatures. Stator core temperatures are monitored by thermocouples (RTDs) embedded between the stator core laminates at strategic locations (see Figs 5.4 and 5.5). In the radial direction, these locations are usually in the tooth center, the core yoke a few inches below the slot bottom, and in the core yoke centered between the slot bottom and the back of the core. In the axial direction, they are positioned at both core ends, generally from the first core packet up to about one-and-a-half feet (45 cm) inward, and in the core center. This provides coverage such that the various core-heating modes may be fully monitored. This includes global overheating, core-end overheating, and localized overheating from local core faults.

Global overheating of the stator core may occur from overfluxing by excessive field current application, either online or offline. An indication that this condition is occurring will also be evident by the stator terminal voltage being above allowable limits. Core overheating can occur during overload, in which case the terminal voltage and field current will almost certainly be above limits. These are inverse time dependent so that a certain amount of overfluxing may be tolerated in a short period of a few seconds. The relationship is exponential with time.

Another possible cause of global core overheating is from the hydrogen gas pressure dropping too low. In this case, the core is simply not getting sufficient cooling. Hydrogen gas purity will also affect core temperature, but the drop in purity has to be

Fig. 5.4 Embedded core thermocouple.

severe for overheating to occur. This will have an effect on most other components in the machine at the same time.

The other cause of global core overheating is high hydrogen gas temperature. This is most likely to occur due to a problem with the hydrogen coolers, such as low flow or loss of raw cooling water. There have also been cases in which the raw water supply is too warm in summer months in southern areas, and even at full raw water flow the cold

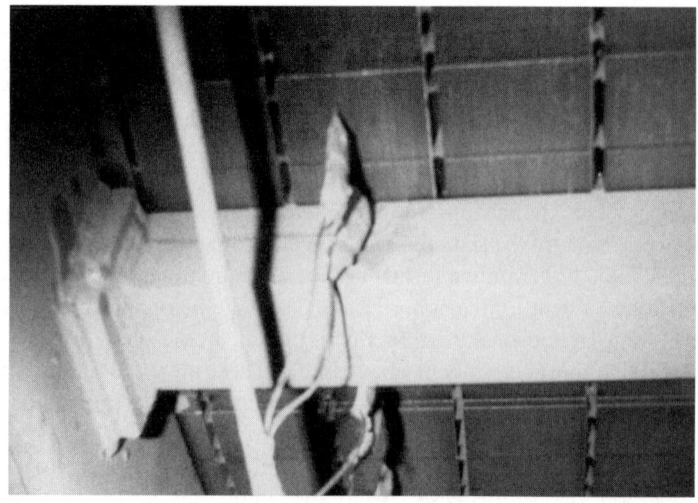

Fig. 5.5 Embedded core thermocouples.

hydrogen gas cannot be maintained below allowable limits. Again, core heating will be higher but overheating is unlikely. The condition should not be left uncorrected though, because the stator core interlaminar insulation life will be adversely affected in the long run by extended operation at higher than rated temperatures.

Core-end overheating is a condition related to a number of factors such as electric loading, airgap and stray flux, and magnetic saturation of the core, which are all affected to some degree by leading power factor operation. The amount of effect in the leading power factor range, for any given machine, will depend on the core-end design in terms of electric and magnetic loading characteristics.

In the core-end overheating mode, the core center is not affected. During leading power factor operation, the interaction of the magnetic fields in the core ends is such that there is a higher degree of axial flux impingement on the core, which tends to saturate the iron at the core ends. As the power factor or field current is reduced, saturation increases and, subsequently, the core-end temperatures rise. The level of temperature rise in the core ends is dependent on a variety of generator design features and, depending on design variations from generator to generator, some machines will see this affect more than others.

Local core overheating is a condition generally related to a small area of the core and is usually due to a localized defect or foreign body present in the stator. It can be associated with a breakdown of the interlaminar insulation or shorting of the core laminates.

Interlaminar insulation breakdown can occur from voltage spikes because of system- or machine-related events, fretting from loose iron, poor space-block welds, and so on. Shorts across laminates occur generally either from foreign objects in the bore or damage to the core from the rotor skid plate or hammer strikes during rewedging activities.

Core-Clamping-Plate Temperatures. Core compression or clamping-plate overheating is also related to the same factors as core-end overheating and somewhat dependent on leading power factor operation. It is affected by the interaction of the magnetic fields at the ends of the generator, the same as the core. TCs or RTDs are attached to the core compression plates to monitor for excessive temperature rise (see Fig. 5.6).

In addition to the end-iron effects from leading power factor operation, the clamping plates are also susceptible to overheating due to abnormalities in the hydrogen gas (i.e., low gas pressure and high gas temperature), since they are also cooled by hydrogen.

Core-End Flux Screen or Shunt Temperatures. Core-end flux screens or shunts are employed to reduce the saturation effect on the stator core-end iron and core compression plates. Again, this occurs primarily during leading power factor operation. The net effect of the flux screens or shunts is to control the stator-end temperatures below the class insulation limit. The flux screens/shunts therefore generate considerable losses and heat from the circulating currents induced in them.

Temperature monitoring in the form of TCs or RTDs is provided in many cases to ensure that the high-temperature limit is not exceeded. In addition to the end-iron ef-

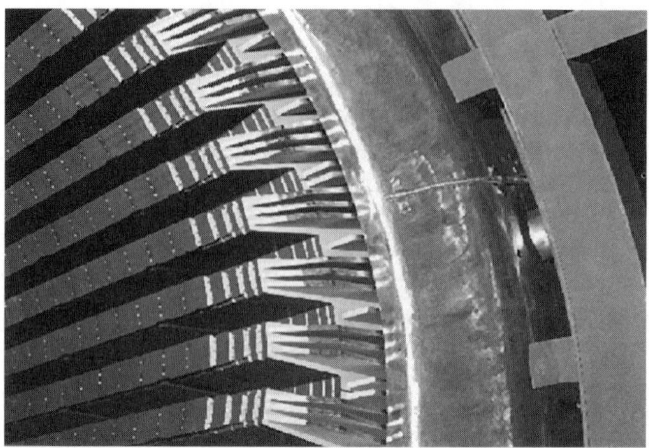

Fig. 5.6 Thermocouple attached to core-end clamping plate in the bore side of the plate where the temperature is usually highest due to stray flux.

fects from leading power factor operation, the clamping plates are also susceptible to overheating due to abnormalities in the hydrogen gas (i.e., low gas pressure and high gas temperature), since they are also cooled by hydrogen. However, there are some designs in existence that employ a water circuit in the flux screens and use stator cooling water for primary heat removal. In such cases, the hydrogen gas has a lesser effect.

Generator Condition Monitor. The generator condition monitor (GCM), commonly known as the *core monitor,* is a device used to detect overheated insulation inside the generator. Although the original intent of the GCM was to provide advance warning of stator core insulation overheating, the fact is that any overheated organic material inside the generator (i.e., insulation) will create pyrolysis products in the form of particulates that can be detected by the GCM (see Figs. 5.7 and 5.8).

The detection of particulates occurs in an ion chamber within the GCM. The ion chamber is designed to work in a hydrogen atmosphere under the high pressures used in large generators. As hydrogen is passed through the ion chamber, the detector inside the chamber produces a constant current through the gas flow. The current is set (usually at 80%) by regulating the gas flow and continuously monitoring it. When particulates (i.e., pyrolysis products from overheated insulation) are present, the current will drop. If the reduction in current goes below a preset alarm limit (generally set at 50%), the GCM signals an alarm.

The alarm tells the operator that a GCM event has occurred in the generator and a check on the validity of the alarm is now required. To assist the operating personnel in this assessment, a filter is employed to check that particulates are in fact present in the hydrogen gas. The operator simply pushes a button to divert the gas flow through the filter assembly before going to the ion chamber. If the alarm is real, the filter will re-

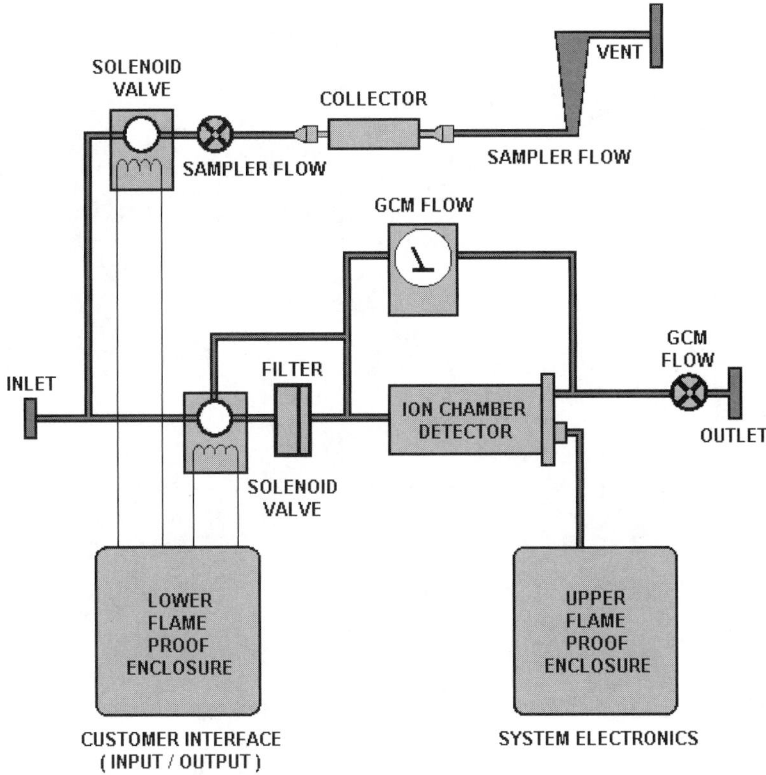

Fig. 5.7 Condition (core) monitoring system. (Courtesy of Environment-One.)

move the particulates and the GCM current will return to normal. Releasing the filter button should bring the alarm back on as long as the particulates are still present.

One of the problems that have been experienced over the years with the GCM is false alarms due to oil mist contamination in the hydrogen gas. The problem can be circumvented by slightly heating the ion chamber, as this causes the oil mist to have no effect on the current signal [3]. For this reason, many GCMs are now equipped with warmed ion chambers.

In the past, ion chambers were heated to the point where they were known to somewhat desensitize real alarms due to overheated insulation. Warming to a lower temperature than previously removes this issue, and the effectiveness of the core monitor is generally good.

To further enhance the use of the GCM, and to aid in determining what the source of any alarm is, an automatic sampling assembly is incorporated in the GCM. When an alarm occurs, a controlled volume of hydrogen gas is passed through the sampler, which has a removable collector. The collector maintains the ability to capture both the particulates and the gas that passes through it. The collector is then sent to a laboratory

Fig. 5.8 Condition (core) monitor installed next to generator.

for analysis of the collected samples. In this way, the source of the overheating or false indication can be determined.

The downside of auto-sampling is that it takes time for the samples to be analyzed while the unit is still operating. A decision can be made to continue running on the chance that the source of overheating will not be fatal to the generator (back-of-core burning, oil mist, etc.) or the unit can be shut down until the analysis results are obtained.

Tagging compounds [4] were developed to assist with identification of the source of internal generator overheating. Tagging compounds are essentially a number of different synthesized chemicals that are chemically and thermally stable. Each of the different "tags" is mixed in trace amounts with the colored epoxy paints that are used to coat the various internal generator components. When a GCM alarm occurs, a sample is taken and the sample is then quickly analyzed with the provided "on-site" equipment. Since the tagging compounds are known, the location of the overheating may be quickly determined (see Fig. 5.9).

Calculation of Available H_2 Flow-Rate Capacity. It is not uncommon for H_2-cooled generator operators to find that their GCMs are not functioning properly due

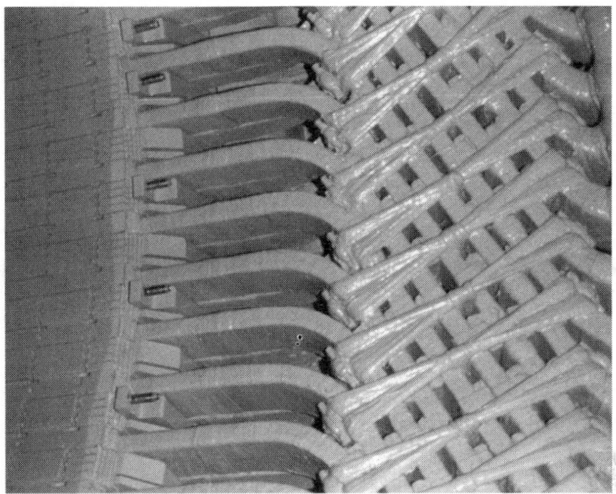

Fig. 5.9 Color-coded tagging compounds. (See color insert.)

to a lack of sufficient H_2 flow rate through the device. Unfortunately, users that discover they are having a problem with hydrogen flow mostly find out about it after the instrument is purchased. In some cases, in particular those involving older units, changing to a new GCM version may be all that is needed, given that newer versions tend to require less hydrogen flow to operate correctly. Sometimes, the solution requires replacing H_2 piping with larger diameter pipes. In other cases, the solution may necessitate the installation of a hydrogen fan external to the generator, placing it next to the GCM. In all cases, it may be advantageous to estimate the available pressure differential before a particular GCM is purchased.

The hydrogen flow via the GCM is obtained by tapping into the generator's casing in two locations with different internal pressure. At first glance, this appears to be a simple issue. However, as indicated above, the resulting pressure head may not be sufficient to overcome the flow resistance in pipes and the GCM's internal flow resistance, resulting in less than adequate flow. To estimate the flow of H_2 thru a given GCM, the operator can perform the simple set of tests described below. The results will be only somewhat accurate, given the nonlinearity (not captured in these calculations) between the pressure drop and the flow rate of a gas flowing through a pipe. However, they should be accurate enough to provide a good ball-park estimate of the available versus required flow leading to the correct operation of a particular GCM.

Figure 5.10 shows a typical GCM attachment to a hydrogen-cooled generator. In the figure, one can see that tapping into the unit's casing is done at different ends of the machine, and on different sides. This arrangement is not necessarily true for every machine in operation. Nevertheless, the figure also clearly shows that the H_2 pipes are of relatively small diameter and quite long. Also shown are some additional valves for H_2 collection. Finally, the figure also shows that in this particular example there are more than two tapped points. In fact, there are three, and one can choose the two providing the maximum hydrogen flow through the GCM.

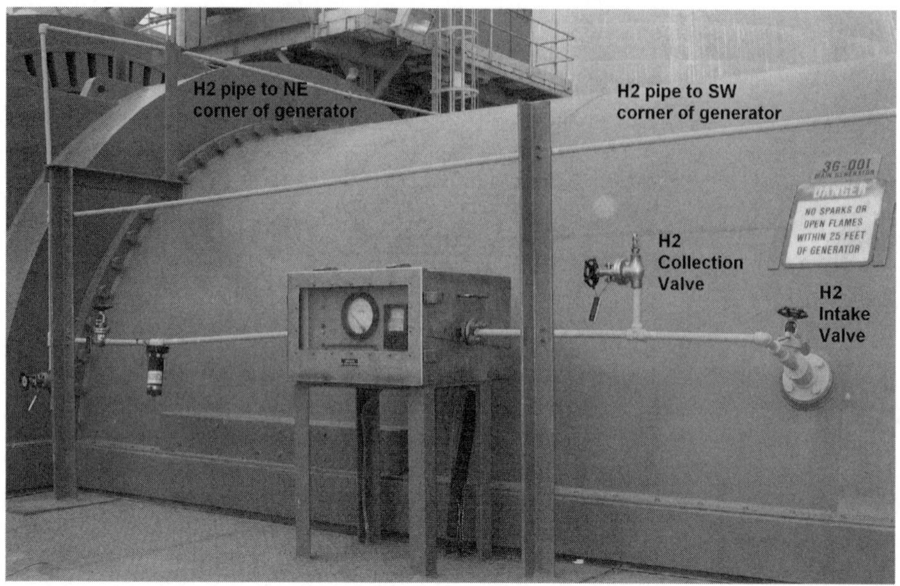

Fig. 5.10 Core monitor and hydrogen feeding pipes.

To perform the required estimation of flow, two points in the piping that will be close to the installed GCM should be chosen for performing the open- and closed-circuit tests described below.

Open-Circuit (OC) Test (see Fig. 5.11). This test is done by plugging the pipes at the chosen location (as close to the GCM as possible) and by measuring the resulting pressure head. In general, one can expect pressures equal to several inches of water. Given that in this situation the gas flow equals zero, the pressure measured is the pressure differential available at the tapped location. Changing this pressure can only be achieved by changing the position of one or both perforations in the generator casings—not a preferable solution.

Closed-Circuit (CC) Test (see Fig. 5.12). The circuit is now completed via an adequate flowmeter (chosen to have a very low internal resistance), and readings are taking of the resulting gas flow rate (for instance, in cubic feet per minute (cfm). In this test, the flow rate is a result of the available pressure head (found in the open-circuit test) and the resistance of all the piping.

Now the actual rate of flow through a given GCM can be calculated (see Fig. 5.13). The calculation requires the value of the GCM flow resistance to be known. This can be obtained from the manufacturer of the instrument.

In this way, users can ascertain if there is enough H_2 head pressure in the existing piping to obtain the minimum required H_2 flow rate through any given GCM. Using

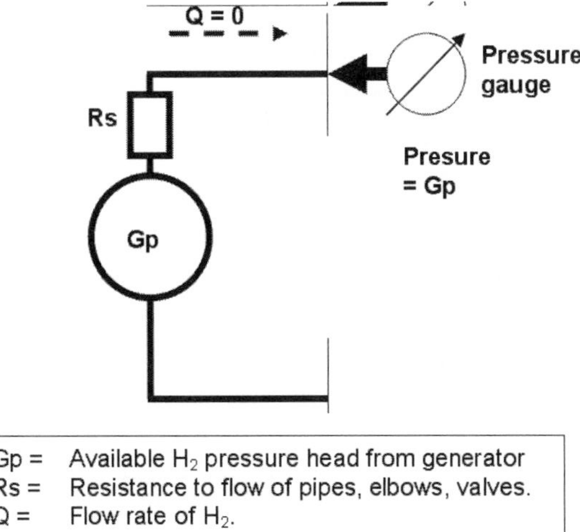

Gp = Available H_2 pressure head from generator
Rs = Resistance to flow of pipes, elbows, valves.
Q = Flow rate of H_2.

Fig. 5.11 Open H_2 circuit test.

these calculations, one could also estimate if changing piping diameter would result in any significant improvement.

Core Vibration. Vibration in the stator core is naturally produced by the magnetic pull in the airgap, originating from the main magnetic field of the rotor, incident on the stator core. The pole or direct axis carries the main flux, while the winding or quadrature axis carries only the leakage and stray fluxes. Therefore, a large difference in magnetic

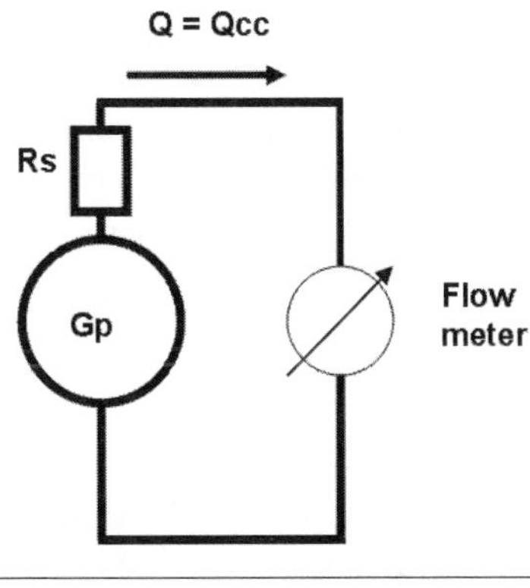

Gp = Available H_2 pressure head from generator
Rs = Resistance to flow from pipes, elbows, valves.
Qcc = Flow rate of H2 with closed circuit.

Fig. 5.12 Closed H_2 circuit test.

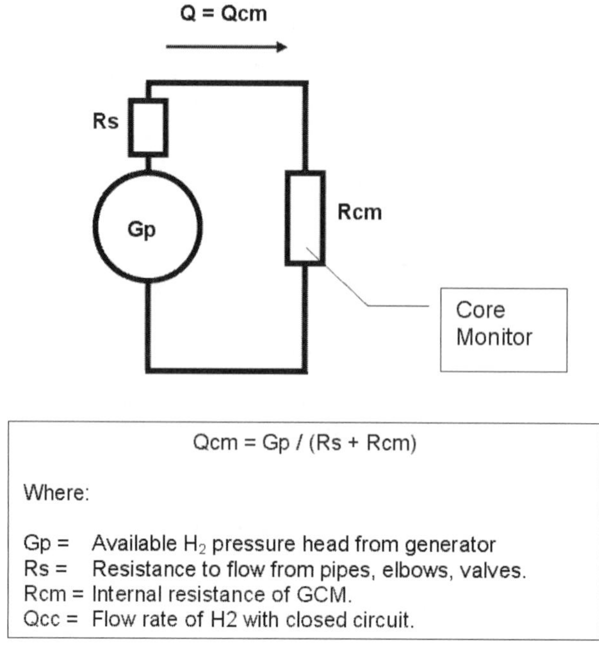

Q = Qcm
Rs
Gp
Rcm
Core Monitor
Qcm = Gp / (Rs + Rcm)
Where:
Gp = Available H2 pressure head from generator
Rs = Resistance to flow from pipes, elbows, valves.
Rcm = Internal resistance of GCM.
Qcc = Flow rate of H2 with closed circuit.

Fig. 5.13 Complete H_2 circuit.

force is inherent between the two axes. A large magnetic force is generated in the pole axes, and a weak magnetic force is present in the winding or quadrature axes. Since each pole has a north and a south associated with it (with a quadrature axis in between each) and there is rotation, the stronger and weaker magnetic pulls generate vibration at twice the line frequency and the result is a twice rotational speed vibration effect.

The core must be maintained tight or fretting will occur between the laminates. Minor fretting will tend to deteriorate the interlaminar insulation, but if the core becomes too loose, the laminates and or the space blocks may even fatigue, with the result being pieces of loose core material breaking off and causing damage. Monitoring of core vibrations can be done with accelerometers mounted on the core back in strategic locations to determine the magnitude and phase of both radial and tangential vibration modes (see Figs. 5.14 and 5.15).

Frame Vibration. Frame vibration is also excited by the same magnetic pull influences on the stator core and, consequently, the resulting vibration produced in the core. There are known cases of vibration resonance occurring on the frame as a result of the frame having a resonant frequency near or twice the line frequency. Resonant frequencies may be corrected by either adding mass to the frame to bring the natural frequency down or by stiffening the frame to drive the natural frequency higher, the object being to move the frame natural frequency away from the exciting frequencies by at least 10%.

Fig. 5.14 Core and frame accelerometers.

Severe damage to the frame can occur by initiating cracks in the frame welds or in the frame members themselves. Residual damage from the high vibrations associated with frame vibration is likely to be transmitted to other components of the generator if the situation becomes severe.

Good core-to-frame coupling is required to ensure that the core and the frame move together. There is evidence of numerous cases in which core frames became "uncou-

Fig. 5.15 Core and frame accelerometers.

pled" from the core and impacting damage is found at the core-to-keybar interface. Such vibrations were corrected in the past by spring mounting the core to the frame or installing a damping arrangement to "detune" the vibration modes.

Monitoring of frame vibrations can also be done with accelerometers mounted on the keybars, frame ribs, or casing structure in strategic locations to determine the magnitude and phase of both radial and tangential vibration modes (see Fig. 5.14 and 5.15).

Liquid Level in Generator Casing. Most large generators are equipped with liquid detectors. As fluids collect in the generator casing they are drained to a tank equipped with a level detector connected to an alarm. The liquid detector does not differentiate between the fluids that collect in it; it only senses that the allowable liquid level in the stator casing has been exceeded.

When the liquid detector alarm is triggered, a sample is required to be taken for determining the source. The possible sources are seal oil, stator cooling water, and hydrogen cooler service water.

5.5.3 Stator Winding

The stator winding is a high-cost component of the stator. Most serious stator problems are statistically found in the stator winding, due to the nature of the construction of the component and the operating duty it must endure with the relatively "soft" materials employed, when compared to the stator core or frame.

To judge the condition of the machine, operating load conditions can normally be correlated to temperature, vibration, partial discharges, and other parameters being monitored. Temperature is the main parameter monitored, and the winding may be monitored for temperature rise of the copper conductors and of the outlet cooling medium. The end winding may be monitored for radial, axial, and tangential vibration at once and twice the operating frequencies, and phase connection rings and terminal bushings may be monitored for both temperature and vibration.

Correlations normally are made of identical quantities at fixed operating conditions, held constant for an adequate time period to establish steady-state conditions. However, some occurrences relate to changing conditions. In these cases, an exact record and methodical variation of conditions may be required for a proper diagnosis.

Stator Cooling Water Inlet Temperature. Pure demineralized water is used as the cooling medium for stator windings in most large generators. It is commonly referred to as the stator cooling water (SCW). Water cooling of stator windings is usually done on generators of 300 MW size and larger, but there are also numerous large machines cooled by application of direct hydrogen cooling in the stator winding. In smaller machines, the cooling is done indirectly by hydrogen gas or air. In this arrangement, the stator core acts as a heat sink to dissipate the losses generated in the stator winding.

The stator cooling water inlet temperature is generally maintained below 50°C [5] but normally operates in the 35 to 40°C range. High inlet water temperature will cause the water outlet temperature from the stator winding to be correspondingly high as well. This increases the possibility of the outlet water temperature exceeding the boiling point. Strict attention must be paid to the inlet stator cooling water temperature, since the outlet stator cooling water temperature will be directly affected. Measurement of the cooling water inlet temperature is done generally by thermocouples or RTDs in the inlet coolant path.

Bulk Stator Cooling Water Outlet Temperature. The stator cooling water outlet temperature is usually maintained below 75°C [5]. The limit is approximately 90°C. Overheating of the stator bars is likely to occur and, possibly, boiling of the stator cooling water if the limit is exceeded.

Monitoring of the bulk stator cooling water outlet temperature provides indication of a range of problems that may be occurring in the stator winding but, generally, is indicative of a global stator winding heating problem rather than a localized winding problem. The other stator winding components affected are the stator terminals and phase connectors if they are water cooled. In many generator designs, the terminals and phase connectors may be cooled by hydrogen while the stator bars are cooled by water.

The main problems indicated by the bulk stator cooling water outlet temperature being high are stator current overload, low flow or a reduction of cooling water to the stator winding, and high inlet cooling water temperature. For all slot-RTDs and/or stator cooling water outlet temperature sensors, high readings also indicate these conditions.

Stator current overload simply means that the stator current is above the maximum allowable limit, and that the I^2R losses in the copper of the winding have exceeded the cooling capacity of the stator cooling water system. Low flow or reduced cooling water in the winding may occur due to plugging, leaks, or reduced flow from the stator cooling water system.

Plugging may occur due to corrosion of the hollow copper conductor bar strands, debris, or dust plugs left in the stator water cooling path after a generator overhaul. Low cooling water flow in the generator can occur if there is a substantial leak. Most likely, the problem will be in the stator cooling water system, external to the generator. Leaks may originate, for example, at piping joints or from cooler tube failure.

Reduced flow from the stator cooling water system may indicate a pump problem or a clogged stator cooling water filter/strainer. High cooling water temperature at the inlet to the generator also implies a stator cooling water system problem. A likely source is the stator cooling water coolers not getting enough raw service water for heat removal from the stator cooling water.

Conductor Bar Hose Outlet Temperatures. The stator cooling water outlet temperature of the individual conductor bars are also generally maintained below 75°C [5]. Again, operating above 90°C will cause a stator bar to overheat and lead to boiling in the individual stator bar if the limit is exceeded. Boiling can cause stator bar strand rupture and an eventual ground fault. Monitoring of individual stator bar hose outlet temperatures is generally done by a thermocouple or RTD in the outlet hose connection. A problem in the particular bar being monitored is indicative of a localized bar problem (see Fig. 5.16).

High stator bar hose outlet temperatures may be caused by strands plugged with debris and corrosion products, hydrogen gas locking, or collapsed strands from magnetic

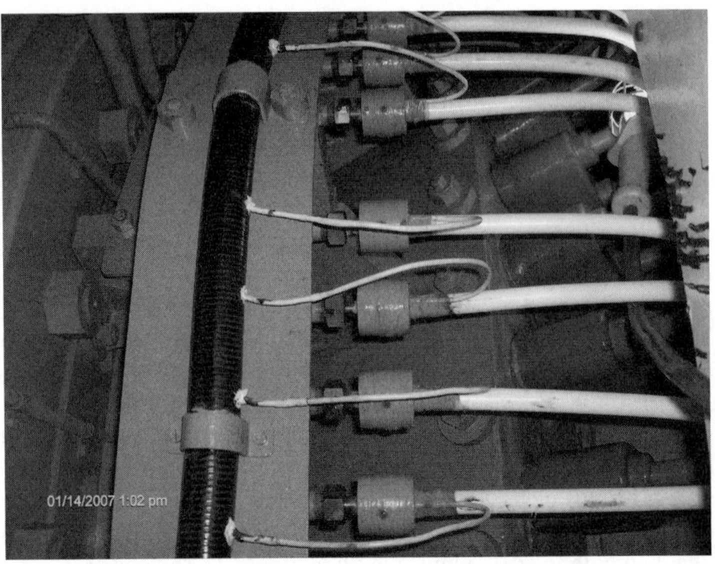

Fig. 5.16 Conductor bar hose outlet thermocouples.

"termites" that collapse the hollow strands by magnetic attraction and vibration effects. In addition, broken strands or electrical connectors and bar bouncing can cause stator bar temperature problems.

Plugging in a conductor bar is a common problem that can be sensed by hose outlet temperature monitoring. It refers to a flow blockage in the stator conductor bars by one of the mechanisms above, with the main result likely being overheated bars due to insufficient cooling water flow.

If the plugging is partial, then stator cooling water flow will continue but not in sufficient quantity to cool the affected stator bar. Therefore, the sensor will pick up the higher temperature of the stator cooling water outlet.

If there is full blockage of coolant flow in a conductor bar for one of the foregoing reasons, the condition may also be indicated by a high-slot RTD reading (discussed in next section) but a lower stator cooling water outlet temperature. This is due to the RTD still reading a high bar temperature while the stator cooling water outlet sensor is detecting the temperature of the cooling gas because there is no cooling water outlet flow. In addition, the core monitor will sound an alarm if the stator winding insulation is burning. To prevent the severe occurrence of the stator cooling water boiling, some manufacturers provide an automatic runback scheme that brings the machine to minimum load and avoids overheating. All manufacturers recommend reducing load during such a condition.

Conductor Bar Slot Temperatures. Conductor bar slot temperatures are monitored by a TC or RTD embedded between the top and bottom bars in the slot.

They are also generally installed toward the coolant outlet end of the generator so that the hottest temperatures in the stator winding slots may be observed.

Well-instrumented generator stator windings employ an RTD or TC embedded between top and bottom conductor bars in each slot (see Fig. 5.18), and a TC installed on the cooling water outlet hose of each individual stator bar, but this is not always the case. In some machines, there may be only one or the other or only a percentage of the slots containing temperature sensors.

In the case in which only partial slot temperature monitoring is installed, a high-temperature reading caused by blockage in either the top or bottom bar in an unmonitored slot will go undetected. With only hose outlet temperature monitoring of the stator bar coolant, total flow blockage in a bar can also cause the overheating to go undetected. The hose outlet sensor temperature will actually revert to the temperature of the hose outlet manifold, which will be at the "bulk" coolant outlet temperature. Since there is no flow of cooling water in this particular case, the true temperature of the stator bar will be unknown.

The optimum monitoring situation is where both slot and hose outlet temperature monitoring are employed. This allows identification of the affected bar/bars and provides a positive means to check the actual temperature of each bar.

In cases in which only one strand in a stator bar is plugged, there may only be a minimal effect on the discharge temperature of the stator bar and its performance. As more strands become plugged, the partial blockage of the coolant flow in one bar will increase and result in a much more identifiable overheating situation.

Slot temperature monitoring allows detection of overload, loss of cooling, and broken conductor strands. But it requires all slots to be instrumented if all bars are to be monitored. Otherwise, slot temperature monitoring is only partial, and trending becomes the useful part of this approach rather than absolute temperature monitoring.

Fig. 5.17 Slot separator with thermocouple installed.

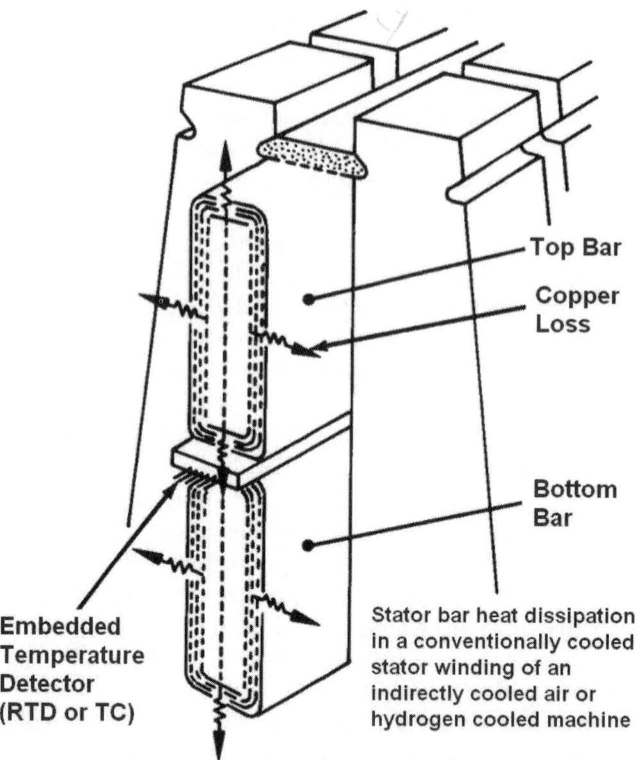

Fig. 5.18 Illustration of a slot RTD installed in separator pad and located in the slot between the top and bottom stator bar (from IEEE Standard 67-2005).

Stator Winding Differential Temperature. Stator winding differential temperature refers to the monitoring of the hottest to coldest operational stator winding temperatures. It further refers to a condition of temperature imbalance between individual stator bars or phases. The condition may be caused by phase imbalance due to a system problem, localized bar-to-bar temperature differences, plugging, other flow restrictions, and high resistance or broken electrical joints. An attempt is generally made to keep the temperature differential to the minimum possible, but it is not uncommon for a machine to have an inherent 10°C difference from hottest to coldest bar temperature. This can be due to differences in the cooling circuit from such things as some bars having the phase connectors in series with the stator bar cooling circuit. Slot RTDs and/or stator cooling water outlet temperature sensors indicate the condition.

Another source of temperature differential in stator bars can be between top and bottom bars, due to differences in the top and bottom bar sizes and magnetic field from the top to the bottom of the slot.

Also, differences in temperature between Roebel-transposed stacks in double Roebel stacked bars may occur within an individual bar due to circulating currents between the two stacks. This, however, causes the overall bar temperature to be higher rather than a difference between bars.

Stator winding differential temperature can also refer to the difference in inlet-to-outlet temperature of the stator winding. This should also be within a characteristic range for any particular machine.

The differential temperature across the stator winding, from the stator cooling water inlet to the outlet of the generator, can be monitored to ascertain that the design temperature difference across the whole stator winding is at the correct level. Correct differential temperature is an indication that the stator cooling water flow rate is adequate for cooling of the stator winding as a whole. This type of monitoring is not indicative of problems in individual stator conductor bars.

When the temperature differential is higher than normal, or higher than recommended by the manufacturer, this may indicate a partial blockage somewhere in the stator winding or stator cooling water system. To determine the source of partial blockage, other testing and monitoring is required to identify the plugging location.

Stator Cooling Water Inlet Pressure. The stator cooling water inlet pressure to the generator is measured by a pressure transmitter in the pipe work on the SCW inlet side and, generally, is kept at a design level to ensure there is correct cooling water flow to the stator winding, but at 5 psi below the hydrogen gas pressure to minimize the possibility of water leakage into the generator.

Stator Winding Differential Pressure. There should be a normal differential pressure that exists across the stator winding, during operation, based on design factors. A higher than normal differential cooling water pressure across the stator winding can indicate that there is a cooling water flow problem. This may be due to some form of plugging or hydrogen gas locking in the stator cooling water system.

The differential pressure across the stator winding from inlet water manifold to outlet water manifold is usually monitored to ensure proper cooling water flow, and maintain the correct operating temperature in the stator winding. If the pressure differential is much higher than expected, a large general obstruction to the flow of the stator cooling water may be present. If the pressure drop is very low across the generator, this may indicate a different problem such as a pump problem, blockage in one of the external system components, or a large leak of stator cooling water out of the external system piping before the generator inlet.

Hydrogen Gas in Stator Cooling Water. On machines with water-cooled stator windings, hydrogen leaking into the stator cooling water can occur at fittings and joints (from a bad or porous braze), corrosion, cracked strands, and from holes due to magnetic termites. Leakage occurs wherever a breach in the stator cooling water system occurs inside the machine, because the hydrogen is maintained at a higher pressure than the stator cooling water. Vibration and temperature cycling further aggravate leakage. It can also cause high conductivity of the stator cooling water as hydrogen contamination increases. Hydrogen gas locking may occur if the leak is large enough, which can also lead to bar overheating.

Hydrogen gas locking is a condition that may occur if the leakage rate of hydrogen into stator cooling water leak is large. Hydrogen gas locking can cause a cooling water flow restriction and eventually overheat the affected stator bar. Another form of gas locking can occur from gas bubbles if the stator cooling water starts to boil in the bar.

Some manufacturers employ a counter system to determine the amount of hydrogen loss from the stator cooling water system during online operation of the generator. The counter is a device used to collect a measured volume of hydrogen gas vented from the stator cooling water storage or makeup tank, and then release the collected gas to the atmosphere. The number of releases, times the volume of the collection chamber, indicates the leakage rate of the hydrogen into the stator cooling water.

Other manufacturers employ a leak monitoring system that is automated to collect and measure the vented gas in the stator cooling water system. It has a secondary purpose of helping to keep the dissolved oxygen content of the stator's cooling water at the correct level. This is a preferable method to vent line bagging and is similar to the vent line counter (see Fig. 3.10 on the gas release alarm tank).

It should be noted that as the hydrogen leaks into the stator cooling water circuit, stator cooling water usually also leaks outward through the leak location because of the capillary action of the stator cooling water. In numerous recorded cases, this has resulted in wet stator bar insulation.

Stator End-Winding Vibration Monitoring. End-winding vibration can damage the mechanical integrity of the stator conductor bars and the other stator winding components that make up the electrical and cooling water delivery portions of the total stator winding. End-winding vibration is usually a symptom of loose end-winding support structures. Checking and retightening of these support structures is required to avoid damage to the stator winding and associated hardware.

Vibration monitoring may be employed to indicate increasing vibration levels and plan outages to make repairs. Damage from vibration can initiate leak problems from cracked strands, broken braze joints, and loosening of fittings. There are many different end-winding support system designs in operation. Some are designed with the intent that they do not need retightening, but loosening may still occur naturally over time due to vibration from the forces induced by the high ac electromagnetic fields inside the machine. Other designs are provided with mechanisms for periodic retightening, and allow for natural loosening. Again, the degree of natural loosening and the time it takes depend on the design of the end-winding support system and the dampening effect of the support system.

Premature loosening may be enhanced by excessive seal-oil ingress, which will also create a greasing effect. Black grease will be seen at ties and blocking interfaces where fretting has occurred. System faults will again affect the integrity of the end-winding support system due to the high forces induced during such events.

In stators that are free of oil, looseness may be indicated by numerous paint cracks at ties and blocking to winding interfaces. In addition, there may be some white or light gray powder from corona discharge at paint cracks, which indicates some looseness.

In some instances, vibration probes are installed to monitor the online vibration characteristics of the stator end winding. Vibration probes are not, however, common instrumentation, and they are not found in the majority of machines in service (see Figs. 5.19 and 5.20).

Measurements are generally done in the tangential, radial, and axial modes.

Stator Winding Ground Alarm/Trip. Stator winding grounds occur on failure of the stator winding ground wall insulation. In the majority of cases, this occurs in the stator slots next to the stator iron. In addition to the stator winding conductor bars, the phase connectors and terminals are covered by this protection. In fact, this is more a protection scheme than a monitoring parameter.

It is generally used to trip the generator on such an occurrence. (This topic is covered fully in Chapter 6.)

However, a ground alarm does not necessarily mean that a winding insulation failure has occurred. It can be caused by instantaneous overvoltages from either machine transients or system spikes, and it only indicates that an event has occurred. Some basic electrical testing is usually carried out to prove that the ground insulation is still viable, and the unit is returned to service if the tests show that no failure has occurred.

Stator Winding Partial Discharge Monitoring. Partial discharge (PD) in large generators is mostly associated with the high-voltage stator conductor bars. It is

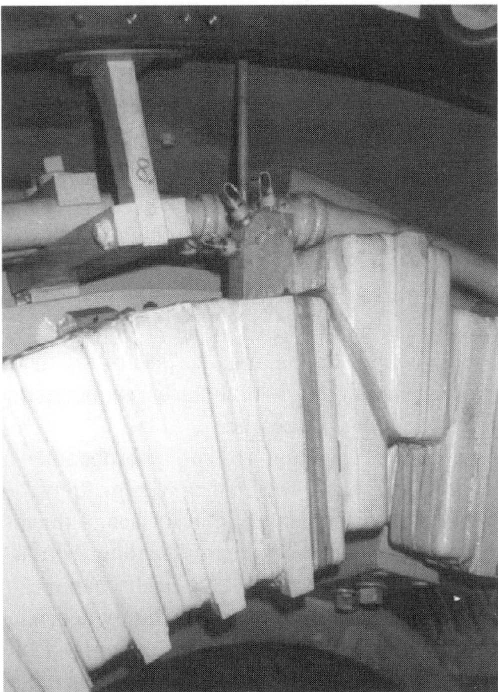

Fig. 5.19 Stator end-winding vibration-monitoring transducers.

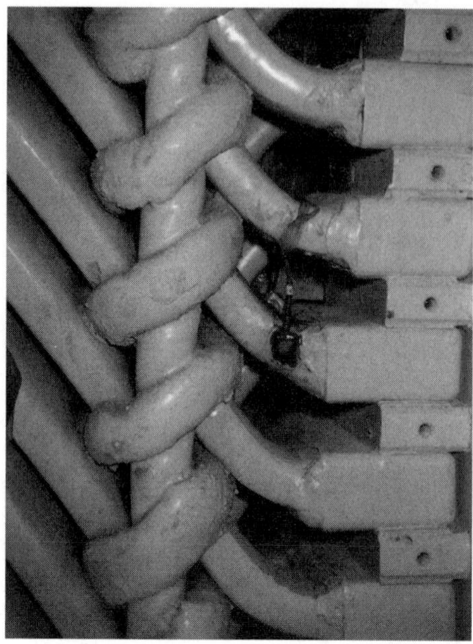

Fig. 5.20 Stator end-winding vibration-monitoring transducers.

the generic term for small electric current pulses flowing from one surface of the generator stator winding to another, or to another component of the generator, usually the stator core. These are not necessarily harmful and are quite normal. Partial discharges are also constantly occurring during operation whenever the stator winding is energized above the *discharge inception* level. Discharge inception is nothing more than the lowest voltage at which continuous partial discharge starts to occur as the applied voltage is increased.

Partial discharge generally occurs across the small airgaps or voids within the main stator winding insulation, and this is referred to as an "internal" type of discharge. Partial discharge can also take place on the outer surface of the stator bar insulation protruding from the end of the stator core slot into the end-winding region, and this is generally referred to as "end-winding" discharge.

The above terms are general in nature and can be broken down even further to describe the different types of discharge within these designations.

Partial discharge that takes place between the surface of the insulation in the stator slot and the slot wall and/or the semiconducting coating (corona-protection layer) is generally referred to as *slot discharge.* This type of discharge can be very destructive to the stator winding groundwall insulation and is highly dependent on design, manufacturing and assembly issues, in order that the insulation system will maintain its integrity for reliability and long service life. The copper conductors within the stator bars must be completely electrically insulated from the stator core to prevent destructive discharge currents flowing to ground. Therefore, the bars should be installed in the slots such that each bar cannot move in its slot and maintains good and continuous electrical contact with the walls of the stator slot, between the top and bottom bars in the slot, and between the bottom bar and the slot bottom. On a high-voltage stator bar, if there is a gap of a critical size between the stator bar and the slot wall, a charge will build up and eventually discharge across the gap to the stator core. This is extremely damaging to the groundwall insulation of the bar.

To avoid such gaps, good surface contact between the stator bars and the stator core must be maintained, and a semiconducting coating is used for corona or discharge suppression. It is usually made of graphite or carbon-loaded varnish, and applied in paint form or within the stator bar armor tape. Also, the semiconducting or corona-protection layer must be of the proper resistance value (typically 500 to 15,000 ohms per square) and applied correctly. If done correctly, this semiconducting layer will maintain the necessary contact between the surface of the stator bar and the core, and prevent damage to the insulation caused by electrical effects.

Restricting bar movement in the slot and proper application of the semiconducting material is critical for good contact and minimizing any gaps that may occur. However, when discharge does occur because one of these has gotten out of control, the mechanism is well known as *slot discharge* and is also of a capacitive nature. It may be recognized by the typical white residue produced.

In the capacitive type of slot discharge, as the semiconductive coating is eroded away, the bar surface resistance increases and the effect simply multiplies. Eventually, the groundwall insulation is electrically pitted and itself erodes away until there is not enough groundwall insulation left to hold the stator voltage, and a ground fault will occur (see Fig. 5.21). This semiconducting layer is extremely important in controlling the capacitive effects of slot discharge by ensuring that the resistance between the bar and the stator core is low enough not to allow voltage build up that will promote discharge. If the semiconducting layer is removed, for whatever reason, the resistance between the bar and core surface can get too high and capacitive effects will overtake.

Sometimes, however, there is slot discharge that originates internally in the stator bar groundwall insulation by delamination of the insulation material or by separation of the groundwall from the copper conductors. Even though the bars may be tight in

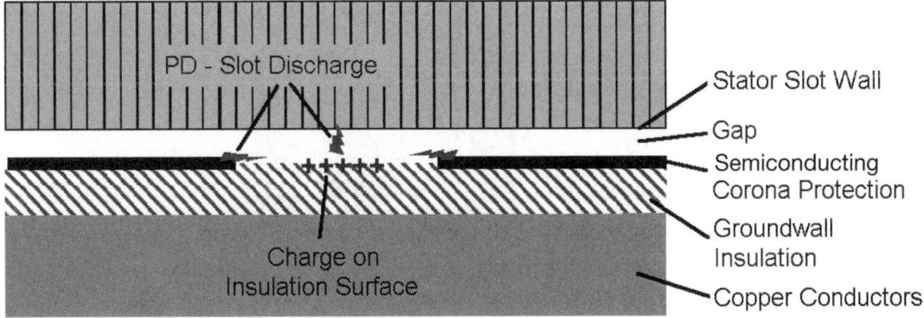

Fig. 5.21 Slot discharge from loss of the semiconducting coating or corona-protection layer, capacitive type discharge. (Courtesy of Alstom Power Inc.)

the slots and the semiconducting coating is intact, voltage breakdown occurs inside the bar insulation and damaging discharge currents result. Therefore, capacitive slot discharge can be broken down even further to surface discharge, internal voids and delamination, and groundwall separation from the copper (see Fig. 5.22), and each of these shows up as a different characteristic of the partial discharge measurements. This will be discussed a little further on in this section under online measurement techniques.

A second form of slot discharge can occur due to the interruption of currents flowing in the corona-protection coating due to movement of the stator bar relative to the slot wall (see Fig. 5.23). This is called bar bouncing (see Section 2.8 for determination of bar bouncing forces in the stator winding). During operation, a voltage is induced along the length of the stator bar by the alternating magnetic fields linking the stator core. This voltage produces a small axial current in the corona-protection coating that flows from one contact with the slot wall to another via the coating. If one of the contacts opens to interrupt the current, then, depending on the distance between contacts and the magnitude of current in the bar, there is a risk of partial discharge. This type of slot discharge may be referred to as *inductive slot discharge* since it arises from the inductive voltage along the stator core.

The key factors for inductive discharge are the magnitude of the bar movement or vibration in the slot, the distance between the points of contact between the bar and the core for current to flow, and the resistance of the semiconducting coating.

Bar movement must be lateral and of significant magnitude to make and break electrical contact between the semiconducting coating on the surface of the bar and the stator core slot wall. The effect of the bar movement is to create a spark gap and is, therefore, sometimes also called vibration sparking.

Once the spark is initiated, the distance over which the current travels between the fixed and intermittent contact points becomes important in terms of the resistance of

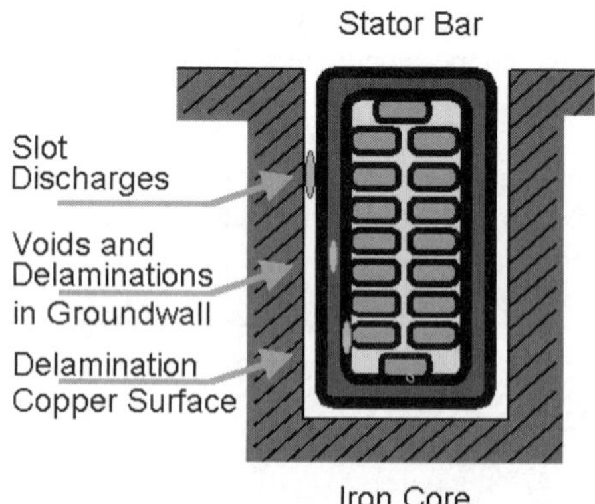

Fig. 5.22 Internal discharges from surface discharge to the core, insulation voids or delamination in the main groundwall insulation, and separation of the groundwall from the copper. (Courtesy of Alstom Power Inc.)

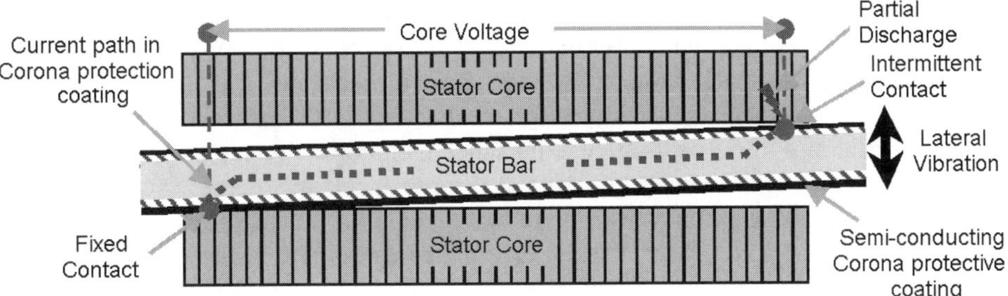

Fig. 5.23 Slot discharges from bar bouncing, inductive type. (Courtesy of Alstom Power Inc.)

the conduction path and the driving voltage behind the discharge. The discharge becomes more intense and, hence, is more destructive in nature as the resistance becomes lower, and as the voltage becomes higher over a longer conduction path.

The resistance of the semiconducting coating has a lower limit (usually 500 ohms) for precisely the purpose of limiting the amount of induced current that can flow along the bar surface as this vibration sparking effect occurs. It is, therefore, easy to see that if good electrical contact can be maintained between the semiconducting layer and the stator slot walls, the current will be harmlessly dissipated by eliminating the source of sparking.

If there is relative movement between the bar and the stator slot wall due to such things as bar bouncing and vibration then, as the bar separates from the stator slot wall, an intense spark will be produced, which is damaging to both the outer corona-protection layer and the insulation on the surface of the bar. The sparking burns away the outer corona-protection layer and the epoxy resin, leaving a carbonized residue and damaging the mica in the groundwall insulation. Carbon is conductive, so the residue lowers the resistance level even further, and the rate of damage increases. As this inductive discharge continues, the danger of finally eroding the semiconducting coating away to produce a surface resistance that is too high becomes likely. The result can be that the inductive discharge now becomes capacitive in nature, as described above. The end result is the same, in that the stator bar is prone to a groundwall failure.

So far, we have discussed slot discharge in some detail, but there is also end-winding discharge to consider as it can also cause severe damage to the insulation system and, at worst, a phase-to-phase flashover under extreme circumstances.

One of the main locations of end-winding partial discharge is at the stress-control voltage-grading coating at the slot exit. Some of the reasons for end-winding discharge are defects such as electric stress concentrations at the interface between the semiconducting slot coating and the stress-control coating, localized mechanical damage on the bar surface, or improper application of the stress-control coating (see Fig. 5.24).

End-winding discharge may also occur further out in the stator end winding, past the stator slot exit, due to chemical contamination, loose conductive particles, vibration, mechanical damage, relative movement of end-winding components, and insufficient spacing between conductor bar involutes in the end winding (see Fig. 5.24). When the discharge is even further out into the phase connections or stator terminals, those components are also susceptible to the above mechanisms.

All of the above-mentioned discharge mechanisms in the end winding require a voltage difference from one location to the other in order for discharge currents to flow along the surface of the bars and across end-winding blockings and ties. Normally, the voltage differences are controlled by the stress-grading coatings, groundwall insulation, and adherence to proper clearances and cleanliness. When there are problems with any of the above issues, severe discharge currents can flow and cause insulation burning and eventual failure. One of the worst cases is when there is discharge between phases, since the voltage difference is large and the fault current is severe when a full breakdown occurs.

All the mechanisms that promote partial discharge create areas of voltage stress where electrical charges build up and discharge. The result is possible damage to the voltage grading systems on the bars, the inter-strand insulation, or the ground-wall insulation.

A failure of the stator winding insulation is costly to fix, in terms of both the capital cost to repair or replace a stator bar and the outage time required to complete the work. Therefore, much effort has been invested over the years in developing techniques to identify the occurrence of PD in the stator winding.

Discharge activity in the stator winding can be measured during an overhaul by energizing the winding with an ac test transformer and checking for the presence of

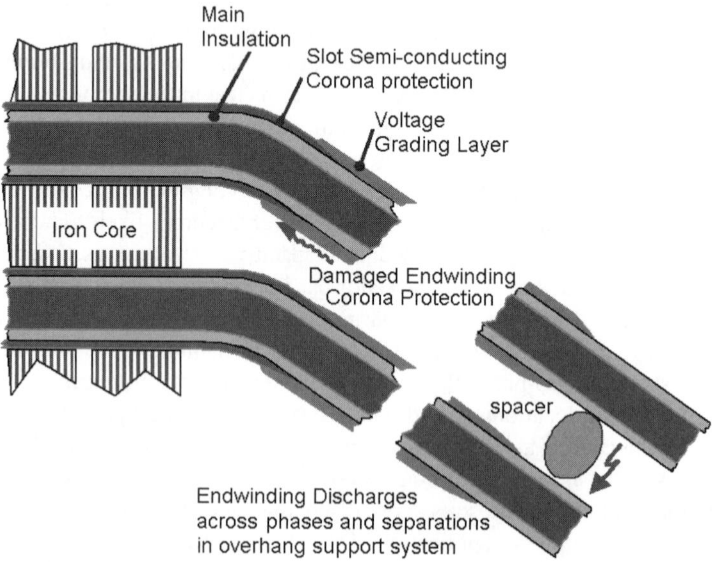

Fig. 5.24 This illustration shows the various locations for surface discharges in the end winding. (Courtesy of Alstom Power Inc.)

the high-frequency currents induced by partial discharge activity. However, although offline testing shows the relative magnitude of partial discharge activity, it is often difficult to identify the cause of any increase in measured levels. Further, offline tests will not readily detect inductive slot discharge since there are no significant electromagnetic or mechanical forces to drive lateral vibration of the bar in the slot. Offline tests are useful, however, in detecting capacitive slot discharge where there is damage to the corona-protection coating, and in detecting some forms of end-winding discharge activity. Basically, the methods for offline detection of PD do not cover the operating effects that also promote PD activity, such as those due to thermal and vibration effects.

To provide the best PD detection, an online method of monitoring is required. Discharge activity in the stator winding can be measured during normal operation (online monitoring) by detecting the high-frequency currents and/or voltages at the connections to the stator terminals, as well as by other means. However, since the generator is ultimately connected to the power system through an isolated phase bus and by excitation systems, and so on, partial discharge measurements are subject to interference from these types of sources external to the stator winding. Various analytical methods and procedures have been developed to try to isolate partial discharge arising in the stator winding and to identify the cause(s) of changes in the levels of activity, as well as to overcome interference signals from external sources. These procedures are more or less successful depending on the design of the stator winding, the magnitude of partial discharge activity, and the relative level of external interference. Unfortunately, there is no technology to date that can interpret the wide variety of partial discharge patterns and make a definitive diagnosis of the exact location of the discharge and the root cause. Some types of discharge do produce easily recognizable patterns, but often there are a number of mechanisms in play that cause multiple patterns to be observed simultaneously. This type of situation makes interpretation complicated and not always fully reliable. Regardless, there are a number of approaches to online PD monitoring, and all are considered viable methods. The following methods are some of those more commonly used for large turbo-generators.

Radio-Frequency Monitoring. Radio-frequency (RF) monitoring is a technique for detecting electrical sparking and arcing or stator winding PD inside the generator. It operates on the premise that arcing in the stator winding will cause radio-frequency currents to flow in the neutral of the winding. The types of stator winding problems or failure mechanisms that will cause these RF currents to flow are conductor bar strand cracking, electrical joint failure, and partial discharges due to insulation problems.

To monitor these currents, a high-frequency current transformer (CT) is placed around the neutral grounding lead, before the neutral grounding transformer, as shown in Figure 5.25.

The output of the CT is fed to a radio-frequency monitor for signal processing and analysis. The signals from the CT are filtered to examine those that are in the correct frequency range for radio-frequency arcing. The monitor generally has a set-point or an alarm limit that is adjusted to a predetermined level for which the RF activity is

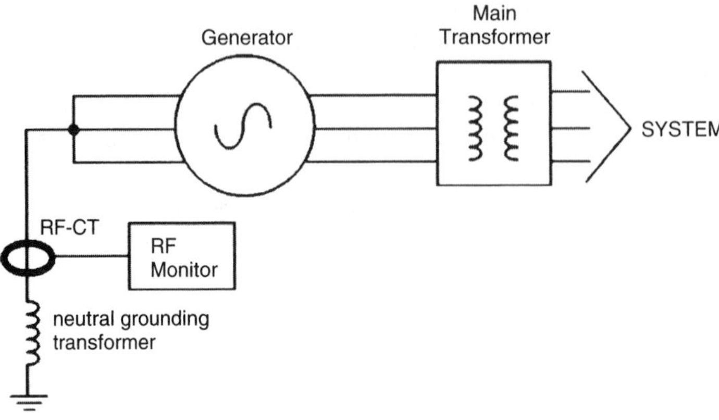

Fig. 5.25 Radio-frequency monitoring.

known to be excessive. When the level of RF activity increases to where the set-point is reached, the operator is notified of an RF problem in the generator.

It is difficult, however, to distinguish between the sources of RF arcing, and it is not always possible to identify the root cause. In most cases, one can only say that arcing is occurring and at a certain level. In addition, filtering does not completely eliminate noise. Therefore, this also becomes a problem in signal discrimination. This is especially true for large turbo-generators in which the noise generated from the slip rings/brush gear and the shaft grounding brushes is considerable. In recent years, however, advances in computer and software analysis have allowed better discrimination between RF signals that are actually PD and those that are noise or from another source.

Capacitive Coupling. Capacitive coupling has been in use since the early 1950s and was developed as an alternative to RF monitoring [4,5,6]. In contrast to RF monitoring, in which detection of PD is at the generator neutral, capacitive coupling is done at the line ends of the generator winding, meaning at the output of each of the phases. The improvement is that PD can be detected on a per-phase basis.

To measure the PD activity on each phase, a tuned capacitor or "capacitive coupler" is connected to each of the generator terminals. The couplers are then connected to an RC band-pass filter for which the output is a signal containing the high-frequency PD pulses. The PD pulses are displayed on a high-speed oscilloscope, where they can be categorized (i.e., negative and positive PD pulses) and quantified (i.e., the magnitudes of the negative and positive PD pulses) by a technician with sufficient skill. Although capacitive coupling has proved to be instrumental in online detection of PD, it does require a great deal of experience in distinguishing the PD pulses from noise in the generator and from the system (see Figs. 5.26 and 5.27). One disadvantage of this test,

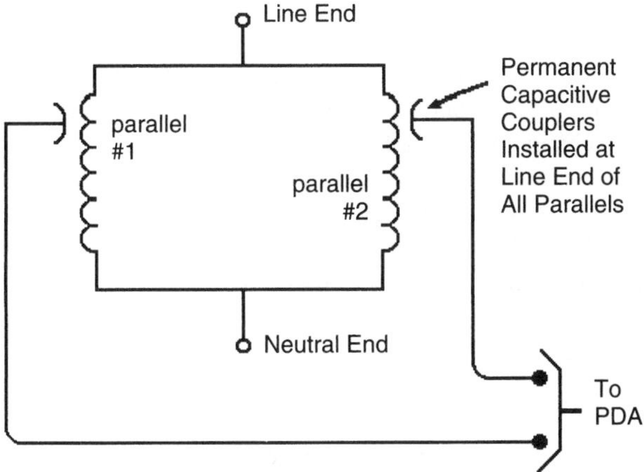

Fig. 5.26 Capacitive coupling.

however, is that the capacitive couplers are generally temporary on steam-turbine generators and require live-line installation—a very dangerous procedure with risk to the test technician.

Once the couplers are installed, three tests are generally done to further identify the type of PD activity occurring. The first is done offline but with the generator on open circuit at rated terminal voltage. The second is done with the generator online at low

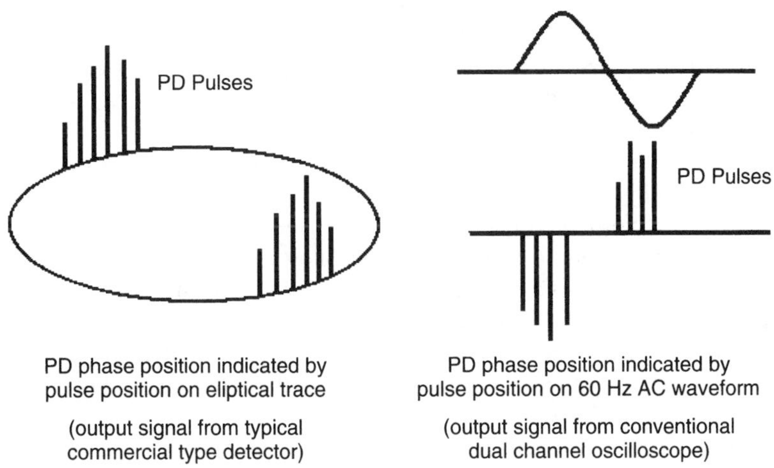

Fig. 5.27 Capacitive coupling signals.

load, and the third, also online, at high load. Measurements are taken on each phase for each of the three conditions.

The loading on the generator is varied to provide distinction between the types of PD activity present. Since temperature and vibration have significant effect on the level of PD activity, changing the loading conditions will cause variations in terminal voltage and stator current. Hence, there will be variation in winding temperature and stator bar bounce forces or bar vibration.

Since positive PD is associated with surface discharges, one could conclude, for instance, that if the magnitude of the measured positive PD pulses increased with load and there were no increases in the magnitude of the negative PD pulses, then there may be slot discharge activity caused by loose stator wedges. Similarly, if there is equal positive and negative PD activity one could conclude that there is discharge in the bulk of the groundwall insulation, and so forth.

In addition, efforts to verify the source of the PD pulses from the generator or from the external system or isolated phase bus (IPB) have also been attempted using two capacitive couplers per phase installed on the IPB, where they can be separated by a number of feet. When a PD pulse is measured on both couplers, its direction may be determined by which coupler is the first to see the pulse. Therefore, if the coupler farthest from the generator picks up the PD pulse first, then its direction is toward the generator, and vice versa. If the coupler closest to the generator is the first to see the pulse, then it had to come from the generator. There has been some success made in the directional capability of capacitive coupling, but again, noise is a problem and often masks the true PD being measured.

Interpretation of PD measurements is not an exact science. It requires a great deal of knowledge on the subject to determine the type of PD activity that a particular machine may be experiencing [9]. See Figures 5.28–5.31 for examples of PD readings obtained by capacitive coupling.

STATOR SLOT COUPLER. The stator slot coupler (SSC) is basically a tuned antenna with two ports (see Fig. 5.32). The antenna is approximately 18 inches long and is embedded in an epoxy/glass laminate with no conducting surfaces exposed. SSCs are installed under the stator wedges at the line ends of the stator winding, such that the highest voltage bars are monitored for best PD detection. Since the SSC is also installed lengthwise in the slot at the core end, its two-port characteristic gives it inherent directional capability [4,5,6].

The problem of noise is virtually eliminated in the SSC. Although the SSC has a very wide frequency response characteristic that allows it to see almost any signal present in the slot in which it is installed, it also has the characteristic of showing the true pulse shape of these signals. This gives it a distinct advantage over other methods that cannot capture the actual nature of the PD pulses. Since PD pulses occur in the 1–5 nanosecond range, and are very distinguishable with the SSC, the level of PD activity can be more closely defined.

In addition, dedicated monitoring devices have been devised to measure the PD activity detected in the SSC. The capability for PD detection using the SSC and its associated monitoring interface is enhanced to include measurement in terms of the

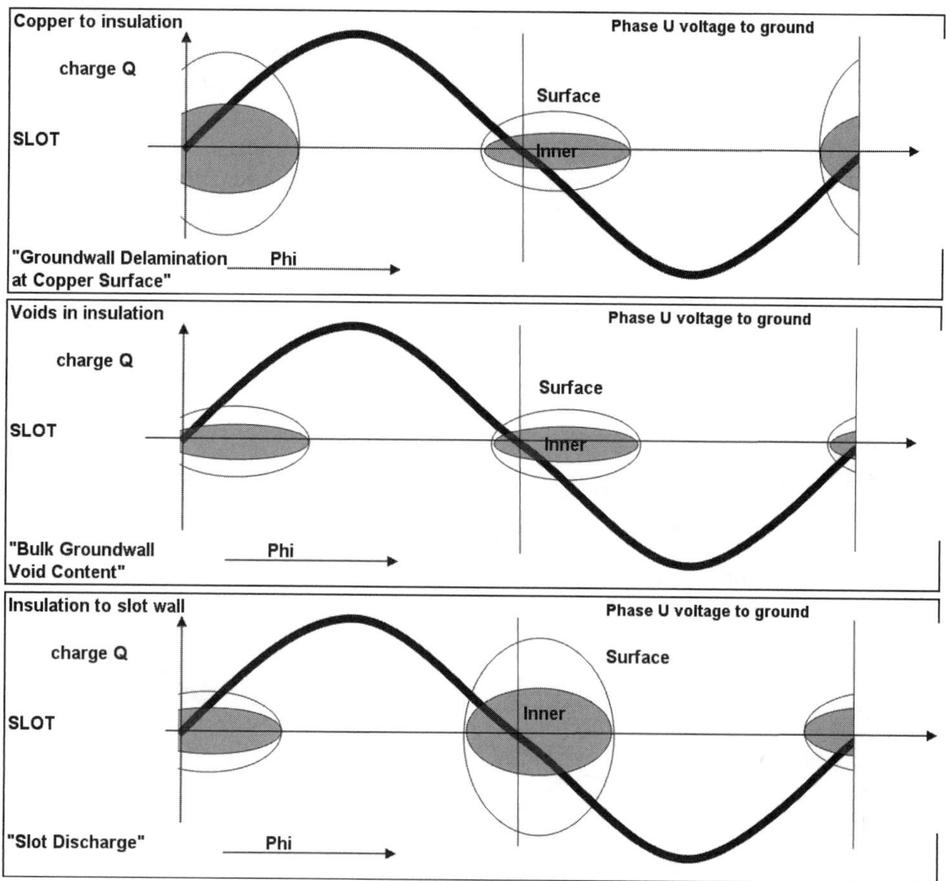

Fig. 5.28 Capacitive coupling, slot discharge patterns. (Courtesy of Alstom Power Inc.)

positive and negative characteristic of the pulses, the number of the pulses, the magnitude of the pulses, the phase relation of the pulses, and the direction of the pulses (i.e., from the slot, from the end winding, or actually under the SSC itself at the end of the slot).

The other advantage of the SSC is that once it is installed, measurements may be taken at any time without the need for exposing live portions of the generator bus work to make connections (see Figures 5.33–36).

5.5.4 Rotor

The rotor has minimal instrumentation due to its dynamic nature. It is a rotating component and it is not possible to attach instruments unless they can transmit information by radio transmission. Regardless, there are numerous ways to monitor the performance of the rotor indirectly. Rotor shaft unbalance, loose rotor parts, and loose retaining rings may show up as increased rotor vibration (amplitude or orbit changes). Degraded blower-blade performance may be detected as loss of acceptable differential fan pressure. Parameters that may be monitored include vibration, field current, field voltage, and vibration for both the exciter and main fields. Electromagnetic characteristics of the rotor can also be used to look for winding problems in the form of shorted turns.

Rotor Winding Temperature. Rotor winding temperature is generally mea-

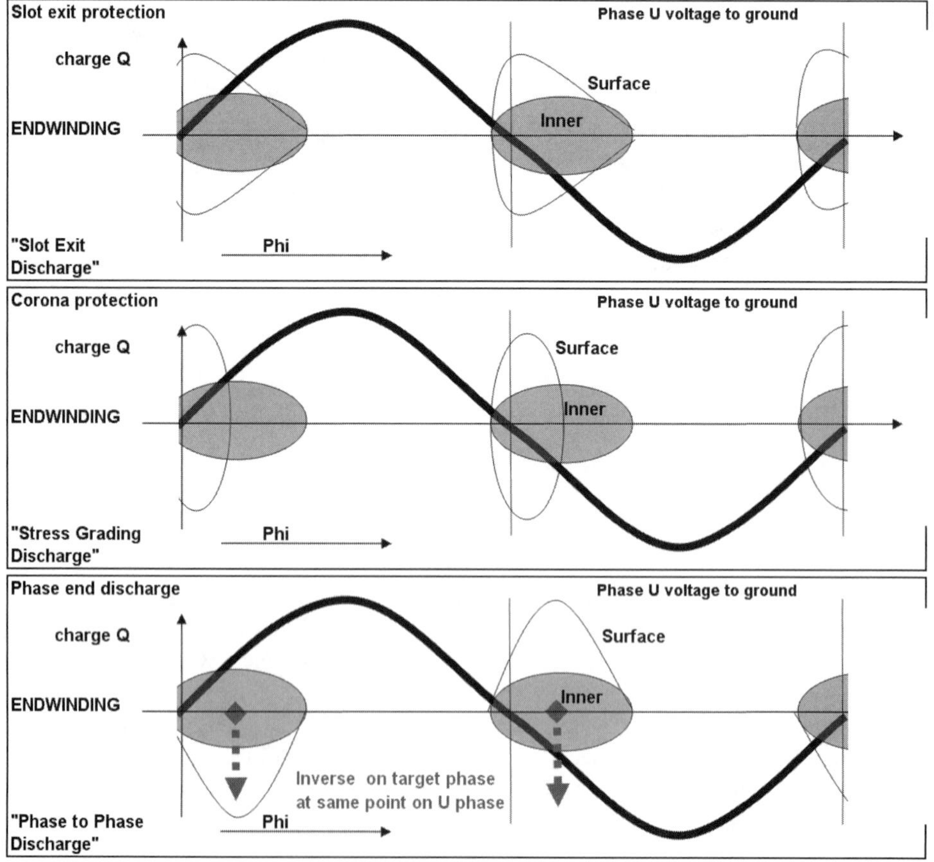

Fig. 5.29 Capacitive coupling, end-winding discharge patterns. (Courtesy of Alstom Power Inc.)

sured as a function of the winding resistance. This produces the average temperature of the winding but does not indicate the temperature of the hottest part of the winding. Locating the hot spot in the rotor winding will be dependent on knowledge of the rotor winding cooling circuit. Generally, the manufacturer can provide a profile of the temperature distribution across the axial length of the rotor winding. The difference between the hot-spot and average winding temperatures produces a multiplier factor for calculation of the hot spot. The multiplier factor is determined as shown in Figure 5.37.

With the hot-spot multiplier known, the rotor-winding hot-spot temperature can be calculated for any load and cold operating gas temperature as follows:

$$T_{hs} = (T_{ave} - T_{H_2cold}) \times M + T_{H2cold}$$

where

T_{ave} $= R_{ave}/R_{ref}(K + T_{ref}) - K$ (IEEE 115) [10]

R_{ave} $= V_f/I_f$, calculated average winding resistance from measured field voltage V_f and field current I_f

R_{ref} = winding resistance measured in factory by manufacturer at known reference temperature T_{ref}

K = 234.5 (copper constant)

T_{H_2cold} = measured value of the cold hydrogen gas temperature

M $= T_{hs}/T_{ave}$, hot-spot multiplier from manufacturer winding temperature profile data

The rotor winding temperature will vary with field current, hydrogen gas temperature, and pressure. Rotor winding high temperatures generally occur as a result of overloading or undercooling. They are caused mainly by excessive field current in the online mode, low hydrogen gas pressure, high hydrogen cold gas temperature, or ventilation problems in the rotor that block the hydrogen cooling gas from flowing to part of the winding. Excessive rotor winding temperature will cause deterioration of the rotor winding interturn and groundwall insulation. Advanced effects of the rotor winding exceeding its temperature limit may be overheating or burning of the insulation or a rotor thermal unbalance.

Reducing field current to remove the energy input to the rotor winding will allow the winding temperature to come down to a tolerable level and stabilize. Failure of the interturn insulation will cause rotor winding shorted turns. Failure of the groundwall or slot-liner insulation will cause rotor grounds. The retaining ring is also insulated from the rotor winding and will be affected by high winding temperatures.

Failure of the retaining-ring insulation also constitutes a rotor winding ground.

Rotor Winding Ground Alarm. Rotor winding grounds are actually a leakage of field current to ground. As previously stated, grounds can occur through or over the slot liner and the retaining-ring insulation. In addition, grounds can occur to the rotor forging through the insulation systems of the sliprings, the radial studs, the up-shaft

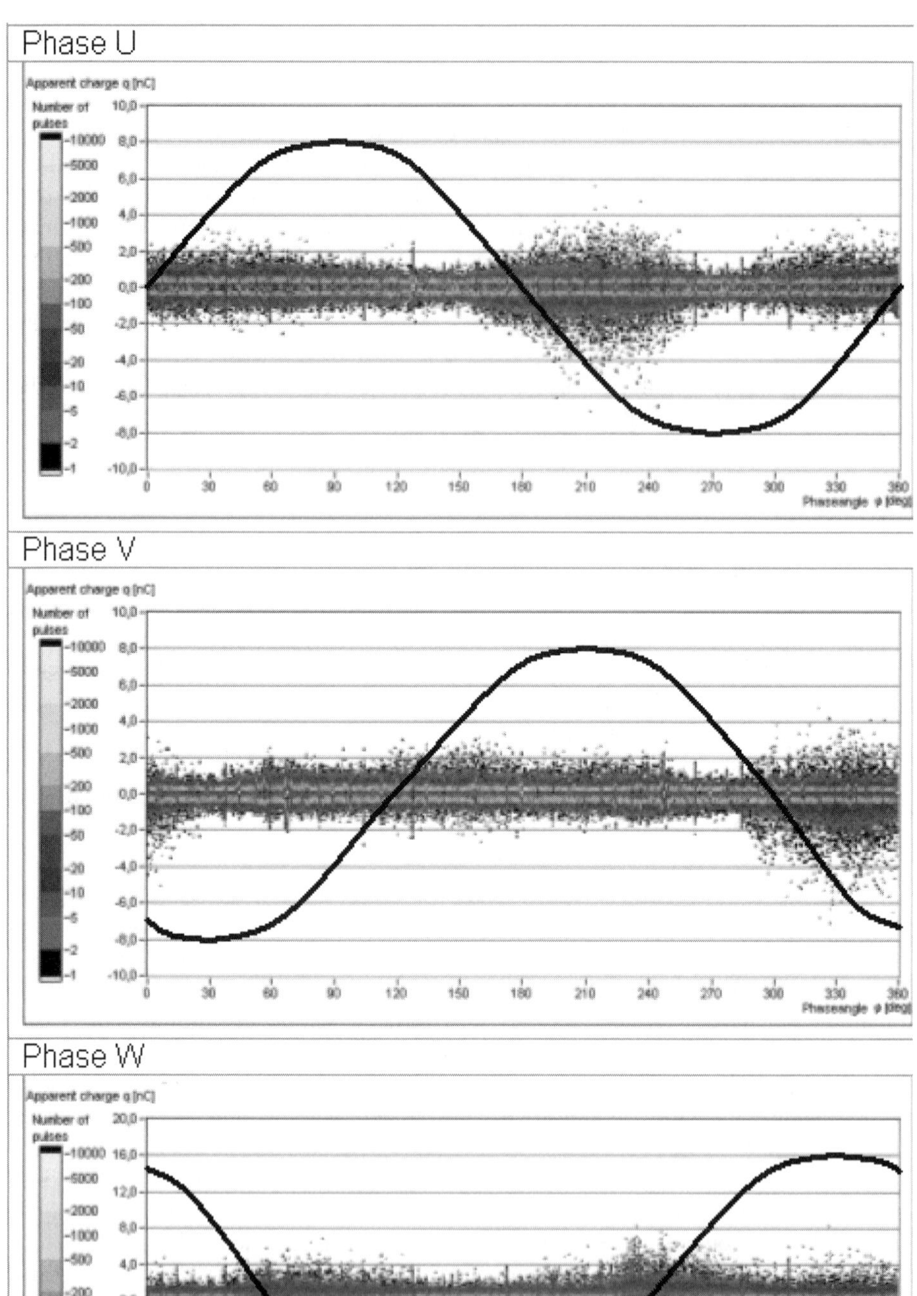

Fig. 5.30 Capacitive coupling, slot discharge example. (See color insert.)

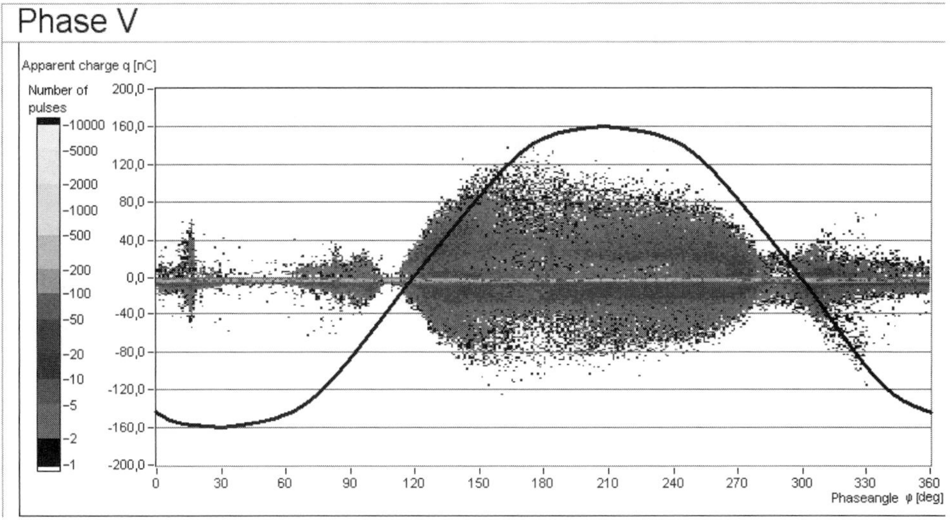

Fig. 5.31 Capacitive coupling, end-winding discharge example, no particular pattern. There is considerable discharge due to a number of mechanisms going on at the same time in this example. (See color insert.)

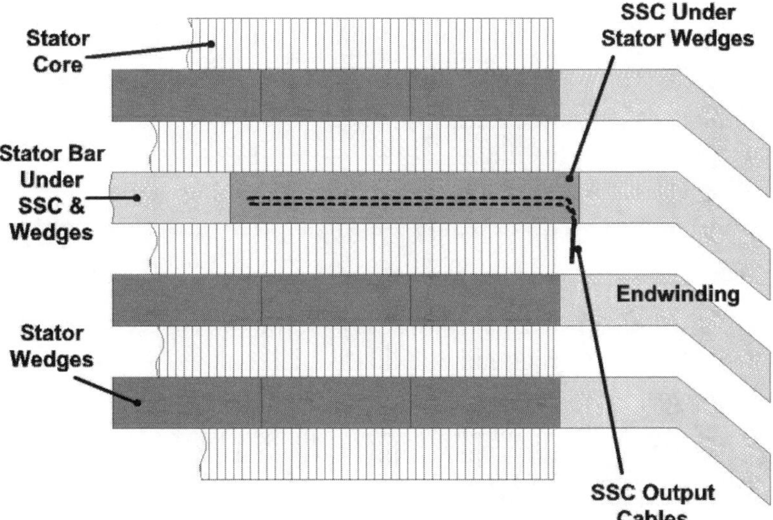

Fig. 5.32 Stator slot coupler plan view.

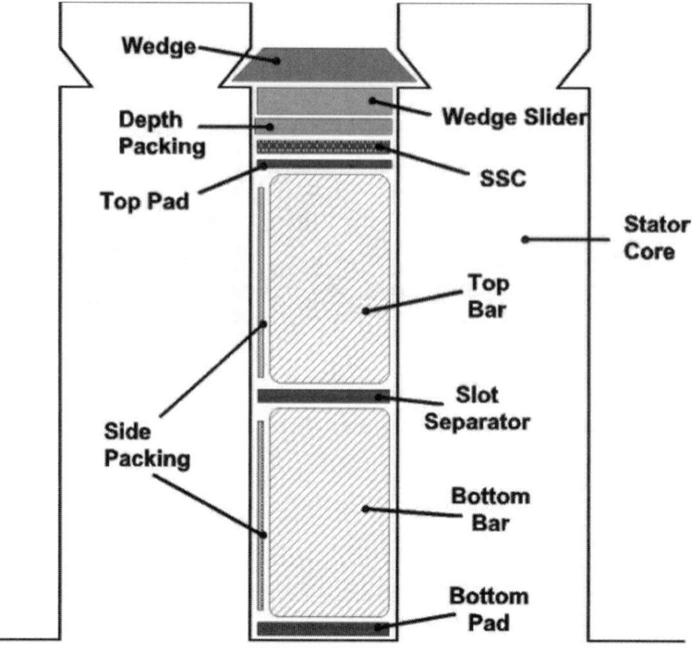

Fig. 5.33 Stator slot coupler end view.

Fig. 5.34 Installed stator slot coupler.

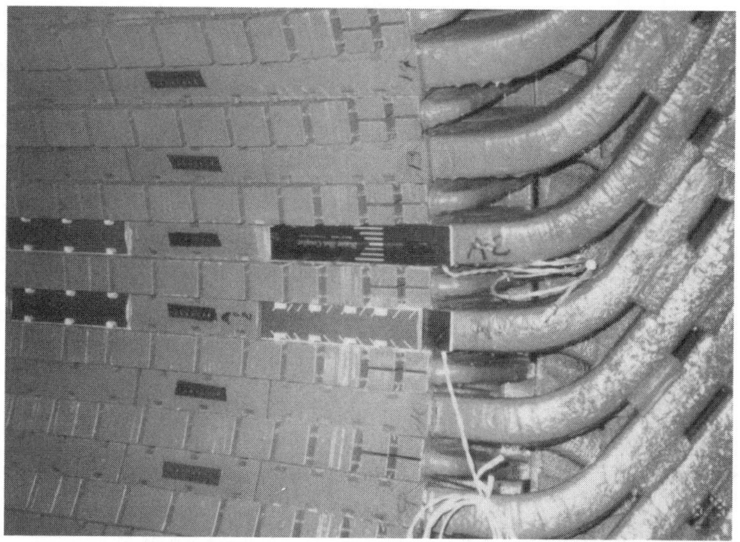

Fig. 5.35 Installed stator slot coupler, another type of stator.

leads or bore copper conductors, and the rotor winding main leads. Grounds can also occur external to the rotor itself, out in the excitation system.

Rotor winding ground protection is applied to the generators to provide warning of a ground. The rotor winding ground alarm is almost always connected as an alarm and not an actual generator trip. It is left up to the operator or suitable authority to decide whether the unit should be taken offline.

Although it is common for utilities to operate with a single rotor ground for short periods of time until a convenient outage, this practice has a high degree of inherent risk. Should a second ground occur anywhere on the rotor, very high currents will circulate through the two ground points, creating overheating in the affected rotor components. The overheating effect is extremely serious and damaging to the machine. A double rotor ground can cause a catastrophic rotor failure. (More about this topic in Chapter 6.)

The discussion so far has basically been about rotor ground monitoring as a "protection" device, but there is also now a device that will provide on- and offline detection of rotor grounds, called a "rotor earth fault resistance monitor" (see Fig. 5.38 and Fig. 5.39). Combining rotor telemetry technology and rotor ground fault measurement techniques, a method of determining the actual ground resistance has been developed. By knowing the actual ground resistance, the progression of any developing fault can be trended and alarm limits set up for operator response. This device further provides fault location diagnostics and allows monitoring of brushless-excitation machines as well.

Rotor-Winding Shorted Turns Detection. Shorted turns in rotor windings

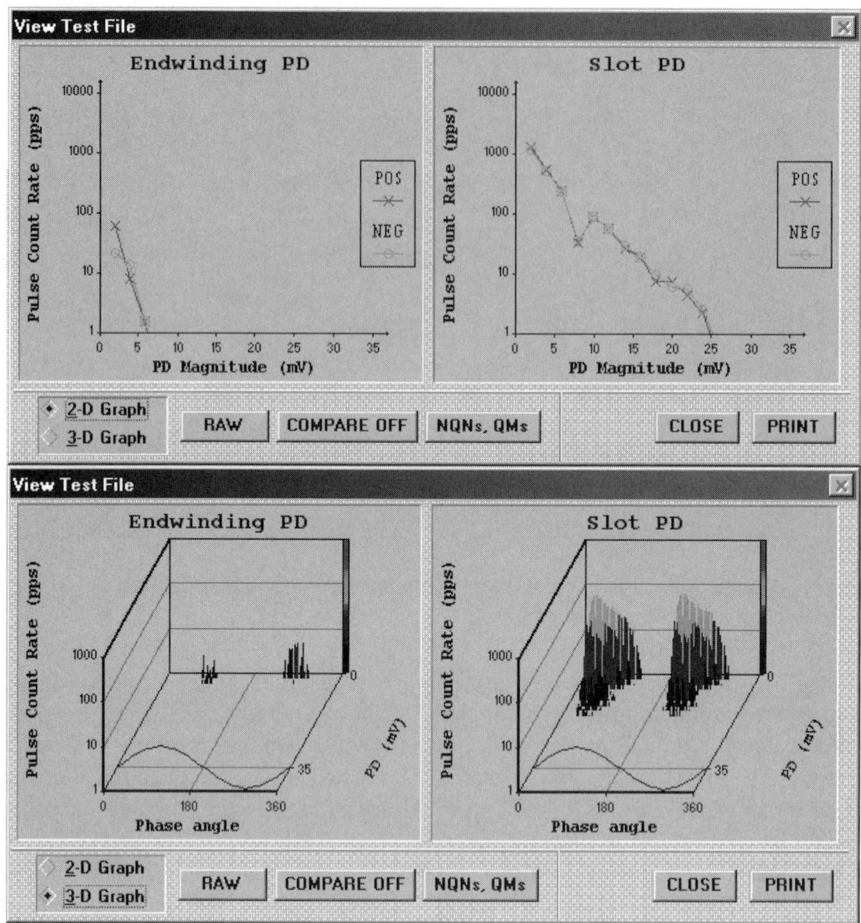

Fig. 5.36 Stator slot coupler, classical slot discharge pattern seen in measurements.

are associated with turn-to-turn shorts on the copper winding as opposed to turn-to-ground faults. Rotor winding shorted turns, or interturn shorts, can occur from an electrical breakdown of the interturn insulation, mechanical damage to the interturn insulation allowing adjacent turn-to-turn contact, or contamination in the slot, which allows leakage currents between turns.

When shorted turns occur, the total ampere-turns produced by the rotor are reduced since the effective number of turns has been reduced by the number of turns shorted. The result is an increase in required field current input to the rotor to maintain the same load point and an increase in rotor-winding temperature.

At the location of the short, there is a high probability of localized heating of the copper winding and arcing damage to the insulation between the turns. This type of

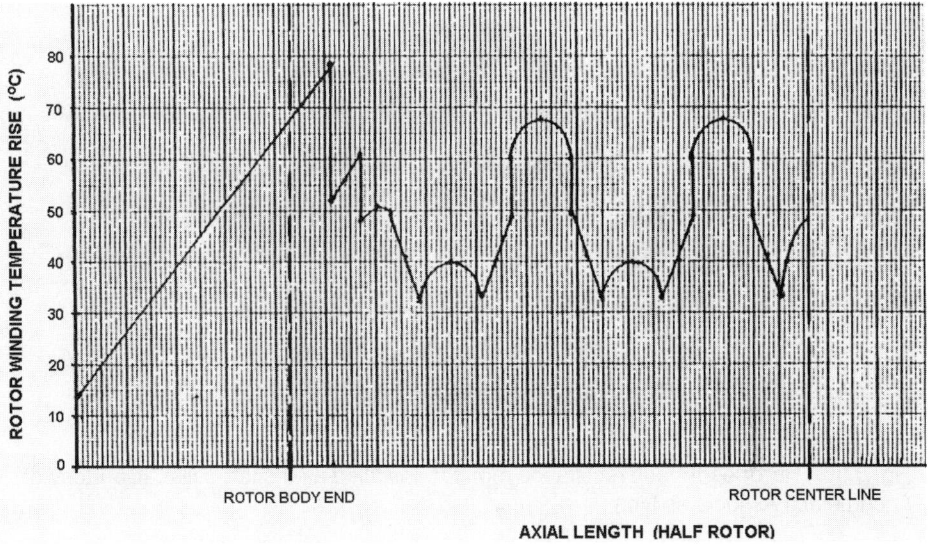

Fig. 5.37 Rotor hot-spot profile.

Fig. 5.38 Rotor earth fault resistance monitor, installed at the shaft end. (Courtesy of Accumetrics Associates Inc.)

Fig. 5.39 Rotor earth fault resistance monitor, installed as a shaft collar. (Courtesy of Accumetrics Associates Inc.)

damage can propagate and worsen the fault such that more turns are affected, or the ground-wall insulation becomes damaged and a rotor-winding ground occurs.

One of the most noticeable effects of shorted turns might be increased rotor vibration due to thermal effects. When a short on one pole of the rotor occurs, a condition of unequal heating in the rotor winding will exist between poles. The unequal heating will cause bowing of the rotor and, hence, vibration. The extent and location of the shorted turns and the heating produced will govern the magnitude of the vibrations produced.

Off-line methods for detecting shorted turns include winding impedance measurements as the rotor speed is varied from zero to rated speed, and RSO (recurrent surge oscillograph) tests based on the principle of time domain reflectometery. In addition, a short of significant magnitude may be identified by producing an open-circuit saturation curve and comparing it to the factory open-circuit saturation curve. If the field current producing the required rated terminal voltage has increased from the original design curve, then a short is likely to be present. The number of shorted turns may be identified by the ratio of the new field-current value over the design field-current value.

All these methods of identifying shorted turns are prone to error and only indicate that a short exists. They do little to help locate which slot the short is in and require special conditions for collecting data or for testing. To better identify shorted turns, and to employ a method that works online, the shorted turns detector (STD) (also called rotor flux monitor or RFM) search coil method has been perfected. Each manufacturer has its own version of the STD, but all work essentially in the same manner.

The STD is actually a search coil (see Fig. 5.40) mounted on the stator core by various methods but located strategically in the airgap. The search coil looks at the variation in magnetic field produced in the airgap by the rotor as it spins. The energized rotor winding and the slotted effect of the winding arc cause a sinusoidal-like signal to be produced at the winding face from the winding slot leakage field of the rotor. The pole face, on the other hand, has no winding, and the signal is more flat since the variation in magnetic field is minimal.

The magnitude of the sinusoidal peaks in the winding face is dependent on the ampere-turns produced by the winding in the various slots. If there is a short in a slot, the peak of the signal for that affected slot, or amount of leakage flux, will be reduced (see Figs. 5.41 and 5.42). The amount of reduction will be dependent on the magnitude of the short. Therefore, by knowing which slot the short is in, an estimate of the number of shorted turns can be made fairly accurately (see Fig. 5.43).

Problems in saturation effects at full load can occur during the data analysis. Most manufacturers now have a dedicated monitor connected to the STD to automate the analysis process. This allows the STD and monitor to act as stand-alone sensor to alarm when a short turn is detected and notify the operator for investigation.

Because of the interference of the main magnetic field with the measurement of the rotor cross-slot or leakage flux for shorted turns detection, it is required to adjust the rotor load angle to the point of minimum interference; this is called the *zero crossing flux density* or ZCFD (see Fig. 5.44). This is merely the center of the stator and rotor main field interaction, where the flux goes from positive to negative and vice versa. At this point, there is the least amount of interference with the leakage field of the rotor slots. Generally, as an online monitoring system, the rotor angle adjustment is not done because the system demands on the unit do not usually allow significant changes to test all ZCFD points for all slots. A preplanned test is usually required to accomplish this, and so it will be discussed further on in Chapter 11, Section 11.7.12.

Fig. 5.40 Installed flux probe.

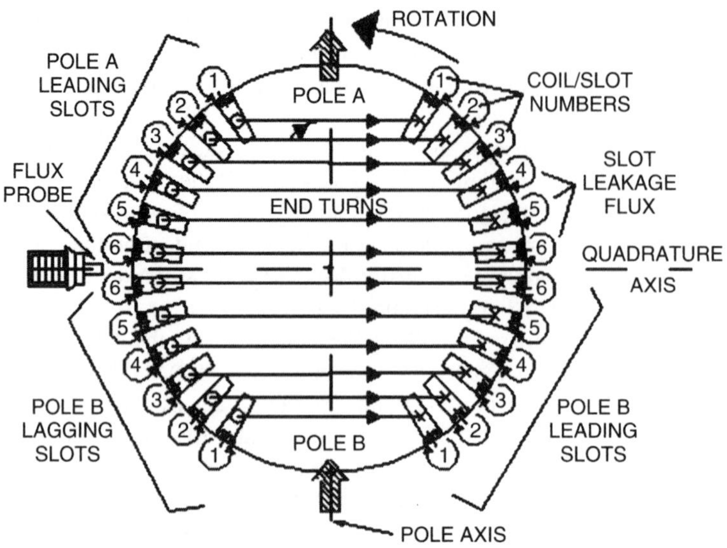

Fig. 5.41 Leakage flux pattern. (Courtesy of GeneratorTech.)

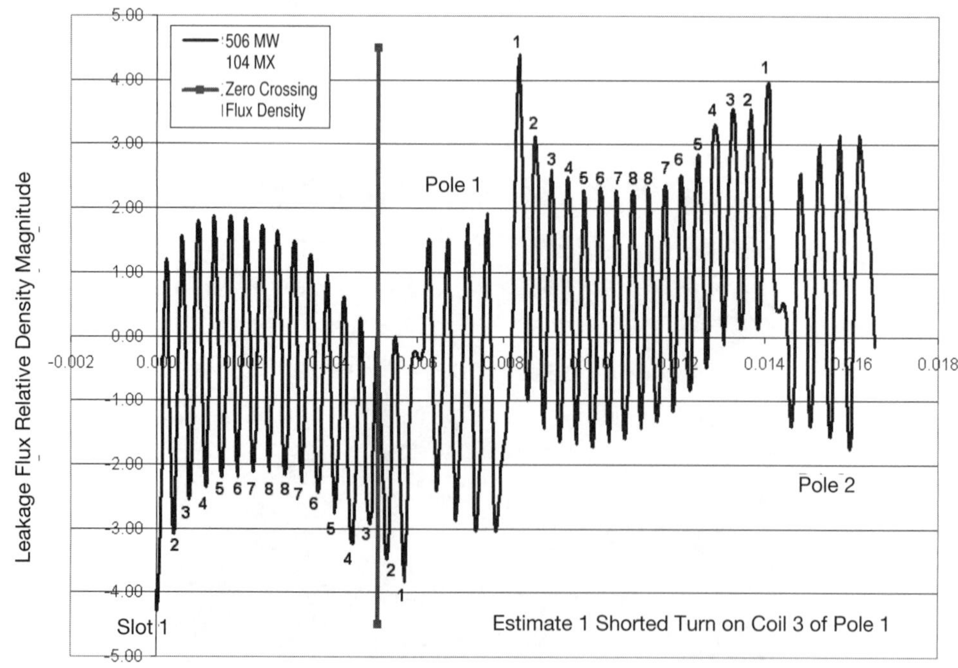

Fig. 5.42 Shorted turns plot. Shorted turn indicated on coil 3 of pole 1.

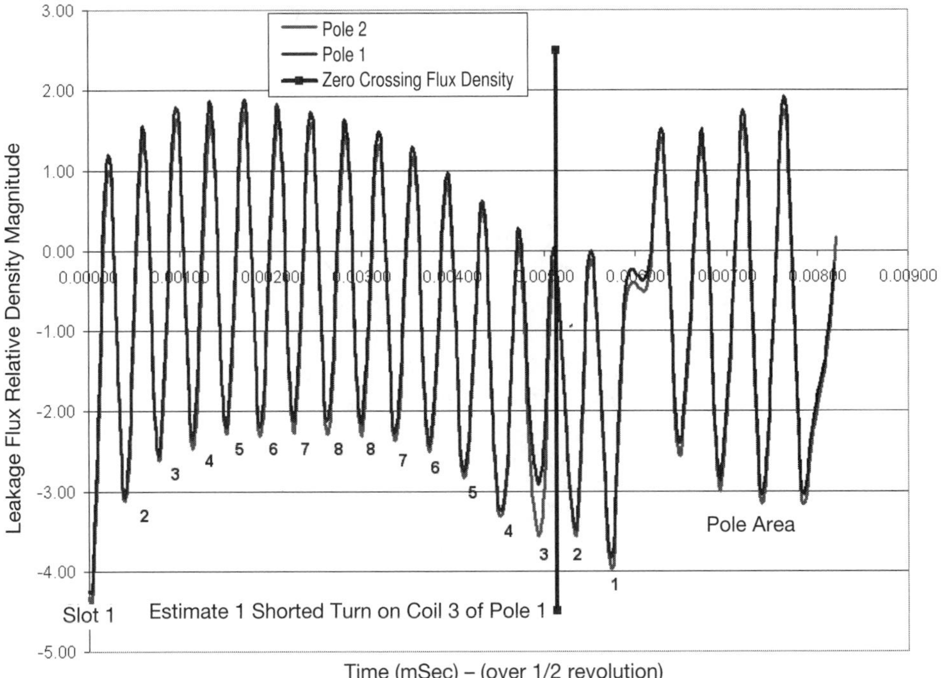

Fig. 5.43 Shorted turns. Pole 2 inverted and superimposed on pole 1 to show the magnitude of the short on coil 3 of pole 1.

Shaft Speed. Shaft speed refers to the shaft's rotational speed. In North America, this is 3600 rpm for two-pole generators and 1800 rpm for four-pole generators. For both two- and four-pole machines, this translates to 60 Hz operation of the generator. For 50 Hz systems, the respective two- and four-pole speeds are 3000 rpm and 1500 rpm.

The speed is generally measured by a probe mounted next to the rotor, looking at a toothed wheel or key phasor on the rotor shaft. The speed signal is used for monitoring, but the frequency of the generator is usually taken from the electrical output of the generator in Hz.

Shaft speed monitoring is particularly useful when looking at the vibration profile of the rotor during run-up and run-down, when the generator goes through its first and second *critical speeds*. Shifts in critical speeds during run-up/run-down can be an indicator of a change in shaft stiffness due to some problem within the rotor, such as a cracked shaft.

Rotor Vibration. Rotor vibration refers to vibration monitoring on the shaft and the bearings. Shaft vibration is the movement of the shaft in relation to the bearing mounts or generator footing, where the vibration probe is normally mounted. Bearing vibration is the movement of the bearings relative to the generator footing.

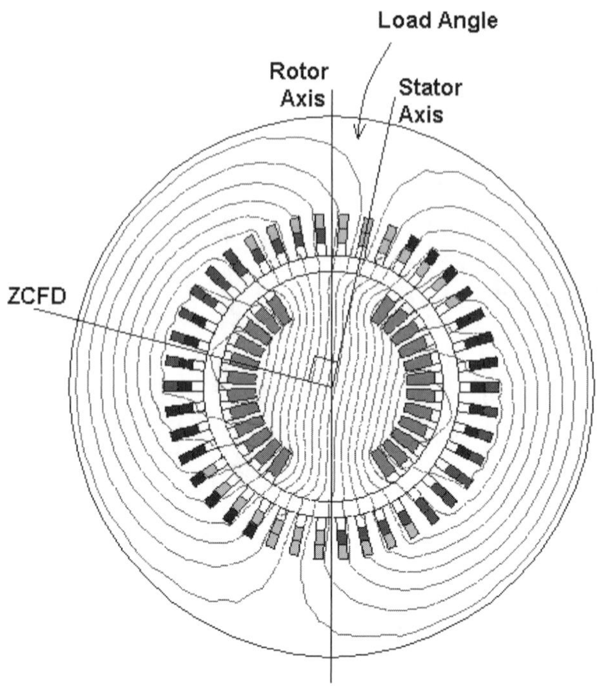

Fig. 5.44 Zero crossing flux density (ZCFD) location in the main magnetic field between stator and rotor. The ZCFD location changes with rotor angle.

Normally, vibration is measured in units of displacement (i.e., thousandths of an inch or micrometers, peak to peak on the displacement signal obtained). Vibration also can be measured as velocity or acceleration. By a simple calculation, these two can be converted to displacement.

All generators operate under strict bearing and shaft vibration guidelines and limits usually set down by the manufacturer, but these are often modified by the experienced operator.

High rotor vibrations can damage the rotor or one of its components, such as the bearings, hydrogen seals, or slip rings. High rotor vibration is caused by a number of mechanical or thermal imbalance problems.

Rotor mechanical imbalance is caused by conditions such as loss of a balance weight, oil whirl, bearing loading, coupling alignment, misaligned retaining rings, hydrogen seal or oil wiper rubs, foundation resonance, a general rotor structural component problem, cracked forging, shorted turns, and thermal effects. The level of severity is usually determined by the magnitude of vibration present, and may require an outage to correct the source of the vibration or to apply balancing weights to offset the imbalance. For mechanical vibration problems, vibration levels generally remain constant regardless of field-current changes but will vary with the shaft rotational speed and, possibly, load.

Rotor thermal imbalance is a condition in which there is uneven heating in the rotor shaft due to a number of possible influences, with the resulting effect being high rotor vibration. Some of the possible causes are rotor shorted turns, negative-sequence heating, or blocked rotor-winding ventilation ducts. Vibrations due to thermal imbalance are usually load dependent; that is, vibrations increase with increases in field current.

Bearing and shaft vibration on both ends of the generator may be monitored to detect any or all of the above abnormalities in terms of the magnitude, phase, and frequency of the vibration at variable load conditions. A frequency analysis can also be applied for detailed analysis of the vibration pattern. Seismic and proximity vibration sensors, usually two sets 90 degrees apart, at each bearing are used for monitoring. The amplitude of the vibration may be monitored by either a shaft-riding accelerometer, bearing-housing-mounted accelerometer, proximity detector, or by velocity detecting sensors (see Figs. 5.45 and 5.46).

Torsional Vibration Monitoring. The torsional monitor is a device or system used to monitor the occurrence of shaft torques, particularly those of a severe nature that can affect the remaining life of the turbo-generator shaft. By measuring the pertinent parameters and entering all the event information into a model of the turbo-generator shaft, the loss of life of the shaft can be calculated.

Torsional events are generally caused by severe system disturbances or power sys-

Fig. 5.45 Bearing *X-Y* proximity, vibration probe.

Fig. 5.46 Shaft vibration proximity probe.

tem resonant frequencies that are inadvertently stimulated. These cause the turbo-generator shaft to respond by oscillating subsynchronously on top of the shaft operating speed. The effect is to cause excessive oscillating torque in the shaft. If not dampened, the oscillation will eventually run away and cause failure of the shaft. As oscillations are damped, they decay to zero. Excess torsional stress on the turbo-generator shaft can lead to loss of life, so it needs to be accounted for. The torsional monitor is a monitoring device used for this purpose.

The parameters required to be measured are stator current, terminal voltage, and shaft speed. A power system stabilizer is used for feedback into the control system, and it helps dampen the oscillations that may be stimulated during system events.

Shaft Voltage and Current Monitoring. During operation, voltages build on the generator rotor shaft. The sources of shaft voltage have been identified as voltage from the excitation system due to unbalanced capacitive coupling, electrostatic voltage from the turbine due to charged water droplets impacting the blades, asymmetric voltage from unsymmetrical stator core stacking, and homopolar voltage from shaft magnetization. If the shaft voltage is not drained to ground, it will rise and break down the various oil films at the bearings, hydrogen seals, turning gear, thrust bearing, and so on. The result will be current discharges and electrical pitting of the critical running surfaces of these components. Mechanical failure could follow.

Inadequate grounding of the rotor shaft will allow these voltages to further build up on the rotor shaft of the generator. Inadequate grounding may be due to a problem with the shaft grounding brushes from wear (requiring replacement brushes), or a problem with the associated shaft grounding circuitry if a monitoring circuit is provided.

In addition to the shaft voltages on the shaft, circulating currents can also flow in the rotor from some of the above-mentioned influences and damage components like H_2 seals and bearings. Therefore, they are insulated to resist this as well as discharge to ground.

High shaft voltages are caused by severe local core faults of large magnitude, which impress voltages back on the shaft from long shorts across the core. Protection against shaft voltage buildup and current discharges is provided in the form of a shaft-grounding device, generally located on the turbine end of the generator rotor shaft. The grounding device consists of a carbon brush(es) or copper braid(s), with one end riding on the rotor shaft and the other connected to ground (see Fig. 5.47).

Shaft voltage and current monitoring schemes are provided, in many cases, to detect the actual shaft voltage level and current flow through the shaft grounding brushes. This has the advantage of providing warning when the shaft-grounding system is no longer functioning properly and requires maintenance. There are numerous monitoring schemes available, and most manufacturers provide their own system with the turbo-generator sets they market. For older machines with only grounding and no monitoring, a monitoring system can usually be retrofited to the existing ground brushes. Figure 5.48 is an example of a simple shaft-grounding circuit that can be retrofited

Fig. 5.47 Shaft grounding brushes.

to a generator, provided there are dual shaft-grounding brushes. The circuit is tunable to match the characteristics of the generator shaft-grounding system and the existing level of operating shaft voltage. However, the manufacturer should be consulted when upgrading the shaft monitoring or attempting to adapt such circuits as in the example given in Figure 5.48. (See also Chapter 6 for a discussion of this topic in terms of generator protection.)

In the circuit of the example shaft-grounding system in Figure 5.48, two brushes are used so that they may be connected in series through the transformer and the rotor shaft. The transformer is used to drop the voltage down and apply a constant potential of 4 volts ac across the brushes. In this way, if there is a problem with the contact of the either of the brushes on the rotor shaft, the circuit will be broken and a drop in current through the CT will be picked up to bring in the *brush problem alarm*. The components are chosen such that they can handle up to 2 amperes flowing in the circuit under all conditions. When a problem with the brushes does occur, the drop in current can be monitored by a small current meter while it triggers the relay to tell the operator that there is a problem with the grounding circuit. It may simply be that the brushes are worn out, fouled, or hung up, and require maintenance or replacement. This is why the circuit is designed as active, so that there is constant monitoring.

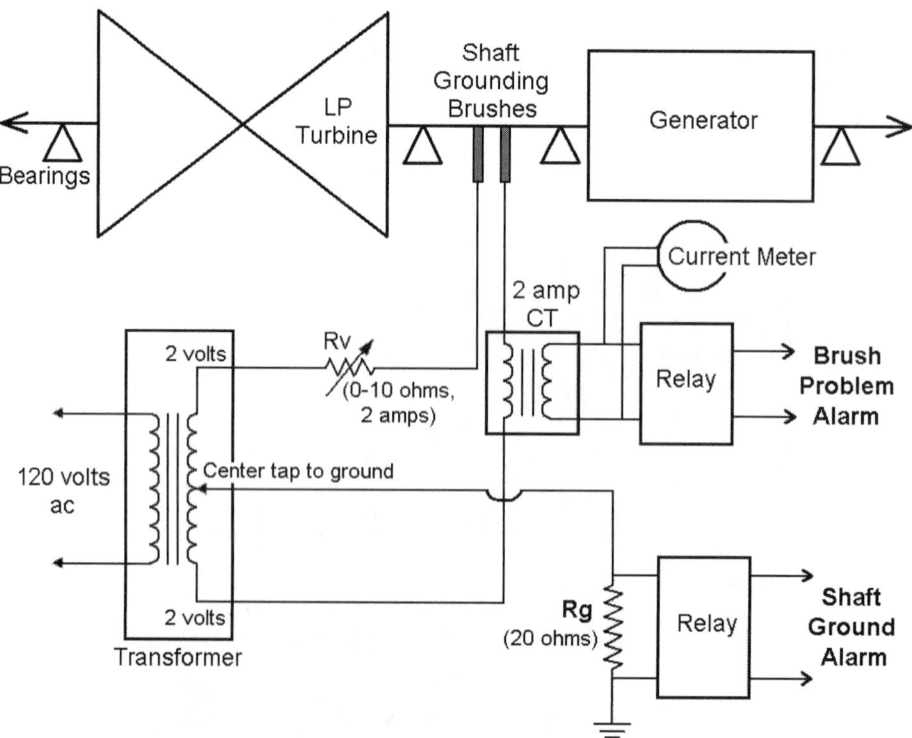

Fig. 5.48 Example of an active shaft-grounding system.

The actual grounding of the shaft is done by both brushes through the center tap of the transformer and a resistor, Rg (see Fig. 5.48), to ground. From the circuit, it can be seen that even if one brush is not working, the other will still ground the shaft through the center tap of the transformer and keep the shaft ground alarm operable. The variable resistor, Rv, and the ground resistor, Rg, are used to tune the circuit such that a consistent drain current is present for monitoring purposes and so both brushes can respond to the *shaft ground alarm*. The resistor, Rg, must be low enough in value so that the grounding circuit does its job and shaft currents can actually drain to ground under normal operating conditions, but it must also have some impedance to help limit currents through the failed component during shaft-ground events. If a ground occurs anywhere on the shaft, current will flow through Rg and the shaft-grounding brushes to complete the circuit. This will trigger the shaft ground alarm by a voltage drop across the resistor Rg, which is sensed by the relay attached. It is possible that if a suitable relay were found, it could suffice as both the drain resistor Rg and the relay.

Note that these voltage and current values are not necessarily the ones that should be used on every machine. These are the values that work on one particular group of machines with a characteristic shaft voltage between 10 and 20 volts. The basic sizing of the components shown should work for most machines, but tuning is required to make the circuit work properly. In addition, the grounding brushes associated with these values are a carbon/copper composite. Adjustments also have to be made when a different brush material is used, due to the different resistance of the brush material. In addition, the circuit depends on the type of relays and components available. This is a basic design that works but needs to be specifically applied to every different machine. Therefore, it requires some testing and experimentation to make it work on any particular machine.

Bearing Metal Temperatures. Rotor bearing metal temperature is a function of the lube-oil inlet temperature to the bearings. In addition, bearing temperatures are affected by vibration, alignment, oil condition, bearing preloads, electrical arcing between the rotor shaft and the bearings, and low lube-oil flow to the bearings. High bearing metal temperatures can cause high bearing-oil outlet temperatures at the affected bearing. Excessive bearing temperatures can result in an overheated bearing and a subsequent rub or wipe.

The bearing metal temperatures of the upper and lower bearing halves, or of the individual pads in a tilting pad bearing, are monitored by thermocouples embedded in the material of the bearing. One or two sensors are typically located in the bearing's lower half, very close to, but not in, the babbitt metal. Because the bearing is generally insulated, the sensor must also be insulated from the bearing to prevent shorting the insulation.

Bearing Inlet Oil Temperature. High inlet-oil temperature will cause the bearing temperatures to rise and thus affect their operation. High inlet-oil temperature is almost always a result of poor operation of the bearing lube-oil system. All generator and turbine bearings will be affected. Should the bearing-oil inlet temperature exceed its limits, investigation of the root cause should begin with the lube-oil system.

Bearing Outlet Oil Temperature. Bearing-oil outlet temperatures are indicated most commonly by a TC or RTD in the oil-outlet stream from each individual bearing. Overheating of the individual bearings is indicated by this parameter, and this should correlate with the bearing metal temperature readings.

Bearing-oil outlet temperatures can also indicate a global problem with the lube-oil system if all bearing-oil outlet temperatures are high. In this case, it is possible that the bearing inlet oil from the lube-oil system has a temperature problem from the lube-oil system. It is also possible that a leak of oil from the system has starved the bearing of oil. The bearing inlet-oil pressure should be checked in such instances to ensure that it is at the correct pressure.

Temperature increase on one bearing can only mean a reduction or loss of lubricating oil to that bearing, pitting of the babbitt metal by excessive shaft currents, bearing loading due to bearing movement, and a rub in the bearing or deterioration of the babbitt material.

Hydrogen-Seal Metal Temperatures. Hydrogen-seal metal temperatures are a function of the seal-oil inlet temperature and the seal friction losses. The hydrogen-seal temperatures are affected by mechanical problems such as seal rubs from misalignment and white-metal degradation, electrical mechanisms such as arcing, and oil-related causes such as low inlet-oil flow to the seals. High hydrogen-seal metal temperatures will cause high hydrogen-seal oil outlet temperatures of the affected seal.

Overheating of the seal metal can cause not only a failure of the seal but subsequent serious damage to other generator components. A hydrogen-seal failure is extremely dangerous since hydrogen is likely to escape from the generator casing at the location of the failure. The hydrogen escaping from the seal can ignite because of the high temperature at the failure point and also because hydrogen leaking from the pressure vessel can self-ignite. Hydrogen-seal metal temperatures are monitored by thermocouples embedded in the seal-ring material.

Hydrogen-Seal Inlet-Oil Temperature. As high inlet-oil temperature raises the hydrogen-seal temperatures, the operation of the oil supply will be affected. High inlet-oil temperature is almost always a result of poor operation of the seal-oil supply system, and will affect both generator hydrogen seals. Should the hydrogen seal-oil inlet temperature exceed its limits, investigation of the root cause should begin with the seal-oil supply system.

Hydrogen-Seal Outlet-Oil Temperature. In general, a TC or RTD in the oil-outlet stream from each individual seal indicates hydrogen-seal oil-outlet temperatures. Overheating of the individual hydrogen seals will be indicated by this parameter, and should correlate with the hydrogen-seal metal temperature readings.

Hydrogen-seal oil-outlet temperatures can also indicate a global problem with the seal-oil supply system if all hydrogen seal-oil outlet temperatures are high. In this case, it is possible that the inlet oil has a temperature problem from the seal-oil supply system.

Hydrogen Seal, Hydrogen/Seal-Oil Differential Pressure. A high degree of hydrogen loss from the generator can cause the hydrogen/seal-oil differential pressure to increase. Similarly, seal-oil inlet pressure to the generator being too high will cause the differential pressure to increase. One of the problems of high hydrogen/seal-oil differential is excess seal oil leaking into the generator. This can overload the drainage system and cause oil to mist and coat the generator's internal components. The problem is that oil has a tendency to promote loosening of components such as the stator wedges.

The differential pressure between the oil inlet to the hydrogen seals and the casing hydrogen gas pressure is continuously monitored. It is always maintained in the range of 5 to 15 psi by a regulating valve to control the differential pressure, with the seal-oil pressure being the higher of the two. The higher seal-oil pressure keeps the hydrogen from escaping the generator casing.

Loss of the hydrogen seal oil at either end of the generator will cause hydrogen leakage at the seal, creating a dangerous condition for both personnel and the internal generator components. When hydrogen under pressure is allowed to leak uncontrolled, it has the capacity to self-ignite. When it does, the flame is invisible to the eye. The only clues to the presence of a hydrogen fire are the emitted heat and the formation of water droplets, since the escaping hydrogen combines with the oxygen in air to make water. The other problem is the reduction in cooling of the generator components due to the reduction in hydrogen pressure.

Rotor Fan Differential Pressure. Rotor fan differential pressure is an indication of the operation of the hydrogen cooling circuit inside the generator. The fan differential pressure is the pressure required to overcome the pressure drop across the entire hydrogen gas cooling circuit inside the generator, such that there is sufficient flow of hydrogen to cool the generator components.

A significant increase in the fan differential pressure is an indication that there may be a flow restriction somewhere in the cooling circuit. A low fan differential may indicate a problem with the fan itself.

5.5.5 Excitation System

ac Power into Exciter. The ac power into the exciter is the power consumed by the excitation system. In static excitation systems, it is the power delivered by the ac supply transformer to the excitation cubicles, and in rotating excitation systems it is the input to the main rotating exciter. For any particular load, the ac power should be at a certain level. A check can be made of the consumption to see if it is in the correct range for the load output.

dc Power out of Exciter. The dc power out of the exciter is the power delivered to the rotor winding after rectification. The difference between the ac power in and dc power out should only be the normal losses. A large differential between the two is an indicator of a possible excitation system fault.

Whether static or rotating, the main exciter can be considered the power stage of the excitation system. When the power stage of the exciter has failed, it may result in a partial or complete loss of excitation to the generator. For a static excitation system, this would include the thyristor bank or transformer. For a rotating exciter, this would include problems with the diode wheel or rotating alternator.

Main Exciter, Cooling Air Inlet Temperature. The cooling air temperature into the main excitation system, whether a rotating exciter or a static exciter, is generally the ambient temperature of the powerhouse. Cooling air is usually taken from the powerhouse, filtered, and directed over the excitation equipment for cooling. It is then discharged back into the powerhouse.

The cooling air inlet temperature will vary with the seasons as the powerhouse general atmosphere temperature changes. Therefore, it is more likely to have temperature problems relating to the cooling air supply for the exciters in the summer.

Main Exciter, Cooling Air Outlet Temperature. The cooling air temperature out of the main exciter (i.e., static or rotating) is an indicator of overheating of the main excitation components. It will also vary on the cooling air inlet temperature. Therefore, the temperature rise of the outlet cooling air above the inlet cooling air of the main exciter should be monitored primarily.

Collectors and Brush Gear, Cooling Air Inlet Temperature. The collector and brush gear inlet air is generally the air from the powerhouse turbine hall at its ambient value. It is filtered and directed over the collectors and brush gear for cooling, and discharged back into the turbine hall.

Collectors and Brush Gear, Cooling Air Outlet Temperature. The cooling air outlet temperature from the collectors and brush gear is an indicator of overheating of these components, since the collector ring temperatures cannot be monitored directly. Overheating is generally caused by current overloading or improper operation of the brush gear.

The outlet air temperature will also vary with changes in the cooling air inlet temperature. Therefore, the temperature rise of the outlet cooling air above the inlet cooling air should be monitored primarily.

Collectors and Brush Gear, Cooling Air Filter Differential Pressure. The collectors and brush gear air filter being clogged or dirty will cause reduced cooling airflow to the collectors and brush gear. This can also cause above-normal temperatures and poor operation of the collectors and brush gear. The pressure drop across the filters is monitored to detect clogging.

5.5.6 Hydrogen Cooling System

Hydrogen is provided in large generators for improved cooling of the internal generator components. Hydrogen is the primary coolant for the stator core and frame, the rotor, and the stator winding in generators with directly cooled stator windings. Hydrogen coolers and a cooling system are provided to remove the heat absorbed by the hydrogen gas.

Bulk Hydrogen Supply Pressure. Hydrogen supply pressure is monitored to indicate if there is a problem with the external hydrogen system pressure controls, or simply if the system's external hydrogen supply is low and requires refreshing. Whatever the reason, the effect of low external hydrogen supply pressure is no available hydrogen makeup because the generator casing loses gas pressure.

Generator Casing Hydrogen Gas Pressure. Hydrogen cooling effectiveness of the generator internal components is proportional to the hydrogen gas pressure. Therefore, the capability curve limits are related to the hydrogen pressure of the generator, and the power output capability of the generator will be affected if the proper pressure is not maintained (see Chapter 4 for an exhaustive treatment of this topic).

The design hydrogen pressure in the generator must be maintained at all times when the generator is operating and there is a regulating valve generally employed in the system to do this. Otherwise, the temperature limits of the various components cooled by hydrogen could be encroached upon. At full load, the hydrogen pressure should be maintained at rated design pressure. At reduced load, the generator is more efficient if the pressure is just slightly higher than needed as determined by the capability curves. A regulator maintains the proper pressure for the purpose of reducing the hydrogen supply pressure to the required operating pressure of the generator.

Low hydrogen gas pressure will result in a temperature rise of all the internal generator components (stator winding, rotor, core, etc.) and eventually cause overheating of the same components if the condition becomes severe enough.

The dielectric strength of the hydrogen gas is also proportional to hydrogen pressure. Therefore, reducing the hydrogen pressure below its rated value will also decrease the insulating capability of the hydrogen gas as well as its cooling capacity.

Low generator casing gas pressure can result if there is a leak of hydrogen gas from the generator. This can occur at casing joints such as the end doors, or by a leak into the stator cooling water, hydrogen cooler service water, and/or into the seal oil. A small amount of hydrogen will continually permeate through any Teflon hoses in the stator water-cooling circuit and provide a slow source of hydrogen loss into the stator cooling water.

There is always some continual loss of hydrogen gas that is made up by the external hydrogen system. Nonetheless, if the external supply is low, the casing pressure cannot be increased as needed.

It is also important to maintain the correct hydrogen pressure, such that it is always lower than the seal-oil pressure. This ensures that hydrogen does not escape from the hydrogen seals.

The casing hydrogen pressure is also maintained above the stator cooling water pressure (usually by about 5 psi above the stator cooling water inlet pressure) and the raw service water pressure of the hydrogen coolers. These systems are designed to handle hydrogen ingress into the water and discharge it safely to atmosphere. It is important to keep the hydrogen pressure higher than the water pressure in these systems so that water does not leak into the generator and create a possible failure mechanism or a high hydrogen dew-point condition.

The hydrogen pressure inside the generator is generally displayed in the control room. In addition, a low-pressure alarm is usually provided and set a few pounds below the normal operating pressure. The existence of a low-pressure alarm indicates that a hydrogen leak exists.

Cold Hydrogen Gas Temperature. The cold hydrogen gas temperature is the ambient operating temperature of the generator, and is controlled by hydrogen cooler operation. It should be maintained at its operating set point, which for most generators is usually between 30 and 40°C. The maximum allowable cold gas temperature by IEEE/ANSI standards is 46°C [5].

The operation of the hydrogen coolers is usually either a once-through system with temperature control throttled by varying the raw water flow to the coolers, or by use of raw water recirculation with temperature control carried out by a temperature control valve, varying the warm raw water recirculation level. Normally the temperature of the cold hydrogen gas supplied by each gas cooler is typically balanced to within 2°C.

The cold gas temperature is measured by RTDs or TCs located in the gas flow path at the discharge of the gas coolers. One sensor is usually provided for each cooler. A temperature indication and high alarm is also provided.

Hot Hydrogen Gas Temperature. The hot hydrogen gas temperature is a measure of the heat absorbed by the cold hydrogen gas from cooling of the generator components. The hot gas temperature should always be maintained with design limits set down by the manufacturer.

In addition to a problem with one or more of the generator components causing the hydrogen gas to become excessively warm, problems with the hydrogen cooling system will also cause high hydrogen gas temperature. This can result from reduced efficiency of the hydrogen coolers, by either high raw water inlet temperature or low raw water flow to the coolers. Excessive hot hydrogen gas temperatures can indicate abnormalities such as low purity or low hydrogen pressure in hydrogen-cooled machines.

The hot gas temperature is usually measured by RTDs or TCs located in the gas flow path at the inlet to each of the hydrogen coolers. A temperature indication and high alarm are also provided.

To measure the hot hydrogen gas temperature directly, generators that have direct hydrogen-cooled stator windings generally have a proportion of the stator conductors with RTDs or TCs at the outlet end of the bars.

Hydrogen Dew-Point Temperature. Hydrogen dew-point temperature is an indicator of the moisture content in the generator casing hydrogen gas. Moisture is undesirable for the stator and rotor insulation systems, since it can initiate insulation failure by electrical tracking, and for various steel components in the generator due to rusting and corrosive effects. Hydrogen dew point may be monitored on a continuous basis by a dew-point indicator and should be maintained at a value much lower than the expected cooling water temperature. It is recommended that the dew point be maintained at less than 0°C [5]. An alarm can be provided if the dew point rises above this set point.

High moisture content in the hydrogen gas is usually due to hydrogen dryer malfunction, contaminated hydrogen supply, poor seal-oil quality, or a malfunction in the seal-oil system on the vacuum treatment plant. Stator cooling water or hydrogen cooler raw water leakage into the generator can also cause high dew point, this is less common.

Moisture is most likely to collect in the generator during shutdown periods, when the internal components begin to cool and the dew-point temperature rises.

Hydrogen Purity. To ensure high cooling efficiency, Hydrogen purity is generally maintained above 98%. If the purity drops too low, an explosive mixture can occur below 74% and above 4%.

A gas analyzer is usually provided to monitor the concentration of hydrogen gas in the generator against CO_2. It is measured against CO_2 because it is CO_2 that is used to purge hydrogen from the generator.

During operation, the hydrogen in the generator is continuously passed through the analyzer. The output from the analyzer may be displayed as hydrogen purity in the control room. If the purity or concentration falls below the manufacturer's recommendation, an alarm can be initiated.

Low hydrogen purity results in increased windage losses and lower efficiency (see Fig. 5.49). Monitoring of purity is extremely important during purging operations for safety reasons, to ensure that the explosive range is not encroached upon.

Hydrogen Makeup or Leakage Rate. The hydrogen makeup or leakage rate is monitored to indicate a loss of hydrogen somewhere in the generator. Hydrogen can leak into the stator cooling water and result in hydrogen gas locking in the stator winding, or leak into the hydrogen coolers. It can also be entrained in the seal oil at an excessive rate or leak from the generator casing at joints or connections. Hydrogen leakage can also occur due to a problematic hydrogen seal, which has dangerous implications due to the temperature of the seal.

The allowable daily leakage is always specified by the manufacturer. However, most generators should be capable of maintaining a total daily leakage rate under 500 cubic feet (14 m^3) per day, with an alarm set point of 1500 cubic feet (42.5 m^3) per day. A gas-totalizing flowmeter on the inlet (makeup) line is usually provided to indicate the total hydrogen consumption of the generator.

Hydrogen gas leakage can also occur, for example, into the stator cooling water system, through the stator bar hollow strands, water hoses, manifolds, O-rings, and terminals. Excessive leakage of hydrogen into stator cooling water can lead to gas locking of the cooling path. This is caused by a gas bubble blocking the stator cooling wa-

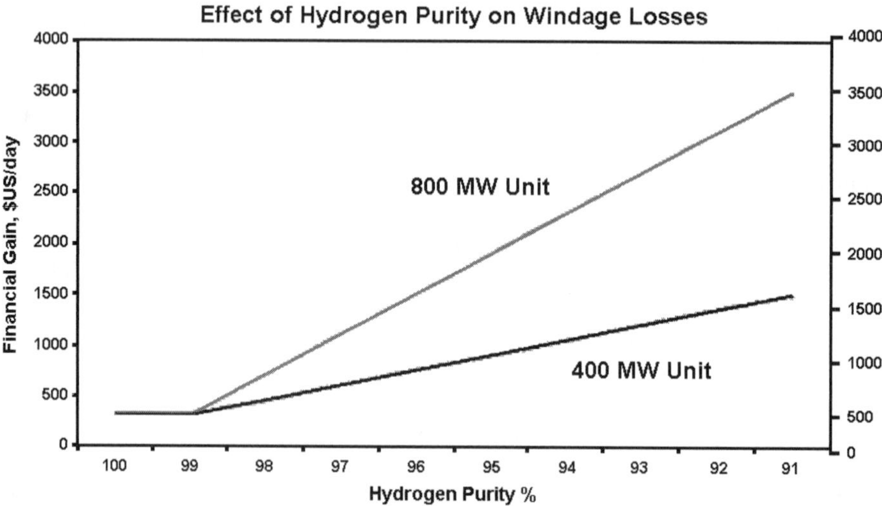

Fig 5.49 This graph shows how a decrease in hydrogen purity affects the generator efficiency by increasing the windage losses (after the Proton Co.).

ter flow and may result in overheating in the stator winding. Venting is often provided for releasing any hydrogen introduced into the stator cooling water system. Also, the amount of vented hydrogen is sometimes measured by a dedicated system for monitoring the leak rate of hydrogen into stator cooling water.

Another source of hydrogen leakage can be the hydrogen coolers. It is not generally possible to monitor the leak rate, but it can be detected by high hydrogen makeup or overall consumption. Hydrogen gas can also leak into the seal-oil system where the hydrogen comes in contact with the seal oil. The hydrogen entrained in the oil is generally not significant in comparison to the other possibilities.

Raw Service Water Inlet Pressure to Hydrogen Coolers. The service water inlet pressure to the hydrogen coolers is monitored to ensure continuous flow of raw cooling water to the hydrogen coolers, and to ensure proper heat removal from the hydrogen gas. Low raw water inlet pressure to the hydrogen coolers is an indication of low flow to the coolers which will affect the cooling of the hydrogen gas. Low service water flow to the hydrogen coolers may be caused by either a failure of the supply pump or a large leak before the generator. Low pressure/flow to the hydrogen coolers will result in reduced cooler efficiency and cause the cold hydrogen gas temperature to rise.

Raw Service Water Inlet Temperature to Hydrogen Coolers. Raw water from a local lake or river is circulated through the hydrogen cooler tubes on one side of the heat exchanger to remove the heat from the hydrogen flowing over the cooler tubes on the other side. The service water inlet temperature to the hydrogen coolers is monitored since it is important in maintaining the correct cold hydrogen gas temperature. It is also monitored for comparison to the outlet water temperature and subsequent determination of the differential temperature. This allows an estimation of the cooler efficiency.

High service water inlet temperature to the hydrogen coolers usually occurs in the summer months. If the temperature of the inlet service water is too high, it can cause a hydrogen cooler reduced-efficiency problem. This will result in the temperature of the cold hydrogen gas rising above its normal set point.

Most generators use a "once-through" raw water system. Nonetheless, hydrogen coolers with a warm service water recirculation system are not uncommon. These provide somewhat more control of the inlet service water temperature to the coolers, but are more useful in times when the service water is too cold.

Raw Service Water Outlet Temperature from Hydrogen Coolers. The raw service water outlet temperature is monitored to assist in detecting problems with the hydrogen coolers. The service water outlet temperature should be proportional to the inlet service water temperature and the generator loading.

Low service water flow to the hydrogen coolers may cause high outlet water temperature. The low-flow condition can be due to silting or fouling of the cooler tubes. This will result in reduced hydrogen cooler efficiency and cause the temperature of the cold hydrogen gas and, subsequently, the generator internal components to rise.

5.5.7 Lube-Oil System

Lubricating oil is required for the generator bearings. Oil flow, temperatures, and pressures are monitored to ensure proper performance of the bearings.

Lube-Oil Cooler Inlet Temperature (Bearing Outlets). Heat exchangers are provided for heat removal from the lube oil. Raw water from a local lake or river is circulated on one side of the cooler to remove the heat from the lube oil circulating on the other side of it.

The inlet water temperature is monitored for comparison to the outlet water temperature and subsequent determination of the differential temperature. This allows an estimation of the cooler efficiency.

Lube-Oil Cooler Outlet Temperature (Bearing Inlets). The temperature of the rotor bearings will vary with the inlet temperature of the lube oil. It is important to keep the inlet lube oil temperature within recommended limits for proper operation of the bearings and to reduce the possibility of bearing overheating.

The lube-oil inlet temperature to the rotor bearings will rise if the lube-oil outlet temperature from the lube-oil system is too high. The outlet temperature will be raised by reduced lube-oil cooler efficiency, due to low service water flow to the lube-oil coolers, or due to high-service water inlet temperature to the lube-oil coolers.

Lube-Oil Pump, Oil Outlet Pressure. The lube-oil pump outlet pressure is monitored to provide warning of the loss of lube-oil flow due to a supply pump problem. A low lube-oil pump outlet pressure implies a problem or failure of the lube-oil pump, and will create a no-flow or low-flow condition of lube oil to the rotor bearings. The result will be a rise in the rotor bearing temperatures if the oil flow is not reestablished immediately.

Failure of the source of lube-oil supply is extremely serious. The unit is usually shut down as soon as possible to avoid bearing damage and possible damage to the rotor shaft.

There are often redundant lube-oil pumps for backup and, in some cases, a dc emergency pump to allow safe unit shutdown should the shaft-driven or ac pumps be unavailable.

Lube-Oil System, Oil Outlet Pressure (Bearing Inlets). The lube-oil system outlet pressure (inlet pressure to bearings) is monitored to ensure that there is lube-oil flow to the bearings. Fouled or plugged lube-oil coolers will cause a flow blockage and reduced lube-oil flow to the rotor bearings. Fouled or plugged lube-oil coolers on the lube-oil side is most likely the result of debris in the lube oil, or lube oil that is in very poor condition, containing sludge that has built up in the lube oil coolers.

A large leak in the lube-oil system before the bearings will also create a low-pressure (low-flow) problem.

Lube-Oil Filter, Differential Oil Pressure. Lube-oil filter differential pressure is used to indicate the operating condition of the filters and the overall lube-oil system. Low differential pressure on the filters is an indication of reduced lube-oil flow, whereas high differential pressure is an indication of high lube-oil flow or a plugged filter.

Fouled or plugged lube-oil filters indicate that the filter has collected an excessive amount of sludge or debris from the lube oil, and that the condition of the lube oil is in question, or that the lube-oil filter has not been cleaned or maintained for some time. A plugged lube-oil filter will cause a lube-oil flow blockage, resulting in reduced lube-oil flow to the rotor bearings.

Lube-Oil Tank Level. The lube-oil tank level is monitored to ensure that there is always a supply of lube-oil for the lube-oil pumps to draw on. Low lube-oil level in the tank implies a significant loss of lube oil in the system. A lube-oil system leak is a possible reason for a low lube-oil tank level.

Lube-Oil Flow. Lube-oil flow to the bearing is monitored to ensure proper operation of the bearings. Lube-oil flow may be indicated by a flow indicator or by low lube-oil pressure at the rotor bearings.

Low lube-oil flow to the rotor bearings can be caused by a flow blockage somewhere in the lube-oil system, a failure of the lube-oil supply pump, or a large lube-oil leak in the lube-oil system before the inlet to the bearings. Low lube-oil flow will result in the bearing temperatures increasing above their normal operating range. The design lube-oil flow should be maintained to avoid damage to the bearings.

Raw Service Water Inlet Pressure to Lube-Oil Coolers. The service water inlet pressure to the lube-oil coolers is monitored to ensure continuous flow of raw cooling water to the coolers and ensure proper heat removal from the lube oil.

Low raw water inlet pressure to the lube-oil coolers is an indication of low flow to the coolers, which will affect the cooling of the lube oil. Low service water flow to the lube-oil coolers may be caused by either a failure of the service water supply pump or a large leak before the coolers. Low pressure/flow to the lube-oil coolers will result in reduced cooler efficiency and cause the lube-oil temperature to rise.

Raw Service Water Inlet Temperature to Lube-Oil Coolers. Raw service water is circulated through the lube-oil coolers on one side of the heat exchanger, to remove the heat from the lube-oil flowing through the other side.

The service water inlet temperature to the lube-oil coolers is monitored to maintain the correct lube-oil inlet temperature to the rotor bearings. It is also monitored for comparison to the service water outlet temperature of the coolers, and subsequent determination of the differential temperature. This allows an estimation of the cooler efficiency.

High service water inlet temperature to the lube-oil coolers usually occurs in the summer months. If the temperature of the inlet service water is too high, it can cause a reduced efficiency problem in the cooler. This will result in the temperature of the lube oil rising above its normal operating level.

Raw Service Water Outlet Temperature from Lube-Oil Coolers. The raw service water outlet temperature is monitored to assist in detecting problems with the lube-oil coolers. The service water outlet temperature should be proportional to the inlet service water temperature.

Low service water flow to or in the lube-oil coolers will cause high outlet water temperature due to reduced efficiency of the coolers. This will result in the temperature of the lube oil rising above its normal operating level. The low-flow condition may originate from silting or fouling of the cooler tubes.

5.5.8 Seal-Oil System

The hydrogen seals require oil for lubrication and for actual sealing of the generator against loss of hydrogen from the casing. In most generators, the seal-oil system is divided into an *air side* and a *hydrogen side*. In some designs, there is also a *vacuum side*.

For proper operation of the seal-oil system, the oil flows, temperatures, and pressures are monitored at various points in the system to ensure correct operation of the hydrogen seals.

Seal-Oil Cooler Inlet Temperature (Hydrogen Seal Outlets). Heat exchangers are provided for heat removal from the seal oil. Raw water from a local lake or river is circulated on one side of the cooler to remove the heat from the seal oil circulating on the other side of the heat exchanger.

The inlet water temperature is monitored for comparison to the outlet water temperature, and subsequent determination of the differential temperature. This allows an estimation of the cooler efficiency.

Seal-Oil Cooler Outlet Temperature (Hydrogen Seal Inlets). The temperature of the hydrogen seals will vary with the inlet temperature of the seal oil. It is important to keep the inlet seal-oil temperature within recommended limits for proper operation of the seals, and to reduce the possibility of hydrogen seal overheating.

High seal oil inlet temperature to the rotor hydrogen seals occurs when the seal-oil outlet temperature from the seal-oil system is too high. This can be caused by reduced seal-oil cooler efficiency due to low service water flow to the seal-oil coolers, or due to high service water inlet temperature to the seal-oil coolers.

Seal-Oil Pump, Oil Outlet Pressure. The seal-oil pump outlet pressure is monitored to provide warning of a loss of seal-oil flow due to a supply pump problem, and to ensure that the pressure is maintained within limits for the hydrogen/seal-oil differential pressure.

Seal-oil pressure that is too high can cause excess seal-oil leakage into the generator, which will cause oil contamination problems with many of the internal generator components.

Low seal-oil pump outlet pressure implies a problem or failure of the seal-oil pump, and will create a no-flow or low-flow condition of seal oil to the rotor hydrogen seals. The result will be a rise in the hydrogen seal temperatures if the oil flow is not reestablished immediately.

Failure of the source of seal-oil supply is extremely serious, since it is required to maintain the generator casing hydrogen pressure. In the event of a serious failure of the hydrogen seals, the unit is usually shut down as soon as possible and purged with CO_2 to avoid a hydrogen fire, seal damage, and possible damage to the rotor shaft.

There are often redundant seal-oil pumps for backup and, in some cases, a dc emergency pump to allow safe unit shutdown should the shaft driven or ac pumps become unavailable for any reason.

Seal-Oil System, Oil Outlet Pressure (Hydrogen Seal Inlets). The seal-oil system outlet pressure (inlet pressure to the hydrogen seals) is monitored to ensure that there is oil flow to the hydrogen seals. The system outlet pressure (inlet pressure to the hydrogen seals) is also monitored to ensure that the pressure is maintained within limits for the hydrogen/seal-oil differential pressure and proper operation of the hydrogen seals.

Seal-oil pressure that is too high can cause excess seal-oil leakage into the generator, which will cause oil contamination problems with many of the internal generator components.

Reduced seal-oil flow to the hydrogen seals is an indicator of fouled or plugged seal-oil coolers. Fouled or plugged seal-oil coolers on the oil side is most likely the result of debris in the seal oil, or seal oil that is in very poor condition containing, sludge that is building up in the seal-oil coolers.

A large leak in the seal-oil system before the seals will also create a low-pressure (low-flow) problem.

Low hydrogen seal inlet pressure is dangerous if it goes below the hydrogen gas pressure in the generator casing. This will allow hydrogen leakage from the generator, creating a safety hazard.

Seal-Oil Filter, Differential Oil Pressure. Seal-oil filter differential pressure is used to indicate the operating condition of the filters and the overall seal-oil system. Low differential pressure on the filters is an indication of reduced seal-oil flow, whereas high differential pressure is an indication of high seal-oil flow or a plugged filter.

Fouled or plugged seal-oil filters indicate that the filter has collected an excessive amount of sludge or debris from the seal oil and that the condition of the seal oil is in question, or that the seal-oil filter has not been cleaned or maintained for some time. A plugged seal-oil filter will cause a seal-oil flow blockage, resulting in reduced seal-oil flow to the hydrogen seals.

Seal-Oil Tank Level. The seal-oil tank level is monitored to ensure that there is always a supply of seal oil for the seal-oil pump to draw on.

Low seal-oil level in the tank implies a significant loss of seal oil in the system. A seal-oil system leak, or excessive oil leakage into the generator, is a possible reason for a low seal-oil tank level.

Seal-Oil Flow. Seal-oil flow to the hydrogen seals is monitored to ensure proper operation of the seals. Seal-oil flow may be indicated by a flow indicator, or by low seal-oil pressure at the hydrogen seal-oil inlets.

Low oil flow to the hydrogen seals can be caused by a flow blockage somewhere in the seal-oil system, a failure of the seal-oil supply pump, or a large seal-oil leak in the seal-oil system before the inlet to the hydrogen seals. Low seal-oil flow will result in the hydrogen seal temperatures increasing above their normal operating range. Very low oil flow to the hydrogen seals can result in loss of the seal, allowing the hydrogen to escape from the generator. The design seal-oil flow should be maintained to avoid damage to the seals.

Excessive oil flow can result in oil entering the machine past the seal oil wipers, through the clearance in the wiper gap. Monitoring the seal-oil pressure and alarming when it deviates from an acceptable range can detect both low- and high-flow conditions.

Seal-Oil Vacuum Tank Pressure. The seal-oil vacuum tank pressure is maintained near vacuum to assist in removing entrapped hydrogen and moisture from the seal oil. Loss of the vacuum will cause improper operation of the seal-oil vacuum system, and may result in oil foaming in the seal-oil vacuum tank. This will allow moisture to carry over into the generator casing, and adversely affect the dew-point temperature of the hydrogen gas.

Failure of the seal-oil vacuum system will require the hydrogen dryers (if provided) to work harder to keep the dew point of the hydrogen gas at design levels. Seal oil vacuum treatment malfunction may occur from failure of the seal-oil vacuum pump or a leak in the vacuum tank.

Raw Service Water Inlet Pressure to Seal-Oil Coolers. The service water inlet pressure to the seal-oil coolers is monitored to ensure continuous flow of raw cooling water to the coolers and ensure proper heat removal from the seal oil. Low raw water inlet pressure to the seal-oil coolers is an indication of low flow to the coolers, which will affect the cooling of the seal oil. Low service water flow to the seal-oil coolers may be caused by either a failure of the service water supply pump or a large leak before the coolers. Low pressure/flow to the seal-oil coolers will result in reduced cooler efficiency and cause the seal-oil temperature to rise.

Raw Service Water Inlet Temperature to Seal-Oil Coolers. Raw service water is circulated through the seal-oil coolers on one side, to remove the heat from the seal oil flowing through the other side of the heat exchanger. The service water inlet temperature to the seal-oil coolers is monitored to maintain the correct seal-oil inlet temperature to the rotor hydrogen seals. It is also monitored for comparison to the service water outlet temperature of the coolers, and subsequent determination of the differential temperature. This allows an estimation of the cooler efficiency.

High service water inlet temperature to the seal-oil coolers usually occurs in the summer months. If the temperature of the inlet service water is too high, it can cause a reduced efficiency problem in the cooler. This will result in the temperature of the seal oil rising above its normal operating level.

Raw Service Water Outlet Temperature from Seal-Oil Coolers. The raw service water outlet temperature is monitored to assist in detecting problems with the seal-oil coolers. The service water outlet temperature should be proportional to the inlet service water temperature.

Low service water flow to or in the seal-oil coolers will cause high outlet water temperature due to reduced efficiency of the coolers. This will result in the temperature of the seal oil to rise above its normal operating level. The low-flow condition may originate from silting or fouling of the cooler tubes.

5.5.9 Stator Cooling Water System

All water-cooled stator windings require an external system to deliver cooling water to the stator winding (and other water-cooled components). The stator cooling water system is regarded as a closed system, and every major generator manufacturer has a specific philosophy regarding its design and the quality of the water used for cooling generator components. Generally, for stator windings, the water must be conditioned by the stator cooling water system to operate in conductor bars that are energized at high voltage and carrying high electrical current. Therefore, the water must be very pure and have minimal conductivity. To accomplish this, a purifier–filtering system is provided with the SCW system.

In all stator cooling water systems, the various parameters involved are monitored at different points in the system. The general parameters monitored are stator cooling water inlet to generator temperature, stator cooling water outlet from generator temperature, stator cooling water inlet pressure to generator, stator cooling water outlet pressure from generator, differential pressure across filters and strainers, system stator cooling water flow, stator cooling water conductivity, and storage and makeup-tank levels. In addition, the raw service water for the stator cooling water coolers is also generally monitored for inlet and outlet temperature of the coolers.

Stator Cooling Water Coolers—Inlet Temperature. Heat exchangers are provided for heat removal from the stator cooling water. Raw water from a local lake or river is circulated on one side of the cooler, to remove the heat from the demineralized stator cooling water circulating on the other side of the heat exchanger.

The inlet water temperature is monitored for comparison to the outlet water temperature, and subsequent determination of the differential temperature. This allows an estimation of the cooler efficiency.

Stator Cooling Water Coolers—Outlet Temperature. Generally, the stator cooling water inlet temperature is maintained below 50°C, but normally operates in the 35 to 40°C range. It is desirable to keep the stator core, winding, and hydrogen gas temperatures in the same range to minimize thermal differentials. Also, the stator cooling water inlet temperature is kept approximately 5°C higher than the hydrogen gas temperature to reduce the possibility of condensation in the machine.

High stator cooling water outlet temperature from the stator cooling water system is usually caused by reduced stator cooling water cooler efficiency, either from low service water flow to the stator cooling water coolers or from high service water inlet temperature to the stator cooling water coolers. High stator cooling water outlet temperature from the stator cooling water system will cause the stator cooling water inlet temperature to the stator winding to be too high and, subsequently, the temperature of the stator winding will rise above its normal operating range.

Stator Cooling Water Pump—Outlet Pressure. The stator cooling water pump outlet pressure is monitored to indicate that there is stator cooling water flow to the stator windings. Generally, there are two ac motor-driven pumps (one in use and the other as backup) used to deliver the cooling water to the windings. In some instances, a dc motor-driven pump is used for emergency shutdown.

A stator cooling water supply pump failure implies loss of stator cooling water supply to the generator. The loss of stator cooling water pump outlet pressure indicates a stator cooling water flow problem. This will result in a rise in temperature above the normal operating range of the stator winding and other components cooled by the stator cooling water system. The generator can usually be operated at some minimum specified load without stator cooling water, but this is generally less than 25% of rated full load (consult the generator manual for manufacturer-set limits). If the unit is operated above this minimum load, the stator winding is likely to overheat.

Stator Cooling Water System—Stator Cooling Water Outlet Pressure. The stator cooling water system outlet pressure (inlet pressure to the stator winding) is monitored to ensure that there is stator cooling water flow to the winding. Loss of stator cooling water to the stator winding will cause loss of coolant to the stator winding and a subsequent rise in the stator winding temperature. There are generally cooling water flow devices employed to control the stator cooling water pumps and the tripping/run-back of the machine.

Fouled or plugged stator cooling water coolers will cause a flow blockage and reduced stator cooling water flow to the winding. Fouled or plugged stator cooling water coolers on the stator cooling water side is most likely the result of debris in the stator cooling water or corrosion of the stator cooling water coolers due to poor stator cooling water chemistry. Backflushing is often done to remove whatever debris or other problem that is causing the flow blockage.

A large leak in the stator cooling water system before the generator will also create a low-pressure (low-flow) problem.

Stator Cooling Water Filter/Strainer—Differential Pressure. Filters and/or strainers, or a combination of both are employed for removal of debris from the stator cooling water. Strainers are generally sized to remove debris in the 20 to 50 micron range and larger, and filters are used for debris in the range of 3 microns and larger. They can be mechanical or organic-type filters and strainers. Strainers are usually mechanical devices and can be cleaned by backflushing procedures outlined in the manufacturer's manual. Filters are usually of the organic type with replaceable cartridges; the cartridges should be replaced with new ones if they are found to be dirty or fouled.

Fouled or plugged stator cooling water filters or strainers are generally caused by the collection of debris removed from the stator cooling water, to a point where the amount of debris is excessive. This is an indication that the filters/strainers may require replacement or backflushing.

It is very important to ensure that the stator cooling water filters and/or strainers are also functioning as required. Any solid particles in the stator cooling water that are not collected by the filters will make their way to the inlets of the stator conductor bars. The hollow strands of the bars represent the narrowest portion of the stator cooling water flow path, and also have numerous turns and angles in the internal flow path due to the end-winding formation of the stator bars and the Roebel twisting employed. The hollow strands become a likely location for debris to build up.

The stator cooling water filter/strainer differential pressure is used to indicate the operating condition of this equipment. Low differential pressure on the filters may indicate reduced stator cooling water flow, whereas high differential pressure may indicate high stator cooling water flow or a plugged filter.

Stator Cooling Water Flow. Stator cooling water flow to the stator winding may be measured directly by a flow measurement device, or indirectly indicated by the stator cooling water inlet pressure to the generator. Continuous cooling water flow is essential to carry away generated heat.

Flow velocities are design specific and are based on such things as heat-carrying capacity of the water, cross-sectional flow area in each bar, and corrosion effects on the copper. It is, therefore, important to have some indication that there is stator cooling water flow, especially during high-load operation.

Low stator cooling water flow may be due to a flow blockage, a large stator cooling water leak in the external stator cooling water system, or a failure of the stator cooling water supply pumps. This will result in stator cooling water starvation of the stator winding, and cause the temperature of the winding to rise above the normal operating levels. The degree of low flow is generally dependent on the root cause. The urgency for corrective action is based on the temperature of the stator winding components.

In addition to debris and copper corrosion buildup, a flow blockage in the stator winding water path inside the generator may be due to hydrogen leaking into stator cooling water. When this occurs, there is a possibility of hydrogen gas locking, which will cause a stator cooling water flow problem. The hydrogen leak into the stator cooling water system usually has to be significant for a flow blockage of this type to occur and the blockage must encompass many hollow strands before bar cooling is significantly affected. The degree of concern will depend on the stator bar design, with regard to such things as whether the bar is of the mixed-strand type or all-hollow-strand type. The manufacturer should be consulted as to the level of concern on this issue for the given bar type in question.

There is almost invariably a certain amount of stator cooling water that will leak out of any opening or fault present in the stator bar, even though the hydrogen is maintained at a higher pressure than the stator cooling water. This is due to the capillary action of water. The leaking stator cooling water is detrimental to the ground-wall insulation and may cause a bar failure if left uncorrected.

Stator Cooling Water Conductivity. A deionizing subsystem is required to maintain low conductivity in the stator cooling water, generally on the order of 0.1 μS/cm. High conductivity of the demineralized cooling water can cause electrical flashover to ground by tracking, particularly at the Teflon hoses where an internal electric tracking path to ground is formed.

High conductivity in the stator cooling water may be the result of a failure of the stator cooling water deionizer, but it can also be caused by internal erosion of the stator conductor bar strands from cavitation or chemical attack on the copper itself. Failure of the stator cooling water system deionizer will allow the conductivity of the stator cooling water flowing through the stator winding to rise.

The deionizer usually polishes a small percentage of the stator cooling water on a continual basis (generally about 5% of the total stator cooling water flow). If left uncorrected, the conductivity of the stator cooling water will continue to rise slowly, risking a possible phase-to-ground leakage current through the increasingly conducting stator cooling water. A ground fault is possible in the worst case.

Stator Cooling Water Oxygen Content. The dissolved oxygen content of the stator cooling water is monitored and controlled to prevent corrosion of the hollow copper strands. Corrosion of hollow copper strands is a common source of plugging in the stator winding, and is mainly due to improper stator cooling water chemistry. In both high- and low-oxygen systems, it is desired to avoid dissolved oxygen contents in the range of 200 to 300 ppb in the stator cooling water. This is the range in which the maximum corrosion rate occurs and, hence, the greatest risk of plugging. In addition, the pH value affects the corrosion rate when it decreases below the recommended levels. Generally, a higher pH will reduce the corrosion rate [11]. Further, a higher pH value in the order of 8.5 (alkaline) or higher decreases the corrosion rate significantly for both high- and low-oxygen-type systems.

High oxygen refers to air-saturated water with dissolved oxygen present in the stator cooling water in the range of greater than 2000 μg/l (ppb) at STP. The high-oxygen system is based on the supposition that the surface of pure copper forms a corrosion-resistant and adherent cupric oxide layer (CuO), which becomes stable in the high-oxygen environment. The oxide layer is generally resistant to corrosion as long as the average water velocity is less than 15 ft/s (4.6 m/s) [11].

Low oxygen refers to stator cooling water with a dissolved oxygen content less than 50 μg/l (ppb). The low-oxygen system is based on the supposition that pure copper does not react with pure water in the absence of dissolved oxygen. The upper limit is set by the corrosion rate that the water cleansing system can handle. The lower limit is set to the level at which copper will not deposit on any insulating surface in the water circuit such as a hose. This is to avoid electrical tracking paths to ground. It has better heat transfer properties at the copper/water interface and a lower copper ion release rate [11].

Hollow steel strands mixed with solid copper strands have been used for the cooling water circuit in the stator bars. The stainless steel allows a better leak seal at the ends of the stator bars, higher mechanical strength due to nature of the materials used, and reduced effects of corrosion.

Corrosion products can also form in the external stator cooling water system when the stator cooling water chemistry is not optimum, since the external piping is also usually made of copper. In some cases, however, the external system piping is composed of stainless steel, and in these instances the corrosion mechanism is reduced. The problem is that the stator cooling water chemistry must still be closely watched, since even stainless steel can corrode if not maintained.

Stator Cooling Water Hydrogen Content. When no leaks are present in the system, hydrogen content is at a minimum. High concentrations of hydrogen may also cause conductivity problems.

High hydrogen leakage into stator cooling water can cause gas locking if the leak rate is too high. Excessive venting of hydrogen is an indication of a high leak rate. Many machines are equipped with a detraining system to collect and release any entrained hydrogen that enters the stator cooling water system. The detraining system further measures the amount of gas collected and the number of releases, so that the leak rate is known. However, the content of the gas released is not fully known. It can only be assumed that the large majority of it is hydrogen, since that is the environment within the generator casing, but small amounts of air may also be present.

Copper/Iron Content. The content of copper and iron in the stator cooling water is normally less than 20 ppb. High concentrations of either can cause conductivity problems.

Stator Cooling Water pH Value. The pH value is manufacturer specific. Generally, there are two modes of operation: neutral and alkaline. Neutral pH refers to a pH value of approximately 7. A neutral pH with low oxygen content (less than 50 μg/l) works best. Oxygen at 200 to 300 μg/l produces the highest corrosion rate, but oxygen over 2000 μg/l will also work [11].

Alkaline pH refers to a high pH value of around 8.5. Again, low oxygen works best, although high oxygen will also work. This method requires an alkalizing subsystem to keep the pH at the proper level [11].

Stator Cooling Water Makeup Tank Level. In the event that stator cooling water is lost, or for filling of the stator cooling water system after shutdown and draining, the system requires replenishing. Therefore, a storage tank to hold sufficient makeup water is required. Some systems are open to the atmosphere, whereas others maintain a hydrogen blanket on top of the water to keep the level of oxygen at a minimum. The storage tank arrangement is manufacturer specific depending on the desired water chemistry.

Stator Cooling Water System—Hydrogen Detraining Tank Level. Since no stator cooling water system is leak-free, there will be some ingress of hydrogen (i.e., permeation of hydrogen through Teflon hoses) and natural collection of other gases such as oxygen in the stator cooling water system. A means for venting off these gases is required. Generally, the excess gases are vented to the atmosphere. In some systems, the venting process is monitored or measured and in other systems there is no monitoring. This is manufacturer specific.

Raw Service Water Inlet Pressure to Stator Cooling Water Coolers. The service water inlet pressure to the stator cooling water coolers is monitored to ensure continuous flow of raw cooling water to the coolers, and ensure proper heat removal from the stator cooling water.

Low raw water inlet pressure to the stator cooling water coolers is an indication of low flow to the coolers, which will affect the cooling of the stator cooling water. Low service water flow to the stator cooling water coolers may be caused by either a failure of the service water supply pump or a large leak before the coolers. Low pressure/flow to the stator cooling water coolers will result in reduced cooler efficiency and cause the stator cooling water temperature to rise.

Raw Service Water Inlet Temperature to Stator Cooling Water Coolers. Raw service water is circulated through the stator cooling water coolers on one side, to remove the heat from the stator cooling water flowing through the other side of the heat exchanger.

The service water inlet temperature to the stator cooling water coolers is monitored to maintain the correct stator cooling water inlet temperature to the stator winding. It is also monitored for comparison to the service water outlet temperature of the coolers and subsequent determination of the differential temperature. This allows an estimation of the cooler efficiency.

High service water inlet temperature to the stator cooling water coolers usually occurs in the summer months. If the temperature of the inlet service water is too high, it can cause a reduced-efficiency problem in the cooler. This will result in the temperature of the stator cooling water rising above its normal operating level.

Raw Service Water Outlet Temperature from Stator Cooling Water Coolers. The raw service water outlet temperature is monitored to assist in detecting problems with the stator cooling water coolers. The service water outlet temperature should be proportional to the inlet service water temperature.

Low service water flow to or in the stator cooling water coolers will cause high outlet water temperature due to reduced efficiency of the coolers. This will result in the temperature of the stator cooling water to rise above its normal operating level. The low-flow condition may originate from silting or fouling of the cooler tubes.

REFERENCES

1. G. S. Klempner, "Expert System Techniques for Monitoring and Diagnostics of Large Steam Turbine Driven Generators," IEEE/PES Winter Power Meeting/Panel Discussion Paper, New York, February 1995.
2. General Electric Company, "Generators for Utility and Industrial Applications," GE Industrial Power Systems, GET-8022, October, 1992.
3. C. C. Carson, S. C. Barton, and R. S. Gill, "The Occurrence and Control of Interference from Oil Mist in the Detection of Overheating in a Generator," presented at IEEE/ASME/ASCE Joint Power Generation Conference, Los Angeles, September 1977.
4. S. C. Barton, C. C. Carson, W. V. Ligon Jr., and J. L. Webb, "Implementation of Pyrolysate Analysis of Materials Employing Tagging Compounds to Locate an Overheated Area in a Generator," presented at GER-3238, IEEE PES Summer Power Meeting, Portland, OR, July 1981.
5. IEEE/ANSI C50.13-1989, "Requirements for Cylindrical-Rotor Synchronous Generators."
6. Electric Power Research Institute, *Handbook to Assess Rotating Machine Insulation Condition,* EPRI Power Plant Series, Vol. 16, November 1988.
7. H. G. Sedding, S. R. Campbell, G. C. Stone, and G. S. Klempner, "A New Sensor for Detecting Partial Discharges in Operating Turbine Generators," presented at IEEE Winter Power Meeting, New York, February 1991.
8. H. G. Sedding, S. R. Campbell, G. C. Stone, G. S. Klempner, W. McDermid, and R. G. Bussey,"Practical On-line Partial Discharge Tests for Turbine Generators and Motors," presented at IEEE 1993 Summer Power Meeting.
9. IEEE Std 1434-2000, "IEEE Trial-Use Guide to Measurement of Partial Discharges in Rotating Machinery."

10. IEEE Std 115-1995, "IEEE Guide: Test Procedures for Synchronous Machines."
11. "Primer on Maintaining the Integrity of Water-Cooled Generator Stator Windings," EPRI TR-105504, Project 2577, September 1995.
12. EPRI-NMAC Tech Note, "Main Generator On-line Monitoring and Diagnostics," Principal Investigator, G. S. Klempner, December 1996.
13. IEEE Std 1-2000, "IEEE Recommended Practice—General Principles for Temperature Limits in the Rating of Electrical Equipment and for the Evaluation of Electrical Insulation."
14. IEEE Std 67-1990, "IEEE Guide for Operation and Maintenance of Turbine Generators."
15. IEEE Std 1129-1992, "IEEE Recommended Practice for Monitoring and Instrumentation of Turbine Generators."
16. ISO 7919-2 (1996), "Mechanical Vibration of Non-reciprocating Machines—Measurements on Rotating Shafts and Evaluation Criteria. Part 2: Large Land-Based Steam Turbine Generator Sets" (First Edition).
17. IEEE Std 1434-2000, "Guide to the Measurement of Partial Discharges in Rotating Machinery."

6

GENERATOR PROTECTION

6.1 BASIC PROTECTION PHILOSOPHY

The *protection system(s)* of any modern electric power grid performs the most crucial function in the system. Protection is a system because it comprises discrete devices (relays, communication means, etc.) and an algorithm that establishes a coordinated method of operation among the protective devices. This is termed *coordination*. Thus, for a protective system to operate correctly, both the settings of the individual relays and the coordination among them must be right. Wrong settings might result in no protection to the protected equipment and systems, and improper coordination might result in unwarranted loss of production.

The key function of any protective system is to minimize the possibility of physical damage to equipment due to a fault anywhere in the system, or from abnormal operation of the equipment (overspeed, undervoltage, etc.). However, the most critical function of any protective scheme is to safeguard those persons who operate the equipment that produces, transmits, and utilizes electricity.

Protective systems are inherently different from other systems in a power plant (or for that matter any other place where electric power is present). They are called to operate infrequently, and when they do, it is crucial they do so flawlessly. One problem that arises from protective systems being activated infrequently is that they are sometimes overlooked. This is a recipe for disaster. The most common reason for cata-

Handbook of Large Turbo-Generator Operations and Maintenance. By Klempner and Kerszenbaum

strophic failure of equipment in power systems is the failure to operate or misoperation of protective systems.

Purchasing, installing, setting/coordinating, and properly maintaining protective systems is not an insignificant expense. Therefore, the extent to which any device or electric circuit is protected depends on the potential cost of not doing so adequately. Electric power generators are most often the most critical electrical apparatus in any power plant. In fact, given the electrical proximity between the generator and the main step-up transformer (SUT), these two most important apparatuses share some of the protective functions. Given the prohibited cost of replacing any of these two—in particular, the generator—significant expense goes into providing the most comprehensive protection coverage.

Protection is considered by many to be an art as much as a science. Although the basic protective components are well known, and the commonly used settings for those devices are spelled out in a number of standards and other widely available literature, the particular combination of protective relays, settings, and coordination schemes are particular to every site. Therefore, it is impossible to describe or prescribe a single protective system for generators. The description we attempt here is of the most commonly encountered protection arrangements and functions.

Protection systems can be divided into systems monitoring current, voltage (at the machine's main terminals and excitation system), windings, and/or cooling-media temperature and pressure, and systems monitoring internal activity, such as partial discharge, decomposition of organic insulation materials, water content, hydrogen impurities, and flux. Protective functions acting on the current, voltage, temperature, and pressure parameters are commonly referred to as primary protection. The others are referred to as secondary protection or *monitoring* devices. Secondary functions tend to be monitored in real time or on demand. For instance, hydrogen purity is monitored online in real time, whereas water content (for water leaks) is not. Temperature detectors (RTDs or thermocouples) on bearings (and sometimes on windings) may be monitored online in real time, or they may not. Furthermore, these functions may more often than not result in an alarm, rather than directly trip the unit (e.g., core monitors). The discussion of where and when to use these monitoring devices and how to set them is provided in Chapter 5. To the primary protective functions monitoring currents, voltages, temperatures, and pressures there can be added the mechanical protective function of vibration. Typically, vibration will alarm, but it can also be set to trip the unit. Protection functions can also be divided into short-circuit protection functions. The short-circuit protection comprises impedance, distance, and current differential protection.

6.2 GENERATOR PROTECTIVE FUNCTIONS

Protection devices are designed to monitor certain conditions and, subsequently, to alarm or trip if a specified condition is detected. The condition is represented by a *function* or protective function code. Thus, there is a relay for every protective function. If a relay only monitors and, thus, protects against a single set of conditions, it is said that the relay is a "single-function device" (see Fig. 6.1). In the past, most relays were single-function devices. With the advent of solid-state electronics, manufacturers have combined several functions in one unit or device. These "multifunction" relays or protective devices (see Fig. 6.2) offer specific protective functions designed for certain types of apparatus. Some multifunction relays are dedicated to transformers, others to motors, and others to generators. Advances in solid-state electronics have led to less costly devices. Today, a multifunction solid-state device with, for instance, five protective functions, is less expensive than five separate relays for five protective functions.

The number of functions covered by different relays and the number of multifunction devices are decided by, among other things, the expected losses of all the protective functions covered by the multifunctional relay if that particular device becomes faulty. A multifunctional relay containing all the protective functions required for the protection of a generator can be combined with a few discrete relays providing backup protection for critical functions. Alternatively, two or more multifunctional relays can be applied, providing partial or comprehensive redundancy. There are many combinations of these discrete and multifunctional relays that can be adopted, depending on when the power plant was build, the size of the units, system conditions, the idiosyncrasy of the designer, and many other factors. Figure 6.3 shows a typical generator protected by a multifunctional digital protection system.

Relays or protection devices are divided into two categories according to how they process data. The first category is that of analog relays; the second is that of numerical

Fig. 6.1 Single-function generator protective device (Basler BE1-25 Synk-check relay).

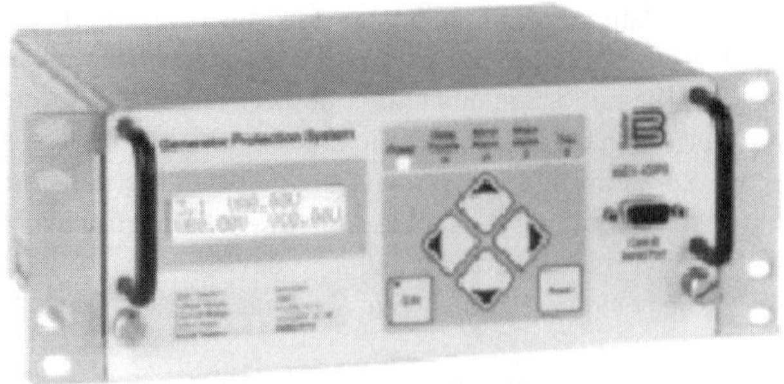

Fig. 6.2 Multifunction generator protective relays from two different manufacturers. Top: Basler's BE1-GPS100 relay. Bottom: GE's DGP digital generator protection relay.

(also called digital) relays. Bear in mind that a relay can be electronic but still process the data in an analog manner. The advantages of numerical processing are various. Accuracy is enhanced. So is flexibility in use. For instance, a numerical relay offers user-shaped protection widows such that the user can change the shape of the *operation/nonoperation* areas for a specific function of the relay. Furthermore, the shape of the region of operation may change according to system conditions (adaptive function).

Finally, there is rather a new—still evolving—approach (from the early 1990s) for protecting large generating units by the so-called expert protection systems. The idea is to protect the unit based not only on the basic protective functions (given in Section 6.3), but also as a combination of protective and monitoring data and built-in expertise in the form of diagnostic prescriptions. Invariably, building the expertise base of these systems consists in expressing probable causes for a particular combination of symptoms, expressed as a probabilistic tree. Southern California Edison, Houston Lighting and Power (with Texas A&M), Ontario Hydro, and EPRI are but a few power producers that have developed these types of software in North America.

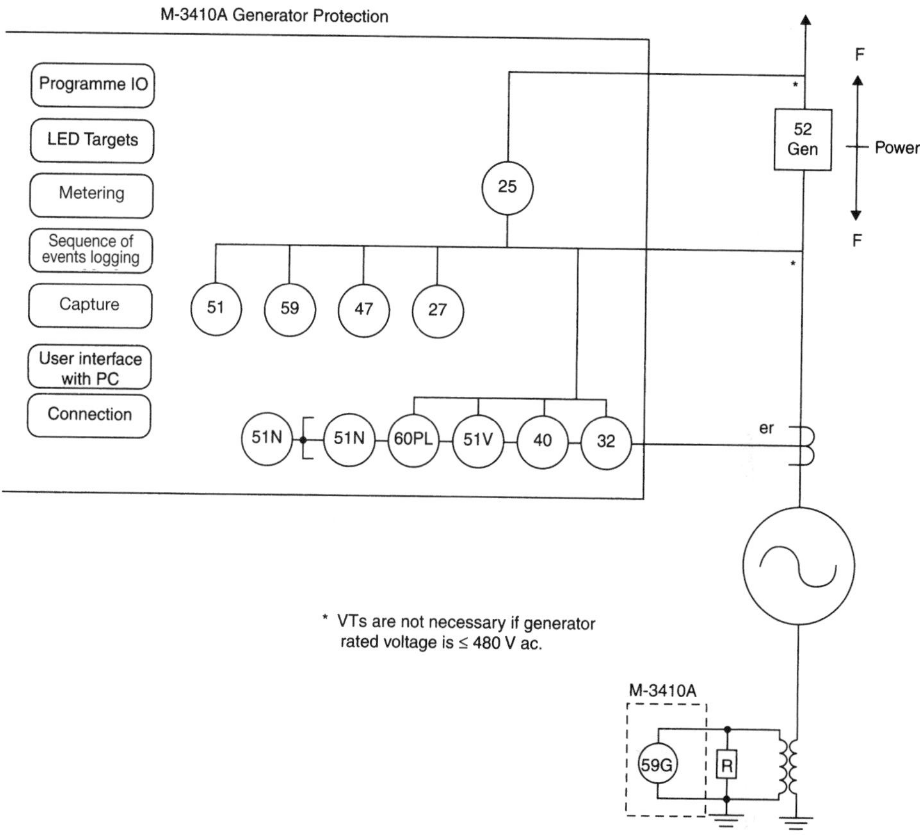

Fig. 6.3 Multifunction generator protection device, Beckwith M-3410A. (Reproduced with permission from Beckwith Electric Co.)

A number, according to a worldwide-accepted nomenclature, identifies protective functions. The functions shown in Table 6.1 are typical of generation protection [1,2].

A number of the functions included in Table 6.1 are so important that they will always find their way into the protection scheme of any generator (e.g., 25, 59, 87). Others may be omitted in some applications (e.g., 49). The larger and more expensive the generator, and the more critical the application, the more comprehensive is the protection applied to protect it from abnormal operating conditions or faults. As explained above, for most large machines, some of the applied protective functions are covered by more than one relay or protective device.

6.3 BRIEF DESCRIPTION OF PROTECTIVE FUNCTIONS

It is beyond the scope and purpose of this book to go into a detailed description of each protective function and the various schemes that incorporate them into a generator's

Table 6.1 Generator protective device function numbers

Device number	Function
15	Synchronizer
21	Distance protection; backup for system generator zone phase faults
24	volts/hertz protection for the generator
25	Sync-check protection
27	Undervoltage
32	Reverse power protection; antimotoring protection for generator (and associated prime mover)
40	Loss-of-field protection
46	Stator unbalanced current protection
49	Stator thermal protection
50B	Instantaneous overcurrent protection used as current detector in a breaker-failure scheme
51GN	Time overcurrent protection; backup for generator ground faults
51TN	Time overcurrent protection; backup for ground faults
51V	Voltage-controlled or voltage-restrained time overcurrent protection; backup for system and generator zone phase faults
59	Overvoltage protection
59BG	Zero-sequence voltage protection; ground fault protection for an ungrounded bus
59GN	Voltage protection; primary ground fault protection for a generator
60	Voltage balance protection; detection of blown potential transformer fuses or otherwise open circuits
61	Time overcurrent protection; detection of turn-to-turn faults in generator windings
62B	Breaker failure protection
64F	Voltage protection; primary protection for rotor ground faults
78	Loss-of-synchronism protection; not commonly used as part of the generator protection package
81	Over- and underfrequency protection
86	Hand-reset lockout auxiliary relay
87B	Differential protection. Primary phase-fault protection for the generator
87GN	Sensitive ground fault protection for the generator
87T	Differential protection for the transformer; may include the generator in some protective schemes
87U	Differential protection for overall unit protection of generator and transformers
94	Self-reset auxiliary tripping relay

protection package. Instead, a basic description of the protective functions and their application will follow. For those readers interested in an in-depth discussion on the topic, a good place to start is reference [1]. See also references [11,12]. For the same reason, no specific values are recommended for setting protective relays. These values oftentimes depend on the particular machine and system to which it is connected. There are numerous sources for information on the setting of protective relays. The manufacturers' manuals are one good place to start. Others are provided in the References section at the end of this chapter.

6.3.1 Synchronizer and Sync-Check Relays (Functions 15 and 25)

The combination of function (15) with function (25) provides the means by which the unit can be brought up to speed automatically and synchronized to the system. Before doing so, the amplitude of the voltages of the system and generator terminal must be within a narrow margin so that the breaker can be closed. So must be the angle of the terminal and system voltages. The *slip,* which is the frequency difference between the machine and the system, must be lower than a given value. Almost always, two relays are provided: the *synchronizer* and the *sync-check.* This division of labor is based on the need to avoid the destructive results of synchronizing a unit out of step due to the failure of a single protective device.

In older installations, mainly those with steam-driven units, it is customary to start and bring the unit up to speed under manual control. Closing the breaker is done manually while the sync-check relay monitors all voltages, vector angles, and frequencies, making sure they are within their prescribed values. Infrequently, some operators close the breaker by keeping the "close" button depressed when the unit is brought to the right speed and voltages, allowing the angle to be taken care by the sync-check relay. This practice has resulted in more than one unit synchronizing out of step due to a failure of the relay (function 25). The failure can be catastrophic. Thus, it is imperative that during manual operation the actual breaker-closing signal be sent when the conditions for synchronization are met, leaving the sync-check system as a backup device, as it is supposed to be. Figure 6.4 shows a discrete sync-check relay.

6.3.2 Short-Circuit Protection (Functions 21, 50, 51, 51V, and 87)

These functions are designed to protect the unit against short circuits in or outside the windings of the alternator. Outside faults can be in the system close to the station's busses, on the main unit transformer or auxiliary transformer(s), on the cable, segregated busses, or insulated phase busses (IPB), between the alternator and the transformers, or on the alternator's windings. In large units, the IPB is designed to reduce any short circuit between the generator and main and auxiliary transformers to a single phase-to-ground fault. This is possible because of the high-impedance grounding of the machine and the fact that all transformers connected to the generator are delta connected on the generator's side, which results in ground faults of very low currents. However, a "benign" single-ground fault inside the generator can develop into a highly destructive phase-to-phase short circuit, and this is the main reason why ground faults inside the generator ought to trip the unit promptly.

Function 51V is a voltage-controlled overcurrent relay that provides voltage control to differentiate between a low-current fault and a normal or abnormally high load condition.

To some extent, most of these functions back each other up. Thus, occasionally some are omitted. Additionally, current-based relays are backed up in the detection of short-circuit events by some voltage-based relays. A typical case is seen in the ground-fault detection scheme of the generator with high-impedance grounding via a

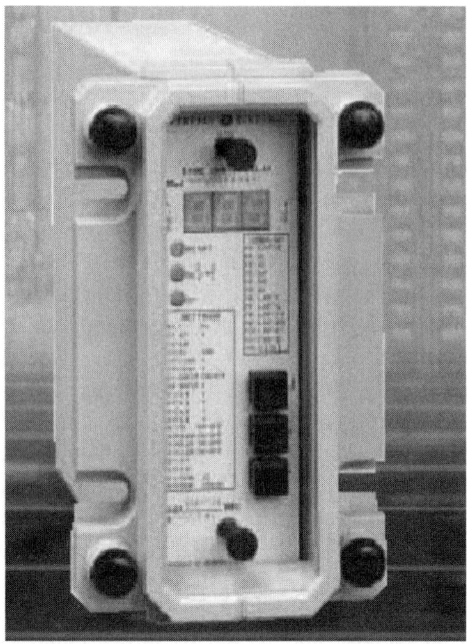

Fig. 6.4 Single-function generator synchronism check device (GE's MLJ digital synchronism check relay).

transformer (see Fig. 6.5). Figure 6.6 shows the type of cubicle that may house this configuration and the cable connection from the generator neutral point to the grounding transformer. The differential protection function (87) is the most critical as it provides protection against the very serious phase-to-phase short circuits. Normally, there are at least three protected areas, each one covered by its own function 87 relay. The first one covers the generator itself, the second covers the auxiliary transformer, and the third covers the main transformer, generator, and low-voltage side of the auxiliary transformer. Each function 87 scheme utilizes a dedicated set of current transformers.

The ground-protection schemes in use today often incorporate a third-harmonic function. This addition to the standard overvoltage and/or overcurrent relays is based on the fact that during normal operation of the generator, a given amount of third-harmonic voltages are present, and during a ground fault these third-harmonic voltages are considerably reduced. This fact is used for protection of the third of the generator's winding close to the neutral, where ground faults tend to generate very small neutral currents (and, hence, may not be detected by the neutral overvoltage or overcurrent protection). Third-harmonic protective devices must be tested periodically, the same as any other protective functions.

In some instances, no overload protection is provided, other than alarming and expected operator intervention. In others, function 51 relays are provided that will alarm,

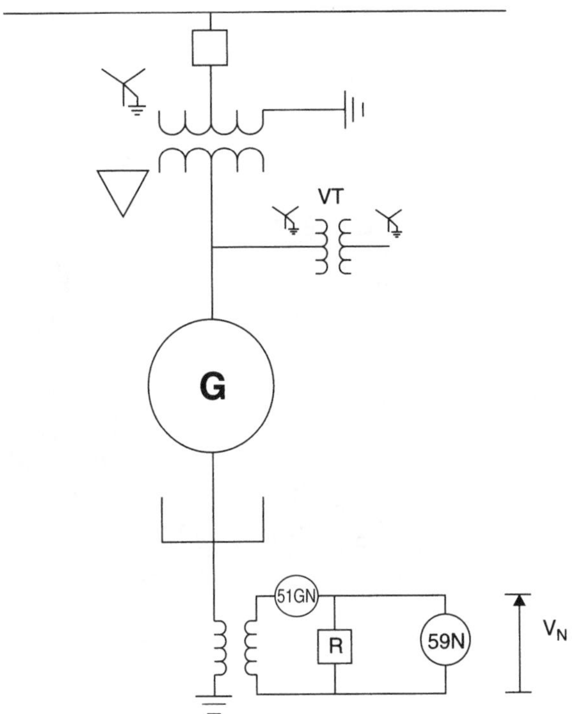

Fig. 6.5 High-impedance grounding showing the main overvoltage protection (function 59N) and the backup overcurrent (function 51GN). (Reproduced with permission from Beckwith Electric Co.)

and then trip the unit under overload conditions. The overload can be extremely onerous if allowed to continue beyond the withstand capabilities of the windings. References [5 and 6] discuss this in detail.

As explained above, the scope of this book does not include protective setting recommendations. For the reader interested in learning about generator ground protection and overload protection, a good place to start is references [3, 5, 6].

6.3.3 Volts/Hertz Protection (Function 24)

As explained in Chapter 4, Sections 4.1.13 and 4.4.8, core damage due to overfluxing is a rare event. However, when severe overfluxing occurs, the most probable result is partial or complete destruction of the core's insulation, with the consequential need to replace it. Therefore, it is critical that V/Hz protection be applied and properly set. Almost invariably, the cases of severe overfluxing occur during run-up, prior to synchronization. One vital component in all V/Hz schemes for any turbo-generator is double-feeding from two independent potential transformers (PTs). Otherwise, loss of a single PT connection may give the excitation system wrong information about the terminal

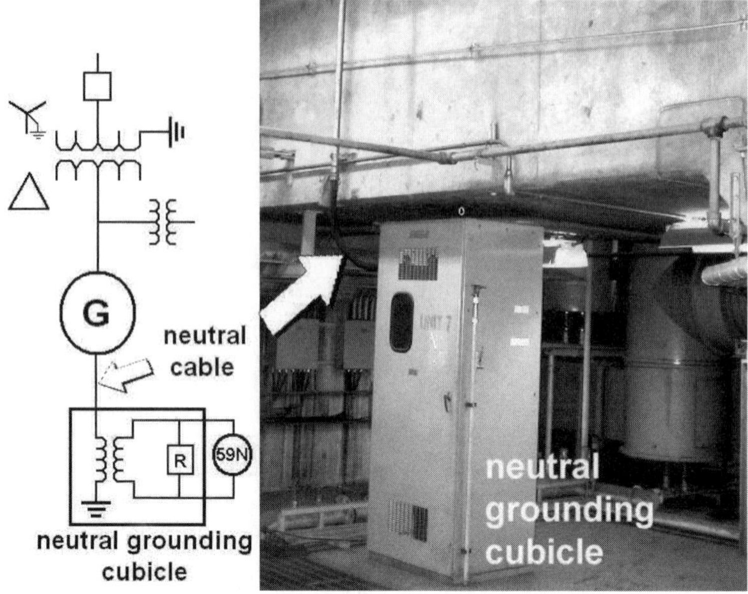

Fig. 6.6 Example of a neutral grounding scheme and actual cubicle where the neutral grounding transformer is located.

voltage, forcing the field current (and terminal voltage) beyond the V/Hz capability of the machine.

It behooves the operator of the turbogenerator to follow OEM instructions in setting the overexcitation relays (function 24). However, a good independent guideline is provided by ANSI/IEEE Standard C.37.106 [4].

Figures 6.7, 6.8, and 6.9 show the damage that can result from an overfluxing event. The unit shown in the figures required restacking of the core with new laminations, rewinding of the stator, and extensive metal removal from the rotor, as well as one new retaining ring. However, in addition to expensive repairs for the damage described, the loss of production was considerable. It took a number of months before the unit could be returned to service.

6.3.4 Over- and Undervoltage Protection (Functions 59 and 27)

As explained in Section 6.3.2, some voltage relays are used for short-circuit protection (on the neutral of the generator, function 59GN). Overvoltage relays are also used as backup to function 24 during normal operation of the machine. During start-up, function 59 will not provide backup to function 24 because a V/Hz condition can readily develop during run-up, even while the terminal voltage is below its rated value.

The undervoltage relays are mainly installed for the purpose of identifying loss of PT voltage or to identify dead-bus conditions for certain alignments.

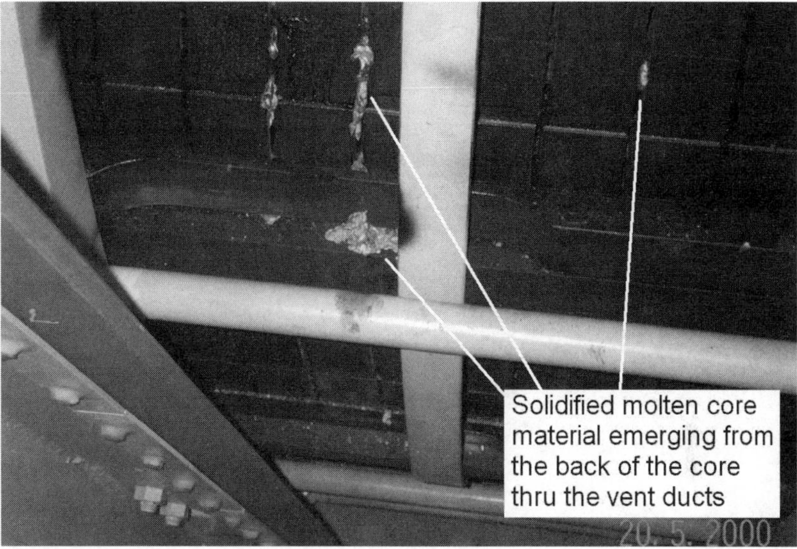

Fig. 6.7 Damage due to an overfluxing event, back-of-core view. This 2-pole, 500 MW, hydrogen-cooled machine underwent a 16-sec overfluxing event of about 130% V/Hz during start-up. After being synchronized to the grid and loaded, it tripped several hours later on stator ground-fault protection. Inspection of the unit revealed extensive amounts of solidified molten core material. The damage to the core due to the overfluxing event started in a number of damaged spots in the core. The molten metal tends to solidify rapidly once it reaches the gas gap due to the flow of cooling hydrogen. Then it is pushed from behind by more iron being melted. The solidified metal in the gap now starts cutting into the rotor components. Thus, the damage to the unit goes well beyond the stator.

6.3.5 Reverse Power Protection (Function 32)

This protective function trips the unit when power flows from the grid to the generator. In this situation, depending on the generator's field condition, the alternator is driven as a synchronous or induction motor. If it is driven as an induction motor, slip-frequency currents will be established in the rotor, potentially damaging damper windings, wedges, retaining rings, and forging. This phenomenon is discussed elsewhere in this book. However, in either case, the reverse power condition may adversely affect the integrity of the prime mover.

Of all the prime movers, steam turbines are the most sensitive to motoring. They also happen to operate on less power input (only a few percent of rated load, compared to combustion turbines requiring up to 50% of rated power). For these reasons, steam-driven generators require sensitive settings for the reverse power relays (function 32), plus some additional protection that may be indicated. A very good discussion of the subject is found in reference [1].

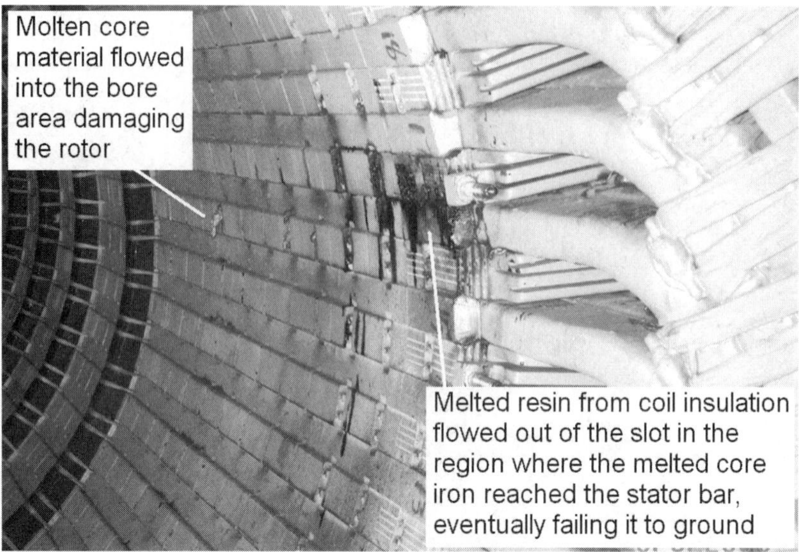

Fig. 6.8 Damage due to an overfluxing event, bore-view. This is the same machine as in Figure 6.7, showing the view from the bore. Molten iron flowed in both directions and in a number of locations along the core. The machine eventually tripped on stator ground fault once the molten iron reached one of the stator conductor bars, overheating the ground-wall insulation and causing its failure.)

Fig. 6.9 Damage to the retaining rings of the machine shown in Figures 6.7 and 6.8.

6.3.6 Loss-of-Field Protection (Function 40)

There are a number of events that may result in an accidental removal of the source of excitation to the generator. This can happen for both brushless and externally excited units. For instance, an unplanned opening of the field breaker, a failure of the exciter, a flashover in the brush rigging, a failure of the automatic voltage regulator (AVR), and a short circuit in the field winding, can all result in a loss-of-excitation condition.

When a generator loses its excitation during normal operation, its speed increases by up to 3 to 5%. The amount of speed increase depends on the generator's load prior to losing its excitation. A lightly loaded unit will experience a much smaller increase in speed than one fully loaded. Additionally, the stator current will normally increase because the generator without its field will operate as an induction machine, receiving its excitation VARs from the network. Accordingly, the stator current may increase by up to 100% of its nominal value.

The onerous effects of the increase in line current will be aggravated by the overheating of rotor components; by the currents induced in the forging and damping winding, if present; and by the overheating of the stator core-end regions. A fully loaded unit that loses its field may experience serious damage very quickly under these conditions. Therefore, the protection against loss-of-field occurrences is set to alarm and trip the unit relatively quickly.

The most widely utilized method of protecting against loss-of-field conditions is that relying on impedance elements. It is based on the fact that the impedance seen from the terminals of the machine follows a distinctive pattern when the field is lost (see Fig. 6.10). Sometimes two relays are used, each looking at the impedance within a different region of operation, so that a loss-of-field condition is captured regardless of the level of prefault loading. Sensing the field current directly or sensing the VAR power flowing *into* the generator is sometimes used for the alarm and trip, but mainly for the alarm and rarely as primary protection.

There are different requirements for designing and setting a protection system against loss of field, depending on the type of machine arrangement (tandem, cross-compound, double winding, etc.). A discussion can be found in references [1, 8, 9, 10].

6.3.7 Stator Unbalanced Current Protection (Function 46)

There are a number of incidents that may result in unbalanced three-phase currents at the terminals of an alternator, for instance, unbalance loads, single-pole opening of a breaker, asymmetrical transmission systems (without or with insufficient transposition), and open circuits. As explained in Chapter 4, Section 4.4.2, unbalanced currents will result in negative-sequence current components flowing on the rotor forging surfaces, retaining rings, rotor wedges, and to some extent in the field windings, in particular, the amortisseur. These rotor negative-sequence currents have the potential of generating high temperatures within seconds, with severe detrimental effects to specific areas of the forging and other rotor components. Figure 6.11 shows one such area in the case of body-mounted retaining rings. However, rotors with spindle-mounted retaining rings are also susceptible to damage by negative-sequence currents.

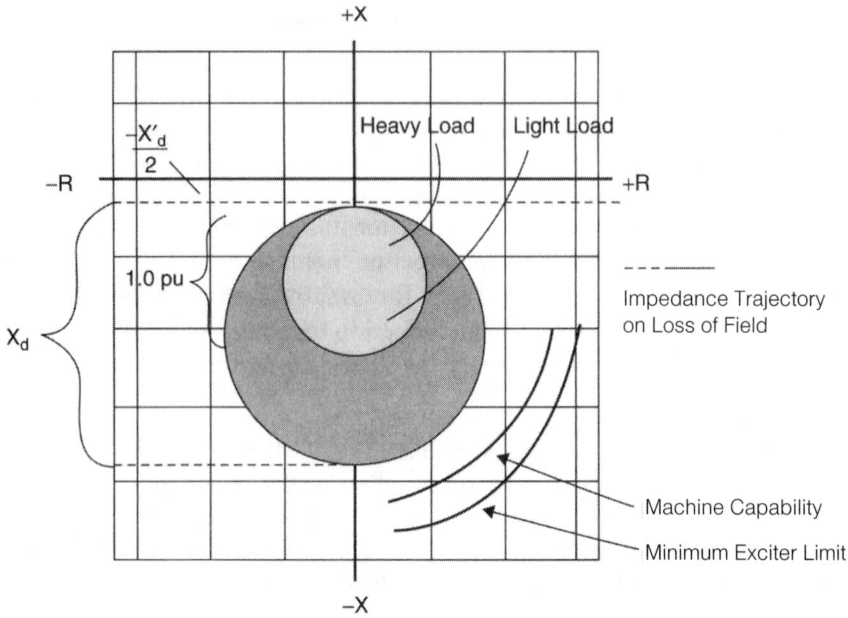

Fig. 6.10 Impedance trajectories followed by a loss-of-field event, depending on the load state of the unit prior to the occurrence, as seen at the terminal of the generator. (From Charles Mozina, Power Plant "Horror Stories," reproduced with permission from Beckwith Electric Co.)

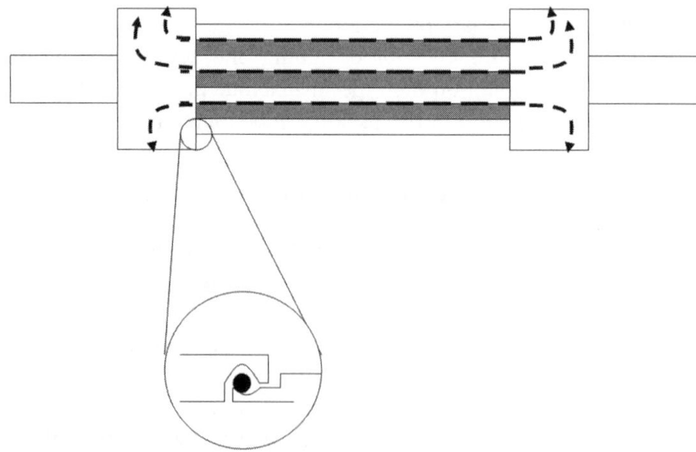

Fig. 6.11 Arrangement of a body-mounted retaining ring and the surface-flowing negative-sequence currents. The detail shows one of the areas susceptible to damage: the locking-ring interface between retaining ring and wedges.

Generators must meet minimal requirements for sustaining unbalanced currents without damage. These requirements are spelled out in ANSI C50.13 [5]. The protection against unbalanced currents is implemented by using overcurrent relays that measure negative-sequence components. Electromechanical relays provide basic protection against most negative-sequence current conditions. However, digital relays allow setting the protected region of operation in such a way that closely matches the withstand capability of the protected generator. This allows a more sensitive and discriminatory approach.

6.3.8 Stator and Rotor Thermal Protection (Function 49)

There are a number of conditions that may result in elevated temperature inside the generator. Presently available techniques allow one to directly monitor temperatures of the stator winding, core, and cooling media. Monitoring of rotor winding temperature is done by measuring field voltage and current, then calculating the rotor-field resistance and comparing the obtained resistance with a known value of ohms at a known temperature. Conditions that may result in higher-than-normal temperatures are overload, core hot spots, bent laminations swelling into vent ducts, winding failures, and cooling failure (clogged filters in air-cooled machines, lack of hydrogen pressure or failure of the hydrogen cooling system in hydrogen-cooled generators, and water blockage or other failure of the water cooling system in water-cooled units). There are other conditions that may result in higher temperatures, such as unbalanced currents; however, these are detected and protected by other protective functions, so they are discussed elsewhere.

In addition to the design limits of each machine (based on such things as temperature rise class and class of insulation), there are ANSI guidelines regarding minimum withstand capability requirements under overload conditions. These requirements are given in ANSI C50-13 [5]. For instance, at 130% overload, the machine should be able to operate without damage for a minimum of 60 seconds. These numbers show that once in an overload condition, the time available to remove the dangerous situation is quickly reduced.

Typically, generators have a number of RTDs (resistance temperature detectors) embedded in their stator windings, with a minimum of two per phase. In some designs, these RTDs are wired to the control room via SCADA or DCS. In manned stations (all large turbogenerators fall in this category, with exception of some "peaking" units), the winding RTDs are used for alarming (overcurrent protection is used for high and sudden overload conditions). In unattended stations (mainly smaller machines), the output from the RTDs may be used to remotely alarm and to control and/or trip the unit. In the United States, the standard RTD has a resistance of 25 ohms at 25°C. When the RTDs are installed during original manufacture, the OEM will place the proper RTD. However, if for any reason RTDs are installed by the operator (e.g., during a partial rewind or any other overhaul), the RTDs must match the operating temperature of the winding. This temperature will most likely be related to the temperature class of the unit and the insulation class.

Some manufacturers of direct hydrogen-cooled stators omit the embedded RTDs. In lieu of them, they install a number of RTDs monitoring the hydrogen paths in the stator bars (in addition to other RTDs monitoring other areas along the machine's gas flow path). The RTDs monitoring the flow of gas in the stator bars are normally installed in the exit boxes at the end windings. If any overheating occurs in a bar or section of the winding, some RTDs will pick up the excess temperature in the gas flowing in that region. During troubleshooting activities carried out on a number of hydrogen-cooled units with directly cooled stators, the authors ascertained that the existence of embedded RTDs, in addition to the gas-flow RTDs, made it easier to determine the location of the faults in the coils. Therefore, it is not a bad idea to specify generators with embedded RTDs (a minimum of two per phase) for those machines with gas-path-flow RTDs. All RTDs should be monitored, including the embedded ones.

Another way to monitor onerous temperatures in the stator is by applying tagging compounds. This technique can also be used to alarm for developing core problems. These tagging components may be used in hydrogen-cooled machines in which a core monitor is installed. More about this can be found in Chapter 5. Core monitors have the ability to monitor and alarm against deterioration of the field winding due to overheating.

Recall that for rotor winding, the only method of estimating the temperature rise of the winding is by measuring the field voltage and current at the collector, and comparing the calculated resistance thereof with a known value at a known temperature. This technique is restricted to those machines with collector rings (external excitation). One method commonly used is to monitor the excitation current. As the excitation current exceeds certain nominal value, relays time the duration of the occurrence and trip the unit as soon as a certain setting is reached. The time–current characteristics followed by the protection try to match the withstand curves for short-time field overloading as contained in ANSI C50.13 [5].

6.3.9 Voltage Balance Protection (Function 60)

The main function of the *voltage balance* relay is to avoid false tripping of other protection relays due to a loss of secondary voltage feed, for instance, by a blown potential transformer (PT) fuse. Voltage balance schemes are possible in most modern and/or large generators because such units have at least two PTs feeding the protection and monitoring systems. The voltage balance relay senses and compares the secondary voltage of different PTs, and when it determines that a "blown-fuse" situation arises, it blocks the operation of certain voltage-controlled relays and alarms. Figure 6.12 shows two PTs being monitored by a voltage balance relay.

In those older alternators (or small units) in which only one PT feeds the protective and excitation systems, it is still possible to sense and alarm for a blown-fuse condition. This is attained by using a scheme that compares negative-sequence voltages in the secondary of the PT (which will arise as a consequence of a primary fault or a blown-fuse condition), with negative currents in the secondary of the current transformer (CT). If negative-sequence currents are not present, it indicates that a fault in the primary system did not occur, and thus it must be a blown-fuse condition. This

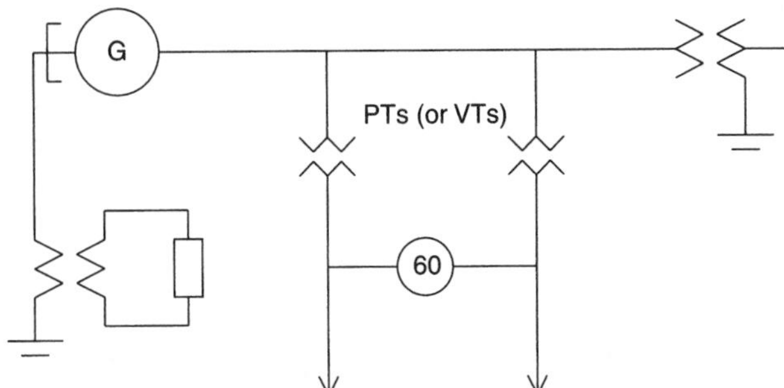

Fig. 6.12 Schematic of two potential transformers (PTs), also called voltage transformers (VTs), feeding two circuits: protection and excitation of a generator. The voltage balance relay (60) monitors the three-phase secondary voltages off the PTs.

voltage/current negative-sequence comparative function can be found in certain modern digital protective packages. Figure 6.13 shows such an arrangement.

6.3.10 Time Overcurrent Protection for Detection of Turn-to-Turn Faults (Function 61)

The most common stator winding design for large turbo-generators is based on a single-turn arrangement. Smaller machines may have multiturn windings. For such machines with at least two parallel circuits, the "split-phase" protective scheme can be used for protection against turn-to-turn short circuits. In this scheme, the circuits in

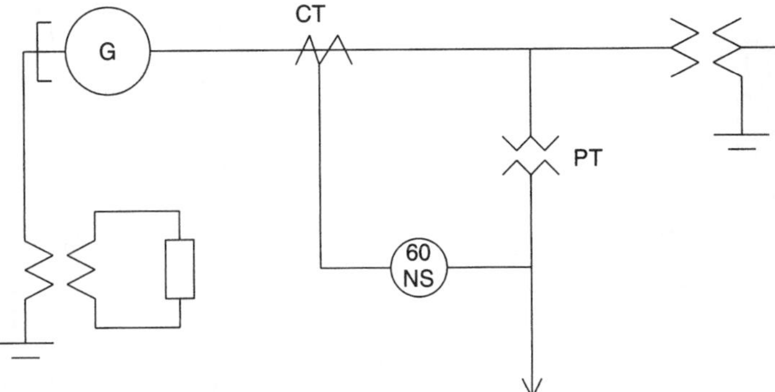

Fig. 6.13 Schematic of a generator with one PT feeding its protection and excitation system. The negative-sequence voltage balance relay (60 NS) monitors the three-phase voltages and currents fed from the PT and a CT. Loss of PT phase ("blown-fuse" condition) is achieved by sensing negative-sequence voltages but not negative-sequence currents (these would exist during a fault).

each phase are split into two equal groups and the currents in each group are compared. Any significant difference would indicate an interturn failure. Reference [5] contains a detailed discussion of this protective scheme and the various arrangements found in industry. The relays used are normally very inverse overcurrent relays and instantaneous trip combinations.

Figure 6.14 shows the construction of a multi-turn stator coil by cross-sectional views.

6.3.11 Breaker Failure Protection (Function 62B)

Most faults involving the generator require tripping the line circuit breakers. Failure of any such circuit breaker to operate properly results in loss of protection and other abnormal conditions, such as motoring. Adverse conditions arise also if only one or two poles of a line circuit breaker operates, for instance, resulting in energization from one line, or single-phase operation with the accompanying negative-sequence currents.

Activation of a breaker failure scheme is carried out by a combination of triggering signals from the generator protective relays, overcurrent relays, and circuit breaker auxiliary switches, via a timer. Some modified schemes also include in their triggering circuit the trip signal from the neutral of the main step-up transformer's overcurrent relay. This change is to protect against circuit breaker *head flashover,* which is when arcing occurs across the circuit breaker contacts due to high voltages. The protection is

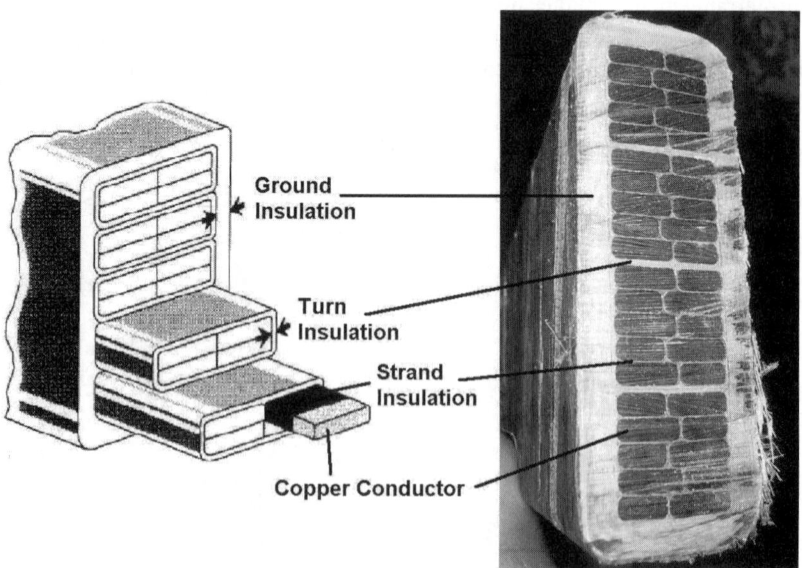

Fig. 6.14 Multiturn stator coil cross-section. The diagram on the left represents a general schematic cross-sectional view of a multiturn coil, whereas the photo on the right is an actual multiturn coil cross section.

designed to operate against the flashover of two poles. Figure 6.15 shows the functional diagram of a simple breaker-failure protective scheme.

6.3.12 Rotor Ground-Fault Protection (Function 64F)

Rotor field windings are designed to operate ungrounded. As a result, a single short to ground, in theory, should not be reason for concern, because it will not interfere with the normal operation of the machine. However, the appearance of a second ground can be very detrimental to the operation of the generator, as well as to its integrity. In fact, the existence of one ground fault will make a second more probable, due to induced field voltages resulting from stator transients. Two coetaneous grounds may result in the following:

- Unbalanced air-/gasgap fluxes with increased rotor vibrations
- Unbalanced thermal heating of the rotor with increased vibrations
- Fluctuating VARs and output voltage
- Major damage to the forging by dc currents (dc currents are known to be able to produce arcs several inches long in the forging of a turbogenerator's rotor during a double ground occurrence). There are enough cases documented of large rotors having to be replaced because of such events.

There are a number of methods in existence for the detection, alarming, and/or tripping of generators due to field ground faults. Some methods use a voltage source, and others use a passive unbalanced bridge. In addition to IEEE C37.102, one good reference is the chapter on protection of the Westinghouse manual on applied protective relaying [13].

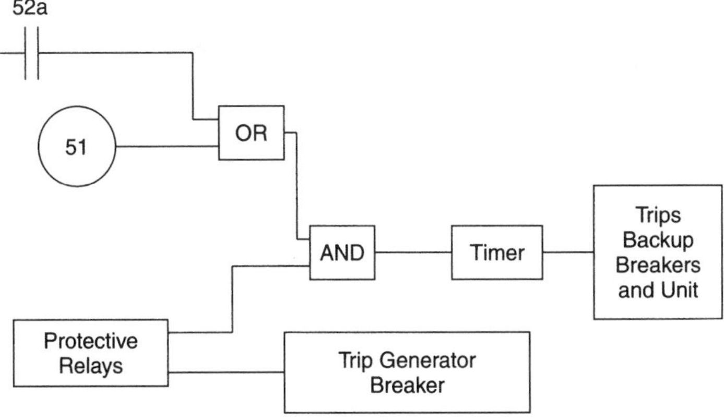

Fig. 6.15 Functional diagram of a generator breaker-failure scheme (from IEEE C37.102).

One can still find an older unit protected by an ac-source field-ground detection method. As shown in Figure 6.16, such methods can damage the noninsulated bearing of the generator if the grounding brush somehow becomes ineffective. To avoid this from happening, the newer designs have the ac source replaced with a dc or any one of the nonlinear resistor-based bridge methods (see Fig. 6.17).

Generators with brushless excitation cannot be directly protected with the type of schemes shown above, due to the lack of collector rings. Many designs of such units use a set of small rings that periodically can be temporarily connected to brushes, via which the tests described above can be performed. For large generators, additional circuits can be found that complement the basic schemes shown above. For instance, a circuit may be added to verify that the brushes are sitting on the rotor.

6.3.13 Over-/Underfrequency Protection (Function 81)

Chapter 4, Section 4.4.13, contains a discussion about the severe consequences to the integrity of the turbo-generator unit while operating even for very short times at

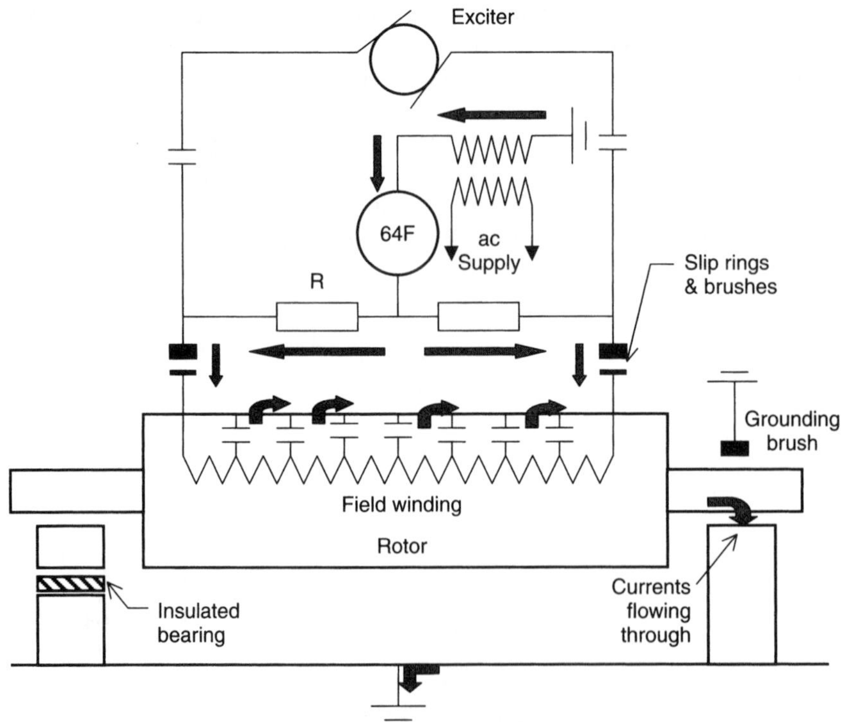

Fig. 6.16 Path of capacitance-coupled currents (shown as thick arrows) flowing in the rotor and through the oil of an noninsulated bearing when the field ground-detection scheme uses an ac supply and the grounding brush is not in operation. These currents, over long periods of time, may result in bearing failure.

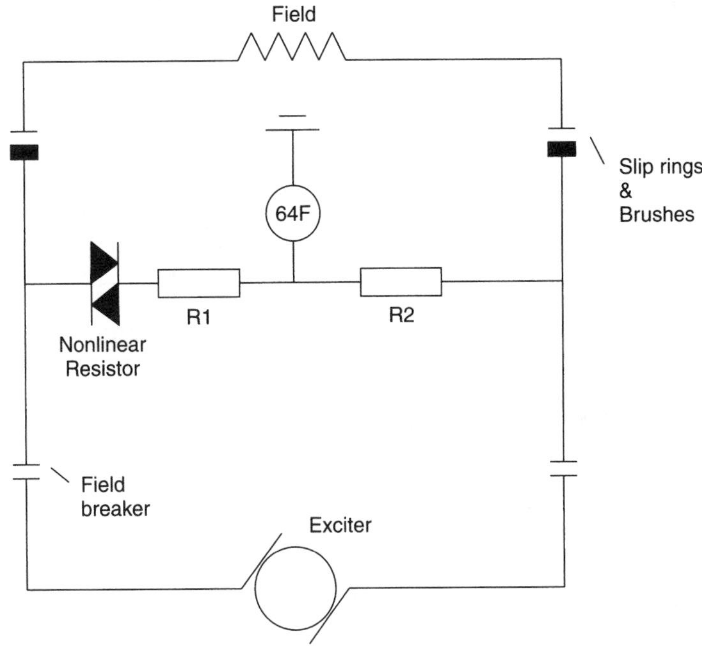

Fig. 6.17 Schematic representation of a ground-detection circuit using voltages developed across a bridge during a field-ground event. The nonlinear resistor is included to artificially shift the neutral point so that a ground anyplace in the winding can be detected.

frequencies higher or lower than nominal. Over- and underfrequency operation generally results from full or partial load rejection or overloading conditions. Load rejection can be caused by a fault in the system or load shedding. Overload conditions may arise from tripping a large generator or a transmission line. What frequency the machine will attain following load rejection or overload is a function of how much load has changed and the governor *droop characteristics*. For instance, a governor with a 5% droop characteristic will cause a 1.5% speed increase for a 30% load rejection (speed change in percentage equals droop in percentage times load change per unit).

As explained in Chapter 4, the manufacturers provide withstand curves that should be used in setting the function 81 relay. A very good discussion of over-/underfrequency operation is given in ANSI/IEEE C37.106 [4]. Figure 6.18 shows an example of over-/underfrequency operating ranges and protective device settings.

6.3.14 Out-of-Step Operation (Loss of Synchronism, Function 78)

There are a number of reasons why a generator may lose synchronization to the system during operation. Regardless of the reason, loss of synchronization (being out of step)

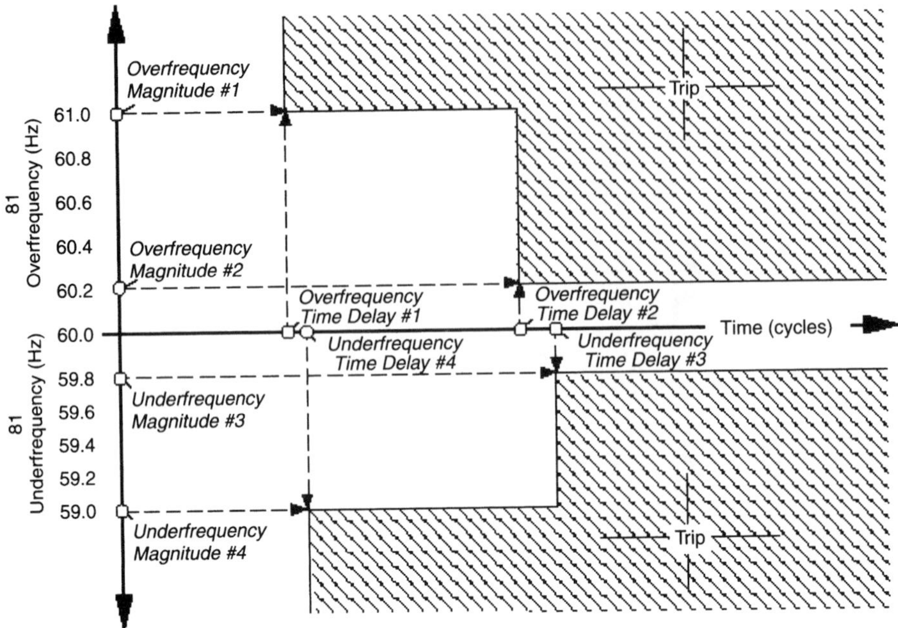

Fig. 6.18 Example of over-/underfrequency region of operation and protective settings. (From Beckwith Intertie/Generator Protection M-3410A Instruction Manual, reproduced with permission from Beckwith Electric Co.)

can have serious detrimental effects on the generator. The end windings and end windings' support are prone to damage and dislocation during such an event. Rotor and coupling damage is also possible. This condition is not too unlike the out-of-step synchronization discussed in Section 4.4.11. To minimize any harmful effects, the protection should separate the generator from the system as soon as possible, preferably during the first half-slip cycle.

Protection against the out-of-step condition is based on the fact that the apparent impedance, as seen at the generator's terminals, changes in a predicted manner during an unstable condition (see Fig. 6.19). This is similar to the loss-of-excitation condition. However, the loss-of-excitation relay will not pick up every occurrence of an out-of-step condition because the apparent-impedance behavior is different for both conditions. Therefore, to fully protect against out-of-step conditions, a dedicated relay (or protective function within a multifunctional device) must be included in the protection package.

Tripping the unit within the first slip cycle has major advantages in the case of an out-of-step event. This fast protective action tends to reduce considerably the very large oscillating shaft torque that can otherwise occur. For those readers interested in an elaborated discussion of this topic, Reference [14] provides a good starting point.

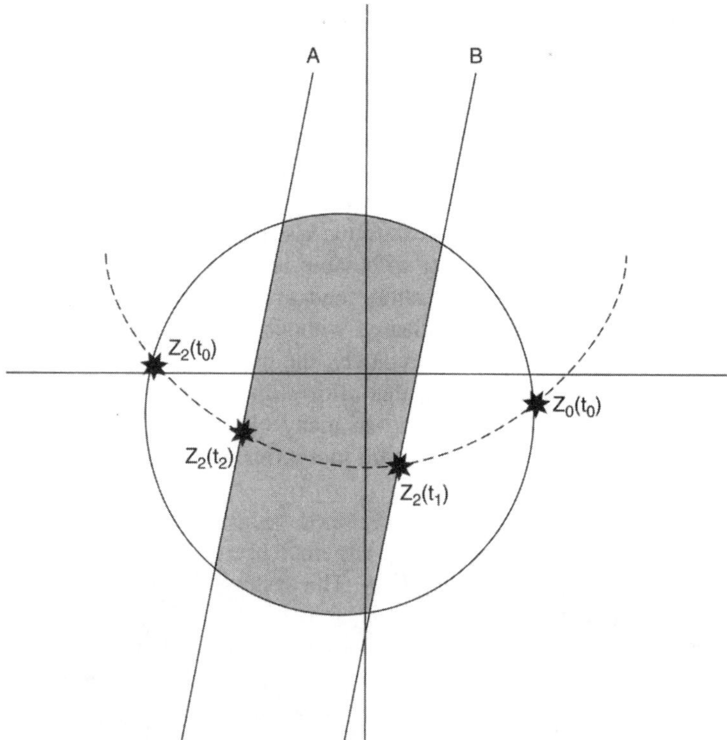

Fig. 6.19 Example of out-of-step region of operation and protective settings. (From Beckwith Generator Protection M-3425 Instruction Manual, reproduced with permission from Beckwith Electric Co.)

6.4 SPECIALIZED PROTECTION SCHEMES

6.4.1 Protection Against Accidental Energization

In many instances, not enough attention is paid to the protection of the generator when the unit is offline. For example, protective devices that are crucial in avoiding disastrous occurrences are left nonoperational. It is important that when the unit is offline, the protective relaying systems are kept operational, and if work is being carried out on them, any risk should be ascertained and mitigating action taken.

One very onerous event that comes to mind is the inadvertent energization of the generator while offline at rest or on the turning gear. When a cylindrical rotor generator (all turbo-generators are of cylindrical rotor construction) is energized while at rest or at very low speeds (e.g., on the turning gear), the equivalent impedance "seen" by the power system is the generator's negative-sequence reactance in series with the negative-sequence resistance. However, it can be shown that the negative-sequence reactance X_2 equals the subtransient reactance X''_d. This low impedance causes currents of up to 500% of nominal to rush into the unit when the main breaker closes. Obviously, these large currents can create havoc in the windings, in addition to the physical risks associated with the turbo-generator unit being accelerated in an unplanned manner (see Fig. 6.20 for damage to a rotor due to a sudden energization).

One point that should be made here is that when a generator rotor is motored from standstill, the consequences for the rotor itself are far more severe than if the rotor is at speed and loses excitation. This is because at standstill the wedges and damper windings are not in good contact with adjacent components and, therefore, the contact resistances that are present are much higher than if the components were at speed and making good contact with lower resistance. The overheating effect is, therefore, greater at zero speed due to higher resistive heating. In addition, the turbo-generator shaft system inertia has to be overcome to pull it up to speed and this further exacerbates the situation by causing the currents flowing to be higher. The effect is like an induction motor starting from standstill; the starting currents in those machines may be as much as six times the running current.

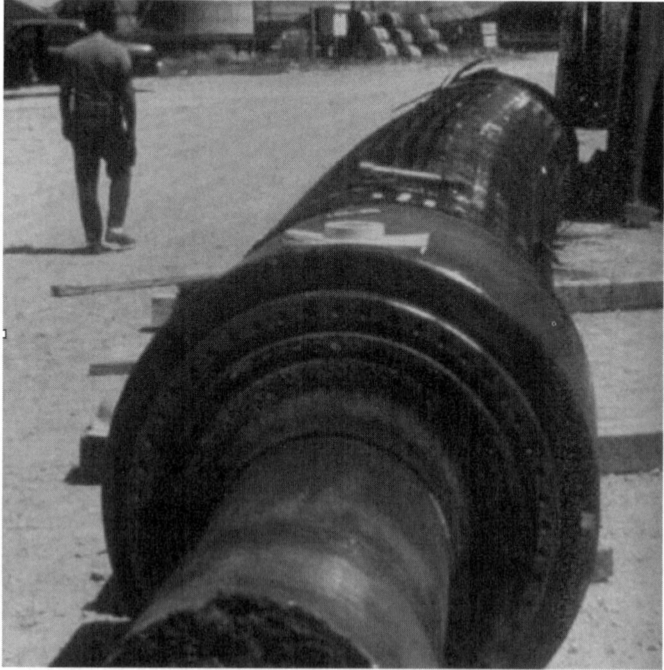

Fig. 6.20 Turbo-generator rotor subject to motoring currents due to sudden energization at standstill. The generator breaker was closed with no excitation on the rotor winding. The rotor was essentially being started as an induction machine as currents flowed in the forging, wedges, retaining rings, and so on. Not being designed for such currents, the rotor and most of its components were severely overheated. The rotor was basically destroyed.

There are several ways that a generator can be protected against this occurrence. The loss-of-field (function 40) and reverse power (function 32) relays can, under certain schemes and settings, provide protection. Other schemes rely on directional overcurrent (O/C) relays, frequency-supervised overcurrent relays, distance relays, and system backup relays. Reference [1] provides a good discussion of the subject.

6.4.2 dc Field Ground Discrimination

Almost everyone involved in the operation of large turbine-driven alternators has had the opportunity to troubleshoot a ground in the dc circuit feeding the rotor field via collector rings. Invariably, these circuits have a type of ground-fault detector. These detect and alarm when one polarity's insulation to ground is breached. These detection systems do not discriminate between a ground inside the rotor and one anywhere in the plant, between generator and excitation source, sometimes located in a different deck. Discrimination can be really important. A ground inside the rotor should result in the unit being brought offline in a short time. This is so because continuous operation risks another ground developing anywhere in the dc system, with the potential of extensive damage to the forging or other critical rotor components due to large dc currents. A ground in the area between the machine and the exciter also has the potential of developing into a second ground in the rotor, with same lamentable results. However, the odds are lower and, therefore, operators tend to keep the unit online until the ground is identified and perhaps cleared without the need of removing the unit from operation immediately. Unfortunately, the overwhelming majority of units in operation do not have a system that allows discrimination between grounds on the rotor or outside the generator. In many older plants, contamination and humidity combined with old cables result in a series of very annoying dc grounds over the years.

One such detection system was developed some 15 years ago by the R&D department of a large U.S. utility, and it has since been successfully installed on a number of units in and outside the utility. The system is based on the installation of a very sensitive Hall-effect current transformer around the dc leads feeding the generator, as close as possible to the collector rings, in addition to an electronic package that simulates grounds every few seconds, not unlike other commercially available devices. The system has proved to be able to detect a difference of about 20 mA between polarities in a bus carrying about 1500 amperes dc. In doing so, it can detect a ground inside the rotor, even a high-resistance ground, providing the operators the opportunity to make rational decisions based on knowledge about the location of the fault.

Figure 6.21 shows the device installed on the dc leads between the excitation breaker and the brush gear of a 400 MVA hydrogen-cooled, two-pole generator. Figures 6.22, 6.23, and 6.24 show a Hall-effect pickup collar being tested on the bench and then installed on two large cross-compound units. Finally, Figure 6.25 shows a pickup installed on a 580 MVA, two-pole unit.

As seen in the figures, the higher-rating units have specially manufactured coaxial copper busses on which the Hall-effect pickup devices are installed. The purpose of this arrangement is to obtain the maximum sensitivity possible.

Fig. 6.21 Photograph of a Hall-effect sensor strapped around the dc-field cables of a 400 MVA, two-pole, 60 Hz, hydrogen-cooled generator.

Fig. 6.22 First dc coaxial bus with a Hall-effect sensor strapped around it, on a test bench. On the front-left of the bench, the electronic package can be seen inside a transparent plastic box.

Fig. 6.23 A dc-coaxial bus and pickup installed on a 800 MVA unit, with directly water-cooled stator. The coaxial bus has been inserted just above the brush gear.

Fig. 6.24 A dc coaxial bus installed on a 750 MVA, directly water-cooled generator. The coaxial bus is inside the brush-gear compartment, very close to the brushes (electrically speaking).

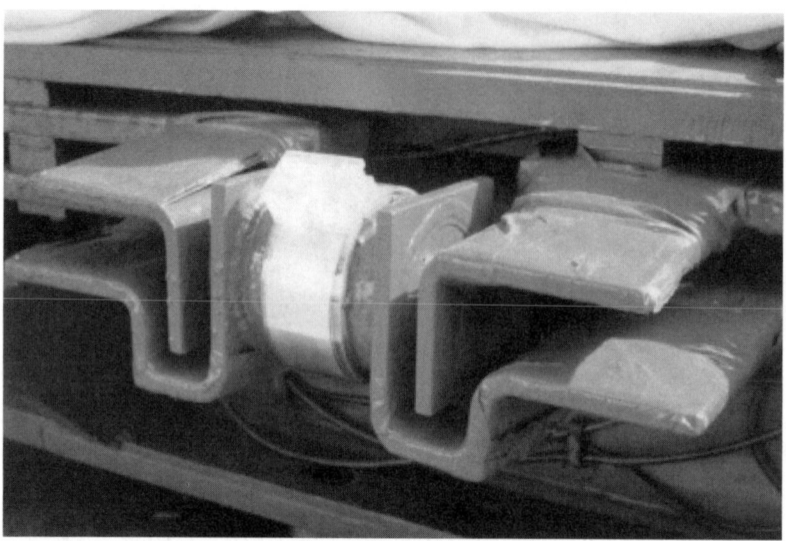

Fig. 6.25 A dc coaxial bus located on the side of the sole plate of a 580 MVA, hydrogen-cooled generator.

6.4.3 Vibration Considerations

Vibration considerations are key to the proper functioning of any rotating machine. In particular, large turbo-generators are very much affected by vibration because the main rotating element—the generator's rotor—is made of a number of critical components. (The detrimental effect of vibrations on the operation of the brushes is explained in Chapter 9.) High vibrations cannot only damage rotating elements but also the armature winding, the core, and the frame. In short, every significant component in a generator is prone to damage by excessive vibrations.

Thus, it is as important to address the protection of the generator against excessive vibration as intensively as for other physical variables. However, as with the rest of the protection functions, no unique philosophy exists. In the United States, many operators follow criteria codify by IRD [15] in a set of curves widely distributed and readily available. Operators of large turbo-machinery may opt to follow ISO 7919-2 [16], written for steam-driven turbo-generators in excess of 50 MW. Gas-turbine sets are covered in Part 4 of ISO 7919.

In all cases, the manufacturer of the generator provides, at the time of delivery of the unit, a set of vibration alarm and trip values. The manufacturer may also recommend following another standard regarding vibration behavior when ramping the unit up or down.

Vibration analysis is a complex topic that requires significant knowledge and elaboration beyond the scope of this book. Thus, it is left to the reader to search additional sources. We provide a general idea of what those criteria measure in Figure 6.26, com-

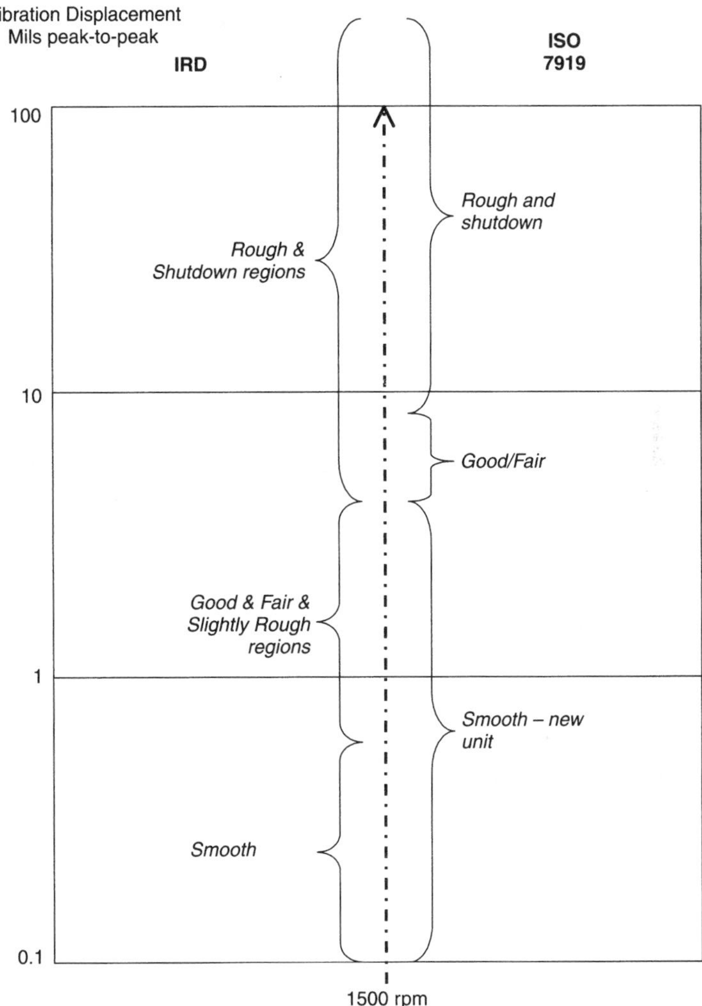

Fig. 6.26 Comparison of the IRD Vibration Severity Chart with ISO 7919-2. The comparison is made for a four-pole, 50 Hz machine (1500 rpm). Note that this is not a strict comparison, because in order to build this graph ISO's peak-to-peak relative shaft displacement values were used. Absolute displacement yields somewhat higher values. IRD's values, on the other hand, are for filtered readings taken on the bearing cap. The comparison is only for illustration purposes, to provide an idea of how the different regions are defined in each case.

paring the IRD chart with ISO 7919-2. Note that the comparison is not entirely fair because each case uses different vibration criteria.

6.4.4 Operation of the Isolated-Phase Bus (IPB) at Reduced Cooling and Risks from H_2 Leaks into the IPB

In Section 2.11, a brief description was introduced about how the terminal boxes of medium- and large-rating generators are connected to the isolated-phase bus (also called iso-phase bus or IPB). It was also noted that in large machines, the IPB is almost always forced cooled by one or, most often, two fans driven by induction motors. The basic need for a bus to carry the generated power from the generator to the main step-up transformer is easily understood by looking at the typical value of stator currents. For example, in the extreme case of a 2000 MVA unit, with terminal voltage of 25 kV, the stator current flowing through the terminals of the generator would be 46,188 amperes per phase. The only practical way of carrying this current is with bus ducts. By isolating the phases, the possibility of sustaining a very deleterious phase-to-phase short circuit between transformers (main step-up and auxiliary) and the generator is made practically nil (this also requires effective deionizers at the generator end of the IPB). Another interesting fact not widely recognized by personnel not directly engaged in the design of IBP systems is that the forces acting upon the busses from the interaction with the other phases are reduced by as much as 90% by the action of each phase enclosure, when compared with the forces developed if the enclosure were not present.

There is no clear-cut MVA rating between units employing IPBs against those with cables. However, most units larger than 300 MVA will deliver their power into isolated-phase busses. Also, there is no magic MVA rating that separates those with forced IPB cooling from those without it. However, most units with output currents of about 25,000 amperes or more will have forced IPB cooling. In many instances, forced cooling may be found in units with IPB capacity as low as 15,000 amperes or so. The actual choice of at which rating to use force-cooled IPB depends on all those considerations mentioned above, plus capital costs and cost of IPB losses. Figure 6.27 shows a typical arrangement for a forced-cooled isolated phase bus.

The cooling fans of a typical IBP are very reliable systems. The squirrel cage induction motors are rugged, the load is constant, and operation is basically continuous. Thus, these motor-fan systems tend to operate reliably for most of the life of the plant. Nevertheless, on occasion, one of these motors can fail. In a single-motor application, this would result in complete loss of forced cooling. Most force-cooled IPBs have two independent motor-driven fans, and each one is sized to provide all the required cooling for full-capacity operation of the IPB. With the loss of the single motor, or the loss of both motors in an IPB with two such fans (for instance due to a loss of power to the motors), a partial derating of the IPB occurs.

In addition to a loss of cooling fans, loss of cooling can be caused by a loss (or reduction) of water flow to the heat exchangers. On the other hand, water flow may have to be stopped due to developing leaks. These leaks may elevate the humidity of the air enclosed in the IPB, increasing the likelihood of developing ground faults.

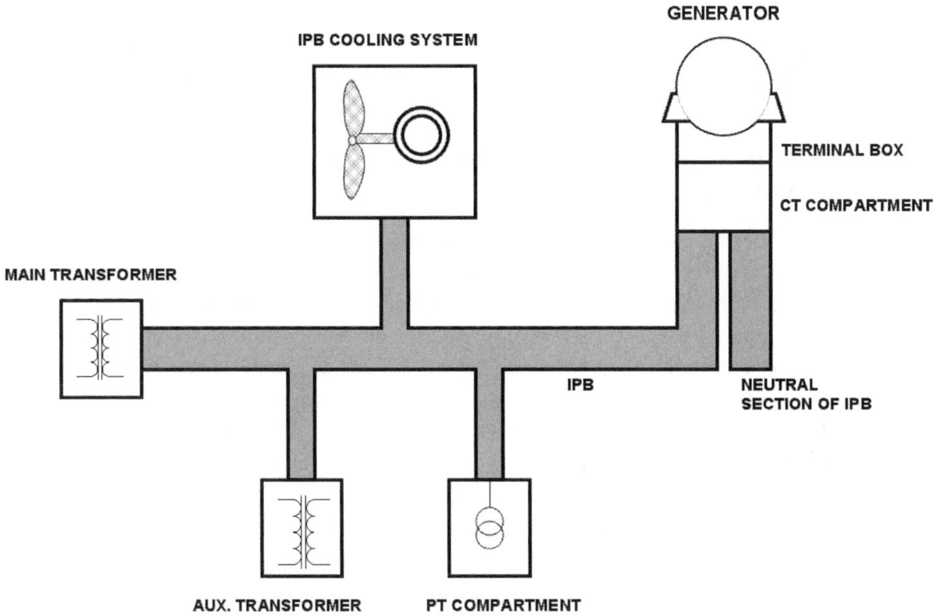

Fig. 6.27 Typical arrangement of a force-cooled IPB.

The output limitations imposed on the generator by any of the aforementioned reasons depend on several factors such as length of IPB, design characteristics, and ambient temperature (summer or winter). In addition, there may be other constraints that do not allow operation at reduced outputs for extended periods. In a typical example for a 22 kV, four-pole unit with a rating of 1300 MVA, the rating of the IPB with all cooling systems in operation is about 34 kA amperes, whereas the rating without forced cooling is about 21 kA. Thus, loss of cooling in this case results in an IPB capacity reduction roughly equal to 38%. The actual reduction in generator apparent output power may be less because, in general, the IPB's maximum capacity is larger than the generator's maximum output by a certain margin, typically around 10%. These values are very specific to a particular design and station; the manufacturer should be consulted to find out what derating applies to in each specific case.

An interesting spin off of forced cooling of IPBs is the issue of accumulation of hydrogen in the IPB due to a leak from the generator into the IPB via the generator bushings. Part of the bushings are in the generator's hydrogen environment, and part protrude into the IPB or CT compartment, which in many cases is an integral part of the IPB. There is no one common policy among stations about how to monitor H_2 concentration in the IPB. Some plants may employ H_2 detectors permanently installed in the IPB, whereas others may just use portable ones when circumstances call for that. In IPBs with forced cooling in operation, the air is quickly mixed along the IPB, reducing the possibility of H_2 accumulating to dangerous levels in any one area. In addition, force-cooled IPBs also maintain certain air exchanges with the environment (allowing air to ingress via filters), so that global accumulation of H_2 in the IPB is basically limited to very low values. In the next section, it is shown how any station can calculate the rate of increase of H_2 in the IPB due to a given leak, when the forced cooling is in operation.

6.4.5 Calculation of the H_2 Mix in the IPB for a Given H_2 Leak from the Generator into the IPB

Before engaging in the calculation of the hydrogen mix, the significance of mixing hydrogen with air in a confined space should be mentioned. As mentioned in other places in this book, a hydrogen–air mixture can be explosive when occurring in certain ratios.

For ignition, the lower explosive volume concentration limit in air (LEL) is 4% for hydrogen. The upper explosive limit (UEL) is 74% for hydrogen. The energy needed to ignite an explosive mixture is very low. To ignite a hydrogen–air mixture near stoichiometric concentration, a spark of only 2 μJ would be sufficient. Open and unprotected electrical equipment could easily cause ignition, and so could static electrical discharges. From the aforementioned, it is obvious that hydrogen leaks, in particular in confined spaces like the iso-phase bus, must be taken very seriously.

At this point, the reader is referred to Figure 6.28 to become acquainted with the various variables.

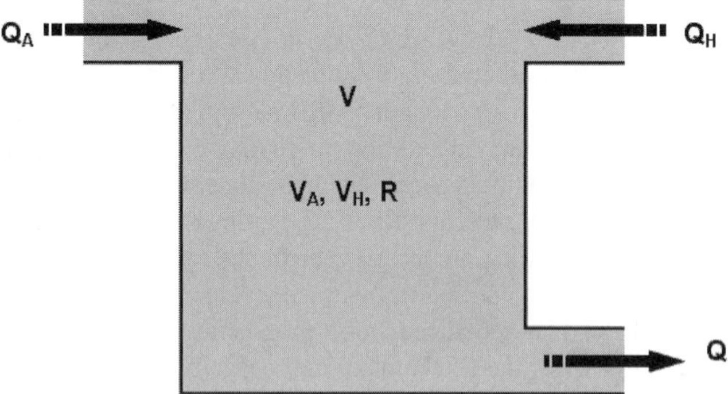

Where:

- Q_A, Q_H, **Q** are given in CFM (cubic feet per minute) or equivalent
- **V**, V_A, V_H, are given in CF or equivalent
- Q_A is the make-up flow of air into the IPB
- Q_H is the leak rate of the H_2 into the IPB
- **Q** is the rate of flow out of the IPB ($Q = Q_A + Q_H$)
- V_A and V_H are the volumes of air and H2 in the IPB in any given time ($V = V_A + V_H$)
- **R** is the volume of H_2 in the IPB given in percent of total volume ($R = V_H / V$)

Fig. 6.28 Model representation of air and H_2 flow in the IPB.

Assumptions:

- The makeup airflow and the rate of H_2 leak into the IPB are constants
- At $t = 0$, the volume of the IPB is filled with air
- The air and H_2 are mixed instantaneously (in fact, this will take at least the time that it takes the gas to circulate the entire IPB for the mixing to happen, on the order of a minute or so)

System Constants:

1. Makeup airflow. The makeup flow thru the IPB filter(s) can be measured with the help of an anemometer. It should be measured with a relatively clean filter. The value obtained can be taken as the highest makeup air flow.
2. Volume of the IPB. The volume of the IPB must include all those areas that the air will flow through.
3. Hydrogen leak rate into the IPB. This is the expected hydrogen leak rate from the generator into the IPB. For the calculation, sensitivity analysis can be made by taking a number of values.
4. Average leak of hydrogen from the generator. This value is the typical H_2 leak rate from the generator in question. This may include the IPB, cooling water, through seals, and so on. Most stations have a good idea of the typical leak rate of the generators. This value could be taken as the highest value of the item above (H_2 leak rate into the IPB), because any higher leak rate becomes noticeable.

From inspection of Figure 6.28, the following equations can be written:

$$dV_{\mathrm{H}} = Q_{\mathrm{H}}dt - Q(V_{\mathrm{H}}/V)\,dt \tag{6.1}$$

$$dV_{\mathrm{A}} = Q_{\mathrm{A}}dt - Q(V_{\mathrm{A}}/V)\,dt \tag{6.2}$$

Equation (6.1) means that the change of hydrogen in the IPP equals the rate by which hydrogen leaks into the IPB, less the rate at which it escapes from the IPB. The rate of escape of the hydrogen depends on the ratio of H_2/V at each instance. In the case of Equation (6.2), the same approach is used, but now for the makeup air.

Equations (6.1) and (6.2) lead to Equations (6.3) and (6.4):

$$V'_{\mathrm{H}} = V_{\mathrm{H}} + [Q_{\mathrm{H}} - Q(V_{\mathrm{H}}/V)]\,dt \tag{6.3}$$

$$V'_{\mathrm{A}} = V_{\mathrm{A}} + [Q_{\mathrm{A}} - Q(V_{\mathrm{A}}/V)]\,d \tag{6.4}$$

where V'_{H} and V'_{A} equal $V_{\mathrm{H}} + dV_{\mathrm{H}}$ and $V_{\mathrm{A}} + dV_{\mathrm{A}}$, respectively.

Additionally, Q_A, Q_H, and Q/V are constant for a given H_2 leak and air makeup rate. Therefore,

$$V'_H + V'_A = V_H + V_A = V$$

$$R'_{H/A} = V'_H/V'_A$$

Replacing V_A with $V - V_H$ and applying simple calculus yields the equation describing the H_2 buildup in the IPB as percentage of total volume:

$$R(\%) = 100\left(\frac{Q_H}{Q}\right)\left[1 - \exp\left(\frac{Q}{V}t\right)\right]$$

The solution has two clear limits: at $t = 0$, $R = 0$ (no hydrogen in the IPB), and at $t = \infty$, $R = 100 \times (Q_H/Q)$.

From the equation above, graphs can be drawn to estimate H_2 buildup in the IPB. The graph in Fig. 6.29 is for a 1300 MVA machine with the following parameters:

- IPB volume (V) = 8018 CF
- Rate of makeup air (Q_A) = 255 CFM
- Daily "normal" H_2 leakage rate = 1500 CF = 10.4 CFM
- Rate of H_2 leak into the IPB (Q_H) = 10 times the daily "normal" leak rate = 10.4 CFM

As shown in Figure 6.29, with a H_2 leak of ten times the "normal" leak of that unit, the LEL is never reached, though it comes close to it after about 66 hours. Please keep in mind this is a very specific example. Other units will certainly exhibit different behavior.

Figure 6.30 shows a more rational scenario than the previous one for the same machine. In this case, the "normal" daily H_2 leak is assumed to go entirely into the IPB. In this case, the maximum mix reached is 0.4%, which is ten times lower than the LEL.

Finally, Figure 6.31 shows the case of a H_2 leak into the IPB equal to five times the "normal" daily leak, together with an IPB filter 75% clogged. In this case, the LEL is reached in less than a day.

Next, Figure 6.32 shows the case of a H_2 leak into the IPB equal to the expected "normal" total leak. However, in this case the makeup air filter is completely clogged. In this hypothetical case, LEL is reached in 23 days. This case is not a realistic scenario, because it is hard to believe that no air will escape and be made up by new air during operation of a force-cooled IPB. However, it provides a good criterion for establishing filter maintenance routines.

By comparing the four cases shown above, it becomes obvious that maintaining good IPB ventilation (i.e., the filters should be unclogged) the risks posed by H_2 leaks into force-cooled IBPs are greatly reduced. Maintenance procedures of the IPB ought to include a periodic check of the filter(s).

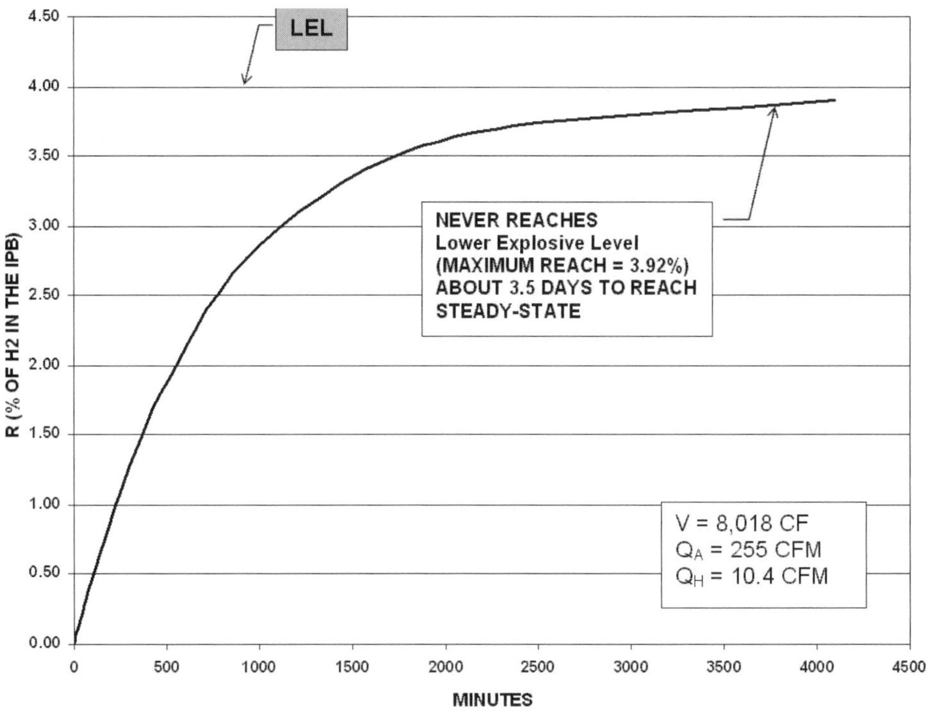

Fig. 6.29 H_2 buildup in IPB with 10 times "normal" daily leak rate.

With the IPB cooling fans out of operation (as during a long outage or due to power failure), the accumulation of H_2 in certain IPB configurations might raise concerns of creating an explosive mixture. One way or another, stations, together with IPB manufacturers, should address this problem, and establish maintenance and monitoring procedures, if deemed necessary, to eliminate these risks.

For illustration purposes, Figures 6.33 and 6.34 are included. Figure 6.33 shows a typical cooling station of a large IPB. Figure 6.34 shows a makeup air filter for the same IPB.

6.5 TRIPPING AND ALARMING METHODS

Earlier in this chapter, we discussed protective functions without reference to how ane actual trip or alarm is implemented. There are many different schemes used to trip a unit as the need arises. There are also many different choices as to when to alarm, when to trip, when to alarm and trip, and so forth. The choices are idiosyncratic; they depend on country, history, age of plant and equipment, and so forth. This book is not geared toward an exhaustive discussion of protection systems and philosophies. There-

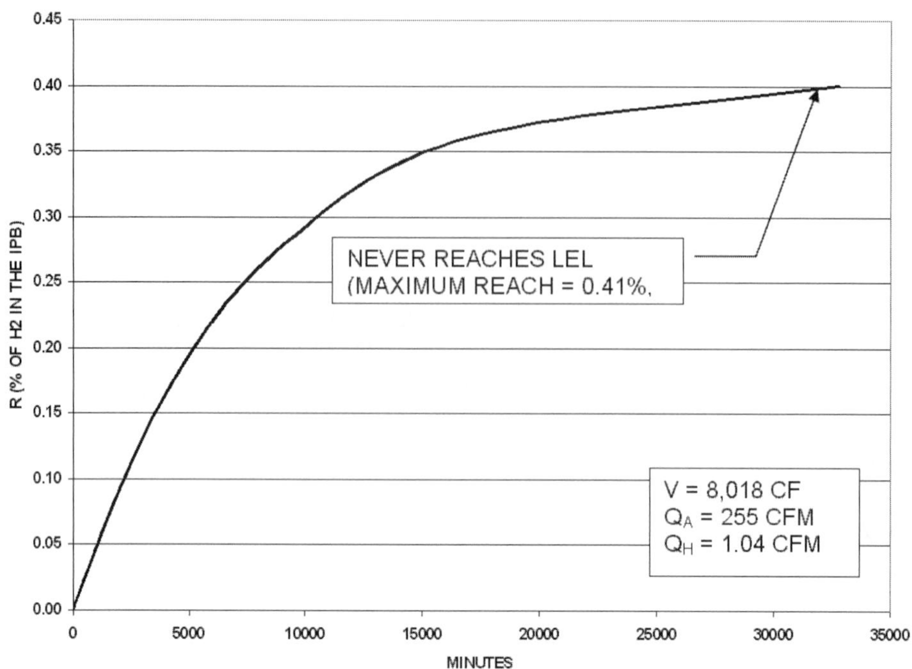

Fig. 6.30 H_2 build-up in IPB with a 1 times "normal" daily leak rate.

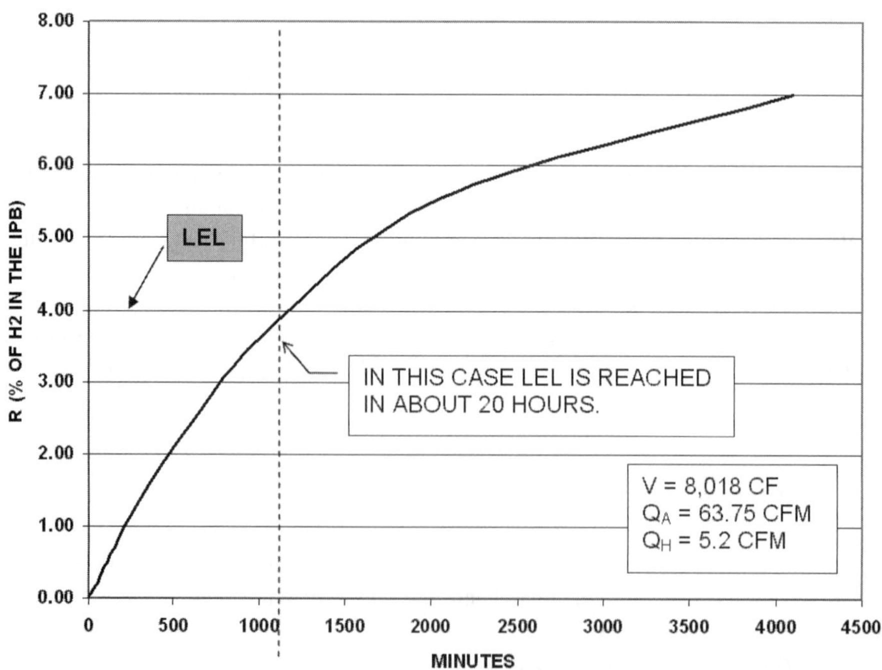

Fig. 6.31 H_2 buildup in the IPB with a 1 times "normal" daily leak rate and makeup filter 75% clogged.

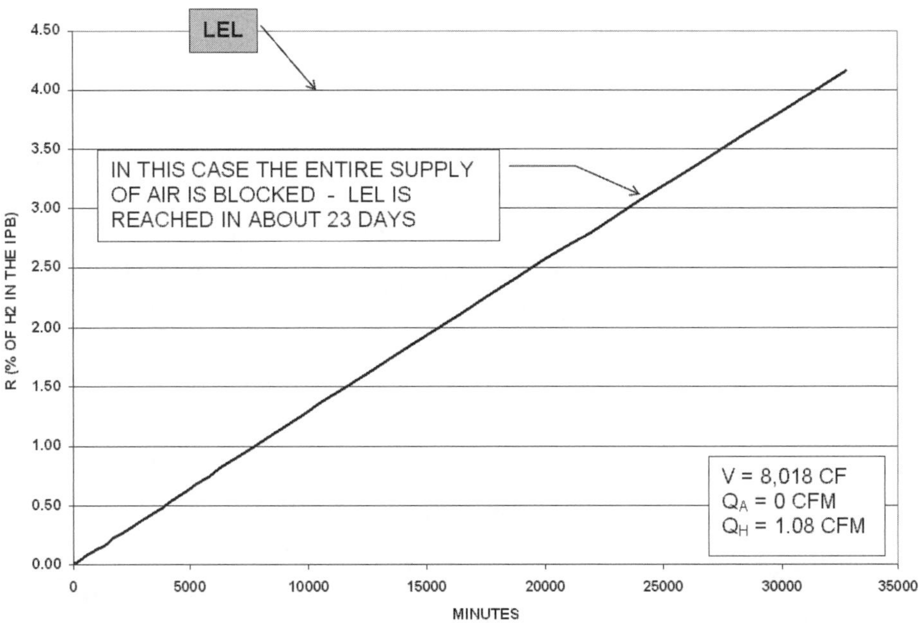

Fig. 6.32 H_2 buildup in IPB with 1 times "normal" daily leak rate and makeup filter 100% clogged.

Fig. 6.33 Cooling station of a 1300 MVA unit IPB.

Fig. 6.34 Filter for makeup air for the IPB of previous figure.

fore, only a nonpartisan listing of some of the elements that go into the decision making will be enumerated.

Factors that could or should be taken into account when determining what type of trip should be implemented are as follows:

- How onerous the fault can be to the generator and turbine
- How onerous the fault can be to the power system
- Overspeed risk posttrip
- Fault spreading to other equipment
- Need for maintaining auxiliary loads post-trip
- Time required for restarting the unit
- Need to trip the excitation
- Need to trip the prime mover

Depending on the answers to those questions, the following tripping decisions may be implemented:

- *Simultaneous Trip.* Trips the turbine and trips the generator's main circuit breaker and excitation circuit breaker. This trip provides the highest degree of protection to the combined unit, but it allows for some overspeed, perhaps even high overspeed.
- *Generator Trip.* The generator main breaker is tripped together with the excitation. This type of trip does not trip the turbine; thus, it must be employed only in those circumstances when overspeed is not expected beyond a very small

amount. The benefit is that it allows the unit to be restored to normal operation in a short time.

- *Circuit Breaker Trip.* Trips the generator main breakers, but neither excitation nor turbine. The advantage of this type of trip is that it keeps the auxiliary transformer(s) energized. It also allows the unit to be reconnected to the system in a short time. Overspeed risks must be taken into account. The fact that the auxiliary load is maintained provides for certain control of overspeed and keeps the turbine in its allowed operating condition (with a minimum of load that most turbines require for maintained operation).
- *Sequential Trip.* Trips the turbine first, then the generator is tripped by its reverse power relay/function. The opening of the line circuit breakers trips the excitation. This trip can prevent overspeed occurrences. It is strongly recommended when implementing this type of trip that backup protection for the reverse power relay be in place.
- *Manual Trip.* The turbine is tripped manually. Then the generator trips by activation of the reverse power relay, followed by the excitation trip, as in the case of the sequential trip.
- *Runback and Trip.* Runback can be manual or automatic, depending on whether the generator monitoring and protection is set up for auto runback. In most cases, it is manual. The operator reduces power to the turbine (steam to the steam unit) to zero power or close to zero, and then follows with a sequential trip. This trip is the normal trip mode, and it is normally employed following an alarm that necessitates the manual tripping of the unit. Some units may also activate an *automatic runback trip* by manual command of the operator.

The following list is a sample of some of the more common protective actions that may be found for the various types of faults. These are not all inclusive and may not be what is used at any given plant. They are merely what is common practice.

Electrical Faults	
Stator overcurrent	Manual or auto runback
Stator ground fault	Simultaneous trip
Stator phase-to-phase fault	Simultaneous trip
Volts/hertz	Simultaneous or generator trip
Stator over-voltage	Voltage restoration by field current runback
Rotor shorted turns	Field current and/or load reduction
Rotor overheating	Simultaneous or sequential trip
Rotor ground	Alarm (but small proportion sequentially tripped)
Loss of excitation	Simultaneous or generator trip
Mechanical and Thermal Faults	
Stator core local overheating	Runback with eventual trip
Bearing vibration excessive	Sequential trip
Synchronizing error	Synchro-check relay protected
Motoring	Simultaneous, generator, or main breaker trip
Loss of SCW flow	Runback (manual or auto), eventual sequential trip

Mechanical and Thermal Faults (cont.)	
Loss of seal-oil pressure	Reduce H_2 pressure and load
Hi SCW conductivity	Alarm and then runback and trip
System Faults	
Unbalanced stator currents	Main breaker, generator, or simultaneous trip
Loss of synchronism	Main breaker, generator, or simultaneous trip
Abnormal frequency operation	Main breaker, generator, or simultaneous trip
Main breaker failure	Main breaker failure protection
Voltage surges	Surge arrestors

REFERENCES

1. ANSI/IEEE C37.102, "AC Generator Protection," 1995-R2001.
2. ANSI/IEEE C37.2, "IEEE Standard Electrical Power System Device Function Numbers," 1996-R2001.
3. ANSI/IEEE C37.101, "IEEE Guide for Generator Ground Protection," 1993-R2000.
4. ANSI/IEEE C37.106, "IEEE Guide for Abnormal Frequency Protection for Power Generating Plants," 2003.
5. IEEE/ANSI C50.13-1989, "Requirements for Cylindrical-Rotor Synchronous Generators."
6. ANSI/IEEE C50.14, "Requirements for Combustion Gas Turbine Driven Cylindrical-Rotor for Synchronous Generators," 1977.
7. ANSI/IEEE Std 67, "IEEE Guide for Operation and Maintenance of Turbine Generators," 2005.
8. Power System Relaying Committee, "Loss-of-Field Relay Operation during System Disturbances," *IEEE Transactions on Power Apparatus and Systems,* September-October 1975.
9. W. P. Gibson, and C. L. Wagner, "Application of Loss-of-Field Relays on Cross Compound Generators," presented at PEA Relay Committee Meeting, Harrisburg, PA, October 24, 1969.
10. J. Berdy, "Loss of Excitation Protection for Modern Synchronous Generators," *IEEE Transactions on Power Apparatus and Systems,* Vol. 94, September-October 1975, p. 1457.
11. GEC, *Protective Relay Application Guide Book.* GEC Measurements, 1975.
12. *Protective Relays, Their Theory and Practice.* Chapman and Hall, 1968.
13. *Applied Protective Relaying,* Chapter 6: Generator Protection, Westinghouse Electric Corporation, Relay and Telecommunications Division, 1982.
14. M. S. Baldwin and R. H. Daugherty, "Tripping of Generators during Asynchronous Operating Conditions," presented at 40th Annual American Power Conference, Chicago, April 24-26, 1978.
15. IRD Mechanalysis, Inc., 1975, *Audio-Visual Customer Training Instruction Manual,* IRD Vibration Severity Chart. (The IRD line of products is now part of the Entek Company, which belongs to Rockwell Automation.)
16. ISO 7919-2 (2nd ed.), 2001, International Standardization Organization, www.iso.ch. (The ISO 7919-2 is part 2 of ISO 7919.)
17. R. W. Smeaton and W. H. Ubert, *Switchgear and Control Handbook,* 3rd ed, McGraw-Hill. (Section 13.2.4 contains a brief description of Isolated Phase Bus.)
18. Delta-Unibus Corporation, Catalog No. 4-200, titled "Delta-Star Isolated Phase Bus."

II

INSPECTION, MAINTENANCE, AND TESTING

7

INSPECTION PRACTICES AND METHODOLOGY

7.1 SITE PREPARATION

Site preparation is the first significant activity that should be carried out before inspecting a generator. Every inspection of a large machine—scheduled or not, long or short—requires a sensible effort toward site preparation. The goal is to minimize the risks of contaminating the machine with any foreign material or object, as well as to ensure a safe environment in which to perform the inspection. Site preparation should be planned ahead of time, and it should be maintained from the moment the generator is opened for inspection until it is sealed and readied for operation. Neglecting to prepare and maintain a proper working environment in and around the generator has the potential for resulting in undue risks to personnel safety and machine integrity.

7.1.1 Foreign Material Exclusion

Foreign material exclusion (FME), a term that originated in the nuclear industry, is the set of procedures geared to minimize the possibility of intrusion into the machine of foreign material before, during, and after the inspection.

In principle, the definition of foreign material is anything not normally present during the operation of the generator that may adversely affect its constituent components if left there. For instance, although ambient air is not necessarily considered a foreign material, the water content of the air is. Water definitely is an extraneous element that should be kept from condensing on the machine's windings, retaining rings, and other parts susceptible to mechanical failure from corrosion, or from electric breakdown of the insulation.

Preventing water from condensing onto the machine's components can be readily accomplished by containing both stator and rotor under protective covers (e.g., tents). Tents or other forms of temporary shelter should be designed to withstand the elements, and due thought must be given to rain and strong winds. It is not uncommon to see a tent with damaged canvas or insufficient retention to the frame allowing rainwater onto the windings, causing unnecessary difficulties later when trying to dry the windings to obtain satisfactory insulation ("megger") readings. In addition to a good cover, maintaining a flow of hot air is critical in preventing humidity from condensing on the windings, forging, retaining rings, and other components. The hot air and the positive pressure differential inside the tent eliminate the condensation of any significant amount of water (Figs. 7.1 and 7.2). Although the flow of hot air is normally suspended during the actual inspection for reasons of personnel comfort, thus allowing some condensation to occur, the subsequent flow of hot air will most probably evaporate the moisture and remove it from the containment area [1]. In lieu of hot air, cool, dry air is nowadays used in many stations, providing a comfortable working environment, while still keeping the moisture out.

It is important to perform any scheduled electric tests with dry windings. Otherwise, results obtained will not be representative of the winding condition under normal operating conditions. While attempting to "dry" a winding with the flow of

Fig. 7.1 Typical hot-air blower.

Fig. 7.2 Lowering a protective tent onto the rotor of a large turbo-generator.

hot air for the purpose of attaining desirable insulation resistance readings, the fact the windings are hot may cause the readings to be too low. It can be helpful at a certain point to remove the hot air, allowing the temperature of the insulation to go down. If the winding has been subjected to hot air for a number of hours, reducing the temperature may result in higher insulation resistance readings. It is important, though, that when the hot air is removed, dehumidifiers be installed in the enclosure, and the last enclosure kept sealed as reasonably as possible. Obviously, when working inside a building, the aforementioned requirements can be relaxed compared to those in stations with open turbine decks where the inspection and lay-down areas are in the open.

It is also important not to inadvertently contaminate the generator with corrosive liquids such as solvents, certain oils, and so forth. Sometimes, extraneous fluids can be introduced by walking over them and then walking into the machine. Therefore, in situations where stringent FME rules are applied, paper booties are worn over the shoes. Some inspectors prefer the use of rubber booties placed over their shoes for a better and safer grip.

Paper, rubber, or cloth booties will go a long way in eliminating the introduction of small pebbles that may be stuck to the sole of the shoes. When pressure is applied to the end winding by walking over it, a small pebble can puncture the insulation, thus creating a region of electric-field concentration. This is worth avoiding. It is good practice not to step on the bare coils. A cloth will suffice to protect the winding from the shoe.

The worst enemies of the windings are any foreign metallic objects. They can become airborne due to the high speed of the cooling gas and break the insulation when striking it. Magnetic particles have been known to cause failures in water-cooled generators by penetrating the insulation over long periods of time, due to the electromagnetic forces acting on the particle. It is unfortunate when magnetic particles (also called magnetic "termites" or "worms") are produced within the generator, resulting in damage to the windings, but it is truly shameful when these particles are introduced by negligent practices. Magnetic as well as nonmagnetic metallic objects may be subject to eddy-current heating, detrimentally affecting the insulation with which they come into contact. Foreign metallic objects such as nails, welding beads, or pens inadvertently left in the bore can short-circuit the laminations of the core (see Fig. 7.3 and 7.4). Continued operation under this condition may result in a winding failure due to localized temperature rise of the core. Precautions should be taken to eliminate the possibility of metallic parts or other foreign objects entering the machine. One step in that direction is masking the vent holes of the rotor while they are kept outside the bore, and covering the rotor altogether when not under inspection or being worked on (see Figs. 7.5 to 7.7).

Metallic objects not required for the examination, or for performing work inside the machine, should be left outside the stator bore. This entails removing any coins and other objects (medallions, beepers, unnecessary pens, pencils, etc.) from pockets prior to entering the machine. Inspection tools should be carried into the machine on an "as needed" basis. When using mirrors or flashlights in otherwise inaccessible areas, these should be secured by strapping them to one's wrist (Fig. 7.8). In partic-

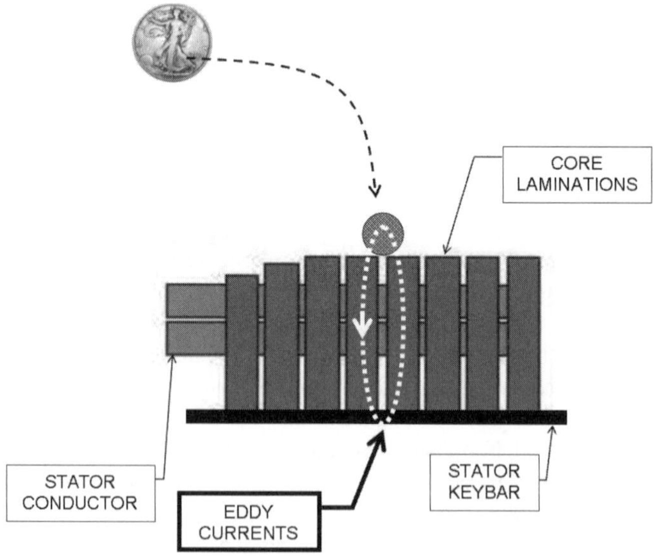

Fig. 7.3 This figure schematically shows how a metal object left inadvertently inside the generator can make its way to the bore of the stator core, shorting a few laminations and effectively creating a path for the flow of large eddy currents. Over time, these currents will overheat the laminations, damaging the interlamination insulation. The end result is a runaway situation that can lead to serious damage to the core. See next figure for such an example.

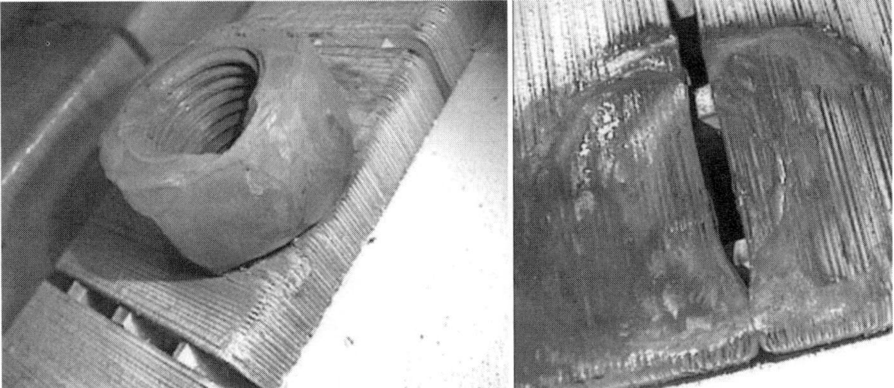

Fig. 7.4 The nut shown in the left photograph was found laying on the stator of a large nuclear-driven generator. The source of the nut is not necessarily due to a breach in FME, but it is presented here to illustrate how a metal object has the potential to damage the laminated core. Laminations were found partially welded to each other under the nut, as shown in the photograph on the right.

ularly compromising situations, such as with very large generators, where loss of production due to an unplanned outage can be extremely expensive, taking an inventory of tools is recommended both before entering the machine and after exiting it. This is a time-consuming practice but recommended for all very large generator inspections (see Figs. 7.9 and 7.10).

Fig. 7.5 Applying masking tape to the vent holes of a large four-pole turbo-generator rotor, with the purpose of eliminating contamination of the rotor winding and other components.

Fig. 7.6 The same rotor as in the previous figure, with the vents covered.

A lack of adequate access control can cause serious damage to the stator or rotor of a large generator, in particular the stator winding (see Figure 7.11). Figure 7.12 shows another case of a metallic object left inside the generator during an outage, finding its way into the end winding of the rotor. This type of occurrence is, unfortunately, not as rare as it should be. Thus, it is critical to account for all tools and parts coming in and out of a generator during an outage.

Fig. 7.7 Rotor body covered to avoid contamination while working on other areas of the rotor (end windings).

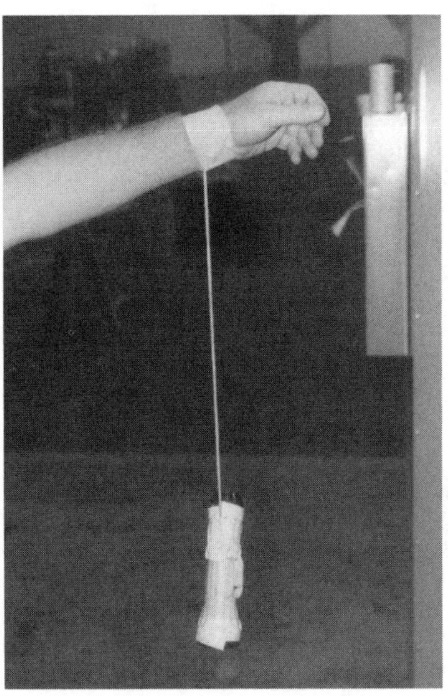

Fig. 7.8 Flashlight strapped to the wrist, to eliminate the possibility of dropping it in inaccessible places during a generator inspection.

Fig. 7.9 Wooden cover with door at the entrance to the bore area (both sides). These covers allow access control of personnel and tools to the machine.

Fig. 7.10 Temporary barrier erected to keep vital components from being contaminated with foreign objects or materials.

Fig. 7.11 The bottom photograph shows damage to the stator of a large generator caused by the bolt shown in the top photograph. The bolt was inadvertently left in the machine due to a breach in FME.

Fig. 7.12 The photograph shows a washer left inside a generator due to a breach of FME. The washer found its way under a retaining ring, becoming lodged between the top turns of a coil and the nonmetallic coil separator. The washer was found during troubleshooting of another problem in the area. If left there, it might have eventually caused a ground fault to the retaining ring, after damaging the insulating liner resting between the retaining ring and the winding.

7.1.2 Foreign Material Exclusion—Procedures

A good FME effort must include clearly written procedures addressing all aspects of training and implementation. Responsibilities must be clearly defined, as well as contingencies for loss of FME control. A typical FME procedure may include the following subjects:

Responsibilities of:

- Planner
- Supervisor
- FME monitor
- FME qualified worker
- Non-FME qualified worker
- Escorts

Issues addressed:

- Definitions
- Controls of FME zones
- FME zone designation

- FME signs
- FME covers
- Appropriate clothing
- Tool controls
- Spare parts controls
- Recovery of lost FME control
- Final FME check

Training

- As required by the level of *responsibility* and the *issues* for each level described above

7.2 EXPERIENCE AND TRAINING

Inspection of large turbo-generators is not a trade that can be solely learned in the classroom. A combination of classroom learning with years of hands-on training, mainly while accompanying an experienced specialist, is what leads to the ability to decipher the root causes of conditions afflicting a generator. Misjudging root causes, or simply not understanding the condition of the machine, can lead to huge monetary losses. Therefore, it is always a good idea to have someone well experienced leading the inspection. When large units are inspected, repaired, or overhauled, most owners will arrange for a manufacturer's specialist to be present, at least through part of the effort.

7.3 SAFETY PROCEDURES—ELECTRICAL CLEARANCES

When carrying out work in an industrial environment, nothing is more important than adhering to all required safety precautions. Large machines opened for inspection often present obstacles in the form of big openings in the floor surface, crevices to crawl thorough, rods and protruding machine members, and so on (see Fig. 7.13). They all demand evaluation of required temporary additions to the site, such as beams over the open floor spaces, handrails (Fig. 7.14), secure ladders, and so forth.

The obstacles just mentioned are all visible to the people engaged in the inspection. However, an invisible and very powerful element to contend with is the voltage potential (or range of voltages) that may be present in a generator. Although rare, electrical accidents can occur when work is performed on large machines.

A comprehensive inspection of a large generator requires direct physical contact with all windings and other elements that are normally energized during the operation of the machine. "Walking the clearance" is jargon used by some to describe the process of inspecting all breakers, cables, and connections that may be sources of electric power to any part of the machine, and making sure they are all deenergized and secure. This means that none of these will be accidentally energized during the inspection. The following is a typical (but by no means all-inclusive) list of safety procedures:

Fig. 7.13 Site of a 1250 MVA unit undergoing overhaul.

- "Personal grounds" (grounding cables) at both ends of the winding of each phase will minimize the possibility of receiving an unexpected electric discharge of dangerous magnitude (Fig. 7.15). Bear in mind that the conductor's cross section must be sized to withstand the available short-circuit currents for the expected duration of the fault, given the protection settings on-site. The connections of the personal grounds must be verified to be tight and secure.

Fig. 7.14 Provisions allowing safe access to the bore of a large generator.

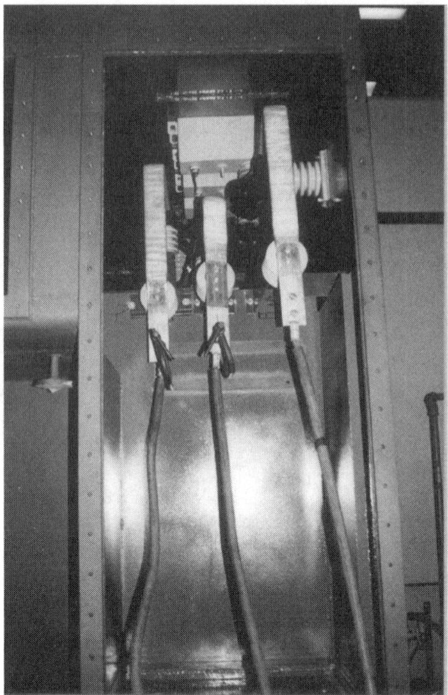

Fig. 7.15 Ground leads applied to a generator being overhauled.

- Phase leads must be open.
- Neutral transformer (if present) must be disconnected or have its leads opened.
- Voltage regulators and other excitation equipment must be disconnected.
- Potential transformers are an additional source of voltage to the main windings and, therefore, they must be disconnected and secured. Space heaters are often overlooked. To keep moisture out, space heaters, when present, are normally left "on" after disassembling the generator; thus, it is important to check that they are disconnected during the inspection.
- All switches that may energize any part of the machine must be clearly tagged. *Only* the person who installed a tag can remove it, after verifying that the inspector(s) have left the machine. If someone else is to remove tags, there should be in place a very clear written procedure that ensures that the next person knows someone is inside the generator.
- In machines with direct gas cooling of the stator windings, discharge resistors are often found installed on the coil knuckles (see Fig. 7.16). When these resistors are faulty or disconnected, the conductors may remain charged for substantial lengths of time. Precaution should be taken when inspecting such windings, in particular if high-voltage tests were performed before disassembling the unit

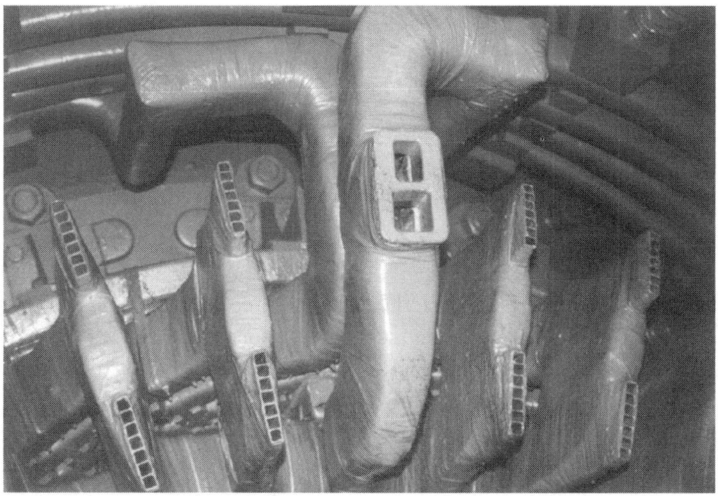

Fig. 7.16 Four knuckle areas (two at each side of the connection lead) of directly gas-cooled stator coils. The discharge resistors are inside the knuckles.

or ahead of the inspection. If in doubt, a grounded *discharge stick* can be used to discharge the suspected coils. *You can never be too safety-conscious when dealing with high-voltage apparatus.*

- Turning gear must be disconnected (fuses removed) and clearly tagged when inspecting with the rotor in place. The last thing you want is for that rotor to inadvertently start turning when you are in there.
- Gas monitors for confined areas must be used to ascertain that the breathing air is of good quality.
- Additional items as each specific case warrants.
- Follow all relevant safety rules and regulations at the site.
- Finally, *you* walk the clearance. Do not take anyone else's word for it. This is even more important when the inspection is performed in a station other than yours.

7.4 INSPECTION FREQUENCY

Certain components in large synchronous machines require inspections (and sometimes maintenance) between scheduled major outages. Other more comprehensive inspections requiring various degrees of machine disassembly are performed during the more lengthy outages.

However, experience shows that a major inspection after one year of operation is highly recommended for new machines. During the initial period, winding-support hardware and other components experience harsher than normal wear. Retightening of core-compression bolts may be required during this first outage. Also, expiration-of-warranty issues may be of consideration.

Subsequent outages and inspections can be performed at longer planned intervals. How long an interval? Minor outages/inspections every 30 months, to major outages/inspections every 60 months, are typical periods recommended by machine manufacturers. The major outages include removal of the rotor, comprehensive electrical and mechanical (nondestructive) tests, and visual inspections. Of course, these intervals tend to be longer for machines spending long periods without operation. Most stations have logs containing the actual number of hours the unit was running and the number of starts/stops. Together with the manufacturer's recommendations, this information forms the basis for scheduling inspections and overhauls.

Large utilities that have many generators in their systems and many years of experience running these machines have formed their own maintenance and inspection criteria and schedules. Although working closely with machine manufacturers, these utilities tend to extend the periods between outages for those machines that experience has shown to have good records of operation, and shorten the periods between outages/inspections for machines that have been characterized by more frequent failures.

Frequently, the major outages during which the opportunity presents itself to carry out a major inspection follow the need to maintain the prime mover more than the generator itself. This is certainly the case with nuclear-powered steam turbines, when the inspection of the turbine and generator is made coincident with the refueling of the reactor. (The subject of frequency of inspection and maintenance is broached in Chapter 12.)

7.5 GENERATOR ACCESSIBILITY

The issue of accessibility is the same for inspection and for maintenance, as both activities often follow each other. The following list summarizes the various levels of access to the generator. The level of access required is commensurate with the purpose of a specific inspection activity, and it will be somewhat different for differently contructed machines.

- Generator casing end-doors (end-brackets) removed
    - Stator end-windings access
    - Access to retaining-ring outer surface
- Rotor removed
    - Access to stator bore
    - Full rotor inspection possible
- Hydrogen coolers removed (vertical coolers; see Fig. 7.17)
    - Access to back of stator core in localized areas
    - Full H_2 cooler inspection and cleaning possible
- Terminal enclosure opened
    - Access to stator terminals
    - Access to back of stator end windings

Fig. 7.17 A vertical H_2 cooler is being removed from a four-pole, 1250 MVA generator. The opening can be used by an inspector to entry the machine for inspecting the end-winding region of the generator, as well as part of the back of the core.

- Belly ports opened

  Access to some back-of-core areas between stator frame ribs

  Access to some key bars

  Access to belly bands

There is often some discussion going on about the need to remove a rotor for a comprehensive inspection, given the availability of some robotic devices to crawl anywhere in the air-/gasgap of large generators, and even probe wedge tightness with some wedging designs. Although the availability of robotic devices, boroscopes, and other miniature cameras alleviate the need for major disassembly for every inspection, there is no replacing the age-old visual inspection inside the bore during major inspections to fully appreciate the condition of a generator.

7.6 INSPECTION TOOLS

Probably the most important item for the inspector is not a gadget or an instrument, but a piece of paper, namely, a record of previous visual inspections and electric tests. Findings from past inspections act like a compass, helping to guide the inspector to areas already proved to be problematic. Most operators of large turbo-generators carry out a minimum set of electrical tests on the machines before disassembling them for a visual inspection. Results from these tests have the salutary advantage of, first, calling attention to problematic areas and, second, allowing comparison with the test results obtained after cleanup and refurbishment performed on the machine. It goes without saying that a comprehensive inspection report should always be written and archived. This report is a helpful reference for the next inspection, often several years later.

An additional source of information that has served the authors of this book well is historical information obtained from identical or similar machines. Still other significant sources of information are the facts or recommendations sheets periodically issued by the manufacturers.

As to the "bag" of tools carried by the inspector, it may include the following (see Fig. 7.18):

- Writing pad attached to a clipboard, and a nonmetallic pen attached to the board with a string
- Safety glasses
- Disposable gloves
- Disposable paper or cloth booties, or multi-use rubber booties
- Disposable paper overall, or cloth overall

Fig. 7.18 Easy-to-carry tool case with the most essential tools for visual inspection of the generator.

- Work shoes with soft sole (rubber)
- Floodlight with an extension cord
- Flashlight and a set of spare batteries. If a flashlight is at risk of falling into inaccessible places, it should be attached to the wrist with string and tape, as shown in Figure 7.8.
- Clean rags.
- A small sealed container with white rags to be used as swabs; useful for taking samples of contamination for later testing, when required.
- A set of mirrors with articulated joints and expandable handles. If the mirrors are at risk of falling to inaccessible places, they should be attached to the wrist with a string and tape.
- A hammer with soft (rubber) and hard (plastic) heads, for probing wedges, insulation blocking, and so on. For a wedge survey of the entire machine, hand-held electromechanical probes are commercially available (Fig. 7.19). After an initial setting, the probe identifies each wedge as *tight, loose,* or *loose/hollow.*
- A set of magnifying glasses or hand-held microscopes to probe for corrosion or electrically originated pitting on retaining rings and on other critical components.
- Charts from manufacturers of commutator brushes depicting observable signs of bad commutation (Fig. 7.20).
- Boroscope, especially suitable for inspecting under the retaining rings, air/gas ducts, air-/gasgaps of machines with the rotor in situ, and other inaccessible spots (Fig. 7.21).
- A good camera capable of taking close shots of small areas.

Fig. 7.19 Commercially available wedge tightness electromechanical tester.

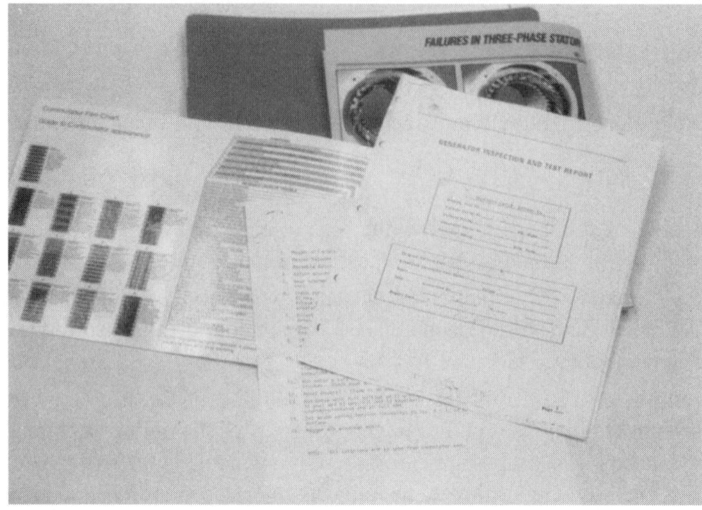

Fig. 7.20 Commutation performance charts from various manufacturers of brushes and other useful information, such as vendor graphs.

- A small magnet for the extraction of loose iron particles.
- A compass to pinpoint magnetized items.

Figures 7.22 to 7.30 show several of the items described above, in addition to other specialized tools for probing different components of the alternator.

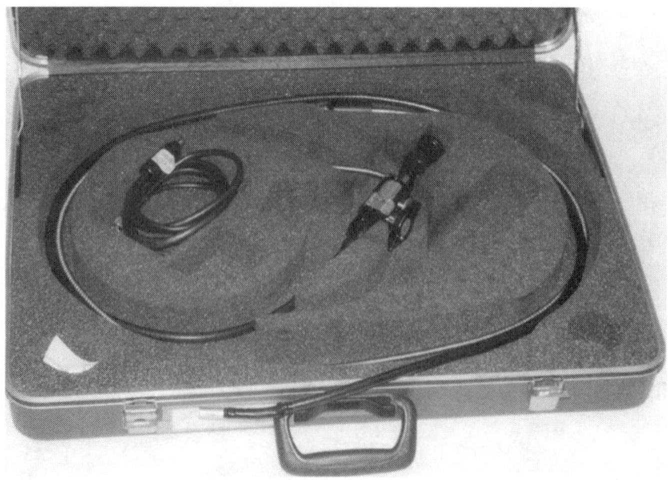

Fig. 7.21 Boroscope used for visual inspection of otherwise inaccessible areas of the machine. Light source and optical video equipment not shown.

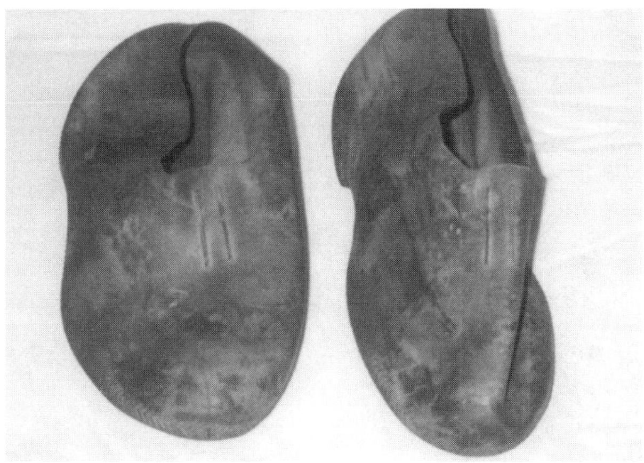

Fig. 7.22 Rubber booties.

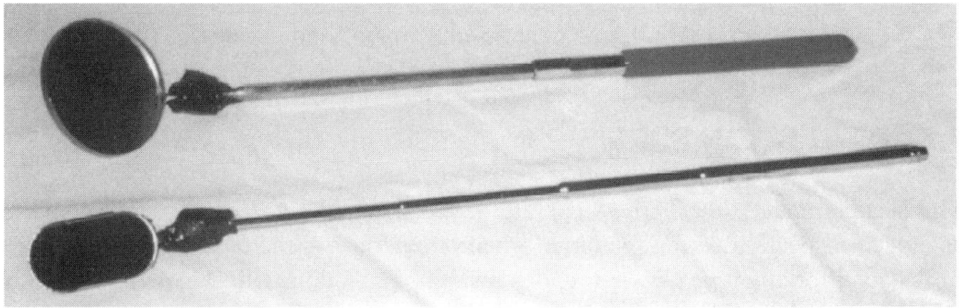

Fig. 7.23 A set of mirrors with articulated joints and expandable handles.

Fig. 7.24 Two styles of hammers for probing wedges and insulation blocking, among other things. Hammers with hard but nonmetallic surfaces are preferable for manual wedge surveying.

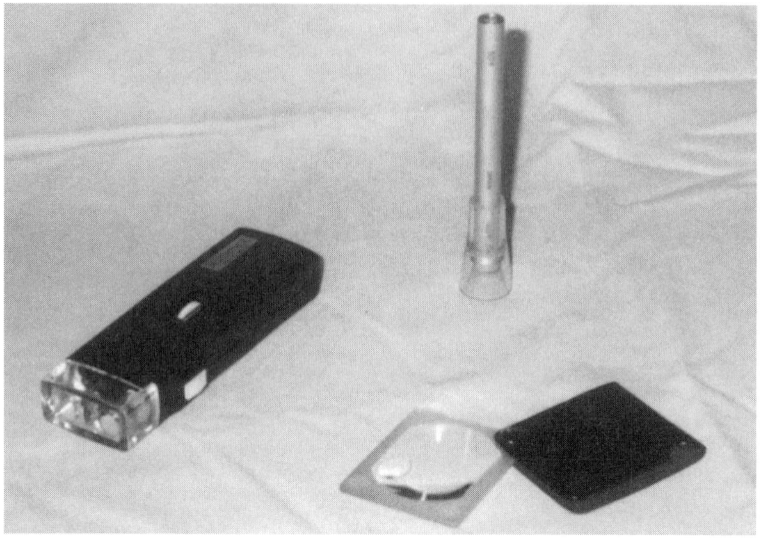

Fig. 7.25 Types of hand-held magnifying glasses.

7.7 INSPECTION FORMS

This section includes the Turbo-Generator Inspection and Test Report, which comprises eleven inspection forms for large synchronous turbo-generators. A similar set of forms good for any type of synchronous machine (including salient-pole units) can be found in Reference [2].

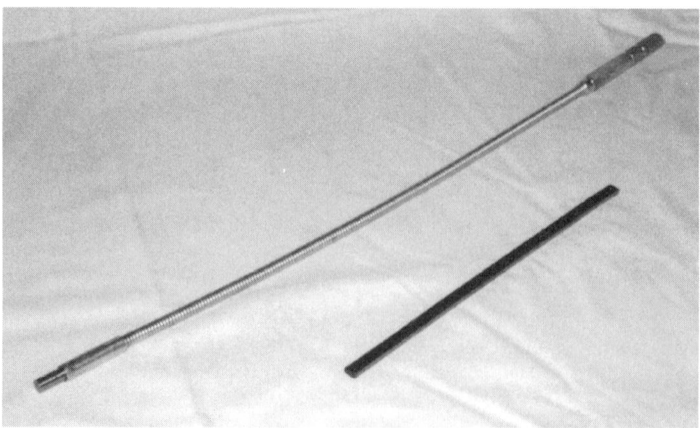

Fig. 7.26 Two types of magnets designed for access to remote locations inside the generator.

Fig. 7.27 Compass to ascertain location of magnetized surfaces, meter, and a range of feeler gauges and rulers for measuring clearances (e.g., between coils and slot).

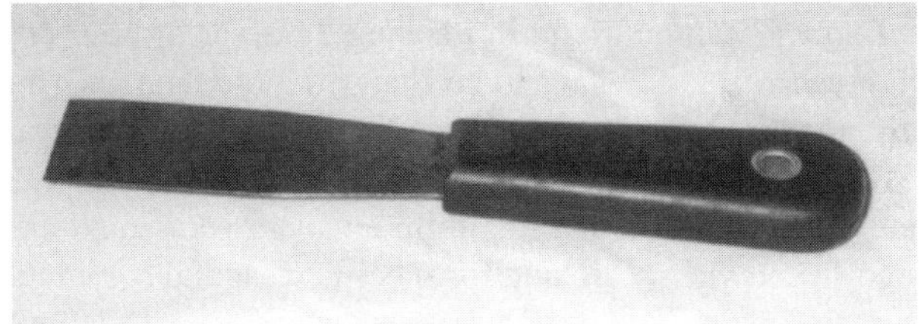

Fig. 7.28 Core tightness checking knife.

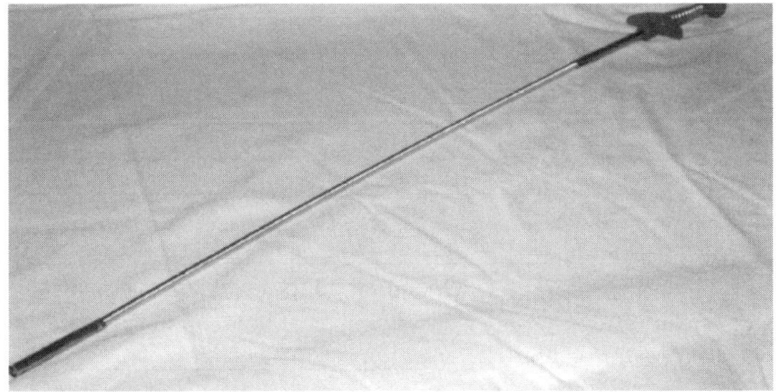

Fig. 7.29 Pincher for collecting loose samples in hard-to-reach places.

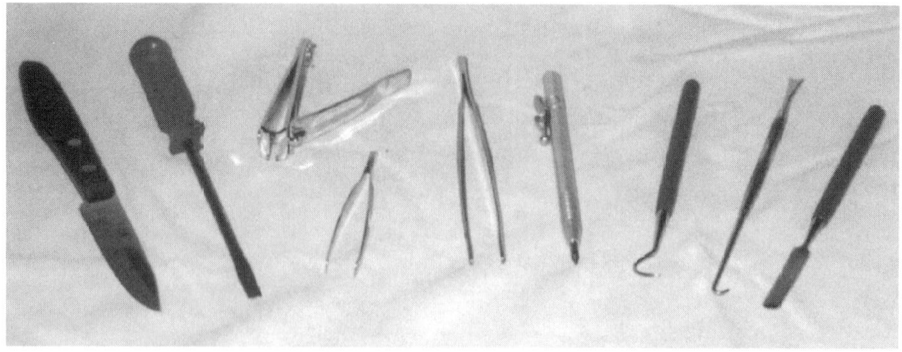

Fig. 7.30 Miscellaneous tools for probing and inspecting.

The inspection forms included in this chapter cover full inspections and routine inspections. They are examples of forms that can be used practically for any turbine-driven generator, of any industrial size. The forms are by nature generic. However, it should take very little effort to recast them in a way that can be quickly adapted to the needs of any machine operator and/or inspector.

Forms 1 to 9 are designed for a full or partial inspection of a machine in various stages of disassembly. Form 10 is for daily, weekly and monthly routine inspections of the brush gear. Form 11 can be useful in tracking the wear of brushes in a generator. Given that each brush will rarely have to be changed more than once a month, using a form for an entire month should suffice. This way, 12 forms—one per month—should be adequate to document brush changes for each brush location and date of replacement for the entire year. Completing this form entails a very small effort, but the rewards can be significant when searching for specific wear and commutation problems.

TURBOGENERATOR INSPECTION AND TEST REPORT

Form 1: Basic Information

Station/Company where installed ______________________________
Unit no. ______________________________
Prime mover type: Steam __________ Gas __________ Diesel __________
Serial no. of generator ______________________________
Manufacturer ____________________ Frame ____________________
Date of manufacture ________________ Year installed ________________
Date of last rewind: Stator ________________ Rotor ________________
Date of last major inspection ______________________________
Total operating hours __________ Operating hours since last overhaul __________
Total number of starts/stops _________ Number of hours in turning gear ________
Present inspection performed by ______________________________
Assisted by ______________________________
Date of Inspection ______________________________

TURBOGENERATOR INSPECTION AND TEST REPORT

Form 2: Nameplate Information

Rated MVA__________ Power Factor__________ Short-circuit ratio__________
Field I_2t ______________
Stator: Line voltage ______________kV Rated current______________ amperes
Field: dc voltage ______________volts dc current______________ amperes
Nominal speed ______________rpm No. of poles______________
Frequency______________Hz
Stator cooling: Open air______________ Air/water______________
Direct Water__________ H_2__________ Direct H_2__________
Rotor cooling: Air____________ Hydrogen____________ Water____________
Stator Insulation: Asphalt______ Epoxy/resin_mica______ VPI (Y/N?)________
Other _______________
Maximum H_2 pressure (psi)_______ Normal H_2 operating pressure (psi) __________
Stator cooling water operating pressure (direct water-cooled units) (psi)____________
Number of stator slots_____________ Number or rotor coils/pole______________

TURBOGENERATOR INSPECTION AND TEST REPORT

Form 3: Inspection Accessibility

	Yes	No
Rotor out of bore		
Inboard top bracket (or full bracket) removed		
Inboard bottom bracket removed		
Outboard top bracket (or full bracket) removed		
Outboard bottom bracket removed		
Bushing well open		
Access via a vertical H_2 cooler (coolers removed___________)		
Inboard retaining ring off		
Outboard retaining ring off		
Exciter's rotor out of exciter's bore		
Stator belly ports opened		

TURBOGENERATOR INSPECTION AND TEST REPORT

Form 4: Stator Inspection Items

Type of blocking: Maple ________; Textolite ________; Conforming ________; Felt ________
End-winding makeup: Z-ring ________; Radius strip ______; Sausages ________; Other ________
Ties: Glass roving _______________; Flax _______________; Other _______________
Type of wedges: Flat _______________; Piggyback _______________; Other _______________
Type of side fillers: Flat ____________; Ripple spring ____________; Nonexisting ____________
Top ripple filler?: Yes _______________; No _______________
Boroscopic inspection of the air/gas-duct area performed? Yes ____________; No____________
If wedge survey performed: % of loose wedges ___________; % of hollow wedges ____________
Number of damaged resistors in nose of water-cooled windings _______________
Number of damaged resistors in nose of directly H_2-cooled windings _______________

"O" for satisfactory
"X" for unsatisfactory

Item	Description	N/A	O/X	Corrective Action
S01	Cleanliness of bore (oil, dust, etc.)			
S02	Air/gas radial ducts clogged?			
S03	Iron oxide deposits?			
S04	End-winding hardware condition (bolts, nuts, etc.)			
S05	HV bushings			
S06	Stand-off insulators			
S07	Bushing vents clogged?			
S08	Greasing/red-oxide deposits on core bolts?			
S09	Space heaters			
S10	Fan-baffle support studs			
S11	Heat exchangers, cleanliness			
S12	Heat exchangers, leaks			
S13	Hydrogen desiccant/dryer			
S14	Core-compression ("belly") belts			
S15	Bearing insulation (at pedestal or Babbitt)			
S16	Coil cleanliness (oil, dust, etc.)			
S17	Blocking condition			
S18	Ties between coils tight?			
S19	Ties between coils too dry?			
S20	Ties to surge rings tight?			
S21	Ties to surge rings too dry?			
S22	Surge rings insulation condition			
S23	Surge rings support assembly			
S24	Additional end-winding support hardware			

Item	Description	N/A	O/X	Corrective Action
S25	RTD and TC wiring hardware			
S26	Asphalt bleeding? Soft spots?			
S27	Tape separation? Girth cracking?			
S28	Insulation galling/necking beyond slot?			
S29	Insulation bulging into air/gas ducts?			
S30	Insulation too dry? Flaking?			
S31	Circumferential bus insulation?			
S32	Corona activity			
S33	Wedges condition (wedge survey below)			
S34	Wedges slipping out at ends of core?			
S35	Side/bottom packing fillers slipping out at ends of core?			
S36	Bars bottomed in slot?			
S37	Laminations bent in bore? Broken? Skid Plate damage?			
S38	Laminations protruding into air/gas ducts?			
S39	Terminal box CTs condition			
S40	Bushing-well insulators and H_2 sealant condition			
S41	Winding support bearing-bolt condition			
S42	Water boxes condition (water-cooled stator)			
S43	PTFE (Teflon) hoses condition (water-cooled stator)			
S44	Manifold leaks/corrosion (water-cooled stators)?			
S45	Magnetic termites found inside machine?			
S46	Back-of-core condition?			
S47	Core tight in the bore? (Knife check for looseness)			
S48	Space block migration or damage in the radial core vent ducts in the bore?			
S49	Back-of-core burning between keybars and core at core ends?			
S50	Core frame or weld cracks?			
S51	Arcing at stator frame ground points?			
S52	Arcing at inner end door joints?			
S53	Condition of flux shields or flux shunts?			
S54	Does flux shield float electrically (i.e., megger value)?			
S55	Condition of internal casing conduits for instrumentation (arcing, breakage, etc.)?			
S56	Condition of core to frame coupling devices (spring bars, any damper pads etc.)?			
S57	Space block migration from radial vents at back of core?			

TURBOGENERATOR INSPECTION AND TEST REPORT

Form 5: Rotor Inspection

Retaining rings type: Magnetic _______; Nonmagnetic _______; % Mn _______; % Cr ________
Spindle-mounted _____________; body-mounted ______________
Collector rings position: Both polarities at same end __________; At opposite ends ___________
Number of collector rings per polarity _______________
Type of conductor: Double copper turn (prone to copper dusting) _____;
Each turn fully insulated in end-winding region _____
Type of wedges: All aluminum ___; All steel ___; Both aluminum and steel ____;
Each wedge full length of rotor ____
Type of shaft-mounted fans: Axial ___; Centrifugal ___; One fan only ___;
Two fans, one per side ___
Type of rotor cooling: In from the ends/out from wedge vents ____;
In from the ends and in/out from wedges _____
Cooling vents on surface of retaining rings _____

"O" for satisfactory
"X" for unsatisfactory

Item	Description	N/A	O/X	Corrective Action
R01	Rotor cleanliness (oil, dust, iron/copper dust)			
R02	Retaining rings' visual appearance			
R03	Centering rings' visual appearance			
R04	Fan rings' visual appearance			
R05	Fretting/movement at rings' fits			
R06	Fan blades and hub condition			
R07	Bearing journals condition (scoring, heat stains?)			
R08	Balance weights/bolts condition			
R09	End wedges (touching end rings? loose?)			
R10	Other wedges (overheated, loose, cracks)?			
R11	End-winding condition (distortion, heat staining, cracked joints, arcing)?			
R12	Collector rings condition (ring surface, groove depth)			
R13	Collector rings insulation condition (megger value at 500 V dc)			
R14	Brush springs' pressure and condition			
R15	Brush rigging condition (clean, damaged, arcing, etc.)?			
R16	Shaft-voltage discharge-brush condition			
R17	Inner/outer H_2 seals condition (pitting, scoring, heat staining, etc.)?			

Item	Description	N/A	O/X	Corrective Action
R18	Pole face cross-cuts for stiffness equalization (heat stains at cut ends, cracks?)			
R19	Venting holes in retaining rings surface condition (pitting, cracks, heat stain, arcing)?			
R20	Venting holes in wedges condition (cracks, blocked)?			
R21	Indication of electric arcing between wedges and retaining rings and/or forging?			
R22	Slot liners slipping at ends under the retaining rings?			
R23	Insulation between conductors fully or partially covering venting holes along slots?			
R24	Condition of coupling and bolt holes (gouging, cracks)?			
R25	Condition of any toothed gears (i.e., turning gear)			
R26	Terminal stud connections to collector rings and end winding under retaining rings			
R27	Pole-face axial slots for stiffness equalization (wedges, balance weights condition)			
R28	End winding coil-to-coil and pole-to-pole connectors (cracked joints, arcing, deformation)?			
R29	Locking tabs secure on bolts etc.			
R30	End-winding packing and blocking condition (missing pieces, cracked, out of place)?			
R31	End-winding interturn insulation condition (insulation migration or other damage)?			

TURBOGENERATOR INSPECTION AND TEST REPORT

Form 6: Excitation Inspection

Type of excitation: Self-excited (brushless) _____; Stand-alone solid state _____;
Stand-alone rotary generator ____ Shaft-driven generator ____;
Other _______________

"O" for satisfactory
"X" for unsatisfactory

Item	Description	N/A	O/X	Corrective Action
E01	Cleanliness (oil, carbon dust, etc.)			
E02	Shaft-mounted diodes' condition			
E03	Diodes' connections and support hardware			
E04	Commutator's condition			
E05	Commutator brushes' condition			
E06	Commutator brush holders' and rigging's condition			
E07	Brush springs pressure and condition			
E08	dc generator's stator condition			
E09	dc generator's armature condition			
E10	Exciter-drive motor cleanliness			
E11	Exciter-drive stator condition			
E12	Exciter-motor's rotor condition			
E13	Field discharge resistor condition			
E14	Solid-state exciter's general appearance			

TURBOGENERATOR INSPECTION AND TEST REPORT

Form 7: Comments

TURBOGENERATOR INSPECTION AND TEST REPORT

Form 8: Wedge Survey

This is a typical table for performing a wedge survey. Enough rows and columns should be included to cover the machine to be inspected. One way to enter information is: "O" for a *tight* wedge, "X" for a *hollow* or *medium* wedge, "L" for a *loose* wedge. Wedges can be tapped at both ends. In that case, the number of notations will be double.

At the end of the survey indicate # of *loose* wedges and % of total, # of *hollow/medium* wedges and % of total, number of *tight* wedges and % of total.

← Turbine End **Wedge Number** Nondrive End →

Slot Number

	1	2	3	4	5	6	7	8	9	10	11	12	13	14	15	16	17	18	19	20	21	.	.	.	.	.	.	N
1																												
2																												
3																												
4																												
5																												
6																												
7																												
.																												
.																												
.																												
.																												
N																												

TURBOGENERATOR INSPECTION AND TEST REPORT

Form 9: Electric Test Data

The following sample of data is from tests performed after machine shutdown (some of these tests are to be performed under hydrogen pressure for hydrogen-cooled generators). Refer to the pertinent standards for the correct test procedures. See references in Chapter 11.

Machine Stator

- Measured conductivity for liquid-cooled generators: __________ micro-ohm/cm
- Measured megger readings of windings to ground. Megger's output voltage: ________V

  30 s ______________, 1 min ______________, 10 min ______________ (MΩ)

  Polarization Index (PI) (10 min/1 min): ______________

  Ambient temperature ______________; Hours after shutdown ______________

  Stator RTDs temperature 1) ___________ 2) ___________ 3) ___________

  4) ___________ 5) ___________ 6) ___________

  (at least 2 per phase)

  Note: For water-cooled stators only 1-minute megger required.

Machine Rotor

- Measured megger readings of windings to ground. Megger's output voltage: ____V

  30 sec ______________; 1 min ______________; 10 min ______________ [MΩ]

  Polarization index (PI) (10 min/1 min): ______________

  Ambient temperature ______________; Hours after shutdown ______________

Alternator Exciter

Stator

- Measured stator megger readings of windings to ground

  30 s ______________, 1 min ______________, 10 min ______________ (MΩ)

  Polarization index (PI) (10 min/1 min): ______________

  Winding temperature ______________

Rotor

- Measured megger readings of windings to ground

  30 s ______________, 1 min ______________, 10 min ______________ (MΩ)

  Polarization index (PI) (10 min/1 min): ______________

  Winding temperature ______________

(*continued*)

Form 9: Electric Test Data *(cont.)*

Alternator Exciter (cont.)

dc Exciter

- Measured megger readings of windings to ground

  30 s ______________, 1 min ______________, 10 min ______________ (MΩ)

  Polarization index (PI) (10 min/1 min): ______________

  Winding temperature ______________

RTDs

- Megger test to ground with 500 V megger
- Measure each RTD's resistance with a bridge
- Compare readings with measured temperature of the winding

Stator Water Outlet Thermocouples

- Measure millivolts and compare readings with measured temperature of water or ambient air (when empty).

Additional Tests

- PD activity readings should be taken before shutdown if PDA sensors installed.
- Air-/gasgap flux waveform should be recorded before shutdown if flux probes installed.

Alarm Checks

The following is a sample of the alarm circuits and activators that require checking (different machines will have different sets of alarms):

- Air filters clogging alarms
- Stator cooling water pressure low
- Water pressure at machine
- Water flow
- Stator water filter
- Stator water-cooling pump
- Water and oil leakage detectors
- Hydrogen seal-oil enlargement detector

TURBOGENERATOR INSPECTION AND TEST REPORT

Form 10: Comprehensive Brush Routine Inspection

Date: ___________ UNIT: ______

Visual Inspection	Chattering Brushes	Sparks	Overheating	Cleanliness	Spring Tension
OK					
Not OK					

Positive Polarity

	Column 1	2	3	4	5	6 etc.
Row 1 Date:	A B C D F	A B C D F	A B C D F	A B C D F	A B C D F	A B C D F
Row 2 Date:	A B C D F	A B C D F	A B C D F	A B C D F	A B C D F	A B C D F
Row 3 Date:	A B C D F	A B C D F	A B C D F	A B C D F	A B C D F	A B C D F
Row 4 Date:	A B C D F	A B C D F	A B C D F	A B C D F	A B C D F	A B C D F
Row 5 Date:	A B C D F	A B C D F	A B C D F	A B C D F	A B C D F	A B C D F
Row etc. Date:	A B C D F	A B C D F	A B C D F	A B C D F	A B C D F	A B C D F

Negative Polarity

	Column 1	2	3	4	5	6 etc.
Row 1 Date:	A B C D F	A B C D F	A B C D F	A B C D F	A B C D F	A B C D F
Row 2 Date:	A B C D F	A B C D F	A B C D F	A B C D F	A B C D F	A B C D F
Row 3 Date:	A B C D F	A B C D F	A B C D F	A B C D F	A B C D F	A B C D F
Row 4 Date:	A B C D F	A B C D F	A B C D F	A B C D F	A B C D F	A B C D F
Row 5 Date:	A B C D F	A B C D F	A B C D F	A B C D F	A B C D F	A B C D F
Row etc. Date:	A B C D F	A B C D F	A B C D F	A B C D F	A B C D F	A B C D F

Form 10: Comprehensive Brush Routine Inspection (*cont.*)

Legend: **A**: Chipped brush, **B**: Cracked brush, **C**: Sticky brush, **C**: Sparking/Arcing, **F**: Replaced with new brush

- Circle the corresponding letter(s) in each box, according to the inspection.
- Inside each box, note the length of the brush.
- If a brush is changed, make a diagonal across the box, and write on top of the length of the old brush, and on the bottom of the length of new brush.

Number of Brushes Replaced: Positive ________ Negative _______ Total ________

Comments:

____

____

____

____

____

Condition of rotor grounding brush: ____________________________________

Shaft voltage reading: ___________________________________

Condition of bearing insulation: NDE Reading: ________ DE Reading: ________

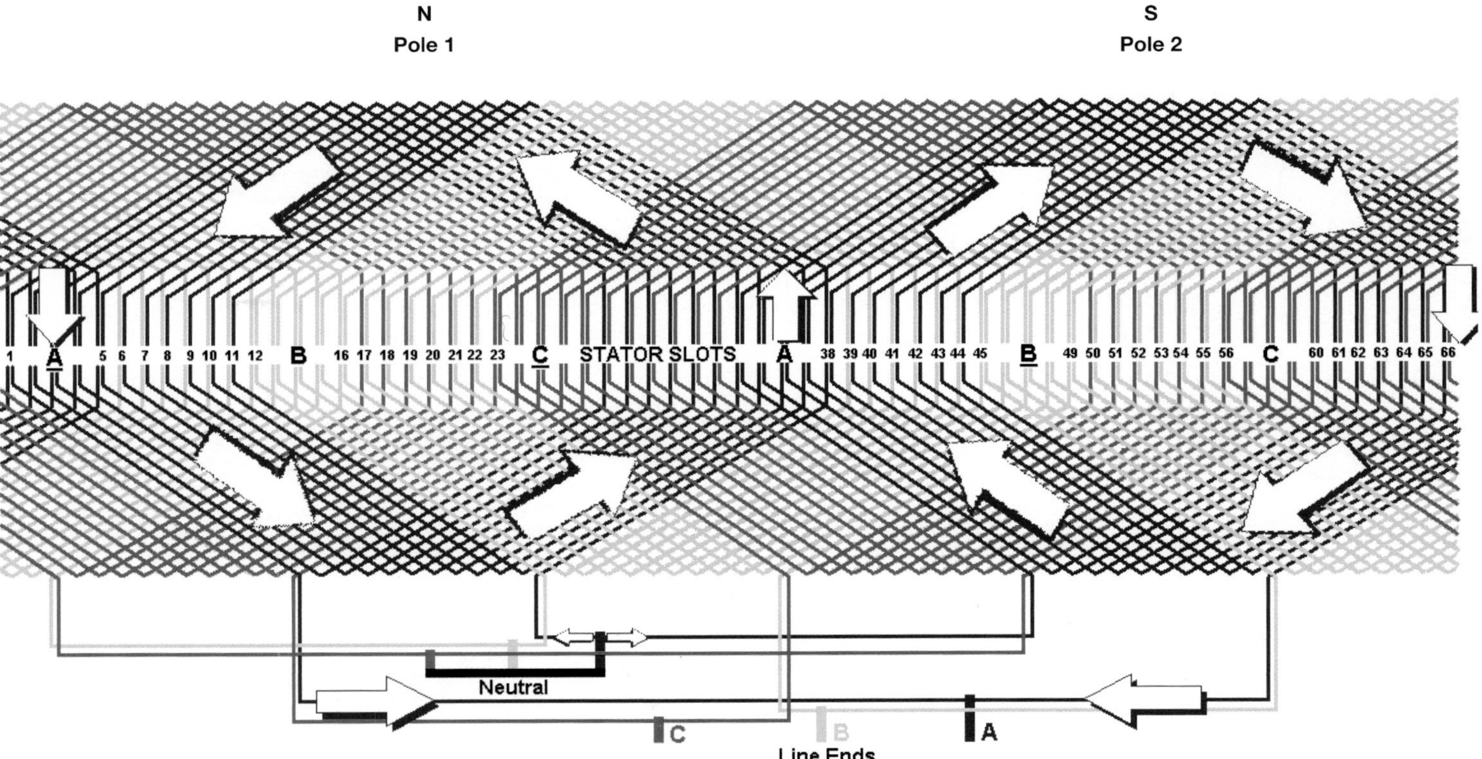

Fig. 1.30 "Developed" view of a two-pole stator, showing the slots, the poles, and the three winding phases. It can be readily seen that the winding runs clockwise under a north pole and counterclockwise under a south pole. A similar pattern is followed by the other two phases, but located 120 electrical degrees apart. See page 29 for text discussion.

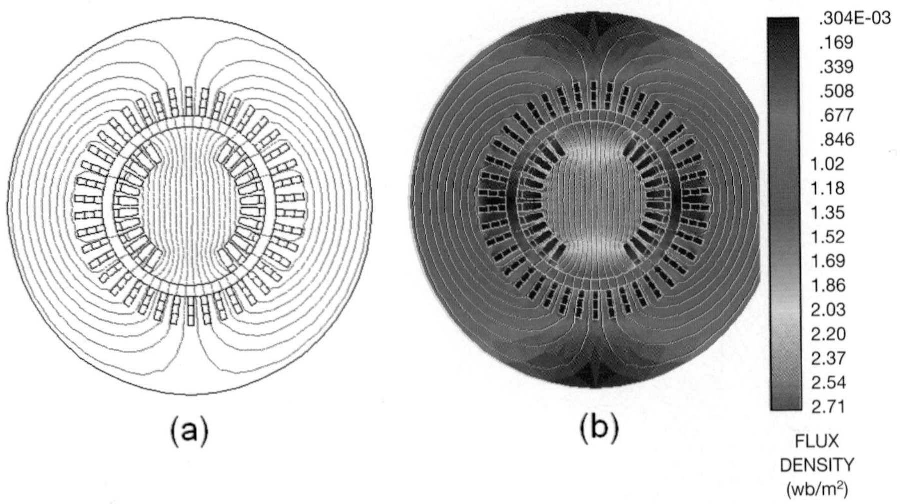

Fig. 2.10 (*a*) Two-pole generator, open-circuit flux distribution. (*b*) Two-pole generator, open-circuit flux density distribution. See page 53 for text discussion.

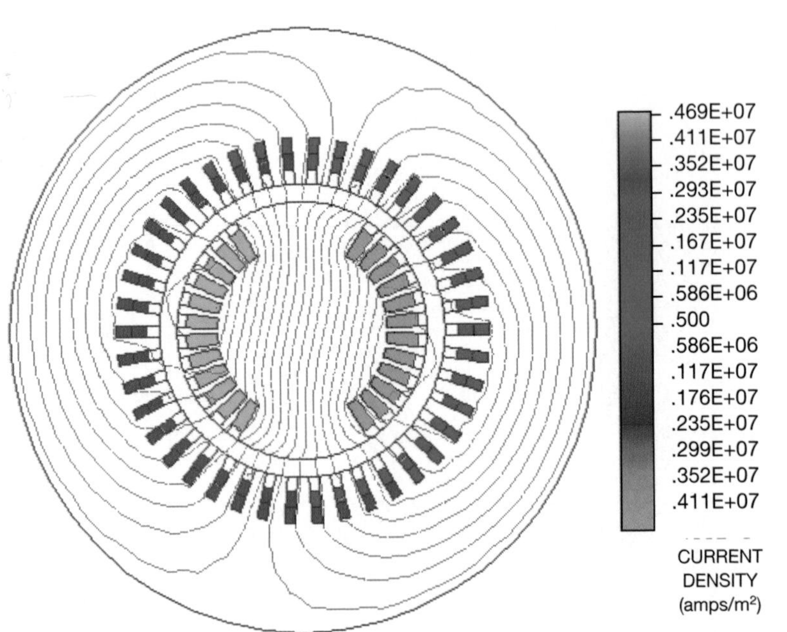

Fig. 2.11 Two-pole generator, full load current density and flux distribution. See page 53 for text discussion.

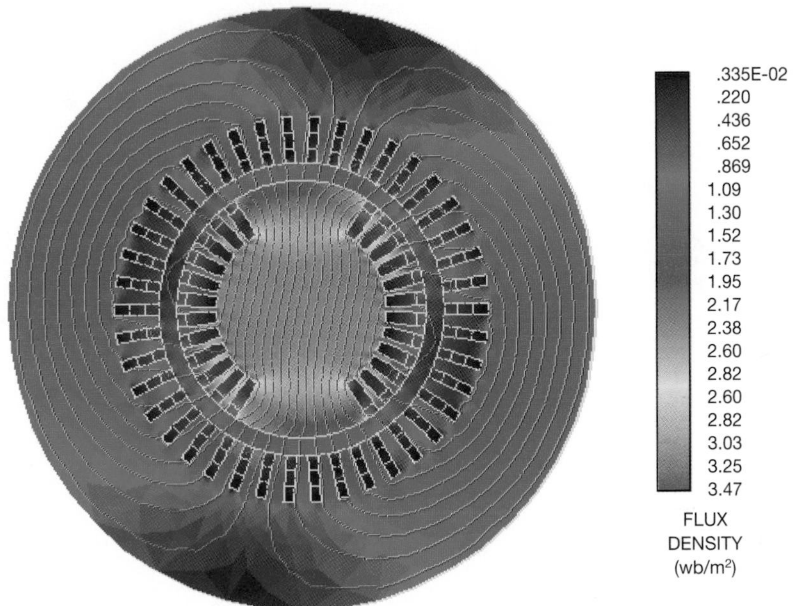

Fig. 2.12 Two-pole generator, full load flux and flux density distribution. See page 54 for text discussion.

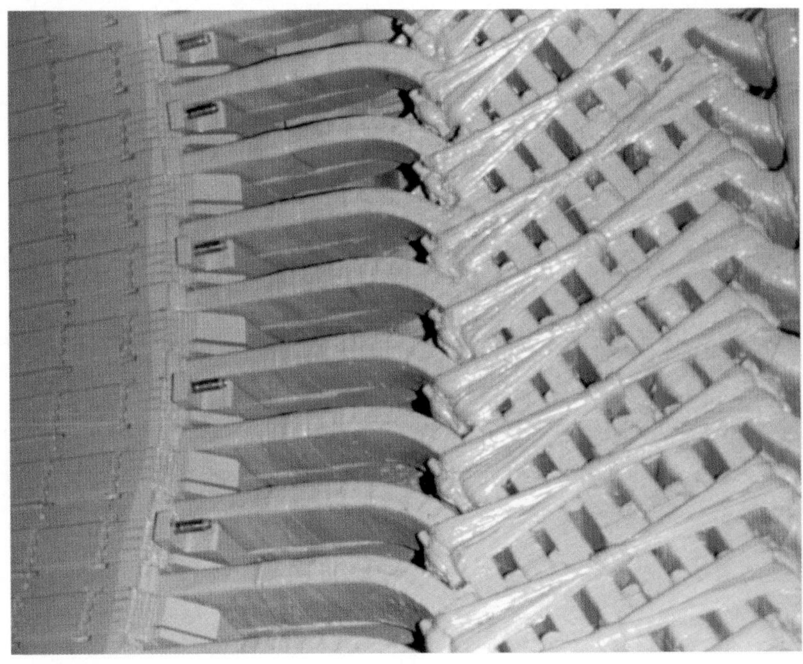

Fig. 5.9 Color-coded tagging compounds. See page 271 for text discussion.

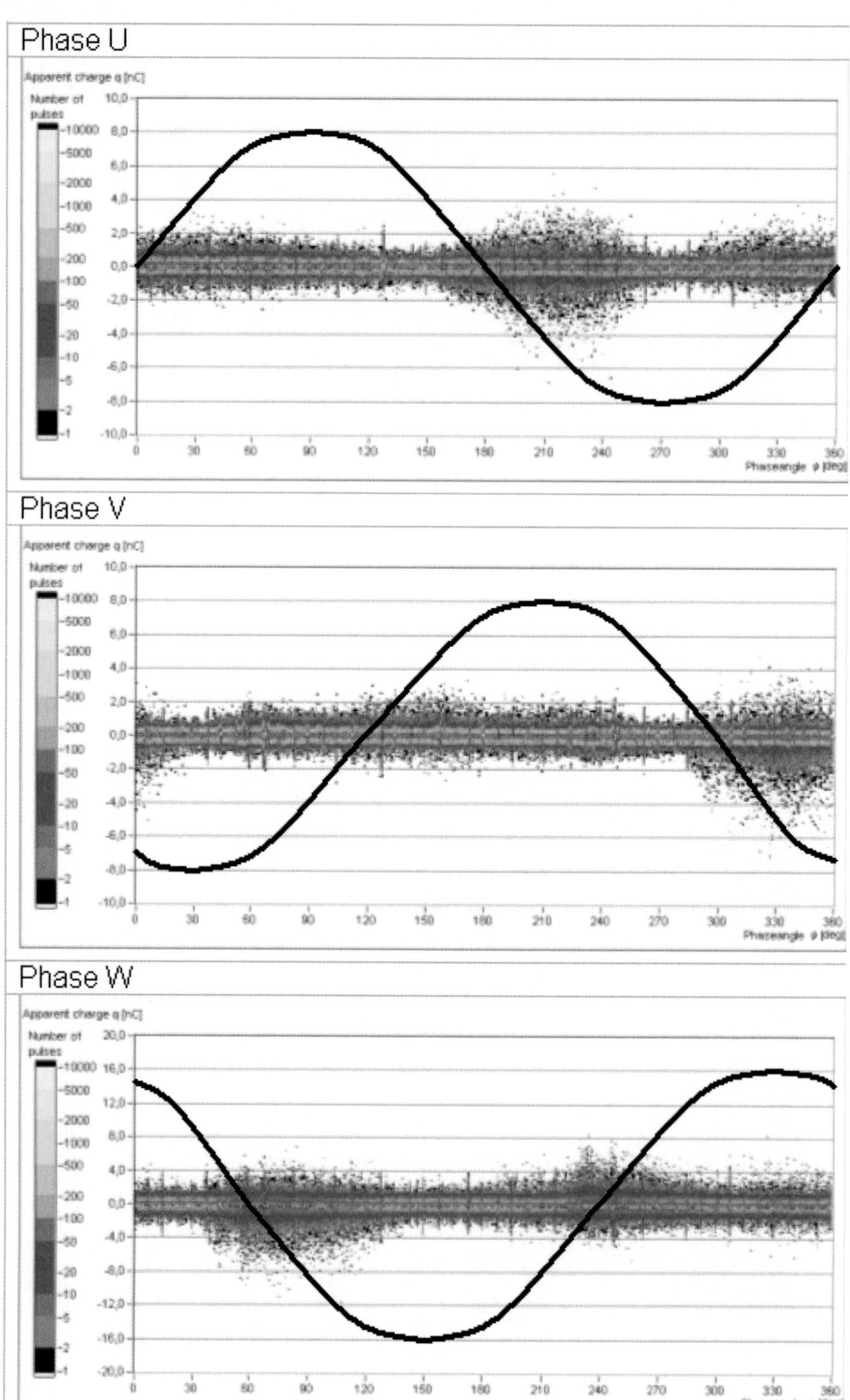

Fig. 5.30 Capacitive coupling, slot discharge example. See page 292 for text discussion.

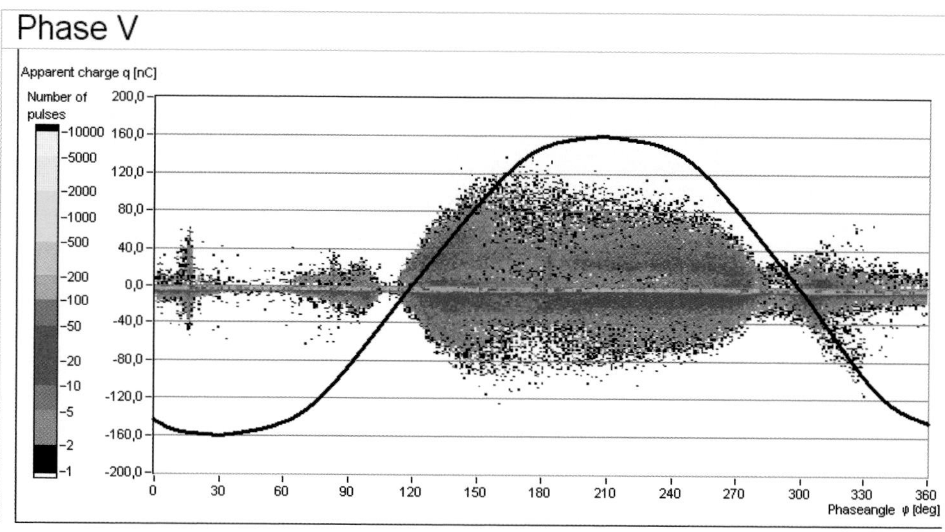

Fig. 5.31 Capacitive coupling, end-winding discharge example, no particular pattern. There is considerable discharge due to a number of mechanisms going on at the same time in this example. See page 292 for text discussion.

Fig. 9.58 Slot liner with embedded copper particles. This situation is prone to leading to a ground fault. See page 583 for text discussion.

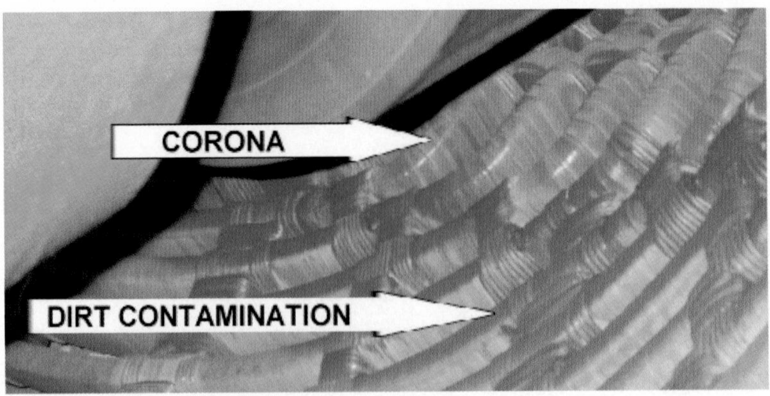

Fig. 8.77 In this photo there is significant dirt contamination shown on the endwinding, but there is also a significant amount of slot-exit partial discharge residue or "corona" discharge present in the form of a white powder. In this particular case, the "corona" discharge residue is easily wiped off and there is no sign of ground-wall insulation deterioration at this point. See pages 473–474 for text discussion.

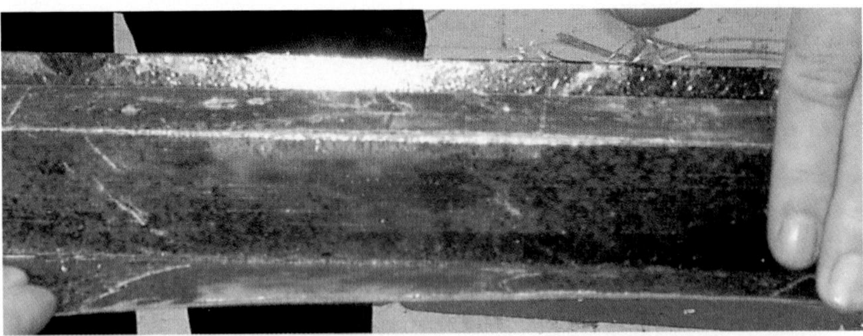

Fig. 8.105 The construction of the ground wall of the stator bar shown is sheet wrapped rather than taped. The ground wall has delaminated from the copper strand stack a few layers out from the copper surface. The voids created in the insulation due to delamination have allowed partial discharge activity to occur, as seen by the yellow-brown patches that are the by-products of the discharge. See page 498 for text discussion.

Fig. 8.106 The ground-wall insulation wrap has been completely removed from the stator bar of Figure 8.105, and the patch of brownish PD residue is clearly seen on the surface of the interstrand insulation of the outer part of the strand stack. See page 498 for text discussion.

Fig. 8.36 This photo shows the "before" and "after" results of dry-ice blasting in a small generator. See pages 439–440 for text discussion.

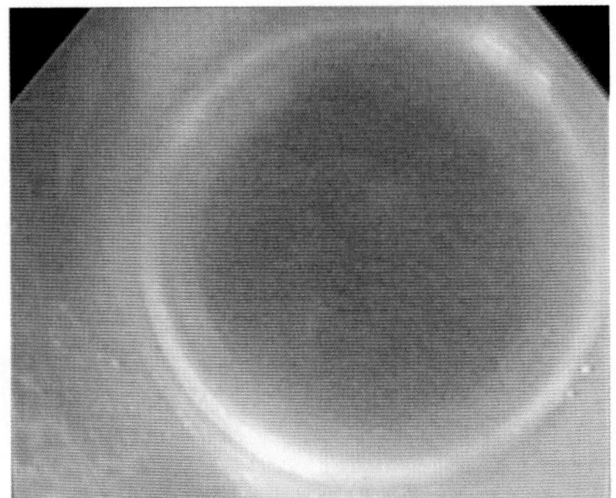

Fig. 8.128 Contaminated hose internals at the fitting attached to the bar nozzle or waterbox. The contamination was introduce after a failure of the stator cooling water cooler tubes that allowed raw service water to leak into the demineralized stator cooling water. See page 519 for text discussion.

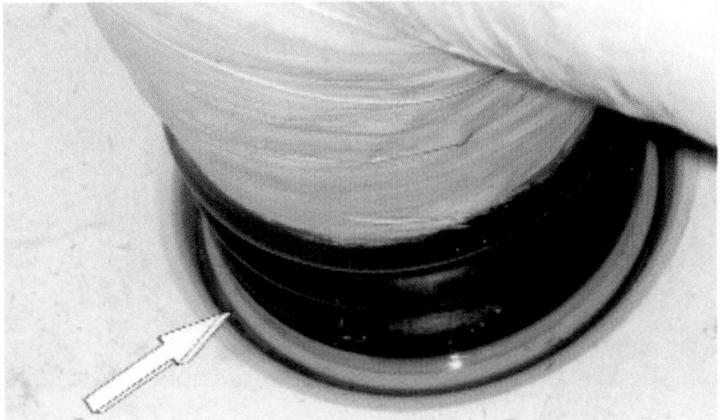

Fig. 8.142 Sealant for HV bushing in a circular reservoir at the casing exit in the bushing well. See pages 529–530 for text discussion.

Fig. 9.104 A different arrangement than previous figure for polishing a slip ring. In addition to the sandpaper, there is a vacuum hose to remove particles during the polishing activity. See pages 615–616 for text discussion.

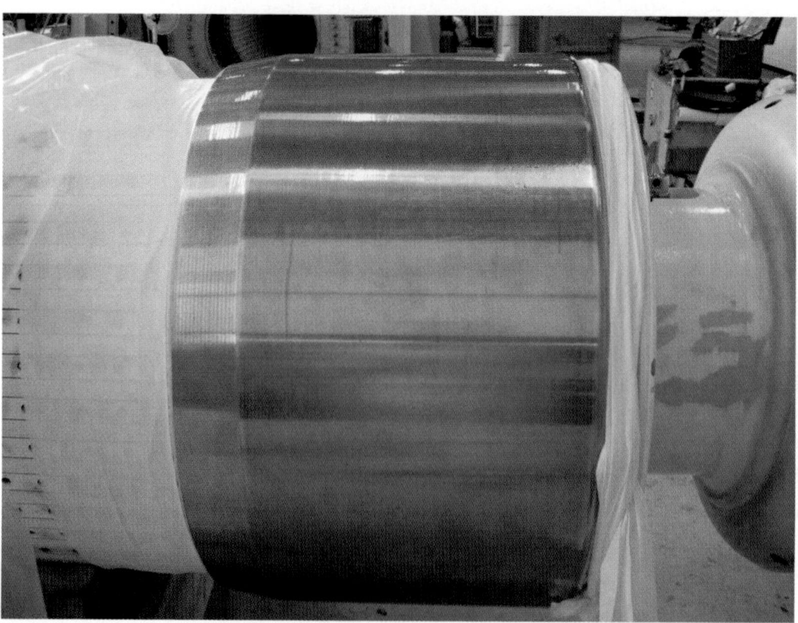

Fig. 11.36 18 Mn–5 Cr retaining ring, fluorescent dye partially applied. See page 722 for text discussion.

TURBOGENERATOR INSPECTION AND TEST REPORT

Form 11: Brush Replacement (to be used for one month and replaced—12 forms per year)

Looking at the brush rigging from the (North/East/West/South) side

Turbine/exciter side

Row 1 (top row)					
Row 2					
Row 3					
Row 4					
Row 5 (bottom row)					

Looking at the brush rigging from the (North/East/West/South) side

Turbine/exciter side

Row 1 (top row)					
Row 2					
Row 3					
Row 4					
Row 5 (bottom row)					

Enter the date of a brush replacement in the respective box.

REFERENCES

1. ANSI/IEEE Std 43-1974 or later, "Recommended Practice for Testing Insulation Resistance of Rotating Machinery."
2. I. Kerszenbaum, *Inspection of Large Synchronous Machines,* IEEE Press, 1996.

8

STATOR INSPECTION

Turbo-generator stators are very resilient major components in spite of being subject to intense vibrations, mechanical forces, and large voltage stresses. In addition to the aforementioned, expansion and contraction of different materials due to changes in temperatures, load shocks, and current and voltage transients also impose high stresses on the various stator components. The stator is a complex system of organic and inorganic materials working together under extreme conditions and they can operate sometimes for many years (between outages or refueling cycles in a nuclear plant) before inspection and electrical testing is performed, or run up to ten years before a detailed visual inspection is performed. Many stators with proper maintenance are known to run for decades before a major refurbishment, such as a rewind, is required. They generally perform reliably for many years and if well maintained can usually outlive their expected design life.

Nevertheless, the criticality of these components requires them to be monitored and well maintained. And due to recent changes in the power industry in terms of deregulation, operational duty on stators has become more onerous because of *two-shifting* and *load following*. Similar to rotors, this has resulted in a corresponding increase in certain modes of stator failures, largely due to thermal cycling effects.

There are a number of statistics published by organizations such as EPRI and INPO about expected life of certain subcomponents (stator windings, cores, cooling water systems, etc.), as well as failure statistics of stators in general. One normally accepted criterion is that stators in general have an expected useful life of at least 25 to 30 years, driven by winding insulation degradation. As stated earlier though, a well-maintained stator may possibly operate well beyond this time frame without a major refurbishment. When refurbishment is required, it is generally a stator winding replacement that is done and revalidation of the stator core condition. Once rewound and refurbished, it is possible for a stator to last an additional 25 to 30 years. Although statistically speaking those numbers may reflect the average of thousands of units, any particular unit may have quite different statistics. Being guided exclusively by industry-wide statistics when deciding on maintenance and life-cycle management of a specific unit in a specific site will certainly result in large errors of judgment. Only intimate knowledge of one's own stator (and overall generator) can lead to the correct maintenance and refurbishment plan, and to "intelligent" life-cycle management decisions. The goal of this chapter is to serve as a guide to learning the specific problems and failure mechanisms and their identification that will make it possible to correctly assess intrinsic risks for a given design, and explicit signs of deterioration and damage and/or impending failure. Although no OEM is specifically identified while discussing a particular issue, the reader should recognize each item discussed as pertinent or not to the specific machine under his/her supervision.

The Stator Inspection form (included in Chapter 7) refers to items comprising the actual stator, as well as the frame, bearings, and other machine-related components. Each item on the form is described below, but not necessarily in the same order as they show up on the form.

8.1 STATOR FRAME AND CASING

8.1.1 External Components

The main external components of the stator frame and casing that require inspection and periodic maintenance are the frame footing and bolts, the generator foundation, the seismic supports, the grounding cables, the piping and connections, the generator end brackets, the end-shield bearing supports, the bearing insulation and pedestals, and the space heaters.

Frame Footing and Bolts. Stator core frames are numerous in design variation, but they all have the basic function of carrying the stator core iron and acting as the main structure of the generator. They are usually bolted down to the turbine deck or power station floor, which may be a floating-slab type of arrangement for seismic purposes. The footings are generally made of steel and very substantial, and so are the bolts used for securing them (Fig. 8.1).

These components must take up the torque loading of the generator for normal and abnormal operating conditions, as well as for the normal operating vibrations. So it is important that they be tightened down securely to ensure that vibrations or torque pulsations are not transmitted to the other generator components in any detrimental way. Particularly, the rotor must be isolated from the stator frame vibrations, and this be-

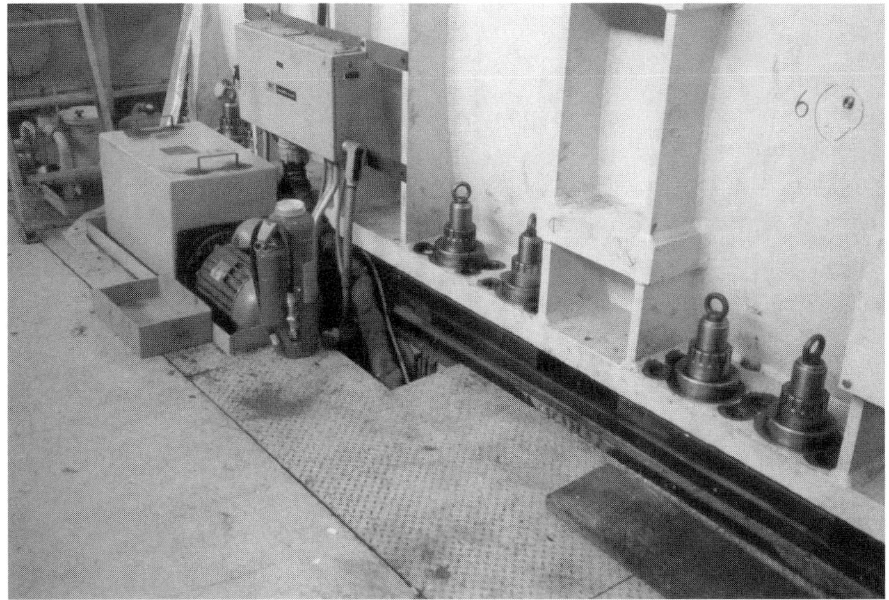

Fig. 8.1 Stator frame footing and bolting arrangement to the turbine deck.

comes a challenge when the bearings are part of the stator end shield, bolted to the outer stator casing or frame.

During outages, the frame support system should always be checked to ensure that there is no loosening or damage present. Bolts may have to be torque checked and retightened if they become loose. When looseness occurs, it is possible in extreme cases for the bolts to be broken during some significant stress event that produces a significant torque reaction. Also, if the stator frame natural frequency is operating too near the nominal once or twice per revolution operating frequency of the machine, then it is also possible for excessive vibration to occur and inflict damage to the footings and bolting arrangement.

Generator Foundation. The generator foundation is a complex structure of concrete and reinforcing steel used to support the entire turbine generator line. It is also tall in nature. In some units it can be up to a few building stories high from the lowest floor of the powerhouse to the operating deck for the turbine generator set. The foundation itself is subject to natural frequencies by the nature of its construction, and to vibrations from operation of the machine. Therefore, it is subjected to significant forces, as mentioned earlier, and from the weight of the machine that the foundation supports. It is important that the structure does not shift in natural frequency and that it remains free from cracks or other structural damage (Fig. 8.2).

The grouting or concrete base under the generator footings of the frame can easily spall or crack; this may be an indication of a foundation problem. Therefore, inspec-

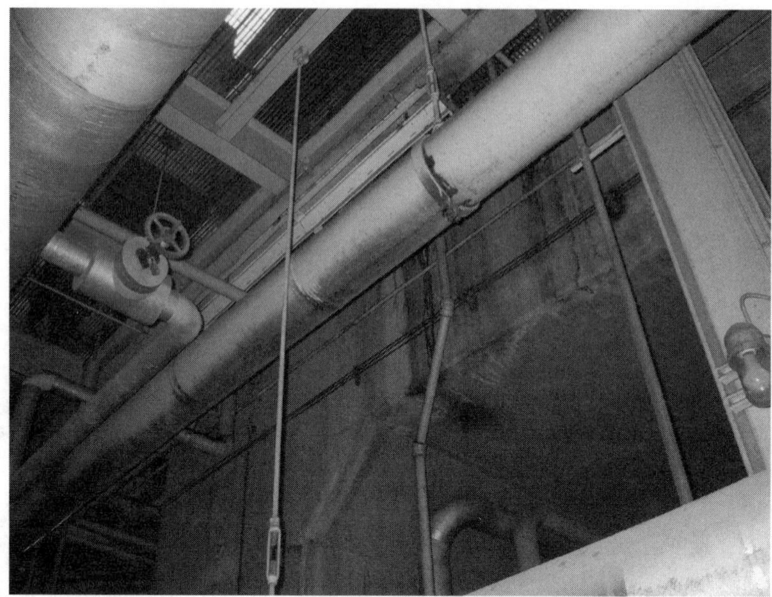

Fig. 8.2 Generator foundation of concrete and embedded reinforcing steel, showing water damage on the concrete surface.

tion of the support structures should be carried out periodically to look for obvious damage such as cracks, deformation, or spalled concrete.

Seismic Supports. Generators are not always built on a fully anchored foundation. A more modern arrangement is to have the generator mounted on an isolating block, with a seismic support structure between it and the main anchored foundation. This way, the generator is supported on a floating slab type of arrangement that is capable of handling extreme seismic events or significant torque reactions in the generating equipment when a system event or turbogenerator line failure of some kind occurs (Fig. 8.3).

The support structure usually also can be adjusted or releveled if it moves out of its specified design settings. This is more an exception than a common occurrence. However, when problems occur with generator frame and casing vibrations, this should be considered as a possible cause of the problem, and the alignment of the seismic supports should be checked against installation and design specifications.

Grounding Cables. All generator frames and casings are grounded to the power plant grounding system. This is usually done at one location only to allow control over the ground current flow and to ensure that there are no circulating currents in the generator frame structure that will cause arcing between components inside the machine (Fig. 8.4).

Fig. 8.3 Seismic support system on a floating slab arrangement under the generator.

Circulating currents in the generator frame can be quite damaging when they occur on a critical surface that seals hydrogen or oil, for example. In addition there is usually considerable heat generated at the arcing location, and this can cause further damage by distortion and deterioration of the component's proper functioning.

The ground cables should be inspected to ensure that they are tight and in generally good condition. Signs of damage would be corrosion, overheating, fraying, or crack-

Fig. 8.4 Grounding cable on the generator casing and frame structure.

ing. Prior to lifting a ground for repairs, if the generator is still in operation, it should be checked to see if it is hot and if there is any current flow. A hot ground cable would indicate significant current flow in it and that there might be a problem in the machine somewhere. An additional temporary ground should be applied at the same location prior to lifting the permanent ground for repairs. This is for personnel safety and for the protection of the generator and its components. Live line procedures should be followed if this type of work is to be done. However, it is better, and recommended, to do this type of work with the machine offline if the grounding is highly energized.

Piping and Connections. In a large generator, there are numerous auxiliary systems carrying gases and fluids to the generator, some very volatile. These include, of course, hydrogen, oil, and water. There is also outlet piping from the generator that carries mixtures of some of these substances in air, where oxygen is present. One can easily see that if air and hydrogen were to mix in a pipe, then there is potential for a hydrogen explosion. Also, the air in an oil pipe provides the oxygen necessary to support combustion if a spark were to occur in one of the piping systems carrying lube or seal oil.

It is, therefore, important that the piping be maintained in good condition. Piping should be grounded so that sparks cannot occur due to any circulating and arcing currents. Likewise, the piping joints and flanges should all be maintained by gasket and O-ring changes at the appropriate maintenance intervals for the particular system. These joints should be kept tight and leak free. In some auxiliary systems, it may be necessary to do periodic pressure testing of the piping to ensure its integrity.

Piping is often overlooked in generator overhauls, but it is an important part of the system. It should be included on the maintenance inspection list for the items mentioned above (Fig. 8.5).

Generator End Brackets. Generator end brackets (also known as end doors or end shields) generally serve multiple functions. They act as the end seal of a pressure vessel at both ends of the generator, since a large generator can be operating under hydrogen pressures up to 90 pisg (620 kPa). The end brackets also act as supports to the hydrogen seals and, in some designs, the rotor, when end-shield bearings are employed (Fig. 8.6).

Because of the nature of the stator construction, the end doors are designed as two halves: a lower and an upper. This is done for ease of installation and removal of the rotor and of the brackets themselves, as well as the other components attached to them. This arrangement of half brackets also presents a requirement to seal them at their half joints, in addition to sealing the face of the stator casing and around the hydrogen seals and bearing housings.

The interface of the end doors and the generator stator casing are steel-on-steel mated surfaces. They require sealing so that hydrogen cannot leak past these close interfaces. There are usually grooves machined into the sealing faces of the end doors on all mating surfaces as well as on the stator casing face, so that a sealant can be injected into the space to prevent hydrogen leakage out of the casing. There must be allowance for imperfection in the fit, due to the flexure of the end doors when under

Fig. 8.5 Piping connections for water, hydrogen, CO_2, oil, and drains under the generator casing and frame structure.

Fig. 8.6 Generator end brackets of a large two-pole turbo-generator. These end brackets are made of two halves with bolted joints at the casing face and the upper and lower mating faces of each bracket.

high pressure. In addition, there are significant vibrations that must also be accounted for in the seal and an injected sealant type of arrangement generally works for this application.

When the generator is opened up, all the sealant must be removed from the grooves in the end brackets and the casing grooves, so that new sealant can be injected on reassembly. During the outage, a thorough inspection should be made of all the sealing surfaces to look for mechanical damage or arcing that may have occurred. If leaks have been a problem, then it may be necessary to carry out some NDE work in the form of surface penetrant application to look for surface roughness or flaws that require correction. This is commonly called *bluing* the surface to look for high and low spots in the mating area.

End-Shield Bearing Supports. End-shield bearing supports are in fact also generator end doors (Figs. 8.7 and 8.8). They are mentioned separately from end doors because not all generators employ this type of arrangement. There are many generators in which external pedestal-mounted bearings are used and the bearings are completely outside the generator (Fig. 8.9).

When a generator does have a bearing-support type of end bracket, it is a substantially stronger component due to the fact that it must also support the rotor. This means that the bolting of the end-shield bearing support to the generator stator casing is also carrying the weight of the rotor. The end-shield bearing support is also then subjected more intensely to the rotor vibration, and there is a closer coupling between rotor and stator vibrations in this type of arrangement. It is a common method for some manu-

Fig. 8.7 Generator end shield with bearing support. This end bracket houses the rotor bearings and hydrogen seals, and supports the rotor in operation.

Fig. 8.8 End-shield bearing arrangement. Bearing, H_2 seal, and oil wiper halves mounted in the lower end shield.

Fig. 8.9 A pedestal bearing arrangement is shown in which the bearings are external to the generator and not part of the end door itself. The collector-end main bearing and steady bearing are mounted on steel plates that are double insulated.

facturers and works quite well, but it requires diligent maintenance during outages to ensure that the bearings are electrically isolated from the end-bracket structure as well as the hydrogen seals that are also carried by the end-shield bearing support arrangement.

The end-shield bearing support structure also has sealant groves and is arranged in halves. In addition to the items mentioned above, all the issues regarding end doors, generally, also apply to end-shield bearing supports.

Bearing Insulation and Pedestals. During the normal operation of electrical rotating machinery, voltages and, hence, currents are induced in the shaft. In the case of large turbo-generators, in addition to those voltages induced in the generator, substantial voltages are created by the rotating elements of the turbine. These shaft voltages and, in particular, the resultant currents, have to be kept to low values; otherwise, subsequent bearing failures can be expected to occur.

The main sources of shaft voltages are as follows [7]:

- Potential applied to the shaft, intentionally or accidentally (e.g., a grounded field conductor and excitation spikes).
- Electrostatic effects due to impinging particles (e.g., shaft-mounted fans in generator turbine blades in steam or gas turbines) or charged lubricants.
- Dissymmetry effects due to change of the reluctance path as a function of the angular position of the rotor and magnetic asymmetries of the core. (This can be due to design, manufacturing details, or core faults of large magnitude.)
- Homopolar flux effects due to axial fluxes originating from magnetized turbine and generator components, especially the generator shaft.
- Movement off magnetic center, due to axial movement of the rotor off the magnetic center, which might result in shaft voltages. (It has been noticed that multipole hydro generators are more sensitive to this phenomenon than two- and four-pole generators.)

In large turbo-generators, voltages of up to 150 V peak to peak from static buildup on the shaft are not uncommon [8]. These voltages have the potential to generate currents that, when allowed to flow freely, will destroy bearing surfaces, oil seals, and other close-tolerance machined surfaces. Shaft-grounding brushes should help mitigate this effect.

As is sometimes assumed, the heat developed by the flow of current does not cause the damage; rather, it is caused by mechanical failure. Shaft-bearing currents will damage the bearing surfaces by pitting resulting from minuscule electric discharges. The pitting will continue until the bearing surfaces lose their low coefficient of friction; then other more dramatic and rapid changes occur, culminating in bearing failure. Bearing currents also adversely effect lubrication oil by altering its chemical properties. Control of shaft voltages is achieved by taking certain precautions when designing the machine and by introducing shaft-grounding elements and/or bearing insulation.

Grounding devices, which can be copper braids or silver or copper graphite brushes, should be inspected often. Bearing insulation can be inspected often in certain bearing insulation designs, with the machine online or offline, and after a certain amount of disassembly of pipes and other attachments has been performed (see Chapter 9 for additional information).

During major inspections, the bearings and the journals are checked for signs of pitting caused by shaft (bearing) currents. These are easily recognized with help of a magnifying glass. They appear as shiny and well-rounded tiny droplets. If the bearing surfaces are found to be damaged by shaft currents, then the reason for the existence of significant shaft currents should be investigated. Strong candidates are bridged or faulty bearing insulation and/or defective or missing grounding devices. Pitted surfaces of the bearings should be rebabbitted. Pitted journals should be polished.

The bearing insulation is commonly made of mechanically strong water- and oil-resistant insulation materials formed from fiberglass or similar bases, laminated and impregnated with resins, polyesters, or epoxies.

In end-shield mounted bearings, the bearing insulation takes the shape of a ring or collar surrounding the bearing or, less commonly, as an insulation layer between bracket and casing, piping, and so forth. In pedestal-type bearings, the insulation is made of one or two plates, normally placed at the bottom of the pedestal. When double insulation plates are used, a floating metal plate is sandwiched between them. This system allows each of the insulation layers to be tested without the necessity of interrupting the operation of the machine, uncoupling it, and taking care of the other arrangements normally required for measuring the bearing insulation (see Chapter 9 for additional information). Machines normally have the nondrive end (outboard end) bearing insulated. However, machines with couplings at both ends or driven by turbines may have both bearings insulated. In some cases, the couplings are also insulated.

It is important to verify during the inspection that the bearing insulation is not contaminated with carbon dust or accidentally bridged with pieces of metal touching the bearing pedestal, temperature or vibration sensors, noninsulated oil piping, and so forth. If electrical testing of the bearing insulation is performed, the measured values should be in the hundreds of thousands of ohms. However, in this application, as in many other areas of insulation practice, a wide range of resistance values can be found in the literature, from 100 KΩ to 10 MΩ or greater. Given the relatively large shaft voltages encountered in large turbine generators, it is preferable to have insulation resistance values in the M(region.

Reliance only on grounding devices is not recommended. A grounding device is required to reduce the level of voltages in the shaft to values compatible with the small clearances encountered in bearing insulation and between shaft and seals. However, the normal contact resistance of grounding devices does not eliminate shaft voltages to the extent that bearing insulation would become redundant [9] and, therefore, bearing and seal insulation cannot be eliminated with shaft grounding systems.

Grounding devices are often taken for granted and, therefore, neglected during normal maintenance procedures. Rather, it is important they be inspected carefully during major inspections.

Space Heaters. It is common with many large generators to have a number of space healers installed. Sometimes, they are located on both sides of the frame, at the end-winding region, and sometimes externally, on the belly of the generator casing (Fig. 8.10).

The function of the space heaters is to keep the moisture out of the machine during shutdown periods, when moisture is most likely to condense onto critical components. By warming the internal air or gas, the water vapor always present in the air or gas does not condense onto any of the critical surfaces. Water condensation is detrimental to insulation and 18 Mn–5 Cr retaining rings. Thus, keeping the space heaters in good operating condition is an important element of correct maintenance when operating large electric rotating machinery.

The inspection should, as a minimum, check for loose connections and integrity of the wires and heating elements. If visual access is not satisfactory, then a resistance test should be taken at the terminal box to ascertain continuity of the circuit.

RTD, Thermocouple, and Miscellaneous Device Connection Panel. The internal instrumentation inside the generator is brought out through penetrations in the stator casing that are specially designed to handle the transition from the internal high pressure atmosphere of the generator to atmospheric pressure outside the casing and frame. (Fig. 8.11) These are generally simple in nature and do not usually require much attention for maintenance. They should however, still be inspected regularly as they can be a source of hydrogen leakage and create significant forced outage time, not to mention the extreme safety hazard due to possible hydrogen fires from a leak.

Fig. 8.10 Externally mounted space heater under the belly of the generator.

Fig. 8.11 External view of a high-pressure penetration for thermocouples and RTDs wired to an external wiring frame.

There are also sometimes special high-pressure penetrations retrofitted for specialized instrumentation that may be introduced after the machine has been in operation. These should also be inspected carefully and any gaskets or O-rings replaced as found necessary (figs. 8.12 and 8.13).

Once the instrumentation is terminated outside of the generator, it must be wired back to specific data collection points, such as the unit computer or dedicated data logger, in order to use the information from the device. Wiring racks mounted in enclosed cabinets are usually located next to or under the generator for this purpose. All of this equipment requires inspection, testing, and maintenance on a regular basis.

8.1.2 Internal Components

The main internal components of the stator frame and casing that require inspection and periodic maintenance are the support structure, wiring, and hardware.

Stator Frame and Support Structure. All parts internal to the generator are exposed to continual vibration, temperature changes, and other mechanical stresses. They may become loose, fractured, or broken. It is, therefore, important to search for these abnormalities during the inspection before they develop into major troubles.

In particular, all components of the winding support assembly are subjected to mechanical stresses from torque reactions due to sudden and large load changes. Such oc-

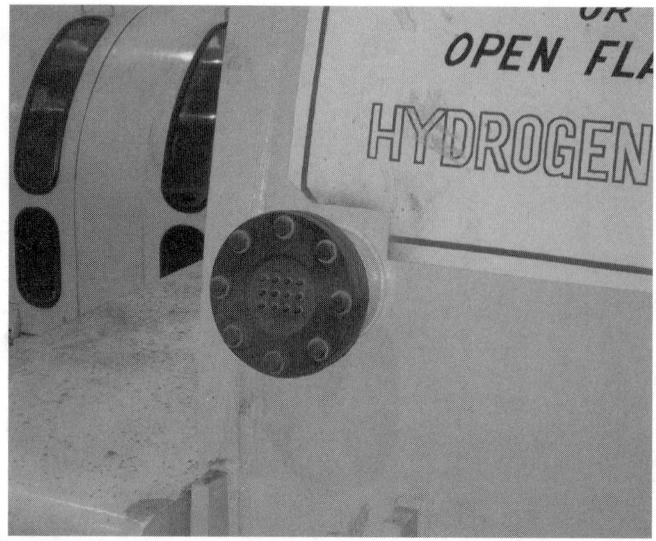

Fig. 8.12 External view of a specialized high-pressure penetration for direct connection of internal partial-discharge monitoring devices.

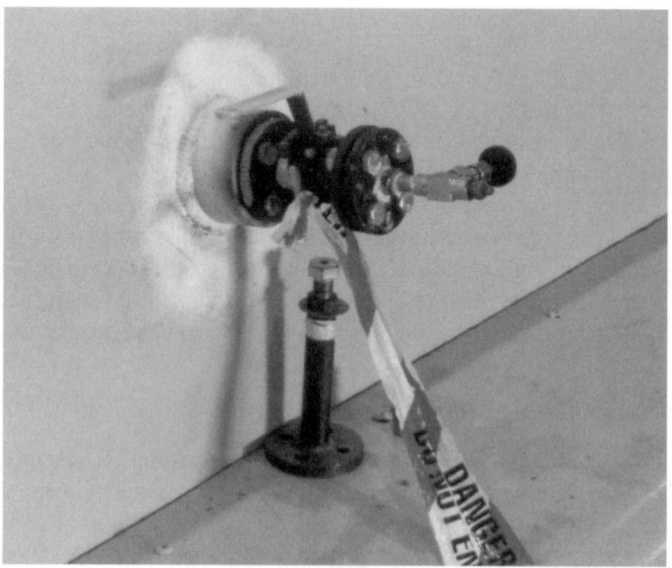

Fig. 8.13 Specialized high-pressure penetration for installation of a flux probe. The probe is not permanently installed for this type of setup but, rather, only for a test and then removed. A valve is provided to ensure that no hydrogen leakage occurs. In modern installations, the flux probe is permanently installed in the stator and hardwired out through a penetration, as described in the text.

Fig. 8.14 Externally mounted instrumentation panel for field wiring of the internal generator instrumentation such as RTDs, thermocouples, and other devices such as vibration probes.

currences can happen during loss of load, short circuits, pole slipping, and closing the generator breaker out of synchronism. Machines subjected to the conditions mentioned above, as well as machines operated with many starts, are more susceptible than others to hardware failure.

Some of the most sensitive components are:

- Compression bolts. Observe if any "greasing" (oil and dust mixed together by the friction of two components vibrating within the machine) is present, indicating loose bolts or core, and integrity of nut-locking device.
- Surge-ring supports. Look for cracking and looseness.
- Fingerplates. Look for cracked or bent fingers.
- Stator core. Look for looseness and mechanical damage.
- Stator windings. Look for looseness, cracking, and greasing.

The main component carrying all of the items mentioned above is the stator frame and its support structure. Vibrations and torques transmitted to the more sensitive components are stimulated by the reaction of the frame. In addition to components used within the stator, there is always the possibility of problems occurring with the frame itself if a resonant frequency exists in the frame structure near the operating or twice-per-operating frequency. Were this situation to occur, vibration of the stator during operation would become very noticeable. In most stators, there is some mechanism employed to dampen vibrations between the stator frame and the iron core, while keeping the two major components well coupled in operation (Fig. 8.15). When vibrations are abnormal, a thorough inspection is required to look for damage to the frame in the form of cracks and fretting between components. This may even require some NDE work in the way of dye penetrant checks if significant cracks are found. Cracks in welds can be repaired by stop-hole drilling where appropriate (Figs. 8.16 and 8.17), or grinding out the crack areas and rewelding. This can be a problem in terms of crack locations and clean conditions when such work is required. It is important not to contaminate the internals of the generator with metallic debris, which can cause additional damage to insulation systems if such debris manages to lodge in the stator windings or other sensitive area.

Vibration dampening components should be inspected for damage when high vibrations are experienced. It is likely in those instances that the vibration control mechanisms will be loosened or damaged in some way and not fulfill their intended function. It is important to keep any vibration-damping device tight so that the core and frame move in synch. Otherwise, there may be impacting between the core and the frame at the once- and twice-per-revolution frequencies, which are stimulated by the rotating magnetic push-and-pull effect of the rotor field.

One of the essential items to inspect for at the back of the core—at the core-to-stator-frame interface—is the keybar area where the core is mounted. If there is high core and frame vibration, the impacting of the two components can initiate cracks in frame

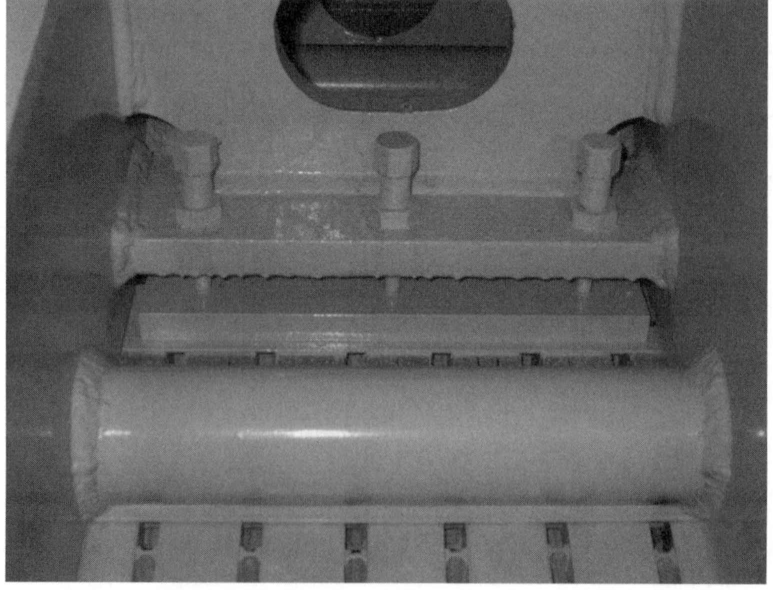

Fig. 8.15 Core-to-frame jacking arrangement keeping the core and frame coupled together to dampen vibrations.

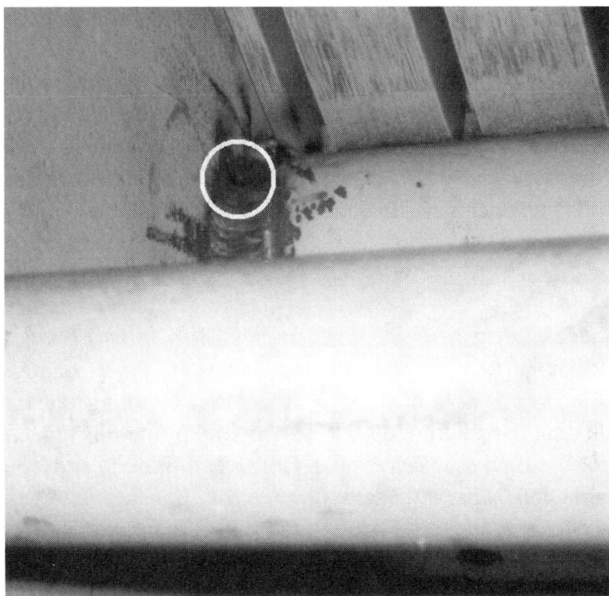

Fig. 8.16 Small weld crack in the frame structure that has been prevented from additional propagation by drilling a stop hole at the leading edge.

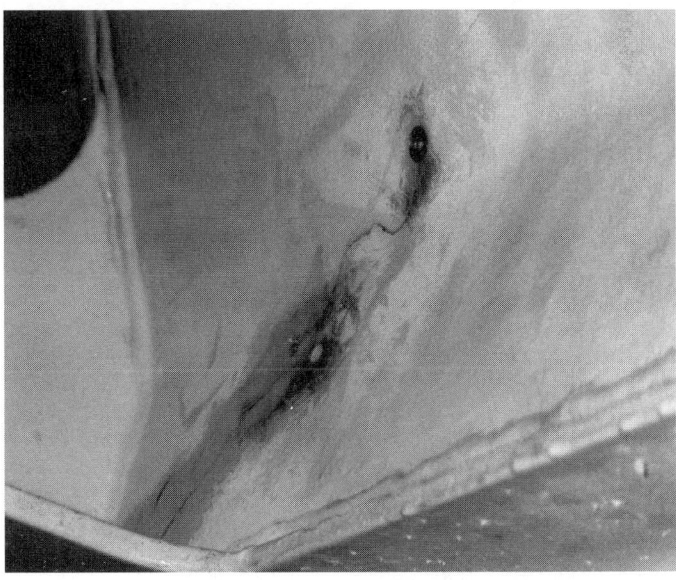

Fig. 8.17 Major frame crack that has been prevented from additional propagation by drilling a stop hole at the leading edge.

welds, as well as in some of the other structural members. The more usual occurrence is fretting debris coming out of the core dovetails where the keybars are located, as shown in Figures 8.18 and 8.19.

When the vibration is due to a resonant frequency, the result can be catastrophic if the vibration is not well damped (Fig. 8.20); even then, well damped vibrations that are near a natural frequency have a tendency to eventually become unstable. This is due to the loosening of components and increasing clearances as the vibrations continue. This is true for all components such as stator end-winding supports as well as the core and frame structure.

RTD and TC Wiring and Miscellaneous Monitoring Devices. Resistance temperature detectors (RTDs) and/or thermocouples (TCs) are mainly found in stator windings, cooling gas flow paths (Figs. 8.21 and 8.22), cooling water paths, and bearings. Stator winding temperature detectors, normally of the RTD type, are located between the stator bars and are inaccessible. However, wiring to and from these devices outside the stator slots is partially accessible for visual inspection. It should be tightly secured along its path to the stator bars, frame, and casing.

Damaged stator winding RTDs or TCs are usually left in place with their wires disconnected or removed, because they are inaccessible. They can only be replaced or re-

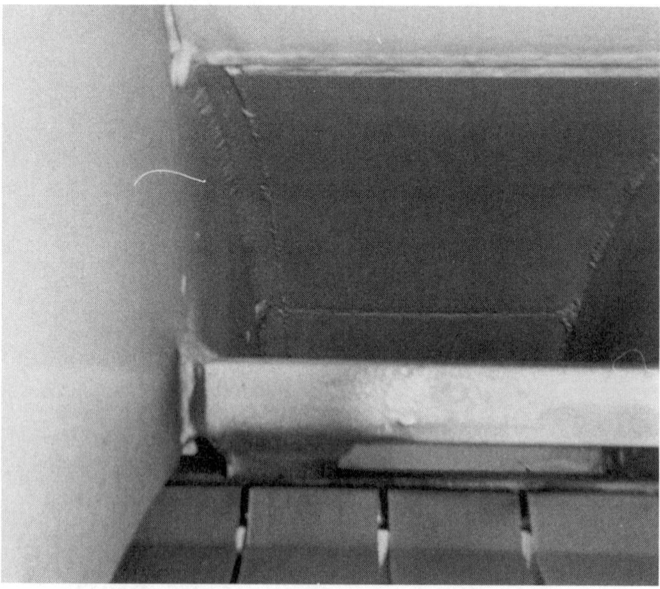

Fig. 8.18 Core-to-keybar interface where looseness between the core and frame has allowed impacting damage; debris can be seen migrating out of the dovetails. The key attached to the keybar component in this machine is aluminum, and the debris is black in color due to the aluminum material and the fact that it is dry from lack of oil mist in the generator.

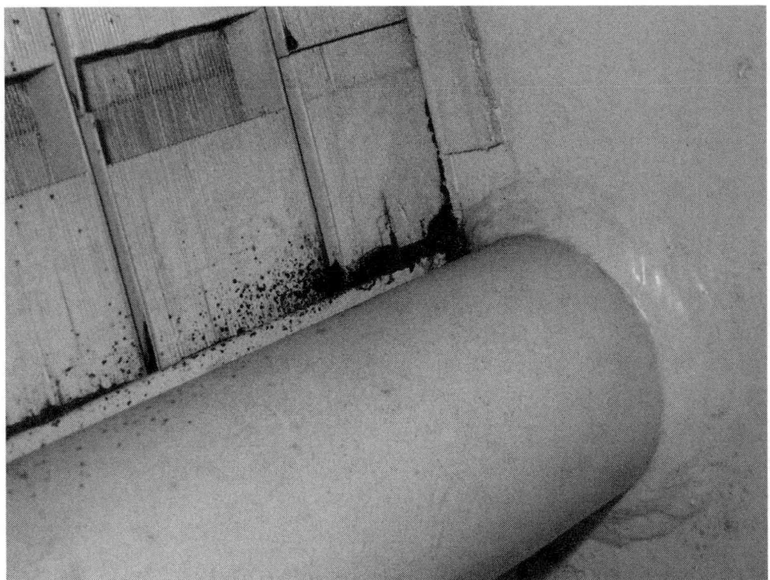

Fig. 8.19 Core-to-keybar interface where looseness between the core and frame has allowed impact damage. Fretting debris can be seen migrating out of the dovetails. The key, attached to the keybar, is made of brass in this machine. In this case there is significant oil in the generator, causing a greasing effect of the fretting debris. The nature of the debris in the fretted material is likely a combination of iron from the core and brass from the key.

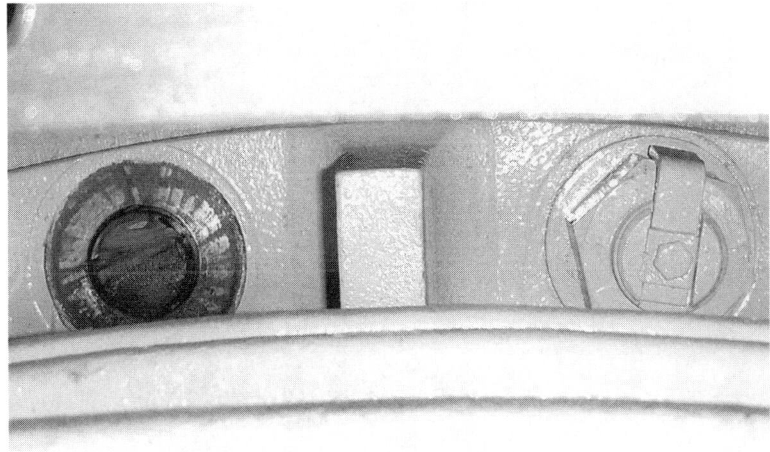

Fig. 8.20 Keybar nut completely severed at the thread due to high vibration at a natural or resonant frequency.

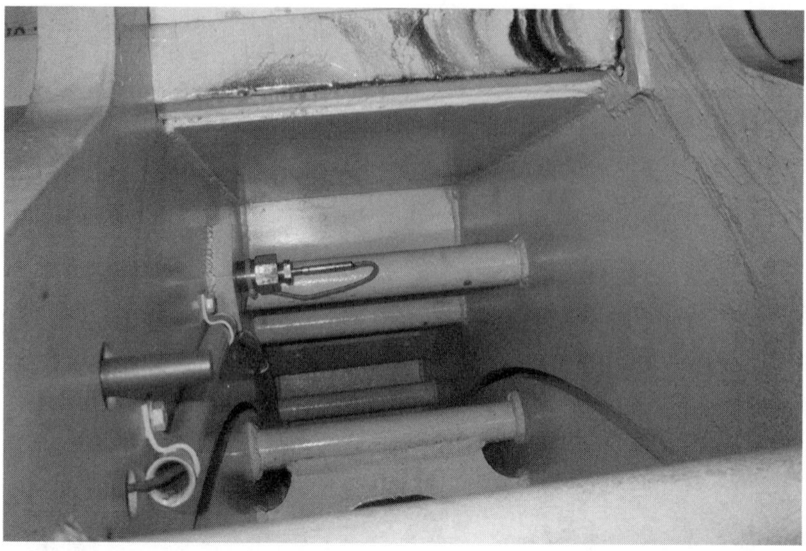

Fig. 8.21 RTD protected in a thermo well for measurement of hydrogen gas temperature.

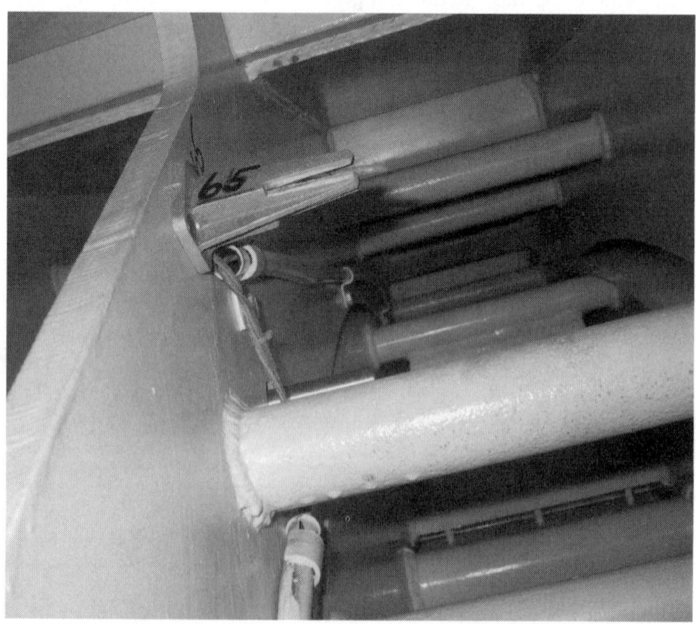

Fig. 8.22 Thermocouple for measurement of hydrogen gas temperature. The protective barrier and conduiting of the wiring is clearly shown in this photo.

paired during a rewind operation. If accessible temperature detectors such as those in gas path bearings are faulty, they can be repaired or replaced during a major inspection or overhaul.

Water-cooled windings with hoses are normally fitted with thermocouples in the outlet end to measure the cooling water outlet temperature from individual stator bars or pairs of bars (Fig. 8.23). These are completely accessible and need to be inspected and checked for accurate operation. They can be replaced during outages if they have been damaged and are no longer functioning. Figure 8.24 shows the damaged sheaths of some hose outlet thermocouples as they enter the main conduit for the thermocouple bundle.

Other types of monitoring devices such as *shorted-turns detectors* (*flux probes*), *partial discharge detectors* and *vibration detection probes* may be found installed in or on a stator and these should all be regularly inspected and repaired or replaced as required.

Monitoring instrumentation is the front line defense during operation; it indicates what is happening to various components inside the machine. The different devices that can be found installed will vary widely in the types and amount of monitoring applied. Regardless, the more monitoring installed, the more that can be diagnosed. Therefore, it is important to inspect and maintain all such devices and all of the conduit and internal connections (Fig. 8.25) that are also present, to ensure that there is a reasonable degree of monitoring available at all times. Of course, only certain portions of

Fig. 8.23 Thermocouples installed in the hose outlets of water-cooled stator bars. The outer protective barrier is removed in this photo so the installation can be more clearly seen.

Fig. 8.24 Damaged thermocouple sheaths.

Fig. 8.25 An example of internal thermocouple connectors with associated wiring runs and securing points shown.

the machine can be instrumented and it is not possible to cover all the internal components, but whatever is instrumented, the devices should at least be kept in good working order.

Miscellaneous Hardware. Observe internal-type space heaters for looseness of bolts and nuts, and integrity of connections. Inspect all conduits and internal attachments of the like for looseness or damage of any kind.

Some generators employ "zone rings" or "gas guides" attached to the stator core, used to more efficiently direct cooling gas flow between the stator and rotor cooling circuits (Figs. 8.26 to 8.28). This hardware is attached by ties and wedging and is susceptible to loosening and damage. They need to be thoroughly inspected and any problems found corrected so that they do not become a potential foreign object in the airgap of the generator when the machine is in operation. The consequences could be severe if debris of that size becomes loose between the stator and rotor when the rotor is at speed.

Most large machines include *air/gas baffles* to direct the cooling air or gas to and from the fans. In this manner, a cooling circuit is established in the machine. So-called fan baffles are subjected to continuous and diversified modes of vibration. Studs or bolts should be inspected for excessive stress and fatigue cracks (Fig. 8.29). If broken during operation, they will very probably cause extensive damage to the windings or rotating elements, particularly the fan blades.

Fig. 8.26 A large two-pole turbine generator showing gas guides in the stator bore. The gas guides are made of insulating material.

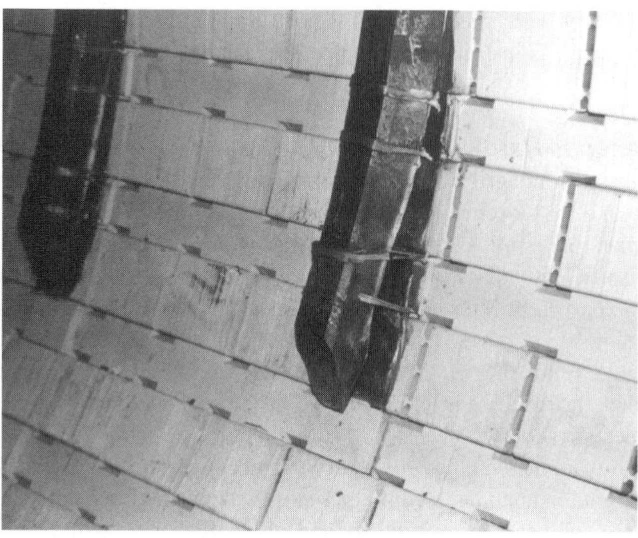

Fig. 8.27 Closer view of the securing method of the gas-guides in the stator bore of another large two-pole turbo-generator. Again, the gas guides are made of insulating material.

Fig. 8.28 Close-up view of the gas deflectors of a large four-pole generator showing damage during the removal of the rotor.

Fig. 8.29 Fan-baffle support threaded stud shown. The stud is mounted on the tip of the core-compression finger. A number of equally spaced studs are placed around the bore.

8.1.3 Caged Stator Cores—Inspection and Removal

By far the most prevalent design for generator stators is one in which the stator core and the frame and wrapper plate that make up the pressure vessel are one structure. This type of construction is termed an "integral" stator. When the stator comes as a separate outer casing with a removable inner core and frame it is termed a "caged core." In earlier times, as the design of machines began to approach 500 MW, overseas manufacturers had difficulty in transporting such large machines across the oceans due both to size but, primarily, weight. The solution to this was to build the stator as a removable "caged core" to allow the separate components to be transported and then assembled on-site.

Although this created a significant amount of additional assembly work, it did solve the problem. But this design has one advantage over an integral stator design in that inner core and frame can be removed and inspected. Thus, the inner frame and back-of-core areas, as well as the inside part of the outer casing, can be worked on if problems occurr. With an integral design, the issue of accessibility comes into play and when there are significant core and frame issues, these can often not be resolved without removal of the entire stator and unstacking the core from the frame, or, in the worst case, replacing the entire stator.

Some of the issues that come up are frame cracks and core and frame vibration. These can be inspected and addressed with the caged core arrangement. The series of photos in Figures 8.30 to 8.33 show a caged core being removed from the outer casing. In this case, a spare stator was available and a simple component exchange was carried out to get the machine back in service while the damaged stator was repaired. When repairs are carried out, it is often possible to install a steel "tire" at each end of the in-

Fig. 8.30 Caged core and frame being removed from the casing. The photo shows how the entire generator stator must be lifted and supported to allow the inner core and frame to be pulled axially out of the casing.

Fig. 8.31 The caged core and frame assembly is almost completely out of the casing in this photo.

Fig. 8.32 The caged core and frame assembly is fully out of the casing and ready for lifting.

Fig. 8.33 This photo shows the empty casing with the inner core and frame removed.

ner core and frame and set the whole structure on a set of rollers so that the stator can be rotated for ease of working. In this way, the stator may be best positioned for whatever job is required (Fig. 8.34).

8.2 STATOR CORE

8.2.1 Stator Bore Contamination

Important information on the condition of the machine may be obtained from a general view of the bore area and frame. Here are some examples:

- Excessive discoloration (and, perhaps, flaking) of paint on the casing, frame, and bore indicates a probable case of overheating. This could be the result of overloading or improper flow of air, gas, or water.
- The presence of large amounts of oil or a dust–oil mixture indicates possible hydrogen seal problems. See Figure 8.35.

Fig. 8.34 The removed caged core and frame are shown with steel tires installed and the entire structure sitting on rollers.

- In certain types of air-cooled machines, large amounts of carbon dust are evidence of deficient sealing between the collector enclosure and main bore areas.
- Excessive amounts of iron powder mixed with oil and dust or found alone in the bore area tend to indicate a loose core or core-to-keybar impacting from high vibration.
- Greasing found in the stator bore is generally a mixture of fretting debris and oil. Care must be taken to determine if the greasing is core interlaminar fretting or fretting from stator-winding wedge loosening.

Regardless of the source of contamination, it is a good idea to remove the contamination material from the stator. In large machines, the stator is often simply dry-wiped until clean. In smaller machines, the cleaning is sometimes done by solvent spraying or high-pressure water, as long as the fluids and residue can be completely removed after application and do not harm the insulation systems in any way. Another method is dry-ice blasting since the dry ice evaporates away and does not leave a residue (Figure 8.36).

Nuts, bolts, small pieces of lamination iron, or other loose objects found inside the machine, oftentimes at the bottom of the casing, should be investigated as to their ori-

Fig. 8.35 Bore area contamination by oil and soot from the rotor and machine internals. The grease-like mixture of oil and debris has been circulating through the rotor winding and stator core cooling paths in this 200 MW hydrogen-cooled turbo-generator.

Fig. 8.36 This photo shows the "before" and "after" results of dry-ice blasting in a small generator. (See color insert.)

gin (Figure 8.37). They may point to loose space heaters, broken laminations or cooler fins, and/or other abnormalities in need of attention. Also, debris may come from threads being stripped when unbolting various components, so great care should be taken to collect any such material that comes loose.

8.2.2 Blocked Cooling Vent Ducts

Blocked cooling vents can derate a machine by restricting the flow of cooling gas through the laminations. This phenomenon is particularly evident in open-air machines, which are more easily contaminated with dirt from the outside surroundings. Therefore, the air filters used to keep this contamination out must be cleaned or replaced regularly to ensure that the cooling air is free from contamination.

Blocked vents are particularly common in open-air machines contaminated with oil, because any dust or debris that does intrude will mix with the oil to form a thick sludge that can plug a vent more easily. Restrictions can show up as hot spots during operation of the machine; that is, temperature readings from one or several temperature detectors will be higher than those obtained from the other detectors. This can also indicate core insulation problems.

Normally, a visual inspection with the aid of a light source will suffice to find this type of blockage. Where possible, a light placed in the back of the core area while observing from the inside can result in good observations of the ducts.

8.2.3 Iron Oxide Deposits

Iron oxide appears as red powder deposited mainly on the bore and in the radial cooling vents of the machine. When mixed with oil, these deposits may be concealed in a mixture of dirt and oil. When iron dust is present (or suspected), this mixture

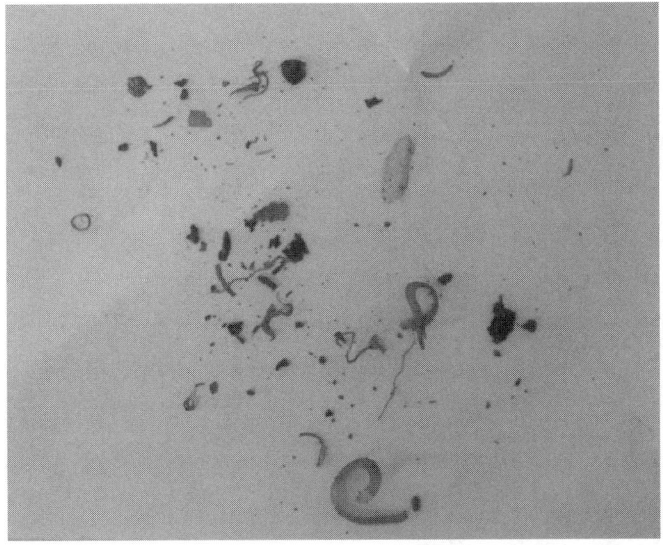

Fig. 8.37 This photo shows the type of debris that may sometimes be found inside the generator casing. The debris shown here is especially dangerous as some of it is metallic and, worse still, magnetic. Magnetic debris can easily be trapped in a magnetic field and bore a hole in soft stator-winding insulation, causing severe damage and a generator ground-fault trip.

should be chemically analyzed for content in proportion to weight. A quick identification during the inspection can be made by subjecting small portions of the mixture to the field of a magnet (one of the desirable inspection tools). If the "dirt" responds to the magnetic attraction, then iron dust is most certainly present in significant proportions.

Iron oxide can result both from loose laminations and loose wedges [1]. In the case of loose wedges, the iron dust is mainly seen in the contact region between wedge and iron. In the case of loose laminations, the iron oxide powder deposits are distributed on larger sections of the machine, on the iron itself. The deposits will tend to concentrate in places adjacent to the cooling vents, having been left there by the flowing gas (see Figs. 8.38 and 8.39). In severe cases, particularly in air-cooled machines, the large amounts of iron oxide powder generated may clog cooling vents.

When the amounts of iron oxide powder present in the machine are substantial, its origins should be thoroughly investigated. If loose wedges are the cause, then they should be tightened by rewedging or another effective method. Loose wedges may abrade themselves to the extent that they come out of the slot. They may also indicate a loose winding condition, with detrimental consequences to stator bar insulation, in particular, the loss of semiconducting paint [2,3]. If the iron oxide originates from the movement of metallic parts, it probably indicates a loose core or loose portions of the core [4].

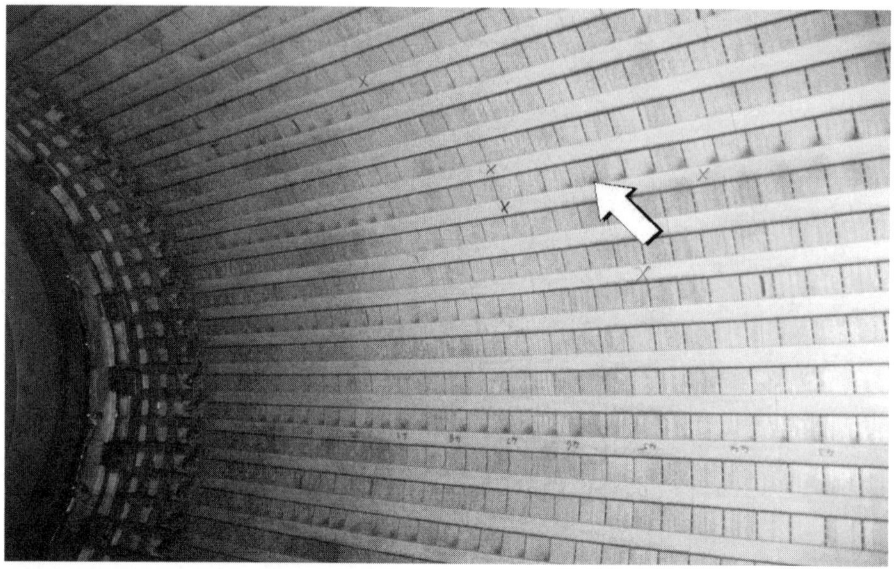

Fig. 8.38 Section of a large turbo-generator bore showing deposits of red iron oxide powder.

Fig. 8.39 Close-up view of Figure 8.38. The accumulation of the red powder depends on the origin of the powder and the pattern of the flow of the cooling gas.

8.2.4 Loose Core Iron/Fretting and Interlaminar Failures

Cores are pressure loaded to given values during the manufacture of the machine. During operation, the core is subjected to continuous thermal expansion and contraction of the laminations (by a phenomenon called *magneto-striction*), at twice the operating frequency. This also affects the supporting structures by thermal cycling and vibration. Machines having properly designed and stacked cores are able to withstand these onerous conditions. In many cases, however, after years of operation, this constant movement of the core components and the accompanying metal fatigue, abrasion, and deformation result in a reduction of the core's loaded pressure. The end result, if not corrected, is abrasion or fretting of the interlaminar insulation. The consequences of such abrasion are spot heating of the core, broken laminations leading to machine contamination with iron particles, and serious failures of core compression bolts, bolt insulation, and core-compression fingers [4]. Other detrimental effects are deterioration of the stator bar insulation due to hot spots in the core, additional core losses, augmented vibrations, and increased audible sound levels.

One test that checks for the presence of a loose core is the insertion of a knife between the laminations at several locations. If a sharp-edged blade penetrates more than a quarter of an inch, such that it goes beyond the "ear" of the stator tooth and into the area that "should" be under pressure from the space blocks, then the core is likely not sufficiently tight. Extreme care should be taken not to break the blade, thus leaving a piece in the laminations. This technique should be used very carefully, especially in any machine with its windings in place. (For a description of the proper procedure, see reference [3].) In many cases, the core may be loose but this may not obvious because there are no signs of damage or fretting debris. In such instances, the only clue to localized looseness is the knife test.

Some utilities check the torque of a sample of the machine's compression bolts at every second or third overhaul, or some other chosen interval. The measured torque value is compared with those recommended by the machine's manufacturer. If these bolts are found to be loose, the entire core is retorqued in accordance with procedures laid down by the manufacturer, but this does not help in determining local looseness in the core.

Looseness in stator cores can show up in a number of ways, from local looseness in the top of the stator teeth, local looseness in the core below slot bottom, and even global looseness of the core in general. In cases of localized looseness in the tooth top of the core iron, the worst case is when it is found at the core end. Core-end looseness, however, is often very noticeable. This is because of the fretting effects from axial flux impingement due to the stray fluxes from the end windings. The axial stray flux impingement creates a push-and-pull effect on the stator core end, due to the ac nature of the rotating magnetic field. This ac flux effect is the actual source of the fretting forces, which can eventually cause accelerated wear between laminations. This tends to erode the interlaminar insulation and allow circulating fault currents to flow, which may eventually overheat the core locally.

If there is sufficient oil present in the machine due to excess seal oil getting by the seals and wipers, it often mixes with the fretting debris from the core and causes a greasing effect (Fig. 8.40). This is also very noticeable but must not be mistaken for

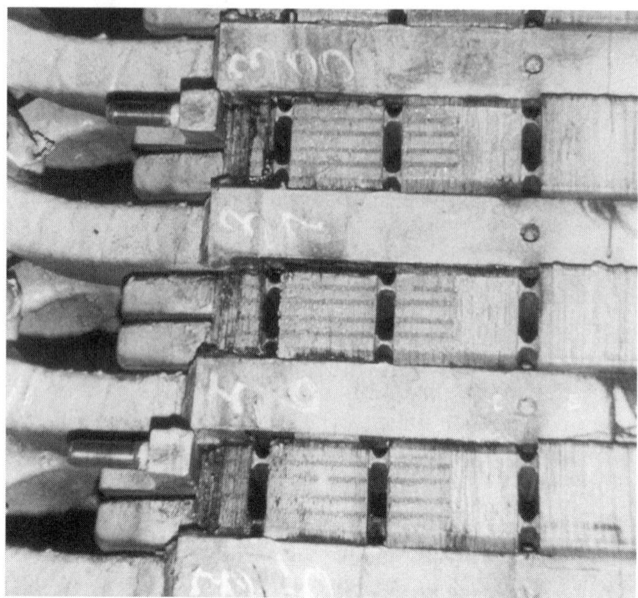

Fig. 8.40 Greasing from loose iron fretting when the core is oil contaminated. Interlaminar insulation deterioration is the result.

greasing that may be occurring from loose wedges. The "knife" test is recommended to confirm the degree of looseness and an EL-CID test to establish if inter-laminar defects exist (Figures 8.41 and 8.42).

In very severe cases of loose cores, the amount of debris produced by the fretting action may be extreme (Figure 8.43) and is often a prelude to an imminent core failure. It should also be noted that this effect is far more prevalent at the core ends, where stray axial flux penetrates the core-end tooth tops and (in addition to the normal radial magnetic pull on the stator core teeth from the rotor) and creates a magnetic push–pull effect due to the rotating north and south of the magnetic poles of the rotor, as mentioned above.

The grease produced is generally not harmful to the generator internal components because it usually consists of nonconducting materials. However, it is a contaminant and may carry conducting particles of iron from the core plate and create a possible flashover potential to the machine. Therefore, this material should always be removed when the opportunity arises. In addition, the grease becomes a lubricant that can work to loosen other components (e.g., winding support systems) if too much is allowed to accumulate. The grease should be removed for this reason as well.

If the fretting damage is not severe, such as in the example of Figure 8.40, then it is often possible to separate the laminations and insert mica sheets (Fig. 8.44) to isolate each successive lamination from the next. This is usually required when EL-CID testing or a flux test that reveals hot spots confirms interlaminar insulation deterioration. It

Fig. 8.41 In this photo the core is clearly loose, as shown by the "knife test" for core tightness and there is only greasing coming from the core iron due to loss of axial pressure, for some reason or another.

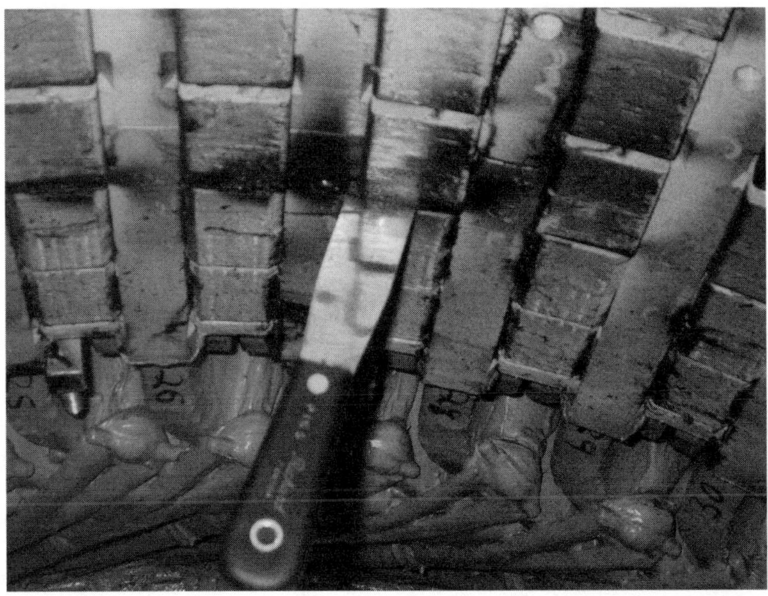

Fig. 8.42 In this photo, the core is shown to be loose by the "knife test," but there is greasing coming from both the core iron and the wedges in the slot. In this case, there is core looseness, due to loss of axial core pressure, and stator-winding wedge looseness. When inspecting for either core or wedge looseness, the inspector must be very careful to make sure which components are actually loose. Wedge tapping along with the "knife test" is required to be sure.

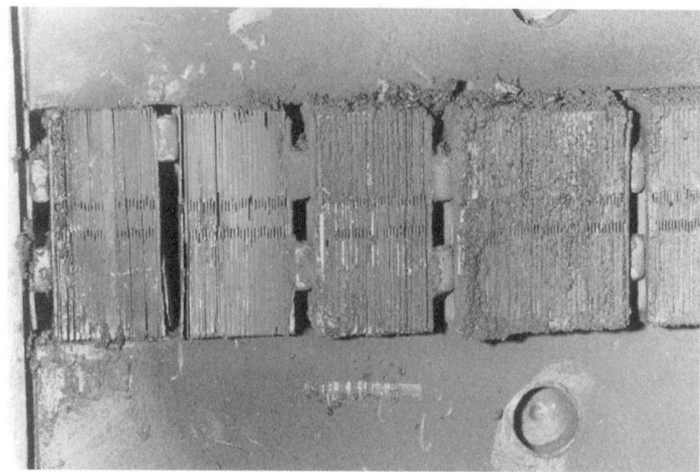

Fig. 8.43 In this photo, the core is clearly in distress without even having to use the "knife test." There is iron oxide in dry form present, from the fretting, as well as overheated oil and fretting debris. Once this type of damage occurs, there is usually excessive heating localized at the damaged area, from loss of interlaminar insulation.

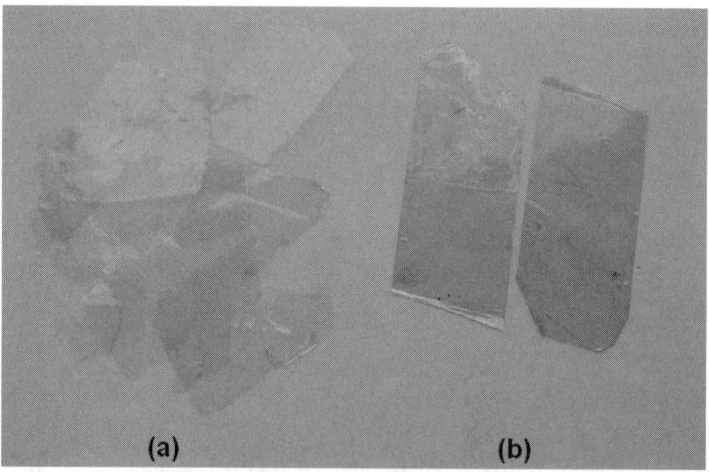

Fig. 8.44 Here can be seen two types of mica. (a) Book mica of the type that is ground up for use in stator-bar insulation systems. It comes in layers but is fairly brittle and easily flakes and breaks into smaller pieces. However, it is good for inserting between laminations because it is very thin and has excellent insulating characteristics. (b) Madras mica, which comes from Madras and is also layered but stiffer than book mica due to the nature of the layering. The layers can be peeled away from one another and used like book mica for lamination separation. The Madras mica is better for defects deeper into the stator core teeth because it is stiffer and will not flake so easily when inserted.

is applicable generally only at tooth tops if the core is significantly loose or it is possible to separate laminations by a core "packet spreader." Once the mica is inserted, the core is then generally treated with a penetrating and insulating epoxy of a suitable type to help solidify and fill any voids remaining. Finally, the core can be released (i.e., the packet spreader removed) and the repair allowed to cure (Fig. 8.45).

In some cases, the core looseness can cause fretting and wearing of the stator-bar insulation as well as the damage to the core itself. In this type of effect, the stator winding may fail before the core, causing a stator ground fault, but more often, the core fails first and overheats the stator bar next to it, initiating the ground fault and subsequent generator trip. (Figures 8.46 to 8.48) The clue as to whether the bar failed first or the core failed first is the degree of damage that occurs. The situation shown in Figures 8.46 to 8.48 is due to a core failure first, because of the significant damage to both the core and the stator bar in that slot. The stator winding ground-wall insulation failure is a result of the overheated core.

In the example shown in Figures 8.49 and 8.50, the looseness is extensive in the core tooth, allowing a very high degree of fretting. When fretting is that severe, pieces of core plate and space blocks will break off and migrate out into the airgap. One might quickly conclude that such a failure requires partial core restacking. However, this is not always the case. If the damage is confined to the upper part of the tooth, the damaged part of the core can sometimes be removed (with the stator winding removed) and replaced with an epoxy/glass false tooth.

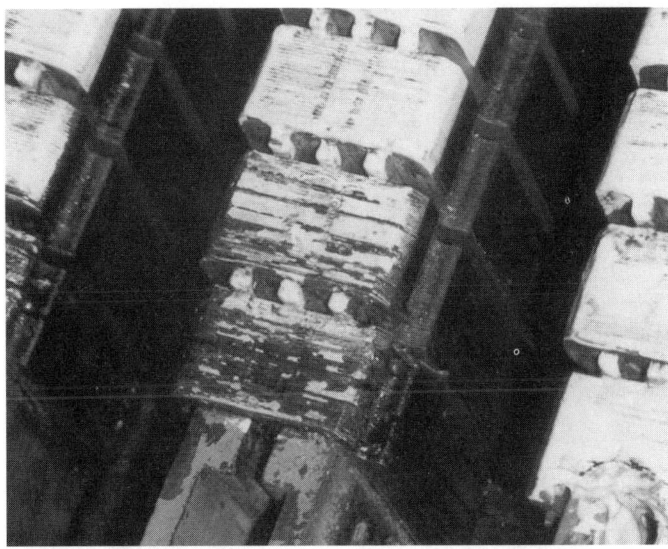

Fig. 8.45 A hot spot from loose iron fretting was repaired by spreading the core teeth and inserting mica to establish interlaminar insulation in the affected area. Penetrating epoxy was finally applied. (Note: At the time of this writing, this core has continued successful operation for over 15 years.)

Fig. 8.46 In this photo, the stator core has failed and melted locally next to a top bar, causing a stator ground fault.

Fig. 8.47 Here the stator bar has been removed from the slot shown in Figure 8.46 so the extent of the core damage can been seen more clearly.

Fig. 8.48 The stator bar that was removed from the slot in two previous figures is shown. The damage to the bar is significant, due to the core failure.

Fig. 8.49 End-packets of broken laminations belonging to a 500 MW turbo-generator.

Fig. 8.50 Packets of broken laminations belonging to the 500 MW turbogenerator shown in the previous figure, repaired with a G10 epoxy/glass false tooth.

A problem that almost invariably occurs with these types of failures is subsequent damage due to the debris now moving in the airgap. These bits of core plate and space blocks are thrown around the airgap by the natural rotation of the rotor, flow of cooling gas, and magnetic influences.

Typical damage is the hammering of tooth tops in the stator bore, which shorts out varying lengths of laminations and creates more core hot spots (Fig. 8.51). In severe instances, loose metallic debris can additionally become lodged in adjacent radial core vents and cause hot spots next to the stator bar (see Fig. 8.52).

In the example of Figure 8.52, the core damage was extensive, covering two full packets and partially the two adjacent packets on each side. The depth of the damage was found to extend to the slot bottom. Initial reaction in circumstances such as the failure described might be to consider restacking the core. However, it is sometimes possible to remove the damaged material and attempt a false-tooth repair.

All the damaged core plate and space blocks must be removed by cutting and grinding. Then the damaged core removal must be terminated at the point where good insulation between laminations can be observed and confirmed by flux testing. The repaired surface is then finished with an electroetching process to ensure that all burrs and edges are removed from the finished surface (see Fig. 8.53). This is done to eliminate the possibility of remaining interlaminar contacts and further hot spots. For additional protection, thermocouples may even be installed from the core back and epoxied to the slot bottom area (Fig. 8.54).

Fig. 8.51 Damaged stator-core tooth tops from loose debris moving through the airgap. Mechanical, thermal, and electrical arcing damage can be observed.

Fig. 8.52 Major core fault caused by loose debris in the airgap lodging in the radial core vents. Subsequently, the core laminations became shorted, and the circulating currents caused a core melt that overheated the stator ground-wall insulation of a conductor bar, leading to failure of the generator after a stator winding ground fault.

Fig. 8.53 Damaged iron has been removed by grinding and the laminations finally separated by electrochemical etching of the core iron surface.

Fig. 8.54 Thermocouples installed on the repaired surface of the fault area and run out the back of the core for termination and permanent online monitoring of the fault area core temperature.

Electroetching is an excellent technique for removing the burrs and rough edges of laminations that have first been ground down to where there is perceived to be good core material and interlaminar insulation between laminates. The authors use the word "perceived" here because it is not always possible to tell if the damaged material has been fully removed. In most instances, using a magnifying glass for close examination will tell whether the grinding process has removed all the damaged material to the point where the interlaminar insulation is visible and the edges of the laminations can be seen. At this point, there are usually rough edges still present and the electroetching process is used to remove the burrs and edges so that there is clear inter-laminar separation and insulation between the laminates. The typical process for electroetching is as follows.

Materials required

- 20% phosphoric acid solution by volume
- Fibreglass wool to hold the acid during the etching process
- Variable and controllable dc current source capable of at least 20 amperes
- Clip-on ammeter to control current level
- Various stainless steel electrodes to suit the contours of the core area where the etching is to be done
- Insulated handle to attach the electrodes to
- Breathing protection, eye protection, and skin protection due to the corrosive nature of the process

Procedure

- Moisten the fiberglass wool with the acid solution. It should not be saturated. (i.e., the acid should not be dripping from the fiberglass).
- Place the moistened fiberglass pad between the electrode and the core where the defect is located, on top of the defect.
- Place the negative polarity of the dc source on the electrode and the positive polarity of the dc source on the core frame (i.e., ground).
- Place the electrode on top of the moistened fiberglass pad and apply approximately 10 amperes per square inch through the electrode. (The amount of current in each instance will have to be a judgment call, and it comes with experience to know when the correct electrolytic action is occurring. This means that material is being removed without actually burning the core iron.)
- Continue the procedure, (replacing the moistened pads as the acid becomes depleted in each) until the laminations become visibly separated.

Postetching Requirements and Cautions

- Do not tighten the core until after etching is completed.
- Ensure that there is interlamination insulation up to the etching point so that no interlaminar contacts are made when tightening is done.
- After tightening the core, check the etched surface with a magnifying glass or other similar device.

- Perform an additional etch of the tightened surface if the magnifier reveals that it is required.
- Verify the repair by an EL-CID and/or flux test with infrared scan.
- Do not drip the acid solution on any other parts of the core or winding, and wipe any spills immediately.
- Most important, the etching should be done by someone experienced in the procedure.

The damaged core material that is removed from the fault area must be replaced with a support structure to keep the stator bars properly supported in the slot and secure the core repair. Figure 8.55 shows a false tooth, constructed of G10 epoxy/glass and constructed in parts, fitted into the space where the damaged core was removed and secured into dowels embedded in the epoxy at the slot bottom, for the example under discussion.

Radial cooling vents were cut into the false tooth to allow cooling of the core below the false tooth. The top of the tooth in the core on either side of the false tooth was chamfered to reduce fringing fluxes and flux concentrations. Final tightening in the axial direction was accomplished by placing 1/16 inch (1.6 mm) G10 epoxy/glass stemming pieces in the chamfered part of the core, on either side of the false tooth. Additional support in the radial direction is provided by dovetails in the false tooth for the winding-slot wedges. In service, the thermocouples have continued to record normal temperatures at the finished surface of the repair.

Fig. 8.55 Epoxy/glass false tooth used to fill the space where the damaged core was removed. Support is provided for the stator bars and remaining repaired core. (Note: This repair, installed almost 30 years ago, is still sound at the time of this writing since installation.)

Finally, on the subject of loose core repair, simple global core retightening by shimming needs to be addressed when there is significant looseness in the teeth tops of the core. Global core retightening is easily accomplished by (local) insertions of numerous "stemming" shims made of epoxy/glass, which are generally about 1/16 inch thick and cut to fit the profile of the stator tooth (see Figs. 8.56 and 8.57).

Each stemming shim is roughed and grooved on both sides, and drilled to create a penetrating epoxy reservoir. When the stemming piece is driven into the core tooth, it is coated with a penetrating and insulating "weeping" epoxy, to secure it in place after it has been driven into position. In fact, it is often a good idea to brush on some "weeping" epoxy in the general area of looseness. This helps consolidate the core once the tightness has been reestablished and the curing cycle begins. It also acts as additional interlaminar insulation.

If the core looseness is significant enough to wear the interlaminar insulation away on successive layers of core plate below the slot bottom (see Fig. 8.58), the danger exists of a runaway core fault that can produce runaway overheating and massive core melting (see Figs. 8.59 and 8.60). This is because the circulating currents are flowing in the core fault area to the keybars as a return path, and the overheating is well away from the stator bars, which would cause the machine to trip on stator ground fault if the winding insulation were breached. In most cases, however, the melted core eventually finds the stator winding as it flows and a ground fault trips the generator (Fig. 8.61). The results of such a failure mode will also cause severe distortion to the core stack (Fig. 8.62), and will most likely require a restack of the stator core with new iron. Also, there may also be distortion of the core frame and this would need to be addressed for any attempted repair by core restacking.

8.2.5 Bent/Broken Laminations in the Bore

Core laminations are often damaged during the removal of the rotor (Figs. 8.63 and 8.64). When this occurs, it is not uncommon for them to become partially short-circuit-

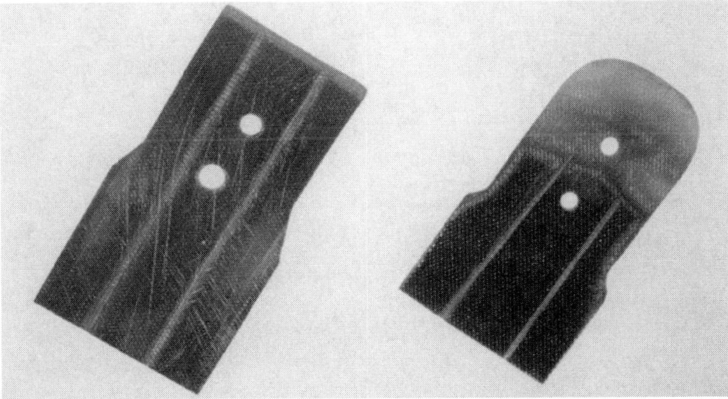

Fig. 8.56 Tooth-top stemming shims made of G10 epoxy/glass laminate.

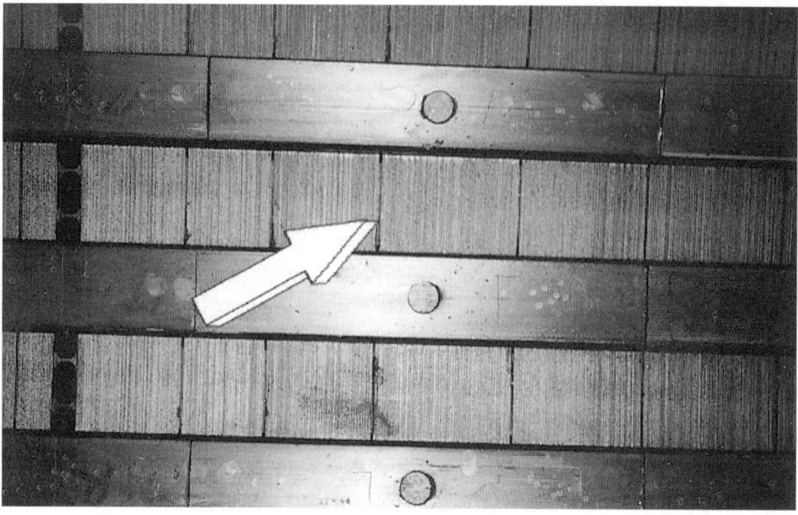

Fig. 8.57 Generally loose core tightened by stemming globally with G10 epoxy/glass shims.

ed. If left in this condition, they may reach excessive temperatures during operation of the machine, resulting in damage to the insulation between laminations. It is, therefore, recommended that laminations or groups of laminations bent during the removal of the rotor be straightened carefully, and their tops separated and impregnated with resin or epoxy so they do not remain short-circuited.

Fig. 8.58 Close-up view of a section of the lamination with the lamination insulation abraded.

Fig. 8.59 Runaway core fault showing molten metal pouring out of the back of the stator core.

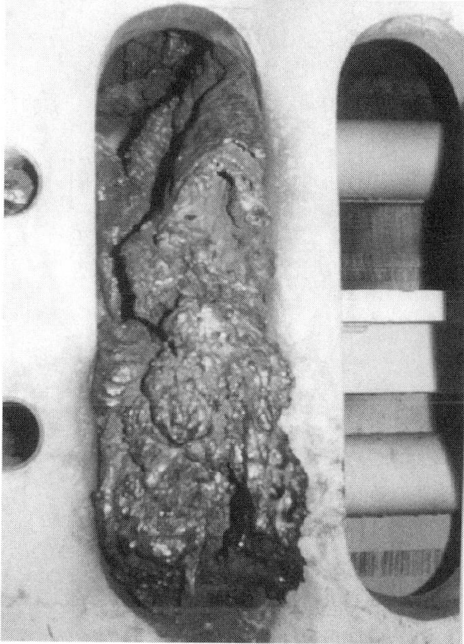

Fig. 8.60 Massive core fault showing molten metal pouring out of the back of the stator core.

Fig. 8.61 Massive core fault in which molten core iron eventually flowed into the stator bore through the radial cooling vents of the core next to the stator winding. The stator winding overheated and failed due to the molten iron.

Fig. 8.62 The stator core iron at the back of the core can been seen to have sustained major distortion from a massive core failure. A complete stator core restack was required for this machine.

Fig. 8.63 Bent laminations from the first packet (closest to the edge of the bore). They were bent during removal of a two-pole turbo-generator rotor.

Fig. 8.64 Same machine as in Figure 8.63. Bent laminations shown at different packets.

It is obvious that care should be taken during replacement of the rotor into the bore to avoid damage to the laminations. Any damaged laminations will remain damaged until the machine is disassembled again, commonly several years later.

As already stated in elsewhere in this book, it is most important to identify laminations that are broken or in the process of breaking up. Broken pieces of lamination can get loose and damage the insulation of the stator bars. Small pieces of lamination can "drill" into the stator bar insulation by the continuous action of the double-line-frequency magnetic forces that the field of the machine exerts over these loose pieces of metal. Some people call these loose small pieces of metal "magnetic termites." Bent laminations are suspected of having experienced excessive metal fatigue. Therefore, the inspector should always assess the source of the bending and the condition of the lamination.

8.2.6 Space Block Support and Migration

Not all lamination problems are the result of a loose core or inflicted damage. As Figure 8.65 shows, entire packets of broken laminations can result from lack of sufficient support by the I-shaped vent separators at the top of the tooth. The core of this particular machine (an air-cooled gas-turbine generator) was otherwise found to be tight. In this case, repairs included the introduction of epoxy glass laminates between laminations, and application of penetrating (low-viscosity) epoxy to the damaged area.

The solution to loose laminations varies, depending on location, severity, and type of problem. It ranges from retightening the core compression bolts to changing com-

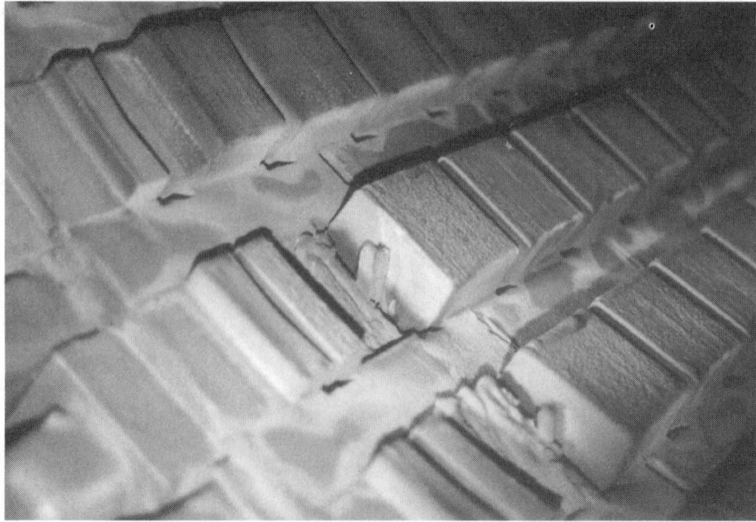

Fig. 8.65 Two packets of broken laminations belonging to an air-cooled gas-turbine generator.

pression plates or parts thereof, to introducing stemming shims, and so forth, as described in Sections 8.2.4.

Another problem that can occur is when a space-block weld breaks, or the space block itself breaks. This may allow the broken part to come loose and cause damage in the stator bore and, locally, to the core teeth, as previously described in Section 8.2.4.

8.2.7 Migration of Broken Core Plate and Space Block Thick Plates

During operation, large turbo-generators may also exhibit migration of broken sections of the core plate or the thicker plates next to the radial vent ducts. The thick plates generally are the same size and shape as the core plate laminates. Further, the thick plate assemblies contain the I-shaped separators that maintain the spacing between packets of laminations, required by the cooling gas to circulate radially within the core. Migration of these components is also usually due to looseness and subsequent metal fatigue.

When migration of the core plate or the vent-spacer assemblies occurs, it is inward toward the bore of the stator. Under these circumstances the bottom of the slot of the lamination damages the bottom conductor bar in the slot by cutting into the groundwall insulation (see Fig. 8.66). The result can be a ground fault or, worse, multiple ground faults from stator bars at different phases, resulting in phase-to-phase short circuits across the laminations. In this instance, the cost of repairs in time, money, and material is generally substantial.

This migration condition can be safely diagnosed by inspection of the bore with the rotor removed. Any migration of these components should be promptly discussed with the machine manufacturer.

8.2.8 Laminations Bulging into Air Vents

On occasion, inspection of the stator bore may reveal that the core laminations are bulging into the air ducts or vent areas, as well as bending away from the core at both ends (see Fig. 8.67). These problems normally occur as a result of weakened lamination support, combined with vibration.

In some designs, the vent spacers are curved toward one corner of the tooth to direct the flow of the cooling gas. In this design, laminations may bulge in the other corner of the tooth.

Laminations with weakened support at both ends of the core are subjected to the axial stray flux impingement of the core end, as described earlier in this chapter. The magnetic forces induced by the twice-operating-frequency axial flux present at those regions of the machine may eventually cause these laminations to crack and break due to metal fatigue.

Repairs to the affected core should be made by the appropriate method, depending on the extent and location of the damage found. It may be necessary to effect repairs as discussed earlier in Sections 8.2.4 and 8.2.5.

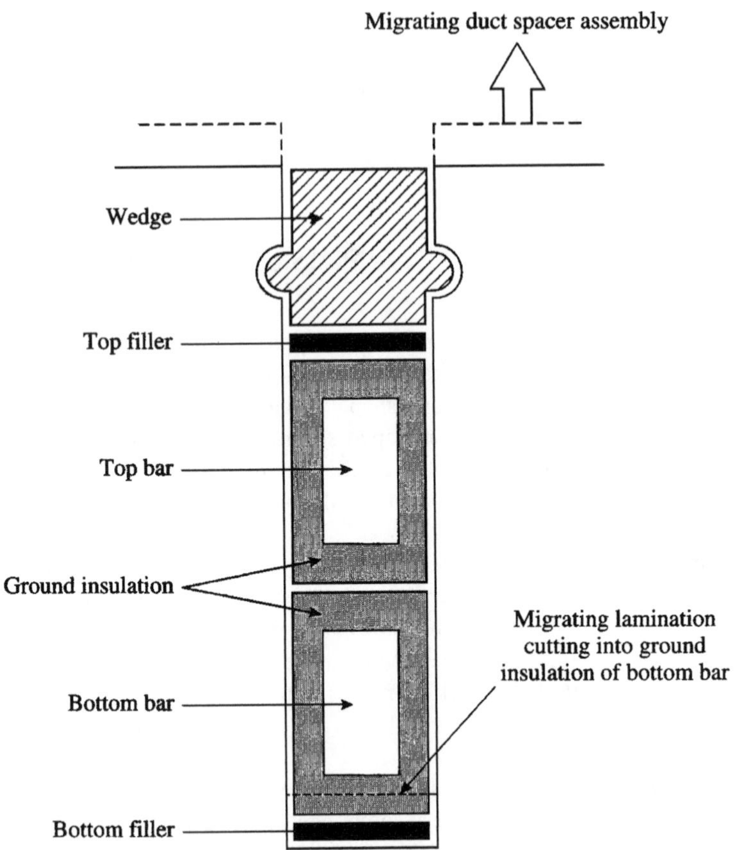

Fig. 8.66 Cross section of slot shows migration of the duct-spacer assembly.

Taking no action will limit the output of the machine due to overheating arising from the restriction on the flow of air (or hydrogen), as well as other heating problems associated with the extra losses generated in the core.

8.2.9 Greasing/Oxide Deposits on Core Bolts

In large machines, the stator core is almost always compressed by core bolts at the stator core back. These are torqued to values specified by the manufacturer during the assembly of the machine. Over time, part of the initial pressure within the core is lost. This looseness of the core is indicated by movement of the nuts and washers on the core-compressing bolts against the compression plates. This movement results in red iron oxide deposits or, when oil is present, in a greasy substance.

If these greasing and/or red iron oxide deposits are encountered during an inspection, the core compression bolts should be tightened to the manufacturer's specified values.

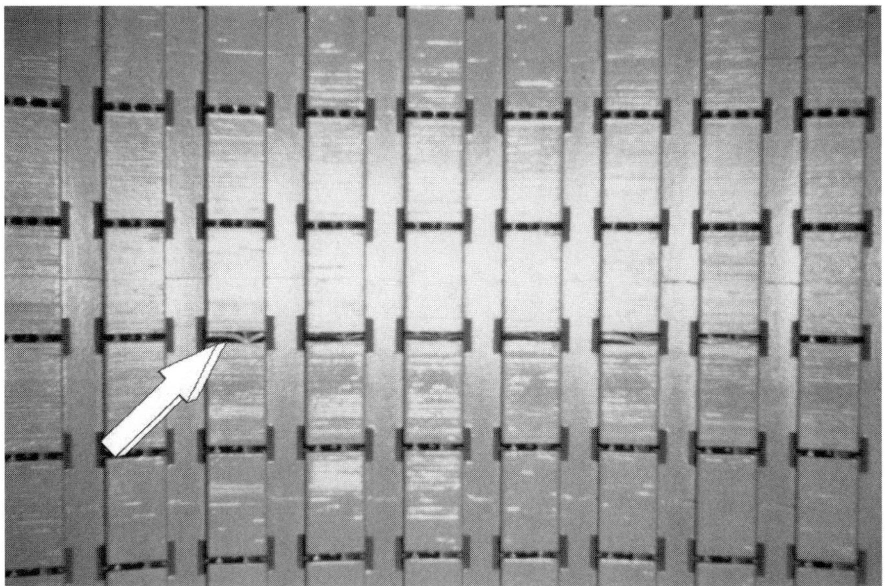

Fig. 8.67 Laminations bulging into a vent duct in a large turbo-generator. In this case, the offset position of the duct spacer provides weak support on one side of the laminations' packet. If this condition is left untreated, continuous vibration of the loose laminations may cause them to break.

In some two- and four-pole machines, the compression bolts travel through the core laminations. In this case, the force per bolt is approximately equal to the total force in the core divided by the number of bolts. For accurate torque values, always contact the respective manufacturer.

It is important to recognize that in machines with expansion-bearing bolts, which allow the end windings to slide (see Section 8.3.5 below), the end windings may have to be relieved of their brackets before the core is compressed. Subsequently, after the core is compressed to the desired pressure, the end windings are tightened again to the supporting brackets. This operation normally entails releasing and tightening several bolts per bracket, which allows the two halves of the brackets to separate. The reason for loosening the bracket halves before compressing the core is to avoid building tension on the end -windings when the core shrinks during the core-compression operation. The same may apply to machines with long cores and without expansion-bearing bolts. Consult with the manufacturer when in doubt.

In certain machine designs, the connection rings are kept tight by means of "bell-type" spring washers. They are conical in shape and inserted at both ends of the core-compression bolts (see Fig. 8.68).

Bell washers have been known to break due to metal fatigue. Being located at the end of the core, in the region of high-velocity moving gas/air driven by the rotor fans, the broken parts can be thrown with force onto the end windings. This can result in im-

Fig. 8.68 Spring washers (also called bell washers) covered with resin-soaked glass tape.

mediate failure or partial damage to the insulation with potential for later development into catastrophic faults of the insulation.

One common practice that eliminates this concern is the use of resin-soaked glass tape to wrap the washers together with the bolt heads and/or nuts. In this manner, all broken parts of the spring washer will remain contained within the hardened glass tape (Fig. 8.68).

During major overhaul inspections, it is good practice to remove the glass tape from the core-compression bolt ends and bell washers to allow their inspection and assess the condition of the washers. There are some users who like to change these washers at intervals of several years (during a major overhaul), even when they do not show visible signs of deterioration

8.2.10 Core-Compression Plates

Core-compression plates or clamping plates are located at both ends of the stator core. They essentially are large and very substantial compression "washers." The core compression plates are mounted over the ends of the stator core building bolts or keybars, and any through-bolt assemblies running the axial length of the stator (Fig. 8.69). Large nuts are used to hold the compression plates to the core, and are pressed in at the factory, as previously described in this chapter in Sections 8.2.4 and 8.2.9. Generally, it is rare to find problems with the compression plates themselves, but they need to be inspected. Loosening of the core pressure and, hence, the compression plates, is the most likely issue to be found.

Fig. 8.69 Core-compression plate on the stator core end.

The keybar and through-bolt nuts of the core-compression plates should be checked for looseness and the manufacturer should consulted regarding retightening. One caution should be noted when retightening, however: if there are known core interlaminar defects found by EL-CID or flux testing, retightening before correcting these may make the defects worse, due to the increased axial pressure exerted and resulting better interlaminar contact.

8.2.11 Core-End Flux Screens and Flux Shunts

One of the components that are often overlooked during stator core inspections are the core-end flux screens (Fig. 8.70), or in some machines the core-end flux shunts (see Chapter 2 regarding flux shunt arrangements) mounted at both ends of the stator core.

Fig. 8.70 Core-end flux screen mounted on a core-end compression plate.

Flux screens are generally made of copper and used to shield the core-compression plates and the core ends from axial stray flux, which can cause core-end overfluxing and subsequent overheating. Flux shunts are used as well to prevent stray flux from entering the core end and to keep core-end heating at acceptable levels.

Inspection of these flux diversion components is required to ensure that they have not overheated and caused damage to adjacent components such as the core-compression plates they are mounted on, or the stator winding that is installed next to them. Flux screens and shunts become very hot in operation, due to the eddy currents induced in them, and are required to be cooled. Sometimes, they are water cooled within the same demineralized cooling water circuit as the stator winding or by the cooling gas of the internal generator.

In addition, flux screens are often electrically isolated from ground to avoid further overheating by additional circulating currents through ground. These should be checked for insulation resistance as well as inspected for thermal and distortion damage. However, there are also flux screen arrangements that call for grounding of the copper screen. In these designs, it is important that the ground connection be checked to ensure it is in good condition and that *it* is the source of grounding of the screen and not some other component that the screen is mounted on, attached to, or touching. Having the screen grounded through a component that is not designed to handle the ground current may cause it to overheat and fail.

8.2.12 Frame-to-Core Compression (Belly) Bands

Many types of two-pole machines include *compression bands,* also known as *belly bands* or *belly belts,* around their spring-supported stator core [6]. These provide control of the vibration levels of the machine. It is not uncommon for additional compression bands to be installed if core vibrations reach unsatisfactory levels (Fig. 8.71).

Belly bands are normally not accessible during an inspection, but many machines do have inspection ports on the underbelly of the stator casing that can be opened up during overhauls to inspect and possibly retighten the bands. However, their inspection is rarely considered other than when stator vibration problems are present and can be attributed to insufficient peripheral compression of the core suspension arrangement. Manufacturers can provide the values to which compression bands should be torqued.

8.2.13 Back-of-Core Burning

Back-of-core burning occurs due to current transfer between the stator core frame keybars and the iron core. Obviously, this occurs at the back of the core, where the keybar system makes contact with the core in the dovetail area. The currents induced can be anywhere from negligible to extremely damaging. The main influences for the current transfer mechanism are the degree of mechanical coupling in the radial and tangential

Fig. 8.71 Stator core and frame compression bands or belly bands.

direction between the core and keybars, and the magnetic saturation effects in the core, especially at the ends. Since the keybar currents are a function of the magnetic saturation levels of the core and, hence, the back-of-core flux density, the higher the flux density and magnetic saturation, the higher the leakage flux and the greater the keybar currents.

First, consider that the keybar structure forms a "squirrel cage" that inherently carries currents due to leakage flux at the back of the core. This leakage flux links the squirrel cage and causes substantial current to flow axially, down the keybars. The current then transfers from keybar to keybar at the core ends via a suitable path, which generally has the lowest impedance. The path can be the core itself, or some part of the core-end clamping structure or the core frame. It most likely occurs in the core when noninsulated keybars are employed, but the current levels are generally not harmful in a machine that has a magnetically well-designed core-end region and good contact between the core and the keybars. "Magnetically well designed" means a core end that does not oversaturate during any mode of operation and a core that is operating at lower saturation levels. In addition, the squirrel cage acts as a damper to minimize, or even negate, flux linkage with the remainder of the core-frame structure.

Generally, back-of-core burning occurs at the core ends. It has been reported that cases of back-of-core burning have been seen away from the ends, but this has undoubtedly been on machines in which a core failure has occurred. In that case, the back-of-core burning occurred directly behind the location of the fault.

Oversaturation of the core ends has been a major contributor to back-of-core burning in many of the cases seen. Therefore, it is essential to prevent this situation. If this cannot be prevented, then current transfer from the keybars to the core must be short-circuited by an alternative and lower-impedance path. But back-of-core burning also requires an intermittent contact or high-resistance joint between the core and the keybars for burning to occur. This has everything to do with the mechanical coupling between the core and keybars, in the radial and tangential directions. Therefore, back-of-core burning can occur without oversaturation of the core ends, but it will be worse in a magnetically poorly designed core end.

Saturation of the core ends is most likely to occur from operation in the leading power factor range. This is due to the stator end-region effects, which produce a higher level of axial flux impingement on the core ends in the leading power factor range. The net effect is that if the core is not protected against or designed properly for this, the end iron will oversaturate and cause excessive leakage flux at the core back, at the ends.

Oversaturation can also occur on a well-designed core end in the leading power factor range, if the particular machine employs magnetic retaining rings. Electromagnetic finite-element analysis on generator end regions has shown a significant difference in effect between the uses of magnetic versus nonmagnetic retaining rings in some machines. In fact, there are cases in which a 40% reduction in core-end flux density at 0.95 pf leading has been achieved when magnetic rings were replaced with nonmagnetic rings.

When there is core-end oversaturation and, hence, higher leakage flux, there is a tendency for the interlaminar voltages at the core ends to increase, and, of course, higher currents to occur in the keybar system due to this localized effect. It is believed that the increased interlaminar voltage at the core ends and the higher currents can create an arcing and sparking effect, which causes the burning to occur at the core/keybar contact areas. In addition, make-and-break (contact) action from stator vibration is suspected to influence the initiation of sparking due to the vibration effects (see Fig. 8.72). If the core ends are oversaturated due to increased axial flux, the circulating currents in the individual laminations will be higher. (Then the core-end temperatures are also raised due to higher losses, which increases the chances of interlaminar core failures in the ends). Therefore, there is a possible contribution to the back-of-core burning simply by the eddy currents transferring and conducting through the keybar/core contact area.

One other interesting fact is that generally not all of the keybars carry the torsional load and, hence, not all the keybars will have good contact with the core. Core-frame-to-core-mechanical coupling analysis by finite element analysis has been done in past, and showed that in some cases only about 20% of the keybars actually take up the torsional load. This may explain why back-of-core burning is not usually seen on all key-

Fig. 8.72 Stator back-of-core burning (shorted laminations where they touch the frame). The severe burning has also created magnetic particles of debris. These become loose and can travel into the critical areas of the stator as "magnetic termites."

bars, even though the leakage flux will be evenly distributed around the back-of-core circumference.

8.2.14 Core-End Overheating

One of the most onerous issues for stator cores is core-end overheating, which is usually most prevalent at the core ends and during leading power factor operation (see Section 2.5 for a detailed description of the mechanism). Higher magnetic saturation and higher flux densities in the core can cause higher interlaminar voltages in the core ends and may have a subsequent effect of arcing across laminations in the axial direction (Fig. 8.73). If there are interlaminar defects present, the situation can become critical much faster. The unfortunate nature of this type of damage is that it cannot be observed directly during an inspection, unless it is at the tooth tops of the stator core. However, if an inspection of the back of the core reveals back-of-core burning, then there is the possibility that some interlaminar breakdown may be occurring, as shown in Figure 8.73 and that the burning at the core back is the return path for the fault. Usually, the defect has to be substantial before core-back burning as the return path can be seen. In any case, it is difficult to determine if the core-back burning is due to a defect near the bore, as the more common occurrence with back-of-core burning is that it is localized at the back of the core only (as described in Section 8.13). When there is uncertainty, the best approach is to carry out EL-CID and flux testing with an infrared scan.

In cases of prolonged operation, when the stator core is experiencing high core-end temperatures, the interlaminar insulation may suffer advanced signs of aging from the excessive heating; this can be seen as "heat staining" on the surface of the laminations

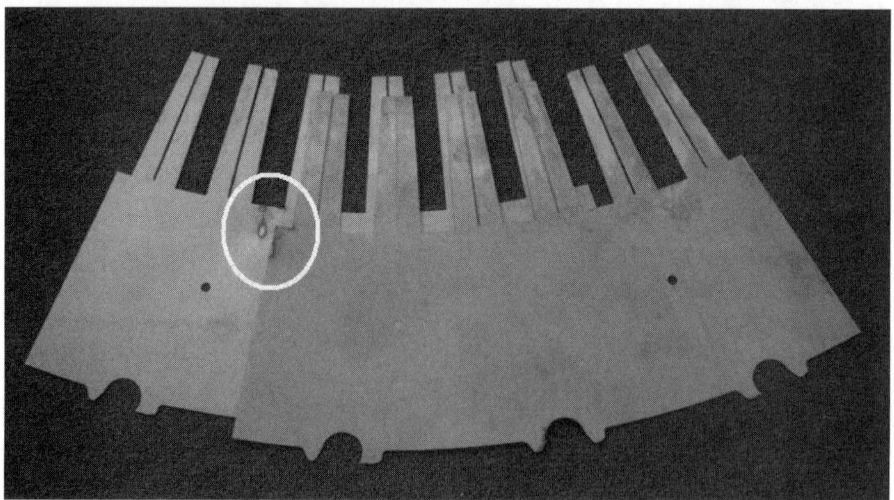

Fig. 8.73 Stator core laminations are shown from two successive layers. The arcing between layers can clearly be seen to occur at a stress point, which is a lamination edge on the upper laminate.

(Fig. 8.74). Again, this is not possible to observe visually during a regular maintenance inspection because it is totally inaccessible. It may be possible to see during a stator rewind with the winding removed, if the heating is severe enough to show discoloration at the bottom of the stator slot. Again, the best approach is to carry out EL-CID and flux testing with an infrared scan. Damage such as that shown in Figure 8.74 will show up with both of these tests if they are done correctly.

If the overheating effect is very severe, then the interlaminar insulation may simply break down and cause a core fault to occur (Figs. 8.75 and 8.76). One clue to a general core-end overheating problem, rather than a localized defect, on a generator stator is that there will be "heat staining" at the slot bottom on every slot. Finite-element analysis of stator cores shows that the hot spot for the flux in the stator yoke behind slot bottom is right at the slot bottom where the heat staining is observed in Figures 8.74 and 8.75. When one adds the effect of end region stray fluxes in the axial direction during leading power factor operation, core end overheating becomes more possible. This is especially prevalent in stator cores in which the core end flux loading is high from high electric loading and there is marginal leading power factor operational capability.

One last point on core-end overheating is that the most common result is subsequent overheating of the stator winding and generator trips on the stator ground fault on a bottom bar, once the ground-wall insulation has broken down enough to allow a discharge to the stator core.

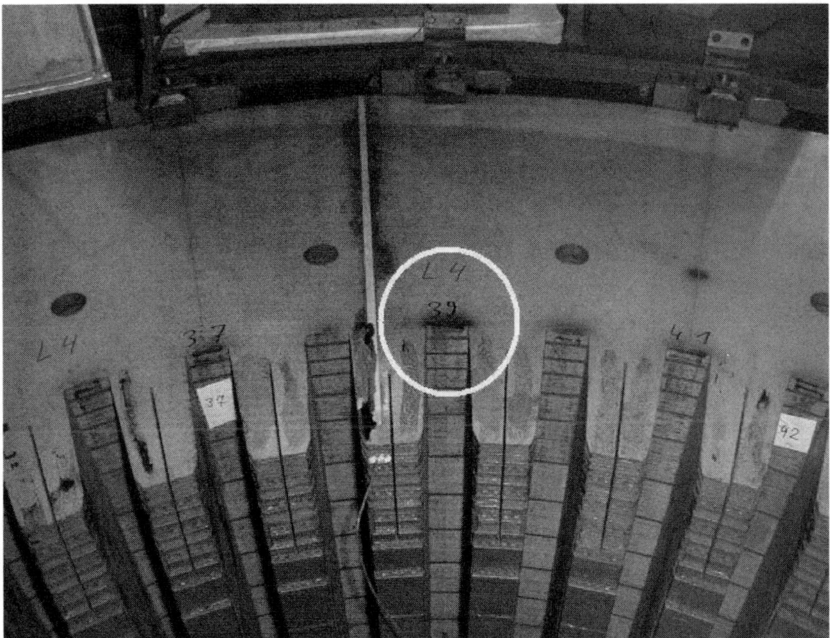

Fig. 8.74 Stator core end section of a core that has been partially unstacked. The light "heat staining" can be seen at the slot bottom on each slot.

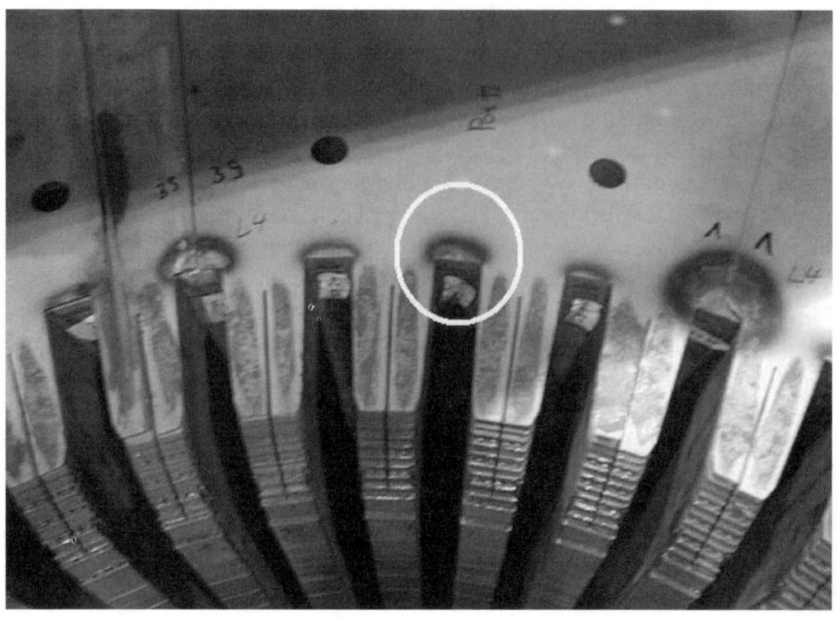

Fig. 8.75 Stator core end that has been partially unstacked. There is heavy "heat staining" visible and the core has suffered a breakdown of the interlaminar insulation and caused a stator ground fault to occur. A close-up of the core melting in one of the slots is shown in the next figure.

Fig. 8.76 A close-up of the core melting at the slot bottom of one of the slots in the machine of Figure 8.75.

8.3 STATOR WINDINGS

8.3.1 Stator Bar/Coil Contamination (Cleanliness)

It is almost certain that during most comprehensive inspections some contamination will be found on the stator bars or coils of the machine. Given that these intensive inspections are performed at intervals of several years, and given the adverse effects contamination has on the integrity of the insulation, it is appropriate to recommend cleaning the stator winding.

Contamination causes degradation of the insulation by providing a medium for currents to flow on the surface of the insulation. This results in tracking and reduction of the insulation properties. In addition, contamination may penetrate the insulation cell via cracks, which occasionally results in tracking followed by a short circuit.

Normally, inspection will determine the measures required for cleaning the insulation based on the degree of contamination found during the inspection. There are several choices; some are mechanical, such as vacuum, compressed air, crushed corn cobs, lime dust, CO_2 pellets, brushes, and cloths. Chemical methods include solvents. Steam cleaning is another option.

The choice of any particular cleaning method should be based primarily on the type of winding insulation, the degree of dirt or contamination encountered, the condition of the windings (how they might be affected by the impact of particles, steam, etc.), and how exposed they are to solvent ingression [10]. Various states and jurisdictions may impose different regulations on the use of the many solvents available. Reference [4 item (7, p. 11)] contains a well-written and detailed description of different methods for cleaning windings.

Air-cooled machines usually sustain atmospheric types of contamination such as moisture and dirt particles (Fig. 8.77), as well as oil contamination. Air-cooled stators should be cleaned regularly due to the more onerous effects of partial discharges in an air environment.

Water-cooled windings present a particularly severe problem regarding oil and/or water contamination. The design of the stator-bar support system within the slots and the higher electromagnetic radial forces present in these machines make these types of windings less tolerant of oil and/or water contamination.

Oil contamination of the *ripple* (*spring*) *fillers* located on one side of the slot reduces the friction designed to retain the stator bars under large radial forces. A winding too saturated by oil contamination could be beyond easy repair and may require a full rewind.

Water ingression is damaging to the wall insulation of the bars. Control of water and oil ingression is primarily an online monitoring activity. However, if excessive oil and/or water contamination is encountered during a major inspection, the inspector has to evaluate the type of response required to return the machine to a reliable operating condition.

Regardless of the type of contamination that may be present, it is always a good idea to remove it if possible. Contamination presents an unpredictable variable in the insulation life of a stator winding, and makes it difficult to predict long-term reliability, since contamination products also affect any electrical test results. The type of

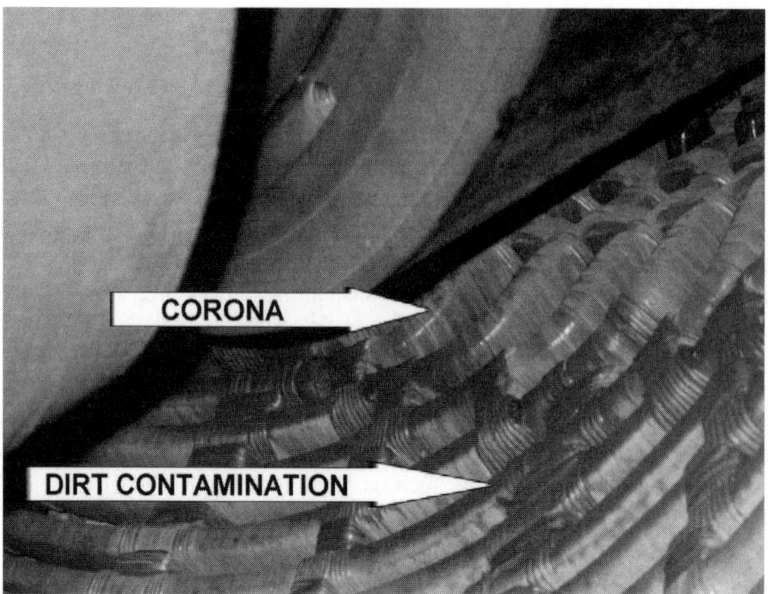

Fig. 8.77 In this photo there is significant dirt contamination shown on the endwinding, but there is also a significant amount of slot-exit partial discharge residue or "corona" discharge present in the form of a white powder. In this particular case, the "corona" discharge residue is easily wiped off and there is no sign of ground-wall insulation deterioration at this point. The dirt contamination in this machine is not responsible for the corona in this particular case. The issue here is more of a slot-exit grading system problem. (See color insert.)

cleaning and cleaning materials to be used should be assessed at the time and in consultation with the manufacturer, since they have the in-depth knowledge of the insulation system and what may or may not be harmful to use.

8.3.2 End-Winding Blocking and Roving

End-winding blocking and roving consist of the material used to separate the sides of the stator bars at the end windings and between end-winding connections. Design trade-offs determine the clearances between the stator bar sides at the end regions, which can generally range from large to relatively small (on the order of a fraction of an inch). Tolerances of the manufacturing process of the stator bars, as well as mechanical stresses arising during their installation, result in many stator bars touching each other at the end windings. Left alone, the stator bars will rub each other at twice the operating frequency during the operation of the machine. Moreover, sudden changes of load or external short circuits can create large movements of the stator bar ends. The continuous movement, combined with the rubbing between stator bars and the large sudden movements that can occur during stress events, as well as different temperature-related expansions and contractions, are strongly detrimental to the integrity of the ground-wall insulation of the stator bars and can severely reduce the expected life of the windings. To eliminate the movement between stator bars, blocking and roving are used.

When the blocking is carried out by inserting separators between the stator bar sides, the blocking materials come in two basic forms: solid or amorphous blocking. Solid blocking can be of rectangular shape or conform to the shape of the stator bars. Both shapes are made of solid insulating materials, such as textolite, maple, and so on. Amorphous blocking is made of felt or felt-like material, which is soaked in resin or an equivalent impregnating material during installation, or as part of a vacuum-pressure impregnation (VPI) process.

Felt-like material is found in most modern machines, whereas older windings are primarily blocked with solid separators. Often, solid separators tend to get loose after long periods of operation. In severe cases, they will even fall from the winding. Operation of the machine with loose or missing blocking will adversely affect the reliability of the winding. Therefore, loose or missing blocking should be treated or replaced as necessary.

Solid blocking material is almost always held in place with *ties* (see Fig. 8.78). In such cases, the condition of the ties determines how effectively the blocking performs. In other cases, no ties are involved and only substantial roving (see Fig. 8.79), impregnated with epoxy and then cured, is employed.

Fig. 8.78 The end-winding of a turbo-generator. Shown are the blockings between the stator bars, held with ties.

Fig. 8.79 Close-up of roving arrangement shows resin-soaked and hardened roving made of glass fiber.

When exposed over a long period of time to the relatively high temperatures encountered in the machine, ties tend to "dry" by evaporation of the solvents and become brittle. Thus, their structure deteriorates and they lose volume. Becoming loose, they allow relative movement between the stator bar sides. Subsequently, the ties and blocking are further deteriorated by abrasion, with negative consequences to the winding integrity.

Ties are, therefore, an item to be inspected carefully. Defective ties may exhibit powder deposits, flaking, tearing, or other telltale signs of deterioration. If relatively few ties show an inadequate condition, they can be treated with lightly viscous epoxies to fill the voids, thereby tightening anew the tie-block structure upon drying out. If a substantial number of ties show signs of degradation, retying the complete winding should be considered.

Inspection of the end windings should include evaluation of the condition of the blocking, ties, and roving. Signs of looseness include greasing, dry/loose/broken ties, powder, abrasion signs on stator bars, cracked paint coating, and missing blocking pieces (Figs. 8.80 to 8.82). Suggested corrections can include one or more of the following: retying, new blocks and ties, amending the stator-bar wall insulation, cleaning, and applying penetrating epoxies or resins. In some cases, though, looseness and high vibration may be due to the natural frequency of the end winding being too close to the twice-per-revolution operating frequency. When such a situation occurs, the best solution is to increase the stiffness and support of the end winding to move the natural fre-

Fig. 8.80 Greasing of stator end-winding ties due to excessive looseness and high vibration.

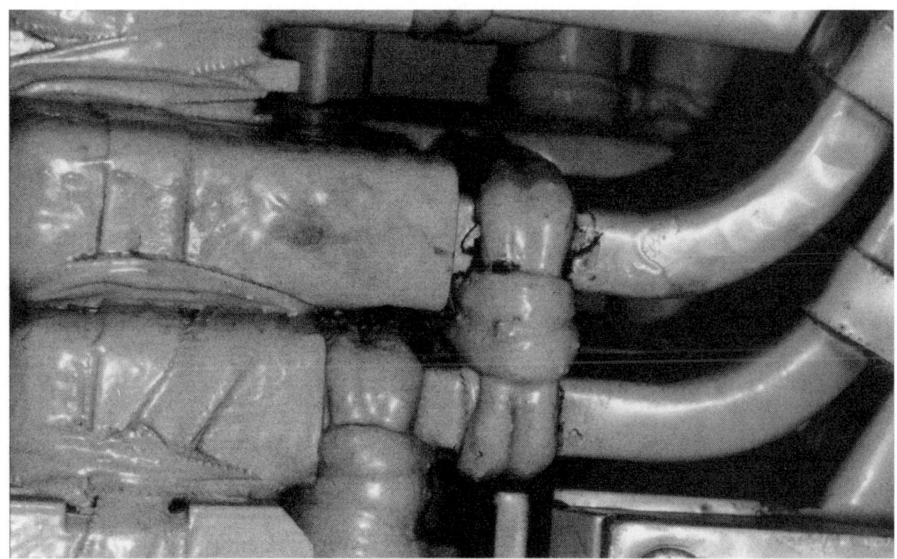

Fig. 8.81 Close-up of localized greasing of a stator end-winding tie from insufficient support.

Fig. 8.82 Greasing of stator end-winding support bracing due to high vibration caused by a resonant frequency in the end winding being too close to the twice-per-revolution frequency.

quency of the end winding higher and further away from the twice-per-revolution frequency of operation. When this is not possible, retightening and increasing the damping effect will help, but this does not usually result in a permanent solution, and such end windings will need to be inspected and maintained more often.

8.3.3 Surge Rings

As explained in the previous section, the end windings of electric machines are subject to substantial movement during sudden changes of load, in particular during short circuits in the "electrical vicinity" of the machine. These movements are detrimental to the integrity of the insulation. Severe movement will deform the windings and possibly crack the insulation of stator bars, lead connections, stator bar connections, and so on. To minimize the damage inflicted by end-winding movements, the stator bars are tied to circular rings commonly called *surge rings* or *support rings* (Figs. 8.83 to 8.85). In large machines, the rings are normally made of steel and sometimes of fiberglass materials. The steel rings are themselves covered with several layers of insulation.

Ties securing the end windings to the surge rings suffer the same type of temperature- and abrasion-dependent degradation as the ties between stator bar sides. When this occurs and is allowed to persist for extended periods of time, abrasion may result

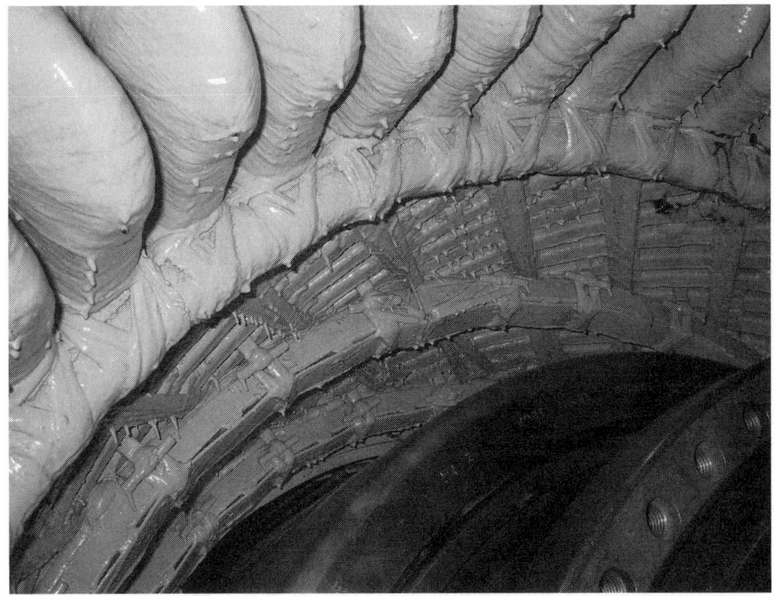

Fig. 8.83 Bore-side multiple surge rings and ties in the end winding of a two-pole stator.

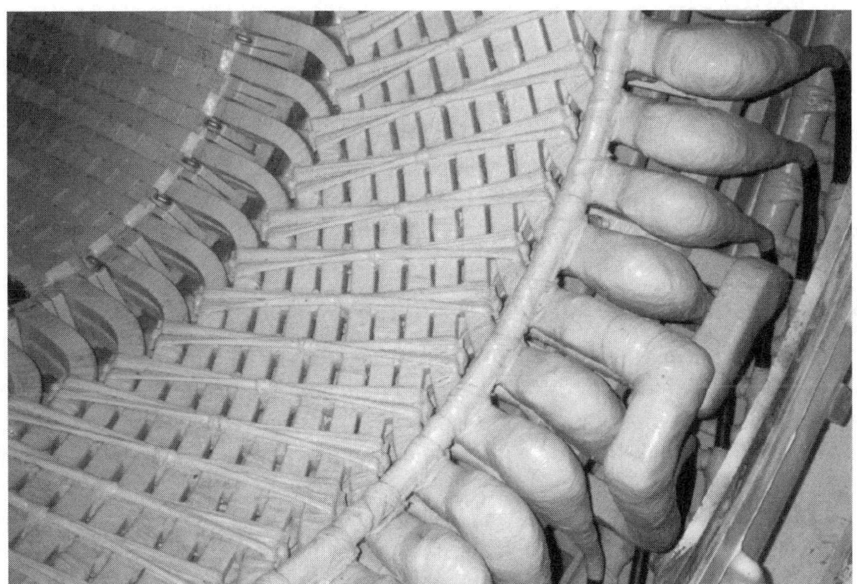

Fig. 8.84 Bore-side single surge ring of a two-pole stator and the ties in the end winding.

Fig. 8.85 Bore-side single surge ring of a four-pole stator and the ties in the end winding. Note the much shorter involute overhang of the end winding due to a much shorter required pitch factor for a four-pole machine. These end windings are usually stiffer and less prone to natural frequency and vibration issues than two-pole stators.

in the wall insulation of the stator bar and the surge-ring insulation. The natural consequence is a phase-to-phase or ground fault at this location.

As with the blocking ties, inspection should include looking for signs of excessive dryness, greasing, deposits of powder (normally of bright color), and so forth. The repairs are similar to those proposed for the blockings and ties (Section 8.3.2).

8.3.4 Surge-Ring Insulation Condition

The purpose of the insulation on surge rings made of steel is to minimize the possibility of a ground fault to the rings. As described in the preceding section, the movement between stator bars and surge rings is minimized by effective ties, but not completely eliminated. The surge-ring insulation provides an additional layer of protection in a critical area of the machine.

Inspection of the surge-ring ties and insulation more often than not requires mirrors, since they are usually in a restricted place, not allowing direct visual access. However it is done, inspection of the windings should always include evaluation of the integrity of the surge-ring insulation, in particular in the areas beneath the ties. Greasing, powder deposits, and other telltale signs should focus the inspector's attention on a probable degraded condition. In addition to deterioration due to movement of the stator bars, the surge-ring insulation can deteriorate due to electrostatic discharges from the stator bars. These appear as electric tracking and/or burnlike marks on the insulation of the surge ring and offending stator bar. If significant, the problem can be taken care of by cleaning the affected area and by adding a few layers of new insulation impregnated with insulating resins and insulating paints.

8.3.5 End-Winding Support Structures

The end-winding support structures constrain the movement of the stator bars by distributing the forces exerted by one stator bar onto other stator bars, and by transmitting them to the frame of the machine. In order to accomplish that, the support structures must remain in sound condition. Inspection should include looking for loose parts, bolts and/or nuts, cracked supports made of solid insulation material, greasing of bolts, and cracked or loose welding.

There are a large variety of support assembly designs, but in most large turbo-generators they are brackets attached to the core-end compression plates (see Fig. 8.86) and, sometimes, to the inner stator frame as well. In some machines, there may even be a surge-type ring of significant size behind the end winding, to which support arms or brackets are attached.

Experience has shown these to be sturdy and reliable structures. Nevertheless, they should be examined, at least during major inspections, to verify that they remain in

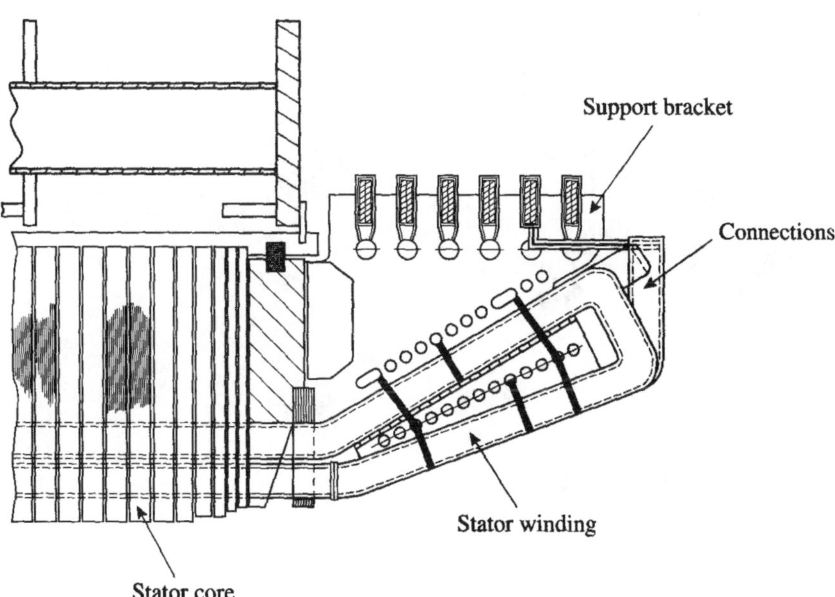

Fig. 8.86 Schematic representation of a typical end-winding support assembly of a small turbo-generator.

good condition. As is the case of the surge-rings, the supporting assembly is not readily accessible for visual inspection. Inspection requires some diligence and agility on the part of the inspector. Mirrors can go a long way in facilitating inspection of the support assembly of these support arrangements.

With some designs of large generators, there is often a means provided to retighten end windings should they come loose due to vibration in operation over time. Such tightening arrangements can be complicated, and are generally found at the back of the end winding. One such mechanism for this purpose is shown in Figures 8.87 and 8.88.

Any type of retightening system must always be checked during an outage and readjusted if necessary. Also, if end-winding vibration monitoring is provided, then this can indicate a need for an outage, to go inside the machine and carry out the retightening. Retightening operations must be done very carefully. The example of Figure 8.89 shows the damage probably caused to the end-winding main support brackets by aggressive tightening operations over the life of this generator. One must seriously evaluate any retightening activity against the strength of materials and other design considerations.

Because of the significant changes in stator winding temperatures that occur inside a generator between operation at high and low loads, as well as when shut down, axial expansion must also be accounted for. In most large turbo-generator designs, the end-winding support assembly is allowed to move in the axial direction to accommodate

Fig. 8.87 End-winding support assembly and retightening mechanism. The retightening rod is locked by an epoxied cord.

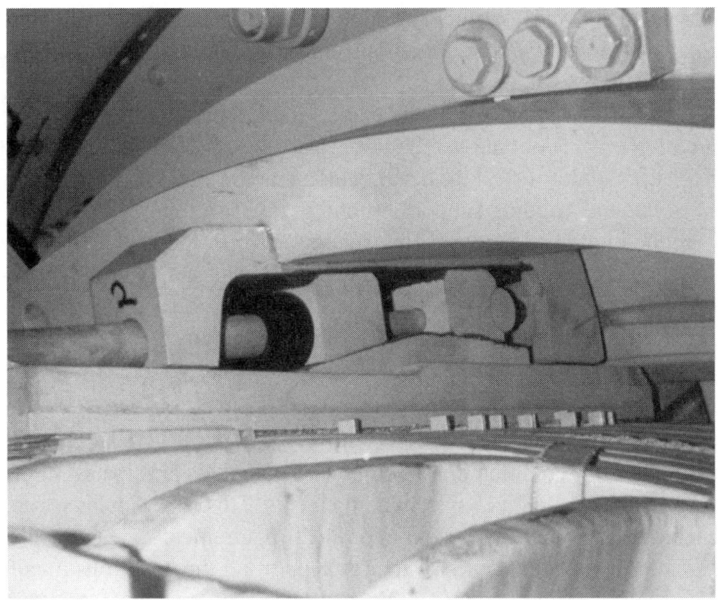

Fig. 8.88 End-winding support assembly and retightening mechanism of Figure 8.87, from a side angle. This photo shows more clearly how the wedging system of the device will increase the pressure between the main support bed and the stator winding when the tightening rod is turned.

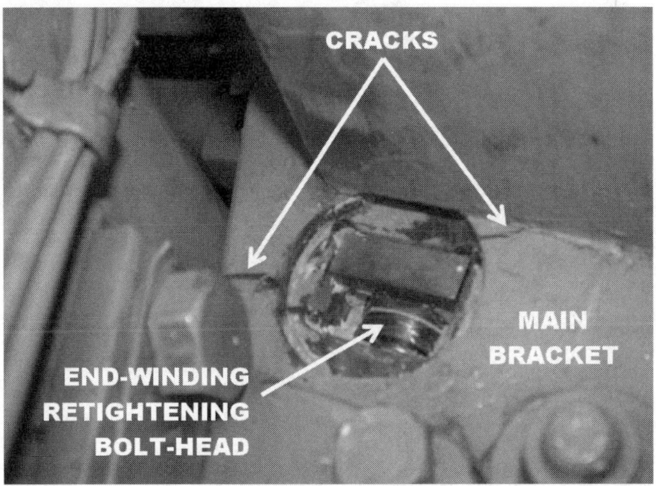

Fig. 8.89 This end-winding main support bracket appears to have been cracked by too much retightening of the end winding. The retightening is carried out during those occasions when the rotor is out of the bore, by means of bolts whose heads are located in specially drilled spaces in the bracket.

the thermal expansion and contraction of the winding. The movement is allowed in the design of all end-winding support structures in one way or another, and there are many variations.

One method is by having the whole end-winding support system mounted on *expansion-bearing bolts.* Continuous vibration may cause fretting and eventually breaking of the bolt assembly (see Fig. 8.90). This situation may encroach upon the free movement of the end-winding support system, giving rise to uneven stress. The stress may damage the integrity of the end-ring support system. Also, some indications exist of abnormal vibration caused by faulty expansion-bearing bolts providing insufficient support to the end windings. Therefore, the expansion-bearing bolts, when present, should undergo close examination during a major inspection.

8.3.6 Ancillary End-Winding Support Hardware

In large machines, it is common to find other hardware that participates in retaining the end-winding structure and other system components. This is particularly true in water-cooled machines, in which there is an additional requirement to support not only the stator bars themselves but also the structures supporting the numerous manifolds, water pipes, and other related hardware. One of the critical areas for the integrity of the water system is the condition of the several O-rings connecting different sections of the manifolds (see Figs. 8.91 to 8.93).

Fig. 8.90 A winding-support bearing bolt. Grease deposits shown on the bolt are an indication of looseness.

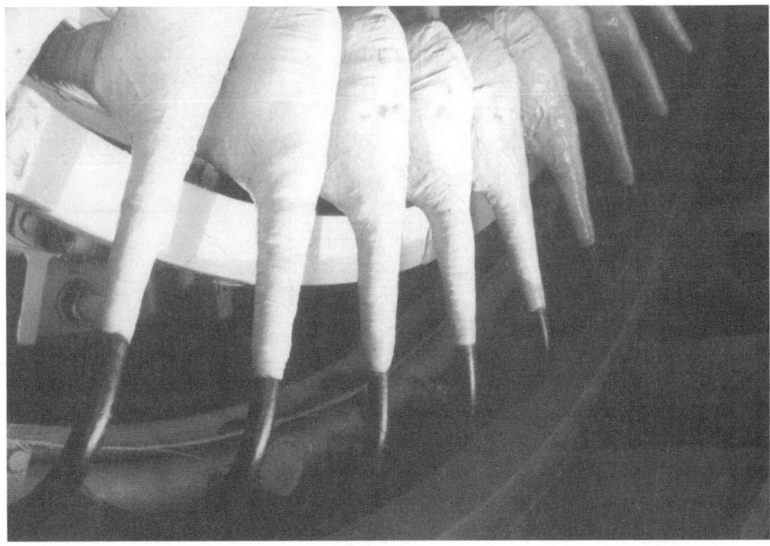

Fig. 8.91 Portion of an end winding of the water-cooled stator of a turbo-generator. The water-carrying hoses connect to the stator bars at the stator bar ends.

Fig. 8.92 Manifolds and other water-carrying piping belonging to a large turbo-generator.

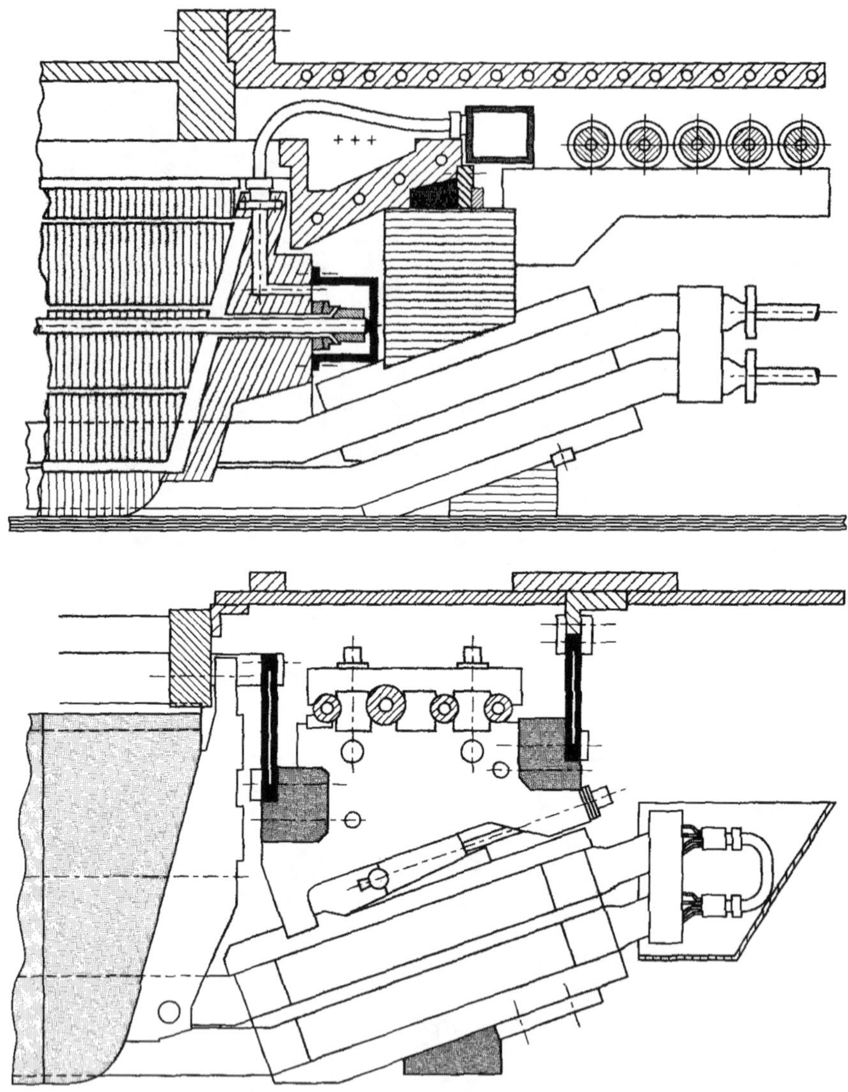

Fig. 8.93 Schematic representations of the end-windings of water-cooled stators of turbine generators. The lower diagram shows a spring-mounted bracket-support system. (Courtesy of ABB.)

It is impractical to describe here the many types of structures found in the industry. However, regardless of the construction, the instructions for the inspector remain the same: Look for indications of looseness; fractured parts; loose parts, bolts, and nuts; and cracked welding. The treatment for any abnormal condition depends on the availability of spare parts and the type of problem.

8.3.7 Asphalt Bleeding/Soft Spots

Sometime during the 1920s and 1930s, the electrical machine manufacturing industry began utilizing asphalt as the bonding agent for the insulation of large synchronous machines. Asphalt was used to bind mica flakes (asphalt micafolium) to form the wall insulation or was used to bond the mica flakes to a tape (asphalt mica tape). In the latter case, the wall insulation was made of a number of layers of tape. The number of layers depended on the tape and the rated voltage of the machine. A final layer of armor tape made of cambric or other materials was commonly used as protection to the cell insulation. In addition to the basic mica and asphalt components, a variety of other elements can be found, such as varnish, asbestos tapes, semiconducting tapes or paints, and Mylar, depending on the type of machine, the manufacturer, and the year of production. The stator bar's fabrication process was completed by a vacuum cycle to remove volatile components, followed by hydraulic pressing. The varnish was applied from a tank or with tape. Asphalt windings are normally rated as class B (130°C).

Windings based on asphalt as the binding element were a great improvement over varnished cambric and cells using shellac as the binding material. Asphalt is less prone to voids due to the evaporation of volatiles and is more resistant to water vapor intrusion. Therefore, it is less susceptible to partial discharge and electric treeing. The insulation in the region of the end windings was a vast improvement over designs made with older insulation systems. Dielectric losses were also reduced. Asphalt allows the ground-wall insulation to become more flexible and less susceptible to cracking and delamination [11]. Asphalt-based stator bars will expand and fill the slot snugly, reducing (in fact, almost eliminating) abrasion of the wall armor due to vibration of the stator bar. Also, a tight fit goes a long way toward minimizing slot discharges, an important mechanism of ground-wall insulation deterioration. Finally, asphalt insulation possesses better heat-transfer capabilities than older insulation systems.

Asphalt-based stator bars are included in the group of *thermoplastic* insulation systems. Their manufacture was gradually replaced during the 1950s and 1960s by an insulation system called *thermosetting.*

Given the popularity of the asphalt-bonded windings before the thermosetting systems were introduced, a very large number of machines still operate with those windings. The most common failure mechanisms of these insulation systems are described in this and subsequent numbered sections.

In conjunction with the aforementioned advantages, asphalt-based insulation systems are prone to develop a number of problems that are very specific to the class of thermoplastic insulation systems. A major disadvantage of this insulation is its poor thermal resilience. When exposed to high temperatures, the asphalt develops a sharp drop in viscosity and thus tends to migrate along the stator bar to areas of less pressure. When allowed by a failure in the armor tape, it can even flow out of the stator bar. Once the thermoplastic element migrates out of a stator bar section, thermal aging is accelerated due to resulting poorer heat-transfer capabilities. In addition, the excess dryness of the area results in increased partial discharge activity within the wall insulation and in the vicinity of the conductors. If the affected area is large enough, magnetically induced movement of the conductors within the insulation will ensue. This creates internal abrasion with subsequent increase of partial discharge activity. All these mechanisms have the potential to develop into interturn and/or turn-to-ground failures.

Asphalt migration might show up as bulging of the insulation in some places. In severe cases, the asphalt will ooze out of the stator bar, running along it and/or dropping onto other surfaces (Figs. 8.94 and 8.95).

Attempts should always be made to treat areas affected by severe bleeding or migration. The area can be patched with armor tape, rebuilt with epoxy-loaded mica tape, or any other procedure deemed proper to the type and severity of the damage.

On occasion, migration of the asphalt is masked from view by the armor tape. The stator bar may appear to be in good condition, without bulging or other visible deformation; however, under the armor tape, a void may be present. Only a spongy or soft yielding to pressure applied by hand on the suspected areas will reveal the presence of a weakened or absent ground-wall insulation. In some cases the location of the soft or spongy area is not directly accessible by hand. In these cases a probe, preferably nonmetallic, should be used. Care should be taken not to further damage the insulation with the probe.

Given the susceptibility of thermoplastic materials to heat, it is important to avoid any abnormal operation resulting in excessive temperature rise of the windings in ma-

Fig. 8.94 Asphalt bleeding from an overheated stator winding of a large turbo-generator. The generator was forced out of service when the stator winding eventually suffered a ground fault. This photo shows the extent of the asphalt bleed after partial disassembly of the endwinding blocking and ties.

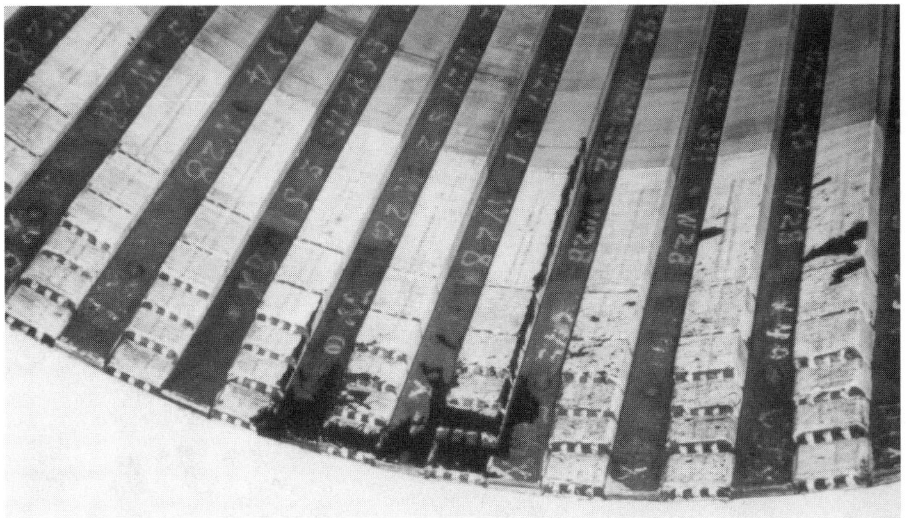

Fig. 8.95 Prior to disassembly of the stator endwinding in Figure 8.94, the asphalt was found flowing from the overheated stator winding through the radial cooling vents in the stator core and the site of the ground fault.

chines with this type of insulation. By the same token, when inspecting machines with these insulation systems that were overheated for a relatively prolonged time by overload or insufficient cooling, one should take into account the possibility of severe winding damage.

8.3.8 Tape Separation/Girth Cracking

A problem common to machines with thermoplastic insulation systems (mainly asphalt) is that the normal thermal cycling undergone by the machine, coupled with the snug fit of the stator bars in their slots, tends to generate cracks within the ground-wall insulation called *tape separation* and *girth cracking*. Although there is some confusion between the two terms (they may mean the same thing in some publications and different things in others), they can be defined as follows.

A tape separation is a separation of the tape covering the wall insulation of the stator bar due to axial expansion and contraction of the conductors and the opposing forces of the slot applied to the wall insulation. In some cases, only the armor tape is separated. In other instances, the mica tape comprising the ground-wall insulation is also affected. When this occurs, a few layers may be affected; in more severe cases, the whole tape will move, creating a "neck" in the stator bar. Sometimes, the ground wall will separate, forming a sharp crack (Fig. 8.96).

A girth crack is the necking of the wall insulation, occasionally reaching all the way to the conductor (see Fig. 8.97). Girth cracking occurs both from thermal cyclic stress-

Fig. 8.96 Tape separation shown in a stator bar of a steam-turbine generator.

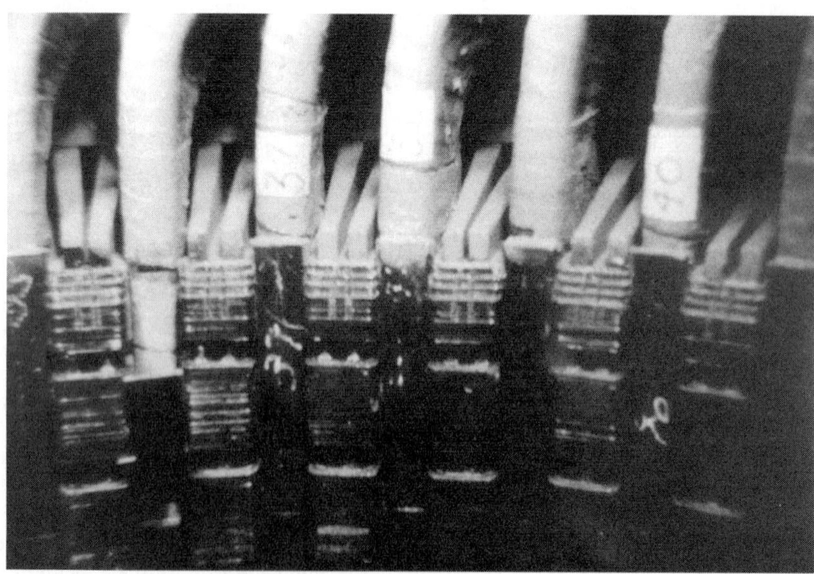

Fig. 8.97 Tape separation and girth cracks shown in the area where the stator bars leave the slot. This is the region of the stator bar most prone to develop this type of problem. Very often, these fissures in the insulation remain hidden underneath the end wedges and/or fillers. One wedge was removed in this case, clearly showing the girth crack on the stator bar.

es in thermoplastic insulation, and in insulation rendered excessively dry and brittle by high temperature. After cracks appear, subsequent humidity or other contaminants allow tracking to take effect, with the probable consequence of a short circuit.

Tape separation and girth cracks are commonly found together. They normally appear in machines with core lengths of about 3 m (10 ft) and up. They appear on the end portion of the stator bars, between the core and the first bend, commonly at about one to two inches from the core. To a lesser extent, tape separation and girth cracks can be found in the cooling vents (Fig. 8.98).

When substantial tape separation and/or girth cracks are found during an inspection, it is recommended that several wedges be removed to allow inspection of the stator bar in the slot next to the affected area. Since tape separation and girth cracks can develop under severe conditions in the slot area, removal of several wedges of the suspected stator bar will help to ascertain if the trouble is localized, or if it affects the slot portion of the stator bar. If the answer is affirmative, more often than not repairs have to be initiated because any degradation of the ground-wall insulation in the slot area has a significant probability of resulting in a short circuit to ground.

Depending on the severity and location of the affected area, the recommended repairs can go from doing nothing to removing a bar. Oftentimes, thin cracks in the insulation are treated with insulating paint. This can help if properly done, with a caveat: The paint can be a carrier for contaminants that will accelerate tracking and possible consequent electric failure. Reference [12] contains an elaborated discussion of the problems described here, with a range of possible actions aimed at repairing them. The fissure shown in Figure 8.99 could be superficial or serious.

Fig. 8.98 This stator bar shows a girth crack in the wall insulation in the area of a cooling vent. The stator bar was removed from the core, and the tape above the girth was peeled off by the inspector to get a better look at the damaged insulation.

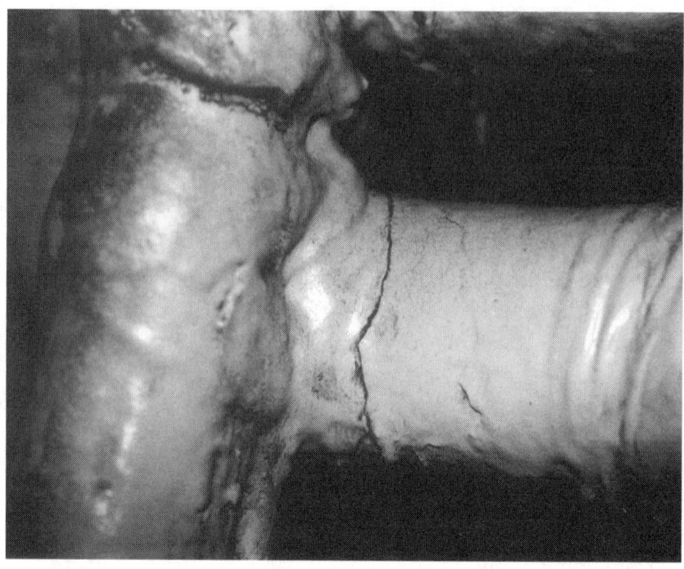

Fig. 8.99 A fissure on the insulation of a stator bar near the surge ring. This could be a superficial crack of the paint or a more serious deep crack.

8.3.9 Insulation Galling/Necking beyond the Slot

Note: For background information on the following discussion, see the two previous sections.

As explained above, *necking* or *galling* of the insulation is the result of thermal cycling in thermoplastic insulation systems such as asphalt. Necking is a lack of insulation, due to cracking and separation. It can also describe the less severe condition of migration of asphalt or other thermoplastic bonding material. Necking is always a sign of a weak point in the ground-wall insulation and should be treated in accordance to its severity and location (see reference [12] for more hints on repair procedures).

Some stator bars are made in the slot portion with a slot wrapper, whereas at the end-winding region a tape is used. The interface between the two regions, close to the end of the core, is called a *scarf joint.* Scarf joints represent a weak mechanical point in the structure of the stator bar, tending to separate under thermal–mechanical cyclic stresses. The indication of a scarf-joint separation is also a necked insulation, or a soft spot. It requires the same type of repairs or attention as previously described for tape separations and girth cracks.

8.3.10 Insulation Bulging into Air Ducts

Note: For background information on the following discussion, see the previous three sections.

Bulging (*balling, puffing*) of the insulation right outside the slot and in the cooling vents is an indication of a soft spot, tape separation, girth cracks, or asphalt migration.

When such a situation is encountered, further inspection of the slot area with the aid of a boroscope is recommended. (See reference [12] for more hints on recommended repair procedures.)

8.3.11 Insulation Condition, Overheating, and Electrical Aging

To correctly evaluate the condition of a particular winding using test and inspection data, one has to have a minimum of information on the composition of the insulation. Different insulation systems react differently to mechanical, electrical, and thermal aging factors.

Unlike the thermoplastic insulation system discussed previously, thermosetting insulation systems, when exposed to elevated temperatures, become dry and brittle. Elevated temperatures may arise from overload conditions, poor cooling (due to lack of coolant pressure, clogged water pipes, clogged vents, etc.); see Fig. 8.100 and 8.101), a damaged core section, negative-sequence currents due to system imbalance, and other causes. Dry and brittle stator bars can also result from many years of normal operation. For example, machines with old shellac or copal resin binders that have been in opera-

Fig. 8.100 The stator bar in this photo was damaged to due operation at full load with no cooling water for an extended period of time (4 hours). The portion of the bar shown is from the stator winding involute or end-winding section of the bar. The ground-wall insulation has been stripped to show the extent of the overheating damage. The material is asphalt–mica and it is evident that the bar was completely destroyed in this failure.

Fig. 8.101 This is a cross section of the same stator bar in Figure 8.100, prior to stripping the ground wall away. Complete delamination and loss of the asphalt binder is observed. The bar has basically been dried out by the overheating.

tion for almost a century will tend to show this type of condition. It has been the industry's experience that with insulation systems introduced prior to the 1950s, the main mechanism for determining the expected life of the insulation is thermal aging.

In the case of epoxy or polyester binders, severe overload conditions also show up as an external discoloration of the insulation. Dry and brittle windings will sometimes show powder accumulation arising from the movement of the shrunken stator bar within the slot. Epoxy–mica insulation systems tend to be even more forgiving than polyester and asphalt–mica systems and, depending on the degree of overheating, they generally are more capable (Figure 8.102).

Like other degradation processes afflicting the insulation of machine windings, once a stator bar becomes too dry and brittle, a positive feedback is established; voids are created within the insulation, further reducing the effectiveness of the heat dissipation from the stator bar. In addition, internal partial discharge is augmented; looseness and internal movement of the conductors and external looseness of the stator bar within the slot result in additional mechanisms of degradation taking over, namely, abrasion and slot discharge.

Internal movement of the conductors and the additional partial discharge activity tend to cause failure in these stator bars in an interturn mode. Consequently, a ground fault develops. In single-bar stator bars, the wall insulation is deteriorated to the extent that a ground fault develops. Besides visual inspections, the condition of an excessive-

Fig. 8.102 This is a cross section of the same stator bar in Figures 8.100 and 8.101 above. However, the portion of the stator bar shown is from the slot and is made of epoxy–mica. The construction of the bar is such that the epoxy–mica portion of the insulation in the slot is joined to the asphalt–mica portion of the insulation in the end winding by an overlapping joint called a *scarf joint*. What is most notable is that the epoxy-mica insulation is virtually unaffected by the loss of coolant event that destroyed the asphalt–mica part of the insulation. The difference in capability is clearly shown in this example and provides the basic understanding of why modern insulation systems have moved to epoxy resin binders.

ly dry and brittle winding can be assessed with a number of electrical tests, such as partial discharge, insulation power factor, and polarization index.

There is little or no remedial procedure for windings that by aging, wrongful operation, or other reasons become too dry and brittle. Nonetheless, knowing the actual condition of the insulation allows for proper planning of a major rewind and/or better assessment of the risks of continued operation.

8.3.12 Corona Activity

Corona activity is defined as the ionization of a gas when exposed to an intense electric field, normally in the vicinity of an electrical conductor. In this form, the definition of corona applies to overhead lines, high-voltage bushings, and other elements producing high concentrations of electric field. In the context of rotating machines, however, the term corona is used interchangeably with partial discharge and slot discharge. This is a high-speed discharge, with a wide range of frequencies (40 kHz to 100 MHz).

Four distinctive types of corona activity can be found within a rotating machine. These are described in the following paragraphs.

1. *Corona Activity on the End Windings*. Ionization of the gas in the end-winding region is present in machines operating with line voltages of several thousand volts. The actual inception voltage (i.e., the voltage at which corona is first observed) depends on the specific geometric configuration of the windings and surrounding structures. Different designs render different concentrations of electric fields for the same voltage, thus resulting in varied levels of inception voltages. Corona activity is directly dependent on the actual electric field concentration. Evidently, proper design practices should be geared to minimize high concentrations of electric field in high-voltage machines. Corona activity can normally be found in machines having voltages of 4 kV or higher.

In general, the highest potential differences exist between phase stator bars (adjacent stator bars belonging to different phases) and line stator bars. It is common, therefore, for the telltale signs of this corona activity to be concentrated in those areas. The most common signs are white or brownish powder deposits on the stator bars (Figure 8.103). In more severe instances, dark burn marks can be found, mainly close to the areas where the stator bars are at close proximity. Some experience is required to distinguish the corona-originated powder deposits from those originated from fretting of the blocking and ties due to the movement of the stator bars (Figure 8.104). Corona-originated powders tend to adhere more tightly to the surfaces and, as mentioned

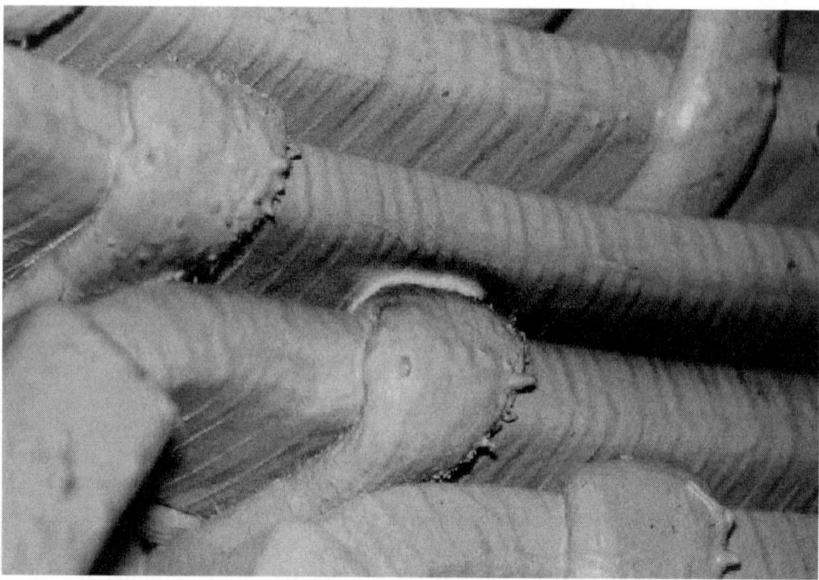

Fig. 8.103 Stator bar phase-to-phase discharge in the end winding at a localized stress point between two bars of different phases.

Fig. 8.104 Stator bar fretting on the surface of the insulation due to mechanical movement (vibration). It could be mistaken for partial discharge activity due to the fretting marks and the heat staining from intense rubbing.

above, tend to be found in areas of high electric field concentration.

Once the machine is in operation, remedial actions will not eliminate the source of corona, as this is design dependent. However, the affected area should be cleaned and the insulation repaired if necessary. These steps will reduce the rate of deterioration of the insulation due to secondary phenomena, such as chemical attack of the insulation by accumulated by-products of the corona activity.

It is interesting to note that partial discharge tests are able under certain conditions to identify if the partial discharge activity is inside the stator bar or if it is between the slot and the stator bar. This subject is covered in more detail below. In the case of activity between slot and stator bar, tightness of the stator bars and wedges should be checked during the visual inspection of the bore.

2. *Internal Partial Discharges.* Partial discharge is defined best as an electric discharge occurring between conductors when the breakdown voltage of the surrounding gas is exceeded. When such a discharge is not followed by the establishment of an arc, it is called *partial discharge* (PD).

PD commonly occurs in voids inside the insulation of the machine. It also occurs between layers of insulation when these are not properly bonded (Figs. 8.105 and

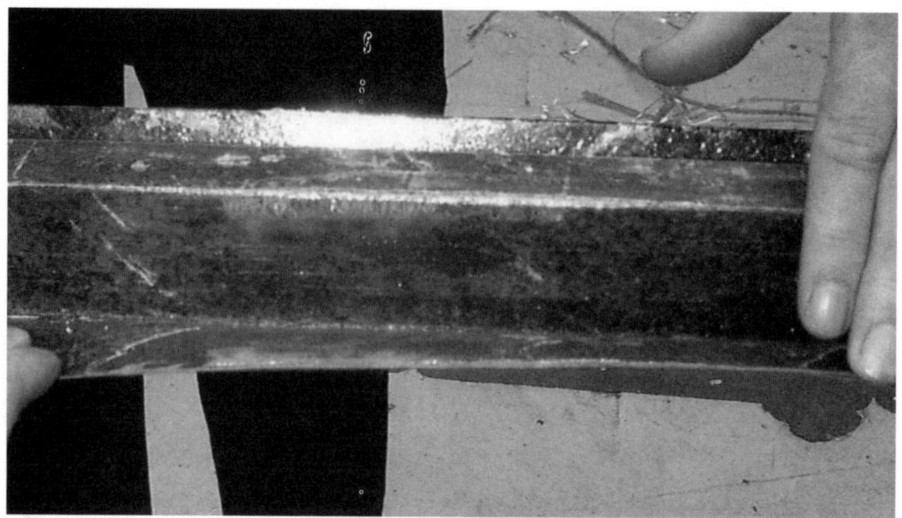

Fig. 8.105 The construction of the ground wall of the stator bar shown is sheet wrapped rather than taped. The ground wall has delaminated from the copper strand stack a few layers out from the copper surface. The voids created in the insulation due to delamination have allowed partial discharge activity to occur, as seen by the yellow-brown patches that are the by-products of the discharge. The indications of the activity in this bar were higher than normal PD measurements in online PD tests, high PD readings, and low corona inception values during offline testing and a subsequent offline TVA probe test that showed numerous areas of high discharge activity in the slots. (See color insert.)

8.106), allowing gaps to remain during the manufacture of the stator bar or to be created during the operation of the machine. The inception voltage of the discharge depends on the size and shape of the voids, as well as on the gas contained within them. Internal voids created during the manufacture of the stator bars contain air. Tape separations created during the operation of the machine will be filled with the coolant medium: air in air-cooled machines and hydrogen in hydrogen-cooled units. Machines operating at about 4000 V or higher are susceptible to PD activity.

PD activity in voids tends to eat away the insulation by electrical erosion (bombardment of the insulation gas by the acceleration of ions), and chemical reaction with the by-products of the electric discharges in the gas. Occasionally a void grows into a "tree" due to continued ionic action within the void. Once established, the tree keeps growing at a fast pace due to PD activity and surface discharges within itself. Trees will grow until the ground-wall insulation is weakened to the extent that a full ground fault is developed. Void augmentation or tree creation due to PD activity may also weaken the strand and conductor insulation, giving rise to shorted turns. This phenomenon is most common in the overhang area of the stator bars, where the manufacturing process does not compress the stator bars to the extent that it does in the slot section.

Fig. 8.106 The ground-wall insulation wrap has been completely removed from the stator bar of Figure 8.105, and the patch of brownish PD residue is clearly seen on the surface of the interstrand insulation of the outer part of the strand stack. What is also notable is that there was no delamination of the strands themselves, only the ground-wall insulation. (See color insert.)

As a result, more voids are created in the overhang section than in the slot section. A particularly troublesome area is the *crossover region* of the stator bars.

Signs of internal PD activity are impossible to detect during a visual inspection of the machine. As will be indicated subsequently, a number of electrical tests are available to assess the extent of PD activity in a particular situation. Nonetheless, visual inspection of the winding and knowledge of the insulation system may provide an indication of the probability of PD activity in the winding. Bloated or puffy windings, indicating internal looseness, will most probably be subject to high PD activity as well. Dry windings are also susceptible to the existence of layer separation and voids, with accompanying PD activity.

Earlier machines that did not use mica were highly susceptible to failure derived from PD activity. However, practically all machines in operation today include inorganic insulation components such as mica and glass. These components are not seriously affected by partial discharges. On the other hand, PD adversely affects the organic materials that make up the bonding structure of the insulation. Old shellac micafolium insulation systems are prone to the formation of voids due to the evaporation of the volatiles in the shellac. The next generation of asphaltic insulation systems went a long way toward reducing voids in the ground-wall insulation. Nevertheless, asphalt-based insulation tends to "swell" or "puff" when a certain temperature limit is exceeded. In addition, weakness of the insulation permits movement between the conductor strands. The consequence of this movement and the generated voids is accelerated PD activity in those spots. All said, thermoplastic (asphaltic) insulation resulted in a great reduction of PD-related faults. Modern epoxy or polyester-based insulation systems offer better bonding. This result in less internal PD activity than in preceding systems.

Aging due to this type of activity can best be avoided by ensuring that no voids are left during the manufacturing process of the stator bars. Vacuum pressure impregnation (VPI) is such a process. Originally applied to smaller machines, in particular, induction motors, VPI windings today can be found in an ever-growing number of synchronous generators. Some European manufacturers use the VPI process in machines with up to about 300 MW ratings, and the upward trend is continuing. However, the repair of stator bars subjected to the VPI process presents some serious challenges.

Other than treating localized areas where a “swell” has occurred, no remedial action exists for a stator bar or winding afflicted with internal PD activity. The only corrective action, if any, recommended due to the severity of the situation is to schedule a winding replacement in accordance with other planned activities for the unit and to evaluate the risk of continued operation in the present situation.

3. *Slot Discharges.* This form of partial discharge is the result of the breakdown of the insulating gas between the stator-bar ground-wall insulation and the iron core inside the slot. In alternating current machines, the stator bar conductors and the opposing slot face act together like a capacitor that is charged and discharged at line frequency. The capacitor “plates” are separated by the ground wall, strand, and turn insulation, as well as by the insulation of the cooling medium, normally air or hydrogen. Given that the breakdown voltage of air is about 1/100th that of the solid insulation (a somewhat lower ratio for hydrogen), the gas tends to break down under the voltage stresses existing in high-voltage machines. The subsequent avalanche of ions abrades the insulation and also attacks it chemically, as observed previously when discussing internal PD modes of failure. Obviously, stator bars close to the high-voltage terminals will be subject to higher slot discharges than stator bars close to the electrical neutral of the machine.

In air-cooled machines, partial discharge is more intensive than in hydrogen-cooled machines. Also, electric discharge in air produces ozone, a very corrosive element. Ozone attacks the organic materials of the wall insulation, accelerating the aging process. The ozone has a characteristic smell. It is not uncommon for air-cooled machines to give away the presence of intensive slot discharge activity by emitting an easily identifiable odor. In these types of machines, it is possible to corroborate the existence of slot discharges by opening inspection plates that allow a view of the slot area of the stator bars through the core vents and watching the light emitted during the discharge. Obviously, all lights in the vicinity of the machine have to be turned off for the duration of the inspection.

As with internal PD activity, slot discharge activity is very difficult to identify through a visual inspection of the machine when it is not in operation. Indirect indicators of possible slot discharge activity are loose wedges and stator bars not snugly fitted in the slots. Asphaltic stator bars tend to sit tightly in their slots, thus showing diminished slot discharge activity. Old shellac and modern thermosetting insulation tend to be more susceptible to slot discharges, unless the stator bars are properly packed and wedged in the slots. *Ripple fillers* (spring-loaded fillers) go a long way toward maintaining a good stator bar-to-slot pressure.

The solution to the problem of localized slot discharge comes in the form of *semiconducting paint* or tape, with which high-voltage stator bars are covered in the slot area. The function of the semiconducting layer is to discharge all of the surface charge through the contact points between the stator bar and the walls of the slot. However, enough points of contact should be established and maintained between the semiconducting paint and the slot to maintain the integrity of the paint or tape. If continuous movement of the stator bar within the slot occurs, the fretting and burning of the few contact points will deteriorate the semiconductor paint or tape to the extent that they cease to protect the ground-wall insulation from the increased slot discharges (see Figs. 8.107–8.111). Eventual failure of the insulation follows. To maintain the integrity of the semiconductor layer, the stator bars have to sit snugly inside the slot.

4. *Surface Discharges* (*Electric Tracking*). Surface discharges are intermittent partial breakdowns of the surface insulation due to high electric fields. The effect of these surface discharges is damage to the organic surface layer of the wall insulation. These discharges, while less damaging than the three types already discussed, do contribute to the general degradation of the insulation.

Surface discharges tend to be concentrated on the overhangs of the windings, in the immediate vicinity of the core (see Fig. 8.77). The grading paint, a feature introduced in high-voltage machines to eliminate high concentrations of electric fields at the end region of the core, also controls the surface discharges generated in this region.

As stated previously, some mechanisms of PD or corona activity cannot be readily identified during visual inspection of the machine. Several tests have been designed to evaluate the presence and intensity of PD activity and the resulting damage to the stator bars.

Fig. 8.107 Stator bars removed from a salient-pole generator. The stator bars show severe loss of their semiconducting paint due to the stator bars not being properly supported in the slots.

Fig. 8.108 Close-up view of one of the affected areas shown in Figure 8.107.

Fig. 8.109 Charring of the semiconducting surface of the stator bar on a *resin-rich* type bar due to excessive discharge activity.

Fig. 8.110 The slot discharge activity on this bar has been significant and caused surface erosion of the semiconducting coating as well as the outer layer of ground-wall insulation.

Fig. 8.111 This is a resin-rich type bar that has suffered severe slot discharge activity and failed on a ground fault. The outer surface is badly burned from the discharge.

All tests designed for the detection of corona or partial discharge activity have to be performed with the machine energized. Some of those (e.g., those requiring the use of a hand-held probe) require the rotor to be removed from the bore.

Some tests, such as the polarization index (PI), dielectric absorption, power factor (PF), power factor tip-up (PF tip-up), and radio-frequency tests (RIV), are general insulation tests that also provide information on PD activity and the damage accumulated in the windings due to PD.

Other tests are specifically designed for the detection of partial discharge; for instance, the integrated discharge energy measurement (Westinghouse), the embedded stator slot coupler (SSC), partial discharge measurements, electrostatic probe tests, and ozone meters. (Reference [10] contains a very good description of the tests. References [13-21] contain additional information on corona/PD activity.)

8.3.13 Stator Wedges

It is very important to inspect the stator wedges. The wedges are one of the main elements controlling the tightness of the stator bars in the slots. Maintaining a positive pressure on the stator bars reduces their movement within the slots, thus minimizing loss of semiconducting coating and wall insulation. Loose wedges may show typical visual clues (Fig. 8.112 and 8.113), but can be readily detected by testing.

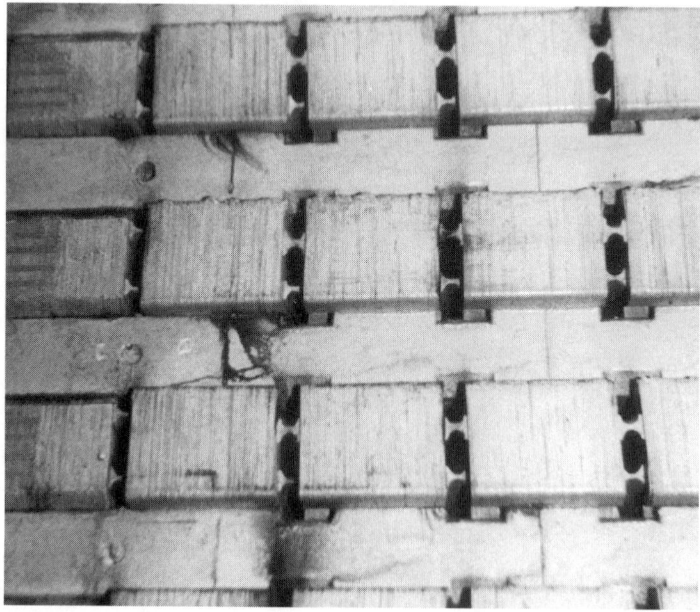

Fig. 8.112 Movement of loose wedges in the presence of oil creates the telltale greasy deposits shown in this photograph. The grease is a mixture of the fretting debris from the wedge and the oil in the bore.

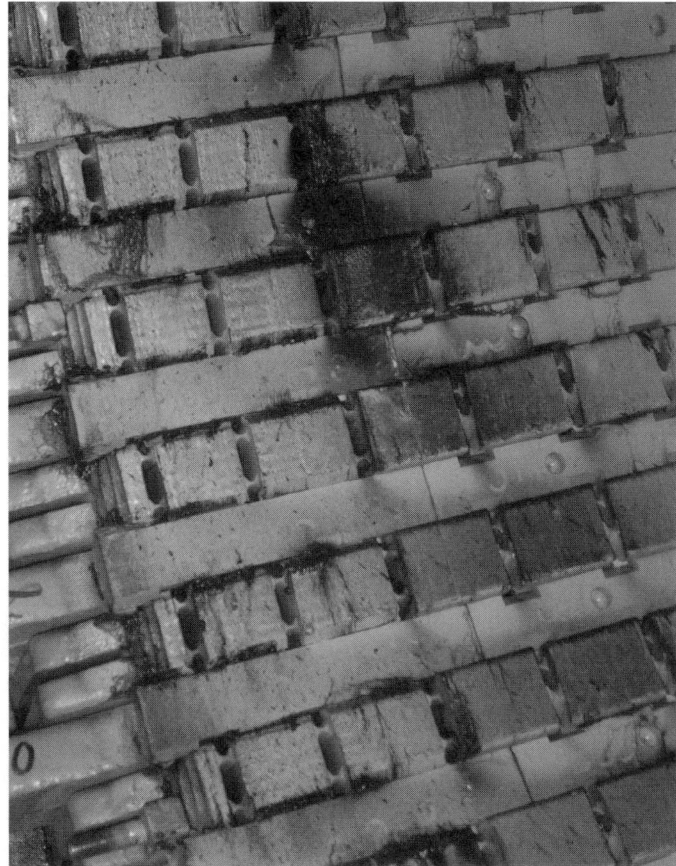

Fig. 8.113 The wedges in this photo are extremely loose and the greasing effect is excessive. Finding this degree of greasing during an inspection would be cause for great concern that the ground-wall insulation of the stator bars may be damaged and a failure mechanism in progress. A check of the wedge tightness must be made and electrical testing must be carried out; some wedges might also have to be removed to enable inspection of the top bar under the wedges.

The common way to inspect wedge condition is to tap on one side of the wedge with a small hammer and sense the amount of movement with the other hand touching the other side of the wedge (see Chapter 11). Given the response, the wedge condition can be classified as *tight, hollow* (medium), or *loose.* A loose wedge is a wedge that responds to the tapping with movement. A hollow wedge is one that does not move but sounds like it is not pressing against the stator bar. It takes some experience to differentiate between hollow and tight wedges; however, this experience can be readily acquired.

A hollow wedge indicates that a clearance exists in the radial direction between the slot and the stator bar. This clearance could be due to poor packing during installation of the stator bars or stator bar shrinkage. Hollow wedges can be found more often in thermosetting stator bars than in thermoplastic stator bars, which tend to fill the slot.

A substantial number of hollow or loose wedges in a row indicate that a stator bar is loose and will probably tend to move within the slot during operation. This is an undesirable condition. If enough wedges are found to be loose or hollow during a partial or random test, a full wedge survey might be desirable. During a wedge survey, all wedges are tapped and the response recorded in a looseness survey form (shown in Chapter 11). The wedges are numbered by their location, both by slot number and position within the slot. If a number of contiguous wedges are not in acceptable condition (hollow or loose), then the slot might be rewedged. Each operator or inspector uses different criteria to decide on the threshold number of unsatisfactory wedges beyond which a "rewedge" is required. A commonly used number is 25% of the total number of wedges. End wedges should always be secured if found loose; application of epoxies and RTV is one common technique of securing these wedges. The decision to rewedge, perform partial repairs, or continue operation "as is" depends also on the importance of the machine to the system, repair costs, and material availability.

A wedge survey performed with the hammer method can be very tedious and time-consuming. New hand-held "tapping" instruments are available. These instruments allow the survey to be conducted in a fraction of the time and with more consistency of readings, in particular when different inspection personnel are involved. In many cases, people tend to interpret differently the response obtained by the "hammer" method. The automatic tapping instrument provides an important measure of consistency between results obtained by various inspectors.

In newly designed machines or recently rewedged machines, having a "ripple filler" under the wedge tends to reduce wedge looseness (Fig. 8.114). In addition, these modern wedges have, on many occasions, been manufactured with small orifices through which a probe can be inserted to measure the tightness of the stator bar against the wedge. Some organizations have recently introduced to the market a robotic method of tapping and/or probing the condition of the wedges in large synchronous machines without the need for removing the rotor from the bore.

One way to identify loose wedges is by the resulting powder deposits along the wedge edges (due to the continuous movement of the wedges) or grease-like contaminants (see Figs. 8.112 and 8.115) if the powder is mixed with oil. The powder deposits, without oil, tend to be yellowish or reddish due to iron oxide resulting from the wedge-core fretting action.

Although it was stated above that a value of 25% of loose and/or hollow wedges could be used as a criterion for rewedging a machine, it has to be recognized that the distribution of those loose/hollow wedges may call for different actions. For instance, a machine may have a relatively small percentage of loose/hollow wedges but have them concentrated in one particular area of the stator. This high concentration of unsatisfactory wedges may require partially rewedging the stator.

In another example, 15 to 20% loose/hollow wedges evenly distributed in one slot may provide enough support to the stator bar; on the other hand, the same 15 to 20% loose/hollow wedges all concentrated in one side of the slot will require rewedging that slot.

Fig. 8.114 Top ripple springs being installed under the stator winding-slot wedges to minimize loosening from vibration effects.

8.3.14 End-Wedge Migration Out of Slot

A not-so-common occurrence is when an end wedge (wedges on both ends of the core) migrates out of the slot end. This can happen if they become loose during operation of the machine, but there are generally mechanisms in place to prevent such migration. It is more common in older, smaller machine designs.

Fig. 8.115 Close-up view of a loose end wedge and "greasing" produced from it (seen along the edge of the wedge).

This condition can be readily detected by inspection with the naked eye or with the aid of mirrors or a boroscope when the rotor is in the bore. A commonly used fix comprises the application of a thick resin, epoxy, or RTV type of material. A substantial number of end wedges moving out of the slot may indicate an overall loose wedge condition, and might warrant a wedge survey (see previous section).

Newly designed machines and most turbine generators have locking designs that do not allow the outward movement of the end wedges.

8.3.15 Side-Packing Fillers

Side-packing fillers may be either flat or rippled in nature. They are generally impregnated with semiconducting material to allow good contact between the stator bars and the core, to provide good contact to ground, and minimize slot discharges. They are inserted into the slot on the side of both the top and bottom stator bars (Fig. 8.116). They are generally tight but may come loose from bar vibration.

Therefore, another indication of loose stator bars, in the radial direction, is the ratcheting movement out of the slot of the top- and bottom-bar side-packing fillers. This phenomenon does not exist in windings processed with global VPI. However, in all other types of windings, it is possible to observe the movement of top and bottom side-packing fillers out of the slots (see Fig. 8.117). This is more commonly found on smaller machines.

Normally, the fillers are driven back (if possible) or broken at the end of the core. In both cases, they are secured with resin, epoxy, or RTV materials.

As with the movement of end wedges, large numbers of fillers slipping out of the core by several inches may indicate a loose winding condition. However, filler movement can also be the result of elongation and contraction of the stator bars due to ther-

Fig. 8.116 Side-packing ripple spring installed in the slot against the side of a top stator bar.

Fig. 8.117 Stator-bar side-packing fillers slipping out of the end of the core by approximately one inch.

mal cycles, even in tight stator bars. A partial wedge survey might provide the required information on the snugness of the stator bars inside the slots.

In larger machines, and in cases of very significant looseness in the slot wedges, the stator bars may "bounce" in the slot and fret against the side-packing fillers. If the side-packing material is of the ripple spring type, the portion of the ripple in the filler that is in contact with the stator bar ground-wall insulation can wear it away and cause a stator winding insulation failure to ground (Fig. 8.118).

8.3.16 Leaks in Water-Cooled Stator Windings

Leaks may occur in water-cooled stator windings from many causes. One of the most common forms is from ordinary plumbing leaks at joints, where O-rings may have become old and dried out, or the components have corroded from age or cracked from vibration (see Fig. 8.119). Also, if there is high stator winding vibration in the slot or end winding, a hollow water-carrying strand may crack.

The possibilities are extensive for this problem, but one of the more recent in the industry is *crevice corrosion leaks* at braze joints of bar-end *water boxes*. A water box, brazed on each end of a stator bar, serves as a cooling-water intake or discharge into

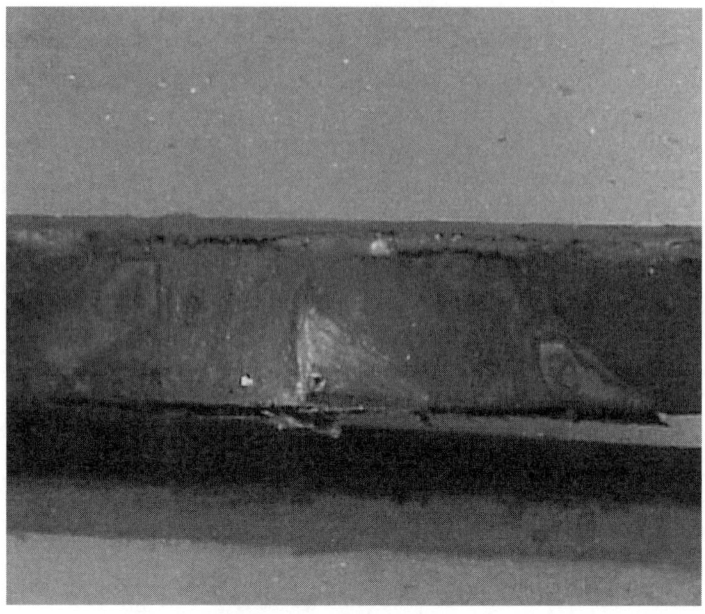

Fig. 8.118 Failed stator bar ground-wall insulation from side-packing ripple-spring fretting against the side of the stator bar.

Fig. 8.119 Plumbing leaks are often found under the end-head insulation in joints.

and from the hollow conductors. A Teflon hose connects the top of the water box with the water manifold. The base of the water box is formed from a clip that fits over the copper strands. All strands are brazed together, and to the copper clip. The brazed portion of the bundle is several centimeters long to seal the water box.

For those machines susceptible to leakage in the brazed joint, the problem appears to exist only in machines that use oxygenated water. The root cause of the problem is crevice corrosion that initiates in the pores of the braze when Cu_3P braze material is used. The brazing material dissolves locally to produce acid phosphate solution. Once this acid pore environment develops, adjacent copper strands may corrode and result in leakage of water into the stator insulation. Repairs for this problem focus on removing the crevice initiation sites, or not exposing them to water. Several individual bar-repair methods are available. They include in-situ replacement of the water boxes, epoxy injection, or laser repair. Partial and full rewinds are also options. Repairs should be performed using qualified inspection processes and operators.

Large leaks in stator windings can be found by visual inspection. However, most leaks start small. Four offline test are recommended: vacuum, pressure, and helium leak tests, and capacitance mapping (refer to Chapter 11 for explanations of these techniques). Figure 8.120 shows such a test in progress.

Since hydrogen pressure is higher than the water pressure in the stator winding, a leak will allow the hydrogen to enter the cooling water and escape through the vent to

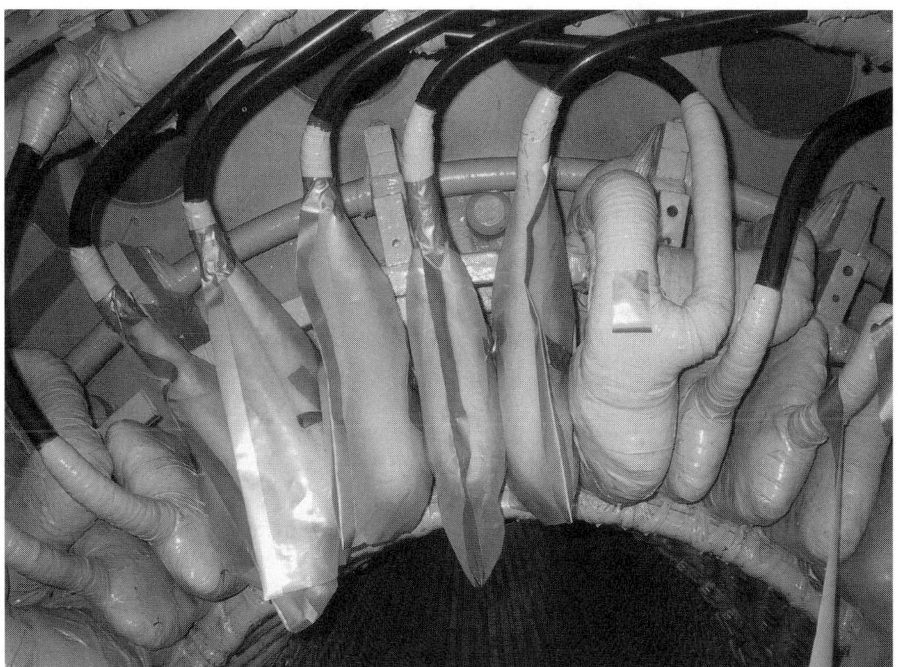

Fig. 8.120 Vent bagging is a method used to detect leaks during a pressure test.

the atmosphere. Measurement and trending of the escaping hydrogen to the atmosphere is used to detect any developing leak.

8.3.17 Magnetic Termites

Particles that cause *wormholes* are commonly known as "magnetic termites." These termites can originate from many possible sources, including metallic debris left in the generator after and overhaul and, possibly, from the molten metal by-products of back-of-core burning.

The concern is that these particles will bore a hole into a critical generator component and cause an expensive failure. The component of main concern, and the one most susceptible, is the stator winding. The stator-winding ground-wall insulation is relatively soft compared to other materials inside the generator, and the copper conductors underneath are at high voltage potential and may carry cooling water (in water-cooled stator windings). Therefore, if a magnetic particle penetrates the ground-wall system, there is a possibility of electrical tracking to an adjacent stator bar or ground, depending on the location of the wormhole and/or hydrogen leakage into the stator cooling water system, and water creepage into the insulation of the affected bar.

The mechanism by which the magnetic particles bore a hole in the stator winding is a combination of magnetic attraction and 60 (or 50) Hz vibration due to eddy currents induced in the magnetic particle, which result in a rotational force on the particle, causing it to rotate in the magnetic field. Because the termite is magnetic, the electromagnetic field in the stator winding pulls the particle toward the center of the magnetic field, namely, the center of the stator bar. The natural 60 (or 50) Hz frequency causes the particle to vibrate at this frequency and, obviously, enhances the boring action of the magnetic particle (Figs. 8.121 and 8.122).

As the hole is bored down to the copper subconductors, the potential for a stator winding failure increases proportionally. However, for electrical tracking to occur, there are a number of other factors involved. If the surface of the exposed copper conductor and the ground-wall insulation are very clean, there may not be enough conducting material on the surface to allow electrical tracking. Hence, the wormhole could remain exposed for an indefinite period without ever failing. This has been seen on numerous occasions on various machines.

The more likely failure mode seen with various wormholes is that of hydrogen leakage into one of the water-carrying, hollow subconductors. One other possibility is that the magnetic termites plug the cooling water flow in the affected subconductor by crushing and wearing it closed. This has also been seen in past occurrences (Figure 8.123).

With regard to termites from back-of-core burning, there is undisputed evidence from a number of machines that back-of-core burning does create small metallic globules that come loose from the back of the core and migrate throughout the generator. The degree of burning determines if molten globules are formed, and to what extent and when they are produced.

The free globules from the back-of-core burning have been know to migrate through the core radial vents into and under the stator wedges, then into the airgap, the

Fig. 8.121 Stator end-winding "worm" or "bore" hole from magnetic termite debris.

Fig. 8.122 A small piece of magnetic debris has lodged between stator-bar end-winding packings and become a magnetic termite, and has started to bore a hole into the end-winding insulation of the bar shown.

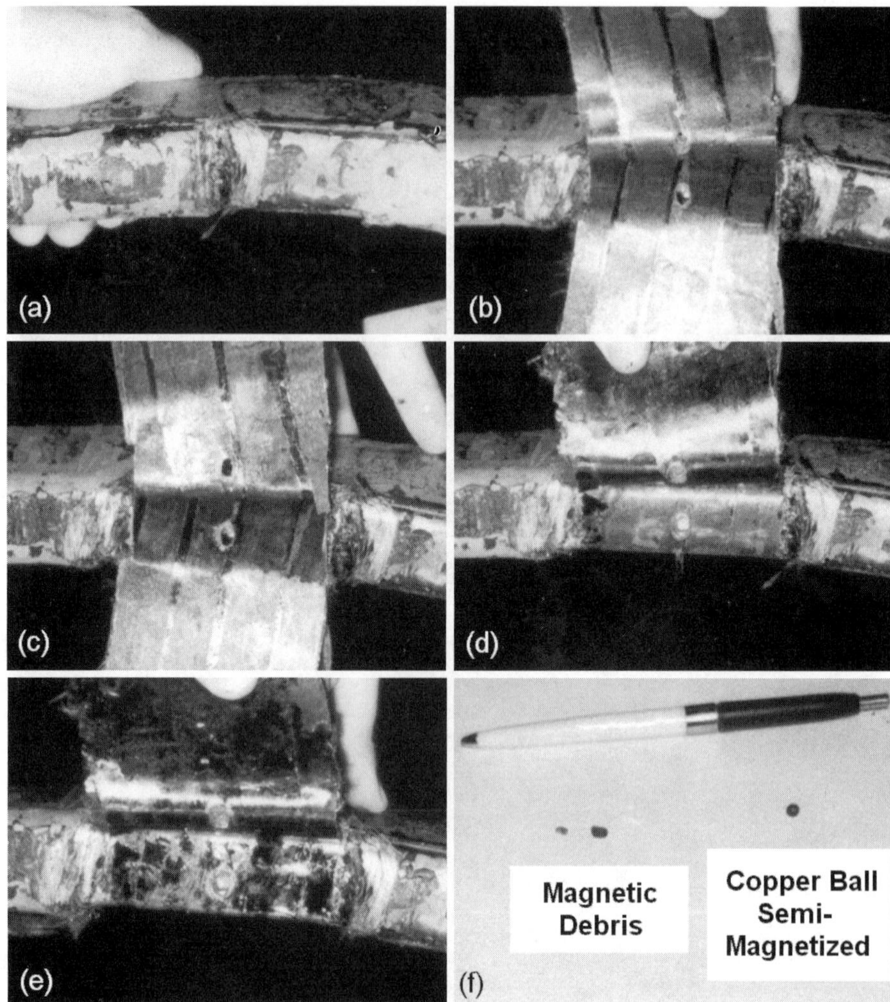

Fig. 8.123 This series of photos shows the progression of the magnetic termite into the stator bar insulation of the bar shown in Figure 8.122 and how it actually crushed the hollow conductor closed, stopping cooling water flow. In this stator bar, all strands are of hollow construction and no leak of hydrogen into stator cooling water occurred, and no electrical failure either. The termite was found on an overhaul and the bar removed and replaced. The interesting thing to note in this incident was that the termite was enhanced in size by picking up copper material as it drilled further into the stator bar. Upon investigation afterward, it was found that the copper material appeared to be slightly magnetized from the effect.

stator end winding, and the rotor. The occurrence of a failure due to one of these particles then depends on where the particle ends up in the machine, and the mechanical, electrical, and magnetic influencing factors around it (Fig. 8.124).

8.3.18 Flow Restriction in Water-Cooled Stator Windings

Flow restrictions in water-cooled windings may occur from debris in the stator cooling-water circuit, hydrogen gas locking from large leaks, crushed hollow strands, or some form of corrosion (Fig. 8.125).

One of the main sources of the corrosion mechanism occurs with any water-cooled generator that is designed to operate with pure water that is either saturated (>2000 ppb dissolved oxygen) or low in oxygen (<50 ppb). The water must be non-conductive (<10 microsiemens/cm) to avoid flashover along the insulating Teflon hoses. An elevated pH renders this effect negligible. However, most generators use neutral pH water. On-line conductivity measurement, thermocouples installed at the water manifold, and pressure drops across the winding, filters, and strainers are used to monitor the system.

The problem of copper corrosion occurs when the stator cooling water is operating off the design point, leaving stagnating water in the winding, or not properly drying or protecting the winding during an outage. This can lead to local buildup of corrosion by-products (cupric and cuprous oxides) that restrict the water flow. The generator output may be curtailed as a result of decreased cooling. The root cause of the problem

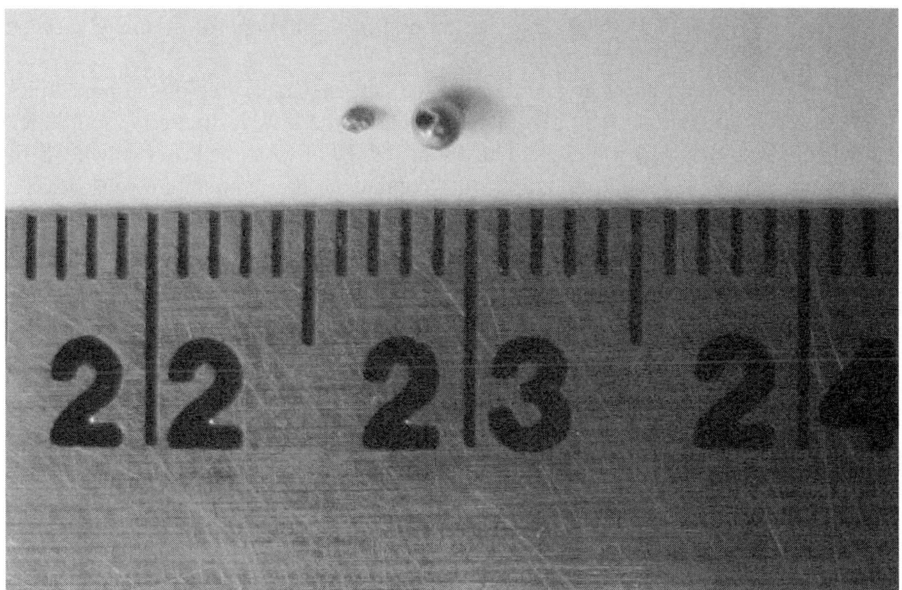

Fig. 8.124 Polished magnetic termite debris removed from a “worm” or “bore” hole.

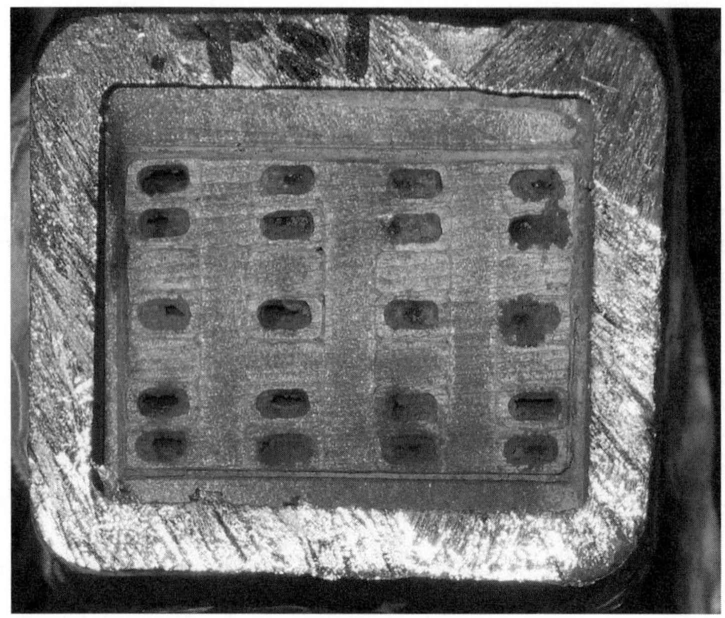

Fig. 8.125 Copper corrosion in the strands of a stator bar due to poor water chemistry.

is operation in the intermediate range between 50 and 2000 ppb of dissolved oxygen levels.

When flow restrictions of any kind occur, temperature (measured by thermocouples and slot RTDs) will increase in the stator winding as the flow restriction in individual strands impacts localized cooling and heat transfer. Totally blocked bars may be difficult to discern from the online temperature data, since the temperature sensor has a tendency to then measure the cooling gas temperature local to it. A water pressure drop across the restricted winding or clogging filters and strainers will also indicate an ongoing problem.

Visual (boroscope) inspection of the waterboxes can reveal partially blocked strands. Ultrasonic flowmeters may also be used to measure flow through individual bars.

Offline cleaning can be done using diluted acid–water backflushing for partially blocked strands. Online cleaning is possible by a patented cleaning process known as Cuproplex, which removes only the oxides while the unit is running. For either offline or online cleaning, the protective oxide layer needs to be restored after any cleaning.

Online monitoring of dissolved oxygen is also recommended to detect conditions that may lead to flow restrictions. Existing conductivity measurement is not uniquely sensitive to dissolved oxygen variations.

The following explains why it is critical that the chemical copper oxides reduction process be carried out in a timely basis. Assuming laminar flow, the volume flow rate is a function of the radius to the fourth power (see Figure 8.126). Therefore, a 10% reduction in radius will cause a 34% reduction in flow. A 20% reduction in diameter will produce a 60% reduction in flow. A 30% reduction in radius will yield a 76% reduction in flow. In addition, as the radius is reduced, turbulence is introduced. Turbulence may further accelerate the copper oxide deposits in restricted areas. Eventually, a number of individual hollow strands become fully blocked. The chemical cleaning process does not unblock a fully blocked strand, so the degradation is permanent. To prevent

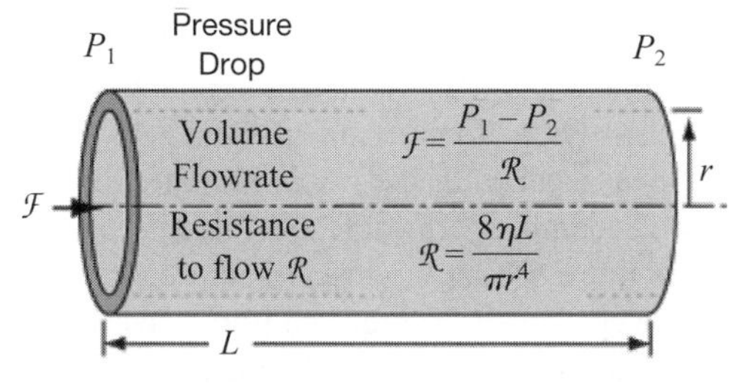

$$\text{VolumeFlowrate} = \frac{\text{Pressure Difference} \times \text{Radius}^4}{8/\pi \text{ Viscosity} \times \text{Length}}$$

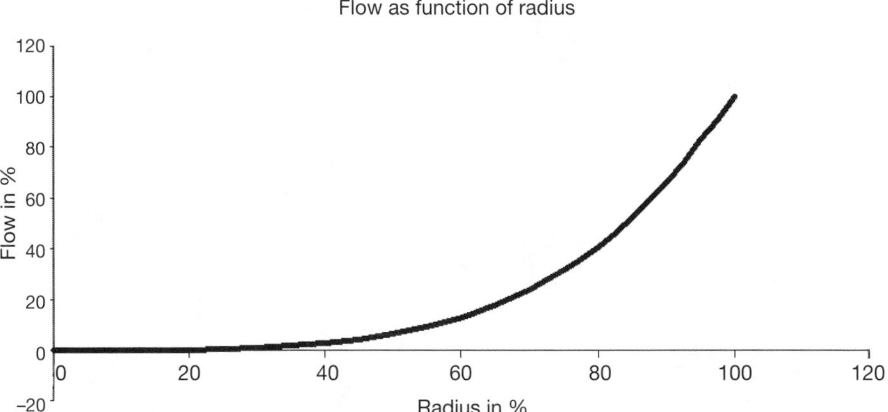

Fig. 8.126 The equation in the figure clearly shows how flow rate through a pipe is a function of the inside radius to the fourth power. This obviously applies to some specific flow conditions and is just an approximation for other cases. Nevertheless, it provides a good view into the very large flow dependency on the radius of the fluid conductor. The relevance of the aforementioned in the case of water-cooled windings is how sensitive the flow is to copper oxide buildup, and how quickly it can go from a good condition to almost no flow. There is a positive feedback in play here: the more copper oxide obstruction there is, the less the flow, and the less the flow, the quicker the copper oxide accumulates.

permanent degradation of the cooling capacity of the winding, it is thus imperative to perform the chemical flushing process on a timely basis.

8.3.19 Hoses, Gaskets, and O-Rings in Water-Cooled Stator Windings

Aside from the stator winding as a major component, there are also numerous small subcomponents of the winding that are often overlooked during maintenance and yet can just as easily create a generator forced outage if they fail. Some of these types of subcomponents are the stator-winding Teflon hoses and fittings, gaskets, and O-rings. The Teflon hoses are flexible and not as resistant to puncture or wear as metal components and can fail more easily due to mechanical causes (Figure 8.127).

Another issue that can occur with the hoses is that over time they may begin to show signs that they are becoming permeable and that hydrogen gas is slowly penetrating the material. This is more difficult to discern and requires special testing to determine if it is happening. Replacement is required if this occurs.

Yet another undesirable problem that can happen is trapped contamination on the inner surface of the hoses from such things as copper oxides and any other material that may intrude (Fig. 8.128). Cleaning of the hoses is often possible and is done by

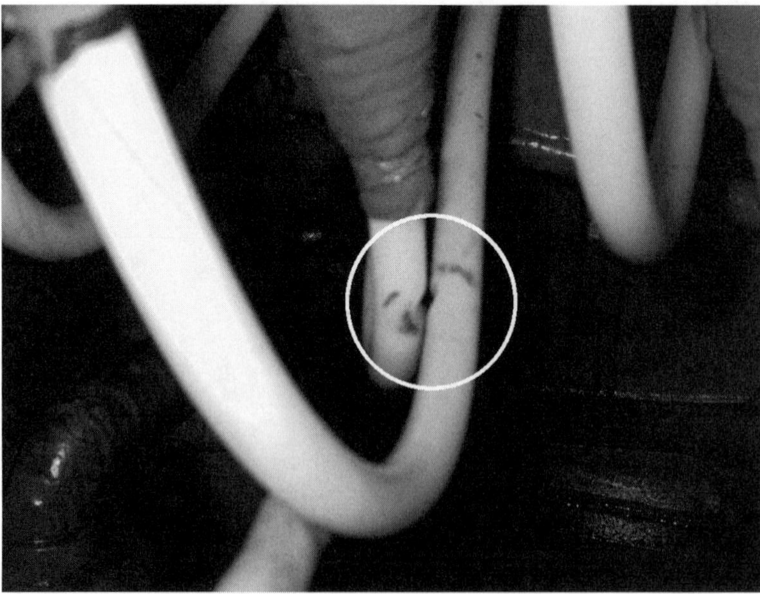

Fig. 8.127 Two hoses were found to be rubbing each other, and due to vibration they incurred fretting damage and are becoming close to wearing through the walls of the hoses. The result would be a high hydrogen leak into the stator cooling water, and if the leak is bad enough, hydrogen gas locking could occur as well.

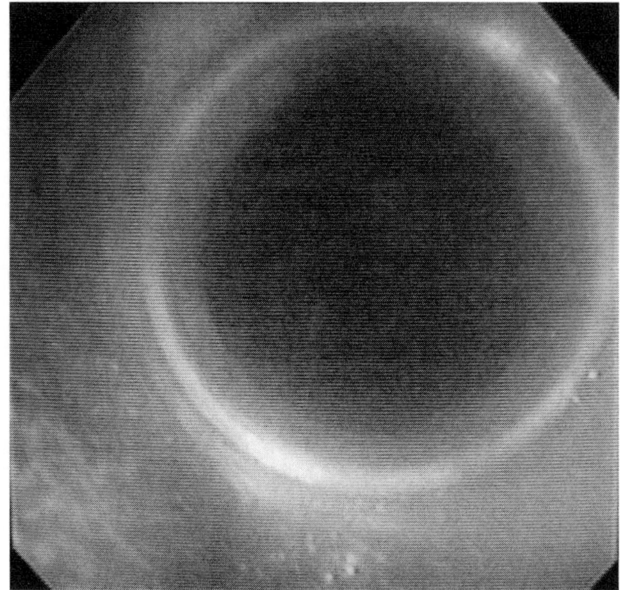

Fig. 8.128 Contaminated hose internals at the fitting attached to the bar nozzle or waterbox. The contamination was introduce after a failure of the stator cooling water cooler tubes that allowed raw service water to leak into the demineralized stator cooling water. A flashover to ground did eventually occur but without any appreciable damage. The hoses were all cleaned, the cooler leak repaired, and the machine put back in service. (See color insert.)

some on a regular basis during major overhauls, as a matter of course. The concern with contamination on the inner surface is the possibility of it allowing electrical tracking and eventual flashover to ground from the high-voltage part of the stator bar. Teflon is used for the hoses to provide an insulation creepage distance for the high-voltage bars to ground and still allow connection to the stator water-cooling circuit for delivery of demineralized water to the bars.

There are also large headers that feed the stator cooling water to the stator winding; these generally contain gaskets and O-rings at the various flanges and fittings to seal them from hydrogen leaking into the stator cooling water. All of these are organic in nature and are commonly called "consumables" because they have a definite life and require replacement from time to time. Some of the contributing factors are simple wear, tearing, drying and embrittlement, disintegration, and deformation (Figs. 8.129–8.131). Depending on the type of material used, a schedule of inspection and replacement is advisable to avoid forced outages from any of the problems that can arise if these subcomponents experience one or more of the aforementioned failure mechanisms. The result may cause such problems as leakage of hydrogen into stator cooling water or bits of the material coming apart and blocking cooling water flow in the stator bars.

Fig. 8.129 Gasket deterioration from overexposure to heat and operation beyond the expected life of the material.

8.4 PHASE CONNECTORS AND TERMINALS

8.4.1 Circumferential Bus Insulation

The term *circumferential bus* is used to describe the circular-shaped *phase-connection buses* found in large turbo-generators, usually at the collector end of the generator. These are required to make the parallel-path connections in the stator winding, as well as the connection to the stator terminals that transfer the power out of the generator (see Fig. 8.132). They are supported by structures made generally of nonconducting

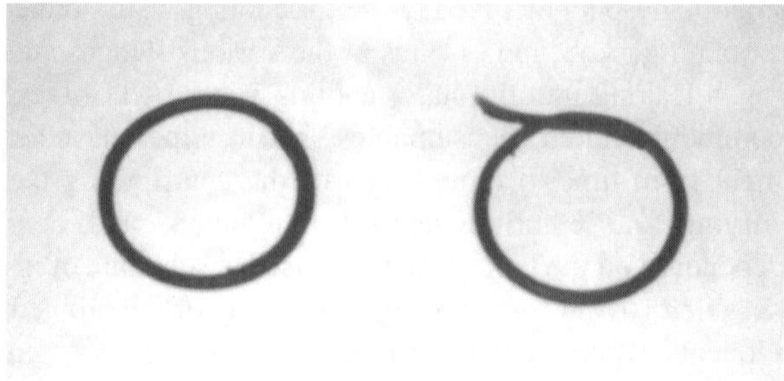

Fig. 8.130 O-rings may be found damaged and torn, such as the one shown, and can easily be the source of a leak if found in the condition shown in the photo.

Fig. 8.131 The O-ring shown has a significantly distorted shape from the roundness it is supposed to have. This type of deformation is termed a *permanent set.* This generally is due to a combination of time, temperature, and pressure, and is also a form of *creep* failure for rubber-type materials.

Fig. 8.132 Circumferential buses in a large four-pole steam-turbine generator.

materials, bolted to the core-end compression plates and the stator frame of the machine. They are generally cooled by the cooling gas in the machine or, in some cases, water cooled as part of the stator cooling water circuit.

Circumferential buses are normally separated from the rest of the winding by a relatively large electric clearance. They are insulated so that there is no conducting path to ground. The most common mode of failure in circumferential busses is a breakdown of the insulation adjacent to the metallic supporting studs, or cracks in the insulation from vibration. The continuous vibration of the heavy copper busses tends to also abrade the insulation in this area

Continuous movement, together with contamination (the busses are located in close proximity to the rotor fan and are subject to the rapid flow of air or gas containing occasional small amounts of seal oil, dust, and other contaminants), tends to produce tracking over the blocking separating the busses, and may eventually result in phase-to-phase failures. Occasionally, low megger readings between the phase under test and the other (grounded) phases can be attributed to the contaminated insulation between circumferential busses.

In cases of excessive vibration, the circumferential conductors may fret within their support structures and abrade the insulation significantly. When oil is present, the greasing effect may also occur on this component (Fig. 8.133).

In some machines, the circumferential bus insulation is painted with a resistive coating to dissipate electric charge. However, if a crack in the paint coating or insu-

Fig. 8.133 Greasing at the support member due to movement between the bus and the support structure.

lation occurs, this can cause the electric charges to concentrate at the sharp corners of the crack, which essentially creates an electric stress point followed by corona discharge. The result of this discharge is a whitish powder forming at the crack location (Fig. 8.134). Discharges of this type, so far removed from the ground plane, are not generally harmful, but they should be inspected and cleaned up in case additional contamination becomes present at some point, which would allow the charges to track to ground.

It is more serious when cracks occur in the phase connectors or support ties. In this case, a "redeye" epoxy is sometimes applied to fill the gaps and help reestablish the insulation (see Fig. 8.135). This is done to minimize the possibility of additional cracking and contamination in the cracks, avoiding electrical tracking failures. This type of repair is generally applied to phase connectors and end windings that have an insulating outer coating only, rather than to the resistive type mentioned earlier.

8.4.2 Phase Droppers

Phase droppers (phase leads) are simply the bus connection to the stator terminals from the circumferential bus, as shown in Figure 8.136. They are also insulated and cooled in the same manner as both the circumferential bus and the terminals. They can be gas or water cooled.

The same problems that exist with the circumferential bus are true for the phase dropper (refer to the previous section). Vibration is sometimes a major concern with

Fig. 8.134 Corona discharge on phase connectors.

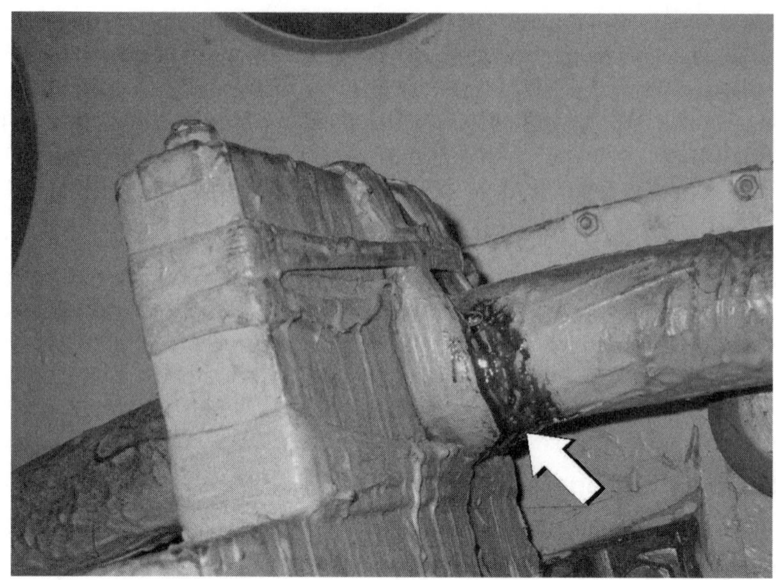

Fig. 8.135 "Redeye" epoxy treatment on phase connector.

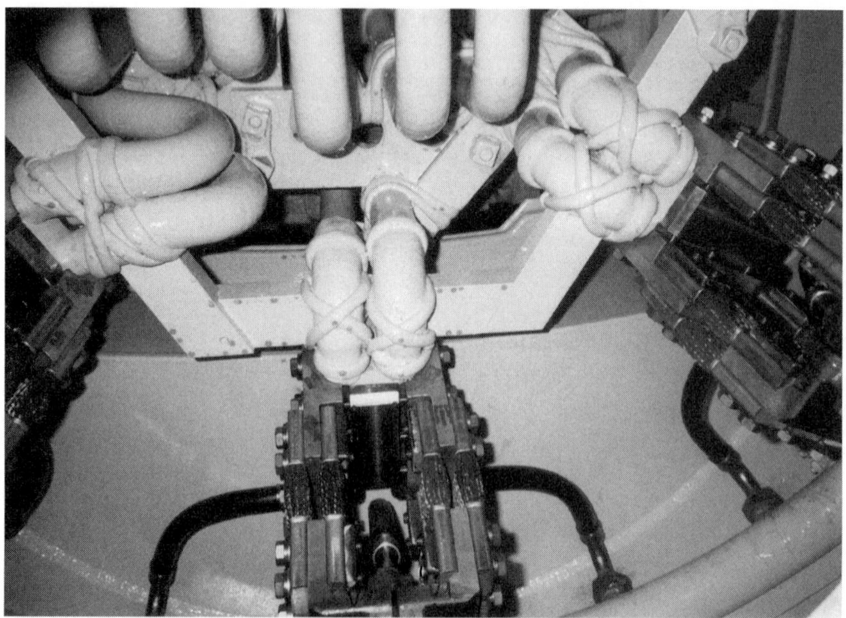

Fig. 8.136 Phase droppers from the circumferential ring buses to the terminals. Clearly seen are the connections between the circumferential buses and the generator terminals.

the droppers, though more so than with the circumferential bus. When excessive vibrations are encountered, as evidenced by greasing or cracked insulation, additional blocking is sometimes added to dampen the vibrations (see Fig. 8.137). Also, redeye epoxy treatments may be applied, as with the circumferential bus or the end windings.

8.4.3 High-Voltage Bushings

There are too many different arrangements of terminal boxes and bushing types in large synchronous machines to describe them all in this book. However, the following general inspection guidelines and comments apply to any arrangement.

Lead bushings are susceptible to damage arising from sudden load changes, excessive vibration, overheating of the leads, and normal vibration over long periods of time. Stator high-voltage bushings should be inspected for evidence of cracks, oil leakage (when oil-filled), and looseness of components. All dirt and tracking residues should be thoroughly cleaned.

In large turbo-generators, the high-voltage bushings are partly contained in sealed bushing wells. Some of the lead bushings have ducts allowing the flow of hydrogen (Fig. 8.138). The ducts should be free of oil, grease, or any foreign elements. Many others are water cooled. In these, connections to the bushings should be inspected for cracks and leaks.

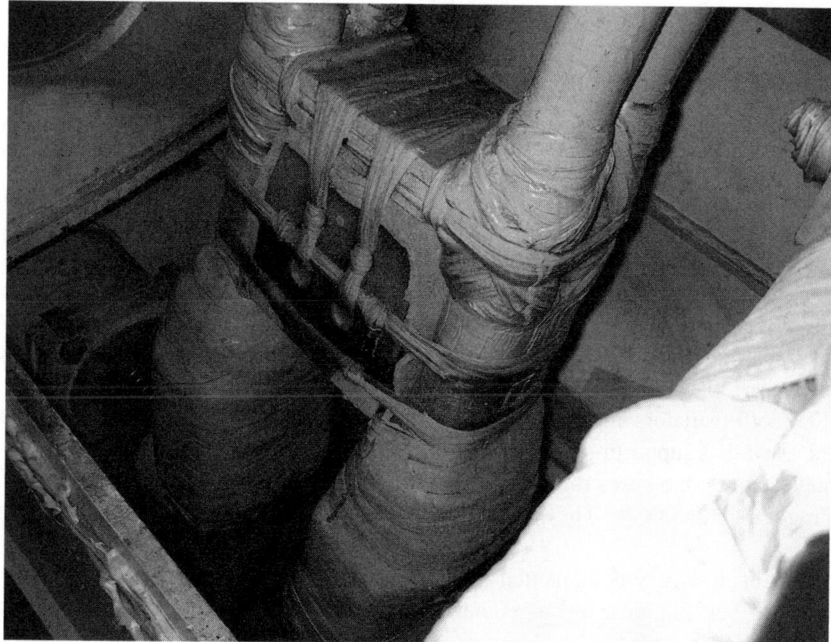

Fig. 8.137 Additional blocking between droppers applied to dampen vibrations.

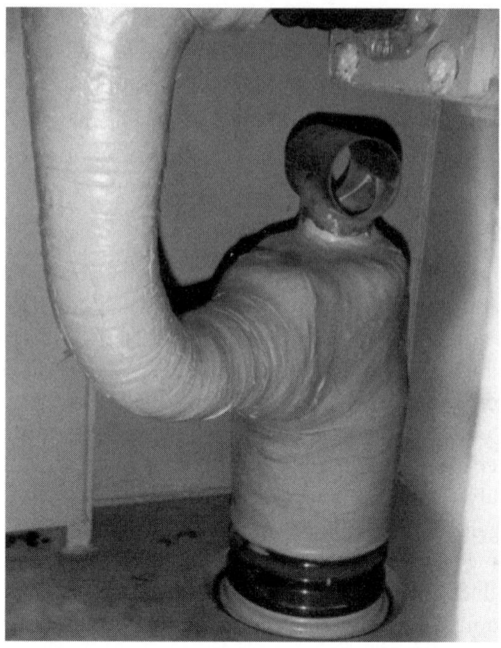

Fig. 8.138 Typical hydrogen-cooled high-voltage bushing inside the bush well of a large 925 MW, four-pole generator.

8.4.4 Standoff Insulators

As with high-voltage bushings, standoff insulators are subjected to continuous forces due to vibration, as well as thermal expansion and contraction of the leads (Fig. 8.139).

Standoff insulators should be inspected for cracks and looseness of their constituent components. External surfaces should be kept clean to avoid extensive tracking, which may result in eventual short circuits to ground.

Old insulators may utilize lead in their construction. In many cases, continuous vibration over a long period of time will deteriorate the insulator, to the extent that the lead-based material will ooze out of the insulator in the region of the porcelain seal. This appears as a gray powder or paste.

Standoff insulators are found inside the case and *bushing well* of large turbo-generators, as well as supporting the lead and neutral buses of all other types of machines. In some instances, the buses may extend to significant distances (sometimes below) from the main machine body. They should be inspected for cleanliness and overall condition.

It has been widely documented that there is a relationship between failures in the connecting leads of large turbo-generators and the condition of the standoff insulators. In many cases in large two- and four-pole machines in which failures of the connection leads occurred, the standoff insulators showed moderate to heavy signs of greasing.

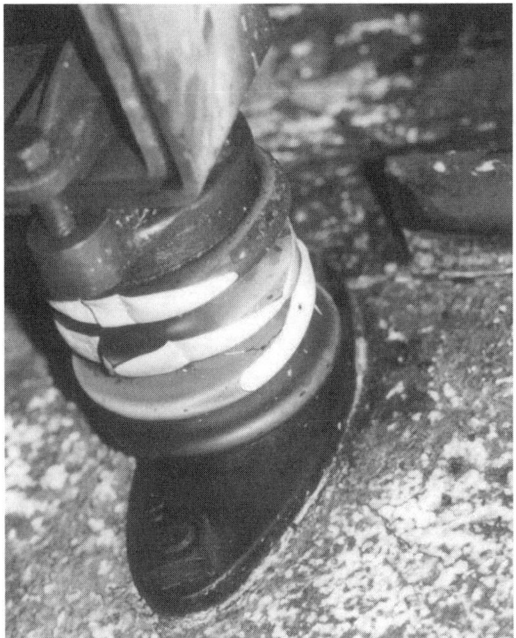

Fig. 8.139 Standoff insulators damaged due to surface contamination.

The greasing is the result of the dust produced by excessive fretting of the seal between the porcelain and the mounting flange and its mixture with oil or oil vapors present in the bushing well.

The normal mode of lead failure is fatigue cracking of the flexible connections due to poor support of a weakened standoff insulator. Partial cracking of the flexible connections results in increased temperature of the conductor. Eventually, the connectors melt, resulting in serious failures.

Additional serious damage can result from the change in the cooling path of the hydrogen due to a shift in the position of the porcelain of the insulator in relation to its supporting flange.

8.4.5 Bushing Vents

As explained earlier, bushings sometimes have passages built inside them to allow the flow of air or hydrogen for cooling purposes. It is important that the vents be inspected to see if they are unclogged. In sealed bushing wells, high-voltage bushings may be flooded by seal oil, and their passages clogged or semiclogged. If clogged, they can be siphoned clean with a venturi pipe or vacuum. Figure 8.140 shows a cross section of a typical bushing well. Figure 8.141 shows the actual bushing and bushing vent inside a bushing well of a hydrogen-cooled generator.

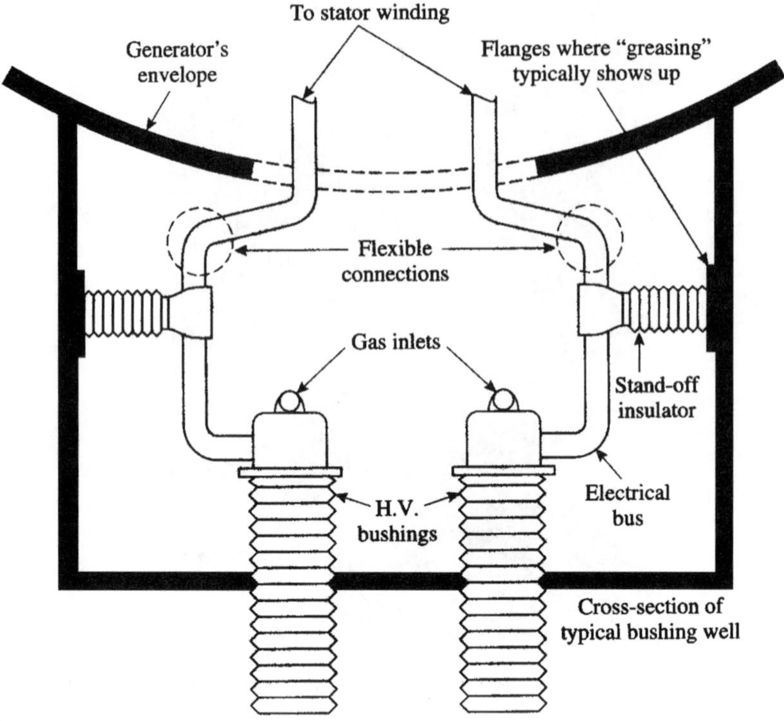

Fig. 8.140 Cross section of a typical bushing well showing standoff insulators and their gas passages.

When possible, a nipple (or nipples) can be placed at the bottom of the bushing well, protruding about ½ inch inside the bushing well. The nipple's protrusion allows the retention within the well of the required viscous-oil seal normally placed in the bushing wells of hydrogen-cooled machines. However, it will allow the flow into a detection and purging pipe of unwanted water and seal oil from the shaft seals that may leak into the machine and overflow the protruding length of the nipple.

8.4.6 Bushing-Well Insulators and Hydrogen Sealant Condition

In large synchronous machines, the connection of the windings to the armature leads is carried out inside a large terminal box (bushing well) underneath the machine. Often, foreign material such as oil, loose bolts, washers, and fragments of insulation become lodged in this compartment. Hence, it is important that the terminal box be opened and inspected, at least during major inspections. In addition, the condition of the bushings and standoff insulators should be inspected (see previous two sections). The resin tape on the terminations should also be inspected for integrity.

These bushing wells normally have small amounts of viscous sealant oil at the bottom of the well. The purpose of this sealant is to eliminate the leak of gas through the

Fig. 8.141 Standoff insulator and high-voltage bushing showing the vent arrangement.

mounting flanges of the high-voltage bushings. The condition of the sealant should be evaluated during an inspection inside the bushing well. Dry or contaminated sealant oil should be removed, and new material should be brought in (Fig. 8.142).

The high-voltage bushing seal should be carefully inspected at every outage and also from the outside as a routine. There are generally gaskets for the bushing flange and these will age with time, temperature, and vibration effects (Fig. 8.143). When they leak, the thick liquid sealant tends to be forced out through the smallest of spaces at the leak point by the higher hydrogen pressure inside the machine (Fig. 8.144). If the sealant is lost altogether, then the hydrogen may leak as well. There is significant danger because of the nature of hydrogen, and the fact that it has a tendency to self-ignite when leaking through a small crevice under high pressure. The flame is invisible and the by-product is water as the hydrogen combines with the oxygen in air. If the hydrogen has the possibility of collecting in a "dead area" where there is no air circulation, then a further danger of an explosion is created if the mixture of air and hydrogen reaches the explosive range.

8.4.7 Generator Current Transformers (CTs)

As part of any major inspection of a large synchronous machine, the condition of the main current transformers (CTs) should be evaluated. In large generators, the CTs are

Fig. 8.142 Sealant for HV bushing in a circular reservoir at the casing exit in the bushing well. (See color insert.)

usually located underneath the machine, just below the terminal box or bushing well. The CTs, which normally come as two or three per phase, are placed around the phase and neutral isophases where these leave the bushing well (Fig. 8.145). In smaller, typically air-cooled, machines, the CTs may be located inside the terminal box. In medium sized hydrogen-cooled machines, there may not be a bushing well or it may be very small, with the major portion of the terminals nonenclosed, outside the machine, and the CTs in the open as well (Fig. 8.146).

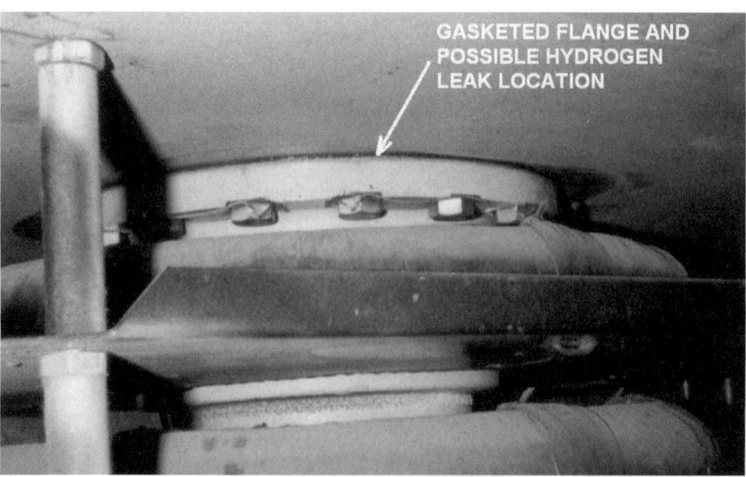

Fig. 8.143 Gasketed flange at the high-voltage bushing penetration.

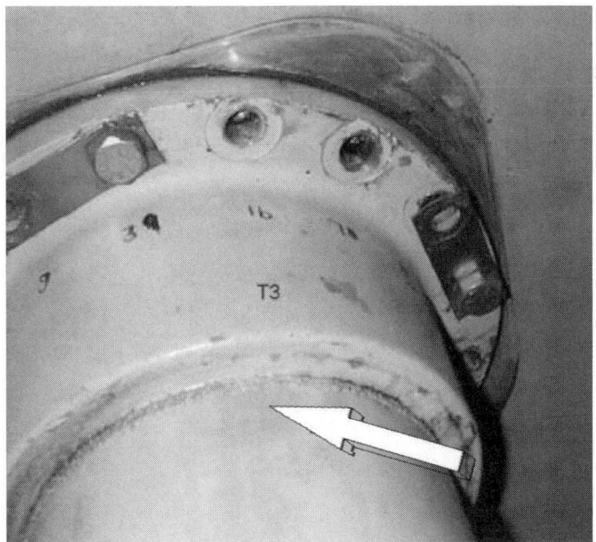

Fig. 8.144 Sealant leak from inside the generator through a leak point at the high-voltage bushing penetration.

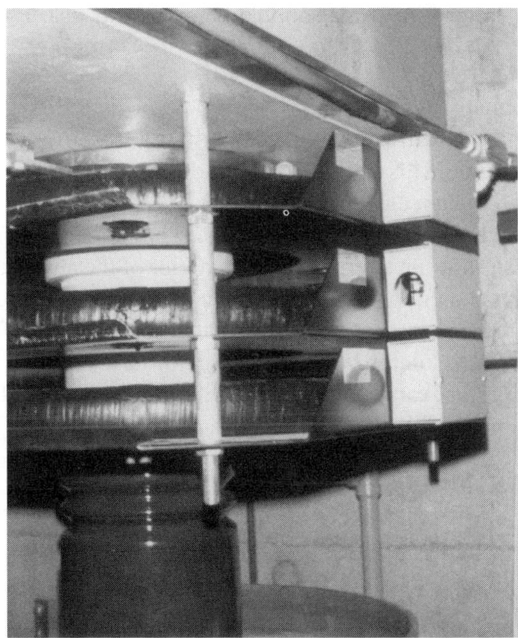

Fig. 8.145 Generator current transformers for metering and protection, installed around the external portion of the generator terminals and supported on trays.

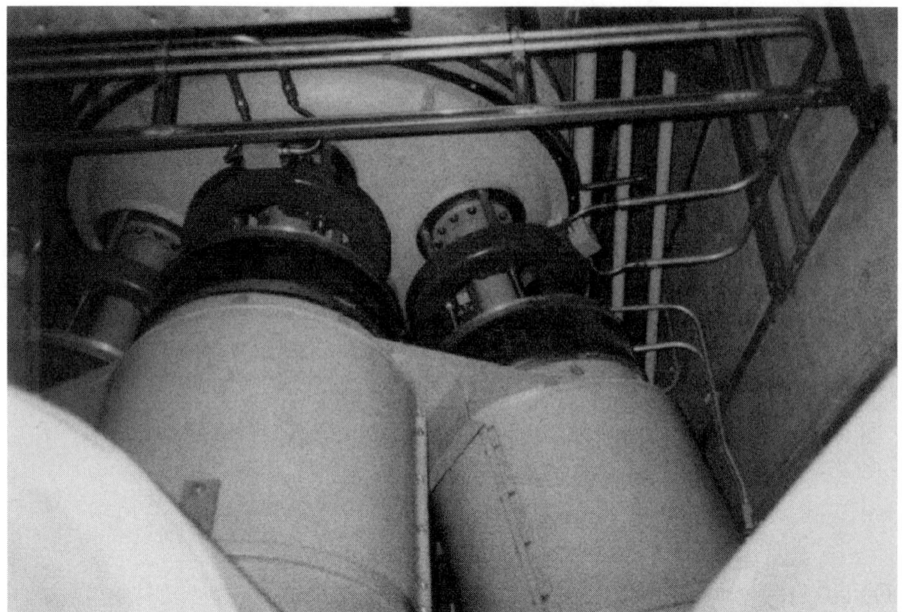

Fig. 8.146 Generator current transformers for metering and protection, installed around the external portion of the generator terminals. This machine is a 200 MW hydrogen-cooled generator with a small bubble-type bushing well.

The CTs are normally mounted on insulated bolts. Grounded metal piping containing the connection leads carries out connection to the CTs. Large machine CTs are commonly manufactured with their windings and core encased in resin-type material. On top of this, an aluminum envelope is typically added.

Internal faults of the current transformer winding that result in substantial heat being generated tend to show up as cracks in the exterior of the aluminum shell, and/or as flow of the encasing resin out of the shell. Signs of overheating, such as leaking resin or discoloration, provide a strong reason for further investigation. The surfaces of the CTs should be cleaned to minimize any possibility of flashover.

Like any other current-carrying element of the power plant, CTs can be scanned with an infrared camera while the machine is in operation to check for abnormal temperatures.

8.5 HYDROGEN COOLERS

Heat exchangers are an integral component of the machine. Typically, large turbo-generators have their hydrogen–water or air–water heat exchangers located inside the machine's casing. In air-cooled machines, the air-to-water heat exchangers may be located in a separate container, often below the machine. In all cases, the maximum allowable rating is directly dependent on satisfactory condition of the heat exchangers.

Hydrogen coolers inside large turbo-generators may be found in many different arrangements, such as dome coolers and vertical and horizontally installed coolers. Figures 8.147 and 8.148 show examples of the vertical and horizontal types. The two major concerns with heat exchangers are clogged water tubes and water leaks.

Through continuous flow, water tends to clog the ducts by causing corrosion and depositing minerals. Once initiated, the clogging process advances at an increasing pace. Not only is the flow of cooling water reduced, but also the heat transfer coefficients are detrimentally affected due to the mineral buildup inside the pipes, resulting in poor heat-transfer properties. It is, therefore, important to visually inspect the tubes and take the recommend corrective action (e.g., descaling) when required.

Heat exchangers, in particular those positioned inside the machine casing, are continuously subjected to vibration. Therefore, it is not uncommon to find leaks developing during the operation of the machine. There are several methods available for the detection of those leaks. Hydrogen sensors are commonly used in hydrogen-cooled machines. The higher pressure of the hydrogen creates a flow of the gas into the coolers, keeping the water from entering the machine. Even in hydrogen-cooled machines, as in other types of machines, water heat exchanger leaks will probably result in water contamination of the windings. A visual inspection can detect a leak by looking at suspicious accumulations of water inside the casing. Pressure tests are also routinely performed, and are recommended for coolers installed inside the machine that have not been inspected for several years. A leak can also be checked by opening the drain tap normally located at the belly of the machine and looking for water. Tubes can also leak

Fig. 8.147 Vertical-type hydrogen coolers.

Fig. 8.148 Horizontal-type hydrogen coolers.

from thinning of their walls by erosion and corrosion. Eddy-current tests can detect excessive thinning of the cooler tube walls.

REFERENCES

1. ANSI/IEEE Std 67-1972, "IEEE Guide for Operation and Maintenance of Turbine Generators," Item 8.6.2, p. 34.
2. H. R. Tomlinson, "Inter-Laminar Insulation Test for Synchronous Machine Stators," *AIEE Transactions,* Vol. 71, Part III, August 1952, pp. 676–677.
3. *EPRI Power Plant Electrical Reference Series,* Vol. 16, Items 5.2.4 and 5.2.5, pp. 5–42, and Item 6.5.4, pp. 6–22, 1991.
4. ANSI/IEEE Std 56-1977, "IEEE Guide for Insulation Maintenance of Large Alternating-Current Rotating Machinery (10,000 kVA and Larger)," Item 6.4.1, p. 10.
5. *EPRI Power Plant Electrical Reference Series,* Vol. 1, pp. 1-11, 1-37, 1-47, 1991.
6. GEI-37081, "Instructions on Hydrogen-Cooled Turbine Generators, Mechanical and Electrical Features," General Electric Co.
7. J. Boyd and H. N. Kaufman, "The Causes and Control of Electrical Currents in Bearings," *Lubrication Engineering,* January 1959, pp. 28–35.
8. G. W. Buckley and R. J. Corkings, "The Importance of Grounding Brushes to the Safe Operation of Large Turbine Generators," *IEEE Transactions on Energy Conversion,* September 1988, pp. 607–612.
9. J. Sparks, "What to Do about Shaft Currents," Plant Operation and Maintenance Section, Power Generation and Transmission, p. 114.
10. The Doble Engineering Company, *Rotating Machinery Insulation—Test Guide,* 1985.

11. *EPRI Power Plant Electrical Reference Series,* Vol. 16, Item 2.1.1, pp. 2–3 and 2–4; Item 2.3.2.7, p. 2–9; Items 3.1.2.1–3, pp. 3-18, 3-20–3-21, 1991.
12. C. A. Duke, T. F. Faulkner, R. C. Price, and C. A. Roberts, "Visual Inspection of Rotating Machinery for Electrical Difficulties," presented at Conference of Doble Clients on Rotating Machinery, 1960.
13. E. H. Povey, "Corona Measurements by the RIV Method," in *Conference of Doble Clients on Rotating Machinery,* p. 3-201, 1958.
14. W. A. Patterson, "Testing for Corona in Generator Stator Windings," in *Conference of Doble Clients on Rotating Machinery,* p. 7-601, 1967.
15. W. A. Rey, "Increased Deterioration of Generator Insulation by Corona Action," in *Conference of Doble Clients on Rotating Machinery,* p. 7-101, 1951.
16. C. A. Duke, "Experience with Slot-Discharge Testing on Generators," in *Conference of Doble Clients on Rotating Machinery,* p. 7-101, 1958.
17. IEEE Std 286-2000, "Recommended Practice for Measurement of Power-Factor Tip-up of Rotating Machinery Stator Bar Insulation."
18. T. W. Dakin, "The Relation of Capacitance Increase with High Voltages to Internal Electric Discharges and Discharging Void Volume," *AIEE Power Apparatus and Systems,* Vol. 78, pp. 790–795, 1959.
19. T. W. Dakin, "A Capacitance Bridge Method for Measuring Integrated Corona-Charge Transfer and Power Loss per Cycle," *AIEE Power Apparatus and Systems,* Vol. 79, pp. 648–653, 1960.
20. W. McDermid, "Review of the Application of the Electromagnetic Probe Method for the Detection of Partial Discharge Activity in Stator Windings," in *Proceedings of the CEA International Symposium on Generator Insulation Tests,* Toronto, 1980.
21. S. R. Campbell, G. C. Stone, H. G. Sedding, G. S. Klempner, W. McDermid, and R. G. Bussey, "Practical On-line Partial Discharge Tests for Turbine Generators and Motors," *IEEE Transactions on Energy Conversion,* Vol. 9, No. 2, 1994.
22. G. Klempner, "Ontario Hydro Experience with Failures in Large Generators due to Loose Stator Core Iron," in *EPRI—Utility Generator Predictive Maintenance and Refurbishment Conference,* December 1–3, 1998, Phoenix, AZ.
23. G. Klempner, "Back of Core Burning," presented at GEC Generators Users Group Meeting, Southern California Edison, San Onofre Nuclear Generating Station, January 10 and 11, 2000.
24. IEEE/ANSI C50.13-1989, "Requirements for Cylindrical-Rotor Synchronous Generators."
25. IEEE Std 1-2000, "IEEE Recommended Practice—General Principles for Temperature Limits in the Rating of Electrical Equipment and for the Evaluation of Electrical Insulation."

ADDITIONAL READING

EPRI Report EL-3564-SR, "Workshop Proceedings: Generator Monitoring and Surveillance," August 1984.

EPRI Report NP-902, "On-line Monitoring and Diagnostic Systems for Generators," September 1979.

9

ROTOR INSPECTION

Turbo-generator rotors are remarkable in that, in spite of being subject to very large stresses, they perform so reliably for many years. The fastest turbo-generator rotors for power generation rotate at 3600 revolutions per minute. At that speed, components experience centrifugal forces of several orders of magnitude larger than gravity. Add to that friction due to the expansion and contraction of different materials in response to changes in temperature, lateral and torsional vibration, load shocks, and current and voltage transients, and one must marvel at the fact these complex systems can operate sometimes for many years (between outages or refueling cycles in a nuclear plant) before cursory electrical testing is performed, or run up to ten years before a detailed visual inspection is performed. Many properly maintained rotors have been known to run for decades before a major refurbishment was required.

Nevertheless, the criticality of these components requires them to be monitored and well maintained. Due to the great stresses under which rotors operate, any unchecked failure has the potential to develop into a catastrophic loss of the entire generator. Even less severe incidents can render the entire unit unavailable for many weeks, with substantial loss of production. Industry's experience with rotors of many manufacturers, sizes, and applications have resulted in shared knowledge about typical weaknesses and specific problems with this or that design. One visible trend is that deregulation has intensified the use of *two-shifting* and *load following,* and this has resulted in a corresponding increase in certain modes of failures, largely due to thermal cycling effects.

There are a number of statistics published by organizations such as EPRI and INPO about the expected life of certain subcomponents (slot liners, collector rings, forgings, etc.), as well as failure statistics of rotors in general. One normally accepted criterion is that rotors in general have an expected useful life of 25 to 30 years, driven by winding insulation degradation, and that once rewound and refurbished they can last an additional 25 to 30 years, at which time the forging may also become a limiting component. Although statistically speaking those numbers may reflect the average of thousands of units, any particular unit may have quite different statistics. Being guided exclusively by industry-wide statistics when deciding on maintenance and life-cycle management of a specific unit in a specific site will certainly result in large errors of judgment. Only intimate knowledge of one's own rotor (and overall generator) can lead to the correct maintenance and refurbishment plan, and to intelligent life-cycle management decisions. The goal of this chapter is to serve as a guide to learning the specific problems and failure mechanisms, and their identification, that will make it possible to correctly assess intrinsic risks for a given design, and to notice explicit signs of deterioration, damage, and/or impending failure. Although no OEM is specifically identified while discussing a particular issue, the reader may recognize each item discussed as pertinent or not to the specific machine under his/her supervision.

This chapter contains a detailed description and discussion of each item mentioned on the Rotor Inspection Form found in Chapter 7, not necessarily in the same order as found on the form.

9.1 ROTOR CLEANLINESS

Cleanliness, or a measure of cleanliness, is important not only to the proper operation of the machine but also to provide the inspector or maintenance crew with clues on the overall condition of the machine.

For instance, a turbo-generator rotor exhibiting numerous deposits of copper particles (copper dust) may indicate excessive movement of the dc field coils when on the turning gear.

Turning gear is the term given to the motor and associated gearbox used to turn the rotor at low speed, when the machine is not in operation. Turning the rotor at low speed keeps it from becoming bowed due to its own weight. In particular, it is important to turn the rotor immediately after bringing the machine offline, when the rotor is still hot; an excessive thermal bow can result if the machine is allowed to rest immediately after operation.

Excessive copper dust migrating out of or simply being present in the winding slot radial vents should alert the inspector to the possibility of the existence or the development of shorted turns and/or ground faults. This can easily be seen when looking into the vent holes in the winding slot wedges. Under those circumstances, the inspector may decide that it would be prudent to carry out one of the available shorted-turn tests. In addition, if copper dusting is found in the rotor end winding, the machine's owner may decide to change the blocking, insulation, and so on. Ground faults that may be due to copper dust can easily be detected with a megger (generic name for an instrument that measures insulation resistance).

Copper dust, iron dust, or any other telltale material may be concealed in a mixture of oil and dirt. In some instances, chemical analysis of a sample is required for assessing the true amount of copper and/or iron dust present. A cotton swab can be used to obtain a sample from the contaminated area, mostly under the retaining rings.

Copper dusting is very specific to those field windings employing turns with two conductors back to back, without insulation between them. Each conductor, which represents half of the turn, is a copper bar that is sometimes flat, or C-shaped, or, more likely E-shaped. When pressed together, this arrangement (with C-shaped and E-shaped conductors) allows the flow of hydrogen cooling gas along the conductor. This cooling method is common for medium- and large-size generators. When at speed, the complete winding is tightly retained in the radial direction, not allowing any significant movement. However, during periods of turning-gear operation, this clearance, and the low turning speed of the rotor, results in a constant pounding of the copper halves. In the long run, copper dust is generated in the winding. This copper dust tends to migrate because of mechanical forces and the flow of the gas. Eventually, the dust can be a source of abrasion of the wall insulation, creating grounds and shorts between turns. Thus, when inspecting the rotor of such a generator, in particular one subjected to long periods of turning-gear operation, one must be attentive to the presence of copper dusting, which may indicate a winding becoming prone to the development of shorted turns and grounds. Figure 9.1 shows two typical examples of the double conductor arrangement. Figure 9.2 shows the copper dust that can migrate from the winding slots in such arrangements as shown in Figure 9.1.

Dirt can mask cracks on the surfaces of critical components such as wedges, fan hubs and blades, retaining rings, and the forging itself. Therefore, critical areas should be cleaned to the full extent of their accessibility. However, it is important that the inspection be carried out both before and after cleaning. Before-cleaning inspections reveal a wealth of information on the rotor or machine condition and operation history, whereas inspection after cleaning allows evaluation of the condition of rotor components.

Besides masking important areas from the inspector, heavy dirt deposits can adversely affect the flow of the cooling gas or air, in effect derating the machine.

9.2 RETAINING RINGS

The retaining rings are the most critical component of the rotor and, normally, the most stressed rotor component. Retaining rings are critical in the sense that their mechanical failure always has catastrophic consequences for the physical integrity of the machine. Therefore, retaining rings deserve the utmost attention from the inspection team and plant personnel engaged in the inspection and maintenance of the unit.

Manufacturers of large synchronous machines usually issue periodic informative technical bulletins. In each bulletin, the attention of the users of a particular type of machine is directed toward problems experienced by other users of machines with similar components and design. Recommended remedial actions necessary to avoid or

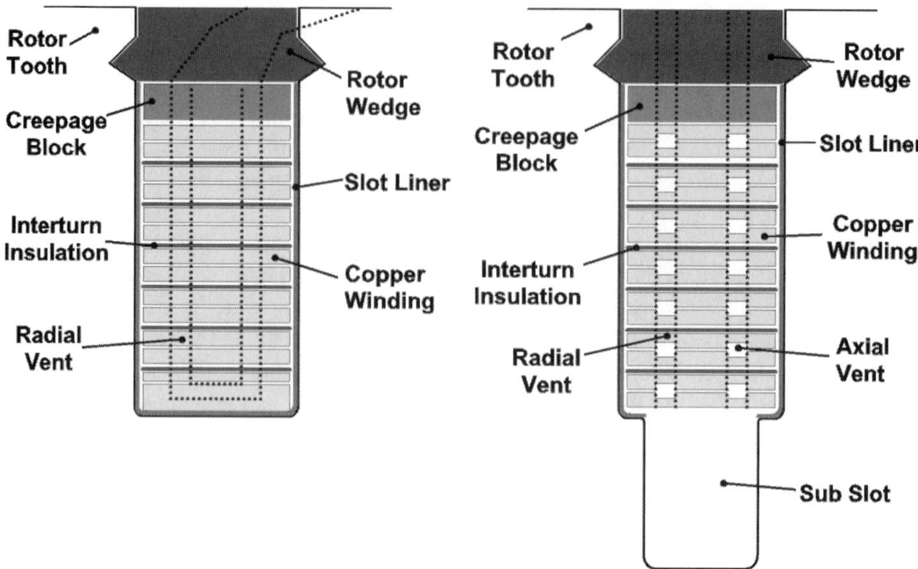

Fig. 9.1 Schematic representations of two types of rotor field windings. (*a*) Each turn is made of two flat copper conductors, with no insulation between each half-turn. This design includes an airgap pickup arrangement with radial vent holes for cooling. (*b*) Each turn is made of two E-shaped copper conductors, with no insulation between each half-turn. This design includes radial vent holes fed from a subslot and axial vents for cooling. Both arrangements allow copper dusting to develop over the years.

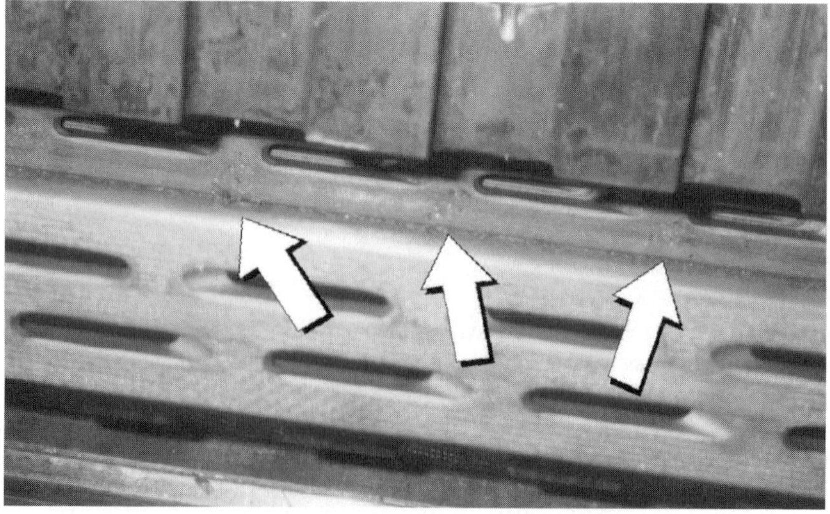

Fig. 9.2 Rotor with winding slot wedges removed and first copper turn lifted. The copper particle dusting can be clearly seen in this photo.

minimize risks of similar problems are presented, oftentimes including a visual or robotic inspection of a specific component(s) in the machine. In the case of retaining rings, a substantial amount of information has been disseminated by the original equipment manufacturers (OEMs), clients, and organizations such as the Electric Power Research Institute (EPRI). In general, the manufacturers' recommendations for the protection of the rings should be followed.

Periodic visual inspection and nondestructive testing (NDT), using eddy-current, ultrasonic, and dye-penetrating techniques, of the retaining rings is highly recommended. The actual length of time between inspections depends on the mode of operation of the machine, which includes total operation-hours and, in particular, whether it is in continuous operation or subject to many start–stop cycles. The successive thermal cycles accompanying a machine subjected to many start–stop operations tend to cause accelerated metal fatigue of the retaining rings.

A major cause for advancing the inspection schedule of the retaining rings is moisture penetration on a machine equipped with the older 18–5 type of ring (18% manganese–5% chromium alloy). The modes of moisture-induced failure of the 18–5 rings are further discussed in Section 9.2.1 and in other chapters.

Another reason for performing at least a partial inspection of the retaining rings is excessive asynchronous "motoring" or generation without the dc field energized. Under these conditions, the machine acts as an induction motor or generator. This can be damaging to rotor components such as wedges and retaining rings, due to the flow of induced currents in those components. Negative-sequence currents also tend to flow in the forging, wedges, and retaining rings. An area prone to damage from arcing is the contact area between the rings and the adjacent rotor wedges (Figs. 9.3 and 9.4). This area should be checked for electric pitting or discoloration indicating the flow of currents. If these symptoms are found, remedial action should be taken. To minimize the risk of current flow between the wedges and the retaining rings, it is good practice to move the wedges away from the rings during overhauls when the rotor is out of the bore. In some cases, the OEM provides instructions of how to prevent these wedges from contacting the retaining rings. Any shift of wedge position mainly occurs during starts, stops, and turning-gear operation. Thus, for units operating mostly at base load, there is little if any shifting of wedges.

Through visual inspection of the retaining rings, a preliminary assessment can be made as to the need to carry out specific tests such as nondestructive examinations (NDE) of several types—eddy current, acoustic, die penetrant—as well as hardness tests, and so forth. If the visual inspection of the surface reveals oxidation traces, pitting, or other warning anomalies, inspection of the inner side of the retaining rings might be advisable (see Fig. 9.5). This entails removing the rings, a costly and somewhat risky operation (see Section 9.2.2).

It goes without saying that for visual inspections to be effective, the paint (if present), oil, and dirt on the surface of the rings have to be removed. Unfortunately, if the paint is removed, the polishing action may conceal otherwise visible trouble spots. Some manufacturers recommend not painting the retaining rings. This depends on the steel type and cooling medium, as well as the manufacturer's own experience and preferences. Some vendors offer technologies that they claim can perform nondestructive

Fig. 9.3 The photo shows a "weld" spot formed between the wedge and the retaining ring (not shown as it has been removed). The "weld" was formed by arcing from surface currents flowing on the surface of the rotor, due to an abnormal operating event.

Fig. 9.4 Excessive asynchronous motoring developed arcs between wedges and between wedges and the rotor teeth, damaging the wedges and the teeth. The excessive heat generated during the abnormal operation also caused the paint to discolor and flake.

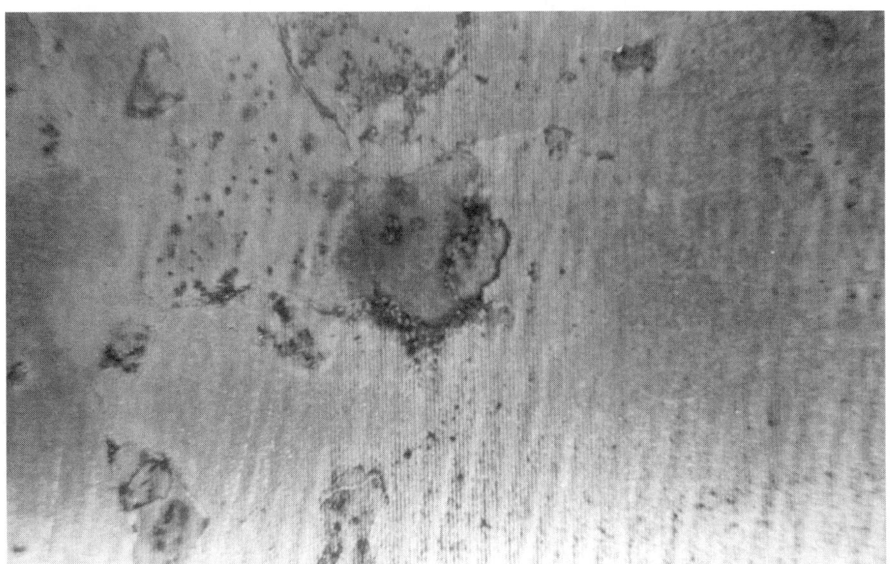

Fig. 9.5 Portion of a retaining ring showing rust deposits where the protective paint has peeled off. Very often, the rust is superficial and easily removed. Any presence of rust should alert the inspector to the possible existence of water-originated pitting. (Reproduced from Kerszenbaum, *Inspection of Large Synchronous Machines,* IEEE Press, with permission.)

examinations without the need to remove the paint. One should make sure that a body of demonstrable experience and adequate references corroborates those statements.

During major inspections, NDE and hardness tests are always recommended. Commonly, these tests require removal of the rotor. Recently, though, some service providers have been able to perform a number of nondestructive tests on retaining rings without removing the rotor from the bore. As stated in the previous paragraph, it is important to ascertain that those remote techniques have been demonstrated to provide meaningful information. Although an inspection technique might work well, it is as important to know what it can miss, as it is to know what it can find.

Several rotor designs have holes drilled in the body of the retaining rings (Figs. 9.6 and 9.7). These allow the flow of gas for cooling purposes. Ring ventilation holes tend to be areas of mechanical stress concentration. The tangential expansion forces cause the areas of maximum stress at the holes to be at the inner diameter (ID) of the ring in the axial direction, namely, facing the ends of the rotor. Some people highly recommend performing eddy-current NDE on all holes drilled in those retaining rings. Other areas of stress concentration that should be closely inspected are *tapped holes* for the locking arrangement on the circumferential ring keys. In general, for any type of retaining ring, EPRI's publication, EL/EM-5117-SR, has valuable information on guidelines for the evaluation of the condition of retaining rings, as well as a list of vendors offering specific inspection tests [1].

Fig. 9.6 Retaining ring of a two-pole turbine generator with vent holes.

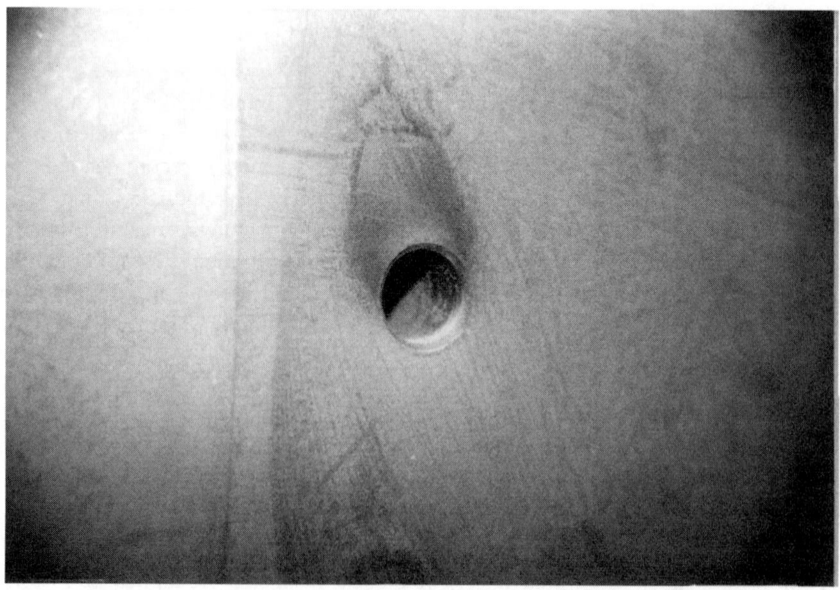

Fig. 9.7 Close-up view of a vent hole from the retaining ring shown in the previous figure. The shadow is the result of deposits and erosion by the continuous high-speed flow of gas leaving the vent in the direction of the shade.

9.2.1 Nonmagnetic 18–5 and 18–18 Retaining Rings

It is extremely important to maintain a dry and clean environment around the rotor, not only during the operation of the machine, but also during the overhaul and inspection activities. Of particular importance is keeping humidity off retaining rings made of 18–5 alloy steel. Corrosion occurring when the rotor is removed from the bore can lead to failure of the ring during future operation (through stress corrosion cracking). The issue of stress corrosion cracking (SCC) of 18–5 retaining rings has resulted over the years in mountains of literature, and as long as some of these rings remain in operation, the topic will remain central to many a plant. Interestingly enough, only very few cases have been documented of catastrophic failures of 18–5 rings due to SCC. This is amazing, considering the many rings with cracks found during routine and/or special inspections. Nevertheless, SCC remains a serious risk due to the overwhelmingly negative effects of any retaining ring failing catastrophically.

Many plants replace their nonmagnetic 18–5 rings with 18–18 rings when the occasion arises due to the improved resistance to SCC of 18–18 rings. Many others keep running with their 18–5 rings. For those plants that do so and for those that plan to replace the 18–5 rings at a later stage, it is important to maintain a rigorous management of these rings. This mainly amounts to keeping moisture away from the rings during operation, during outages, and at standstill, and carrying out careful NDE examinations of the rings when the rotor is out of the bore. Some operators also remove the rings for NDE of the rings' internal-diameter (ID) surfaces, where most moisture tends to accumulate while the rotor is exposed to the environment. Figure 9.8 shows areas particularly prone to corrosion in the ID of a retaining ring.

Some utilities have dual-tower hydrogen dryers to allow regeneration of the desiccant in one tower while the other is in operation. Most also maintain effective monitor-

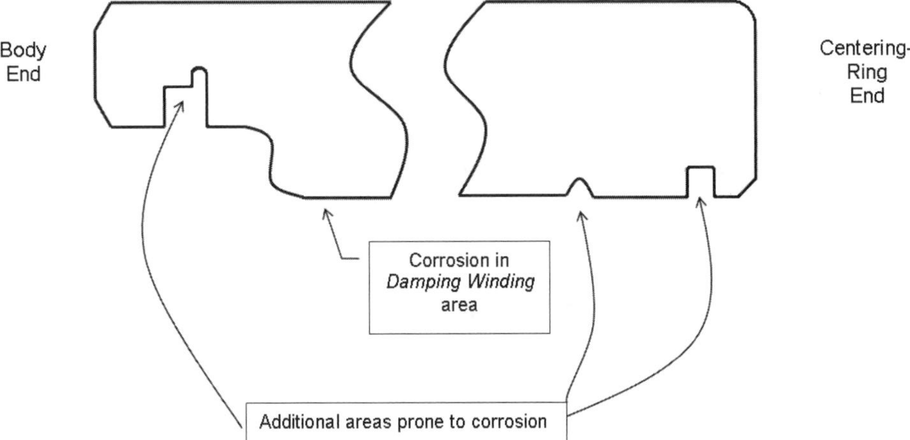

Fig. 9.8 Schematic representation of areas prone to corrosion on the ID surface of a body-mounted retaining ring, in particular, nonmagnetic 18–5 rings.

ing of the hydrogen in the unit while it is not in operation. Air-cooled units can take advantage of space heaters to maintain the air temperature above the dew point inside the machine while the unit is not running.

Given that retaining rings remain stressed while the unit is idle (while the rings are mounted on the rotor), stress corrosion cracking will occur under conditions where moisture is present. Thus, operators of units with nonmagnetic retaining rings have developed various techniques to keep moisture out of the rings during outages as well as during operation. Among the techniques or activities available are:

- Always placing the rotor under a tent while it remains removed from the bore. The tent should be subjected to heat lamps, hot-air blowers, or dry cool air.
- Not allowing rags soaked with water or any other liquid to rest on the retaining rings.
- Not allowing the tent canvas to partially rest on the rotor (points of contact are prone to condensation).
- Keeping dehumidifiers in the enclosure.
- Maintaining dry hydrogen (or CO_2 when present) in the unit while not in operation (e.g., long periods on the turning gear or just idle).

As stated above, there is a much literature on the subject. In particular, EPRI has followed this issue closely over the years and produced a significant number of informative publications. For those operators and inspectors who have in their plants or are involved with nonmagnetic 18–5 rings, References [2,3,4] contain a wealth of information, and are a good point of entry to a number of other EPRI and non-EPRI publications.

It should also be understood that the use of 18–18 retaining-rings does not negate the need to maintain and inspect them [5]. Although 18–18 rings have better resistance to the aqueous stress corrosion mechanism, they are still susceptible to it under conditions in which *halides* and *copper ions* are present. There have been cases of such cracks found in 18–18 rings [6,7,8,9].

Finally, one must remember that certain rotor designs include other components that are made of nonmagnetic 18–5 materials, for instance, *airgap separating rings,* also called *zone rings* and *airgap clearance rings.* These rings and the other 18–5 components, when present, must receive the same attention and care regarding inspections and moisture control as the larger retaining rings.

9.2.2 Removal of Retaining Rings

The removal of retaining rings and their reassembly onto the forging is one of the most critical maintenance activities in turbo-generators. During the process of removing or reassembling the retaining rings, a number of things can go wrong. First is the ring itself. As stated in other sections of this book, the retaining rings are the most highly stressed mechanical component of the generator. Thus, any damage to the ring, such as cracks, overheating, and galling in the area of the ring's shrink seating, can cause the ring to fail catastrophically during subsequent operation. Also, overheating of the ring in situ can damage the liner insulation under the ring, and may severely damage the winding insulation in the end winding. In fact, it has happened many times that an otherwise good winding had to be completely removed, cleaned, reinsulated, and reassembled because of excessive damage to the end-winding insulation during removal or reassembly of a retaining ring. Another concern is heating the ring too quickly, which can alter the metallurgical properties of the ring. Finally, the shrink-seating area of the rotor's forging can also be damaged by galling.

Given those risks, it is a very serious effort for a company, especially a utility, to remove the retaining rings. Some large utilities do have shops where these rings are removed, by their own personnel, by representatives of the OEM, or by third-party companies experienced in this type of operation.

Historically, the first method to heat and expand a retaining ring for the purpose of removing it or installing it was by using two single gas-flame torches at the same time, having one person on each side of the ring. This method caused a lot of aggravation. Many such operations have been mishandled because of the reasons stated above, due to uneven heating of the ring, and because of overheating. Eventually, the technique evolved into placing a number of torches on a ring that fits around the retaining ring. In this manner, a much more uniform heating was achieved. Figure 9.9 shows a sketch of this technique.

Following the torch ring, a better technique was developed by applying a heating blanket around the retaining ring. The system consists of an insulation covering on top

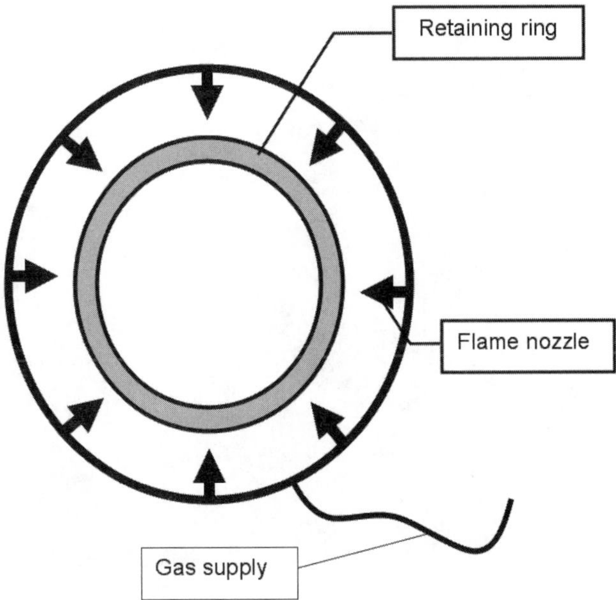

Fig. 9.9 Flame removal of retaining rings by gas torch ring.

of a ring of electrical heating elements that are wrapped around the retaining rings (see Fig. 9.10).

This technique allows a much higher level of temperature control than by the use of gas torches. Thermostats control the temperature of the retaining ring within narrow margins.

Finally, still a more efficient technique has been developed that uses induction heating. This method is based on an insulated conductor wrapped around the retaining ring, carrying ac current. An insulating blanket is also placed over the cables, so that the temperature of the ring is maintained. The application of an ac magnetic flux to the bulky retaining ring causes significant eddy losses that translate into a temperature rise. The benefits of induction heating are fine control of temperature, fast response, and no damage to the surface of the ring by applying direct heating. The weight of the cables wrapped on the ring is considerably less than that of the thermal elements; therefore, the ring can be manipulated with the cables and wrapping left on it while installing it onto the forging, or while removing it. Thus, the temperature of the ring decays slowly, giving more time to the personnel involved with the operation to do it right. Figure 9.11 shows a retaining ring being removed by induction heating.

It has been pointed out above that the removal or installation of a retaining ring is best left to the professionals. In fact, although both removal and installation of retaining rings carry significant risks to the components, the most difficult activity is that of installing the ring over the end winding packed with insulation, amortisseurs, and the insulation liner. The trick is to heat the ring enough so that a manageable

Fig. 9.10 Retaining ring of a turbo-generator's rotor being prepared for its removal by electrical heating.

Fig. 9.11 Retaining ring of a turbo-generator rotor being removed by induction heating. At present, this is probably the most effective technique to safely remove retaining rings without damaging the ring or the insulation of the field winding.

clearance is maintained between the shrink surfaces of ring and forging while avoiding overheating it. That clearance is generally about 0.030″. As the temperature decays, once the source of heat is removed, the operation must be done carefully but relatively quickly. Otherwise, the heating process might have to be repeated, including reinstalling the temporary steel bands that hold the end winding tightly together. This is true in all cases. Nevertheless, there are techniques that make the operation simpler. For instance, specialized shops (such as OEMs or other large outfits) may have a large apparatus that holds the retaining rings in the right position and at some axial distance from the rotor forging. As the ring reaches its proper temperature for mounting, the apparatus moves on rails so that ring comes exactly onto the end-winding and forging sitting areas.

Other techniques, based on special articulated clamps, certainly required experienced personnel to avoiding reheating the ring several times. Figure 9.12 shows a set of articulated clamps, and Fig. 9.13 shows the ring hanging by the clamps.

One additional problem that can happen when removing a retaining ring is causing unforeseen damage from some defect already present but unobservable without the ring being removed. Such a situation could be due to previous damage on the rotor forging shrink-fit surface (Fig. 9.14) or negative-sequence or motoring current flowing from the shrink fit of the forging to the retaining ring, causing arcing and fusing of the two components. When the retaining ring is pulled axially to remove it, the fused point

Fig. 9.12 Retaining ring mounted on the forging showing the articulated set of clamps. The clamp mounted on the forging is used as an anchor to push the ring off the forging. The clamp mounted on the retaining ring is used to lift the ring for re-assembly onto the forging or to hold it during removal.

Fig. 9.13 Retaining ring held with articulated clamps as it hangs from the crane.

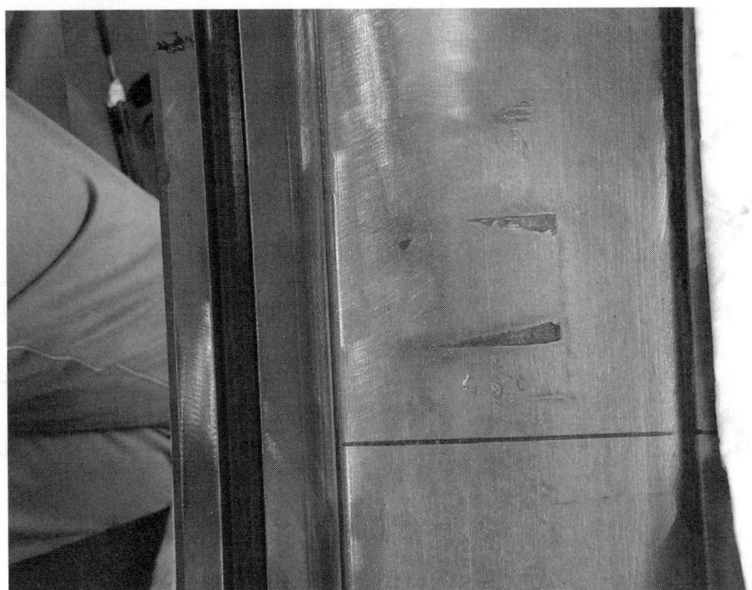

Fig. 9.14 Retaining ring with galling damage that occurred due to removal. The rotor had previous damage to the shrink-fit surface, and removal of the ring made the situation worse. The remedy in this case was to machine the inner surface to clear the galling marks. The amount of machining allowable is that amount that still allows the rotor to be capable of 15% overspeed capability. Anything less and the ability to do overspeed tests would be compromised and could lead to the danger of the ring losing its shrink seating, and floating off the forging during operation.

breaks and creates a jagged surface and debris. This can cause galling of the retaining-ring shrink-fit surface, as well as damage to the forging shrink-fit surface.

Although removing the retaining ring safely has its difficulties, reinstalling it is a far more complex operation because now the end winding must be secured with steel bands that are removed one at a time, as the ring slides over it. The steel bands are temporarily installed so that the end winding is compressed to its operational dimension. Without these bands, the retaining ring cannot be installed.

Some designs are simple; the retaining-ring liner and the damper winding are secured under the steel bands and the retaining ring is a single piece that must be moved over the end winding. In other designs, part of the damper winding must be first placed inside the retaining ring, and held there by some type of device until the ring goes over the end winding. There may be several damper winding parts inside the ring. The entire device is now more difficult to assemble onto the forging and great dexterity and experience is required from those engaged in this operation. Figures 9.15 and 9.16 show such a retaining ring and its attachments.

Figure 9.17 provides a good look at what can be seen when the retaining rings of the previous figures are removed.

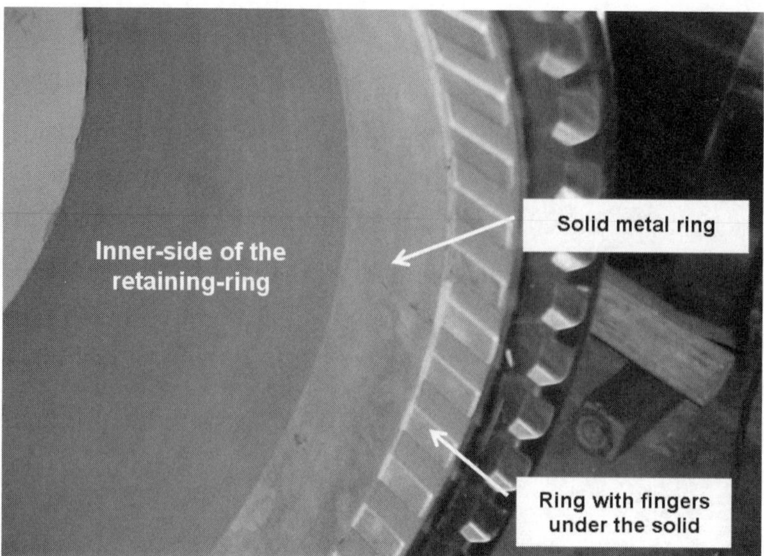

Fig. 9.15 Retaining ring being readied for reassembly onto the rotor. The photo shows the ring with fingers (when the retaining ring is mounted, the fingers are in contact with the top turns of the winding, providing a closed electrical circuit from side to side of the rotor). The ring with fingers actually is made of a number of segments held together by the other ring shown in the figure. This contraption must be held securely when reassembling the retaining ring onto the forging, as it goes over the end winding. The next figure shows how it is held together.

Fig. 9.16 The same retaining ring of the previous figure, showing how the damper winding is kept together during reassembly of the retaining ring onto the rotor body.

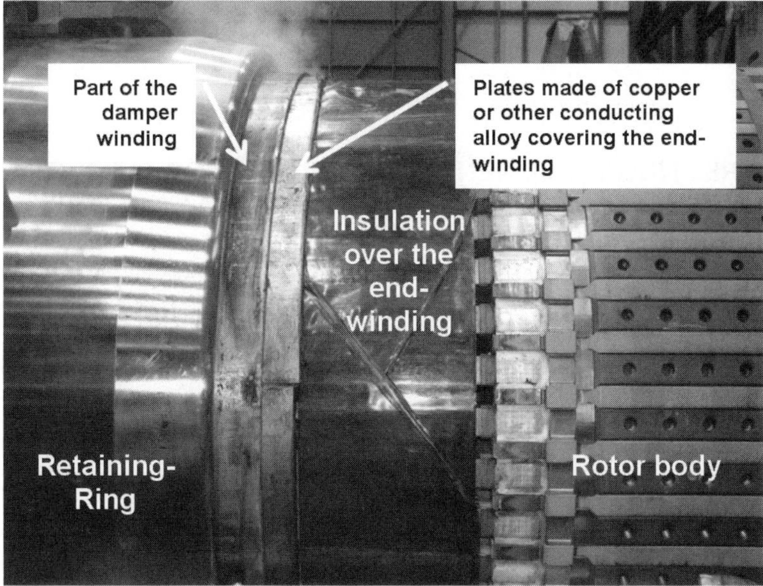

Fig. 9.17 View of the retaining ring of the previous figure being removed. The photograph shows some of the components that go under the retaining ring. These components are shown not in their original position (covering the end winding), but shifted toward the direction in which the retaining ring is removed.

Formula for Retaining-Ring Growth with Temperature. Regardless of the method of removing a retaining ring, heat is either applied to or induced in the ring to make the ring diameter grow such that it will fit onto the forging over the shrink-fit area of the teeth. The problem is that not all rings are the same size and not all rings use the same shrink-fit dimensions. To install or remove a retaining ring, one has to know the actual shrink-fit dimension for the rotor in question; then the temperature to heat the ring to such that it will go on or come off can be calculated. The equation to calculate this is as follows:

$$\text{Temperature, } T = \text{ambient} + \Delta T = \text{ambient} + \frac{\Delta l + 0.030''}{l \times k}, \text{ degrees C}$$

where

ΔT = additional temperature above ambient to achieve shrink-fit growth size

Δl = shrink fit of retaining ring

l = inside diameter of retaining-ring shrink-fit area

k = thermal expansion constant

= 13×10^{-6} for 18–18 rings

= 11×10^{-6} for 18–4 or 18–5 rings

Note in the formula that an additional 0.030″ is added to the Δl (shrink fit) of the ring to allow for space to install or remove rings without causing retaining-ring liner damage.

9.3 FRETTING/MOVEMENT AT INTERFERENCE-FIT SURFACES OF WEDGES AND RINGS

Look for fretting or other movement signs between the contact surfaces of shrink-fit components. Signs of fretting or movement could indicate excessive heating of the shrunk member or abnormal forces experienced during the operation of the machine. Examples are cracks in the shrunk member and insufficient interference fit. Given the serious consequences of failure of any rotating element, prompt attention should be given to any such sign of distress.

9.3.1 Tooth Cracking

There are two basic designs for retaining rings: the spindle-mounted design and the body-mounted design. Each design has characteristic weak points demanding close examination during an inspection. Figure 9.18 shows the basic retaining-ring designs and their weak points.

It is important to note that when removing a body-mounted retaining ring, the inspector should look for the tooth top-cracking phenomenon. This alludes to when the top part of a rotor tooth directly subjected to the pressure exercised by the shrunk retaining ring cracks or breaks due to the mechanical stress to which it is subjected. In severe cases, parts of the tooth have been known to come loose during removal of the retaining ring (see next section). Some designs are more prone to this phenomenon than others. Whenever a body-mounted retaining ring is removed, however, the inspector should consider looking for cracked rotor teeth.

Inspections for tooth cracking require removing the retaining rings and partial removal of the wedges. Given the costs and risks to the rotor associated with these activities, they are mainly reserved for those units that are in one way or another prone to these phenomena, as learned from experience within and without the plant, or from OEM technical information bulletins and/or recommendations.

Tooth-Top Cracking. With certain older designs of body-mounted retaining rings, the area of the rotor teeth between the groove for the ring key and the end of the rotor was prone to the development of cracks (Figs. 9.19 and 9.20). Many times, the upper sections of the teeth broke off and came apart when the rings where removed for inspection. The causes of this type failure are attributed to a number of weaknesses in the design. In particular, the design of the key groove and shrink-fit area stresses. Many rotors in which these types of cracks were found could be fixed and returned to operation. The fix is not trivial and requires a substantial outage and expense. Reference [4] provides a good entry point to literature on this subject. Bear in mind that those vendors whose rotors have experienced these problems are well versed by now

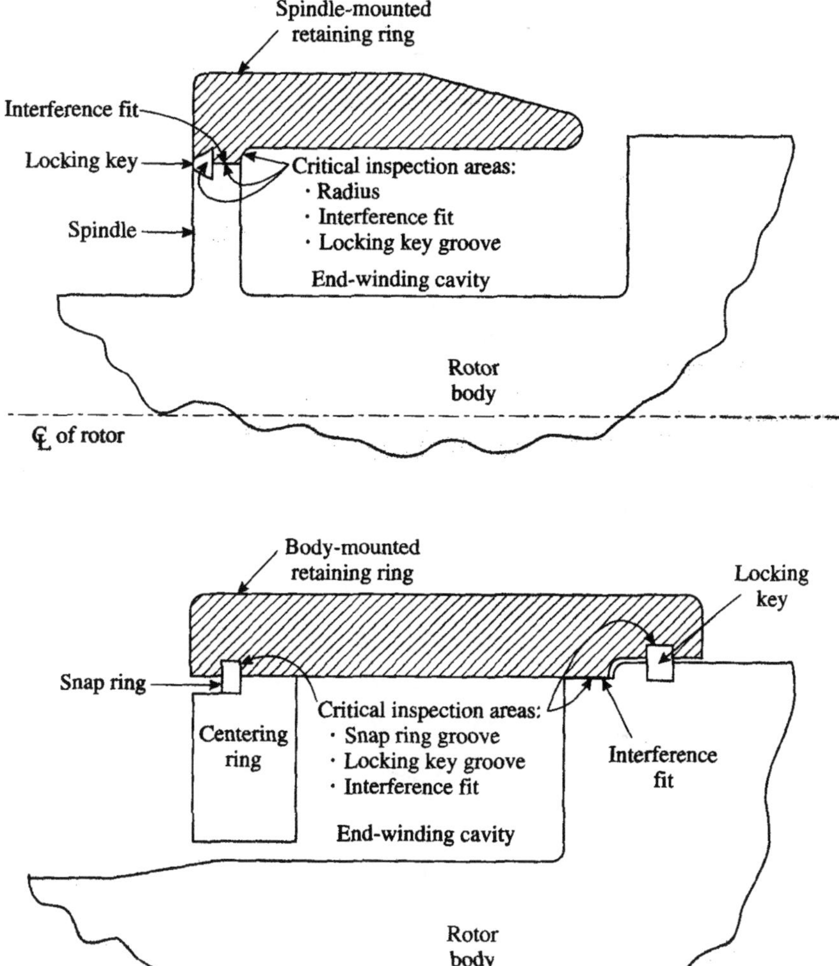

Fig. 9.18 Schematic representation of typical spindle-mounted and body-mounted retaining rings.

on how (if possible) to repair them. Thus, it is highly recommended to work together with the OEM in addressing such a problem.

Fretting-Fatigue Tooth Cracking. Another phenomenon that results in damage to the contact area between rotor wedges and the slot area is due to cumulative fretting fatigue. In general, the cracks are on the forging in the area close to the wedge ends. This phenomenon is practically exclusive to short steel wedges. The main mechanism iden-

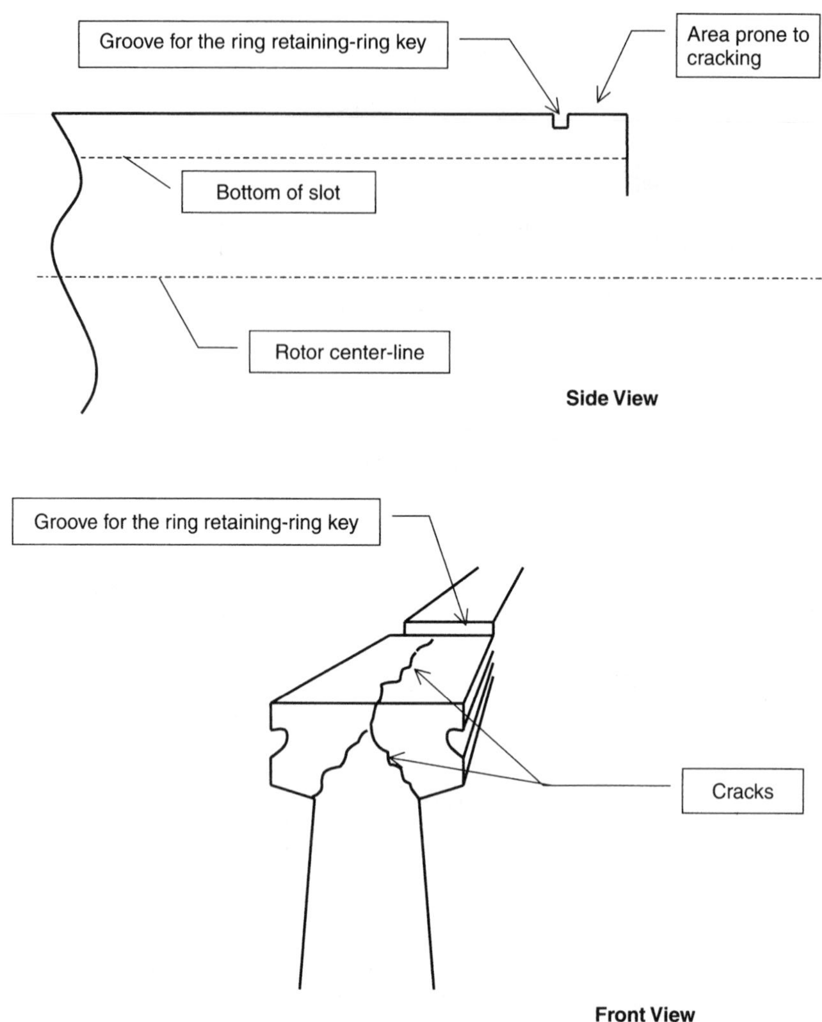

Fig. 9.19 Schematic representation of typical tooth-top cracking.

tified as responsible for the development of this type of failure is the cumulative effects of fretting of the short steel wedges against the slot due to start/stop operations, turning-gear operation, and cyclic loading. For reasons of weight, elastic modulus, and other physical characteristics, aluminum wedges do not result in a similar problem.

Fretting-fatigue-originated cracks are normally repairable by grinding of the affected area. Replacing steel wedges with aluminum wedges, when possible, basically eliminates further development of new cracks.

In many designs, the first slot next to the pole face (see Fig. 9.21) has steel wedges, whereas the rest have aluminum wedges. The steel wedges are introduced there to ef-

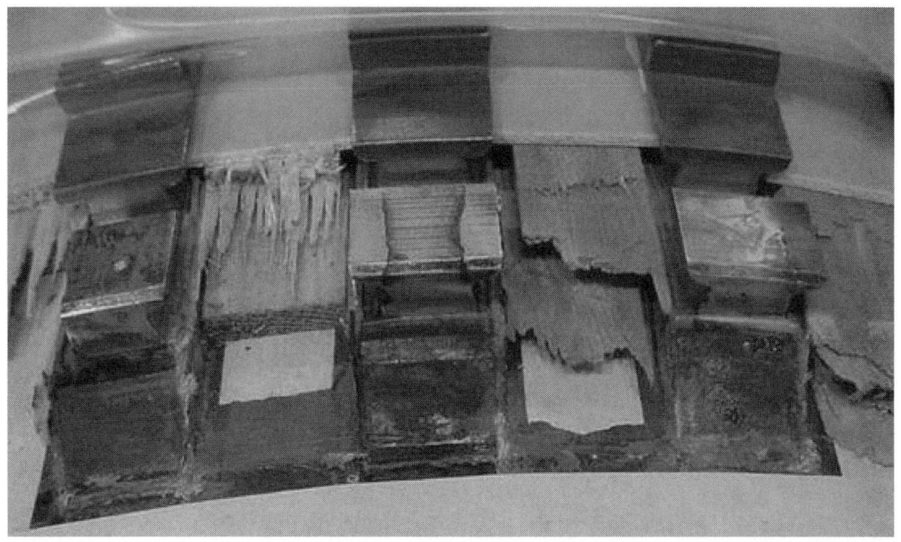

Fig. 9.20 Tooth-top cracking in a two-pole, 60 Hz, hydrogen-cooled generator.

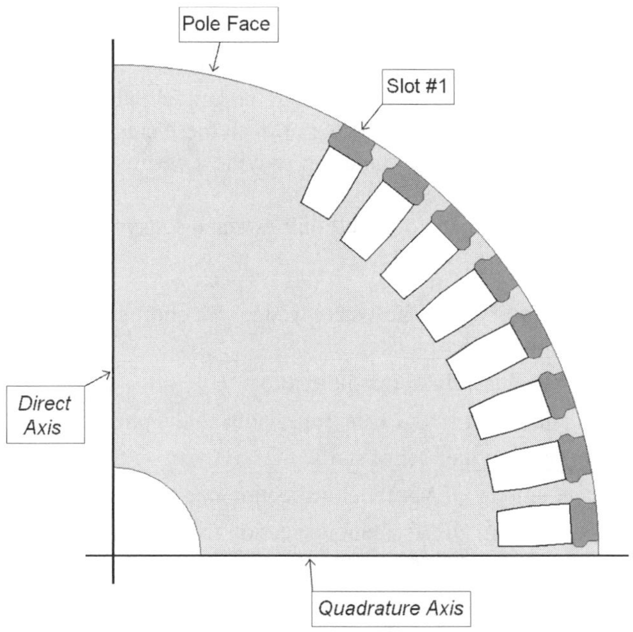

Fig. 9.21 Rotor quadrant showing position of slot #1.

fectively increase the width of the pole face, for magnetic reasons. Fretting fatigue between slot and wedge will happen then on the first slot, the one with the steel wedges. One change targeting the elimination of fretting-fatigue crack initiation is the one that replaces the steel wedges in the first slot next to the pole face with aluminum wedges. Studies done by several OEMs show that replacing the steel wedges with aluminum wedges increases the excitation requirements by only up to 1% while introducing negligible changes to the performance of the unit. An alternative approach is to replace all short steel wedges with long, full-length aluminum wedges. This approach also eliminates the fretting-fatigue issue while introducing very small changes to the originally designed performance characteristics.

This type of problem has not normally posed a great risk of catastrophic failure for most rotors. It has been mainly discovered and taken care during rewinds of the field winding, or when other winding repairs require the removal of the retaining rings and wedges. One is also usually alerted to it by high rotor mechanical vibration once the crack becomes significant. However, for those machines that are subjected to many start/stop operations, cyclic loading, and long periods on the turning gear, plant operators should consult with the OEM about establishing an inspection plan addressing these concerns. For instance, during one of the major planned outages, the rotor could be partially dismantled to allow an inspection of the areas of the slot prone to fretting-fatigue cracking.

The in-depth research of the fretting-fatigue-cracking phenomenon was originally carried out in the United Kingdom during the 1970s. During that time, it was found that some fretting-fatigue cracks grew so big that the cracks were almost halfway through the rotor forging in a few rotors. The subject rotors that were affected so dramatically all had short steel components in use in a rotor slot, and they had a very high length-to-diameter ratio, with high inherent bending stress in the rotor body. Nowadays, research on this subject is also pursued in the United States by some vendors because of the large number of units showing up with these cracks. References [5-8] are a good introduction to the problem, and can provide a wealth of additional references for those interested in this particular problem.

The following is a summary of where things stand today regarding wedge-to-slot fretting fatigue:

- Found only on rotors with short steel wedges (or short steel components such as magnetic pole-face filler blocks).
- Found almost exclusively in two-pole rotors.
- Rotors with aluminum wedges are impervious to the phenomenon.
- Depends strongly on number of starts.
- Disagreement among OEMs about inspection intervals.
- Disagreement among OEMs about inspection techniques.
- Disagreement among OEMs about crack propagation.
- Phenomenon first studied in detail in the U.K. (at Parsons) many years ago.
- Fretting-fatigue threshold sharply decreases with decreasing wedge-to-wedge contact gap (on the dovetail load area).

- Filleted or tapered contact surfaces of wedges at their ends reduces fretting.
- Full-length wedges eliminate pressure points and relative movement, eliminating fretting.
- Fretting-fatigue threshold decreases with increasing hardness of wedge material (therefore, aluminum wedges are less prone to create fretting than steel wedges).
- Better forging manufacturing and slot machining result in lower residual stresses and less crack growth after initiation.

Figures 9.22 and 9.23 show two examples of slot fretting-fatigue indications.

There is a well-developed body of theoretical analysis explaining this phenomenon. The original work, done by one OEM in the U.K. in the 1970s, has been enhanced and supported by recent studies carried out with modern finite-element analysis (FEA) and fracture-mechanics analysis. Although these analyses are quite complex, placing them outside the scope of this book, it is interesting to note that there is a common denominator among the OEMs regarding some of the main mechanisms of crack initiation, yet there are disagreements about crack propagation. Most agree that most fretting occurs with the unit accelerating or decelerating, when there is more movement between wedge and slot. Also, there is agreement that negative-sequence currents can harden the steel in the area of contact and be a factor, in some instances, of crack initiation due to fretting fatigue. However, the well-documented initial number of cases of fretting fatigue in which the rotors sustained a catastrophic failure had no initiation other than pure fretting fatigue of short steel components on a steel forging. For rotors that are deemed susceptible to fretting fatigue, one must be careful to also inspect those rotors

Fig. 9.22 Fretting-fatigue marks in area of slot dovetail.

Fig. 9.23 Fretting-fatigue marks in area of slot dovetail.

for this phenomenon, even when there are no operating events that may influence the problem. Temperature differentials in the rotor forging are different when the unit ramps up in speed than when it coasts down. Some speculate that most of the stress for crack initiation and growth happens when the unit is on its way down. All agree that residual stress from the manufacturing process in the rotor teeth is critical for crack propagation. Figures 9.24, 9.25, and 9.26 schematically show some of the strain introduced into the slot dovetail area during upload and download of the unit.

When fretting marks, cracks, or other indications are found, there is always the issue of what to do about them. The reality is that the large majority of indications and cracks found do not propagate or they do so very slowly and may not become a life-limiting factor for the forging. However, most agree that it is prudent to remove them, regardless if they appear to be dormant and not propagating. A couple of approaches have been taken to remove any damage found. One is to simply blend or grind out the indication if it is small or to make a series of machining cuts to remove the cracks (Figure 9.27). Obviously if the crack is very large, then further evaluation is required to determine if it can be removed so that the rotor forging will still be operable, or if the forging has to be retired.

One final issue is the additional excitation required due to removal of any steel wedges that are made of magnetic material. Another concern is the effect on the waveform, but it is unaffected by this change. Replacement of steel wedges with aluminum wedges has been done on many rotors since the late 1970s and operation has continued

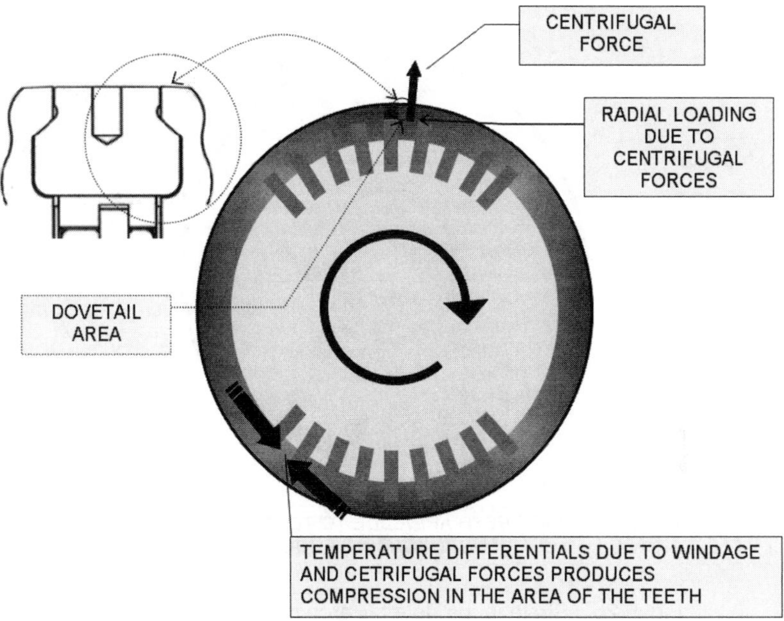

Fig. 9.24 Strains in the rotor during acceleration.

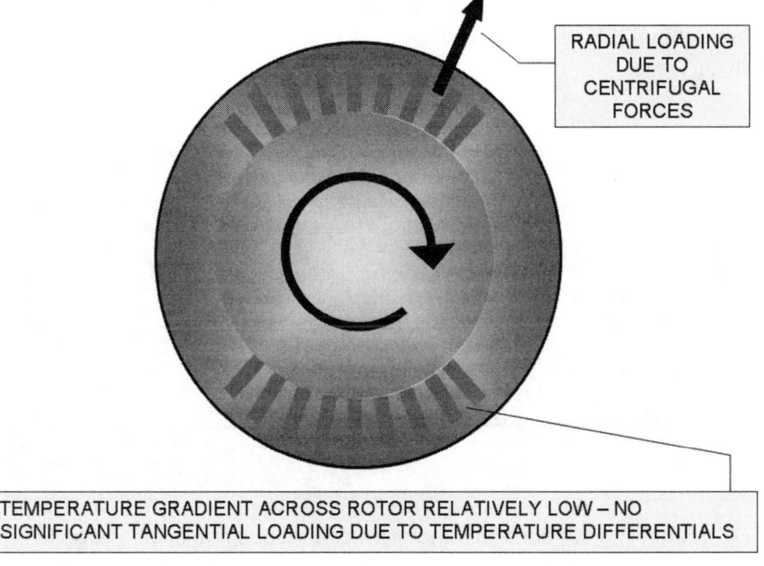

Fig. 9.25 Strain in the rotor tooth during steady-state operation.

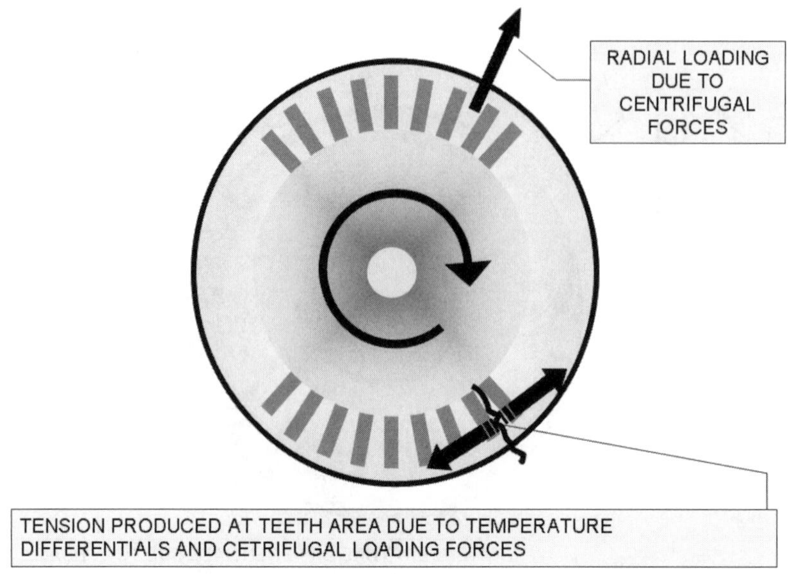

Fig. 9.26 Strain in the dovetail area during deceleration.

since then with no problems. The main effect is an increase in excitation of less than 1% on load and up to about 3% on open circuit. The open-circuit condition is no issue because the excitation level is always well within exciter capability, and the on-load condition, which requires an additional 1% excitation, only comes into play if one operates out to the full lagging power factor. Most exciters have this margin and it is not generally an issue.

Fig. 9.27 Machining cut done to remove fretting fatigue marks on a large two-pole rotor that had numerous indications found. None were deemed to be life limiting but all the indications found were machined out. The steel wedges in the affected slot were replaced with aluminum wedges.

9.4 CENTERING (BALANCE) RINGS

Retaining rings are shrunk onto *centering* or *balance rings,* or onto the forging (spindle mounted; see Fig. 9.13 and the bottom panel of Figure 9.18). The interference surfaces include arrangements for the presence of *locking keys* and *locking-key grooves*. These are areas of concentrated stress. They should be inspected for stress-related cracks and other anomalies. The shrink area of the centering ring should be subject to NDE tests when the retaining rings happen to be removed.

9.5 FAN RINGS OR HUBS

In two- or four-pole machines, one arrangement of the rotor-mounted fans has the fan blades attached to fan rings or hubs shrunk onto the shaft (Figs. 9.28 and 9.29). In other arrangements, the fans are of the centrifugal type of design (Fig. 9.30). Either type requires attention to possible cracks forming in the shrink areas where the fan hub attaches to the shaft. Occasionally, the need for removing a retaining ring requires removing the rotor-mounted fans. These events provide an opportunity for a good visual and NDT inspection of the shrink areas and shrunk components. Also, when centrifugal fans are employed, the fan vanes are sometimes riveted to the fan plates and these areas should be NDE inspected for cracks. Vanes have been known to come off from

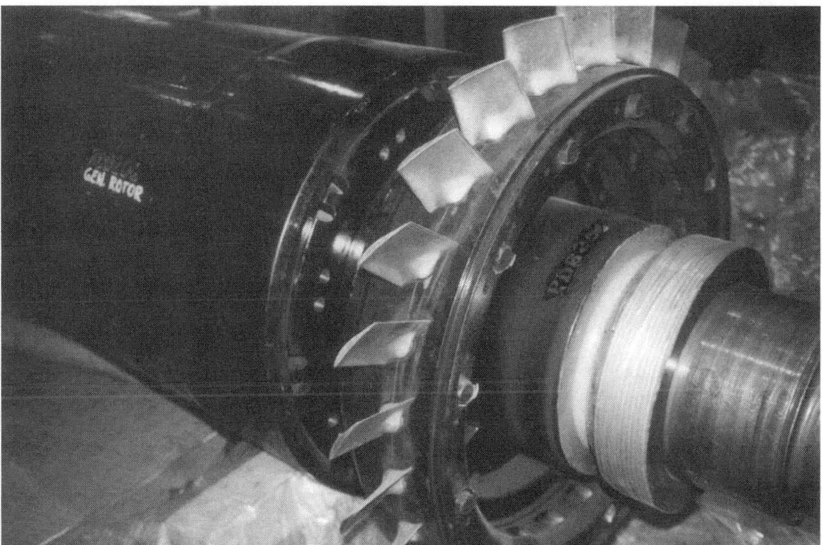

Fig. 9.28 Fan ring and fan blades of a two-pole cylindrical rotor. In this design, the diameters of the fan ring and blades are larger than the diameter of the bore. Thus, the rotor can only be removed in the direction of the fan unless the fan is disassembled prior to removal of the rotor. Compare this design with the one shown in the following figure.

Fig. 9.29 Two-pole cylindrical rotor with its axial fan having an overall diameter smaller than the diameter of the retaining rings.

Fig. 9.30 Two-pole rotor sporting a centrifugal fan.

fatigue of the rivets or tenons used to hold them in place. Figure 9.31 shows a typical fan ring of a two-pole, 180 MVA, hydrogen-cooled unit.

9.6 FAN BLADES

Two-pole and four-pole machines tend to have axial fans or centrifugal fans on their rotors, as stated above. In either case, the fan blades might be an integral part of the fan hub or, as with most axial fans, be either welded or bolted to the fan hub (Fig. 9.32). Axial-fan blades must be inspected in the area where they attach to the hub. Centrifugal-fan blades or vanes must be inspected for cracks in the area where they are welded or riveted to the lateral plates, as mentioned in the previous section. In all cases, interference-fit stresses and/or centrifugal forces induce mechanical stresses in those elements (Fig. 9.33).

It is important to keep in mind that in any of the many versions of fan blade attachment, it is very important to examine the condition of the blades and the soundness of their attachment. Cracks at the root of axial-fan blades are not uncommon. If bolted, attention should be given to snugness and nut-lock condition. It is, therefore, important during a major inspection to assess their mechanical integrity visually and, when the situation requires, by NDE techniques.

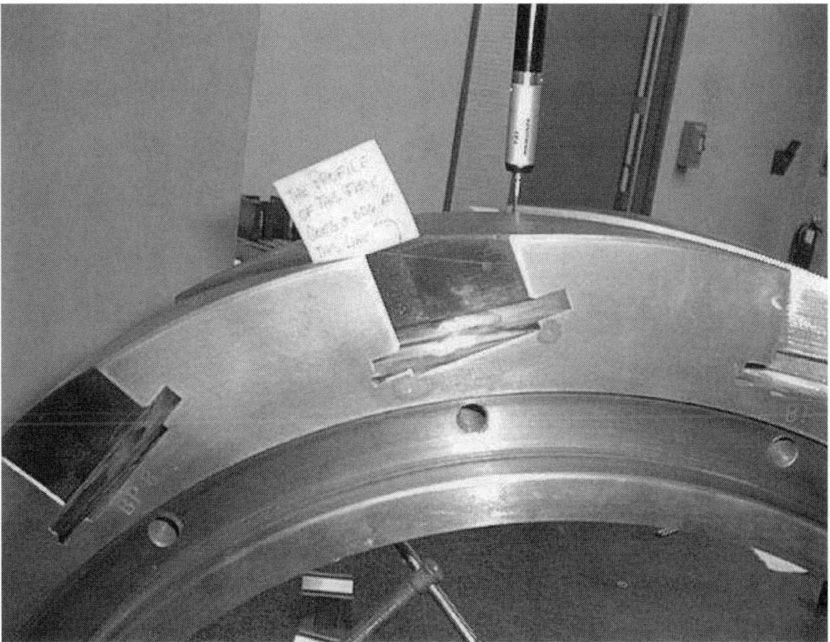

Fig. 9.31 Fan ring of a two-pole, 180 MVA, hydrogen-cooled alternator.

Fig. 9.32 Axial-fan blades removed from the fan ring or hub for inspection and NDE. Note that the blade attachment is a circular fit, allowing the blade angle to be adjusted if required.

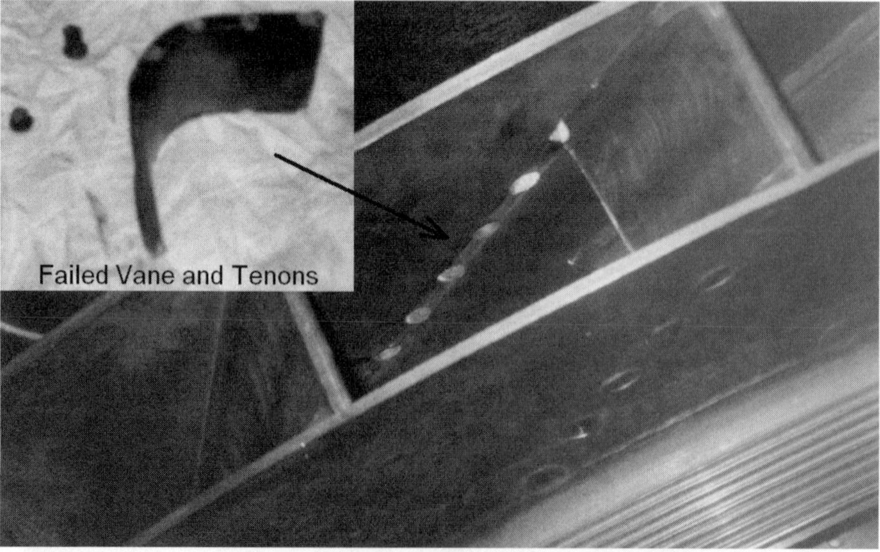

Fig. 9.33 The fan in this photo shows a vane that has failed at the rivets or tenons and come off the main assembly. The vane became a missile and struck the stator end-winding supports, causing significant damage to the stator as well.

Figure 9.34 shows the unusual double-axial rotor-mounted fan arrangement and the stationary guide fins of a two-pole unit built in the 1960s. This type of design is known for presenting challenging stresses to the larger-diameter row of blades. A unit with this type of arrangement ought to have its larger-diameter row of blades carefully inspected, specifically at the roots.

9.7 BEARINGS AND JOURNALS

All bearings for the size of generators discussed in this book are of the *friction* (*sleeve*) type. However, the actual design of the bearing can be very different for a number of characteristics. This diversity makes it impossible to describe them all here. Nevertheless, the following comments apply to any type of bearing.

Inspection of the bearing journals should be an integral part of any extensive machine inspection. Such inspection is, of course, combined with inspections of the babbitt metal and other bearing components, as well as the inspection of the oil-baffle labyrinth, and measurement of oil-seal ring clearances and bearing clearances.

Inspection of the used oil may shed some light on the condition of the bearing. Dirt, discoloration, and acidity are reasons for a change in the oil. Any excessive abnormal condition should be investigated for its root causes. Metal particles in the oil are an indication of bearing problems.

Fig. 9.34 Rotor-mounted double-axial-fan arrangement plus stator-mounted stationary row of gas guides, on a two-pole, hydrogen-cooled turbo-generator. The stationary gas guides can be seen behind the larger-diameter row of moving blades.

When operated under design temperature and load conditions, sleeve bearings are a very reliable component of the machine. However, severe vibration, lack of sufficient flow of oil, deficient cooling, and external pollutants such as foreign materials and shaft currents can result in bearing failure. In Chapter 8 covering stator components, it was noted that shaft voltages and currents can be detrimental to bearings. The existence of shaft currents tends to show up as minuscule spots of melted material. When viewed with the aid of a magnifying glass, depending on the angle of the light, they may appear as tiny rounded pits. It is, therefore, recommended to look for signs of electric pitting of the bearing-journal surfaces at every opportunity when visual access to that area is available. Foreign materials tend to become embedded in the babbitt. Figure 9.35 shows the babbitt metal and half-bearing of a two-pole unit. Figure 9.36 shows same babbitt metal being "scraped" prior to its installation in the unit. The need for this operation may arise from severe heat staining if a bearing has been overloaded or misaligned (Fig. 9.37).

Polishing the journals and resurfacing (rebabbitting) the bearings are common operations carried out during major overhauls. If very severe damage has been incurred, complete rebabitting may be required.

An additional activity related to bearing inspection is checking the integrity of *bearing insulation* and *grounding brushes.* Figure 9.38 depicts a full view of a bracket-mounted bearing.

Working with large friction bearings is a delicate task requiring adequate expertise. It is thus highly recommended that only individuals with the proper experience and

Fig. 9.35 Section of a bracket-mounted bearing from a two-pole, 180 MVA, hydrogen-cooled generator.

Fig. 9.36 Section of a bracket-mounted bearing from a two-pole, 180 MVA, hydrogen-cooled generator having its babbitt metal scrapped in preparation for installation after overhaul.

Fig. 9.37 Heat-stained lower half of a main generator bearing.

Fig. 9.38 Full view of a bracket-mounted bearing.

knowledge be assigned the task of inspecting and refurbishing the bearings of a large machine.

For the reader interested in state-of-the-art developments in the realm of bearing applications for large machinery, EPRI has been sponsoring projects conducive to the application of magnetic bearings in large electric rotating machinery. When (if ever) this technology becomes widely implemented, things will never be the same in the areas of bearing operation and maintenance.

9.8 BALANCE WEIGHTS AND BOLTS

In large machines, balance weights, bolts, nuts, and any other rotor attachments are subjected to intensive centrifugal forces. For example, a 2 ounce nut will exert a centrifugal force of over 1000 pounds when attached to the rotor belonging to a typical two-pole machine. In addition, the constant vibrations and thermal cycles tend to work these loose. A weight, nut, or bolt that has broken loose under these conditions may result in a major failure of critical components of the machine. For these reasons, these elements are secured with many types of locking mechanisms or staking arrangements. It is important during the inspection of the unit to verify that all such nuts, bolts, and weights are secure and all locking devices are in order. This is particularly necessary when work has been performed on the rotating member during an overhaul.

Balancing weights may come in many shapes and be placed in a number of locations on the rotor body and/or components. For instance, balancing weights may be located on the circumferential ring, just underneath the shrink area of the retaining rings. Additional weights may be located on the collector-ring structure (see Fig. 9.39).

Balancing bolts can also be found along the forging on the pole faces. These bolts are part of the factory balancing, and it is best to leave them undisturbed during major overhauls (e.g., a rewind) done in a setting other than a shop with a balancing pit. The inspector should ascertain that those bolts are secured in their original position (by looking at the pinlocking or staking points placed during balancing). See Figures 9.40 and 9.41 for examples of balancing bolts on forging.

9.9 END WEDGES AND DAMPER WINDINGS

During unbalanced load conditions, as well as during system oscillations or other types of abnormal operation, alternating currents are established in the body of the rotor. These currents tend to flow along the rotor body (poles and teeth), along the wedges, and in the end bells (retaining rings). This was discussed in Chapter 4 (see Fig. 4.19).

During the bridging of high-resistance contact areas, the currents can give rise to very localized pitting of the metal. These high-resistance areas are mainly found in the contact surfaces between the wedges and the slot, between different wedges, and between wedges and retaining rings. Figure 9.42 shows one example of a wedge that

Fig. 9.39 Balancing weights on the structure of a collector ring.

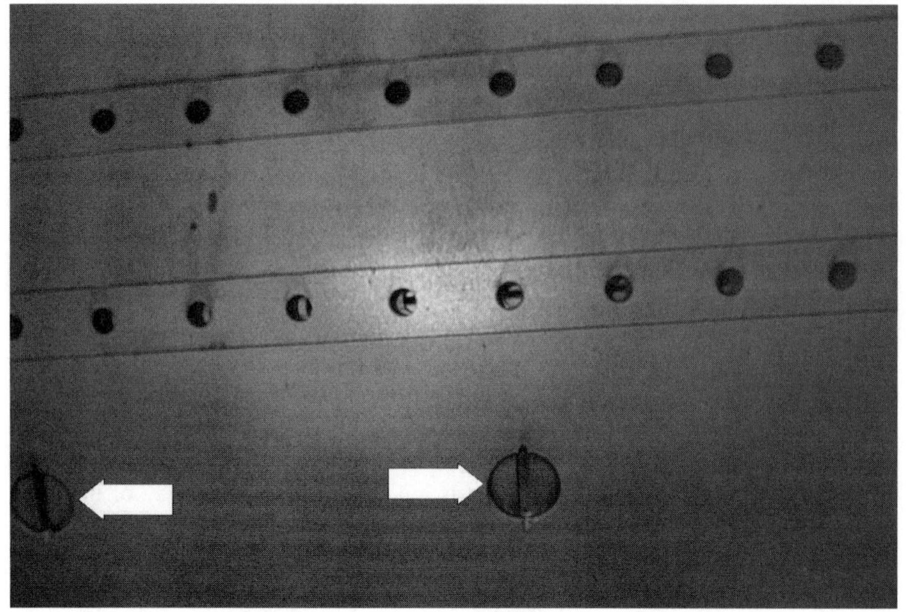

Fig. 9.40 Balancing weights on the pole face of a two-pole rotor.

Fig. 9.41 Balancing weights on the pole face of a two-pole rotor.

Fig. 9.42 Wedge that migrated toward the retaining ring until it touched it. Skin currents due to abnormal operation events or power swings resulted in electric pitting of the contact area. Both the wedge and the retaining ring need to be inspected for damage and an evaluation of the required repairs made.

has migrated toward the end bell; rotor-body currents resulted in pitting of the wedge.

It is possible that severe pitting of the end bells can lead to catastrophic failure due to crack growth initiated from pitting; it is for this reason that the end wedges are kept from touching the end rings in most machines. Although normally pin locked, wedges tend to migrate toward the end rings during the operation of the machine. If this situation is encountered during inspections with the rotor removed, the offending wedges should be driven back away from the rings. Figure 9.43 presents a view of wedges kept about 2 millimeters from touching the retaining rings. Figure 9.44 shows the resulting gap between wedges when the end wedges migrate toward the retaining rings. Figure 9.45 is a view of a wedge with a full length of the slot designed to touch the retaining ring.

After severe motoring or generation with the field off (induction-mode operation), inspection of the rotor should always include inspection of the wedge-ring contact area as well as between the wedges themselves and between wedges and the rotor body. It is important to note that any sign of burning should be carefully investigated. (See Figs. 9.3, 9.42, and 9.49 for damage due to electric currents between wedges and retaining rings, and see Fig. 9.46 for damage to wedges and forging due to electric currents between the wedges and the forging.)

Fig. 9.43 Narrow strips in the figure (two-pole, 80 MVA unit) are the end wedges. The wedges are touching the end bell. (From Kerszenbaum, *Inspection of Large Synchronous Machines,* 1996, IEEE Press, reprinted with permission.)

Fig. 9.44 Two-pole generator in which the narrow strips are the wedges (with numbers on them). The various gaps indicate the amount of migration toward the end rings by the end wedges. (From Kerszenbaum, *Inspection of Large Synchronous Machines,* 1996, IEEE Press, reprinted with permission.)

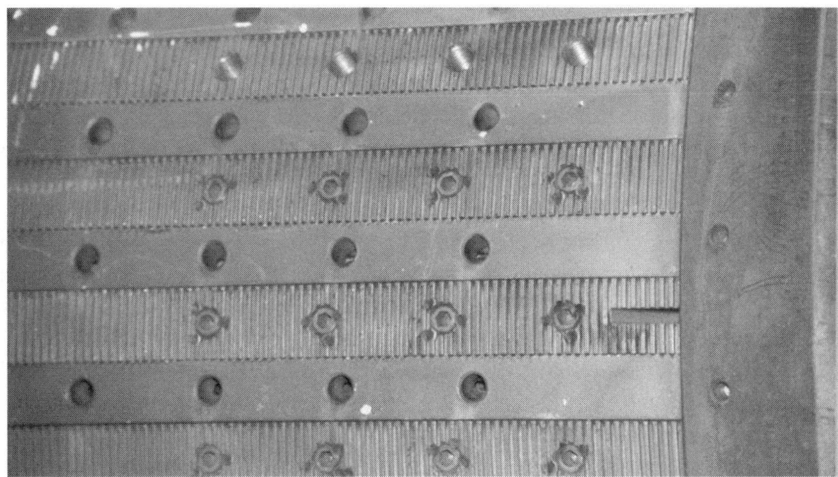

Fig. 9.45 Turbo-generator rotor with one aluminum wedge per slot. In this machine, the wedges are designed to touch the retaining rings.

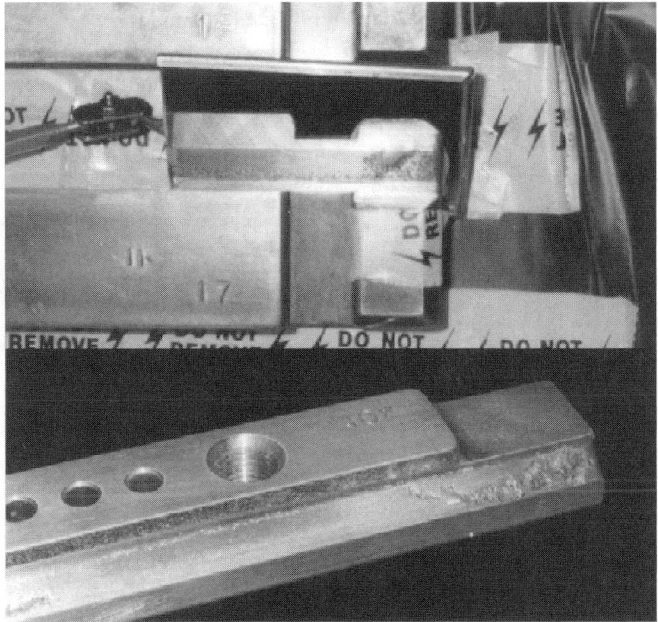

Fig. 9.46 Arcing damage incurred between the rotor forging and the end of a full-length aluminum wedge due to low-speed motoring of a cross-compound machine during an attempt to pull the low-speed rotor to speed by magnetic coupling to the high-speed machine. A mismatch in turning gear speed between the two rotor lines caused low "slip" frequency currents to flow in the easiest path available.

Some manufacturers include a damper winding in two- and four-pole turbo-generators. These are designed to minimize the adverse effects of induced currents in the rotor. Damper windings (amortisseurs) were already discussed in Chapter 2. Figure 9.47 shows an amortisseur winding of the copper sheet type and the contact between the amortisseur and the end wedges designed to carry the skin currents. Despite the existence of a damper winding, these rotors should also be inspected for signs of induced currents. These may damage the forging or other components due to failure of the amortisseur winding to carry all the induced currents (e.g., due to higher-than-normal contact resistance to the end wedges). Figure 9.48 shows the finger area of the amortisseur with the wedge removed. This area ought to be carefully inspected for cracks or other damage whenever the end rings are removed for whatever reason.

Other manufacturers design their rotor wedges to be equal to the full length between the end bells and to be in contact with them (see Fig. 9.45). These wedges, made of aluminum and short-circuited by the retaining rings, are designed to conduct the negative-sequence currents (when present), providing adequate damping of rotor oscillations.

9.10 OTHER WEDGES

Check all wedges for discoloration, which indicates that the machine has been operated under abnormal conditions. Check for cracks and excessive looseness. If the clearances between wedge and slot appear to be excessive, contact the OEM for recommen-

Fig. 9.47 Two-pole, 160 MVA rotor with a copper-sheet amortisseur. The fingers of the amortisseur are in contact with the end wedges.

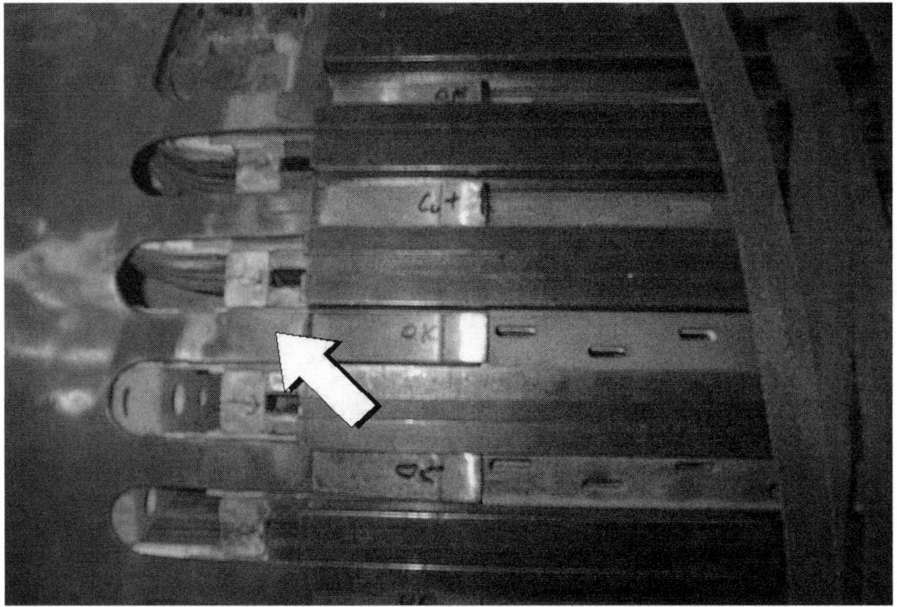

Fig. 9.48 Damping winding of the previous figure. Its fingers are shown in the slot area with the end wedges removed. This is required for a good visual inspection of the fingers. Cracks in the fingers can develop into full loss of contact between amortisseur and wedges.

dations. The various designs and manufacturers differ in the value of the clearance allowed between rotor wedge and slot.

Overheated wedges, in particular those made from aluminum alloys, might have become distorted or lost their original mechanical properties. Consult the OEM for recommendations for each particular design. Figure 9.49 shows a wedge discolored by excessive flow of skin current during an abnormal event.

In the case of aluminum and brass wedges, particular attention should be given to points of greatest mechanical stress concentration. Inspection of these can only be carried out with the wedge (and hence at least one end ring) removed. Obviously, this costly operation is only justified under very special circumstances (see Fig. 9.50).

9.11 WINDINGS—GENERAL

In Section 2.13 of Chapter 2, a general description of rotor windings is presented. In the present section, the authors attempt to review the most common problems that one may find in rotor windings during operation of the machine.

In terms of stochastic data, field windings tend to have expected lives shorter than those of stator windings (25 to 30 years vs. 30 to 40 years). This is so, even though sta-

Fig. 9.49 A discolored end wedge. The discoloration is the result of excessive skin-current flow between wedge and retaining ring. As can be seen by the location of the pinlocks, the wedge migrated toward the ring. The wedge below (number 19) remained in its original place and does not show signs of overheating. The damage to the wedge touching the retaining ring indicates that this area should be inspected for damage. The wedge below number 19 also shows signs of discoloration due to overheating. This wedge also migrated toward the retaining ring until full contact has was established.

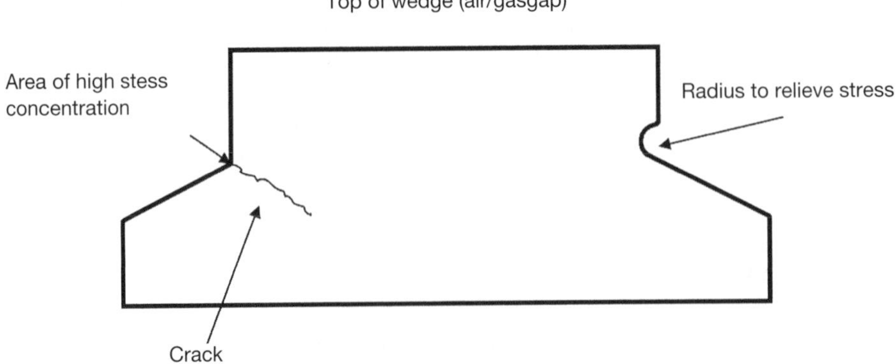

Fig. 9.50 Cross section of a wedge showing typical stress areas.

tor windings are subject to much higher voltage stresses. The reason for this disparity is rooted in the much more intense mechanical and thermal stresses that rotor windings endure, in addition to having a thin layer of interturn insulation. As the rotor accelerates from standstill to full speed, the winding experiences a positional adjustment due to the large centrifugal forces acting upon it. This, coupled with the elongation of the conductor due to heating, introduces bending and sliding between turns, and between turns and slot-wall insulation, between end-winding turns and retaining-ring lining insulation. A similar process occurs as the unit deloads and it is brought to standstill. In addition, the winding elongates and contracts due to load (real and reactive) changes during operation. Finally, torsional oscillations due to grid events or operational activities (such as a "rough" synchronization) add additional strain to the winding. Design peculiarities coupled with turning-gear operation is another source of stress, as is copper dusting, covered in detail elsewhere in the book. Thus, it should be no surprise that these highly stressed components tend to experience more aging and damage than others. The job of the inspector is to understand operational (e.g., double-shifting) and design (e.g., consolidated vs. unconsolidated coils, type of interturn insulation, or type of blocking) sources of strain on the rotor winding, and look for telling signs of damage or impending damage for each specific type of design.

9.11.1 Conductor Material

Field windings are almost exclusively made from high-conductivity copper. In some special cases (such as in units affected by top-turn cracking), the top turns in some slots may be made of a copper alloy of lower conductivity. There are a few odd designs in operation in which rotor windings are made of aluminum. There are a number of interesting trade-offs between copper and aluminum as conductor material in electrical apparatus. For instance, it is quite common to find transformers—in particular dry-type units or oil-filled units of low ratings—that use aluminum conductors. Aluminum conductors also are found in many squirrel cages of induction machines. Table 9.1 lists some of the trade-offs between copper and aluminum as electrical conductors.

As shown in Table 9.1, copper has an advantage over aluminum in almost every physical characteristic when it comes to application in a highly stressed environment such as a turbo-generator rotor. However, when comparing the effectiveness of copper versus aluminum as electrical conductors when the specific weight is included, aluminum is about twice as effective. At the time of this writing, the price of copper is about twice that of aluminum, and that may add some incentive to using aluminum. So, why then do the vast majority of rotors have copper for their rotor winding conductors? It is because of copper's physical properties versus aluminum, and the harsh conditions of the application more than make up for the price and weight advantage held by aluminum. Also, there is one critical design constraint that favors copper: the fact that an aluminum winding would require more slot cross section (all else being equal), and that means a bigger rotor and a bigger machine.

In spite of the above, some turbo-generators were manufactured with aluminum windings, and if the reader has the opportunity to inspect/troubleshoot such a unit,

Table 9.1 Comparison of several characteristics of copper and aluminum for conductor use.

Characteristic	Copper	Aluminum
Electrical conductivity	**100**	63
Specific weight	100	**30**
Heat transfer	**100**	59
Thermal elongation	**100**	134
Melting point	**100**	62
Effectiveness as electrical conductor per unit of weight	100/100 = 1	63/30 ≅ **2**
Toughness	**Cu** more *tough* than Al	
Brittleness	**Cu** less *brittle* than Al	
Ductility	**Cu** has higher *ductility* than Al	
Malleability	Cu less *malleable* than **Al**	
Corrosion resistance	**Cu** more resistant to *corrosion* than Al	

*All values are normalized against copper being equal to 100%.

he/she should be aware of some very specific issues associated with those windings. Figure 9.51 shows a rotor with aluminum windings headed for a rewind.

One major challenge for the OEMs of rotors with aluminum windings is the joints between the copper and the aluminum portions of the winding (due to the same advantages that copper has over aluminum regarding their physical characteristics, copper may be found in sections of aluminum windings, such as top and bottom turns, and leads). Figures 9.52 and 9.53 show some details of the copper and aluminum arrangement.

9.12 WINDINGS—SLOT REGION

9.12.1 Slot Liner

The first subcomponent of interest in the slot area is the *slot liner* (also called *wall insulation*). Made usually of Nomex, resin-impregnated glass fiber, or other similar compound, its main function is to provide electrical insulation between the conductors and the forging, by "wrapping" the coil inside the slot. Figure 2.54 shows the slot liner and its position in the slot area. Slot liners can entirely wrap the coil or, in designs with a cooling subslot, can be made of two sections, one against each of the slot walls (see Fig. 2.54 and Fig. 9.1). The slot liner must provide good electrical insulation while withstanding severe thermal and mechanical forces. It is, therefore, not surprising that the main mechanisms of failure are mechanical in nature, being tearing and/or abrasion. Abrasion arises from copper dust present between turns and the wall insulation (see sections on copper dusting), or/and from axial and side forces introduced by coil deformations, centrifugal forces, and, mainly, elongation and contraction forces due to load changes and starts and stops. Two major differing designs can be found affecting slot liners: *consolidated* coils and *unconsolidated* coils. The vast majority of designs

Fig. 9.51 Two-pole, 60 Hz, 183 MVA, hydrogen-cooled turbo-generator rotor with aluminum windings headed for rewind. The whitish aluminum conductors are clearly visible in the end windings.

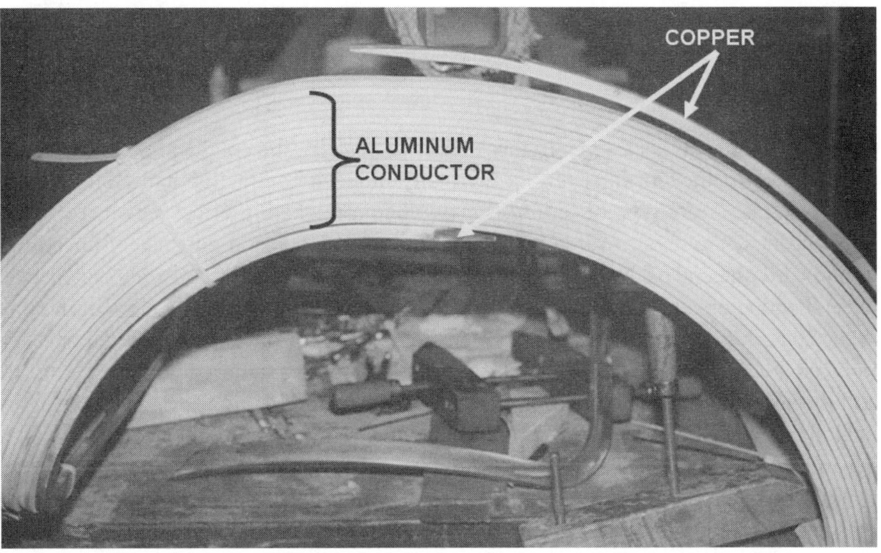

Fig. 9.52 New coil for the 183 MVA rotor shown in the previous figure. The photo clearly shows that the bottom and top turns in this coil are made of copper. This Cu–Al combination is typical of these aluminum-winding rotors. The joint between the Al and the Cu portion of the windings is a challenging area in the manufacturing of these coils. Using rivets has been a preferred method, and past experience indicates problems with these types of joints.

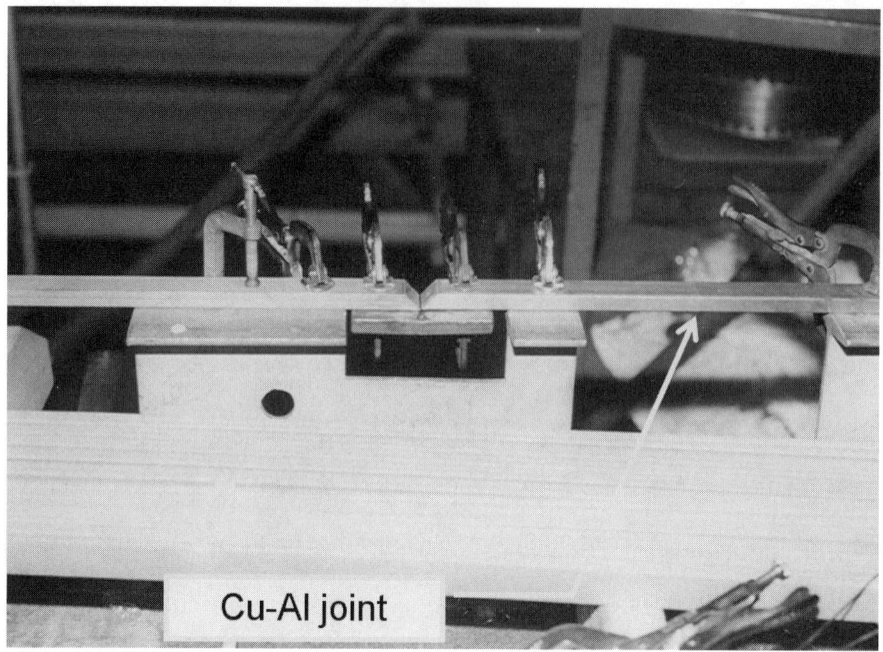

Fig. 9.53 The same rotor shown in Figure 9.51. The photo shows a copper–aluminum joint already made, next to an aluminum–aluminum weld ready to be made.

are of the unconsolidated type. In these types of designs, the conductors are rather free to slide against the slot liner during thermally induced contraction and elongation. In these designs, a rare type of failure occurs when a ratcheting effect between coil and liner "pushes" the liner toward one side of the forging. The consequence of this migration toward one end of the rotor can eventually result in a lack of *creepage* distance between conductors and forging. In same designs, about 10 to 20 centimeters of the liner in the center of the machine is glued to the slot to keep it from shifting toward one side. In consolidated coils, the slot liner is completely bonded and cured with the rest of the coil during manufacturing, making it an integral part of the coil. The liner is designed to elongate with the copper turns without damage. In this type of design, there is no independent migration of the liner toward one side, unless the entire coil does it. These coils have designed into them features that allow them to contract and elongate along the slot, with remaining pinned down at the center of the slot. Figure 9.54 shows a migrated slot liner.

As with any other component of the coil in the slot area, inspection with the naked eye is extremely difficult unless at least one retaining ring and the slot wedges are removed. A poor alternative is to look with mirrors or, better still, with a borescope, under the retaining ring, looking for unusual protrusion of the slot liner toward one end of the unit. In a few cases, a tear in the liner can be found by visual inspection after, at least, wedges are removed and part of the coil lifted, but mostly when significant coil

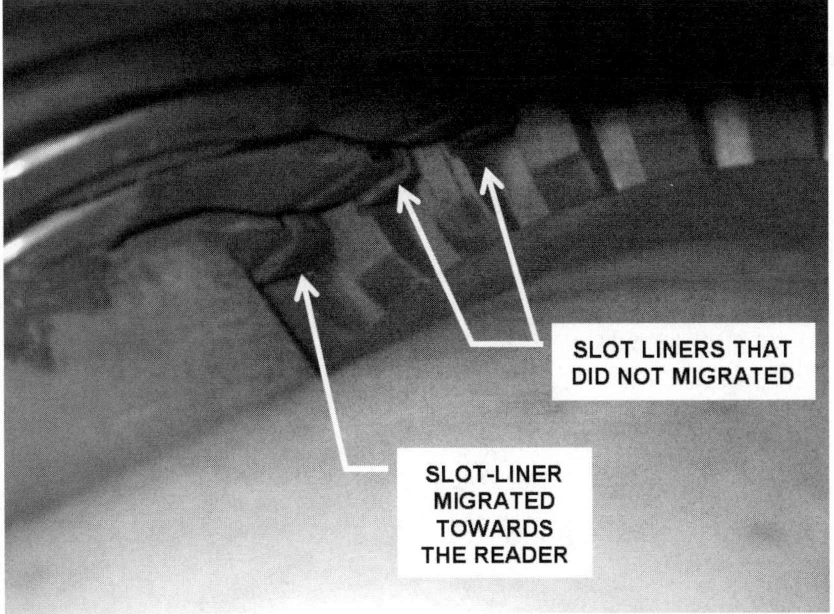

Fig. 9.54 The figure shows a number of coils (the bottom turn can be seen) as they leave the slot. The slot liner in one of the coils (coil #1) has migrated about 2 to 3 centimeters toward the reader.

disassembly has been carried out for whatever reason, or because a ground occurred during operation. Obviously, a megger test or, certainly, a hi-pot test will uncover a damaged liner.

Figures 9.55 to 9.59 show different cases of slot-liner failure or degradation. Each one of these cases was caused by a different failure mechanism.

9.12.2 Turn Insulation

Insulation strips are included between the conductors. These strips form the *turn insulation.* Turn insulation can be made of Nomex, resin-impregnated glass fiber, or similar material. It is rather thin, usually about 10 mils (1/4 mm) or so. From the point of view of providing adequate electrical insulation, one must keep in mind that the interturn voltage during normal operating conditions is about 1 to 3 volts for most machines, and about 10 times that for the most intense transient condition; thus, a few mils will suffice. As with the slot liner, usually the failure of the turn insulation in the slot region is caused by similar mechanisms as those causing slot-liner damage: abrasion, tearing, and overheating. In unconsolidated coils with two copper conductors per turn, copper dusting can be a major source of turn insulation distress in units subject to significant turning-gear operation. Double-shifting or many start–stop operations also result in significant mechanical stress to the turn insulation due to many cycles of elon-

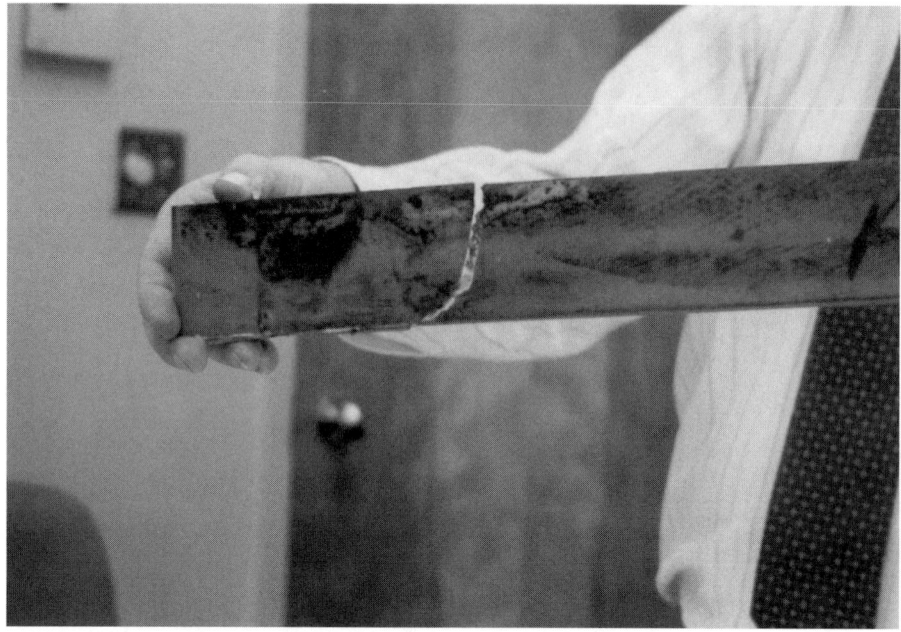

Fig. 9.55 Slot liner torn by elongation and contraction of the coil and high friction forces between the coil and the liner.

gation and contraction of the coils. As with any other component inside the slot, only partial or total disassembly of the coil allows visual inspection of the interturn insulation. Megger tests, hi-pot tests, and RSO tests, as well as other tests, can uncover damaged interturn insulation without disassembly (see Chapter 11 for details). Figure 9.59 shows a strip of interturn insulation damaged by a shorted turn. Figure 9.60 shows a strip of overheated interturn insulation.

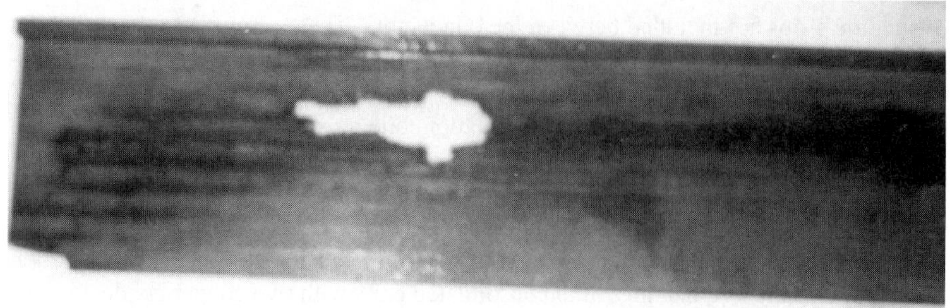

Fig. 9.56 Damage to a slot liner due to extensive abrasion and heating.

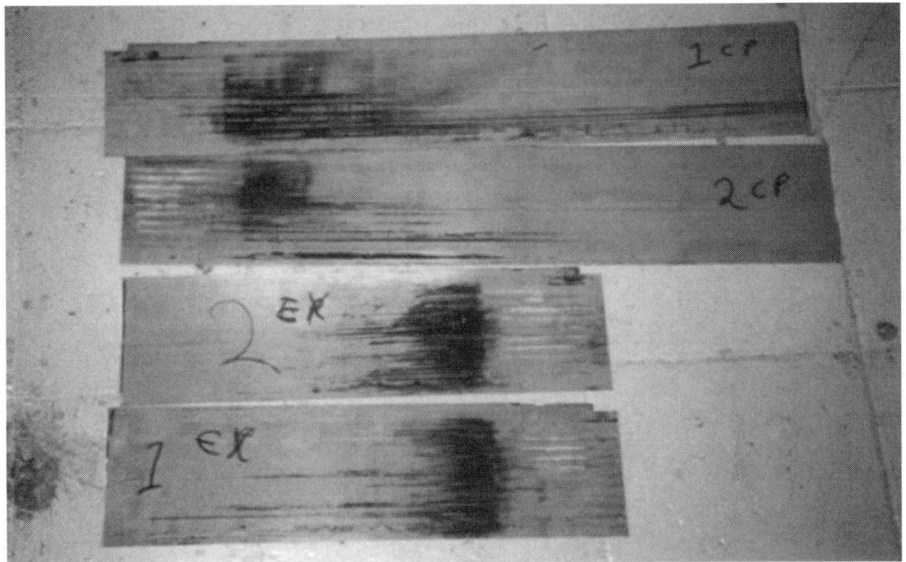

Fig. 9.57 Slot liner wear due to abrasion.

Fig. 9.58 Slot liner with embedded copper particles. This situation is prone to leading to a ground fault. (See color insert.)

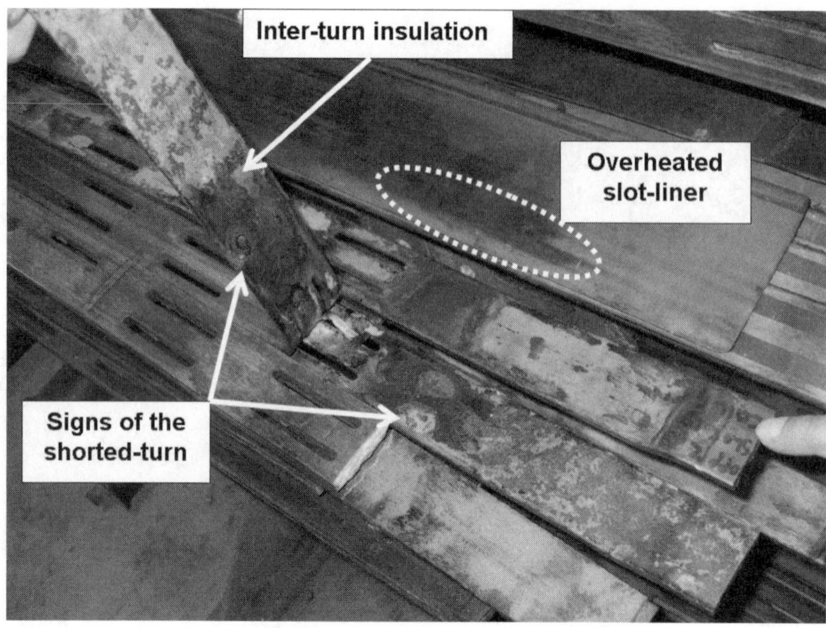

Fig. 9.59 Heavily blackened slot liner in a hydrogen-cooled, two-pole generator. The overheating of the insulation occurred in close proximity to a shorted turn.

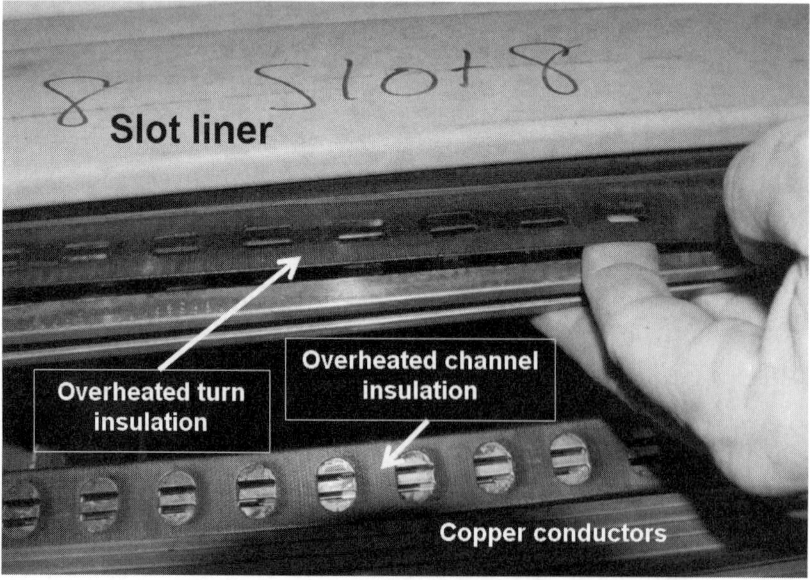

Fig. 9.60 The figure shows the components of a coil lifted from the slot. The coil was subjected to overheating due to partial blockage of the rotor cooling vents.

9.12.3 Creepage Block and Top Channel

Figure 9.1 shows, among other insulation components, the *creepage block* (or *creepage strip*) of insulation, that goes between the last turn and the wedge. In some designs, between the last turn and the creepage strip there is a C-shaped strip of hard insulation called the "top channel." These components are the top-most insulation in the slot. The creepage block (so-called because its dimensions provide creepage electrical clearance between top turn and forging) also serves as the top "slip plane" between coil and wedges, that is, as the coil elongates and contracts it slides across the creepage strip. Therefore, when for some reason or another the sliding becomes too harsh, creepage strips may tear or break, weakening the insulation and coil support systems in that slot. Figure 9.60 presents a view of both the top channel and the interturn insulation strip. Figure 9.61 shows the top channel and the creepage block. Figure 9.62 depicts a cracked top channel.

As already stated, there are a number of failure and degradation mechanisms happening in the slot area of the rotor winding. However, direct visual examination of this region of the machine is almost impossible. The most that can be achieved without having to remove at least a retaining ring and some or all wedges in a slot is to "peek" through the radial vents (if available), or to slip a borescope under the re-

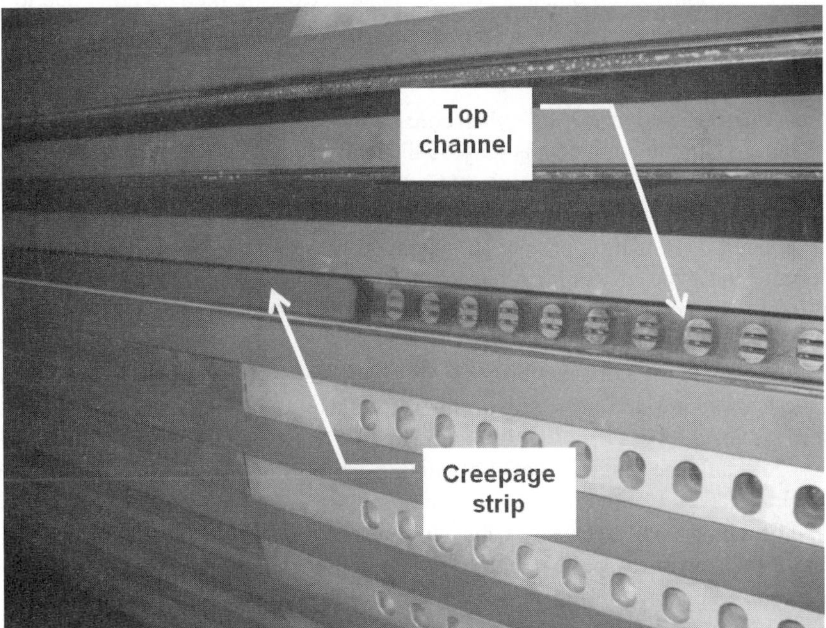

Fig. 9.61 This photograph shows the rotor of a two-pole, hydrogen-cooled generator with an output of about 200 MVA. Shown in the figure is a slot with the wedges removed and a section of the creepage strip also removed. The top conductor's cooling "slits" can be seen through the top insulation channel.

Fig. 9.62 Top insulation channel shows cracks (inside the marked area) due to mechanical strain.

taining rings. Experience shows that borescopic examination is not as simple as it may appear. Simple components may be difficult to identify properly via the distorted and narrow view of the borescope. Certainly, faults and degradation are not easily identifiable. Many things shine in the borescope's light, and even a small speck of dust or shred of insulation or varnish may appear like a major problem. Only experience with this type of examination and a good understanding of the geometry and physics of the rotor will facilitate making sense of those things seen through the borescope.

Going back to the slot area, the borescope will allow some view of the coils coming out the forging and, in rotors with a subslot, the inspector may be able to "sneak" the borescope inside the subslot and obtain a partial view of the bottom turn (see Fig. 9.63). All said, the condition of the winding in the slot area is mainly ascertained through electrical testing and removal of at least one retaining ring and wedge, and partial lifting of turns. Because of this, close inspection of the slot area is done mainly when, for whatever reason, a retaining ring is removed, or because test results, a failure, or some concrete information about a deteriorating condition requires it. For example, Figure 9.64 shows a shorted turn found with the aid of an airgap flux probe. Troubleshooting this type of short, deep inside the slot area, requires a significant amount of rotor disassembly. Normally, this repair would not be attempted for a single shorted turn, but as part of a larger troubleshooting effort (e.g., removal of a number of shorts).

Fig. 9.63 View of the subslot (bottom turn shown at bottom of figure) through a borescope inserted under the retaining ring.

9.12.4 C-Channel Subslot

In the right panel of Figure 9.1, the slot is shown with a subslot, and the conductors resting on a shoulder of the slot; the subslot is machined out of the forging as a continuation of the slot but with a smaller width. This arrangement provides for strong support of the coil above the subslot. On the other hand, in some small rotors, the bottom conductor rests above a C-shaped insulation channel, forming a subslot for providing room for axial flow of cooling air. Figure 9.65 schematically shows such an arrangement. Figure 9.66 shows a real example.

Experience has shown that over time the C-shaped insulation channel may deteriorate or otherwise tear apart, blocking the passage of air through the subslot, with subsequent detrimental thermal effects, but also weakening the coil support. Figure 9.67 shows this type of damage.

9.13 END WINDINGS AND MAIN LEADS

The rotor end windings in a turbo-generator are subject to intense centrifugal forces, as well as forces deriving from the cyclic changes in temperature. The end rings (retaining rings) contain the radial displacement of the copper end windings and the support-

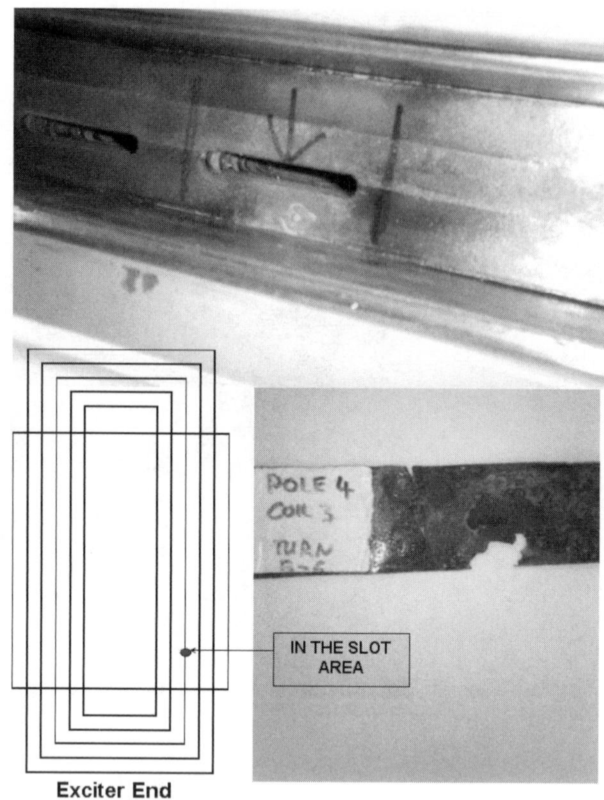

Fig. 9.64 Shorted turn found inside the slot. The bottom-right photo shows the damaged turn insulation. The bottom-left sketch shows the location of the fault. (Courtesy of Alstom Power Inc.)

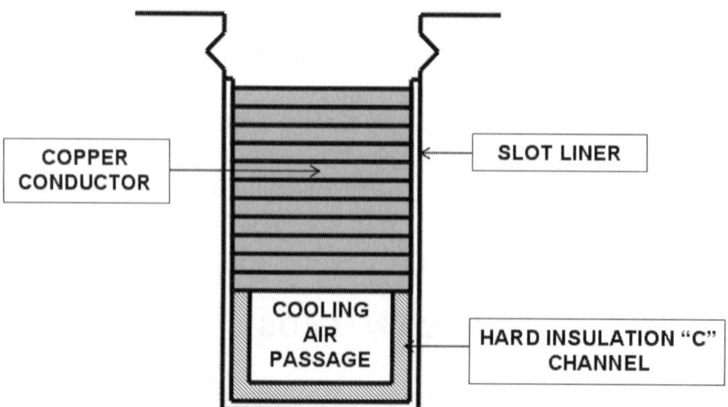

Fig. 9.65 Schematic representation of a rotor coil resting on top of a C-channel made of hard insulation material.

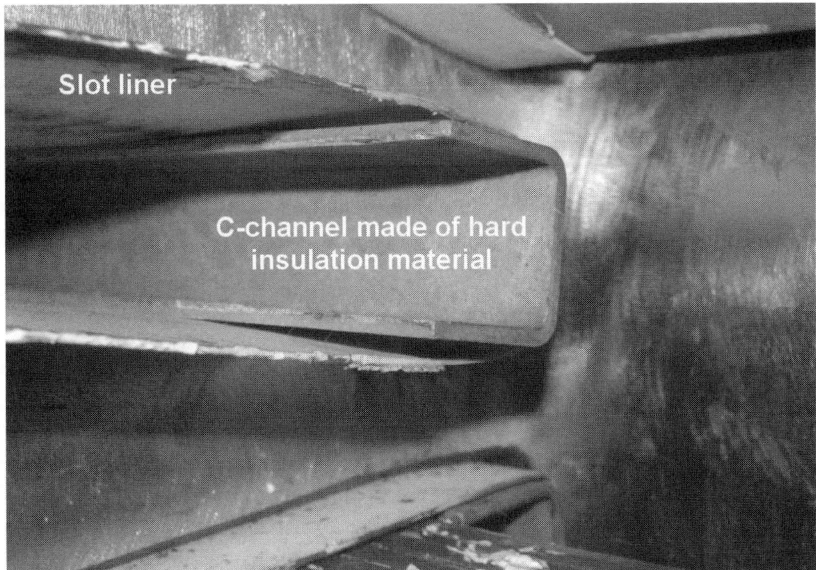

Fig. 9.66 The slot in the center of the photo clearly shows the C-channel made of insulation resting inside the slot liner in the bottom of the slot. The bottom slot shows the copper conductors resting on top of the C-channel forming a subslot for the axial flow of cooling air.

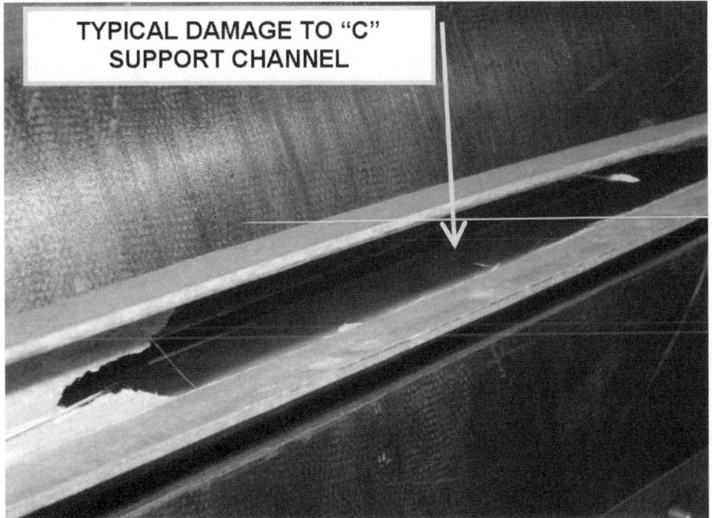

Fig. 9.67 The C-channel made of hard insulating material shown inside the slot of the previous figure can now be seen after it was lifted from the slot. The bottom of this channel came apart from the rest of the channel, blocking the passage of cooling air. Similar damage was found in other slots of this rotor.

ing blocks. However, a well-designed rotor allows for relatively free movement of the entire winding in the axial direction. It is important that regardless of the axial movement of the whole end winding, the distances between the individual coils underneath the end rings are maintained. The distances between individual coils are maintained by the introduction of preformed insulating blocks. Figures 9.68, 9.69, and 9.70 show different views of end windings of two-pole cylindrical rotors. The reader can refer to the chapter on design and construction for a more elaborated discussion of these components and their designed function.

9.13.1 Retaining Ring Liners

Between the retaining ring and the rest of the end winding, a wrap of insulation is included. The purpose of this wrap is to insulate between the electrified rotor field and the retaining ring at ground level. This wrap of insulation may, depending on the OEM and machine design, be solid insulation in the form of cylinder, or of a number of Nomex or Nomex-like insulation layers, or of a combination of layers of Nomex, Kapton, and/or similar insulating components. G10 epoxy–glass is also widely used, often in conjunction with Nomex. In almost all arrangements, the Nomex liner must be replaced each time the retaining ring is removed and before it is reinstalled, and, in some designs, where the liner is actually a solid-insulation cylinder, the liner can be reused if

Fig. 9.68 Front view of the end winding of a two-pole cylindrical rotor having the retaining ring removed. This rotor has seven coils per pole. Some blocking can be seen between the coils. This rotor is in the final stages of a rewind operation.

it was not damaged during the removal of the retaining ring. In some designs, the liner

Fig. 9.69 The same rotor shown in the previous figure, but before the "old" winding has been stripped away. The angle of the photo shows the point of transition from one coil to another (top-left of the end winding).

Fig. 9.70 Top view of an older and lower-rated two-pole rotor compared to the one shown in the previous figure. The blocking between coils has been removed. As seen in the photo, there are also seven coils per pole in this example. This field winding has been in operation for many years. (from Kerszenbaum, *Inspection of Large Synchronous Machines,* 1996, IEEE Press, reprinted with permission.)

is directly in contact with the retaining ring, and in others, the liner is under the damper winding (amortisseur). In Figures 2.46 and 9.47, the retaining-ring liner can be seen under the amortisseur. In Figures 9.71, the liner is under a set of plates that entirely wrap it.

Inspection of the liner, after it is removed from under the retaining ring, might provide a clue about the behavior of the rotor winding during operation. For instance, if a rotor turn gets "caught" by the retaining-ring liner, it will leave a mark. When there is no distortion damage, the markings left by the coils on the liner will show a smooth "picture" of the coils. Figure 9.72 clearly shows this situation.

Figures 9.73, 9.74, and 9.75 show two examples of failure of the retaining-ring liner, resulting in a ground fault from the energized field winding through the liner's insulation.

9.13.2 End Turns and Blocking

However well designed, the end windings of turbo-generators tend to deform. The degree depends on the machine's operating age, the type of loading (cyclic or continuous), any abnormal operating conditions encountered (short circuits, power swings, etc.), and the actual physical design of the coils. If the distortion is excessive (i.e., the

Fig. 9.71 This rotor had its retaining ring removed. Under the retaining ring there are a number of copper or titanium-alloy segments that entirely cover the liner. In the outside of these segments rest toothed short segments that make up the amortisseur. In this design then, the liner is on top of the end winding but not in direct contact with the retaining ring.

Fig. 9.72 The coils of the end winding of a turbo-generator rotor left on the retaining-ring liner. The image shows that here was no excessive friction between the coils and the liner.

Fig. 9.73 Damaged retaining-ring liner. The damage occurred during operation and resulted in a short circuit from coils 6 and 7 to the retaining ring. The damaged region is clearly seen in the figure (marked by the white broken line).

Fig. 9.74 Damage to the retaining ring from the fault described in Figure 9.73.

Fig. 9.75 The figure shows how a ground fault bridged between one of the coils (bottom left) through a Nomex layer of the liner (top left) and a Kapton layer of the liner (top right) and to the titanium-alloy plate under the retaining ring. The metallic layer over the liner and under the retaining ring protects the latter from ground faults. See Figure 9.71 for an overview of the end winding in this machine.

individual turns are not aligned in each coil), then they will become prone to developing shorted-turn faults. This type of fault is more likely to occur in dc fields in which the individual turns in the end-winding section are not fully insulated but are protected only by a single layer of insulation between the turns. Excessive deformation can also result in cracks or breaks in the conductors, in particular in those coils having a sharp 90-degree angle in each corner. Figure 4.22 in Chapter 4 is a prime example of a deformed end-winding conductor resulting in shorted turns. This example indeed presents a very large winding distortion, but it is not a unique or rare event. Figure 9.76 shows another case of an end winding deformed by cyclic load changes and copper ratcheting. A similar situation happened in the case of Figure 9.77.

There are many other types of deformations that may afflict an end winding. For example, Figures 9.78, 9.79, and 9.80 show turn deformation due to a combination of thermally induced expansion and contraction forces, and the reaction forces from the blocking and the liner.

Every rotor with some accumulated operational time will exhibit some deformation of its rotor winding to some degree. In most cases, the deformations are small and do not introduce any impediment to the normal expansion and contraction of the coils

Fig. 9.76 Potentially undesirable consequences of insufficient blocking while a rotor is undergoing cyclic loading. It is important to carefully inspect the alignment of coils and single turns in the end winding during rotor inspections. This condition can easily be detected using a mirror and flashlight under the retaining ring.

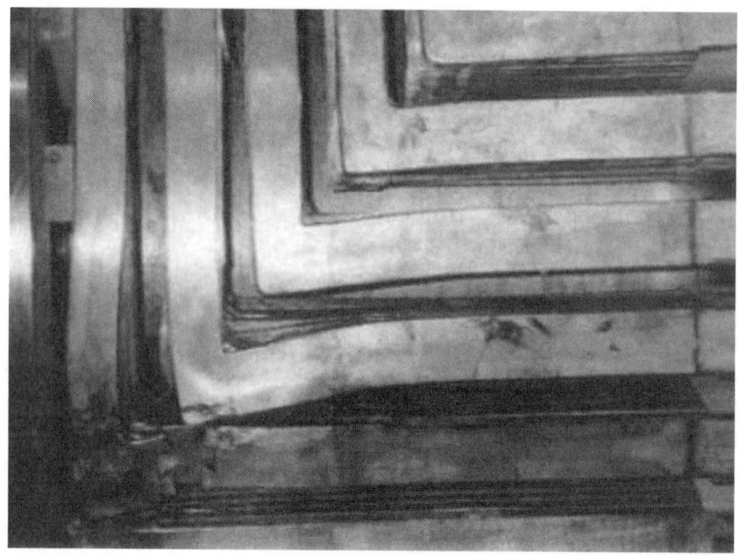

Fig. 9.77 The figure shows a square-corner coil with the upper turns distorted due to poor blocking. The top turn made contact with the top turn of the neighboring coil.

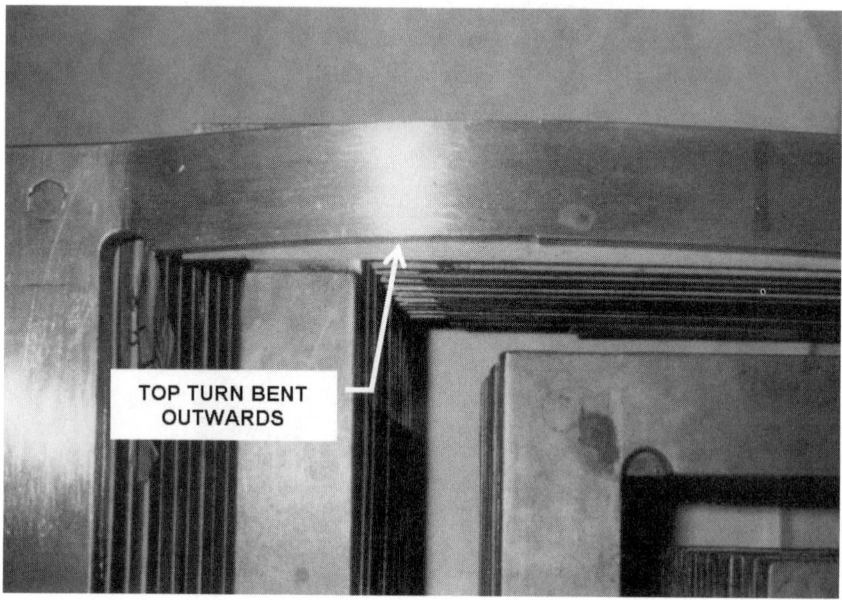

Fig. 9.78 Deformation of the top turn due to circumferential thermal expansion and rigid blocking that does not allow any expansion. The turn then bends outward and into the retaining-ring liner.

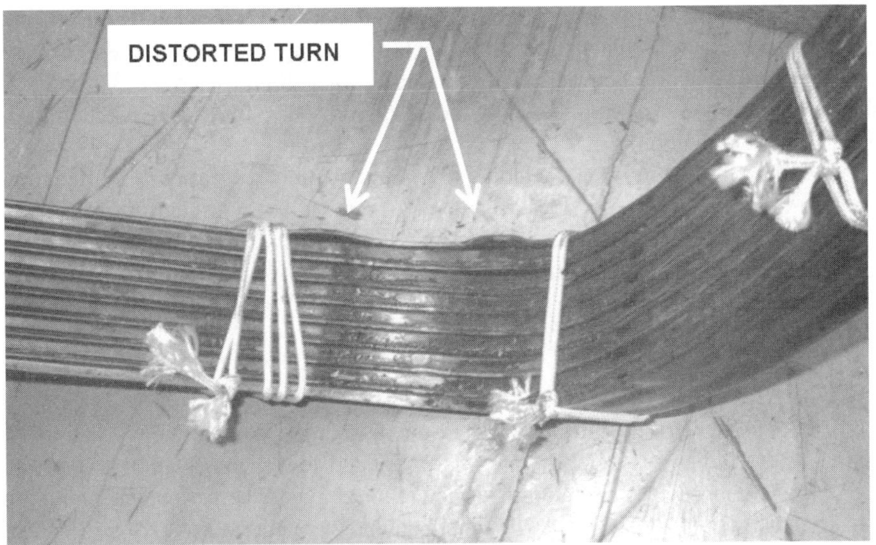

Fig. 9.79 The figure shows a complete coil lifted from the slot, with the top copper strip warped.

Fig. 9.80 The figure shows a rotor in which each turn is made of two bars side by side inside the slot area, and separated by about 1 inch in the end winding. The area marked with the dotted line shows top turns deformed by crushing forces.

along their designed slip planes. In other cases, a section of the winding becomes so warped that it creates ratcheting against the retaining-ring liner or slot liner, or causes shorted turns or shorts between two contiguous coils. Figures 9.81, 9.82, and 9.83 show major turn migration in the end-winding region. The photos were taken during a borescopic examination.

Migration of turns in the end winding in most large units will result in shorted turns, because these types of units do not have, for the most, every second turn insulated. Figures 9.84 and 9.85 show two cases of migration and disintegration of the interturn insulation. Obviously, a situation like this requires serious evaluation for repair and refurbishment.

9.13.3 Shorted Turns

Experience shows that one of the modes of operation most detrimental to the integrity of the field winding in turbogenerators is prolonged low-speed rotation of the rotor ("on turning gear"). This operation is designed to eliminate the permanent bending of the shaft that would otherwise occur if the machine were allowed to cool down standing still when coming offline. However, the low-speed rotation results in continuous pounding of the windings due to their own weight, in particular the end-winding regions of the field winding. This continuous pounding is exacerbated if the clearances between the winding and the insulation are excessive. The result of this continuous pounding is the creation of copper dust that may result in short-circuiting turns of the

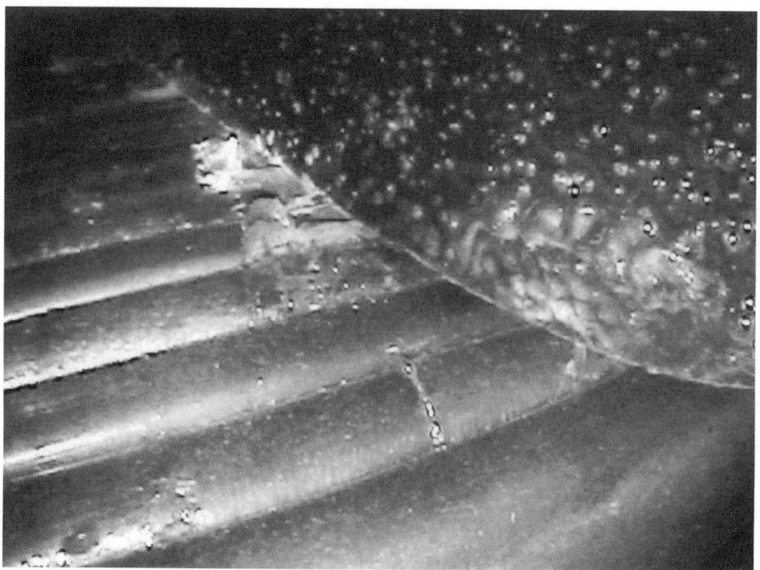

Fig. 9.81 Borescopic view of a coil with its turns properly aligned in an air-cooled, two-pole rotor. Compare this view with the one in next figure.

Fig. 9.82 Same rotor as the one in the previous figure, but a different rotor coil, also viewed via a borescope. Note the drastic misalignment of some of the turns. Interestingly, this major migration did not result in shorted turns due to the fact each second turn is insulated in the end-winding region.

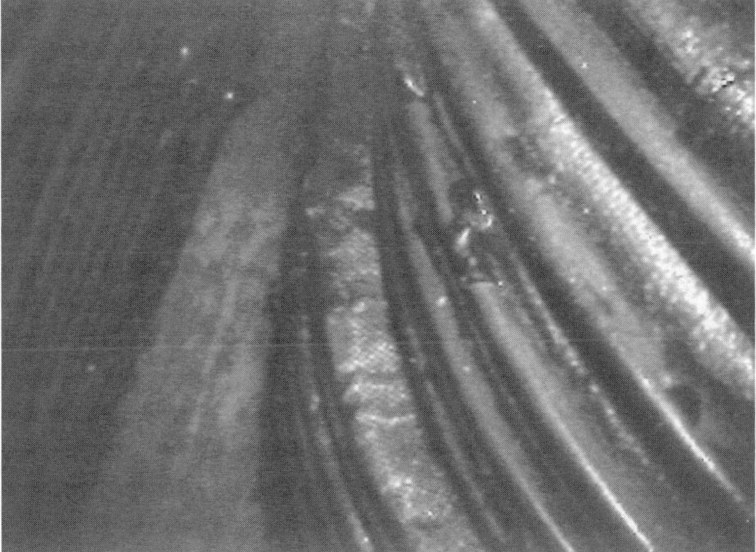

Fig. 9.83 Same rotor as in the previous figure. This photo taken through a borescope is from a third coil showing major turn migration in the end winding.

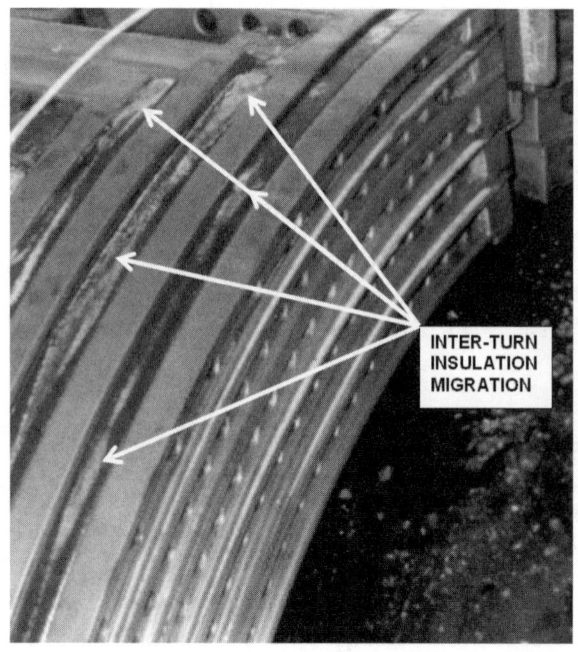

Fig. 9.84 Interturn insulation migration in a large two-pole, hydrogen-cooled rotor. Axial cooling vents can be seen in the conductors. Continuing migration of the interturn insulation is conducive to shorted turns.

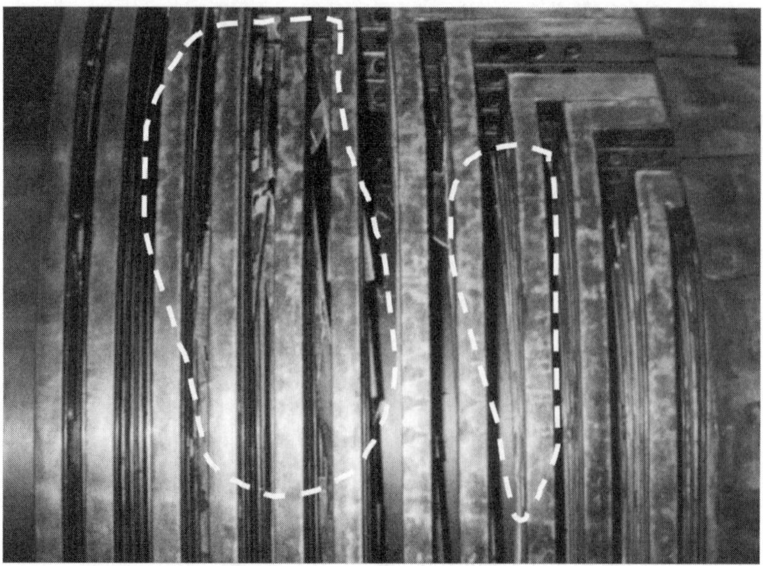

Fig. 9.85 The marked areas show interturn insulation migration and disintegration.

winding or even provide a path to ground in the slot region of the winding. Some designs are more prone than others to the development of copper dusting. This issue is discussed in depth in other sections of this book (refer to the Index). Figures 9.1 and 9.2 in this chapter schematically show conductor arrangements that are prone to develop copper dusting under long turning-gear operation. The key determinant is the existence of two or more conductors per turn. The movement of the copper on copper after substantial periods of turning-gear operation will result in copper dusting if the coils are relatively loose. Obviously, generators with these types of rotor designs and operating under double-shifting conditions, or with long periods on the turning gear, should be inspected carefully for this type of phenomenon.

Relatively new coil designs (preformed and cured outside the slot) are supposed to be very resistant to the formation of copper dusting. Coils fabricated in this way are called "consolidated."

9.13.4 Top-Tooth Cracking

In certain designs and under certain operating conditions, the top turns of the field-winding tend to crack and eventually break in the region just outside the slot. This type of failure can result in excessive damage to the rotor because of major ground faults developing as a result of the broken conductors. This area of the windings should be carefully inspected when the retaining rings are removed for any other reason, in particular if the operator is aware that the machine in question belongs to the family of machines that have experienced these types of problems. Rotors with spindle-mounted retaining rings are prone to develop top-turn cracking. Figure 9.86 provides an elementary view of this phenomenon. There are a number of possible fixes to this problem. One such solution is to replace the top turn with one made of a nonconducting harder alloy (in fact, this removes one active turn from the winding, but this is not a problem, given the margin encountered in excitation systems). There are no large turbo-generators being manufactured these days with spindle-mounted retaining rings. Body mounting of retaining rings is the preferred method.

9.13.5 dc Main Leads

The weakest point on the copper main dc leads connecting the collector rings to the end windings appears to be at the elbow of the lead, where it emerges from the shaft and rises to connect to the coils (Fig. 9.87). To accommodate the axial movement of the end winding as it warms or cools down, sections of this lead are often made of many thin copper sheets braced together, with solid conductors at both ends. The most common main lead configuration is one that is solid for most of its length and is "leaved" at the ends to allow installation over the radial studs (Fig. 9.88). If the lead end connection to the radial stud were not flexible, it would be impossible to slip the connection end over the radial stud. Continuous vibrations of the unit as well as load-cyclic operation will weaken these copper sheets, especially at any bends. These can crack and, in some extreme cases, result in open circuit of the field under operation. Thus, it is important to inspect these leads when the rotor is out of the bore for a major

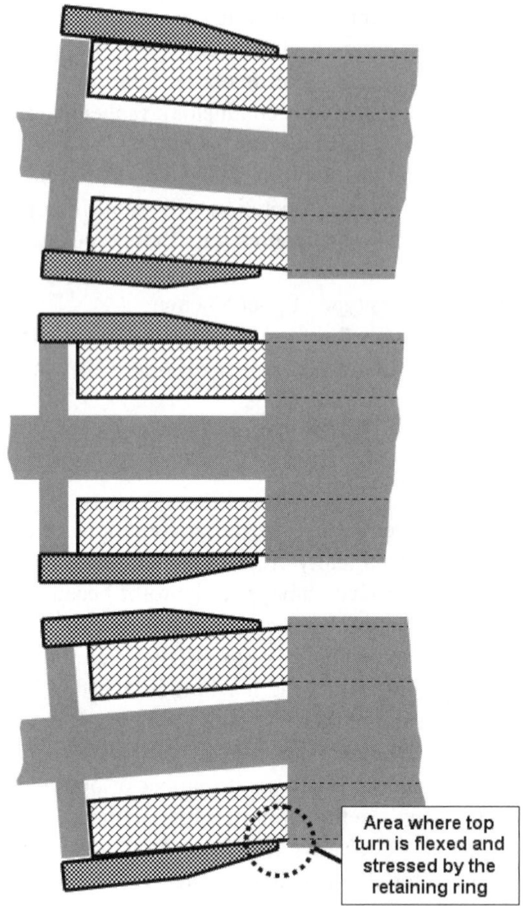

Fig. 9.86 In rotors with spindle-mounted retaining rings, the movement of the ring due to vibrations, frequent starts–stops, and sudden changes of load tend to flex the rotor in such a way that the top turns are stressed, in some cases resulting in cracked or broken conductors, and even ground faults. This area should be inspected carefully when the retaining rings come off. A number of solutions have been implemented by the OEMs.

overhaul/inspection of the unit. With the retaining rings on, direct visual accessibility is nonexistent. On occasion, a mirror may allow a peek at that area. A borescope will almost always allow inspection of the area. It is common during major overhauls to remove the wedges retaining those leads (partially under the retaining ring) to inspect the leads more carefully. Figure 9.89 shows a main lead slot opened for inspection by removal of the main lead wedges (Figure 9.90).

Many large turbo-generators use main leads made of thin sheets of copper, as explained above, to provide flexibility to the lead. Although this arrangement is still

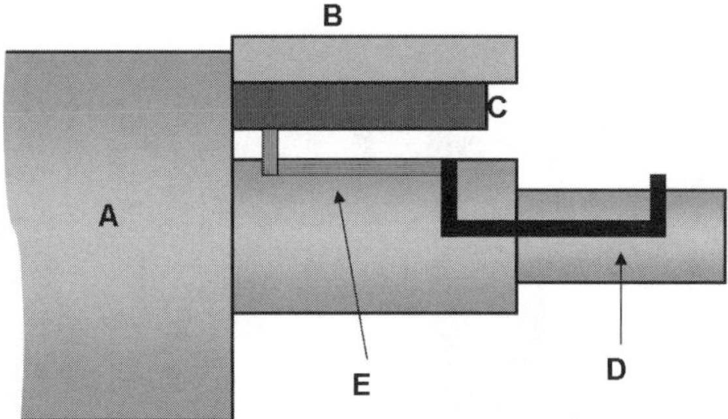

Fig. 9.87 Main lead connecting the radial terminal studs from the bore of the shaft to the end winding under the end bells. In the figure, A is the rotor forging, B is the retaining ring, C is the end winding, D represents the vertical studs and the horizontal solid copper bars carrying the dc field current from the collectors to the field winding, and **E** represents the copper main lead connecting the inner radial stud to the end winding. The 90-degree bend of the main lead is generally prone to cracking that may, in extreme cases, result in open field circuits. This is an area that should be inspected carefully when the rotor is out of the bore. The inspection can be combined with the inspection of the inner hydrogen seal, if one exists.

Fig. 9.88 This photo shows a new main lead of a combined solid and flexible configuration prior to installation of a rotor winding, during a rewind operation.

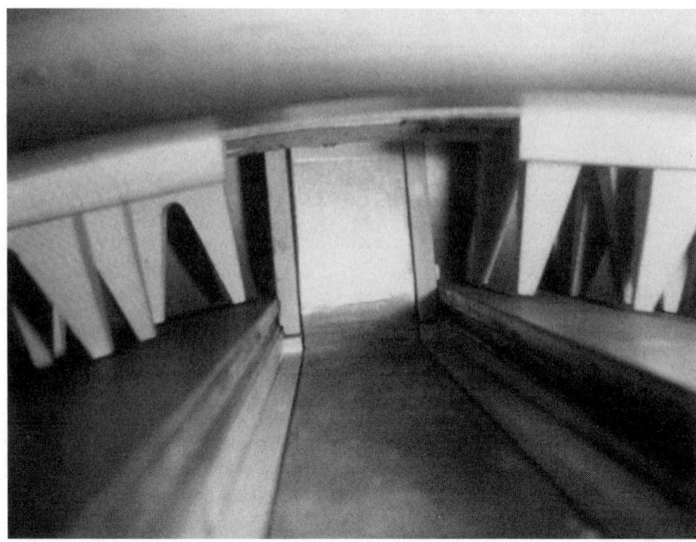

Fig. 9.89 This photo shows a main lead as viewed under the retaining ring, with the main lead wedges and top insulation strip removed. This method of inspection is restrictive, but is the best that can be done without major dismantling. It is sufficient to find main lead cracks in the 90 degree bend where most main leads fail.

prone to long-term failure, one should be much more concerned if the generator has a solid copper lead. Figure 9.91 shows such a lead and the crack found on it. A lead made of several sheets will crack and degrade slowly over many hours of operation. However, in a solid lead, like the one in Figure 9.91, a crack can quickly grow and lead to failure. If troubleshooting is required, one may want to consider replacing the solid lead with a flexible one, if the rest of the arrangement of the end winding accommodates this change.

Fig. 9.90 This photo shows a main lead wedge removed from the main lead slot during an overhaul. These are generally NDE inspected for damage.

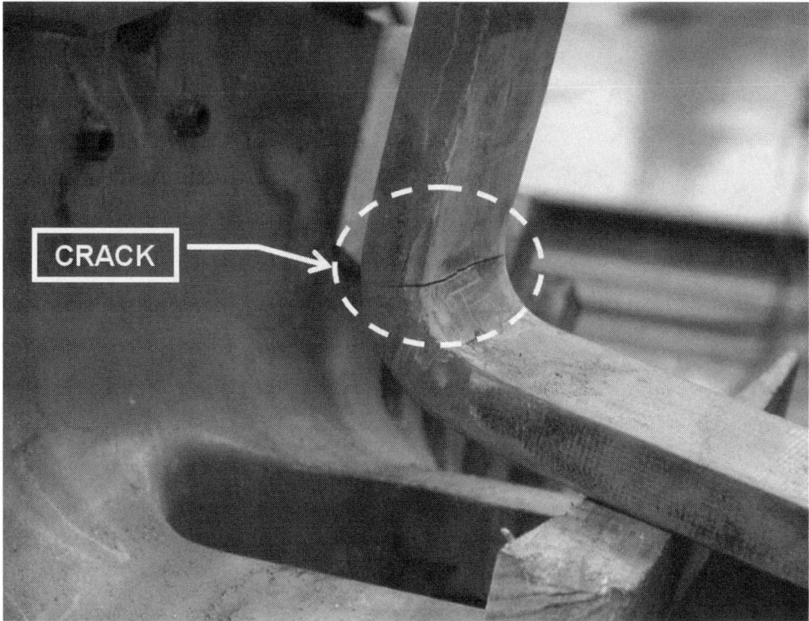

Fig. 9.91 Solid main lead of a rotor field winding of a two-pole turbo-generator. The lead is shown outside its slot, and a serious crack can be seen in the elbow area.

As well as damage to the main leads, the main lead wedges are also subject to high dynamic loading, fretting, and vibration. These have been known to crack and need to be inspected by NDE, usually dye-penetrant examination (Fig. 9.92).

9.13.6 Coil and Pole Connections

Connections between coils in the same pole and between coils in different poles tend to be stressed by centrifugal and thermally induced forces, and well as by lateral and torsional vibrations. Special care is normally taken during the design process to make these connections able to withstand the rigorous environment they operate in, for the expected life of the rotor winding. Nevertheless, careful inspection of these zones is highly recommended when access is available. Figures 9.93–9.97 present different cases of failures of the joints between coils and/or poles.

In summary, inspection of the end windings is not simple. They are quite inaccessible. They are confined between the retaining rings, shaft, and rotor body. For the most part, insulation blocks surround them. However difficult, a good and patient effort will most likely yield enough information on the condition of the winding to determine whether the machine requires additional inspections or if it is fit for operation. The most useful tools are the mirror and the borescope. Check for alignment of the coils as well as turns within the individual coils. Check for loose blocking (Fig. 9.98) and cracked turns. In particular, try to concentrate on the top turns close to the slot. Also,

Fig. 9.92 This main lead wedge was removed from a 575 MW, two-pole generator, and found to be severely cracked. NDE was not needed to find this degree of cracking.

pay attention to the condition of the leads at their elbows. Look for dislocated, bent, cracked, or otherwise damaged blocking (Figs. 9.99, 9.100, and 9.101 show various examples of end-winding blocks being distorted, broken, or both.

Look for excessive copper dust. If this is present, the inspector may recommend an electric test to check for the presence of shorted turns in the field. Chapter 11 on maintenance testing sheds some light on available methods for detecting shorted turns. If

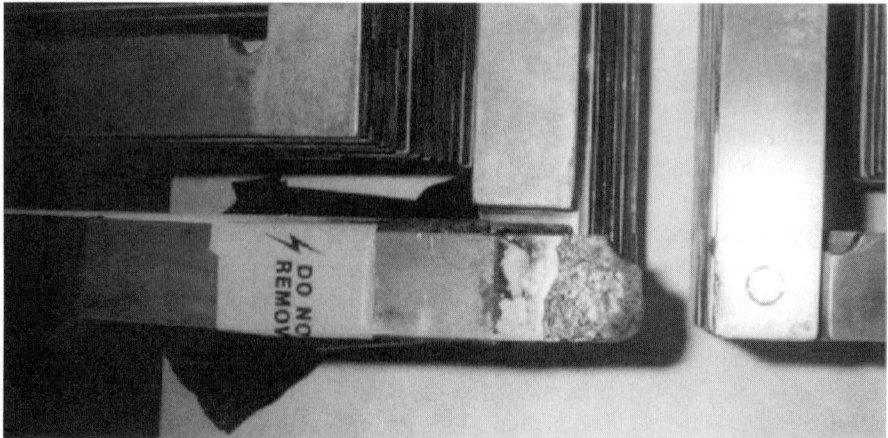

Fig. 9.93 Photo of a burned out pole-to-pole connector of a two-pole, hydrogen-cooled cylindrical rotor.

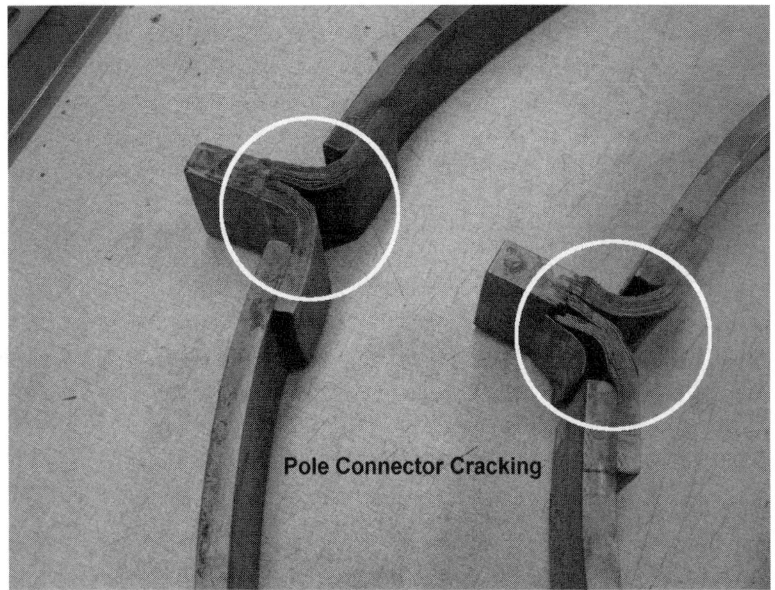

Fig. 9.94 Cracking of the flexible leads of pole connectors.

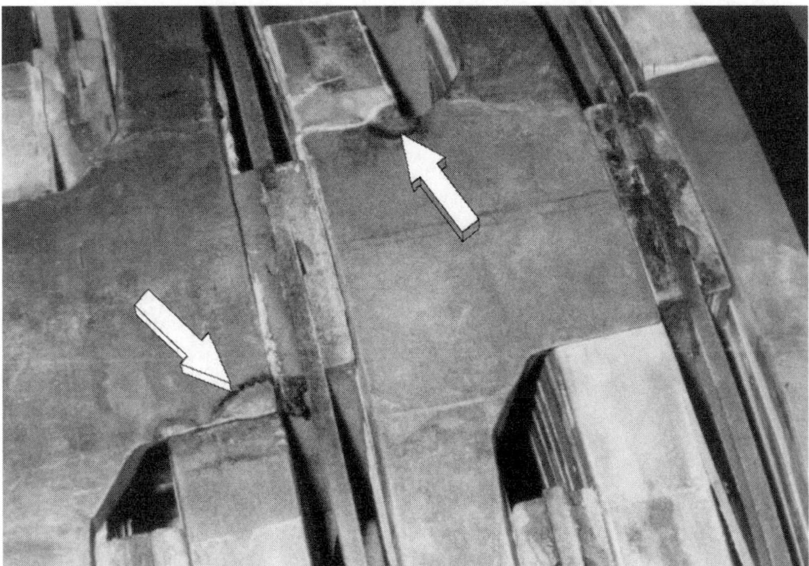

Fig. 9.95 Arcing of the joints between coil crossover connectors. The arcing is the result of high stresses on the braze joints and separation that had been initiated.

Fig. 9.96 Massive failure of a joint between coils led to significant melting of conductor copper.

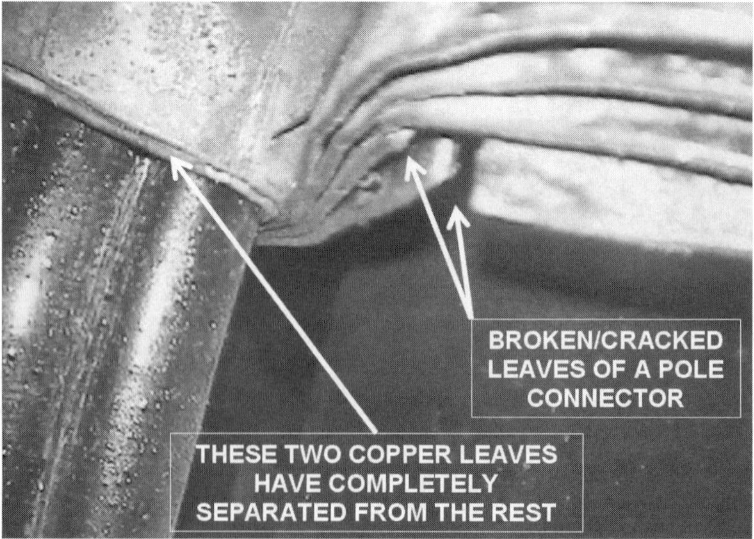

Fig. 9.97 View through the borescope of a pole joint. Some of the sheets in the joint are seen to be cracked. This interesting design is based on a double-joint arrangement, so loosing one joint due to cracking of the copper sheets still leaves another to carry the dc current.

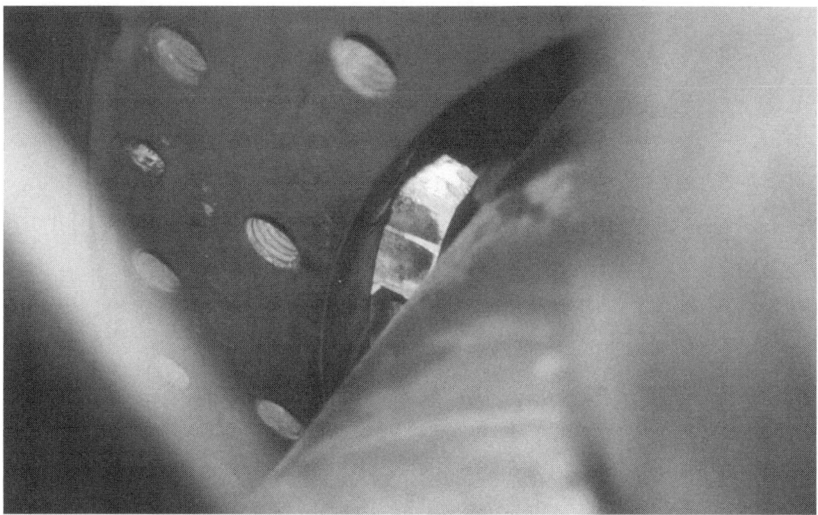

Fig. 9.98 Loose insulating blocking of the end winding of an air-cooled gas-turbine generator. The blocking skidded from its position between the coils under the retaining ring, and it is seen resting on the shaft. The loose blocking made itself known by a repetitive sound during turning-gear operation. (From Kerszenbaum, *Inspection of Large Synchronous Machines,* 1996, IEEE Press, reprinted with permission.)

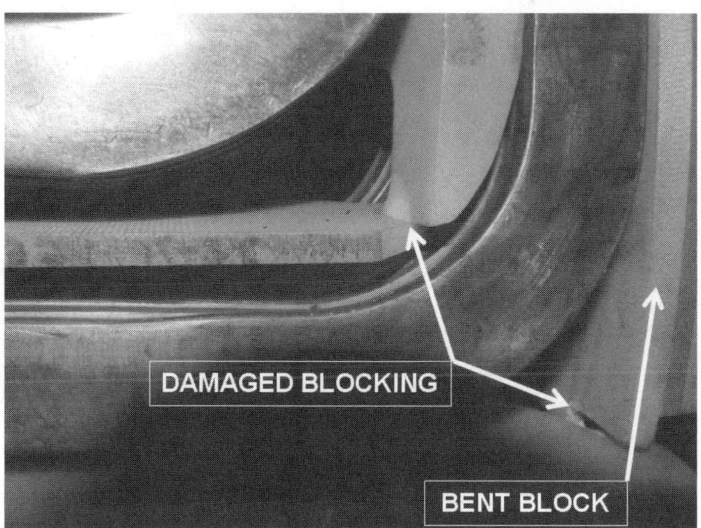

Fig. 9.99 A number of coils in this 3600 rpm, hydrogen-cooler, 180 MW rotor experienced severe coil migration by ratcheting on the retaining-ring liner. The very large forces acting on the coils due to ratcheting can be seen in the broken and bent blocking.

Fig. 9.100 Photo an end-winding block with a large crack. Depending on location, this can be epoxied back to one piece or may require replacement.

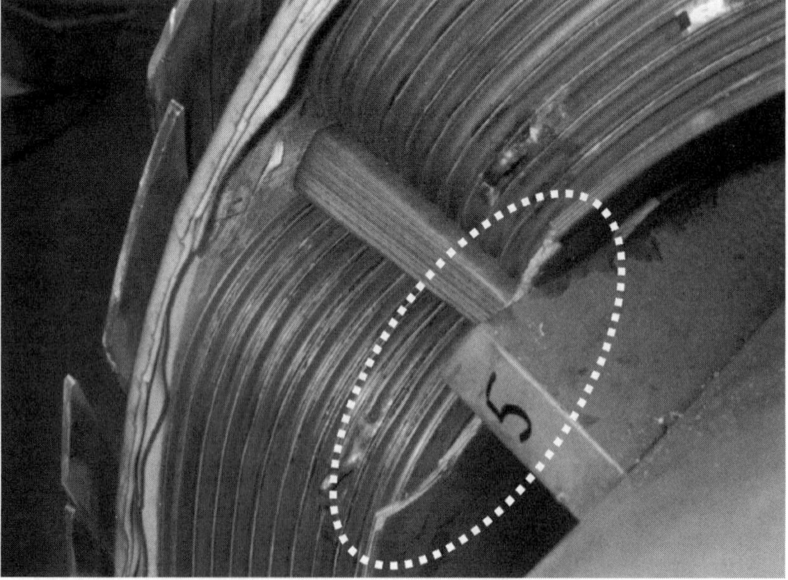

Fig. 9.101 A piece of gas baffle was broken away in this 3600 rpm, hydrogen-cooled rotor. Discovery of this situation requires the inspector to look for the lost piece and evaluate the possibility of it being lodged in one of rotor's cooling passages.

deemed necessary, close inspection of the leads might require the removal of the lead wedges, as explained above. This entails some additional disassembly.

When excessive copper dust is encountered, removal may be tried by blowing, or sometimes by washing with steam or liquids. Although the practice of using liquids or steam in turbo rotors exists, some experts are wary of applying it. The same, to a lesser degree, applies to dry blowing. Some experts believe that these procedures result in carrying the dust back into the slot region, where it could end up causing more damage than before. An alternative—vacuuming—may not remove all of the copper dust, but it will probably minimize the risk of contamination of other vital areas.

9.14 COLLECTOR RINGS

The brush ring and the collectors or slip rings are probably where most of the wear and tear on a synchronous generator occurs (Fig. 9.102). Although simple in construction and appearance, the physics of the transfer of electric current at the brush-collector contact surface is rather complex. Under some conditions, the brushes and the collec-

Fig. 9.102 One collector ring (slip ring) of a large turbo-generator. There are two collectors like this one in the generator, one for each polarity. The width of the ring indicates that it belongs to a unit of relatively high rating. The helical groove and vent holes are for reduction of air pressure under the brushes so that they are held firmly in contact with the collector surface for cooling purposes and for moving carbon deposits away from the ring. (From Kerszenbaum, *Inspection of Large Synchronous Machines,* 1996, IEEE Press, reprinted with permission.)

tor wear very little over many hours of operation, whereas under other environmental conditions, the same brushes and/or collector may wear overnight. The conditions that affect brush-collector performance are as follows:

- Level of humidity
- Contamination (gases, solids, or liquids, particularly carbon and/or metal dust)
- Current density
- Ambient temperature (too low or too high) and cooling of the collector and brushes
- Change of brush grade, use of wrong grade or multiple grades
- Altitude (barometric pressure)
- Unevenness of collector surface
- Run-out of the collector surface
- Brush pressure
- Brush angle to the collector surface
- Brush contact surface requires "bedding" to the collector ring
- Correct clearance of the brush inside brush holder
- Insulation condition at bottom of collectors
- Poor shunt connections

In addition to these elements that directly affect the operation of the brushes, there are other ways in which the performance of the brush collector is influenced. For instance, a solid-state source of field current with poor commutation characteristics (i.e., severe harmonic content) results in accelerated wear of the brushes. Similar problems might arise from poor commutation in rotating exciters. Machine vibration also affects the performance of the brushes. Although vibration levels in electric machinery are kept within limits due to other considerations, it is convenient to remember that elevated vibration levels will tend to have a detrimental effect on the operation of the brushes. Existing literature on the subject states that vibration levels higher than about 1.5 mils peak-to-peak displacement for 3600 rpm machines and 2.5 mils for 1800 rpm units will tend to have a negative effect on the operation of the brushes.

Another cause of bad brush-collector performance is axial misalignment of the brushes against the collector. This misalignment can arise from a combination of thermal expansion of the rotor, rotor axial movement off its magnetic center during operation, and wrong initial positioning of the brushes. To avoid this situation, proper axial positioning of the brushes against the collector rings should be visually corroborated when the machine is both cold (after start-up) and hot (after several hours of operation at full load).

Good or bad performance of the brushes depends on the condition of the contact film (called the *patina*). This film is made of carbon and copper and/or metal-oxide particles, elements present in the surrounding atmosphere, and water vapor. The film is primarily dependent on the collector metal being vaporized by the flow of electricity, then oxidized and combined with water. This conducting film lubricates the collector, permitting the brushes to slide with a minimum of wear. However, the integrity of the film is very fragile. Thus, the conditions required for a good film should be maintained as much as possible. Very low temperatures, contaminants, and other parameters detailed previously are likely to affect the condition of the film. Attention should be paid to substantial changes of any one of these conditions, and their possible effect on the performance of the film, should be evaluated.

Given the volatility of the brushes' performance, in particular for machines in which the brushes and collectors are exposed to the environment, a cursory inspection should be made every day the machine is running. More detailed inspections should be carried out in weekly or biweekly periods, or in particular when the machine is standing still or on turning gear.

Close attention should be paid to the condition of the brush rigging and collectors during a major overhaul inspection. At this time, evaluation of the collector's condition will indicate if "turning" or "polishing" of the collector's surfaces is required (see Figs. 9.103 and 9.104 showing collector-ring polishing in situ). Also, attention should be given to any action that may be required to improve the insulation of the collectors, or to replace the collectors due to insufficient thickness or insufficient groove depth (for collectors with cooling grooves), or if rotation of the polarities is indicated. It is normal for brushes to wear faster on the positive collector ring and slower on the nega-

Fig. 9.103 Polishing the collector or slip rings of a 1250 MVA four-pole generator rotor. A grinding stone is used in situ while the rotor is on turning gear, to achieve the desired surface.

Fig. 9.104 A different arrangement than previous figure for polishing a slip ring. In addition to the sandpaper, there is a vacuum hose to remove particles during the polishing activity. (See color insert.)

tive ring. This property is used to defer the replacement of the slip rings by changing polarities periodically (e.g., during major overhauls).

This polarity dependence is more accentuated in *synchronous condensers* of the salient pole variety, where the collectors are also located (in many cases) within the hydrogen medium. Therefore, changing polarities is very common in these types of machines. (Very few cylindrical rotor generators are used as condensers; those used are very often generators used as condensers after the prime mover is decommissioned and replaced with a *pony motor* for starting up the unit.)

In some designs of large turbo-generators, the resulting savings from the collectors may not justify the expense incurred in changing polarities. When polishing the collectors, enough thickness of metal must be maintained to avoid disintegration by the centrifugal forces encountered during operation. The minimum thickness depends on such parameters as type of metal, diameter of the ring, speed of the machine, and calculated surface temperature. We are not aware of a published formula applicable to every type of machine design. The customary and proper action is to obtain data from the machine manufacturer regarding minimum recommended thickness, as well as minimum recommended depth of grooves (if called for in the design). When polishing or "turning" the rings, it is important that the finished surfaces do not exceed the recommended run-out limit of 1 to 2 mils for 3600 rpm machines. Although higher runouts may be acceptable for lower-speed machines, it is recommended to remain within the 2 mils limit.

During extended periods in which the machine is inoperative, the brushes should be lifted from the slip rings and the slip-rings should be protected from humidity or other contamination. The same protection is necessary when the rotor is removed from the machine during large overhauls. In any case, the condition of the collector surfaces should be checked before returning the machine to operation.

During overhauls, the condition of the brush-pressure springs should be evaluated. Discoloration of the springs can be a sign of overheating. Overheating deteriorates the performance of the spring. Normal brush pressures are in the range of 1.75 to 2.25 psi of brush cross-sectional area. These, like any other values given in this book, are for reference purposes only: actual values should be obtained from the brush manufacturer based on the grade of the brushes, the speed of the machine, current densities, and so on. For machines in which overhauls occur at long intervals (5, 6, 7, or more years), the cost of replacing the brush springs could be more than compensated for by the consequent savings in brush and collector material. The constant-force type of spring is preferable.

The dependence on so many factors often means that the search for reasons for bad brush performance may prove to be extremely elusive. The combinations are too numerous to be described here. However, over the years, manufacturers of brushes have developed graphs and procedures as aids to solve problems in a step-by-step approach.

Figures 9.105, 9.106, and 9.107 show the slip-ring assembly in various stages of assembly on the rotor of a two-pole, 60 Hz, 180 MVA unit.

Fig. 9.105 Slip-ring assembly on a 180 MVA, two-pole, 60 Hz rotor. It can be seen that each polarity is made of two rings. Each polarity's connection to the solid copper riser is made in the gap between the same polarity rings (see next figure).

Fig. 9.106 Slip-ring assembly of the previous figure being installed. In the photo, the insulation under the ring is clearly marked by the arrows. The large hole between the rings of the inner pair of rings is where the connection to the solid copper radial lead is made. A similar hole and connection exists on the other pair of rings, 180 degrees away from the viewer.

Fig. 9.107 The pair of rings for same polarity of the previous photo now shown with the connection made to the radial stud. A final brazing of the nut ensures perfect contact, so that it will not come loose during operation.

9.15 COLLECTOR RING INSULATION

The collectors (slip rings) are mounted on a layer of insulation material. This insulation is subject to contamination by oil, water, carbon dust, copper, iron and other metal particles, as well as other chemicals. If those contaminants are not periodically removed, insulation breakdown may occur. This can show up as a grounded field or, worse, may result in a severe short circuit of the dc field current. Figures 9.108 and 9.109 show two instances in which substantial sections of shaft have been "eaten away" by a severe dc ground fault at that location, combined with another short in the brush-gear area. The main reason why the areas under the collector ring shown in the figures suffered such extensive damage was accumulation of carbon dust compromising the insulation under the collectors.

Any severe short circuit at the area of the collector rings is to be avoided at all costs. The resulting damage to the shaft in this particular area can be such that its future use may be compromised. The reason for this is that the area in question is intrinsically weak due to the presence of the vertical holes for the vertical studs that serve to establish the electrical connection between the collectors and the horizontal copper bars running inside the bore of the shaft leading to the end winding. Figure 9.110 clearly demonstrates the weakness of this region of the shaft, though the catastrophic failure

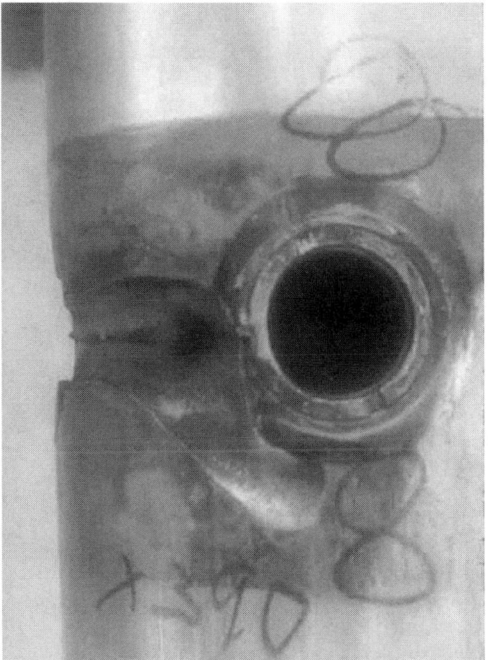

Fig. 9.108 Major dc current short circuit to ground under the collector ring. Faults of this severity have the potential to result in scrapping the entire rotor.

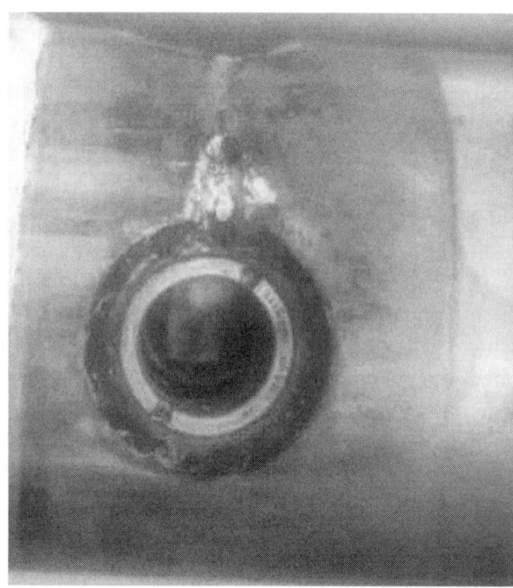

Fig. 9.109 Another severe dc current short circuit under the slip rings, resulting in major damage to the shaft.

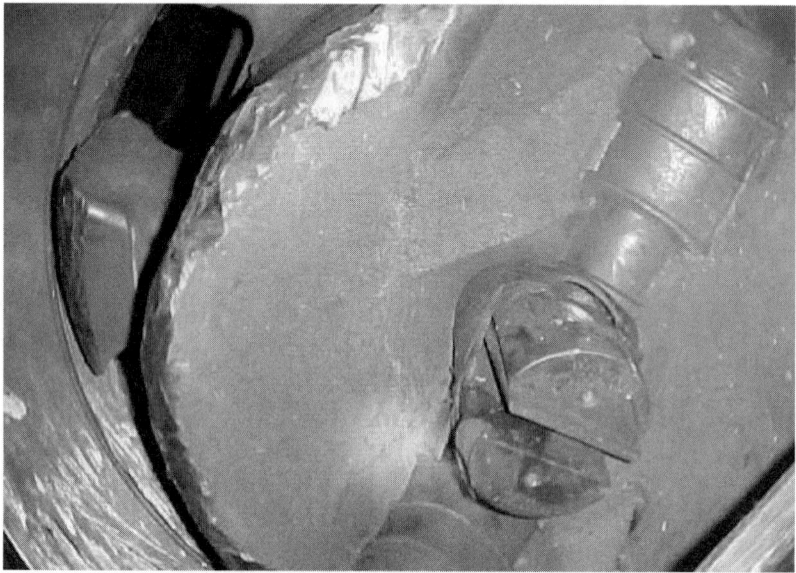

Fig. 9.110 The shaft of this two-pole generator sheared where the vertical holes for the dc current-carrying studs are located. The failure was due to an overspeed condition. The photo affords a rare view of the bore copper bars and the vertical studs. Parts of the insulation can be readily seen.

of this unit was due to an overspeeding condition and not to a short-circuiting of the dc field system. Figure 9.111 shows the same area of a shaft without damage.

During major inspections, it is important to spend time assessing the integrity of this insulation. Figures 9.112 and 9.113 show photos of collector rings being removed for replacement of the solid cylindrical insulation placed between ring and shaft.

9.16 BORE COPPER AND RADIAL (VERTICAL) TERMINAL STUD CONNECTORS

In the previous section, the vertical (radial) copper studs were mentioned several times. Also, the solid copper bars running inside the bore of the shaft from the outer to the inner radial studs were mentioned. Figure 9.110 shows the end of the bore-bars in situ (albeit due to an unfortunate event). Figure 9.114 shows a bore copper conductor outside the bore. Figure 9.115 shows the bore bars installed and viewed from the end of the shaft.

Access to these conductors and to the radial studs is only possible after significant disassembly of the rotor. During major overhauls, it is not uncommon for the connection to the studs to be removed for inspection of the hydrogen seals (chevron seals). At that time, it is possible to inspect to a certain extent the condition of the copper bars. If no unusual field-related events occurred during major inspection periods, it is safe to assume that the condition of these conductors is acceptable. The insulation between

Fig. 9.111 View of a healthy shaft in the vicinity of the vertical hole where the solid vertical studs are located.

Fig. 9.112 Single-polarity double slip ring on a 180 MVA, two-pole, 60 Hz rotor being readied for removal.

Fig. 9.113 Double slip ring from the previous figure shown removed after heating it. The rings are mounted by shrinking them onto the shaft. Removal and installation requires careful heating of the rings until the interference fit is relieved.

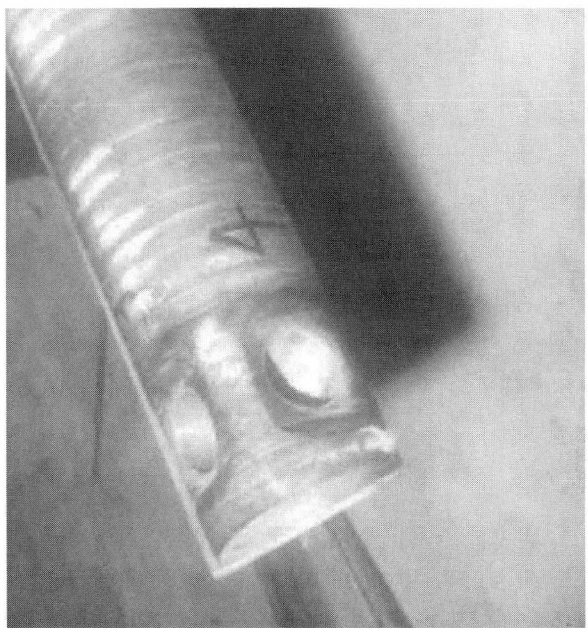

Fig. 9.114 One of a pair of solid copper bars normally located in the bore of a turbogenerator rotor. The function of the bar is to carry the dc field current from the collector rings or diode wheel (for self-excited units) to the end winding. The connections are made through the vertical (radial) studs. The bar shows kinks on its surfaces left there by the forces and thermal–mechanical stresses encountered during operation.

the bore bars can only be directly checked by "meggering" each one with the other grounded and disconnected from the winding. This can be done if the inner radial studs are disconnected for inspection of that area as well as the main flexible lead (discussed earlier in this chapter). If an unusual event occurred (e.g., a ground fault of the dc field system near or on the brush-gear/collector areas), it would be wise to inspect these areas for damage. The determining factors will be the severity of the fault, general condition of the rotor and rotor winding (age, cleanliness, etc.), and the machine's operational history.

Figures 9.116 and 9.117 show the inner and outer connections, respectively, of the main lead to the radial stud. As can be seen in Figure 9.116, in the inner case, the vertical stud is just underneath the centrifugal fan. In this area, it is possible to open the connection without removing the fan. Also seen are the wedges covering the main lead. Inspection of this area requires removal of the wedges. This is not an uncommon activity during long overhauls. In some hydrogen-cooled machines with double sealing, the inner studs as well as the outer studs have seals to prevent the hydrogen from escaping the unit via the bore of the shaft. One of the tests to the rotor during a major overhaul includes pressure testing of these seals (see Chapter 11 on testing).

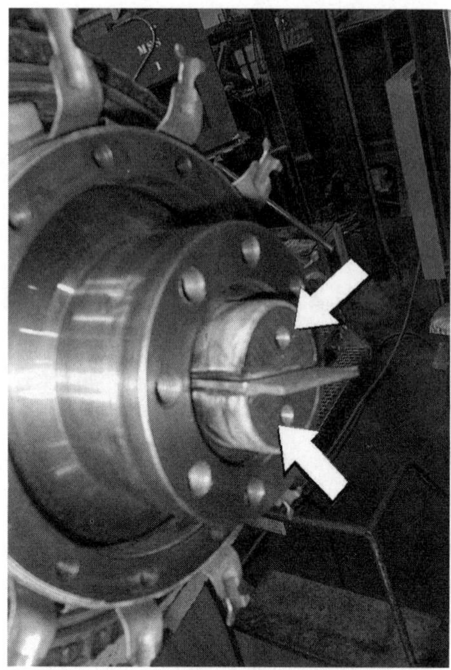

Fig. 9.115 Good view of the bore conductors that run axially inside the shaft. The insulation between both halves (one of each polarity) is clearly visible between them.

One of the major concerns during assembly of the radial connectors or studs is that they should be installed so that they are properly aligned and that no undue or abnormal stress is put on threaded connections or the stud itself (Fig. 9.118). When these are not installed properly, they can fail in a disastrous manner causing missile damage to the stator as well as severe dc arcing damage to the rotor forging (Fig. 9.119).

9.17 BRUSH-SPRING PRESSURE AND GENERAL CONDITION

As stated in Section 9.14, normal brush pressures are in the range of 1.75 to 2.25 psi of brush cross-sectional area. However, actual required values should be obtained from the brush manufacturer or the machine's OEM. The required pressure depends on the brush grade, humidity, current density, peripheral speed, and so forth. Once the desired pressure is known, it is important to make certain that the springs are providing continuous pressure. Constant-force springs are preferable. Discoloration of the springs may indicate a weakened brush and the need for spring replacement.

The condition of the brush springs, like that of brushes, slip rings, and brush gear, must be checked on a regular basis. The overhaul stage should provide the opportunity for brush-spring replacements, if necessary, in accord with operational experience

Fig. 9.116 View of the connection between the inner radial (vertical) studs and the main lead running under the wedges, seen on a two-pole, 60 Hz, 20 MVA generator. The arrow points to the connection nut.

Fig. 9.117 Lead connection, 160 MW unit.

and/or vendor advice. In older machines, a long overhaul could include modifying the brush rig to switch to constant-pressure spring brush holders, if these components were not installed when the machine was originally assembled. Figure 9.120 shows a pair of brush holders with the brushes installed and in good condition. The photo in Figure 9.121 shows a brush contact surface that is damaged and would require resurfacing of the brush. The surface shown is not yet at the failure point but indicates there has been very poor performance and that there is likely some surface damage on the collector rings as well that most likely would require some resurfacing repairs. The condition of the brushes generally indicates what the condition of the collector rings is likely to be during an inspection. Figure 9.122 shows the catastrophic impact on the brush holders from the event shown in Figure 9.123, due to lack of proper brush maintenance and general lack of cleaning of the brush-rigging area.

9.18 BRUSH RIGGING

The brush rig is insulated from ground. During major inspections, this insulation should be examined. A type of contamination similar to that which affects the insulation of the slip rings, discussed in Section 9.15, can be found in the insulation between

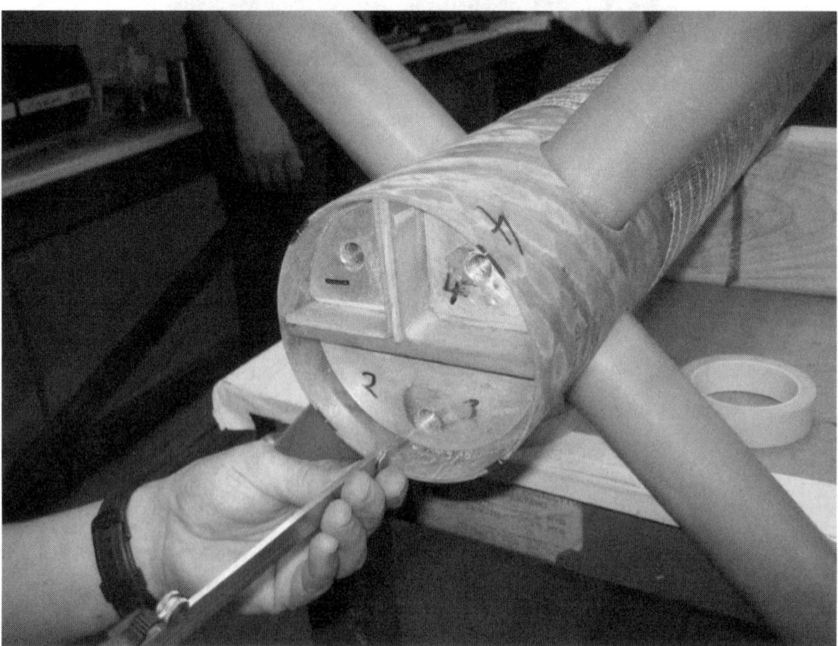

Fig. 9.118 The bore copper and radial connectors of a four-pole rotor, on the bench, before installation in the rotor. The components are temporarily assembled outside the rotor to check that all parts fit together, and to perform electrical insulation testing.

Fig. 9.119 One of the radial studs of the rotor shown had a screw-thread failure and the stud was thrown completely out of the forging. The forging was found severely damaged in the bore-hole area upon closer inspection.

Fig. 9.120 Brush holders with brushes installed and ready for further installation in

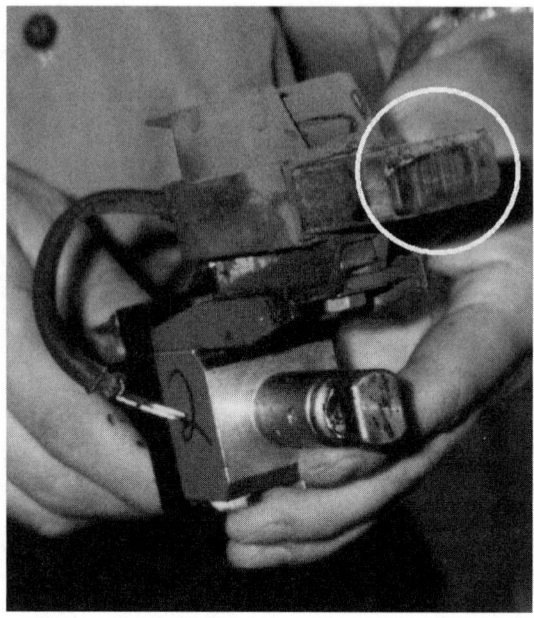

Fig. 9.121 Damaged brush running surface from poor operation of the collector ring surface.

Fig. 9.122 The brushes on these brush holders were left to completely run out. Eventually, the spring behind the brush made contact with the collector's surface, resulting in the catastrophic failure shown in Figure 9.123.

Fig. 9.123 View of excitation base (two-pole, ~800 MVA unit) destroyed by a major short circuit, due to poor maintenance of brushes and lack of general cleanliness of brush-rigging area.

the brush rig and frame. During overhauls it is often necessary to remove the brush -rig in order to disassemble the machine. Thus, the integrity of the brush rig and its insulation is easy to evaluate. Figure 9.123 shows a compromised insulation due to accumulation of carbon dust (from the brushes).

9.19 SHAFT VOLTAGE DISCHARGE (GROUNDING) BRUSHES

In Chapter 8, the dangers to the integrity of the bearings caused by shaft voltages were discussed. In large turbo-generators, discharge brushes are installed whose function is preventing the buildup of excessive shaft voltages. The condition of these devices (normally brushes of different types, e.g., copper braid and carbon) should be carefully and periodically inspected. In some units, the grounding brushes can be readily accessed and, thus, standard procedures call for a cursory inspection during the daily rounds of maintenance electricians. In other machines, visual access requires some disassembly and, thus, special procedures are laid out to ensure that the brushes are inspected on a periodic basis.

In some large units, monitoring systems are installed that send an alarm when higher than normal shaft voltages are present (indicating a possible failure of the ground brushes). There are a number of commercial devices available. They connect to the discharge circuit (i.e., the circuit connecting the grounding brush to ground). Some monitor the current flowing in the circuit. Any significant change on the current's magnitude or spectrum may indicate a problem (see Section 5.5.4 for a detailed description of such an online monitoring device). For instance, if the current goes to zero, it indicates that the grounding circuit is open (due, e.g., to a broken brush or brush spring). Other devices utilize additional "monitoring" brushes. See Figure 9.124 for such a device.

If a ground does occur on the shaft through a bearing or hydrogen seal, there will be considerable arcing from the rotor shaft to the component in which the breakdown has occurred. At the point where breakdown occurs, the shaft grounding brush actually becomes the return path in which circulating currents flow. Figure 9.127 shows the type of electrical pitting damage on the rotor shaft caused by ground currents and arcing. Figure 9.128 shows the electrical pitting and white metal melting that has occurred on a hydrogen seal face due to ground currents and arcing. Figures 9.125 and 9.126 show the components of a shaft grounding system.

9.20 ROTOR WINDING MAIN LEAD HYDROGEN SEALING—INNER AND OUTER

In hydrogen-cooled turbo-generators with brush-fed excitation, the leads between the dc field winding and the collector rings are routed through the interior of the shaft (i.e.,

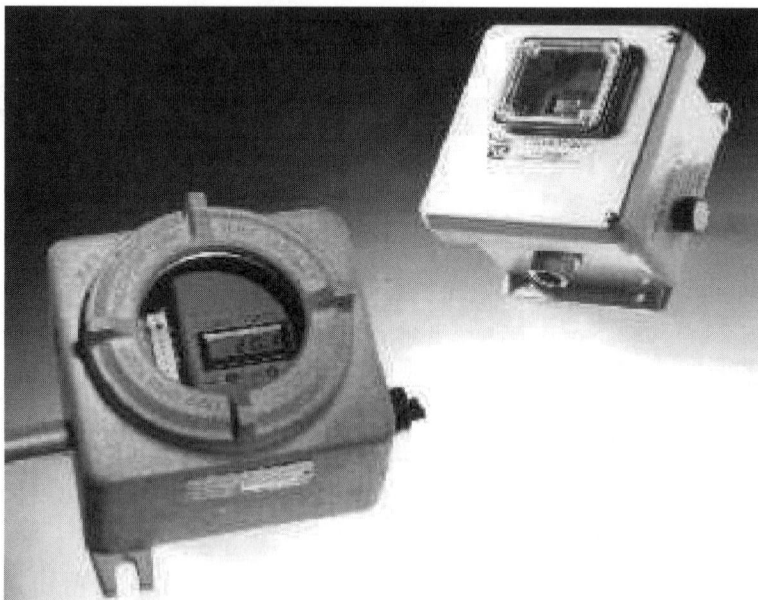

Fig. 9.124 Commercially available shaft-grounding monitoring device. (Reproduced with permission from the MPS Company.)

Fig. 9.125 Double grounding-brush arrangement on a 1300 MVA, 1800 rpm unit. This arrangement allows one to change one brush at a time, maintaining continuous grounding on the shaft.

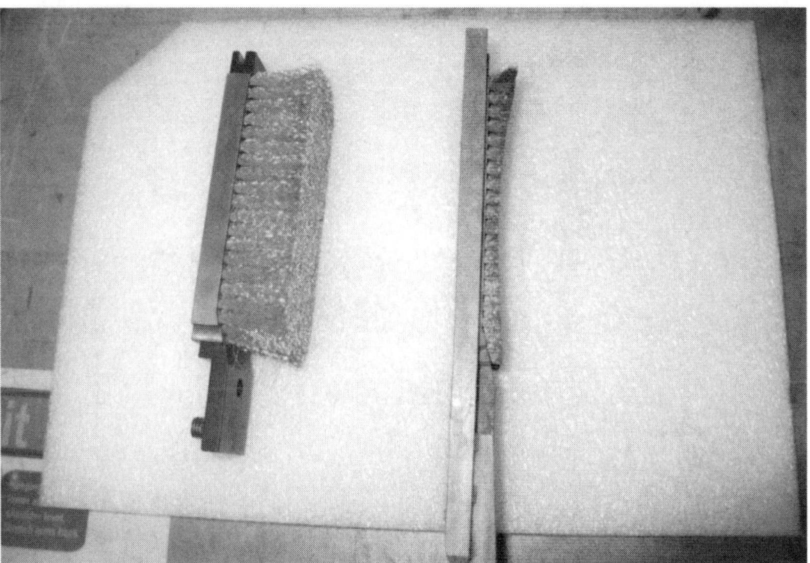

Fig. 9.126 One of the two grounding brushes of the shaft-grounding system shown in the previous figure. Two brushes are shown: one new and ready to be installed, and one removed for replacement.

Fig. 9.127 The rotor shaft in this photo shows an extreme case of electrical arcing from very high voltage built up on the shaft and then breaking down the bearing-oil film. The actual root cause was a stator core fault of significant magnitude that caused up to 1000 volts to be present on the rotor shaft. The grounding system alarmed in this particular case, but the voltage was excessively high and electrical arcing caused significant damage to the shaft journal. There was similar damage to the bearings, although no photo is available.

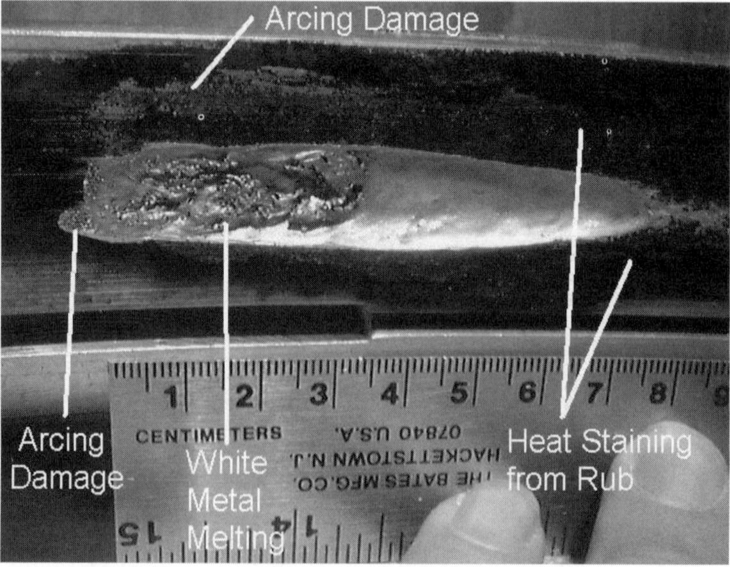

Fig. 9.128 This photo shows the arcing damage found on the face of a journal-type hydrogen seal after a rub, significant arcing, and ground currents flowing. As well as electrical pitting, melting of the white metal is seen and heat staining from the shaft rub.

portions of the shaft are hollow). (See Figs. 9.87, 9.114, and 9.115, as well as Chapter 2 for a comprehensive description of these components.) Were specific precautions not taken, the same space would allow the pressurized hydrogen in the machine to escape. Escaped hydrogen has been known to mix with oxygen and cause explosions in the brush-rig compartment due to arcing of the brushes or other igniting source. This directly poses a potential safety hazard to personnel.

To eliminate the possibility of hydrogen leaks, so-called hydrogen seals are installed. Often, two sets are installed, one in the inner part of the rotor, where the leads leave the shaft and connect to the field winding, and another outside the casing, beneath the collector rings, where the connection between those and the leads is located. There are two seals in each location, one per lead. However, in some machines there is only one set of hydrogen seals, normally in the inner side of the shaft.

It is practically impossible to assess the condition of these seals visually, unless a significant dismantling of rotor parts is carried out. Even then, the final and definitive test is a pressure test carried out by means of specially provided nipples connecting to the bore, or by fabricating a specially designed cylindrical can, which fits snugly over the shaft extension and seals the area of the slip rings. This can is then pressurized, exceeding the maximum operating pressure, and the pressure is monitored for a number of hours. If the pressure is maintained, then the integrity of the hydrogen seals is demonstrated. This type of testing is carried out during extended outages, mostly with the rotor is outside the bore. The test can be done with the rotor in situ, if required. Obviously, air-cooled units do not have the same sealing requirements as hydrogen-cooled units. Nevertheless, sealing of the bore area in the shaft is always important to keep the contamination away from there, for the purpose of preventing failure of the insulation that keeps the bore conductors from failing to ground or between polarities.

Figure 9.129 shows the hydrogen seals in the end of the shaft of a machine designed with one of the collector rings connected at the end of the shaft to the bore conductors.

It is also important to seal the turbine end of the shaft if it has a full-length hollow borehole, and most rotors pre-1990s machines have such full-length hollow bores. Sealing is generally done by a screwed-in faceplate with an inner rubber gasket on the end of the shaft. In addition, a test plug is often installed to allow a rotor bore pressure test to ensure that the seal is leak tight (Figure 9.130).

9.21 CIRCUMFERENTIAL POLE SLOTS (BODY FLEX SLOTS)

In certain round-rotor designs, a number of circumferential slots are machined in the poles of the forging, a few inches apart from each other (see Fig. 9.131). The purpose of these slots is to introduce a measure of flexing freedom to the rotor by stiffness equalization in the pole axis to minimize vibration. (See Chapter 2 for more information about the function of body flex slots.) The tips of the slots, where they reach the outside diameter (OD) of the rotor in the face of the pole, are areas prone to develop stress cracks under certain abnormal operating conditions. For instance, excessive asynchronous operation (without the dc field current) might result in currents flowing through the body of the rotor for extended periods. In this case, the tips of the slots be-

Fig. 9.129 Collector end-of-shaft hydrogen seal.

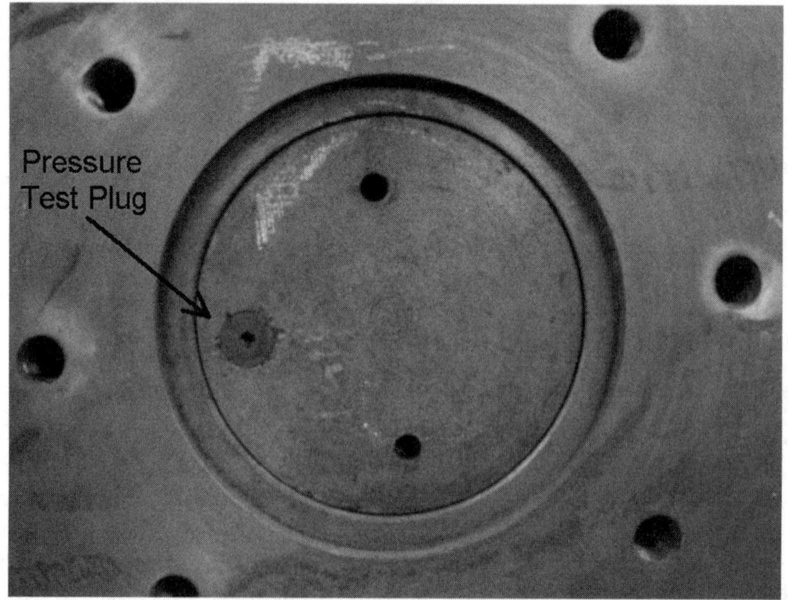

Fig. 9.130 Turbine end-of-shaft hydrogen seal plate with pressure-test plug.

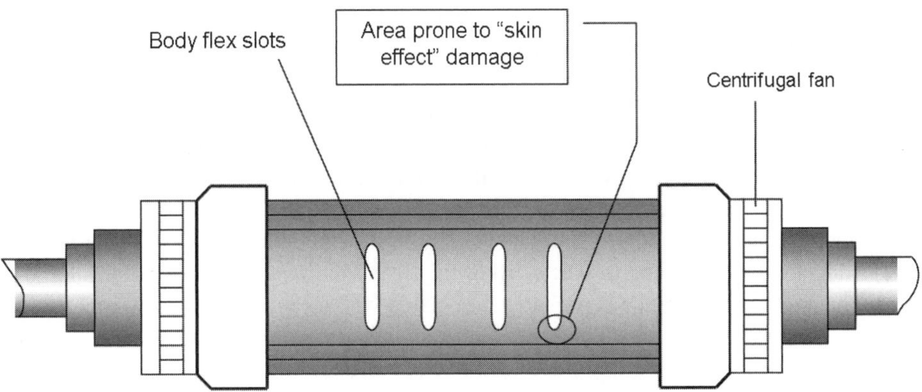

Fig. 9.131 This sketch shows the location of the flex slots and the areas requiring careful inspection for signs of overheating due to skin currents.

come areas carrying high current densities. The highly localized heat generated by the high-density current flowing while these abnormal conditions exist may result in localized changes to the metallurgic properties of the steel. Consequently, the more brittle metal may develop cracks due to metal fatigue as the rotor flexes at twice the supply frequency.

To avoid the possibility of catastrophic crack growth, the inspection team should be aware of any signs of overheating in those areas (normally appearing as discoloration of the metal). If encountered, a nondestructive test or examination (NDT or NDE) should be performed. Any crack found should be cause for further investigation and the initiation of adequate remedial measures.

9.22 BLOCKED ROTOR RADIAL VENT HOLES—SHIFTING OF WINDING AND/OR INSULATION

Rotor designs may or may not include radial vent holes in the retaining rings. Figures 9.6 and 9.7 show such a ring. However, all large turbo-generators have radial vent holes of one shape or another on their body. It is essential when performing an inspection of the rotor to verify that the vents offer unobstructed passage for the cooling gas, be it hydrogen or air. It is easy, with a mirror and a flashlight, to look inside the vent hole and see any obstruction when the rotor is outside the bore.

Obstructions can come in several forms. However, by and large, the most common are introduced by shifting of the insulation between conductors (or/and the top insulating filler), or by shifting of the entire winding. Insulation may shift as a whole or in pieces broken away from the main insulating strip between turns.

Small amounts of coil and insulation shifting is not uncommon. Most rotors with many hours of operation will exhibit some blockage of the vents. There is no magic number as to what amounts to the beginning of a problem. It is also dependent on the general pattern exhibited (i.e., the shift may be only in one slot, or a small number of slots, or all around the circumference). However, in general, when the combined blockage of the radial vent holes in any slot (as seen from the wedges) is around 15% or higher, it is important to consult with the OEM about remedial actions, if so indicated. Keep in mind that any substantial blocking of the vents results in increased operating temperature of the rotor winding, to the point that the design margin may be eroded, reducing the expected life of the insulation.

In some cases, blocked or semiblocked vent holes can be the result of a more ominous development: shifting of part or all of the field winding toward one end of the rotor. This type of occurrence is probably associated with a rotor exhibiting thermal sensitivity (see Chapter 11). This problem requires partial or total rewind and/or reblocking of the winding. Figures 9.132 and 9.133 show examples of blocked vents due to this type of winding shift.

9.23 COUPLINGS AND COUPLING BOLTS

During major overhaul inspections, the couplings (turbine to generator and generator to exciter, if existing) should be checked for signs of distress. The condition of the coupling is important to maintaining good mechanical alignment and, thus, low vibration levels. Additionally, signs of chafing, fretting, or other physical anomalies may indicate the occurrence of an abnormal operational event in which large torque swings were involved. This type of event may be detrimental to other components of the rotating equipment, and further investigation and perhaps inspection and testing activities may be warranted. Figure 9.134 shows a coupling removed from the rotor. Figure 9.135 shows damage found on the coupling.

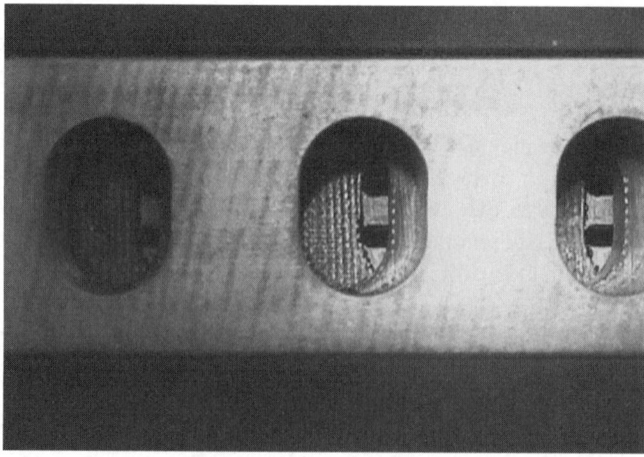

Fig. 9.132 Radial vent of a rotor about 80% blocked due to shifting of the winding and interturn insulation (two-pole, 60 Hz, 180 MVA unit).

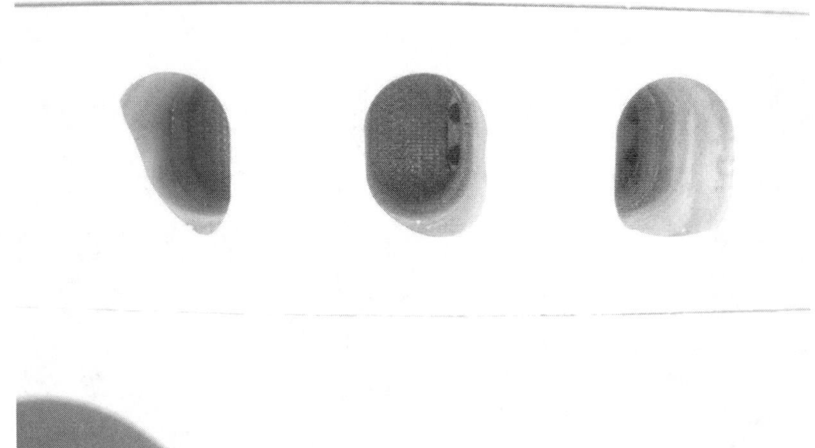

Fig. 9.133 Same rotor as in previous figure. The vent is almost completely blocked.

Fig. 9.134 Coupling from a two-pole generator removed as part of the rewinding of the rotor. The keybars can be seen in situ.

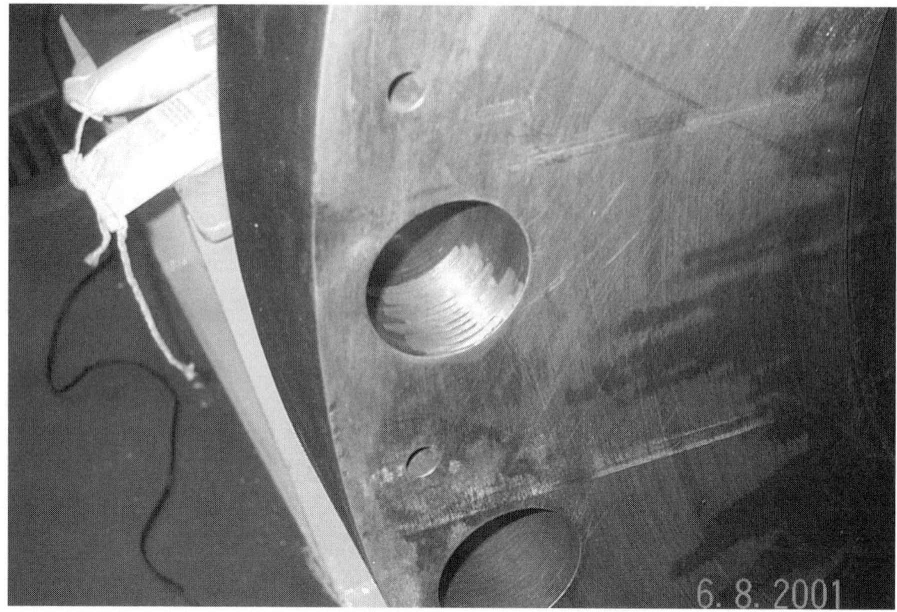

Fig. 9.135 Damage to the threads in one hole belonging to the coupling in the previous figure.

9.24 BEARING INSULATION

Bearings of large turbo-generators are of the *friction* type, also called *sleeve* bearings. It goes without saying that the bearings of large alternators must be in excellent condition to operate reliably at low vibration levels over long periods of time (some units overhauls as many as eight years go between). Thus, the bearings must be carefully inspected and refurbished during a long overhaul by knowledgeable and experienced personnel. The babbitt metal must be inspected for anomalies and proper dimensions, and, if required, the bearing must be rebabbitted. The shell must be inspected for cracks or any signs of distress. The same applies to the vibration probes and other instrumentation connected to the bearings.

One of the most important operational and maintenance issues related to bearings in high-voltage, high-power rotating machinery is control of shaft voltages and currents. During the normal operation of the generator, voltages and, hence, currents are induced in the shaft. In the case of large turbo-generators, in addition to voltages induced in the generator, substantial voltages are created by the rotating elements of the turbine. These shaft voltages and, in particular, the resultant currents, have to be kept to low values; otherwise, subsequent bearing failures can be expected to occur. The causes for the presence of shaft voltages and currents are discussed in detail in Chapter 11.

In large turbo-generators, peak-to-peak voltages in excess of 150 V are not uncommon [14]. Such voltages are largely due to electrostatic build-up from the turbine, caused by water droplets impacting on the blades. These voltages have the potential to generate currents that, when allowed to flow freely, will destroy bearing surfaces, oil seals, and other close-tolerance machined surfaces.

The heat developed by the flow of current does not, as is sometimes assumed, cause damage. Instead, the damage is caused by mechanical action. Shaft bearing currents will damage the bearing surfaces by pitting resulting from minuscule electric discharges. The pitting will continue until the bearing surfaces lose their low coefficient of friction; then other more dramatic and rapid changes occur, culminating in bearing failure. Bearing currents also have adverse effects on the lubrication oil by altering its chemical properties. Control of shaft voltages is achieved by taking certain precautions when designing the machine and by introducing shaft-grounding elements and/or bearing insulation.

Grounding devices, which can be copper braids or silver graphite or copper graphite brushes, should be inspected often. Bearing insulation can be inspected often in certain bearing-insulation designs with the machine online or offline, and after a certain amount of disassembly of pipes and other attachments has been performed.

During major inspections, the bearings are inspected for signs of pitting caused by shaft (bearing) currents. These are easily recognized with help of a magnifying glass. They appear as shiny and well-rounded little droplets. If the bearing surfaces are found to be damaged by shaft currents, then the reason for the existence of significant shaft currents should be investigated. Strong candidates are a bridged or faulty bearing insulation, and/or defective or missing grounding devices. Pitted surfaces of the bearings should be rebabbitted. Pitted journals should be polished.

The bearing insulation is commonly made of mechanically strong water- and oil-resistant insulation materials formed from fiberglass or similar bases, laminated and impregnated with resins, polyesters, or epoxies.

In bracket-mounted (end-shield-mounted) bearings, the bearing insulation takes the shape of a ring or collar surrounding the bearing or, less commonly, as an insulation layer between bracket and casing, piping, and so forth. In pedestal-type bearings, the insulation is made of one or two plates, normally placed at the bottom of the pedestal. When double plates are used, they have "sandwiched" between them a grounded metal plate (Fig. 9.136). This system allows each of the insulation layers to be tested without the necessity of interrupting the operation of the machine, uncoupling it, and taking care of the other arrangements normally required to measure the bearing insulation.

Machines normally have the non-drive-end (outboard-end) bearing insulated. However, machines with couplings at both ends or those driven by turbines may have both bearings insulated. In some cases, the couplings are also insulated.

It is important to verify during the inspection that the bearing insulation is not contaminated with carbon dust or accidentally bridged with chunks of metal touching the bearing pedestal, temperature or vibration sensors, noninsulated oil piping, and so forth.

If electrical testing of the bearings' insulation is performed, the measured values should be in the hundreds of thousands of ohms. However, in this application, as in many other areas of insulation practice, a wide range of resistance values can be found in the literature: from 100 kΩ to 10 MΩ or greater. Given the relatively large shaft

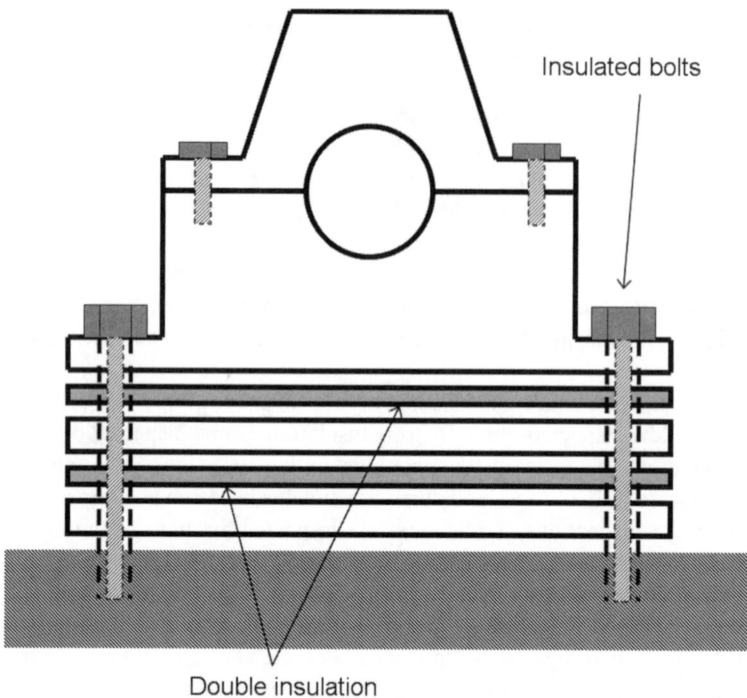

Fig. 9.136 Schematic representation of a pedestal bearing with two layers of insulation. This arrangement allows testing the value of the insulation in each layer with the generator in operation. Otherwise, testing the insulation of a bearing with one layer of insulation only can be done with the unit shut down, and making the measurement will actually ground the rotor.

voltages encountered in large turbine generators, it is preferable to have insulation-resistance values in the megaohm region.

Reliance only on grounding devices is not recommended. A grounding device is required to reduce the level of voltages in the shaft to values compatible with the small clearances encountered in bearing insulation and between shaft and seals. However, the normal contact resistance of grounding devices does not eliminate shaft voltages to the extent that the bearing insulation would become redundant [15]. Grounding devices are often taken for granted—and therefore neglected—during normal maintenance procedures. Thus, it is important they be inspected carefully during major inspections (see Section 11.9 on bearing testing in Chapter 11).

9.25 HYDROGEN SEALS

All hydrogen-cooled generators have in common the need to maintain the hydrogen pressure inside the machine at anywhere from 30 to 90 psi (about 2 to 6 atmospheres) without allowing a significant leak to the exterior. As explained elsewhere in this book, hydrogen leaks can pose significant safety risks, because certain concentration levels of hydrogen and air can become highly explosive. Accidents have occurred due to hydrogen leaking and concentrating in areas next to the generator (for instance, the excitation housing).

Hydrogen can leak out of the machine in a number of ways. For instance, hydrogen can leak into the water cooling system of generators with direct water-cooled stators. This type of problem is addressed elsewhere in this book. Hydrogen can also leak through cracks in the envelope of the generator, as well as through inadequate sitting of end-shields, bolts, wires, covers, and any other plates covering penetrations through the casings. Hydrogen can also leak through the bore of the shaft, when the upshaft solid seals are dry or otherwise aged, and through testing holes tapped to the shaft bore. These are also covered elsewhere in this book. The main concern is hydrogen leaking at the shaft-to-stator casing interface of the machine, where a specific clearance is required to allow unhindered rotation. To deal with this issue, *hydrogen seals* are installed.

Hydrogen seals are systems that use oil at pressures higher than the hydrogen that is injected, filling the gap between the shaft and the seals. Chapter 2 fully describes the *journal* and *thrust-collar* type of seals. The following is a list of items to be inspected and tested during a generator inspection.

9.25.1 Journal Seals

Inspection:

Babbit baterial and steel shell
Bronze or brass seal rings
Seal housing
Seal-oil wipers
Springs and pressure components
Gaskets and O-rings

NDE:

Liquid-penetrating inspection (LPI) for cracks
Ultrasonic (UT) for babbit bonding to shell

Insulation Resistance:

Megger check of the seal insulation to ground, to ensure that the rotor shaft is not grounded through the hydrogen seals. Usually done at no higher than 500 volts.

Typical Maintenance:

Seal rebabbitting
Babbit scraping
Bronze-seal repairs
Consumable component replacement
Oil flushing

Figures 9.137 to 9.140 show various types of damage to journal-type hydrogen seals and seal wipers.

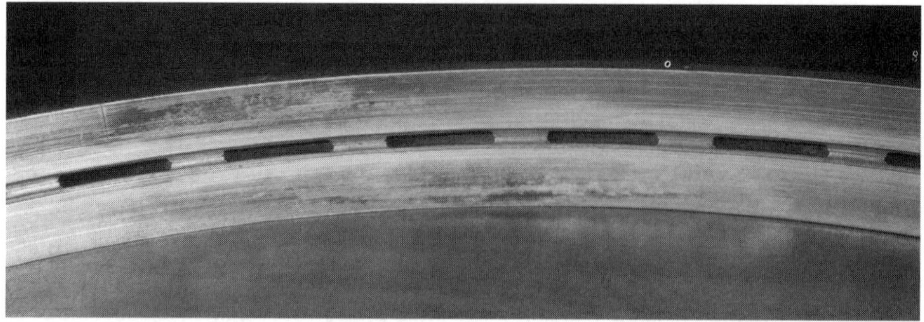

Fig. 9.137 Mild heat-staining damage on a journal-type hydrogen seal from a "light" shaft-to-seal rub.

9.25.2 Thrust-Collar Seals

These types of seals are less common than journal seals. They are somewhat more difficult to align and maintain, but, in general, the same type of inspection activities that were listed for journal seals are also carried out on thrust-collar seals.

Figures 9.142 to 9.144 show various types of damage on thrust-collar-type hydrogen seals.

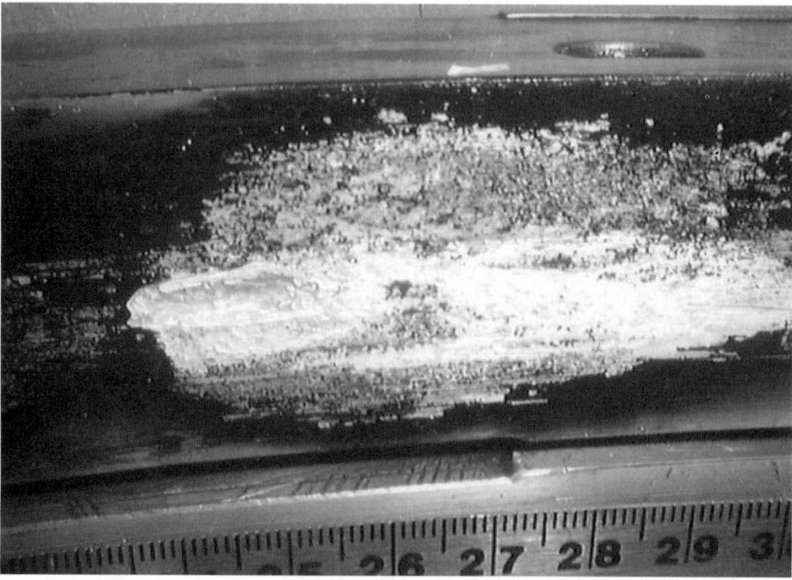

Fig. 9.138 Massive damage on a journal-type hydrogen seal from loss of oil and a severe rub. The combined effects of severe heat staining, electrical pitting, melting, and debris pick-up, with resulting white-metal smudging, can be seen.

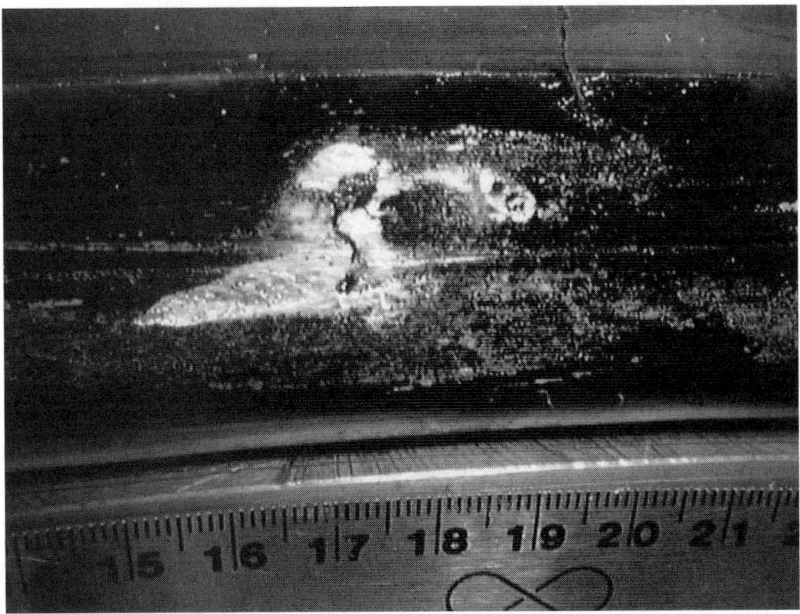

Fig. 9.139 This photo shows the same damage as in Figure 9.138, but further advanced such that the white metal has hardened and cracked beyond repair.

Fig. 9.140 Damaged journal-type hydrogen-seal wipers are shown here. The wiper strips have been deformed by a rub and allowed the seal oil to blow past the wipers, causing a very dangerous operating situation. The possibility of hydrogen leakage from the machine is a concern with failures such as this.

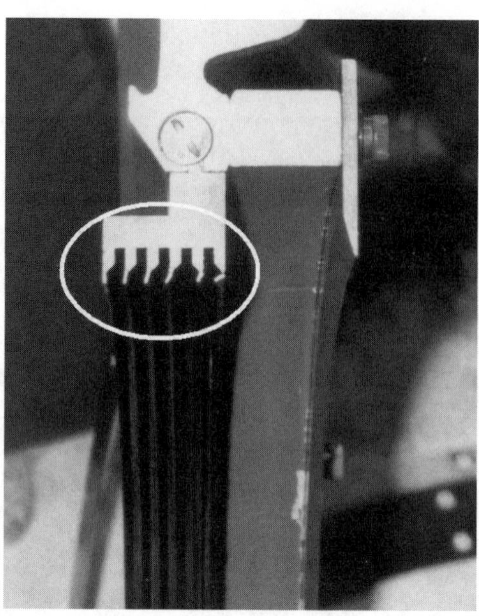

Fig. 9.141 The journal-type hydrogen-seal wipers shown in this photo have been severely rubbed such that some are bent over and cracked. Extensive seal-oil leakage past the wipers again creates a very dangerous operating situation for oil and hydrogen fires occuring if hydrogen were to leak at this location.

Fig. 9.142 Damage on a thrust-collar-type hydrogen seal. The continuous seal face on the hydrogen side of the seal is heat stained from a rub, as is the tapered land section on the air side of the seal. In addition, the tapered land portion has cracked and hardened white-metal damage.

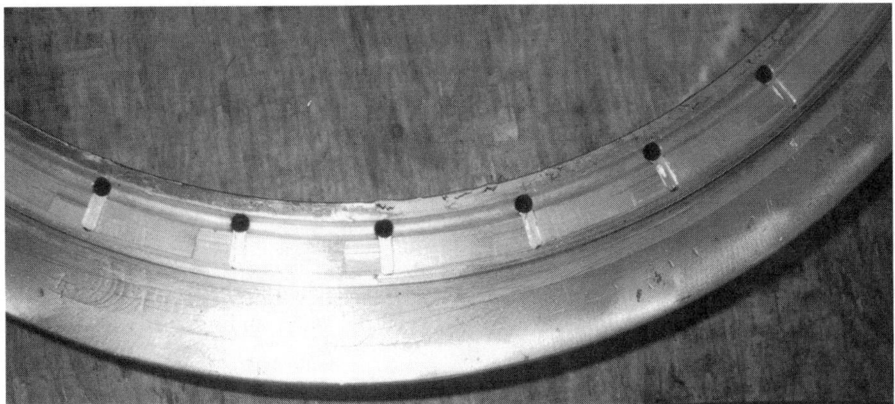

Fig. 9.143 This thrust-collar-type hydrogen seal shows cracked white metal on the continuous seal surface on the hydrogen side of the seal.

9.25.3 Carbon Seals

In recent years, a completely new hydrogen seal has been developed. Made of a carbon wiping surface, it appears to exhibit a number of advantages over the conventional babbit seal. It uses less oil compared to conventional seals, is smaller, and can last significantly longer under oil starvation without causing major damage to the shaft. Finally, the carbon seal virtually eliminates seal rubs due to the large shaft-to-seal housing clearance afforded by its design.

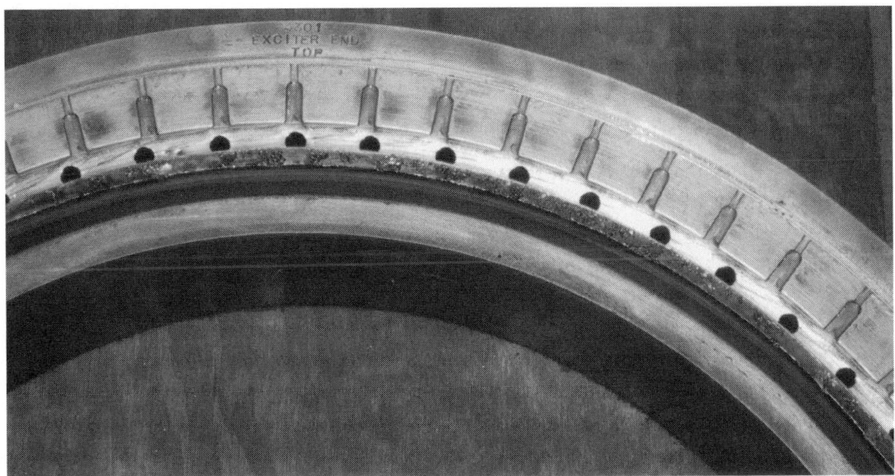

Fig. 9.144 This thrust-collar-type hydrogen seal shows eroded white metal on the continuous seal surface from electrical arcing and mechanical pickup of debris in the oil.

As a retrofit, the new seal will probably require considerable expense. When installed in new machines, its long-term advantages may be a good trade-off against any additional initial cost.

Figures 9.145, 9.146, and 9.147 show the carbon hydrogen seal and some of its performance characteristics.

9.26 ROTOR-BODY ZONE RINGS

Some generators employ *zone rings* or *gas guides* attached to the rotor body, used to more efficiently direct cooling gas flow between the stator and rotor cooling circuits

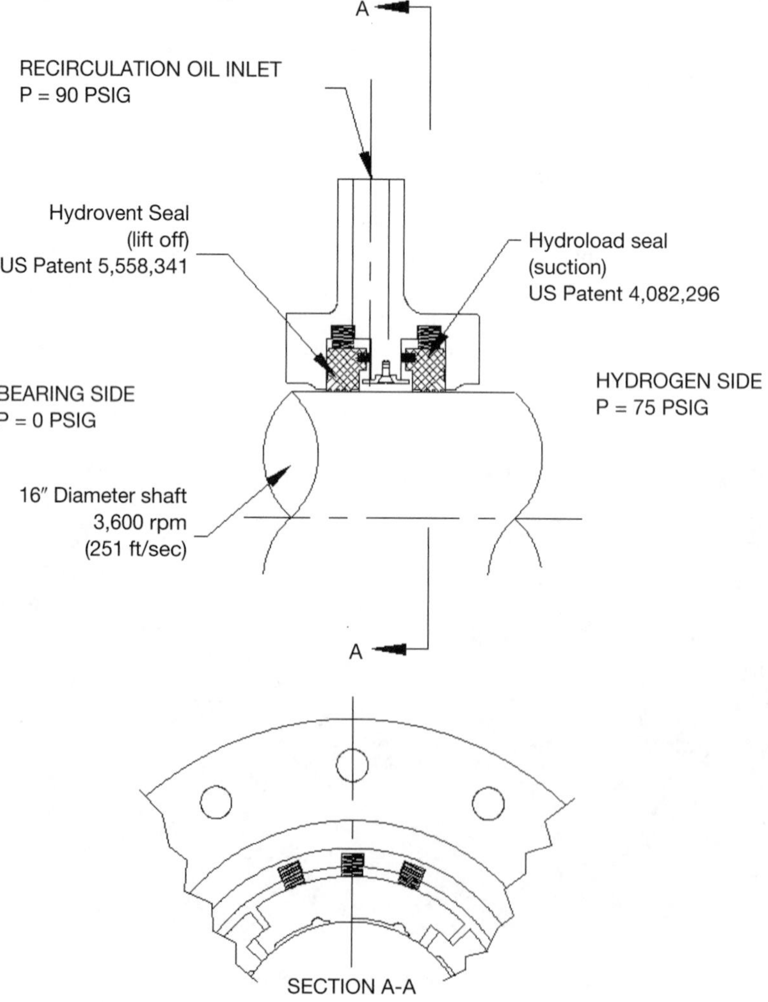

Fig. 9.145 Carbon journal-type oil seal (courtesy of Siemens & Stein Seal Co.).

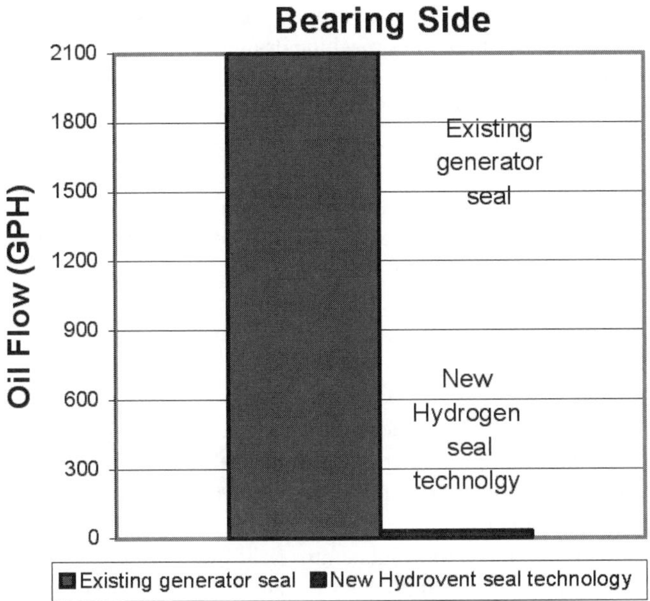

Fig. 9.146 Oil flow of carbon seal versus existing technology—bearing side (courtesy of Siemens).

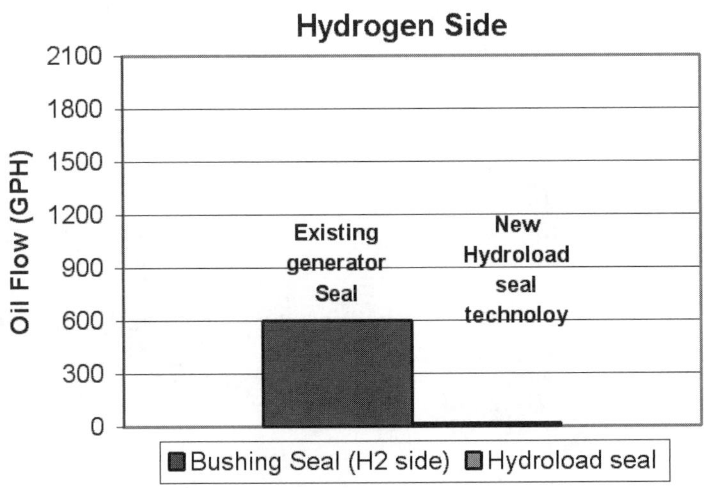

Fig. 9.147 Oil flow of carbon seal versus existing technology—hydrogen side (courtesy of Siemens).

(Figure 9.148). The purpose of the rotor-body zone rings is basically the same as the stator zone rings employed on other machine designs.

Rotor zone rings are made of nonmagnetic steel (similar to retaining-ring materials) and are shrunk fit to the rotor body. They are a highly stressed component and susceptible to the same basic mechanisms as retaining rings. They need to be thoroughly inspected periodically and any problems found corrected so that they do not become a potential foreign object in the airgap of the generator when the machine is in operation. The consequences can be severe if large debris comes loose between the stator and rotor when the rotor is at speed.

9.27 ROTOR REMOVAL

This chapter has been all about inspection of the rotor, mostly while removed from the stator. Removing the rotor poses some significant risks to the integrity of the stator and the rotor itself. Like any other critical activity, it must be carried out by knowledgeable and experienced personnel.

The operation to remove a turbo-generator field from the stator is very similar in all situations, though there are dissimilarities depending on the size of the rotor; the tools available, such as crane and clamps; and the location of the machine. Figures 9.149 and 9.150 show a smallish rotor being removed.

Figure 9.151 shows the shoe of a very large four-pole rotor firmly secured to the shaft.

Once the trailing end of the rotor is firmly supported by the sliding shoe and the skid plate is greased and tightly held by steel wires so it does not slide with the rotor, the *pulling* stage starts. Pulling is carried out by different means, depending on the size of the machine and the tools available. Figure 9.152 shows a small rotor being pulled with a forklift. Figure 9.153 shows a very large rotor being pulled by a dedicated trolley, which slides by means of two hydraulic pistons (Fig. 9.154).

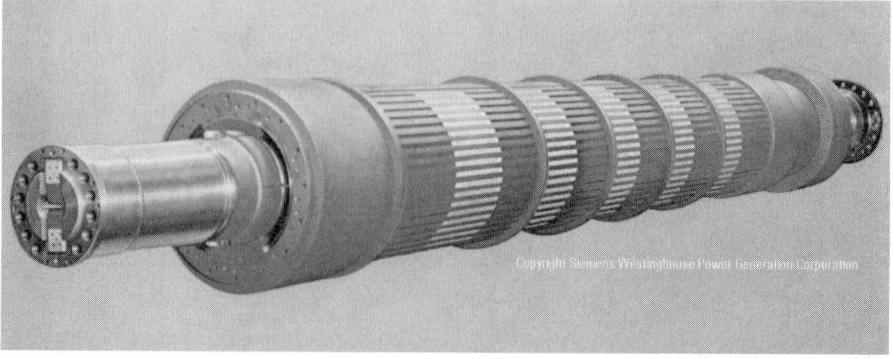

Fig. 9.148 Rotor-body zone rings installed on a large two-pole rotor. (Reprinted with permission of Siemens Westinghouse Power Corporation.)

One side of the rotor
held by a sling from
a crane. The journal
surface where the
sling is attached
must be protected
from damage

Fig. 9.149 A small two-pole rotor being removed from the stator. The sling used at the end of the rotor to be removed first is visible (direction of removal indicated by the thick arrow).

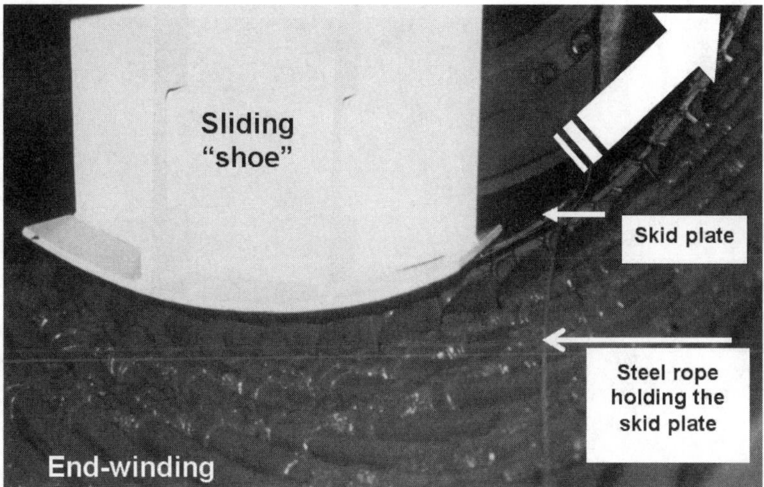

Fig. 9.150 The photograph shows the other end of the rotor in Figure 9.149. The direction of removal is away from the viewer, as indicated by the thick arrow. This end of the rotor rests on a so-called "shoe." The shoe is firmly bolted to the rotor shaft, and it slides on a skid plate made of steel and supported from one side by steel wires. Grease is applied to the skid plate so that the shoe slides easily.

Fig. 9.151 Massive "shoe" supporting on end of a 200 metric ton rotor being removed. The shoe is strongly held to the shaft by clamps and sits on the greased skid plate.

Fig. 9.152 Small rotor being pulled out of the stator by a forklift.

Fig. 9.153 Large four-pole rotor leading end attached to a hydraulic trolley used to pull the rotor out without a tilt.

Fig. 9.154 The trolley of the previous figure is pulled by two hydraulic pistons.

The common denominator of the pulling activity is that it must be slow and very well controlled at all times. Otherwise, damage to the stator, the rotor, or both can occur. Common risks are tilting the rotor unevenly in the bore, causing a rub against the stator laminations, or against zone rings, when present; or having the rotor slide too far, damaging the end winding. Figures 8.63 and 8.64 in Chapter 8 show an example of damage to the stator laminations while removing a two-pole rotor. Figure 9.155 shows core damage while removing another two-pole rotor. Figure 8.28 in Chapter 8 shows a stator rubber zone ring damaged (ties torn) by a rotor during removal or installation (it was not clear to the plant personnel when the ring was damaged).

One thing is certain: every effort carried out in removing the rotor carefully must be redoubled when reinstalling it into the bore. The reason for this is that whereas damages done during the removal operation can be seen with the naked eye and fixed during the outage, damage done during the installation procedure may not be seen for an entire cycle (perhaps up to 10 years) until the rotor comes out again. In many cases, no further deterioration may occur, but in some, lamination damage can over time and create core hot spots (in the teeth) that eventually require a much larger repair job.

The next stage happens when the center of gravity of the rotor is somewhat past the end of the unit and the trailing end of the rotor is still resting close to the straight part of the core on the shoe. Then, the leading end of the rotor must be held securely in position while a sling is wrapped around the forging. In very large units, using dedicated pull trolleys such as that shown in Figure 9.135 allows the rotor to remain on the trol-

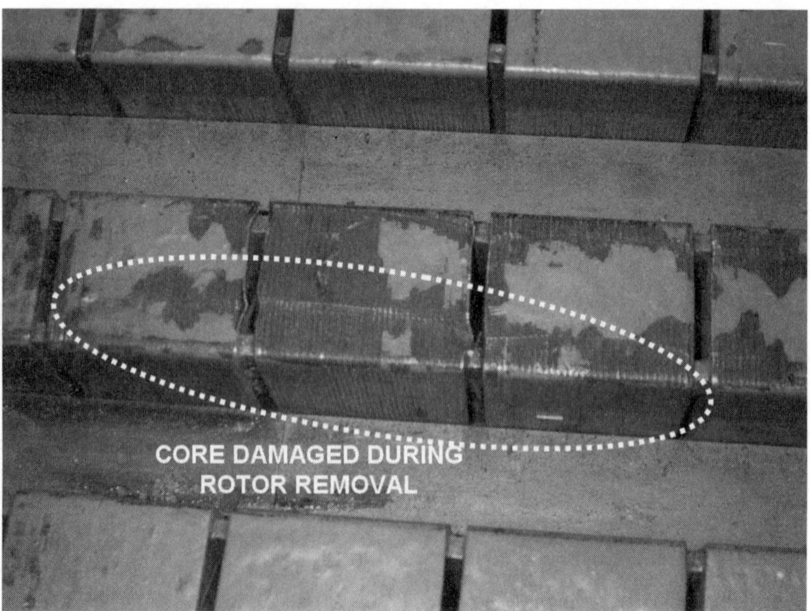

Fig. 9.155 Damaged core during a rotor removal operation. This is a two-pole machine with deep stator-tooth bridges.

ley until the newly attached sling very slightly lifts it and moves it out the bore. In locations where a trolley is not available and the unit's leading edge is supported by a sling as it moves out the bore (Figure 9.152), the rotor must be temporarily cribbed on blocks of timber. It is of the utmost importance that the cribbing not be placed under the retaining ring or under the forging in the area where the wedges are located to avoid serious damage. The cribbing must only be placed under the pole face. Thus, it is important to take this into consideration before the shoe is attached to the rotor at the beginning of the operation. The rotor must be rotated such that one of the pole faces is "looking" down. Figures 9.156 and 9.157 show a small rotor before the cribbing is erected, and then after it is placed on the cribbing. Figure 9.158 shows a cribbed rotor of a large unit. In both cases, the same criteria are followed.

Next, after the sling is moved to the center of the rotor—making sure that the load is balanced so that no tilting will occur that can damage the core laminations—the rotor is slowly completely pulled out the bore. The final resting place of the rotor can be on cribs or on rollers. Rollers are most often preferred because the rotor can be easily rotated for inspection or other activities. Figures 9.159, 9.160, and 9.161 show small, large, and very large rotors being lifted during removal from the stator. Figure 9.162 shows a rotor set outside the stator on rollers. The rollers can be placed under the shaft or, in some cases, under the forging if the wedges are in. Figure 9.163 shows a skid plate still in the stator bore after the rotor was removed.

The procedure for reinstalling the rotor in the stator bore includes all the steps delineated above but in reverse.

Fig. 9.156 Rotor being supported at the leading end by a sling attached to an overhead crane.

Fig. 9.157 Rotor of previous figure shown with the leading edge resting on a timber crib. The crib is under a pole face close to the retaining ring, but without touching it.

Fig. 9.158 A 300 MW, two-pole rotor with its leading end temporarily resting on a timber crib. The crib is supporting the rotor at its pole face.

Fig. 9.159 Small rotor hanging from a crane with the shoe still attached.

Fig. 9.160 Slings being attached to a large two-pole rotor for removal from the stator.

Fig. 9.161 Very large and heavy (200 tons) four-pole rotor being lifted by slings attached to a dedicated lifting beam weighting about 5 tons.

Fig. 9.162 Rotor placed on rollers after being removed from the stator.

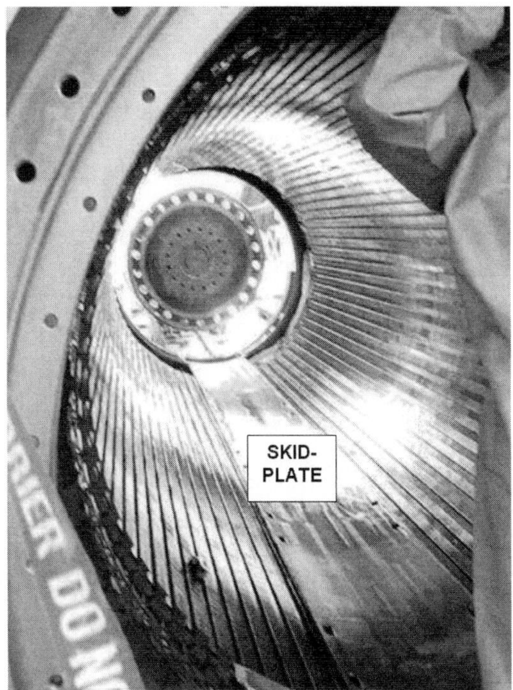

Fig. 9.163 A skid plate inside the stator bore after the rotor was removed. Between the skid plate and the core, a soft material is placed to protect the laminations from damage. The top surface of the skid plate is greased so that the rotor's shoe will slide without too much resistance.

REFERENCES

1. EPRI Special Report EL/EM-5117-SR, "Guidelines for Evaluation of Generator Retaining-rings," April 1987.
2. EPRI Report TR-10209, "Evaluation of Nonmagnetic Generator Retaining-rings," October 1994.
3. EPRI Report TR-102949, "Generator Retaining-Ring Moisture Protection Guide," September 1993.
4. EPRI Report EL-5825, Proceedings: Generator Retaining-Ring Workshop, May 1988.
5. "Turbine Generator Retaining-rings—Supplementary Information on the Inspection of Rings Manufactured in 18Mn-18Cr Steel," *Electra Magazine,* October 1998.
6. H. Feichtinger, G. Stein, and I. Hucklenbroich, "Case History of a 18Mn–18Cr Retaining-ring Affected by Stress Corrosion Cracks," presented at EPRI-Generator Retaining-ring Workshop, December 1997, Miami, FL.
7. M. O. Speidel and R. Magdowski, "Stress Corrosion Service Experience with Generator Retaining-rings," presented at EPRI-Generator Retaining-ring Workshop, December 1997, Miami, FL.

8. G. Stein, M. Wagner, and L. Hentz, "Investigation of Crack Formation on the Inner Diameter of a Retaining-ring," presented at EPRI-Generator Retaining-ring Workshop, December 1997, Miami, FL.
9. A. G. Seidel, "Surface Indications on 18Mn–18Cr Retaining-ring," presented at EPRI-Generator Retaining-ring Workshop, December 1997, Miami.
10. M. C. Murphy, "The Development of Fretting Fatigue Cracks in the Drax Generator Rotor," NEI Parsons, CEGB Symposium, 1979.
11. S. Akkaram, M. Chati, S. Sundaram, and G. Barnes, "Fretting Fatigue Modeling and Repair Evaluation for Generator Rotors," GT2003-38827 (General Electric), in *Proceedings of ASME Turbo Expo 2003, Power for Land, Sea, and Air,* June 16–19, 2003, Atlanta, GA.
12. T. Hattori, T. Nakamura, and T. Ishizuka, "Fretting Fatigue Analysis Strength Improvement Models with Grooving or Knurling on a Contact Surface," in *Standardization of Fretting Fatigue Test Methods and Equipment, ASTM STP 1159,* M. Helmi Attia and R. B. Waterhouse (Eds.), American Society for Testing and Materials, Philadelphia, 1992, pp. 101–114.
13. P. A. McVeigh, G. Harish, T. N. Farris, and M. P. Szolwinski, "Modeling Interfacial Conditions in Nominally Flat Contacts for Application to Fretting Fatigue of Turbine Engine Components," *International Journal of Fatigue,* Vol. 21, pp. 157–165, 1999.
14. G. W. Buckley and R. J. Corkings, "The Importance of grounding Brushes to the Safe Operation of Large Turbine Generators," *IEEE Transactions on Energy Conversion,* September 1988, pp. 607–612.
15. J. Sparks, "What to Do about Shaft Currents," Plant Operation and Maintenance Section, Power Generation and Transmission, p. 114.

10

AUXILIARIES INSPECTION

Some generator auxiliary systems are required to be in service even when the generator itself is not in operation. For instance, when the generator is offline but on turning gear while cooling off, the lube-oil system is still circulating oil to the bearing and the seal-oil system is still circulating oil to the hydrogen seals to maintain the generator hydrogen pressure. This means that the hydrogen system is operating for the described condition. Thus, it is easily seen that the auxiliaries operate more than the generator and that they require maintenance. The following sections provide a description of the things that should be inspected and checked during a generator outage for inspection and maintenance.

10.1 LUBE-OIL SYSTEM

Inspection of the lube-oil system is usually done at every major outage. It is usually done as a turbine maintenance item and so will only be touched on briefly here.

Prior to shutdown, a complete set of oil pressure, temperature, and bearing-metal temperature readings should be obtained to look for any abnormal readings of these parameters. Abnormal readings can help to identify components for inspection in the lube-oil system for the purpose of determining what the cause of the abnormality is.

Handbook of Large Turbo-Generator Operations and Maintenance. By Klempner and Kerszenbaum

For major outages, the main oil tank should be drained and cleaned, and the following items checked and inspected:

- Tank condition and cleanliness
- Piping connections for looseness and defects
- Transfer and check valves for ease of operation and overall condition
- Oil coolers—both water and oil sides (a leak test may be done to ensure that there is no passage of one fluid of the system into the other)
- Filters cleaned or replaced, depending on the type used
- Pressure switches and gauges overhauled and recalibrated
- Bearings and other components of all pumps properly lubricated
- Vapor extractor and oil-pump regulator operation
- Monitoring instrumentation
- Oil purifier or centrifuge

Major maintenance activities in addition to the above may consist of flushing and backflushing of the lube-oil system, and a complete change of the oil.

10.2 HYDROGEN COOLING SYSTEM

Inspection of the hydrogen cooling system is usually done at every major outage. The major components of the hydrogen cooling system should all be checked for general condition and operational functionality. The main components are as follows:

- Hydrogen dryers and control components
- Hydrogen bulk supply or generation plant
- Hydrogen control cabinet and instrumentation

Major maintenance activities during an outage may consist of the following:

- Desiccant recharging in the hydrogen dryers
- Hydrogen supply replenishing

The large majority of minor maintenance activities on the hydrogen cooling system, outside of the coolers themselves, are instrumentation, piping and valve issues. Individually, the maintenance of these types of items is not a large effort, but it may become large if all instrumentation requires recalibration and operational checks. Time should be allotted during major outages to ensure that these items are attended to so they do not become operation problems while the machine is in service. During maintenance of the hydrogen system, great care must be taken due to the danger of hydrogen explosions or fires. Sparkless tools made of brass are usually used to avoid ignition of the hydrogen, but the system should not be worked on until the hydrogen is safely purged.

One of the continual maintenance items is making sure there is always a good supply of hydrogen and CO_2 available. This activity does not require an outage. Generally, the two methods of hydrogen supply are purchasing bottles of hydrogen and simply replacing them as necessary, or having an in-house hydrogen production facility. The in-house facility represents yet another subsystem and will require periodic maintenance as necessary.

Probably the most worked on pieces of equipment in the hydrogen system are the dryers, and these also require the use of sparkless tools. These can often be worked on while in operation since they usually come as twin tower units. One side is operational while the other regenerates its desiccant. This is discussed in detail in the following section.

10.2.1 Hydrogen Desiccant/Dryer

A typical maximum dew point for hydrogen in a machine is 30°F (–1.1°C). Periodic purity checks should suffice to maintain the dew point low enough to avoid condensation during operation of the machine. Nevertheless, if purity cannot be ascertained, it is important to ensure that only dry hydrogen enters the machine. Moisture carried by the hydrogen into the machine, if allowed to condense, may adversely affect critical components.

Moisture in the windings can cause tracking, with subsequent failure of the insulation. Moisture on the retaining rings can be a serious source of problems, in particular, if the rings are made of austenitic steel (18% manganese–5% chromium alloy), commonly known as 18–5 rings. In the presence of moisture, this material tends to develop stress-corrosion cracking, which grows relatively fast. This can result in catastrophic failure of the ring. Retaining rings made out of 18–5 alloys are being changed throughout the industry to 18–18 alloy rings, following recommendations by manufacturers. Some machine operators keep their 18–5 rings but subject them to periodic NDE, and are very careful to avoid ingression of moisture to the machine. The new 18–18 rings have proved to be less prone to stress-corrosion cracking but are still susceptible to water-initiated corrosion when halides or copper ions are present in the water. Moisture carried by hydrogen can also result in corrosion of the iron laminations, zone-rings, fan-rings, nonmagnetic rotor wedges, bolts, studs, brakes, and other structural members. Carbonate by-products from welding fluxes used in hydrogen coolers have been known to result in winding failures.

Manufacturers are divided as to the best methods to maintain dry hydrogen. However, most base their decision for installing or not installing hydrogen dryers on the bearing seal-oil arrangement they employ. Some manufacturers do not recommend the use of hydrogen dryers and rely on seal-oil vacuum treatment to extract the moisture from the hydrogen.

If hydrogen dryers are installed, they should be inspected to ascertain that they are in good working condition. Desiccants are often neglected, particularly in places where scheduled maintenance procedures are lacking; therefore, they should be visually inspected as part of the machine inspection procedure. Plant personnel should make sure that hydrogen dryers are in good working condition at the time of shutdown. This is the most critical time, as it is when the moisture will condense on critical components of the machine.

10.3 SEAL-OIL SYSTEM

Inspection of the seal-oil system is usually done at every major outage. The major components of the seal-oil system should be all checked for general condition and operating function. The main components to be inspected are as follows:

- Coolers. The coolers need to be free from leaks between the raw water and the oil side. A leak of raw water can contaminate the oil and cause operational problems with the seals. Periodic NDE of the cooler tubes is required to check for wall thinning and avoid leaks. Also, cleaning should be done of the water and oil sides at regular intervals to maintain proper flows and heat removal capability.
- Filters. Filters should be periodically checked and cleaned to ensure that they do not become clogged and impede flow. There are many different types and it is best to follow the manufacturer's instructions for maintenance.
- Pumps. Normal maintenance usually carried out on pumps should be done during the maintenance outages.
- Oil reservoir condition and cleanliness. The seal oil supply is usually taken from the main lube-oil supply and separated off. The oil can become affected over time and needs to be checked for proper chemical composition, dryness, and entrained impurities. The oil quality should be monitored on a regular basis and chemically cleaned or replaced with new oil as required.
- Seal-oil vacuum tank. Always inspect the tank sight glass for foaming in or on the surface of the seal oil in the tank. There should be minimal foaming observable. Excessive foaming indicates too much moisture or hydrogen entrainment in the seal oil (fig. 10.1)
- Hydrogen detraining system. Requires general maintenance of tanks and reservoirs and associated instrumentation.
- Piping connections. Normal maintenance of piping and bolted connections.
- Pressure switches and gauges. Should be overhauled and recalibrated if necessary.
- Monitoring instrumentation. During overhauls, monitoring instrumentation and devices should be recalibrated and manufacturer's recommended maintenance carried out.
- Inspect hydrogen seal components. Hydrogen seal components should be inspected and NDE tested as described in Chapter 9 regarding hydrogen seal maintenance.

Major maintenance activities may consist of the following:

- Flushing and backflushing of the system. This is to clear the system of any debris collected.

Fig. 10.1 Seal oil tank with sight glass window showing a high degree of foaming in the oil, indicating excessive moisture.

- Oil changes. Oil quality may deteriorate due to a number of reasons. When the oil cannot be cleaned up by additives or filtering, it should be replaced.
- Maintenance of the seal surfaces. Should be carried out as per Chapter 9 regarding hydrogen seal maintenance.
- Replacement of gaskets and "O" rings
- Realignment of the seals and housings. Done during reassembly of the hydrogen seals.
- Filter cleaning or replacement.

10.4 STATOR COOLING WATER SYSTEM

Inspection of the stator cooling water system is usually done at every major outage. The major components of the stator cooling water system should all be checked for general condition and operating function. The main components are as follows:

- Demineralized water makeup or storage tank. Generally, there are two basic arrangements for the stator cooling water makeup tank. They are either open to atmosphere, creating a "high-oxygen" stator cooling water environment, or they have a hydrogen blanket or atmosphere covering the surface of the water in the tank, creating a "low-oxygen" stator cooling water environment. The manufacturer's recommendations for maintenance of the stator cooling water and the atmospheric environment intended should be followed. It is not usually good practice to alter the manufacturer's operational intent regarding "high" or "low" oxygen content stator cooling water schemes. The reason for this is that the incorrect oxygen content can quickly cause corrosion plugging in the hollow conductors of the stator winding, leading to overheating. All instrumentation associated with the tank, particularly the level detectors and any venting systems, should be maintained in good working order.

- Pipework. The stator cooling water pipework is usually either stainless steel or copper, to avoid contaminants in the stator cooling water that would create corrosion effects or cause the conductivity to deviate from design.
- Automatic and manual valves. The portions of the valves that come in contact with the demineralized stator cooling water are also generally made from the same materials as the pipework, for the same reasons of maintaining low conductivity and minimizing corrosion products.
- Heat exchangers. The heat exchangers need to be free from leaks between the raw water and the demineralized water side. A leak of raw water can contaminate the demineralized cooling water side and cause a ground fault or flashover in the stator winding if the conductivity gets too high from contamination. Periodic NDE of the cooler tubes or plates is required to avoid leaks. Also, cleaning of the raw water side should be done on regular intervals to maintain proper flow and heat-removal capability.
- Filters and strainers. Mechanical-type strainers are usually used for larger debris and often have the capability to carry out back-flushing to clear any debris. This needs to be done on a regular maintenance basis. Also, organic or other type filters should be periodically checked and replaced when they become clogged and flow is reduced. Most often, one will see blackish copper oxide deposits or sometimes minute copper particles or even bits of gasket materials.
- Pumps. Normal maintenance usually carried out on pumps should be done during the maintenance outages.
- Pump motors. Normal maintenance usually carried out on pump motors should be done during the maintenance outages.
- Deionizer. The deionizer requires monitoring and replacement of the cation and anion resins so that they maintain their ability to keep the stator cooling water conductivity at the required level. This type of maintenance is also normally done during the general maintenance outage for the generator.
- Alkalizer. The alkalizer is a specific piece of equipment employed in conjunction with the deionizer to keep the pH level at about 8.5 (alkaline). The chemicals used in the system need to be monitored and the supply maintained for proper operation.
- Hydrogen detraining system. The hydrogen detraining system is basically a tank used to remove any hydrogen or other gases that normally get entrained in the

stator cooling water. The detraining system vents and tank-level indicators and other instrumentation need to be monitored continuously and periodically maintained during outages. All vents need to be monitored to ensure they are clear and venting properly at all times.

- Gas release and leak monitoring systems, if provided. Some stator cooling water systems are equipped with leak detection systems that work by collecting the hydrogen gas that has leaked into the stator cooling water in abnormally excessive quantities. The leak detection system accounts for the amount of leakage in to stator cooling water and then vents it off through the venting system. The leak detection chamber and all the instrumentation associated needs to be monitored and maintained at regular intervals according to the manufacturer's recommendations.
- Monitoring instrumentation. Required to be periodically calibrated and maintained at regular intervals.

Major maintenance activities may consist of the following:

- Stator cooling water system flushing and backflushing. Flushing and back-flushing are generally done to remove debris and help clean out anything that may be lodged in the conductor bar ends and strand openings. The method is to switch over the piping from the inlet and outlet of the stator winding, with the intent of flowing the stator cooling water in the opposite direction to normal flow. The debris is then picked up in the strainers and filters.
- Chemical flushing. Sometimes, debris is not in loose form, but rather a product of corrosion of the copper conductors. This has been discussed earlier on in Chapter 3 in the sections on stator cooling water. When copper corrosion does occur from such problems as poor stator cooling water chemistry, the corrosion can take place at a fast rate. Chemical cleaning methods are available to reduce and often remove the corrosion products so that operation can continue as normal. The methods and chemicals used vary from manufacturer to manufacturer and cleaning can be done both on- and offline. When doing it online, the effort usually requires more time because the flow of chemicals must be kept to a low rate so that conductivity limits are not exceeded. This is desirable for the obvious reason of maintaining production, especially in a nuclear environment, but it requires extreme caution during the operation. It is highly advisable to do any chemical cleaning with the assistance of the manufacturer or under his direction.
- Filter replacement. Filter replacement is a fairly common practice and simple task. It is usually done during an outage and takes very little time if only replacement is required and no other issues are present.
- Deionizing system resin recharging. This is done as a normal course of action during maintenance outages when the resins are found to be depleted and no longer able to maintain low conductivity of the stator cooling water.

10.5 EXCITERS

Inspection and maintenance is generally performed on a five- to eight-year basis for major items, and at weekly or monthly intervals for brushes, commutators, and collectors. The inspection of the excitation system depends on the type of excitation:

- Rotary (generator driven or independent)
- Static (solid state)
- Brushless (shaft mounted)

In general, the work is performed at the planned intervals mentioned above.

10.5.1 Rotating Systems Inspection

Electrical Components:

- Commutator surface
- Insulating mica
- Brush surface
- Brush clearance to brush holder
- Holder clearance to commutator
- Brush spacing
- Pigtail condition
- Armature winding condition
- Field-winding condition
- Buswork connections
- RTDs and/or thermocouples

Mechanical Components:

- Cleanliness
- Bearings
- Deflectors
- Insulated couplings
- Airgap clearances
- Alignment to the generator's rotor (when driven by main generator)
- Filters

10.5.2 Static Systems Inspection

General Components:

- Cleanliness
- Ventilation system

- Connections
- Rectifier short/open tests
- Other, as required per manufacturer and type of system

10.5.3 Brushless Systems Inspection

Electrical Components:

- Cleanliness
- Buswork connections
- Diodes

Mechanical Components:

- Bearings
- Deflectors
- Airgap clearances
- Alignment to the generator's rotor
- Filters
- Cleanliness

10.5.4 Specific Inspection Items

Excitation Cleanliness and General Appearance. As described in preceding sections on generator auxiliaries, there are various types of excitation systems that are typically employed in large synchronous machines, whether self-excited, stand-alone, solid-state, shaft-driven generator, stand-alone generators, or some other system. It is impossible to address in detail so many different types of systems and their components. Therefore, when discussing the inspection of excitation components, we will mainly cover the general issues mentioned in the Excitation Inspection Form (see Chapter 7).

In Chapter 9, we stressed the sensitivity of the operation of the collector brushes to the brush-spring collector condition, as well as to the suitability of the environment, which includes the surrounding atmosphere and dirt or other contaminants. Therefore, excessive presence of these contaminants is a plausible indication of less-than-effective operation. If excessive carbon dust or copper dust is present, it can indicate too much wear on the brushes and collector. If left untreated, a large amount of dust can restrict the movement of the brushes in their brush holders, resulting in a quickly deteriorating situation. Excessive copper and/or carbon dust may also result in a ground fault on the collectors of the brush rig (see Fig. 10.2 for such an occurrence).

In short, the accumulated contaminants may provide a clue to the operation of the machine. However, contaminants should be removed as often as necessary to allow for effective brush operation.

Cleanliness obviously also applies to other equipment associated with the excitation system. For instance, when free-standing, solid-state-controlled rectifiers provide the

Fig. 10.2 Major failure of a two-pole, 800 MVA unit due to a multiple short circuit to ground. The failure was attributed to maintenance deficiencies in the area of periodic inspection of the brushes and brush-holder conditions, as well as general cleanliness of the brush rigging. In the photo, the outer brush gear has already been disassembled. The inner one, where the fault occurred, is shown partially disassembled.

field current, they should also be opened and, as a minimum, given a cursory inspection. The soundness of the main connections should be verified. Filters, if present, should be cleaned or replaced, and the condition of the cooling fans, if installed, should be evaluated.

Shaft-Mounted Diodes. Many smaller synchronous motors and generators, and an increasing number of larger machines, are being supplied with shaft-mounted excitation systems. In this arrangement, a shaft-mounted auxiliary winding is energized by a controlled flux created by a stationary winding. The rotating winding produces alternating currents, which are then rectified by a shaft-mounted, solid-state rectifier bridge. The rectified currents then flow into the dc field winding through a set of leads. The bridge consists of diodes mounted around the perimeter of the rotor in a diode-supporting rig, also called a "diode wheel." Figure 10.3 shows a diode wheel belonging to a 20 MVA, two-pole unit.

Experience shows that these diodes are very reliable. Testing the diodes may not be required in most cases. However, it is important to disconnect both leads from the diode bridge if any potential test is to be carried out on the field winding. Otherwise,

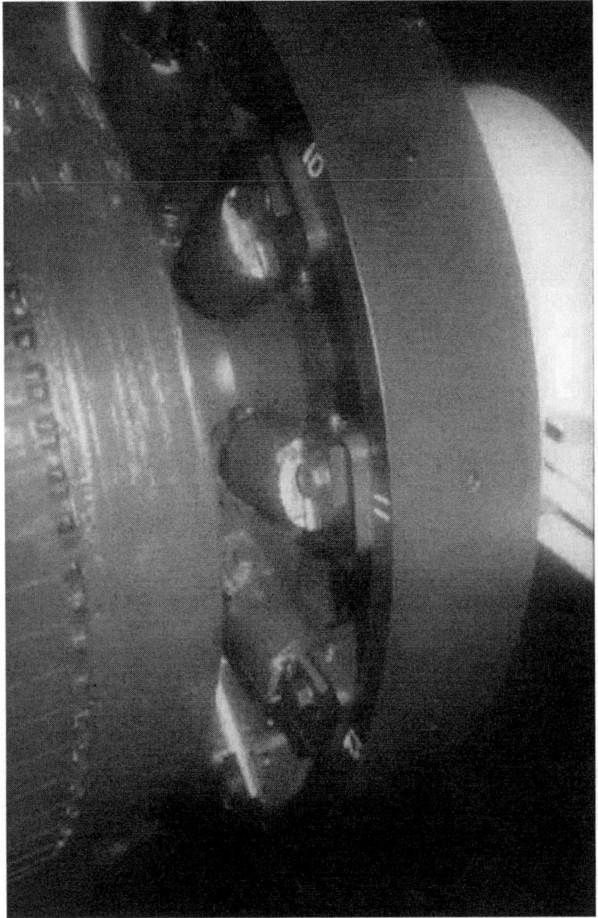

Fig. 10.3 Shaft-mounted exciter and diode wheel in a two-pole, 20 MVA generator.

the voltages generated by equipment such as hi-pot and/or megger devices will destroy the solid-state junction of the diodes. The diodes' heat sinks should be cleaned, if necessary, to improve their cooling during operation.

Diode Connections and Support Hardware. Although the diodes themselves are very reliable and may not need to be electrically tested, the connections between the diodes and leads, as well as the soundness of the supporting rig and its attachment to the rotor, are subjected to continuous vibration and mechanical centrifugal forces. Therefore, their soundness should be checked during a major inspection while the rotor is out of the bore.

Fuses commonly protect the excitation diodes. These fuses normally have a pin that is released upon activation of the fuse. If such fuses are present, the inspector should ascertain that they are all operational by making sure that none have been activated.

Commutator, Brushes, Springs, and Brush Rigging. The commutator of the a machine (in a rotating excitation system) should be inspected with the same scrutiny as the brushes and collectors in the main generator. Most of the topics discussed in regard to collector rings in Chapter 9 also apply to the commutator. In addition, special care should be taken to remove excessive accumulation of brush carbon and other dust in undercut mica grooves. This accumulation is intensified by the existence of oil contamination. Signs of arcing across mica, between adjacent segments, or between commutator and shaft indicate the need for cleaning. Hot spots should be noted and their causes investigated. Loosened and/or shifted segments should be noted and, if necessary, corrective measures should be taken. Unequal spacing of the brushes around the commutator may lead to defective commutation. If this situation is suspected, the position of the brushes can be measured by tracing their contour on a strip of thin paper placed between brushes and commutator. Subsequently, the paper is re-

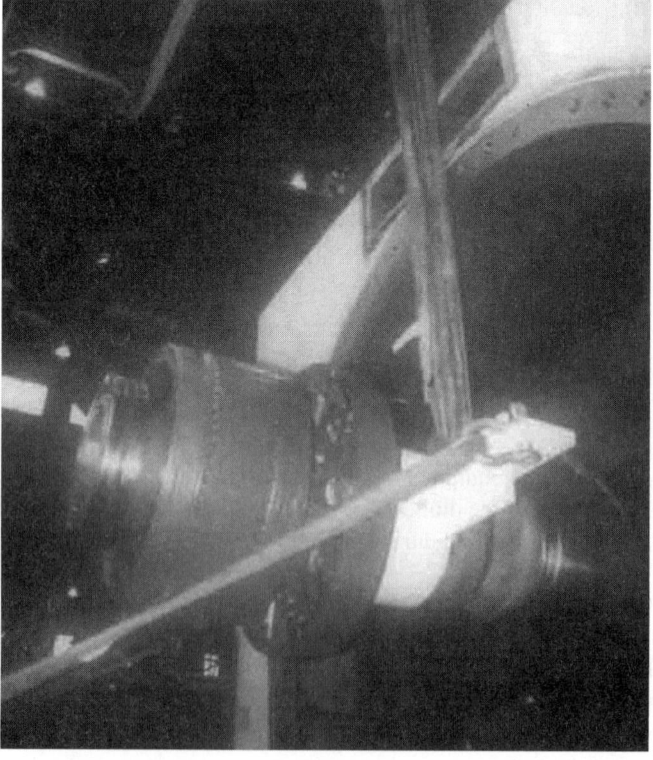

Fig. 10.4 View of the shaft-mounted exciter armature and diode wheel of the rotor of the previous figure.

trieved and the distances are measured. If necessary, corrections can be made by shifting the brush holders on the brush rig.

All the topics discussed in Chapter 9 regarding commutation film, brushes, springs, environment, and brush-rig support apply here entirely.

dc Generator Stator. If a rotating exciter is used, whether shaft-mounted or independently driven, it should be subjected to the same close inspection as that of the rest of the generator. It is an unfortunate situation when an undermaintained motor-exciter system fails, paralyzing a very large generator in good operating condition. In general, stators and rotors of rotating exciters should undergo inspection procedures similar to those that apply to the main generator. Electric tests are also similar to those performed on the main generator. All major items for inspection described in this book for the generator's stator (Chapter 8) apply to some degree to the stator of the exciter.

dc Generator Armature. Apart from general appearance, a visual inspection should assess the soundness of the connections to the commutator segments. The insulation should be analyzed for an indication of overheating, excessive dryness, movement, and so on. Figure 10.4 shows the armature of a shaft-mounted exciter of a turbo-generator.

Exciter-Drive Motor Cleanliness and Stator Condition. On some large turbo-generators, the excitation is drawn from independently driven dc generators. The

Fig. 10.5 Squirrel-cage rotor belonging to the induction motor driving the exciter of a large turbo-generator.

drivers of these machines are induction motors with ratings up to several thousand horsepower. These high-rating motors are not available as off-the-shelf purchased equipment. Therefore, a catastrophic failure of such a motor may paralyze a major generating unit for long periods of time. To avoid these costly outages, the exciter's driver should receive the same attention as the main unit, although this is seldom the case. Inspection of the driver's stator is made along the lines of the items discussed in Chapter 8 for the main unit.

Exciter-Motor Rotor. Practically all exciter electric drivers are of squirrel-cage construction. This is a very rugged and reliable rotor. The major point for inspection is the integrity of the joints between the rotor bars and the short-circuiting rings. Any signs of discoloration due to high temperature should be investigated. See Figure 10.5 for such a squirrel-cage rotor.

Field-Discharge Resistor. In many turbine-driven generators equipped with breakers in the main field leads, the field is discharged across an external resistor when the main field breaker is opened. The resistor eliminates the high voltages induced in any large inductance when the current flowing through it is abruptly interrupted. From time to time, the integrity of this resistor should be confirmed by inspection. A bad resistor can be the cause of expensive field-winding repairs.

11

GENERATOR MAINTENANCE TESTING

For the purpose of this chapter, generator maintenance testing refers to tests that are done generally offline or for some special condition, as opposed to online testing, which is actually a form of monitoring for diagnostic purposes while the generator is producing power. However, there are certain online tests that are done and classified as testing rather than monitoring, because they are not done on a full-time basis, meaning continuously monitored. Those types are discussed in this chapter as well.

Tests on large turbo-generators are a very serious business. Improper testing or test preparation can cause expensive unnecessary losses to the machine and expose the personnel to lethal dangers. Hence, tests must be carried out only by well-trained professionals, following all relevant and applicable rules and standards.

11.1 STATOR CORE MECHANICAL TESTS

11.1.1 Core Tightness

Stator cores can become loose from vibration and thermal cycling. If they do, then there is concern for interlaminar fretting of the insulation coating on the individual laminations. Loss of the interlaminar insulation can cause shorts and local overheating of the core. To ensure that the core is tight, a stator core tightness test can be done when the rotor is removed. The test is done by inserting a thin, tapered, and hardened steel knife blade between laminations to determine the degree of looseness in the stator iron (Fig. 11.1).

This test requires some experience to get a "feel" for what is or is not loose. However, in general, one should only be able to get the knife blade just into the core, if at all, and not be able to push it any distance inward, in the radial direction.

The testing should be repeated all around the circumference of the stator bore, and over the full length of the core, in random locations. Particular attention should be paid to the core ends, as this is where the majority of loosening generally occurs.

When looseness is found in the core, there are numerous methods available for remedying the situation. These include localized core stemming with shaped epoxy–glass inserts, retorquing of the core, and so on, up to restacking of the core iron.

11.1.2 Core and Frame Vibration Testing

The maximum vibration of the stator core and core frame should be less than 50 μm (about 2 mils) peak to peak (unfiltered), with no natural resonance within the frequency ranges of 50–75 Hz and 100–140 Hz for 60 Hz systems, and about 40–65 Hz and 80–120 Hz for 50 Hz systems. The problems associated with high vibrations are premature stator core interlaminar and stator winding insulation wear, and structural problems of the core and frame. To maintain low vibration and avoid problems, it is best to

Fig. 11.1 Stator core tightness testing by the knife insertion method.

have a tight core and to have good mechanical coupling between the core and frame. This ensures that no wear occurs between the two components at the keybars. Low absolute vibrations and low relative vibration between the core and frame, with the two components in phase, is a good indicator that the core and frame structure is sound.

Testing can be done offline or online. However, for either type of test, vibration transducers (accelerometers) must be mounted on the core and frame, inside the machine. This includes the stator center, both ends, and locations on the circumference based on the nodal vibration patterns of the stator (four nodes for two-pole machines and eight nodes for four-pole machines). The vibrations are generally measured in the radial and tangential directions when looking toward the end of the machine. In addition, it is desirable to use a number of portable, magnetic-based transducers, which may be moved around the outside of the generator casing, to ensure complete analysis of the machine.

Offline testing is not generally done unless there is a known problem with core and frame vibration. The excitation source must be artificially applied in this method, and there are a couple of ways to accomplish this. One is to simply strike the frame with a heavy rubber hammer and measure the frequencies at which the vibrations peak. However, this does not usually produce a significant result because the stimulus is so small. The other method is to attach a *shaker* device to the frame to stimulate the stator at a fixed frequency, and then measure the frequencies at which the vibrations peak. The shaker method generally produces good results because there is a significant stimulus and it can be controlled. The problem with this type of testing is that it does not give an accurate picture of the true vibration of the core and frame in operation.

Online testing is generally the best way to get a complete vibration analysis of the core and frame. With online testing it is possible to look at all operating modes of the stator and determine which parameter has the most influence on vibration. During testing, the variable parameters are stator current, field current, hydrogen pressure, and hydrogen temperature. In some cases, it has also been useful to valve out individual hydrogen coolers, successively, to change the cooling pattern and determine its effect on the core and frame vibration.

When core and/or frame vibrations are present, it is necessary to determine if the vibrations are most prevalent on the core or frame, and if the two components require better coupling to each other. The possibilities of what may be found on any individual machine are too vast to cover here. However, as a rule of thumb, it is often found that the core and frame will have a natural frequency that is too close to the forbidden zones, and that they do require some artificial means to ensure better mechanical coupling between the two. In such cases, vibration damping is also usually required. When trying to dampen excessive vibrations, there are two methods employed. The first method requires adding mass to lower the natural frequency. The second method entails stiffening the structure to raise the natural frequency. Adding mass can be difficult if there is no good place to attach it, and stiffening can sometimes cause problems with overstressing the frame welds, causing them to crack.

Stator core and frame vibration problems are very complex to analyze, and even more complex and expensive to solve. They should be addressed on a machine-to-machine basis, as each case will be unique.

11.2 STATOR CORE ELECTRICAL TESTS

11.2.1 EL CID Testing

The EL CID (ELectromagnetic Core Imperfection Detector) technique was originally devised as portable test equipment for inspection and repair of rotating electric machine stator cores. It was devised as a low excitation power alternative to the high power level stator core flux test, for looking for stator core interlaminar insulation problems. Its application has been shown to be applicable to turbine generators, hydraulic generators, small generators, and large motors. The subject of this book, however, is confined to the class of large two and four pole, round-rotor machines, commonly referred to as turbine-driven generators. The information contained in this section is a brief discussion of the EL CID test technique and basic interpretation of results.

Traditionally, stator core interlaminar insulation testing has been done using the “ring” or “loop” flux test method, in which rated or near-rated flux is induced in the stator core yoke. This in turn induces circulating currents from the faulted area, usually to the back of the core at the core-to-keybar interface (Fig. 11.2). These circulating currents cause excessive heating in areas where the stator iron is damaged. The heat produced is generally detected and quantified using established infrared techniques. This method has been proved to be successful over the years, but it requires a large power source and considerable time, manpower, and resources to complete.

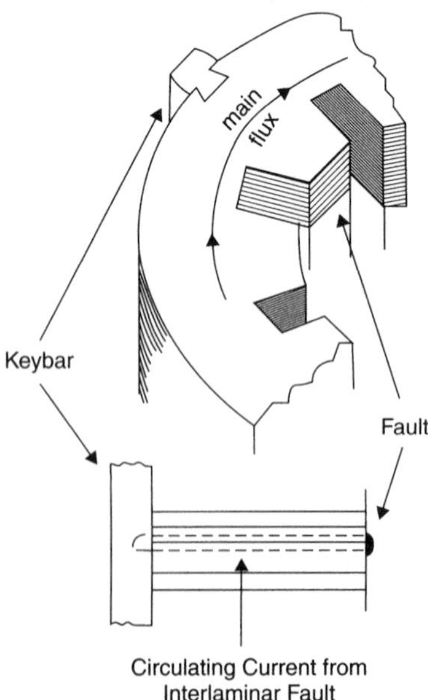

Fig. 11.2 Stator core fault current path from fault to core back at the keybars, by induced flux. (Courtesy of ADWEL International.)

Starting in the early 1980s, the EL CID test has been developed as an alternative to the ring flux test. The technique is based on the detection of core faults by measuring the magnetic flux resulting from the current flowing in the fault area, at only 3 to 4% of rated flux in the core. Furthermore, the test usually requires only two or even one man to complete (using the latest version) in less than one eight-hour shift.

EL CID Test Procedure. The level of excitation needed to produce the desired flux in the stator core-back area is generally determined by a combination of the stator design parameters and the power supply available to achieve the required flux level. For most generators, the standard 120 V ac (North America, etc.) or the 230 V ac (Europe, etc.) outlet, with a current capacity of 15 to 20 amp, is usually adequate.

The characteristics of most stators are such that four to seven turns of a #10 AWG insulated wire (2.5 mm^2) can be used to carry the excitation current for the test. The winding is then energized to the required volts per turn to produce approximately 3 to 4% of rated flux, usually corresponding to around 5 volts per meter across the stator iron. A Powerstat or Variac is best used for voltage and supply current control.

The signal-processing unit of the EL CID test equipment measures detected fault current (in quadrature mode) in milliamperes. By theory and experimentation, a measurement of at least (100 mA is required at 4% excitation of the core before it is considered that the core has significant damage affecting the interlaminar insulation.

The excitation winding and power supply are set up during the test as shown in Figure 11.3. The EL CID equipment is set up as shown in Figure 11.4 (original analog set) and Figure 11.5 (newer digital set).

In the older analog sets, a separate coil is placed in the bore over undamaged iron, as shown in Figure 11.4, to supply the reference signal. In the newer digital version, a CT (shown in Fig. 11.5) is placed around the excitation winding to reference the supply signal. The CT was also an option on later analog sets. The digital equipment uses a laptop computer to store the axial traces, whereas a plotter was used in the original version.

The sensor head (*chattock potentiometer*) is pulled axially along the core at a speed slower than one meter every twenty seconds and always bridging two stator teeth, as shown in Figure 11.6. (The slower speed is important, as the standard chattock coil has a magnetic sensing area of only 4 mm diameter, and both the digital and analog systems have a definite time needed to record the phase/quad signals to sufficient resolution. The digital set records the phase and quad values every 2 mm. Any faster testing results in some missed test points with the digital system or potential inaccuracy due to settling time with the analog system).

The fault current signal is read directly off the signal processor meter and input to a computer or chart recorder to trace out the readings as a function of the axial position along the stator core. When the sensor head is over undamaged iron, the meter should read zero if it was calibrated previous to the test for a condition in which no fault current is circulating. In actual practice, no insulation system is perfect, and some background signal is usually detected. In addition, the contact resistance of the core-to-keybar interface is not zero and can be found to vary between near 0 to 2 ohms. This also affects the EL CID signal that is measured. Usually, anywhere from

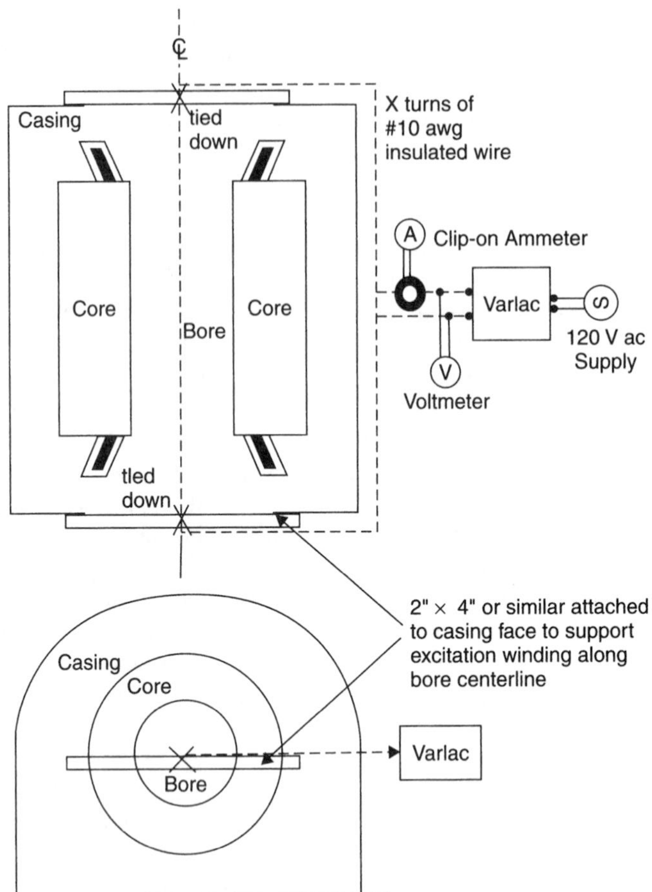

Fig. 11.3 Excitation winding and power supply setup for EL CID testing.

a 0 to (20 mA EL CID signal (in quad mode) is found to be normal when a good core is tested.

The EL CID test is somewhat similar to the rated flux test in which the undamaged iron slowly heats up, producing a background level due to eddy current losses in each laminations. During flux testing, this is recorded as the ambient core temperature rise. Where there is damaged or deteriorated core insulation, the core overheats and is detected as a hot spot above the core ambient temperature due to high fault currents circulating locally.

In the EL CID test, when the sensor head is placed over damaged core areas, the primary indication of a fault is obtained by detecting the flux produced by a current flowing in phase quadrature with respect to the excitation magnetizing current (the phase current). This flux is then converted back to an indicated current (the quad current) assumed to be flowing in the fault (Fig. 11.7). For this reason, the quad current

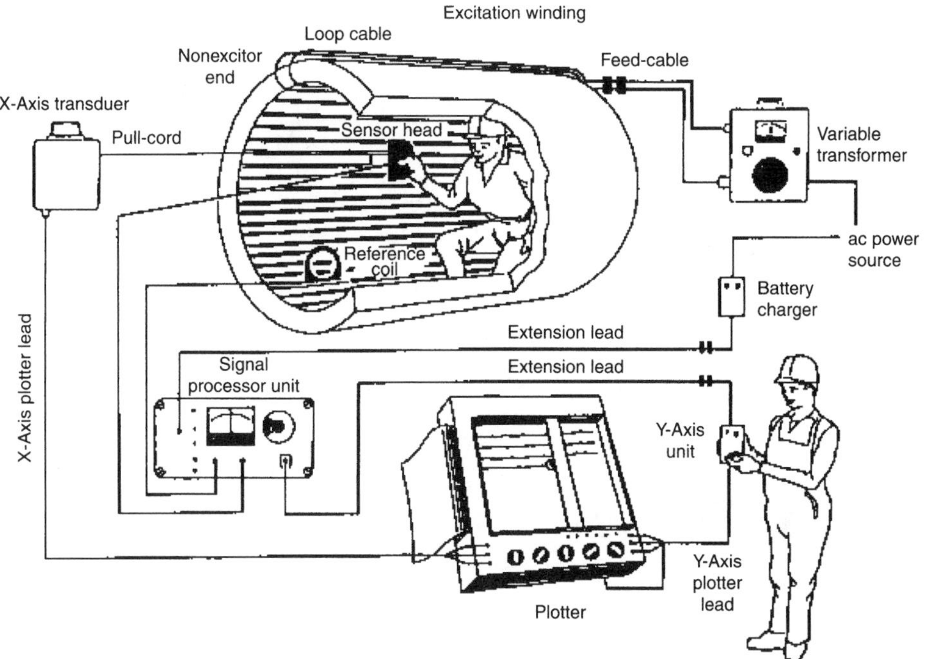

Fig. 11.4 Original analog EL CID equipment setup for testing large turbogenerators. (Courtesy of ADWEL International.)

detected by the EL CID processor is frequently referred to as the fault current (although for large faults the phase current may be affected as well, especially when the fault current path is highly inductive). The quad current is indicated on the signal processor meter and the traces recorded on the plotter (original analog EL CID equipment) or computer (newer digital EL CID equipment).

EL CID Experience. The EL CID test has proved to be extremely reliable in detecting and locating core problems. It can cut the time and manpower required for core testing to within one eight-hour shift, whereas a flux test may have taken a few days to set up, and then a day to test, and another day to dismantle the test equipment.

In the large majority of the EL CID tests on turbo-generator machines, experience has shown that EL CID is very reliable in determining that actual core faults or interlaminar insulation deterioration exist. In other words, if a core defect exists, then EL CID is likely to find it. Then, if the core is indicated to be defect-free by an EL CID test, there is a very high probability that it actually is free of defects.

Large signals may be found at tooth-top locations on the core and only indicate a significant surface fault. Local surface faults are generally indicated by faults that show very localized signals, either high or low in magnitude and positive in polarity, if within the test coil span (assuming the standard EL CID test setup). Deeper faults can

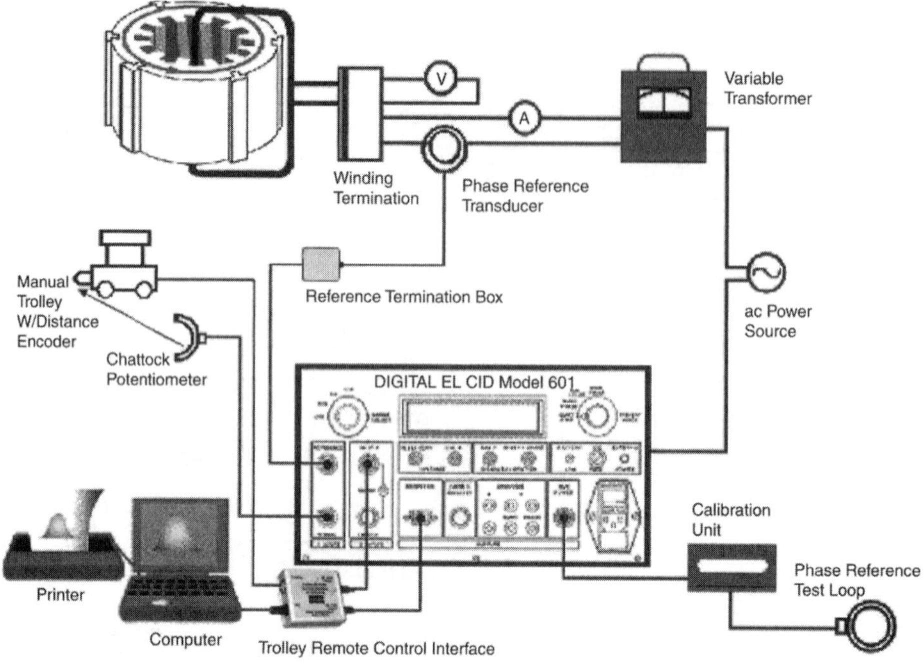

Fig. 11.5 Digital EL CID equipment setup for testing large turbo-generators. (Courtesy of ADWEL International.)

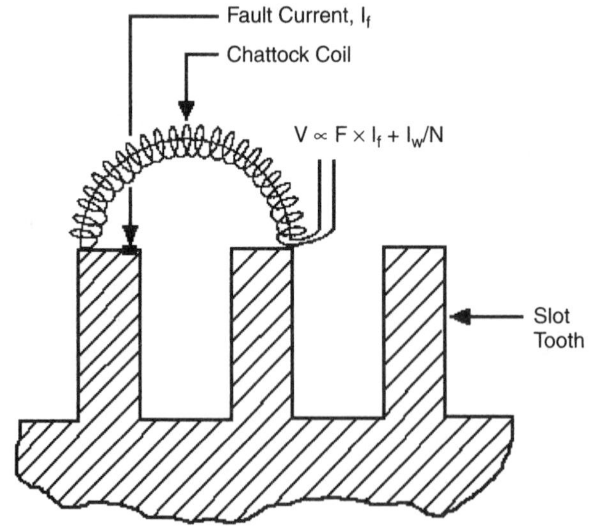

Fig. 11.6 EL CID sensor head (chattock potentiometer) basic circuit. (Courtesy of ADWEL International.)

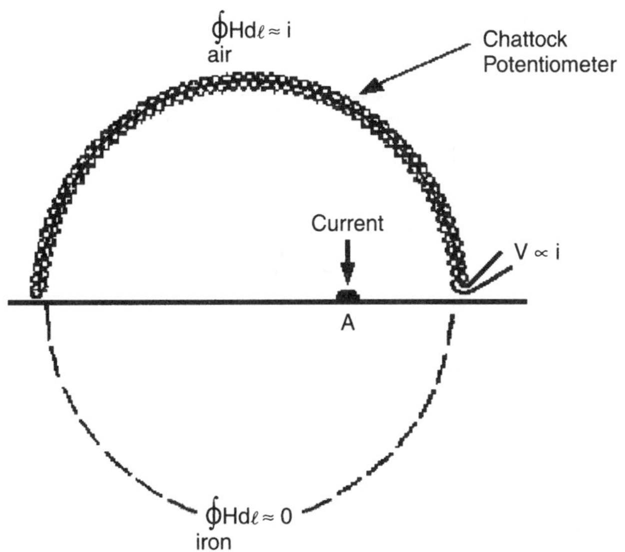

Fig. 11.7 EL CID sensor coil (chattock potentiometer) winding voltage and induced current is proportional to the current flowing in the fault. The iron portion of the circuit is neglected due its low magnetic reluctance compared to air. (Courtesy of ADWEL International.)

generally be seen over a larger scanning area, and also often become opposite in polarity as the sensor head gets away from the fault area. This is because the fault is outside the flux path of the chattock coil sensing the fault current, and the magnetic potential difference is reversed.

Figure 11.8 shows a general basic interpretation of the EL CID signals that can be expected to be seen based on fault location. The magnitudes in Figure 11.8 are only relative to one another to give an idea of what might be expected for faults of roughly the same severity at different locations. The peaks, and widths of the peaks, will vary from fault to fault as their size varies, and as they are more or less severe.

Attempts have been made to correlate EL CID signal readings to temperatures that would be created in the defect area during a flux test. The basic premise of the EL CID test significance level of ±100 mA is that this level represents a 5 to 10°C temperature rise, as seen in flux tests, and, therefore, it is just at the level of temperature rise at which most OEMs and experienced stator core experts should carry out repairs to the core iron.

There are a number of issues that make questionable the assumption that correlates EL CID's signal to temperature. First, many core testers carry out flux tests at widely differing flux levels. Some prefer to test at 100% of rated flux level, whereas others test at about the 80% level. Different operators also apply the flux test over widely varying time periods. Yet, by all indications, all seem to work on the same temperature rise criteria. Obviously, a 10°C rise at 100% of rated flux is much less significant than at 80% of rated flux. This is because the 80% level is generally at the knee of the B–H

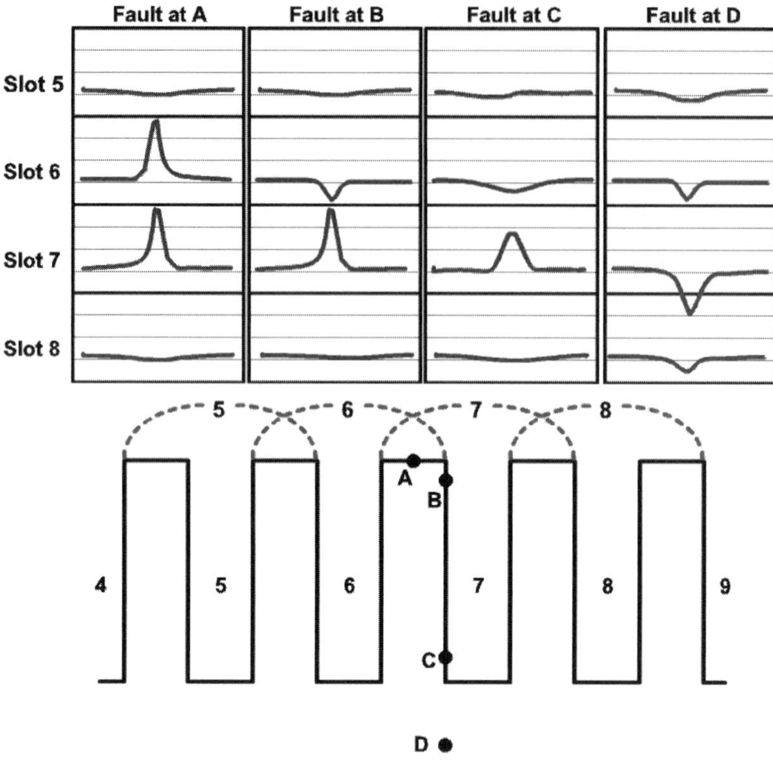

Fig. 11.8 EL CID signal interpretation of surface and deep faults.

curve (relationship between flux density and ampere-turns per meter) and the curve is exponential. Increasing to 100% when the temperature rise is already 10°C will increase the temperature in the fault.

There is also the influence and undetermined significance of core-to-keybar contact resistance at the back of the core. The concern here is that the resistance can be generally measured to vary from nearly 0 to 2 ohms, which will affect the EL CID signal as well as the temperature measured during the flux test.

For the low flux levels of the EL CID test, the core-to-keybar contact resistance may be quite significant. This is one of the unknowns for which there is little data one way or another. But it should be noted that on ungrounded cores (i.e., cores with insulated keybars and infinite core-to-keybar contact resistance), the EL CID and the flux test are both ineffective unless there are two faults in proximity that allow circulating current to flow and hence be detected by either test. This has been proven by tests done on ungrounded cores, where single faults do not show up until the core is artificially grounded at the back near the keybar behind the location of the fault.

Generally, there is only one keybar that is grounded on such core arrangements (for purposes of the stator ground fault relay), and the testing for single faults is effective in that lamination region only. To test such a core arrangement by either EL CID or flux test, the core must be purposely grounded at the back, between every keybar circumferentially, and on every lamination axially. One can easily see the difficulty in this, as this arrangement is not even possible on most stator designs, because of the frame construction. One has to consider this possibility when purchasing a new stator. If a single core fault occurs, then the insulated core may allow operation where the grounded core would fail based on the progression of the fault. However, if a second fault occurs in the same proximity, then a much more severe fault may occur and go undetected until the effect of the failure causes additional collateral damage, which takes the machine out of service. This is generally a matter of user preference and both philosophies are sound, each in its own way.

In addition, there is the problem that core faults can manifest themselves in many forms and levels of severity. It is not uncommon for a surface iron smudge to show a very high EL CID signal and yet not produce much heat when looked at under infrared light in a flux test. The opposite is true as well. It is not uncommon for an EL CID signal that is not much higher than the manufacturer's recommended significance level to produce significant heat. In particular, with very small faults on the surface where EL CID does not produce a significant signal, there are sometimes high-spot temperatures detected by infrared light, but there is insufficient power to cause damage.

One advantage is that EL CID can detect deep-seated faults, which may often not show as particularly large temperature rises on the surface but can be damaging to the body of the core or adjacent conductor bar insulation. This is due to the fact that the attenuation of EL CID signals is generally less than the attenuation of temperature rises with depth of fault.

The difficulty in correlating EL CID and flux test temperatures, therefore, comes from many issues as stated above and a number of other possible influences as listed below:

- Core-plate grade (i.e., grain oriented vs. nonoriented steels)
- Lamination insulation grades
- Axial length of the fault
- Total size of the fault
- Electrical resistance of the interlaminar fault (i.e., deteriorated or fretted insulation type damage as opposed to hard-contact, low-resistance type faults)
- Geometry differences in core structure from one machine to another
- Limitations of the earlier EL CID test equipment, in relation to the size of the chattock coil itself and the relative size of any fault being measured. (Current standard Chattock coils have only a 4 mm diameter magnetic sense area and, thus, are able to detect very small faults, particularly, if the suspected fault area is investigated/scanned slowly enough.)

All these issues can significantly affect both the observed EL CID signal and the temperature produced during a flux test. Some are better known and quantified than others. Trying to correlate temperatures to EL CID signals under so many variables is difficult, unless all of these parameters can be taken into account. In other words, the core under test must be well known to be able to make such a correlation.

There is one other factor regarding EL CID signal interpretation that has to do with readings taken in the phase mode, as opposed to the normal quad mode reading that Figure 11.8 is based on. Basically when a stator has, for example, four turns of an excitation winding and is carrying 12 amperes, then it has an excitation level of 48 ampere-turns (A-T). When the EL CID signal processor is set to phase mode and a reading is taken from tooth center to tooth center across one slot, a signal of 1 ampere (48 A-T divided by 48 slots) should be read. Generally, for most fault areas this is the reading that will be evident. However, in some cases, much higher current is read in the phase mode than the simple magnetic potential based on excitation and slot geometry. One of the things that has been seen when this type of situation occurs is that very high quad readings are often also present, and the fault is usually at the bottom of the slot or in the core yoke area. Correspondingly, there is not always much heat given off during flux testing, and the two tests do not always correlate when this occurs. There is very little experimental data on this point and, again, it shows that some uncertainties remain in interpretation of EL CID test results. It is believed that phase readings are also significant and should be factored into the test interpretation. Just what that interpretation should be is unclear to date, due to the difference in faults from case to case.

For the test interpreter, probably the main difficulty is knowing nothing about the core under test or the type of fault found. The core defect is often not visible and, thus, the tester is confounded in trying to determine is how deep it is and how severe it is. The general consensus of the people surveyed on this issue is that, more often than not, they cannot tell how severe a detected fault is and must depend on a flux test with an infrared scan in making that determination. The ring flux test remains the best test to determine the actual temperature rise of any fault, and whether repairs are required. If the fault is suspected to be deep seated from the EL CID test result, the ring flux thresholds should be appropriately adjusted. Once the core is repaired, an EL CID test can usually show that the repair is successful by the absence of a defect signal. This is perhaps the real benefit of EL CID testing.

There seems to be general consensus that if an EL CID test is performed and no damage is found, then the core is defect free. EL CID has gained good credibility in its ability to determine and locate the presence of faults, and to verify successful repairs of faults. The general consensus also is that more work is required on EL CID signal correlation with temperature rise in fault locations.

The general opinion to date is that both the EL CID and flux testing together are still required to provide the best information on any core defect found.

11.2.2 Rated Flux Test with Infrared Scan

The rated flux test is a high-energy test used to check the integrity of the insulation between the laminations in the stator core. It is also commonly referred to as the ring flux test, in which near-rated flux (normally about 80%) is induced in the stator core yoke. This in turn induces circulating currents and excessive heating in areas where the stator iron is damaged (Fig. 11.2). The heat produced is detected and quantified using established infrared techniques.

It should be noted that there are differing rationales about the level of flux that should be used for stator core testing. Some test at 100%, whereas others test as low as 60%. The lower value requires more time for the core to heat up and, thus, show a fault, but it is thought to be less likely for a fault to run away to failure at lower levels due to lower excitation voltage and less heat induced. The higher flux level rationale is that some faults may not be seen at lower excitation levels and that there is reasonable degree of control so that a fault in unlikely to run away during the testing. And further, if a fault were to progress to a major failure during testing, it would probably be near failure in service. The 80% level is a middle ground between the two schools of thought.

Flux is produced in the iron by looping a cable around the core in toroidal fashion (Fig. 11.9) and circulating a current at operating frequency. The flux required for the flux test is half the normal operating flux due to the difference in the way the flux is induced in operation (Fig. 11.10) from that of the flux test (Fig. 11.11).

The power supply for the cable is usually taken from two phases of one of the high-voltage breakers (e.g., 4 kV) in the plant, or a portable motor generator set. The correct number of turns are looped around the core to produce the required level of flux. IEEE

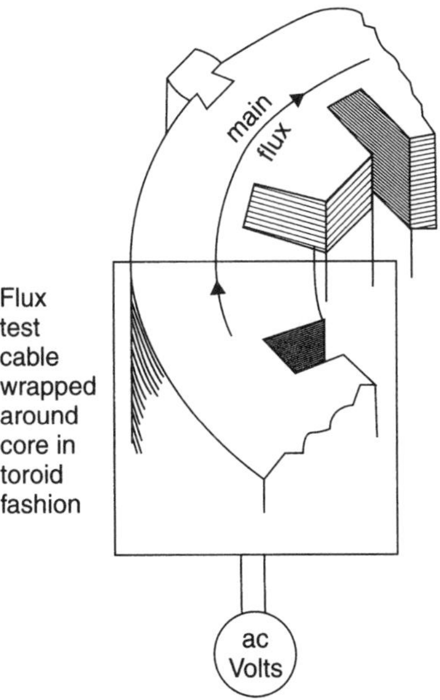

Fig. 11.9 Flux path during flux testing.

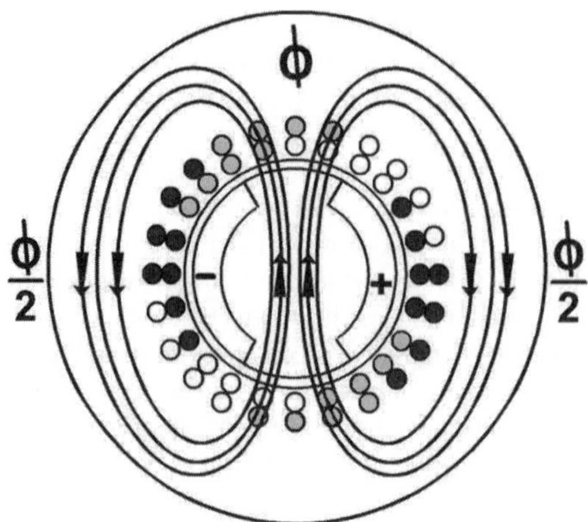

Fig. 11.10 Operational flux path.

Std 432 [11] provides the following expression to find the rated volts per turn required on the stator core:

$$\text{Voltage per turn of test coil} = \frac{1.05V_{\text{LL}}}{2\sqrt{3}K_{\text{w}}N}$$

where 1.05 has the +5% terminal voltage factored in and
V_{LL} = line-to-line voltage
K_{w} = winding factor
N = number of turns per phase in series in the stator winding

As the equation shows, the available power supply voltage is divided by the calculated volts per turn above to give the correct number of turns to loop through the core. If the number of turns includes a fraction, then the next highest number is used, to reduce the flux level to below 100% or below rated. Using too high of a flux level can create core damage, since there is no cooling on the core during the test.

The voltage per turn required for a flux test can also be determined once the machine flux per pole pair is know from the equation given in Section 2.4. This is calculated as follows:

$$\text{Flux volts per turn} = \frac{2\pi}{\sqrt{2}}f\left(\frac{\Phi}{2}\right)$$

Once the number of turns is known, the current capacity is required to size the cable and ensure the power source can handle the current that will be drawn. Knowledge of the specific B–H characteristic of the subject core being tested is required for this. It cannot be stressed enough that the exact B–H characteristic of the stator core should be

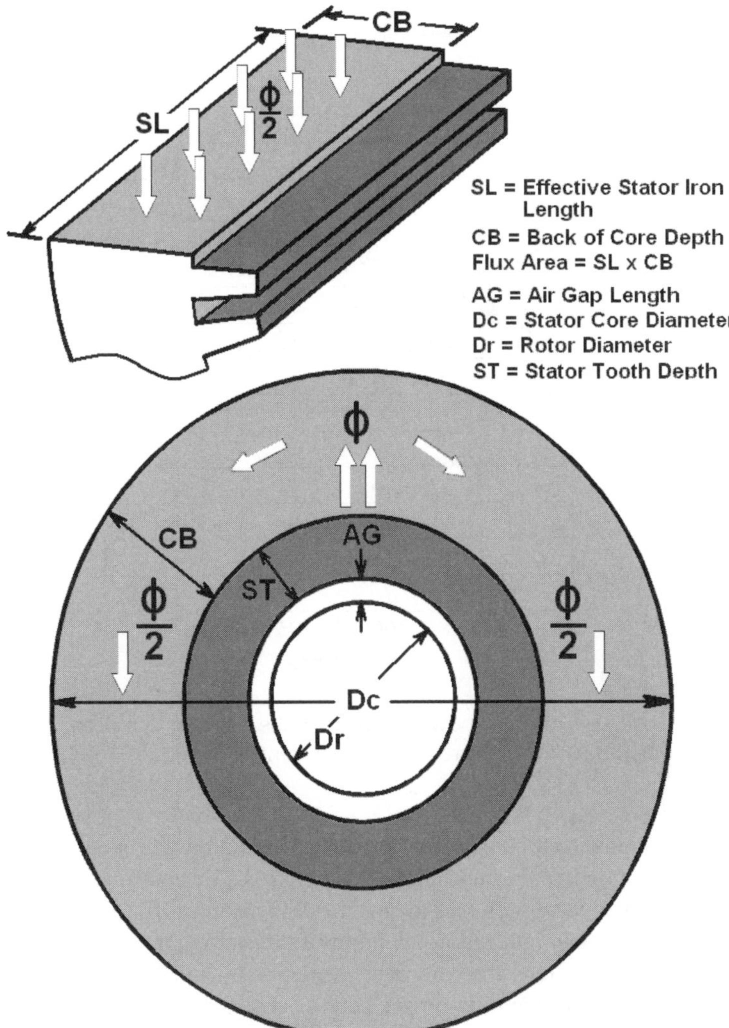

Fig. 11.11 Flux density in the core behind the slot during a flux test is a function of flux per pole pair divided by 2 (from Fig. 11.10) over the effective core iron area.

known in relation to the flux volts per turn as well as the current required from the power source (Fig. 11.12). In many instances it is unknown and, therefore, the number of ampere-turns required must be estimated based on industry curves for the most likely grade of core plate that will be used in the machine under test. A higher-end and lower-end core-plate grade are usually selected to provide a range of possible operating characteristics for the subject core. These are selected to provide a range of possible excitation requirements, based on B–H curves taken from small and large turbo-

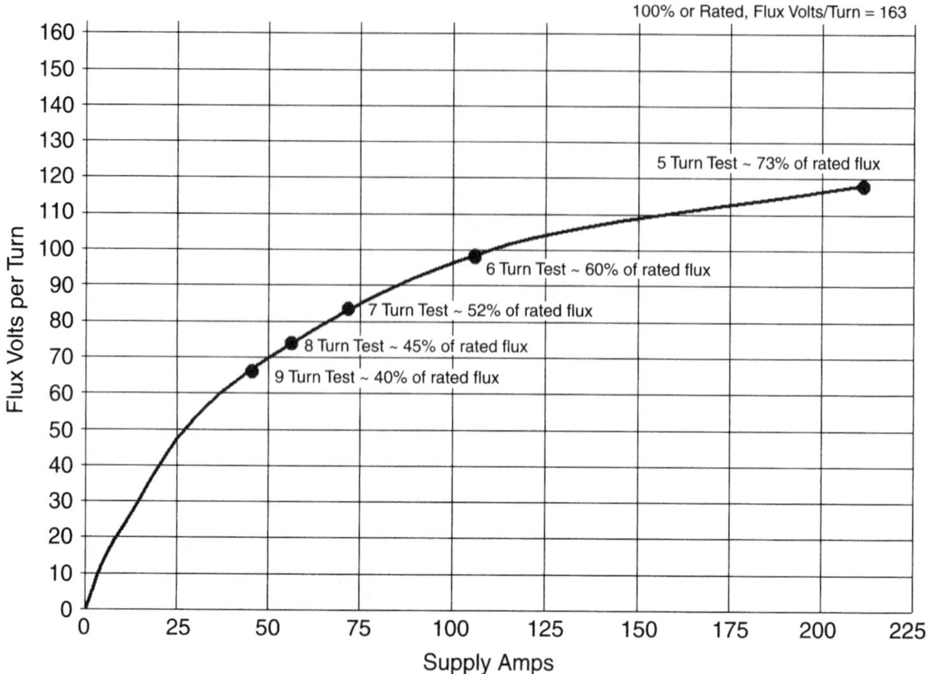

Fig. 11.12 B–H curve example showing the flux volts per turn, the percentage of rated flux, and the number of turns versus supply current.

generator applications. It should also be noted that it is not always enough to know the B–H curve characteristics of a machine. The overall B–H characteristic of the stator core as an assembly is more preferred to have for flux testing as this takes into account the specific reluctances introduced in the build of the stator core from such things as air space between laminations and core-plate insulation thickness, etc.

From the winding configuration for the subject generator, the power supply available, and the B–H curves, an estimate can be made of the number of turns required to achieve the required level of flux for the test. This is generally in the 70 to 90% range of rated flux. The current that would be flowing in the flux cable will depend on the actual B–H characteristic of the stator iron as an assembled stator core (as mentioned above) and, therefore, this must be carefully estimated for safety of both personnel and the equipment. When the B–H curve of the machine is in doubt, adding a higher number of turns will reduce the level of flux. Then one successively removes turns and records the current attained as the flux volts per turn increases. Successive voltage application in this manner can be made until a B–H curve is created and the proper number of turns found (Fig. 11.12).

At this point, an example for the flux test calculation is useful due to the dangerous nature of the test. It is critical to get the excitation level correct because if the wrong level is calculated and it is above 100% of rated flux, damage to the stator core iron is likely. Also, personnel safety is a concern if the cable is not sized correctly for the current that will be drawn from the voltage supply. Generally, when the excitation cable is energized there is usually no voltage control and a simple breaker closure is used as the switch to start and stop the test. When the breaker is closed to start a flux test, the inrush can cause voltage spikes up to 20 times steady state, in addition to the high current from the source.

Flux Test Example. The following example is for a 500 MW, two-pole stator, using the machine flux and winding factor formulas from Section 2.4, where (for the example stator core):

V_{LL} = 22,000 line-to-line terminal voltage
k = two parallel paths per phase
f = 60 Hz
N_{ph} = 14 stator winding turns per phase
slots = 42
phases = 3

For the winding factor k_w, the parameters are as follows:

$\beta = \pi/3$ phases $= \pi/3$
$\eta = 42$ slots/2 poles/3 phases $= 7$
$\gamma = \pi/(42$ slots/2 poles$) = \pi/21$
ρ = stator winding pitch (from winding diagram) = 17/21

Working out the winding factor,

$$k_w = \text{distribution factor, } k_d \times \text{pitch factor, } k_p$$

$$\begin{aligned} k_w &= \frac{\sin(\beta/2)}{\eta \sin(\gamma/2)} \times (\rho\pi/2) \\ &= \frac{\sin(\pi/6)}{7 \sin(\pi/42)} \times \sin(17\pi/42) \\ &= 0.9134 \end{aligned}$$

Plugging all the above parameters into the machine flux formula then gives

$$\text{Machine Flux, } \Phi = \frac{V_{LL}k}{7.7\, f k_w N_{ph}} = \frac{22{,}000 \times 2}{7.7 \times 60 \times 0.9134 \times 14} = 7.46 \text{ webers}$$

Once the level of machine flux is known, the flux volts per turn required can be calculated as follows:

$$\text{Rated flux volts per turn} = \frac{2\pi}{\sqrt{2}} f\left(\frac{\Phi}{2}\right) = \frac{\pi}{\sqrt{2}} \times 60 \times 7.46 = 944 \text{ volts per turn}$$

Once the flux volts per turn is known, the power supply must be chosen such that it can supply both the required voltage and the amperes to provide an approximately 80% rated flux test. In most cases, this is usually a 4 kV circuit breaker with high ampere capacity. Given that the supply available in the example is a 4 kV circuit breaker, the number of turns of cable to wrap around the core is calculated by dividing the supply voltage (4,160 volts) by the flux volts per turn (994 volts per turn in this example), which results in approximately 4.2 turns for a 100% rated flux test. Since the number of turns cannot be a fraction, the next-highest single digit number is used. This would be 5 in the example. If the supply voltage of 4,160 volts is now divided by the number of turns, 5, then the flux volts per turn for a 5 turn test would be 832 volts. To now determine the level of flux for a 5 turn test, divide 832 volts by the rated flux volts per turn of 994 volts and the flux level is 83.7% of rated flux. This level of flux is considered ideal for test purposes.

The next issue is to determine the amount of current that will be drawn from the 4 kV supply so that an adequate 4 kV supply breaker can be selected in the plant as well as now being able to determine the size of the excitation cable for the test. To determine the amperes required from a B–H curve such as the one shown in Figure 11.12, it is first required to determine the *flux density B*. The flux density is calculated as the machine flux divided by 2 (F/2), all divided by the area of the back of the core (SL × CB) from Figure 11.11:

Therefore, the flux density for the example is,

$$\text{Flux density, } B = \frac{\Phi/2}{\text{SL} \times \text{CB}} = \frac{7.46/2}{6.38 \times 0.46} = 1.27 \text{ webers/m}^2 \text{ (tesla)}$$

where SL = effective stator iron length (total length minus all vents)
= 6.38 meters (for the example stator)

CB = length from slot bottom to core back
= 0.46 meters (for the example stator)

DS_{sb} = diameter of stator at slot bottom
= Dr + 2AG + 2ST (see Fig. 11.13)
= 1.1 + (2 × 0.079) + (2 × 0.197)
= 1.652 meters (for the example stator)

The next step in the calculation is to look up the ampere-turns per meter (*H*) on the B–H curve, that corresponds to the calculated value of flux density (*B*), 1.27 webers/m^2. In the particular example, the B–H curve gives a value of about 690 AT/m (Fig. 11.13).

Since flux takes the path of least resistance in the iron circuit, the distance that the flux travels in a flux test is basically the circumference of the stator at the slot bottom. This distance is the sum of the rotor diameter plus twice the airgap length plus twice

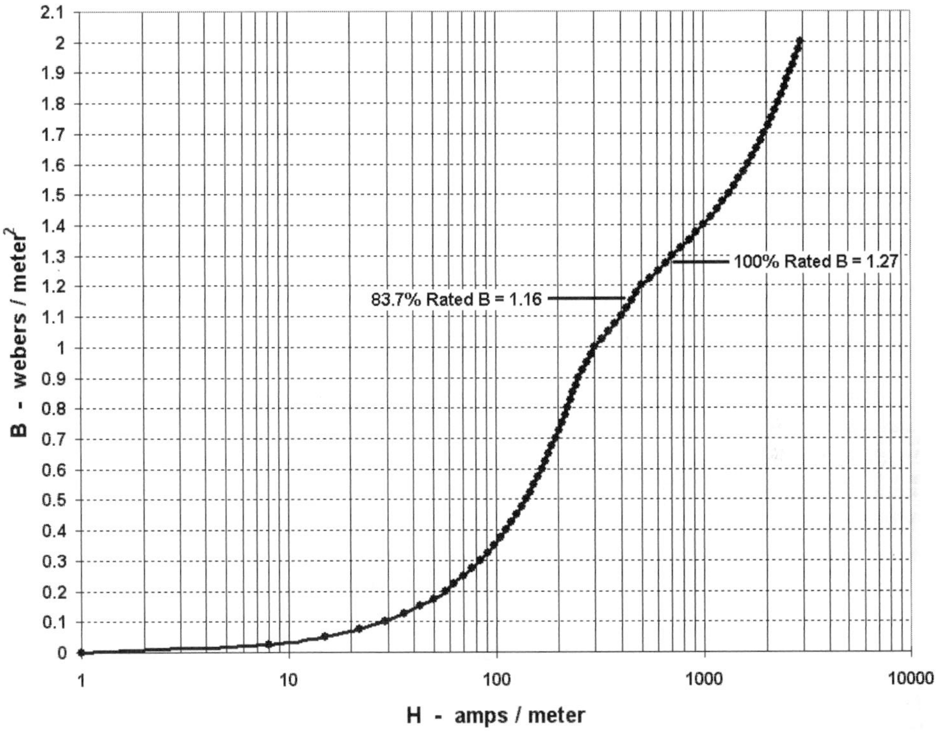

Fig. 11.13 B–H curve for example 500 MW, two pole stator.

the length of the stator tooth, times π. From Figure 11.11, this length is equal to (Dr + 2Ag + 2St)π = πDS_{sb}. The ampere-turns per meter from the B–H curve are approximately 690 AT/m. Therefore, the *magnetomotive force* required to push the rated flux around the stator circumference at the slot bottom is ***MMF*** = $H \times \pi DS_{sb}$ = 690 AT/m $\times \pi \times$ 1.652 = 3,581.0 AT. And subsequently dividing this result by 4.2 (calculated rated turns) gives 852.6 amperes required from the 4 kV supply to produce a 100% rated flux test. However, the 5 turn test produces an 83.7% flux test and, therefore, the flux density for the test is B = 1.27 (rated) × 83.7% = 1.16 wb/m^2. Looking this up on the B–H curve gives a new H of 430 AT/m. The magnetomotive force required to push flux around the stator circumference at the slot bottom at 83.7% of rated flux is ***MMF*** = 430 AT/m × π × 1.652 = 2,231.7 AT. Dividing the ***MMF*** by 5 turns results in 446.34 amperes from the 4 kV supply. This means that the supply must be capable of delivering this level of current continuously for the test. Also, the cable to wrap around the stator core may now be selected and the calculated value indicates that a 500 MCM cable would do the job safely.

The flux test is set up as basically shown in Figure 11.14. The power supply is selected as per the above calculation and connected as shown. The cable is wound through the stator bore the correct number of times and connected back to the power supply. Protection for the test cable is set up to provide "ground fault" and "overcurrent." The stator core, frame, and windings were all grounded for their protection and that of the test personnel. The CTs should also shorted at the terminals and grounded.

Metering is set up to provide measurements of supply voltage and current. A single loop of cable is installed additionally to measure the actual flux volts on the stator core during the test. This is done to provide an accurate measurement of the induced voltage across the core and the level of flux as well.

In some cases an infrared-nonreflecting mirror is used to monitor the temperature of the stator core when angled viewing from outside the stator bore is difficult (Fig. 11.15). The mirror provides a known surface to accurately measure the temperatures so that the absolute and relative rise of temperatures in the core defect areas can be recorded.

Once the flux is established in the core, it is maintained for at least 30 minutes to 1 hour. The temperature of the core should be maintained within values not significantly higher than those encountered during operation. Under these conditions, the temperature rises in the core are monitored and recorded while the existence of hot spots is investigated with infrared monitoring equipment (and, possibly, a nonreflecting mirror) (Figs. 11.16 to 11.18).

The temperature rises of the "good" core areas (ambient core-temperature rise) are

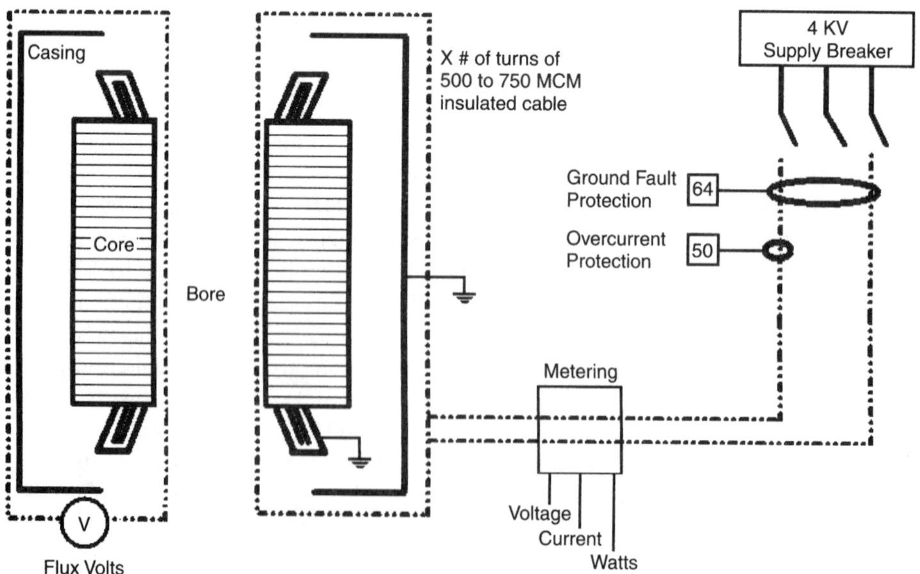

Fig. 11.14 Stator core, excitation, and protection setup for a flux test of a large turbogenerator stator core.

Fig. 11.15 Stator core excitation setup for a flux test of a large turbo-generator stator core, with nonreflecting mirror for infrared applications.

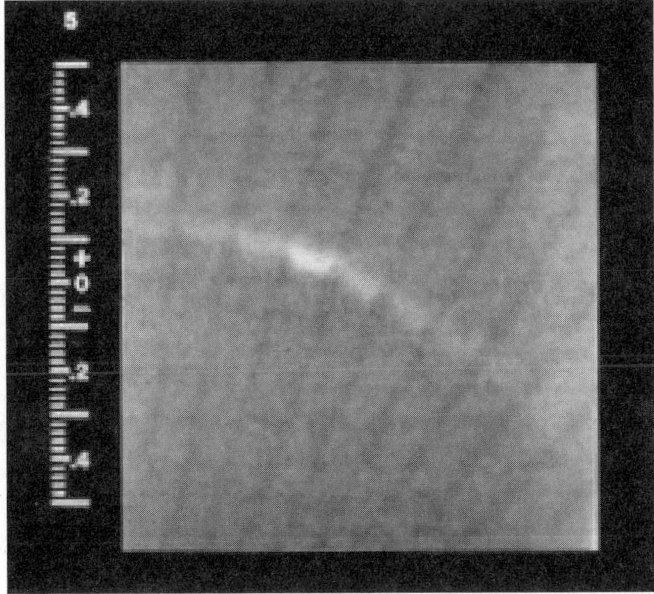

Fig. 11.16 Flux testing with direct infrared scanning of a core hot spot in the stator bore.

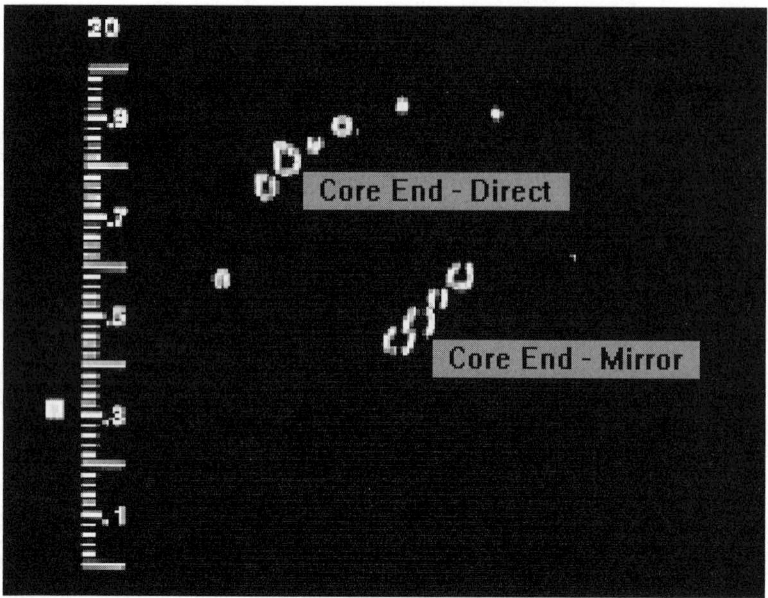

Fig. 11.17 Four core-end hot spots as seen in the nonreflecting mirror and directly, using an isotherm level to measure the temperature of the defects.

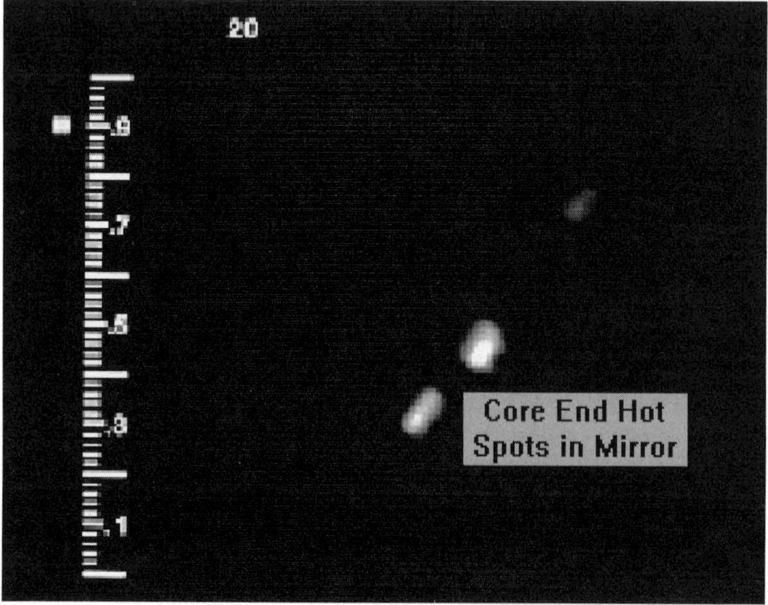

Fig. 11.18 Same core-end hot spots seen by zooming in with the infrared camera on the nonreflecting mirror.

then compared to the temperature rise profile of any defective locations found. Once the defects are located and characterized, repair solutions can then be addressed.

Figure 11.19 shows the relative experience in the industry with the typical types of core faults encountered, and how they appear during flux testing. It should be noted that these are general examples and may not be the case for a particular core tested. However, they show the general trend that the large majority of faults seem to follow.

11.2.3 Core Loss Test

Stator core loss is a function of terminal voltage. The open-circuit saturation curve for the core iron determines it. The core loss for any particular generator is always determined at the factory by the manufacturer and is not a test that is generally done at on-site.

The core loss is determined by the generator being coupled to, and driven by, a calibrated motor. The friction and windage (mechanical) losses are calculated and separated out from the electrical losses to provide a value of core loss for the stator [7].

If there is suspected wear of the interlaminar insulation in the core, on a large scale, it may be possible that a core loss test could be done to compare the present value to the "as new" value to determine the extent of deterioration occurring. However, the serious challenge of driving the generator on-site with a calibrated motor for all practical purposes limits this test to the OEM's factory.

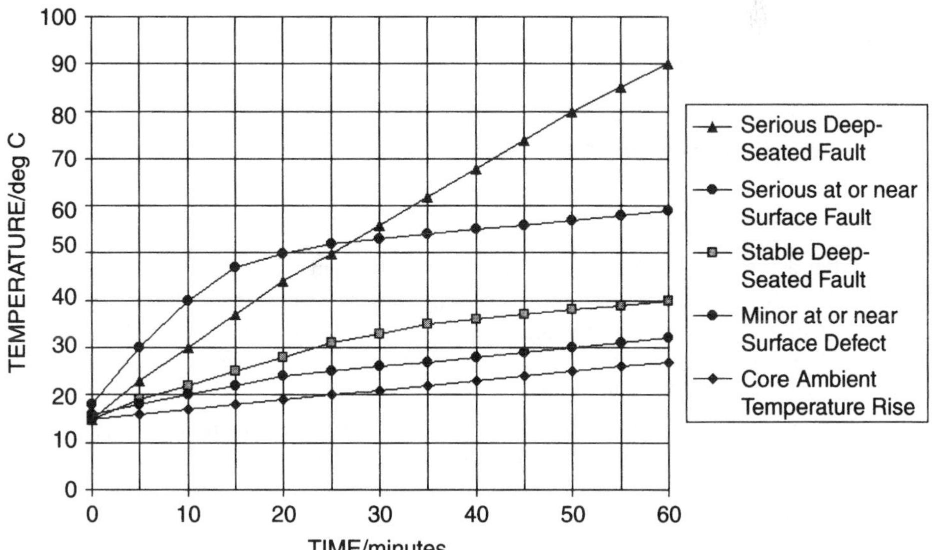

Fig. 11.19 Temperature rise profiles of typical core fault types during flux testing with infrared scanning.

11.2.4 Through-Bolt Insulation Resistance

There are a few manufacturers that provide through-bolts in their stators to pull the cores tight. These through-bolts are full-length bolts inserted axially through the core through holes in the core iron. There are many of them located symmetrically around the circumference of the core, a few inches below the stator winding slots. The ends are threaded and terminated at each end through a pressure plate, where a nut is installed to maintain compression after the core is pressed to a few hundred psi.

The entire through-bolt assembly is insulated generally by a cured epoxy–glass tape wrap or a phenolic tube through the core, and an arrangement of insulators at the pressure plates and nuts. This is done to ensure that the through-bolts do not create any short circuits across the stator core laminations and cause a core failure due to circulating currents.

To ensure that the insulation is in good condition, the insulation resistance of the through-bolts is checked by meggering at 500 V dc. A good reading should be in the hundreds of megaohms range.

11.2.5 Insulation Resistance of Flux Screens

Most large generators are provided with some form of flux screening for the stator core end. This is to prevent overheating in the core ends due to stray flux from the stator end winding. When flux screens are used, they are either well ground to the stator frame or fully insulated from the core end to ensure that no additional circulating currents flow between the core and the flux screens, which would create additional unwanted heating in the core end.

To ensure that the insulation is in good condition on isolated-type flux screens, the insulation resistance of the flux screens is checked by meggering at 500 V dc. A good reading should be in the hundreds of megaohms range.

11.3 STATOR WINDING MECHANICAL TESTS

11.3.1 Wedge Tightness

Loose stator winding wedges allow the stator winding in the core slots to move because of electromagnetic bar bounce forces and vibration. This can, in turn, cause the ground-wall insulation of the winding to wear and fret against the stator core iron and eventually fail by ground fault. Therefore, it is necessary to maintain the stator winding wedges tight to avoid such problems.

Checking wedge tightness is done by tapping with a suitable hammer (e.g., an 8 oz ball-peen) and recording the degree of looseness on a chart (Figs. 11.20 and 11.21). During the tapping, the wedge should be checked by tapping along its full length and feeling for the vibration as well as listening to the sound produced. Tight wedges will vibrate very little and emit a hard ringing sound. Loose wedges will vibrate noticeably and emit a very hollow sound. Wedges with vibrations and sounds in between these indicate that the wedge is in the process of becoming loose. It requires a trained ear and

Fig. 11.20 Stator winding wedge tightness testing by the manual tapping method.

feel to determine the degree of looseness in wedges of various machines because not all wedging systems are alike and they do not all sound and feel the same.

The criteria for determining when rewedging is required is well established, with some variation from one manufacturer to another, and from one operator to another. However, the following is the accepted rule of thumb. Re-wedging in a particular slot is required if:

1. Less than 75% of the wedges are tight in the slot
2. Three or more adjacent wedges are fully loose
3. Fully loose end wedges must be retightened.

Since the early 1980s, there have also been automated systems developed for wedge-tightness testing (Fig. 7.19). Such devices mechanically stimulate the wedge being tested and record the vibration produced from it. With some modern devices (see Fig. 11.22), the vibration results are input to a computer and a map is produced.

Once a wedge-tightness survey has been performed, it is always a good practice to record the tightness for successive surveys of the machine, so that an estimate can be made of when tightening should be done. The individual surveys of a machine should also be compared to other identical machines in a multiunit station, since they do not always loosen at the same rate. This can be seen in an actual example shown in Figure 11.23, taken from a four-unit station with identical machines; three of four generators tended to loosen at the same rate, whereas one loosened much quicker.

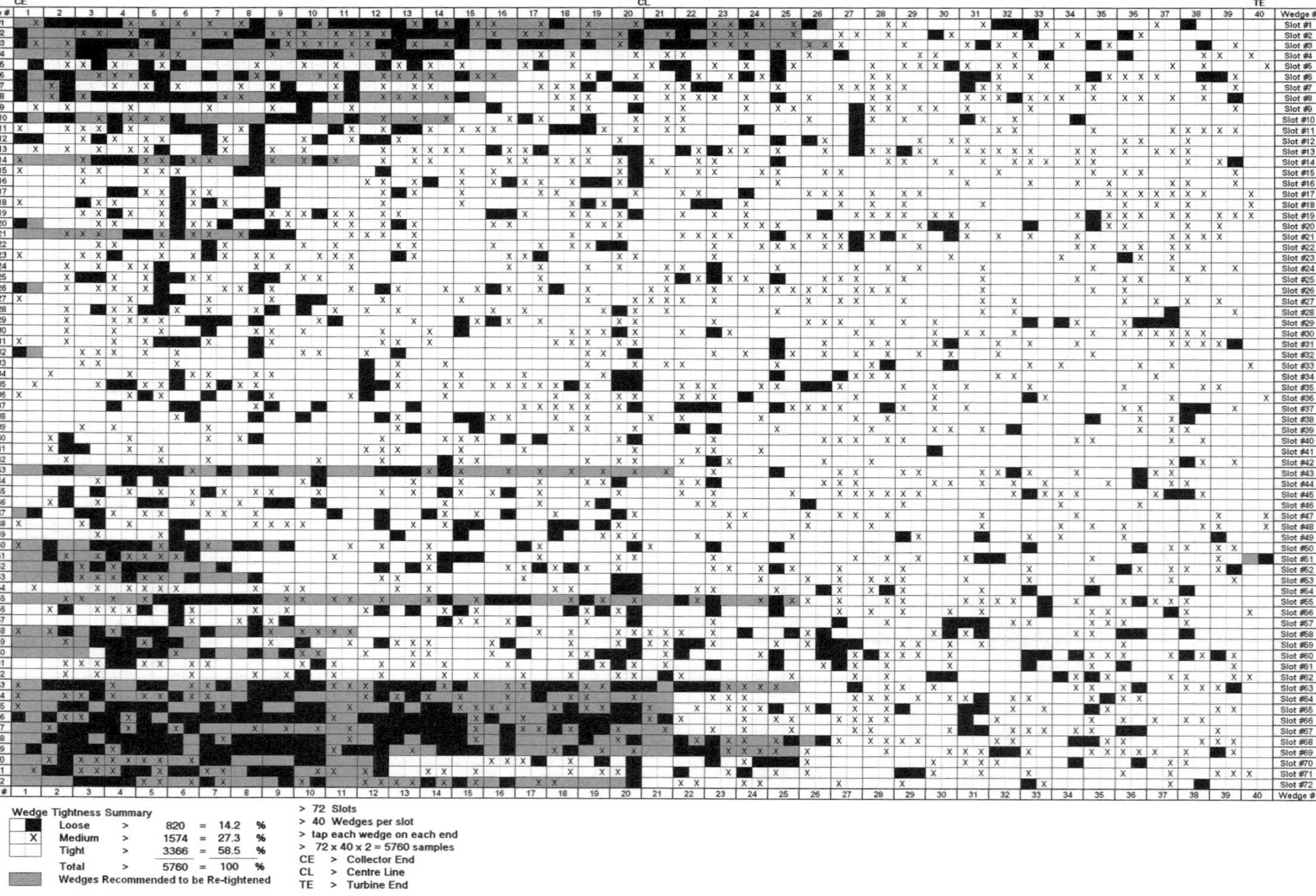
CE
CL
TE
Wedge Tightness Summary
Loose > 820 = 14.2 %
Medium > 1574 = 27.3 %
Tight > 3366 = 58.5 %
Total > 5760 = 100 %
Wedges Recommended to be Re-tightened
> 72 Slots
> 40 Wedges per slot
> tap each wedge on each end
> 72 x 40 x 2 = 5760 samples
CE > Collector End
CL > Centre Line
TE > Turbine End

Fig. 11.21 Real example of a stator-winding wedge-tightness survey map showing tight, medium, and loose wedges. Also, the dark areas of loose wedges are shadowed over where rewedging during an outage was recommended and carried out.

Fig. 11.22 Automated testing and mapping system. (Courtesy of ADWEL International.)

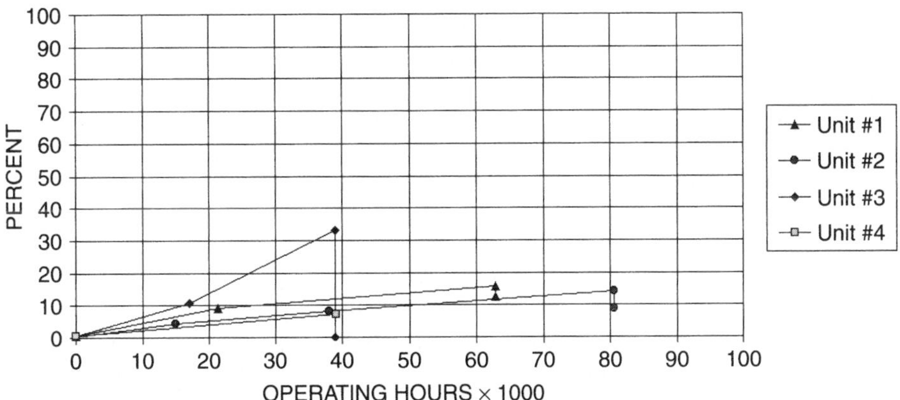

Fig. 11.23 Stator-winding wedge-looseness profile record over time. This shows the difference between four supposedly identical machines that did not loosen at the same rate.

11.3.2 Stator End-Winding Vibration

High vibration on the stator end winding can cause premature winding failures. Acceptable vibration limits may vary from one manufacturer to another. Generally, however, a good rule is to keep the maximum vibration to less than 50 μm (~2 mil) peak to peak (unfiltered), with no natural resonance within the frequency ranges 50–75 Hz and 100–140 Hz for 60 Hz units or about 40–65 Hz and 80–120 Hz for 50 Hz units (similar to the stator core and frame limits). When in doubt, it is best to contact the OEM for a recommendation on vibration limits.

Measurements are accomplished by vibration transducers mounted on the coil heads or stator bar connections at each end of the machine. Mounting is done in such a way that the transducers are insulated from, and noninvasive to, the electrical circuit of the winding. The transducers should be capable of radial and tangential vibration measurements.

If permanent mounting is not possible (as discussed in Chapter 5, Section 5.5.3), another method is to carry out an *impact frequency spectrum analysis* of the end winding. This is done with temporary vibration transducers mounted on the end winding and a calibrated impact hammer to measure the vibration signatures. This test is also known as a "bump" test.

As in the stator core and frame, end-winding vibration problems are very difficult to diagnose and expensive to fix. They should also be addressed on a machine-to-machine basis, as each case will present a unique situation.

11.4 WATER-COOLED STATOR WINDING TESTS

11.4.1 Air Pressure Decay

The pressure decay test is probably the oldest test method for detecting leaks. The stator winding is first completely dried out internally, and then pressurized up to the level of the generator operating hydrogen gas pressure. This provides a very high pressure differential and a leak direction that is normal to the operating leak direction of water.

When pressurized to levels as high as the operating hydrogen gas pressure, a 1 psi pressure drop will only correspond to about 1 cu ft loss of volume. This test is, therefore, not sensitive to small leaks and highly dependent on atmospheric conditions of temperature and barometric pressure. It is carried out by pressurizing the stator winding to about 30 psi and watching that the drop in pressure over a 6 hour period does not exceed 0.1 psi. To recognize smaller leaks, the test must be carried out over a period as long as 24 hours using very accurate and calibrated pressure gauges.

During pressurization, "snooping" is also used to identify leaks. This technique entails application of a soapy water solution to the winding in selected areas. It is generally used at joints, fittings, and points that are likely sources of leaks. If a leak is present, the soapy water will bubble, indicating that the tracer gas is escaping at the bubble location.

Another method of detecting leaks is to use an ultrasonic probe to listen for the sound created by the leaks (Fig. 11.24).

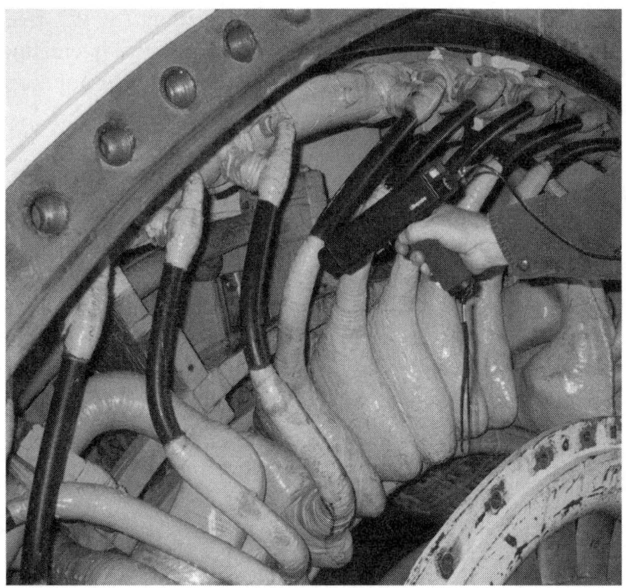

Fig. 11.24 Ultrasonic probe used to detect sound from air pressure test leaks.

11.4.2 Tracer Gases

Freon gas was, at one time, the preferred gas for leak tracing and snooping. Due to environmental factors, Freon has been superseded by the use of helium gas. Helium is a light gas that is molecularly small in size, making it ideal for this application. In addition, it is nontoxic and nonhazardous.

The trace method involves pressurizing the stator winding with helium, or a mixture of air and helium. A portable helium gas detector is then used to "sniff" over the external surfaces of the exposed portions of the stator winding for sources of helium. Obviously, this method requires access to the generator internals, and even so, it is still not be possible to cover the entire winding by this method. Therefore, it is only one of the tests that are required to fully identify leak problems and locate the source.

The helium tracer technique has been shown to be very successful in pinpointing stator winding leaks.

11.4.3 Vacuum Decay

Vacuum decay testing is very sensitive compared to pressure decay testing. It is generally used to determine the leak sensitivity of the entire winding. This test has the advantage that it is relatively insensitive to atmospheric conditions of temperature and barometric pressure. Further, it only requires a few hours of testing to determine the leak rate, since vacuum gauges are far more sensitive than pressure gauges. The test leak direction is opposite to the operating leak direction of water but normal to the operating leak direction of hydrogen into stator cooling water (SCW). Even the smallest leaks will be detected by this method. The test is carried out by creating a vacuum of 29 inches of mercury and watching that the vacuum gauge does not drop by more than 0.2 inches of mercury over 6 hours.

11.4.4 Pressure Drop

The differential pressure across the stator winding from the stator cooling water inlet to the outlet of the generator can also be tested in the offline mode to ensure that the design pressure drop across the whole stator winding is at the correct level. Alternatively, each individual stator bar can be pressure-drop tested to estimate the stator cooling water flow rate in each bar, and ensure that it is above the minimum recommended level for adequate cooling of the stator winding.

11.4.5 Flow Testing

Flow checks are generally done on individual bars to determine if there are restrictions in the cooling water path. Both air and water have been used to accommodate this test, but it is usually done with water. The actual flow in each stator bar is measured to compare it to the design level of flow required for the particular machine being tested. Some flow checks are made using ultrasound techniques to monitor the flow rates. Low-flow bars are then noted for corrective action such as backflushing or chemical flushing.

11.4.6 Capacitance Mapping

Capacitance mapping is a nondestructive electrical test method used to identify the presence of moisture in the ground-wall insulation due to stator winding leaks. The technique is based on the difference in dielectric constant of the ground-wall insulation and that of water. All top and bottom bars are tested at both ends of the machine, providing an average measurement of capacitance. The measurement is made by a capacitance probe placed on the side of each bar. Good insulation is indicated by measurements falling within (2 standard deviations from the average. Measurements of capacitance greater than +3 standard deviations are considered suspect, and those greater than +5 are generally found to be wet and to fail their subsequent electrical tests. Failure of this test tends to indicate that the stator bar is not suitable for long-term service.

Care must be taken when performing this test, since the capacitance measurements are also sensitive to the application of the capacitance probe, bar surface condition and coatings, grading paints, and the condition of the ground-wall insulation itself.

11.5 STATOR WINDING ELECTRICAL TESTS

Stator windings are comprised of materials with specific resistive and dielectric qualities. The materials used include mica, Dacron tapes, glass tapes, asphalt binders, polyester resins, and epoxy resin binders. There are also insulating, resistive, and stress grading paints applied to various portions of the winding to ensure controlled distribution of the voltage on the individual stator conductor bars. All of the materials are used and applied in such a manner as to ensure proper functioning and a reasonable degree of long-term reliability of the winding. The stator winding insulation system is complex and requires a variety of tests to establish its present condition and expected long-term reliability. Therefore, to fully test the stator winding, so that the best possible determination of the winding condition can be made, it is desirable to perform both ac and dc tests.

The dc tests are generally sensitive to the presence of cracks, moisture, particle contamination, or a general degradation of the electrical creepage path. During dc application, the voltage is divided according to the dc leakage resistance. Basically, dc is used to test the conductivity of the insulation system. The dc testing has the advantage that it less damaging to the insulation due to the absence of corona and partial discharges associated with ac.

The ac testing, on the other hand, applies a more realistic electrical stress to the winding, since it operates on ac when in service. When the ac test voltage is applied, it is actually applied across several dielectric components of the winding insulation, which are effectively in series. Therefore the leakage current must go through each of the dielectrics until it reaches ground potential. Under ac, therefore, the voltage is divided according to the relative permittivity of each of the dielectric materials.

The ac testing is in fact far more searching than dc. In addition to the conducting properties of the insulation, ac testing is also capable of determining the loss or power factor characteristics and the dielectric properties. In addition, the mechanical integrity of the insulation can be also be determined by the capacitance characteristics of the winding in terms of insulation delamination.

As with many other issues about testing the insulation of large electric machines, experts and operators have different opinions about whether dc and/or ac is more convenient. Some only prefer dc tests, others prefer ac testing and others like to use both.

11.5.1 Pretesting Requirements

If the stator winding is water cooled, it must be completely dried prior to all testing to obtain meaningful results. If there is stator cooling water left in the winding, it will alter the test results and give a distorted picture of the insulation condition.

All three phases must be isolated to ensure that all testing is carried out on the stator winding only. This means that each phase should be completely separated at the neutral point and floated from ground. The line ends of the stator winding should be separated from the isolated phase bus (or cables in smaller units) just outside the generator, at the stator terminals.

The generator current transformer windings should be shorted and grounded to avoid induced high voltage and possible discharge failure of the insulation. All instrumentation leads should be grounded to also avoid induced high voltage and possible discharge failure of the insulation.

Before conducting any high-voltage testing of the unit, consult vendor and/or pertinent standards.

11.5.2 Series Winding Resistance

This test is used to measure the ohmic resistance of the copper in each phase of the stator winding. Given the relatively low dc series resistance of windings of large machines, the measurement accuracy requires significance to a minimum of four decimal places.

The purpose of the test is to detect shorted turns, bad connections, wrong connections, and open circuits. Acceptable test results consist of the three resistance values (one per phase) to be balanced within a 0.5% deviation from the average.

The test is very sensitive to differences in of temperature between sections of the winding. The machine should be at room temperature when the test is performed.

As with any other electrical test, the results should be compared with original factory data, if available. This test can be performed on stator and rotor windings.

11.5.3 Insulation Resistance (IR)

The purpose of this test is to measure the ohmic resistance between the conductors in each of the three phases and ground (i.e., the stator core). This test is generally regarded as an initial test to look for gross problems with the insulation system, and to ensure that further high-voltage electrical testing may "relatively" safely continue, in terms of danger of failing the insulation.

Normally, the measurements of IR will be in the megaohm range for good insulation after the winding is subjected to a dc test voltage, usually anywhere from 500 to 5000 V, for one minute. The minimum acceptable reading per IEEE Standard 43 [4] is (VLL in kV + 1) MΩ. The test is carried out with a megger device. However, resistance bridges may also be employed.

The dc test voltage level is usually specified based on the operating voltage range of the machine, the particular component of the generator being tested, operator's policy and previous experience, and knowledge of the present condition of the insulation in the machine. Although the readings obtained will be somewhat voltage dependent, this dependency becomes insignificant for machines in which the insulation is dry and in good condition. It is essential that water-cooled stator windings be completely dried before any testing so that any poor readings will be due to a "real" problem and not to residual moisture from the stator cooling water.

The readings are also sensitive to factors like humidity, surface contamination of the coils, and temperature. Readings should be corrected to a base temperature of 40°C by the following equation:

$$R\,(40°\mathrm{C}) = K \times R_{\mathrm{measured}}(°\mathrm{C})$$

Where K is a temperature-dependent coefficient that can be obtained from IEEE Std. 43 (Fig. 11.25).

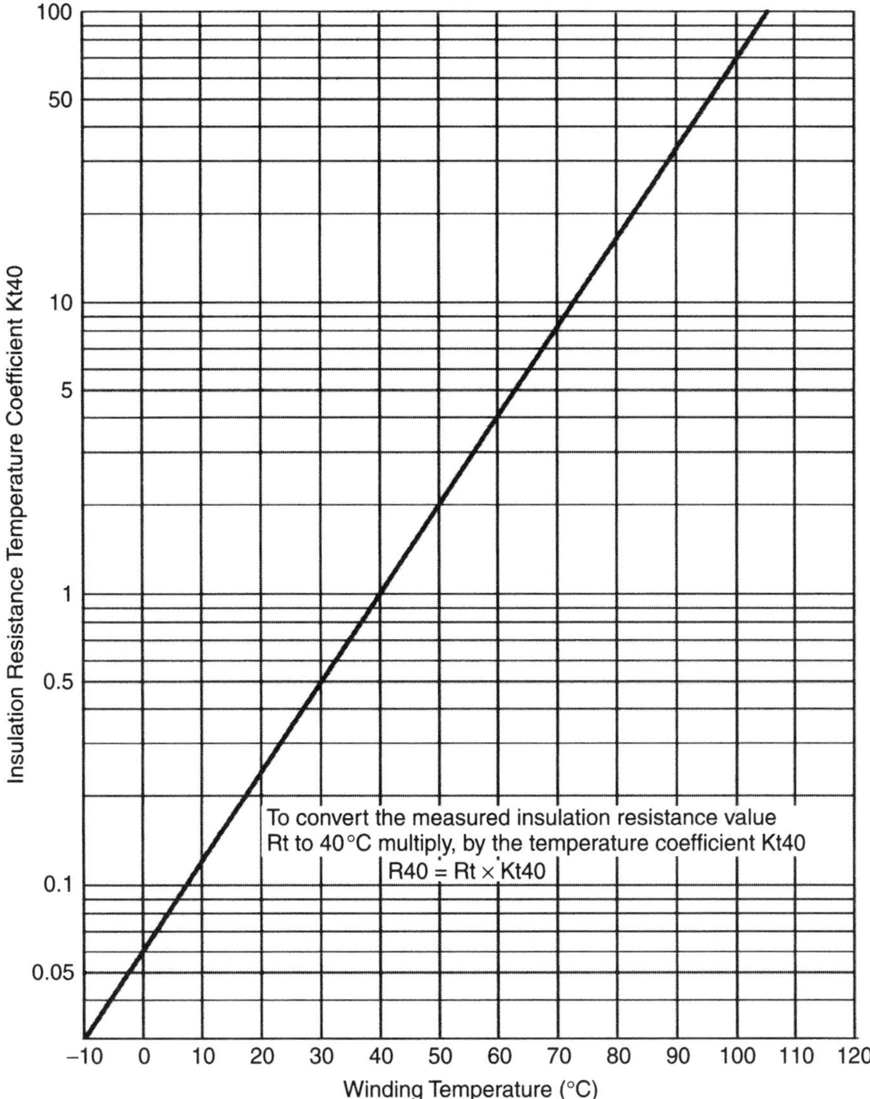

Fig. 11.25 Insulation resistance versus winding temperature curve (courtesy of EPRI).

The following equation can be used to obtain K to some degree of accuracy in lieu of the graph:

$$K = 0.0635 \times \exp[0.06895 \times T_{\text{measurement}} \text{ (in °C)}]$$

Insulation resistance tests are performed on both stator and rotor windings, core-end flux screens, and core compression or through-bolts as mentioned previously.

The insulation between the core-end flux screen and the stator core-end iron ensures that the flux screen maintains its capability to shield the core-end from axial flux, and keep the resulting circulating currents within the flux screen, without providing a current path to the core.

The insulation between core-compression bolts and the iron keeps the through-bolts from short-circulating the insulation between the core laminations. Otherwise, large eddy currents generated within the core would produce heat and temperatures, which could further damage the interlaminar insulation, as well as the insulation of the windings. Figure 11.26 shows typical IR behavior as a function of time.

11.5.4 Polarization Index (PI)

Insulation resistance is time dependent as well as being a function of dryness. The amount of change in the IR measured during the first few minutes depends on the insulation condition and the amount of contamination and moisture present. Therefore, when the insulation system is clean and dry, the IR value tends to increase as the charge is absorbed by the dielectric material in the insulation. When the insulation is dirty or wet or a gross insulation problem is present, the charge does not hold, and the

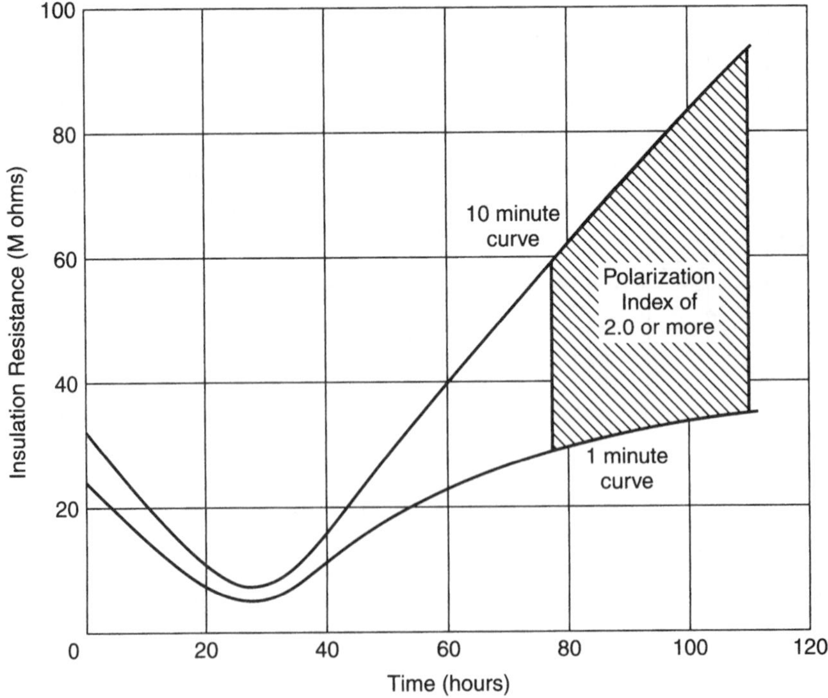

Fig. 11.26 Insulation resistance as a function of time and dryness. (Courtesy of EPRI.)

IR value will not increase because of constant leakage current at the problem area. Therefore, the ratio between the resistance reading at 10 minutes and the reading at 1 minute produces a number or "polarization index," which is essentially used to determine how clean and dry the winding is (Fig. 11.27).

Class B and F windings tend to show higher PI values than windings made of class A insulation. The PI value is also dependent on the existence of a semiconducting layer.

The recommended minimum PI values are as follows:

Class A insulation: 1.5
Class B insulation: 2.0
Class F insulation: 2.0

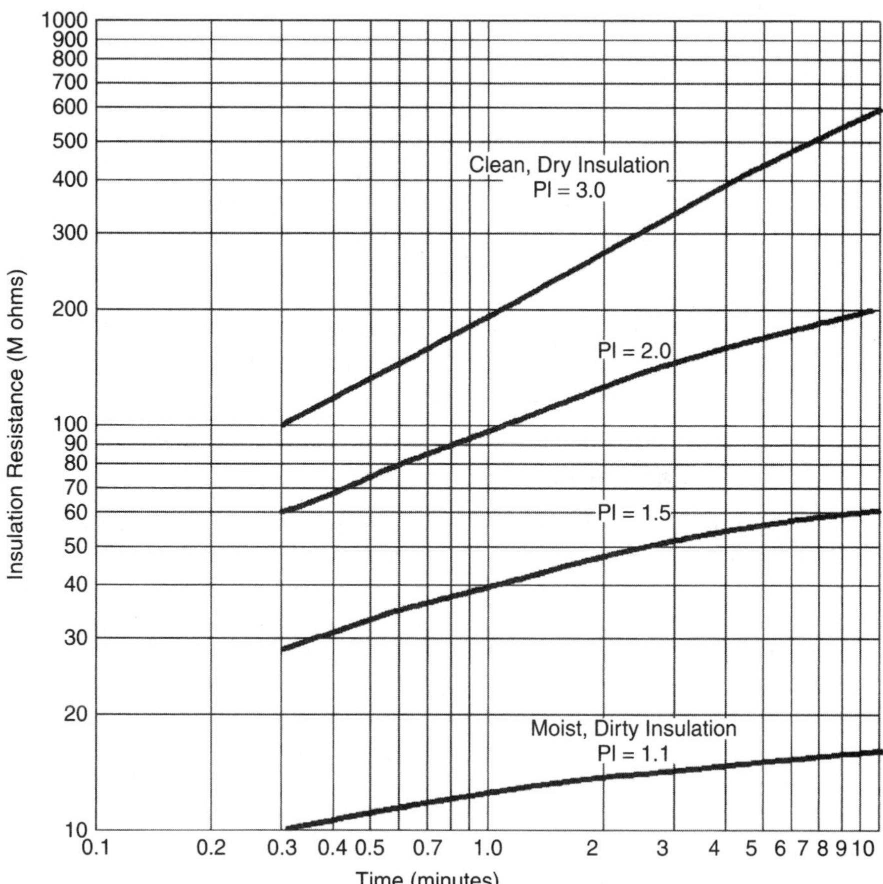

Fig. 11.27 Polarization index curves for insulation resistance as a function of time. (Courtesy of EPRI.)

The same megger used for the IR readings should be used to determine the PI. The PI readings should be done on a per-phase basis at the same voltage as the IR test, and can be used as a go/no-go test before subjecting the machine to subsequent high-voltage tests, either ac or dc. The IR readings for the PI test should also be corrected to 40°C as in the IR test. Performing the high-voltage tests on wet insulation may result in unnecessary failure of the insulation.

11.5.5 Dielectric Absorption during dc Voltage Application

Dielectric absorption current characteristics can be used to measure the aging of the resin binder in the ground-wall insulation. When applying dc voltage to insulation material, a time-dependent flow of current is established. This current has a constant component, called the conduction current or leakage current, and a transient component, called the absorption current.

Absorption current is a function of the polarization of the molecules in the binding material. The older the binding material, the more polarized it becomes, and the more absorption current flows. Therefore, this test is best used as a comparison test between the winding condition at different times, and between similar windings.

Absorption current is also temperature dependent. This fact should be taken into consideration when performing the test and interpreting the results.

Absorption current is also dependent on the amount of voids in the insulation. The dependence is inverse, meaning that an increase in the number of voids in the insulation will tend to reduce the magnitude of absorption current. The contradictory effects regarding void density and aging of the binding material renders this test difficult to interpret. It is best when used in conjunction with other dielectric tests, such as partial discharge and dissipation factor tip-up tests.

11.5.6 dc Leakage or Ramped Voltage

The dc leakage or ramped voltage test is a controlled dc voltage application designed to test the winding in such a manner as to monitor the dc leakage current at the same time that the dc voltage is increased. The leakage current is plotted against the dc voltage applied to give early warning of any impending insulation breakdown. This helps in limiting damage by shutting down the test before a full breakdown occurs (Fig. 11.28).

When dc voltage is applied to the winding, a time-dependent flow of current is established. This current has a constant component, called the *conduction* or *leakage current,* and an initial component, called the *harging* or *absorption current.* Therefore, it is advisable to raise the voltage to the first level of the kV/min rate, hold it for 10 minutes to get beyond the charging phase of the voltage application, and perform the test while dealing primarily with the leakage current. This way, the charging current influence on the leakage current rate of rise will be minimized.

The final dc test voltage level is generally in the range of 125 to 150% of (V_{LL} × 1.7) kV dc (IEEE Std C50.13). The value actually chosen between 125% up to 150% of the test voltage is dependent on the age of the machine insulation and knowledge of its general condition.

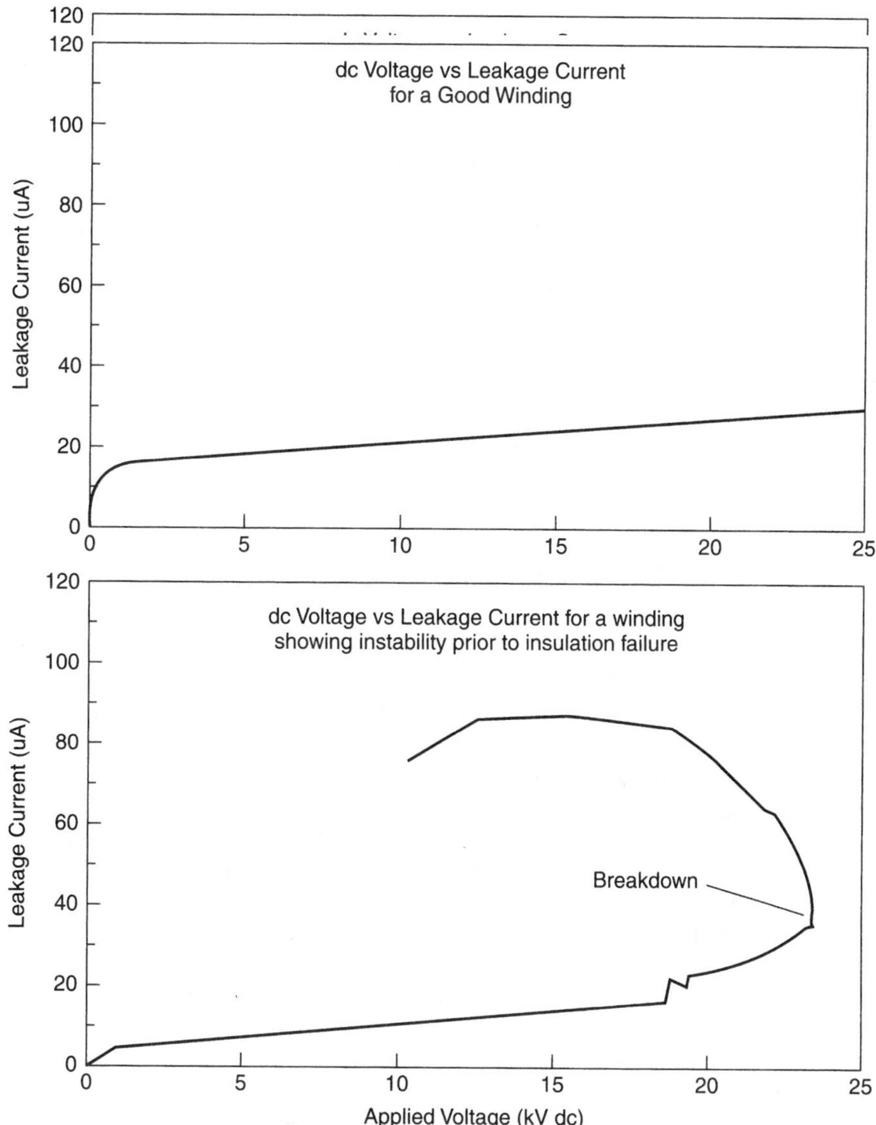

Fig. 11.28 dc Ramp test plot. (Courtesy of EPRI.)

The ramp rate is selected at 3% of the final test voltage level in kV dc/minute (IEEE Std 95). The ramp rate usually is in the range of 1.5 to 2 KV per minute.

Generally, the ramping portion of the test is automated to allow a steady increase in the voltage. A dc hi-pot set with the capability of a timed and steady voltage increase is required. If this is not possible with the equipment available, then a basic dc hi-pot set can be used to raise the voltage in the predetermined 3% voltage steps, holding each step for 1 minute.

11.5.7 dc Hi-Pot

The dc hi-pot test is used to ascertain if the winding is capable of sustaining the required rated voltage levels (without a breakdown of the insulation), with a reasonable degree of assurance for capability to withstand overvoltages and transients, and maintain an acceptable insulation life. The test consists of applying high voltage to the winding (the three phases together, or one at a time, with the other two grounded) for one minute.

The recommended test voltage level is $[(2 \times V_{LL} + 1000) \times 1.7]$ kV dc for new windings (IEEE Std C50.13). The recommended test voltage level for field testing and maintenance purposes is 125 to 150% of $(V_{LL} \times 1.7)$ kV dc (IEEE Std C50.13). The value actually chosen for the test voltage is dependent on the age of the machine insulation, knowledge of its general condition, and the specific situation calling for a test.

11.5.8 ac Hi-Pot

The ac hi-pot test is also used to ascertain if the winding is capable of sustaining the required rated voltage levels (without a breakdown of the insulation), with a reasonable degree of assurance for capability to withstand overvoltages and transients, and maintain an acceptable insulation life. The test consists of applying high voltage to the winding (the three phases together, or one at a time with the other two grounded) for one minute.

The recommended test voltage level is $(2 \times V_{LL} + 1000)$ kV ac for new windings (IEEE Std C50.13). The recommended test voltage level for field-testing and maintenance purposes is 125 to 150% of V_{LL} kV ac (IEEE Std C50.13). The value actually chosen for the test voltage is dependent on the age of the machine insulation, knowledge of its general condition, and the specific reasons for the calling for a test.

The ac testing is generally done at a power frequency of 60 Hz but may also be carried out at a low frequency of 0.1 Hz, which is the accepted industry standard. Generally, the ac hi-pot is a "pass" or "fail" type of test. However, this is not always the case. There are often times when arcing can be heard and even seen (Fig. 11.29), and the test can be stopped until the problem area is repaired. Then retesting may be carried out to prove the repairs.

When there is a true problem with the ground-wall insulation itself and an ac hi-pot is carried out, there are always the unfortunate occasions in which the insulation is not able to withstand the applied voltage and no advanced warning occurs, such as the arcing shown in Figure 11.29. The only warning may be an increased audible crackling due to the leakage discharge increasing. The end result would be a complete breakdown and flashover to ground (Figs. 11.30 and 11.31). When the failure point is small and confined, it is not always easy to locate even during the actual breakdown. Generally, the failure occurs and a dull thud-like noise is heard, rather than a sharp crack. The bar then has to be verified as failed by a megger test if the failure point is not obvious. In some cases, the actual location of the breakdown may only be found upon disassembly and removal of the stator bar.

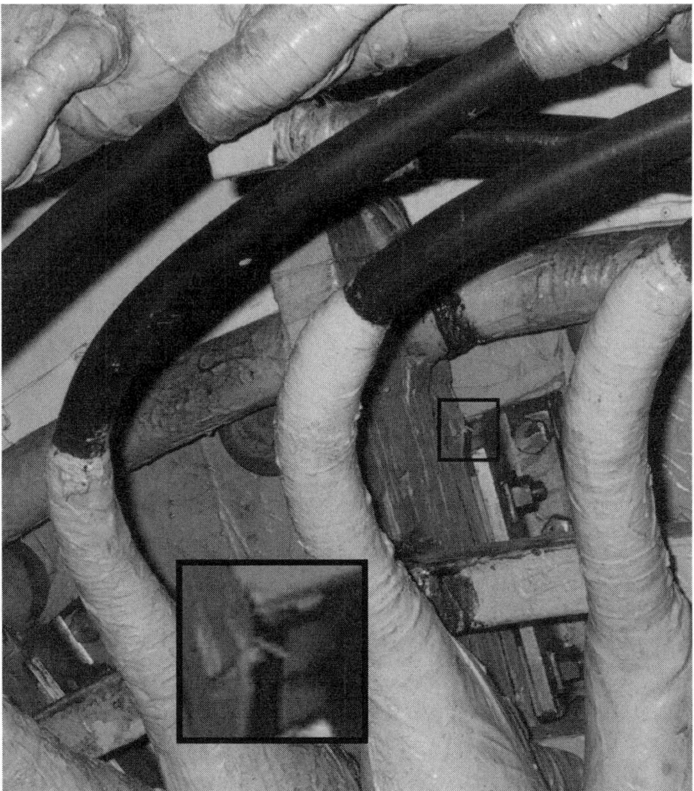

Fig. 11.29 Arcing occurring between a phase connector and ground during an ac hi-pot test.

Testing is usually done on water-cooled windings with the system drained and vacuum dried before voltage application. However, there are instances where good dc measurements have been recorded on two phases while one appears grounded and the winding is technically "wet." In such an instance, ac testing has been done as a next step, but this is a rare occasion. It is not recommended to proceed with this type of testing unless an expert is present who knows how to handle this type of situation. A "wet" ac test should be performed when no failure point can be found and yet the winding will not hold dc voltage and internal contamination of the stator winding hoses is suspected. Under dry conditions, the winding will pass high-voltage testing, and under wet conditions, the contamination will be conducting. Depending on the type of contamination and its conductivity, the hoses may glow under high-voltage ac (Fig. 11.32).

One other issue that always seems to arise when trying to accomplish an ac hi-pot test at a power station is the ability to energize a complete stator winding phase with whatever power source is available, to the required voltage level. Many times, the leakage current of the hi-pot test set is a limiting factor in the ability of the hi-pot set to

Fig. 11.30 Failure point during an ac hi-pot test of a stator bar of a 500 MW water-cooled machine. The location was a few inches into the slot on a top bar and was not obvious.

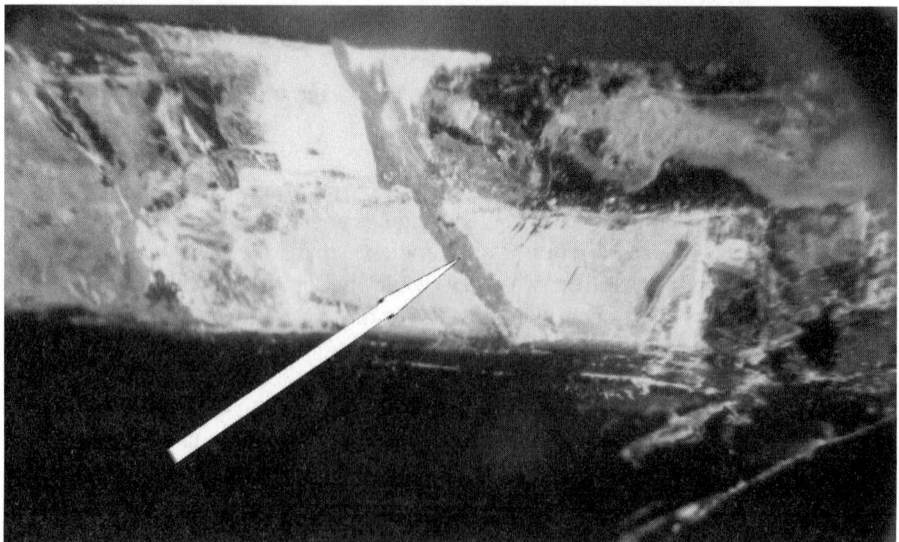

Fig. 11.31 Failed stator bar of Figure 11.30 after stripping off the outer groundwall insulation layer. The electrical tracking path can be seen next to the copper stack on the inside of the ground-wall layer, extending from the top of the bar to the bottom.

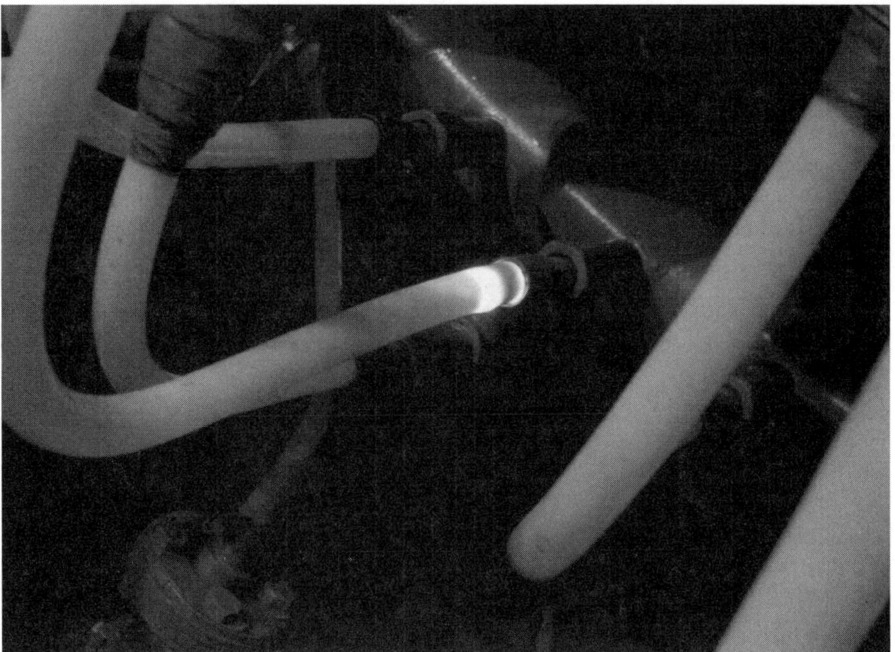

Fig. 11.32 Glowing hoses during a "wet" ac hi-pot test where contamination of the hoses is present.

energize an entire phase to the desired level. Since the power frequency transformers used for energizing the winding are usually large and heavy, it is often difficult to arrange for them to be transported and located on-site for the testing. This is why some opt for low-frequency, resonant ac hi-pot or even dc hi-pot testing, so that the equipment is easier to find and transport. dc hi-pot testing is the easiest because the equipment is very lightweight in comparison to ac equipment, and can be transported easily. The problem with this approach, however, is that only the resistive characteristics of the insulation system can be tested with dc. If one requires some knowledge of the dielectric qualities of the insulation or needs to do such tests as *partial discharge* or *TVA probe,* then ac is required.

To estimate if the local power supply can achieve the required voltage level for the ac hi-pot test, one needs to know the total capacitance of the stator winding phase as a whole. Knowing the capacitance, the power frequency and the voltage level to be tested to, one can determine the leakage current that will occur and, hence, if the power source for the hi-pot equipment will accommodate the voltage level required. This can be estimated as follows:

$$\text{Leakage current} = \frac{\text{Applied test voltage}}{jXc}, \qquad \text{amperes}$$

where
jXc = $\frac{1}{2}\pi fC$
f = power frequency (60 Hz)
C = stator winding capacitance per phase

At this point, an example is useful. Consider a machine with a per-phase stator winding capacitance of 0.25 μF. If the test voltage is 33,000 volts ac, then the leakage current would be as follows:

$$\text{Leakage current} = \frac{\text{applied test voltage}}{j/2\pi fC}, \qquad \text{amperes}$$

$$= \text{applied test voltage} \times 2\pi fC$$

$$= 33{,}000 \times 2\pi \times 60 \times 0.25 \times 10^{-6}$$

$$= 3.11 \text{ amperes}$$

11.5.9 Partial Discharge (PD) Off-line Testing

In principle, PD measurements are based on direct measurement of the pulses of high-frequency current discharges created during the occurrence of partial discharges. Some offline methods are based on a capacitive link between the whole of the winding and the measurement equipment. These setups allow the measurement of PD activity in whole windings, or one phase at a time.

To measure the PD activity in smaller sections of the winding, methods based on an electromagnetic probe or pickup (which is mounted on a hand-held, electrically insulated stick) has been developed. One such probe is known as the TVA probe and is used to traverse the entire length of a slot in the stator bore to search for localized sources of PD. Therefore, each slot is probed over its full length.

The partial discharge tests are carried out from voltages below the inception voltage up to rated voltage.

On-line partial discharge analysis can be performed by modern instrumentation and methods described in the following sections.

PD Monitoring by Capacitive Coupling. Partial discharge monitoring by capacitive coupling is thoroughly described in Chapter 5, Section 5.5.3. It is generally an online test nowadays, but offline measurements are also done on a regular basis. The test setup for offline capacitive coupling is generally as shown in Figure 11.33.

During operation of a generator, the voltage on the stator winding is graded according to the line-to-neutral connection. Thus, when an online test is performed, the bars near the neutral end of the winding are not subjected to high voltage, which represents the actual operating condition. In the offline test, all stator bars are energized to the level of the test voltage applied and, therefore, all may show PD activity. However, in the offline test, the effects of vibration and bar forces are not in play. These issues are important and must be taken into consideration when analyzing the test results and their implications for the condition of the unit.

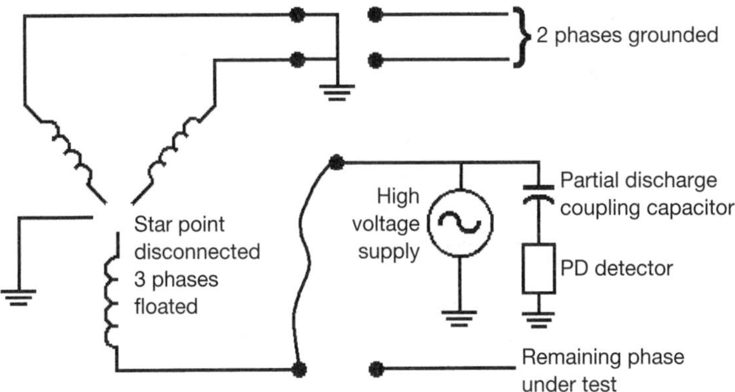

Fig. 11.33 PD testing by capacitive coupling, offline (courtesy of EPRI).

PD Monitoring by Stator Slot Coupler. The stator slot coupler (SSC) is basically a tuned antenna with two ports. The antenna is approximately 18 inches (46 cm) long and is embedded in an epoxy–glass laminate with no conducting surfaces exposed. SSCs are installed under the stator wedges at the line ends of the stator winding, such that the highest voltage bars are monitored for best PD detection. Since the SSC is also installed lengthwise in the slot at the core end, its two-port characteristic gives it inherent directional capability (see Fig. 5.19).

The problem of noise is virtually eliminated in the SSC. Although the SSC has a very wide frequency response characteristic that allows it to see almost any signal present in the slot in which it is installed, it also has the characteristic of showing the true pulse shape of these signals. This gives it a distinct advantage over other methods, which cannot capture the actual nature of the PD pulses. Since PD pulses occur in the 1 to 5 nanoseconds range and are very distinguishable with the SSC, the level of PD activity can be more closely defined.

In addition, dedicated monitoring devices have been devised to measure the PD activity detected in the SSC. The capability for PD detection using the SSC and its associated monitoring interface is enhanced to include measurement in terms of the positive and negative characteristics of the pulses, the number of the pulses, the magnitude of the pulses, the phase relation of the pulses, and the direction of the pulses (i.e., now from the slot or from the end winding or actually under the SSC itself at the end of the slot).

The other advantage of the SSC is that once it is installed, measurements may be taken at any time without the need for exposing live portions of the generator bus work, for the purpose of making connections to the test equipment.

Corona Probing for PD. Partial discharge tests in general determine only the relative condition of the stator winding from the generator terminals. They do not locate specific sites of deterioration or damage in the winding. To do this, the winding must be locally scanned with special probes designed to detect localized sources of PD while the winding is energized to the level of line-to-neutral voltage. There are a couple of variations of probe types, one based on radio-frequency noise and the other on acoustical noise. (SSCs do provide some information about the location of the PD activity. The more SSCs installed in a particular winding, the higher the accuracy in determining the location of the offending bar.)

The "TVA probe" gets its name from the Tennessee Valley Authority, where it was first popularized (Fig. 11.34). It is based on an earlier Westinghouse probe design that was sensitive to RF signals produced by PD in the winding. It functions by picking up the RF energy radiated from active PD sites in the winding. The greater the PD, the greater the RF energy produced. The tip of the TVA probe employs a loop antenna similar to that used in an AM radio. The TVA antenna is tuned to about 5 MHz so that it is sensitive to near-field RF discharge. The output of the antenna is directed by a coaxial cable to a tuned RF amplifier and a peak-reading ammeter that is sensitive to peak PD pulses. The closer the antenna is brought to an RF (or PD) source, the higher the output on the meter.

The "ultrasonic probe" is based on acoustic noise produced by localized PD sites. The noise is similar to a crackling sound that one might hear when next to a high-voltage overhead transmission line on a wet day. This noise is loudest in the ultrasonic frequency range around 40 kHz. A highly directional microphone, sensitive to the 40 kHz noise, is used to locate the site of the PD discharges. Given that ultrasonic noise does not easily penetrate insulation, the ultrasonic probe test is primarily sensitive to surface PD, namely, the sites of slot discharge and surface end-winding PD. See Figure 11.24.

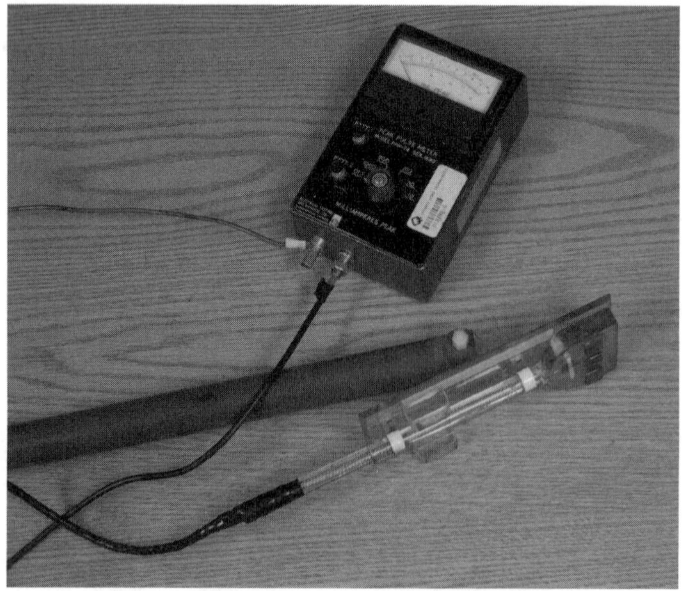

Fig. 11.34 TVA probe for corona probing in the stator slots.

11.5.10 Capacitance Measurements

Capacitance measurements are a method of measurement by which the quality of the insulation can be indicated. The measurements are, of course, done with ac voltage, and generally on a per-phase basis.

Each phase of the stator winding is energized to line-to-neutral voltage, while the other two phases are grounded. The power factor of the winding is measured with a capacitance bridge to determine the value of the per-phase winding capacitance. Comparison of the measured capacitance to the factory measured values, and then successive capacitance readings, can aid in showing deterioration of the ground-wall insulation over time.

11.5.11 Dissipation/Power Factor Testing

The dissipation factor (or tan δ) is an ac test used to measure the bulk quality of the ground-wall insulation by measuring the dielectric loss (primarily due to partial discharges) per unit of volume of the insulation:

$$\text{Dissipation Factor, DF} = \tan \delta$$

$$\text{Insulation Power Factor, IPF} = \frac{\text{DF}}{\sqrt{1 + \text{DF}^2}} = \sin \delta$$

Results are generally dependent on the type of the dielectric material in the insulation system. An increase in DF over the life of the winding can be attributed to an increase in internal voids, delaminations, and/or increased slot-coil contact resistance (i.e., deterioration of the semiconducting paint in the slot).

The readings are dimensionless quantities expressed as percentages. The absolute values obtained are, again, a function of the type of insulation system being measured and are also directly affected by the temperature of the winding. Therefore, it is important that insulation power factor readings be taken at similar temperatures. The results are even more useful, however, in relative terms by comparison of present readings to past readings. Successive measurements provide a scale of the deterioration rate of the insulation system over time.

Therefore, when using dissipation factor as a function of time, it is important to maintain constant conditions during testing. The DF readings are directly affected by the temperature of the winding and are also a function of the applied voltage. Therefore, comparisons with previous readings should be made on tests done at similar temperatures and the same voltage levels.

Since dissipation factor readings are somewhat void dependent, the *dissipation/insulation power factor* ratio will increase with an increase in the amount of voids present in the insulation. This phenomenon is the basis of the DF/IPF tip-up test.

11.5.12 Dissipation/Power Factor Tip-up Test

The dissipation factor tip-up or Δ tan δ test looks at the void content in the insulation. That is to say, the dissipation factor will increase with an increase in the amount of voids or delamination present in the insulation. In addition, it provides information on other ionizing losses in the form of partial and slot discharges.

The test is done by taking DF (or insulation power factor) measurements at different voltages. A set of readings is therefore obtained, which forms an ascending curve. A fast change of insulation power factor with increasing voltage tends to indicate a coil with many voids. The test is based on the fact that ionization, both internal and external to the insulation, is voltage dependent.

The test is done generally at 25 and 100% of the rated phase-to-neutral voltage. The tip-up value is the DF measurement at the higher voltage, minus the DF measurement at the lower voltage (IEEE Std 286). Good readings for an epoxy–mica system, indicating minimal void content in the insulation, are typically less than 1%. Good readings for an asphalt system are generally in the 3% range (Fig. 11.35).

This test will give a good evaluation of the winding as a group. However, any bad coil that deviates greatly from the rest will not be discerned by this test. To ferret out individual bars that exhibit higher discharges, a partial discharge test is done with the addition of manual probing for the location of the discharges if high levels are found to exist.

11.6 ROTOR MECHANICAL TESTING

11.6.1 Rotor Vibration

Vibration measurement is one the most important online measurements taken on the machine. Each manufacturer gives its own recommendations for alarm and trip levels.

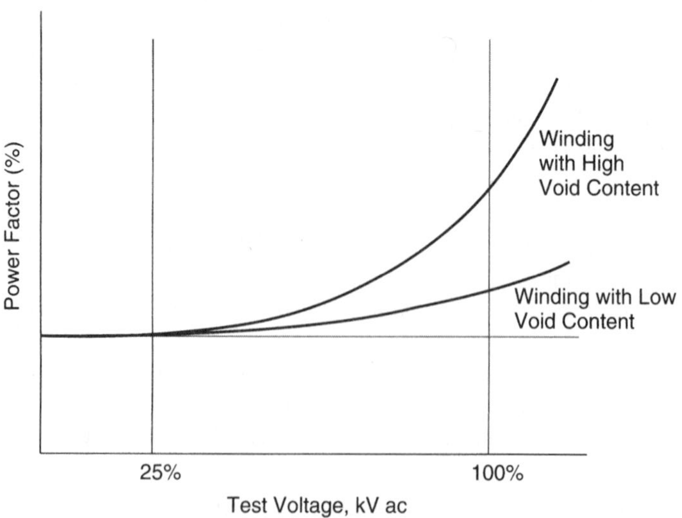

Fig. 11.35 Dissipation factor tip-up test profile comparison for good and poor results.

The information in the following table is from an O&M engineering specification for "peak" vibration at the "bearing pedestal":

Machine speed	Maximum amplitude
0–999 rpm	3 mils
1000–1499 rpm	2.5 mils
1500–2999 rpm	2 mils
3000 rpm and above	1 mil

Vibration is monitored continuously, and vibration charts are normally available int the control room. Vibration is monitored in all turbine and generator bearings.

Although vibration monitoring is generally considered an online monitoring function, there are many occasions where it is necessary to carry out additional and specific vibration testing to look for such problems as component rubs or rotor-winding shorted turns to determine the correct course of action.

This type of detailed vibration testing is very specialized and requires additional equipment to be connected to the vibration probes installed on the generator. The type of additional testing inferred would be to allow characterization of the vibration measurements with regard to both magnitude and phase relation, and to allow frequency spectrum analysis during cold and hot run-ups to speed and run-downs from speed. In addition, load changes and field current changes allow the differentiation between mechanically and thermally induced vibrations.

This is a very detailed topic for which entire books have been written and the literature that can be found is substantial. Refer to Chapter 6 for an additional discussion of the subject.

11.6.2 Rotor Nondestructive Examination Inspection Techniques

Nondestructive examination (NDE) of generator rotors is usually done to look for cracks and inclusions in highly stressed dynamic components. There are generally two types of NDE testing:

1. Surface (visual, magnetic particle, liquid penetrant, and eddy current)
2. Volumetric (radiographic and ultrasonic)

There are generally six different test methods:

1. Visual (VI)
2. Radiographic (RT)
3. Magnetic particle (MPI)
4. Liquid penetrant (LPI)
5. Ultrasonic (UT)
6. Eddy current (ECT)

Visual Inspection. Good detailed visual inspection of generator components may require any or all of the following conditions and equipment:

- Adequate lighting
- Magnifying glass
- Mirrors with rotating heads
- Digital camera
- Borescope (with video recorder)
- Feeler gauges
- Pocket ruler
- Straightedge
- Magnet

The effectiveness of visual inspection is surprisingly good if the inspector is knowledgeable and thorough. For example, surface flaws—cracks, porosity, surface finish, rusting, and corrosion—can usually be easily seen, and much can be deduced from this type of inspection. From the visual inspection, one can sometimes determine as much as needed to make the right decision on maintenance action.

The advantages of visual inspection are that it is low cost, simple, and quick. It can be done during any type of generator work that is in progress, permitting correction of faults.

Visual inspection does, of course, have limitations. It is applicable to surface defects only and provides no permanent record, unless recorded by photographs or video.

Radiographic. Radiographic testing is usually done by commercial X-ray or gamma units made especially for welds, castings, and forgings. It requires film and processing of the film when completed. It is not generally done for generator components, but it can be used if there are major frame weld problems. It is based on the differences in density or discontinuities in the item being examined.

The major defects that radiography can detect are interior macroscopic flaws such as cracks, porosity, and inclusions. Poor welds can be easily seen by this method.

This technique has the advantage of being applicable to most materials and once carried out, the defect areas are recorded on film to provide a permanent record.

The difficulties with radiography are that it requires skill in getting a good angle of exposure, use of the equipment, and interpreting the indications found. It is also somewhat hazardous to health as any radiographic method and requires safety precautions to be taken to protect the equipment operator and any personnel in the area of the testing.

Magnetic Particle. Magnetic particle inspection for surface flaws requires no special equipment but it does require high-current application to align the magnetic powders that show any flaws that are observable. This is in the range of 200 to 20,000 amperes. The magnetic powders are in dry or wet form and may be fluorescent for viewing under ultraviolet light. Pigments are often added to dry metallic powders to make them more visible (yellow, red, black, or gray). In the wet method, iron particles are coated with fluorescent pigment in water or an oil-based suspension. Antifreeze may be added to water-based solutions to impede particle mobility.

Magnetization is done by passing current through a multi-turned coil, looped through or around the part to be examined, with no electrical contact. The magnetization field is parallel to the axis of coil-longitudinal magnetization. The flux density is proportional to the current times the number of turns in the coil. With more turns, less current is required. A flux density of 1 tesla is satisfactory for most generator applications. Defects that are perpendicular to the magnetic field produce the most pronounced indications. The magnetic particles become attracted to discontinuities due to the high flux concentrations at the affected areas.

For low-carbon steels (with little or no retentiveness), a continuous application of magnetic particles is done while the magnetizing current is on. The magnetizing current is usually dc, from rectified ac. It is more penetrating than ac—up to half an inch below the surface.

For high-carbon steels, (with high retentiveness), magnetizing current is applied, followed by the magnetic particles, after the current is switched off. The magnetizing current is generally ac stepped down from single-phase ac voltage. It is effective to approximately one one thousand of an inch (1 mil) below the surface.

Magnetic particle inspection, or "mag-particle" as it is commonly called, is common for use in rotor component inspection of forgings, couplings, and steel wedges. It is excellent for detecting surface or near-surface defects such as cracks in ferromagnetic materials. It is simple to perform and cost-effective.

It is somewhat limited in that it is applicable to ferromagnetic materials only, and does require some skill in interpretation of defect indications and recognition of irregular patterns.

Liquid Penetrant. Liquid-penetrating inspection is widely used for generators, specifically on most rotor components, since it can be applied to both ferrous and nonferrous materials. It is primarily used for retaining-ring inspection. It is considered to have greater sensitivity than magnetic particle inspection and is also a very cost-effective method of NDE.

The equipment required generally consists of the following:

- Dye penetrating fluid, fluorescent or otherwise
- Developer
- Developer application equipment
- Ultraviolet light source, if the fluorescent method is used

Dye-penetrating examination is able to detect surface cracks and pitting or any discontinuity open to the surface and not readily visible to the eye.

The technique requires a clean surface, so preparation of the surface by cleaning and polishing is first required. Then the liquid penetrant is applied and allowed time to weep into discontinuities by capillary action. Once penetration is complete, the excess penetrant is removed by wiping but not completely cleaned away. Next developer is applied, which absorbs the penetrant and acts as a type of blotter. Defects are then easily seen by eye.

The limitations of LPI are that it only detects surface defects and it is also temperature sensitive. It should only be used in the range of 60 to 90(F. This is due to viscosity issues with the penetrant as temperature varies outside the above-mentioned range.

Figures 11.36 to 11.38 show some of the uses of LPI for defect detection on generators.

Ultrasonic. Ultrasonic examination is probably the second most widely used NDE method for generator components. It is highly specialized and requires special commercial ultrasonic equipment, either of the pulse-echo or transmission type (Fig. 11.39).

UT is able to detect surface and subsurface flaws, including those too small to be detected by other methods. The size range in subsurface flaws is generally about 2 to 3 mm. Ultrasonic NDE is used for rotor forging bores, retaining rings, and specialized subsurface defects that can occur in high-stress areas of the rotor. It is a very sensitive technique and can look at areas that are difficult access by other NDE methods. It does have the disadvantage of being difficult to obtain good interpretable signals in areas where the specimen geometry is complex, such as the portion of a retaining ring in

Fig. 11.36 18 Mn–5 Cr retaining ring, fluorescent dye partially applied. (See color insert.)

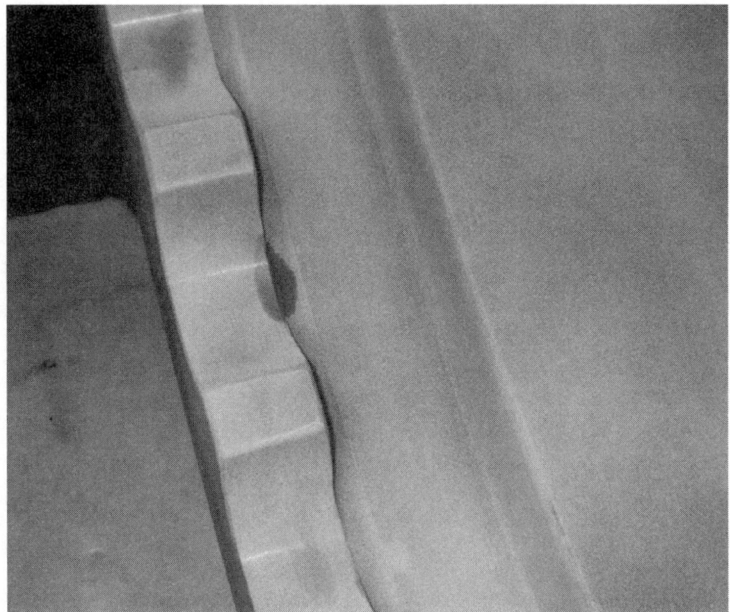

Fig. 11.37 18 Mn–5 Cr retaining ring, developer applied and indication shown in the darkened spot at the castellated fitting.

Fig. 11.38 Rotor forging tooth-root cracks seen by fluorescent dye with developer applied.

Fig. 11.39 Rotor forging tooth-root crack NDE by ultrasonic measurement with a 45° angle probe.

which castellated fitting of the shrink-fit area is employed. It also requires an operator with a high degree of training and skill to carry out the testing and interpret the results.

The two basic methods are described below.

Pulse-Echo Method. The pulse-echo method is the most widely used method. The ultrasonic pulse echo instrument generates high-voltage electrical pulses of short, evenly timed duration. These pulses are applied to the transducer, which converts them into mechanical vibrations that are applied to the material being inspected. The sound reflected back to the transducer is converted back to electrical pulses, which are amplified and displayed on the cathode ray tube (CRT) as vertical pulses. The same transducer is used to transmit and to receive.

Through-Transmission Method. In the through-transmission method, two transducers are required: one for sending and the other for receiving. Either short pulses or continuous waves are transmitted into the material. The quality of the material being tested is measured in terms of energy lost by a sound beam as it travels through the material.

To determine the location of discontinuities within a test specimen, the visual display unit (VDU) horizontal display is divided into convenient increments such as centimeters or inches. At a given sensitivity (gain) setting, the amplitude is determined by the strength of the signal generated by the reflected sound wave. The VDU thus displays two types of information: the distance (time) of the discontinuity from the transducer and the relative magnitude of the reflected energy.

There are generally two types of wave applications to the test specimen: longitudinal (compression) and shear (transverse) waves. Longitudinal waves have back-and-forth particle vibrations in the direction of the wave propagation. Shear waves have particle vibrations perpendicular to the direction of wave motion. Shear waves will not travel through liquids or gasses. In some materials, the velocity of a shear wave is about half that of longitudinal waves. Therefore, the wavelength is shorter (by about half), permitting smaller discontinuities to be located.

Grain structure has a great influence on the acoustical properties of a material. A steel forging generally has a fine-grain structure and has a low damping effect on the sound beam. A casting, however, generally has a coarser grain structure, which is more difficult to get sound through.

When a discontinuity is not normal (at 90°) to the incident wave, the reflected wave will be at an angle. The result is a reduction in the amplitude of the discontinuity indication displayed on the CRT.

Transducers come in many shapes, sizes, and physical characteristics. Some common types include paintbrush, dual element, single element, angle beam, focused, mosaic, contact, and immersion. Single-element transducers may be transmitters only, receivers only, or both. Double-element transducers may be single transducers mounted either side by side or stacked. In a double-element transducer, one is a transmitter and the other is a receiver. Double-element transducers have better near-surface resolution because the receiver can receive discontinuity signals before the transmitter completes its transmission.

A newer UT technique that is now more commonly used is called *phased array*. The application for this is becoming more widely used and has recently been introduced to inspect generator rotors and some of their components. Existing UT examination is done as described above using a single probe with a single fixed-beam angle and it must be moved physically to obtain readings over the surface of the specimen being tested. Generally, there are many different probes used for various applications. The advantage of the phased array technique is the use of a single array probe with the ability to generate many sound-beam angles in sequence and inspect a larger cross section of the component being examined. Beam angle adjustments can be made in the horizontal and vertical directions as appropriate to the geometry of the specimen under examination. The use of phased array UT improves on the performance of conventional UT inspection by reducing the inspection time and data analysis time, increasing the reliability of data analysis, and improving in the detection of small defects. Also, the complexity of the overall equipment is reduced.

Eddy Current. Eddy-current examination requires very special commercial-type equipment, consisting of the following components:

- Vector voltmeter
- Differential bridge

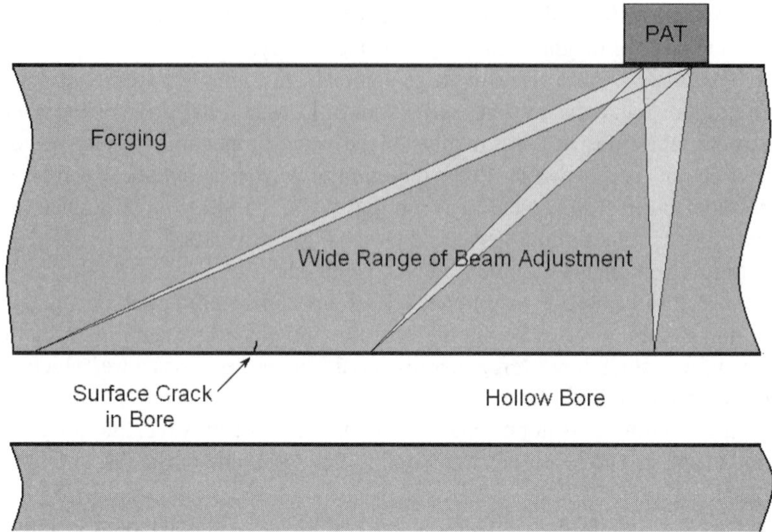

Fig. 11.40 Generator rotor example for the phased array technique to find bore cracks in a rotor forging. The phased array transducer is stationary and the beam is adjusted to find any cracks.

- Sensor coils
- Frequency generator
- Phase-sensitive detector
- Impedance-plane display instrument

Eddy-current examination is not widely used for generator applications, but it has been useful on occasions when tight cracks in retaining rings cannot be seen by LPI, MPI, or UT. It is able to penetrate layers of good material to detect hidden flaws in metals such as inclusions and tight cracks. Examples of the types of defects that can be seen by eddy-current examination are surface and near-surface defects, delamination of multilayered components, and inclusions.

The basic principle is that of an alternating current flowing through a coil, producing an alternating magnetic field in the specimen being tested. When the coil is then placed near a test specimen that is conductive, the magnetic field causes eddy currents to flow. This flow of eddy currents depends on the physical and electrical characteristics of the test specimen. The eddy currents will avoid cracks and seek higher-conducting regions in the specimen. As the eddy currents flow in the test specimen, they generate their own magnetic field. This field interacts with the magnetic field produced by the coil and changes the coil's impedance. Specialized instruments then measure and display these changes in the impedance, allowing the test technician to interpret information about the test specimen, specifically, the presence and size of flaws in it.

There are numerous types of eddy-current probes used, and the type of probe chosen depends on the material being tested and the application. The types of probes are absolute, differential, reflection, unshielded, and shielded. The main advantages to this method in general are low cost, ease and speed of use, high sensitivity to microscopic defects, and automation.

11.6.3 Some Additional Rotor NDE Specifics

NDE tests are done generally performed on the following generator rotor components:

- Forging
- Retaining rings
- Fans

Typical NDE tests include magnetic particle, liquid penetrant, ultrasonic (and boresonic), and, sometimes, eddy current. An additional test hardness test is sometimes performed.

Rotor Forgings. NDE is most often done during rewinding of the rotor, when revalidation of the forging is required. This may include the following inspections:

- Magnetic particle inspection (MPI) of the rotor fillet radii inboard of the journals, shaft radius at the slot exit interface, fan seating and keyways, pole-face crosscuts, pole and winding slots, and so on
- Liquid penetrant inspection (LPI) of the exciter-end lead-slot radii, inboard and outboard radial pinholes, and coupling-bolt holes.
- Ultrasonic examination of the rotor bore (boresonic)

In addition, the rotor forging generally undergoes full-dimensional checks and runout or truth checks.

Fans. Rotor fans are highly stressed components of the rotor. Fan blades and vanes are generally tested by LPI and UT.

Retaining Rings. Retaining rings are by far the component subjected to the highest scrutiny during major overhauls. One of the most comprehensive studies on this topic is given in the EPRI Special Report EL/EM-5117-SR, "Guidelines for Evaluation of Generator Retaining-Rings," April 1987 [18].

Generally, generator rotors employ three different types of retaining-ring material. These are Ni–Cr–Mo–Va magnetic steel (ferritic), 18 Mn–5 Cr nonmagnetic steel, and 18 Mn–18 Cr nonmagnetic steel.

One of the major station concerns is the ability to perform NDE on large generator rotor retaining rings without removing the rotors from the generator, let alone the retaining rings from the rotors. This is driven by the high cost of rotor removal and subsequent removal of the retaining rings from the rotors, and the time consumption involved. In addition, removing retaining rings is not desirable because of the risk of damaging the rings, the rotor forging, and/or end windings in the process.

Retaining rings constructed from magnetic material are generally sound in terms of strength, resistance to corrosion, and so on. However, they are undesirable with regard to their electromagnetic characteristics and resulting effect on the overall machine losses.

Generator rotor retaining rings made from 18 Mn–5 Cr austenitic, nonmagnetic steel are, on the other hand, highly susceptible to the mechanism of stress corrosion cracking when exposed to a moisture laden environment during operation or at standstill. This has become an industrywide problem without regard to a specific manufacturer.

Several ring failures have occurred in the industry, resulting in considerable damage to the overall generators and their support structures. These incidents were attributed to stress corrosion cracking caused by exposure to moisture. These rings had apparently not been adequately protected from the environment.

Once a crack has been initiated, the propagation rate of the crack will depend on the ring design (geometry, stress levels, etc.), environment, and the operating conditions to which it is subjected. Also, the ring material is a factor as the critical crack size for the 18 Mn–5 Cr material is smaller than the 18 Mn–18 Cr material. It is extremely difficult to predict when small cracks will reach the critical crack size at which failure is imminent. An attempt to determine this has been made and is published in [18]. Inspection is the only means available to assess retaining-ring integrity and provide an opportunity to correct problems encountered.

With the advent of a new retaining-ring material known as 18–18 (18 Mn–18 Cr), which is highly resistant to the stress-corrosion cracking mechanism, all manufacturers are advocating that the 18 Mn–5 Cr retaining rings be replaced with new 18–18 rings because of their superior properties. 18 Mn–18 Cr steel is claimed to be very resistant to corrosion and stress-corrosion cracking in the operating environments of large generators. As a result, fewer inspections will be needed for the retaining rings [20]. (For additional information on the subject, see the sections covering retaining rings in Chapter 9.)

TESTING FREQUENCY. Depending on availability, it is desirable to conduct retaining-ring inspections during a period of one to two years following the in-service date, and again six to eight years after that. Subsequent inspections are prescribed based on a combination of information gained during previous inspections and adverse operating conditions since the last inspection.

Adverse operating conditions would include water leakage into the generator, high or sustained dew-point excursions, local system faults, synchronizing errors, high rotor temperature excursions, continuously changing load requirements or two-shifting, overspeeds, rotor at standstill in uncontrolled environmental conditions, and so on.

INSPECTION PROCEDURE. The recommended process for an acceptable assessment of retaining-ring integrity requires removal of both rings from the rotor body completely. A thorough cleaning of all ring surfaces prior to inspection follows this. The cleaning is achieved by hand polishing of shrink-fit areas and by using electrically nonconductive abrasives on all surfaces to achieve bare-metal conditions. The inspection is carried out during a detailed dye-penetrant process, using a highly sensitive fluorescent liquid penetrant to expose suspect areas containing cracks, pits, or defects.

Local grinding and/or polishing will normally remove any flaws, but there are limitations on the amount of corrective action allowed. It is essential that the limits not be exceeded during the corrective action as the suspect areas are removed.

Removal of material from the shrink-fit area is limited to the amount at which 15% overspeed capability is maintained. The design overspeed capability is 20% above the normal operating level. Overspeed trips are set at 10%, thus allowing a minimum 5% margin for the material to be removed. The determination of the depth of the skim to be taken off the shrink-fit are, when required, should be referred back to the OEM.

Acceptance Criteria. Acceptance criteria are normally determined by, and are the responsibility of, the manufacturer/suppler of the item being assessed. However, acceptance criteria are usually based on detection of defects and not crack/pit/defect sizing. It is much easier to establish an acceptable situation by first establishing "not-acceptable" criteria.

Reasonable not-acceptable criteria adopted by some OEMs and users are:

1. No linear defects greater than 1 mm in length.
2. No linear array of three or more point defects with < 3 mm between points.
3. No random groups of three or more point defects within a circle of < 5 mm diameter.
4. No random groups conforming to criterion three but < 5 mm from a similar group.

Note: A point defect is a surface defect < 0.5 mm in diameter.

If a defect is found and sized, it is usually desirable to determine the critical crack size, regardless of the fact that one generally does not leave even the smallest indication in a retaining ring. This is because they can quickly grow in size if the internal generator conditions are not good. Estimating the critical crack size depends largely on the dimensions of the individual retaining ring and the total stress on the ring, as well as the actual material, its *fracture toughness* characteristics and *stress intensity factor.* We have already seen in Section 2.16 how an estimate of total stress may be obtained by the summation of ring hoop stress and the copper loading. From this, an estimate of critical crack size is possible.

Carrying on from Section 2.16, the well known equation for the stress intensity factor K1c is a function of the operating stress on the ring, which is the calculated total stress from Section 2.16, and the critical crack size as per the following basic formula:

$$K1_c = 1.128\,\sigma\,\sqrt{a_{cr}}$$

where
Kl_c = fracture toughness of the ring material
σ = nominal stress on the ring at the operating speed
a_{cr} = critical crack depth

Since it is the critical crack size that is required to be determined, the formula can be rewritten as follows:

$$a_{cr} = \left(\frac{Kl_c}{1.128\sigma} \right)^2, \qquad \text{inches}$$

Initiation of runaway crack growth and eventual ring failure occurs when the tip of the crack a_{cr} reaches the critical size for the ring or, more precisely, the fracture toughness of the ring material.

Starting with a ring made of 18 Mn–18 Cr and the configuration of the ring shown in Fig. 2.68, the total stress on the ring at rated speed has previously been calculated as 526 MPa (75 Ksi). At 20% overpseed, the stress on the ring worked out to be 758 MPa (110 Ksi). We also can find the K1c fracture toughness of the 18 Mn–18 Cr material from a materials handbook by looking up the value for K1c that corresponds to the 0.2% *yield strength* (see note below) of the material. In doing so, one would find that it gives a lower-bound stress-intensity factor of 50 Ksi for a 0.2% yield strength of 175 Ksi (1200 MPa). Therefore, if we now wish to determine the critical crack size for the example ring operating at rated speed, we find the following:

$$a_{cr} = \left(\frac{50}{1.128 \times 75} \right)^2 = 0.349 \text{ inches or } 8.9 \text{ mm}$$

And for the 20% overspeed case we obtain the following critical crack size, which is much less, as stands to reason, due to the higher stress on the ring at overspeed:

$$a_{cr} = \left(\frac{50}{1.128 \times 110} \right)^2 = 0.162 \text{ inches or } 4.1 \text{ mm}$$

From the above simple estimation, one can see the need for such close and accurate examination of retaining rings for even the smallest indication. Given poor internal generator atmospheric conditions, it is easy to see how a small initiation will not take much to become critical.

Note: As a good rule of thumb, the following may be considered.

The factor of safety of any part of the rotor and attachments should not be less than 1.5 at 20% overspeed, based on the calculated nominal stress on the cross section under consideration, and based on the material yield point or 0.2% proof stress, when there is no defined yield point for the material.

This should be checked specifically on the retaining rings because of their highly stressed nature. The calculated local maximum stress concentration in the retaining rings should not exceed the 0.2% proof stress of the material, considering the effective friction factor between the ring and the rotor.

Stress concentrations in the rotor teeth under the retaining ring shrink-fit area should be avoided. The maximum compressive stress at any stress concentration should not exceed the 0.2% proof strength. The calculated local maximum stress concentration in the rotor teeth at synchronous speed should not exceed 85% of the 0.2% proof stress of the material.

New rotor components should be designed for at least 10,000 start-up cycles and 500 overspeeds to 20% above rated over a life of at least 25 years. Reused rotor components should be revalidated for at least 7,500 start-up cycles and 500 overspeeds to 20% above rated over a remaining life of at least 20 years.

Detailed stress calculations should be carried out to ensure the above and an in-depth review done for any component having a safety factor less than the above specified values.

IN-SITU TESTING. At the present time, ultrasonic examination of retaining rings assembled on a rotor body is being debated by users and testers in the industry. There are some who are staunch proponents of the in-situ technique, whereas others insist it is not sensitive enough.

At last report to the authors, the smallest cracks that can be detected are on the order of 2 to 3 mm in the shrink area and only as small as 1 mm in the smooth bore areas. On rings with castellated fittings, it may not be possible to resolve indications identified due to the complex geometry involved.

It should also be understood that these crack sizes quoted above are large in terms of significance for retaining rings. Even pitting marks will propagate into cracks when the internal generator hydrogen conditions are not good. The critical crack size is different for each ring design and must be evaluated independently.

The bottom line is that even the smallest cracks cannot be tolerated, especially in the shrink area. It is the modest opinion of the authors that the only way to know for sure if the ring is free of defects is to remove it for detailed NDE surface-type inspection methods and ultrasonic or eddy current testing for detection of inclusions.

Finally, there are some in the industry who consider it desirable to install the new 18 Mn–18 Cr retaining rings so that inspection is unnecessary. Figure 11.41 shows an in-situ testing device in operation.

11.6.4 Air Pressure Test of Rotor Bore

The vast majority of large generator rotors in existence have hollow bores, which extend the full length of the rotor forging. These boreholes are generally in the range of 3 to 5 inches ($\approx$ 7.5–13 cm) in diameter, and when the contents of the bore are removed one can see from one end right through to the other.

Boreholes were used mostly in the past due to impurities and porosity in the forgings, which tended to concentrate in the center. They serve two purposes: the first is to remove the material defects, and the second is to provide an access for performing boresonic (ultrasonic) inspection of the rotor bore. In modern forgings, the material

Fig. 11.41 In-situ robotic-rotor retaining-ring ultrasonic testing. (Courtesy of Alstom Power Inc.)

manufacturing processes are so improved that a full rotor length borehole is not generally required. Only a short borehole at the exciter end of the rotor is provided to accommodate the up-shaft lead, which connects the rotor winding to the slip rings via the radial terminal studs.

In hydrogen cooled machines, rotor seals are installed in the shaft, sealing the hole in the shaft where the up-shaft lead or bore copper is installed. This is between the slip rings and the field winding. The seals are generally made of rubbery material, which under pressure (sometimes from a nut) expands, filling the area between the bolt connecting the bus/conductor in the hollow of the shaft to the slip rings. On the excitation end, some rotors have only one set of seals close to the collectors, whereas other rotors have a second set of seals, where the leads exit the shaft, under the retaining rings. On rotors with a full-length borehole, the turbine end of the rotor is also sealed by a face plate arrangement on the coupling face.

The integrity of the rotor seals is normally checked during major overhauls. Depending on the design of the rotor, some can be pressure tested through a nipple permanently installed on the shaft, or by placing a can over the shaft extension and collector assembly, tightly sealed against the rotor forging, and pressurizing the can.

11.7 ROTOR ELECTRICAL TESTING

11.7.1 Winding Resistance

The field-winding series resistance is measured to determine the ohmic resistance of the total copper winding in the rotor. Given the relatively low dc series resistance of windings of large machines, the measurement accuracy requires significance to a minimum of four decimal places.

The purpose of the test is to detect shorted turns, bad connections, wrong connections, and open circuits. The machine should be at room temperature when the test is performed. As with most other electrical tests, the results should be compared with original factory data, if available.

11.7.2 Insulation Resistance (IR)

The purpose of the IR test is to measure the ohmic resistance between the total rotor winding insulation and ground (i.e., the rotor forging). This test is generally regarded as an initial test to look for gross problems of the insulation system and to ensure that further high-voltage electrical testing may (relatively) safely continue, in terms of danger of failing the insulation.

Normally, the measurements of IR will be in the megaohm range for good insulation, after the winding is subjected to a dc test voltage, usually anywhere from 500 to 1000 V, for one minute. The minimum acceptable reading per IEEE Standard 43 is (V_f in kV + 1) MΩ. The test is carried out with a megger device. The dc test voltage level is usually specified based on the operating and field forcing voltage of the rotor, utility policy and previous experience, and knowledge of the present condition of the insulation in the rotor.

It is essential that the rotor winding be completely dried before any testing so that any poor readings will be due to a "real" problem and not residual moisture. The readings are also sensitive to factors such as humidity, surface contamination of the coils, and temperature. Readings should be corrected to a base temperature of 40°C (Fig. 11.26).

All of the above also applies to the rotor bore copper and collector rings.

11.7.3 Polarization Index (PI)

Insulation resistance is time dependent as well as being a function of dryness for rotor insulation, just as in the stator. The amount of change in the IR measured during the first few minutes depends on the insulation condition and the amount of contamination and moisture present. Therefore, when the insulation system is clean and dry, the IR value tends to increase as the dielectric material in the insulation absorbs the charge. When the insulation is dirty or wet or a gross insulation problem is present, the charge does not hold. The IR value will not increase because of a constant leakage current at the problem area. Thus, the ratio between the resistance reading at 10 minutes and the reading at 1 minute produces a number or "polarization index," which is essentially used to determine how clean and dry the winding is.

The recommended minimum PI values are as follows:

- Class B insulation: 2.0
- Class F insulation: 2.0

The same megger used for the IR readings should be used to determine the PI. The PI readings should be done at the same voltage as the IR test and can be used as a go/no-go test before subjecting the rotor to subsequent high-voltage tests, either ac or dc. The IR readings for the PI test should also be corrected to 40°C as in the IR test (Fig. 11.25). Performing the high-voltage tests on wet insulation can cause needless failure of the insulation.

The use of PI and its acceptable values for field windings is often in dispute, because most rotors have no insulation on significant sections of their winding. This tends to skew the PI readings. The reader is referred to IEEE Sdandard 43, Section 12 for further elaboration on this topic. As stated therein, PI for field windings of large turbo-generators with significant portions of the winding without insulation should be used mainly for trending purposes.

11.7.4 dc Hi-Pot

The dc hi-pot test is used to ascertain if the winding is capable of sustaining the required rated voltage levels (without a breakdown of the insulation), with a reasonable degree of assurance for capability to withstand overvoltages and transients, and maintain an acceptable insulation life. The test consists of applying high voltage to the rotor winding for one minute.

The dc hi-pot testing on rotor windings is normally done between 1500 V up to approximately 10 times the rated field voltage.

11.7.5 ac Hi-Pot

The ac hi-pot test is also used to ascertain if the winding is capable of sustaining the required rated voltage levels (without a breakdown of the insulation) with a reasonable degree of assurance for capability to withstand overvoltages and transients, and maintain an acceptable insulation life. The test consists of applying ac high voltage to the rotor winding for one minute.

The ac hi-pot testing on rotor windings is also normally done at to 10 times the rated field voltage at line frequency of 60 Hz.

11.7.6 Shorted Turns Detection—General

Shorted turns in rotor windings are associated with turn-to-turn shorts on the copper winding, as opposed to turn-to-ground faults. Rotor winding shorted turns, or interturn shorts, can occur from an electrical breakdown of the interturn insulation, mechanical damage to the interturn insulation allowing adjacent turn-to-turn contact, or contamination in the slot, which allows leakage currents between turns.

When shorted turns occur, the total ampere-turns produced by the rotor are reduced, since the effective number of turns has been reduced by the number of turns shorted. The result is an increase in required field current input to the rotor to maintain the same load point, and an increase in rotor winding temperature.

At the location of the short, there is a high probability of localized heating of the copper winding and arcing damage to the insulation between the turns. This type of damage can propagate and worsen the fault, such that more turns are affected or the ground-wall insulation becomes damaged, and a rotor-winding ground occurs.

One of the most noticeable effects of shorted turns is increased rotor vibration due to thermal and magnetic effects. When a short on one pole of the rotor occurs, a condition of unequal heating in the rotor winding will exist between poles. The unequal heating may cause bowing of the rotor and, hence, vibration. The extent and location of the shorted turns and the heating produced will govern the magnitude of the vibrations produced. Unbalanced magnetic pull (UMP) occurs (mainly in four-pole machines) because shorted turns may create asymmetries in the distribution of the radial flux around the circumference of the airgap. This is somewhat different than the classical definition of UMP, which generally refers to the force acting on a rotor due to asymmetry of the airgap.

The general relationship between the location of the shorted turn/turns and vibration due to thermal effects is as follows:

- Lower vibration is generally experienced when the short is on the Q axis.
- Higher vibration is generally experienced when the short is nearer the pole or D axis.

Stated differently, the rotor is more prone to vibration due to shorted turns if the shorts are located in the "small coils" rather than in the "large coils," the "small coils" being those located closer to the pole faces.

The reasoning for the above is the lack of symmetry of faults nearer the pole face. There is an inherent imbalance in the geometry and heating effect on the rotor forging.

Offline methods for detecting shorted turns include winding impedance measurements as the rotor speed is varied from zero to rated speed, and RSO (recurrent surge oscillation) tests based on the principle of time-domain reflectometry. In addition, a short of significant magnitude may be identified by producing an open-circuit (OC) saturation curve and comparing it to the design OC saturation curve. If the field current required to produce rated terminal voltage has increased from the original design curve, then a short is likely present. The number of shorted turns may be identified by the ratio of the new field current value over the design field current value.

All of these methods of identifying shorted turns are prone to error and only indicate that a short exists. They do little to help locate which slot the short is in and require special conditions for collecting the data or for testing. To better identify shorted turns, and to employ a method that works online, the *search coil method* has been perfected. Each OEM has their own version of a search coil method, but all work essentially in the same manner.

11.7.7 Shorted Turns Detection by Recurrent Surge Oscillation (RSO)

In the RSO method, a low-voltage (a few volts), high-frequency (kHz range) surge wave is injected at each one of the collector rings. The two signals are then compared to determine if the same waveform is observed at each collector ring. If the waveform is identical, then no shorts are present. Variations in the two waveforms indicates that shorts are present. This method is based on the principle of time-domain reflectometry.

This method has the advantage of allowing the rotor to be spun while doing the measurements, to determine if the shorts are also speed sensitive. The advantage of this is that the mechanical loading effects can be taken into consideration. In the spinning RSO, there may be shorts that reveal themselves that are not seen when the rotor is at rest, since at rest there is no mechanical load on the winding turns other than their own weight.

Because the RSO also works on a time-of-flight principle, the location of the coil number where the shorts are, as well as which pole, becomes discernable by this method. Shorts nearer the slip rings show up as blips in the RSO pulse nearer the left side of the traces. The magnitude of the blip increases for the number of turns shorted at the particular location (i.e., the particular coil) as more turns are shorted.

In the "at-rest" test, the RSO is connected directly to the winding via the collector rings and is testing generally done when the brushes have all been lifted from the ring surface or, alternatively, the rotor is removed from the stator completely. Also, the RSO requires a ground connection, to the forging in this case, since it is the ground point for the rotor winding. Thus, only the winding impedance is seen by the high-frequency, low-voltage pulses sent by the RSO.

In the "spinning" RSO test, to accommodate the moving rotor, the leads of the RSO must be connected to the brush rigging, and the connection to the winding is then implemented via the brushes' collector rings. For best results, all the brushes should be lifted, except one per collector ring for the connection point. The ground connection to the shaft is made by attaching the RSO ground connection to a piece of copper braid that can be held on the rotating shaft similar to the way shaft voltage measurements or shaft ground connections are made. However, with this connection setup anything connected to the excitation equipment is "seen" by the pulses (e.g., leads, contacts, field breaker, field resistor, and excitation equipment). The principle of operation of the RSO is comparing the pulses inserted in each polarity terminal of the winding and their reflections. The test is extremely sensitive to any asymmetry in the path of the pulses. From a point of view of the wave impedance seen by the high-frequency pulses, the field winding is by nature very symmetrical, but the excitation system is anything but that. Therefore, in order to obtain any significant signature on the condition of the field winding, the "noise" originating in the path toward the excitation must be reduced as much as possible. This is achieved by opening the excitation leads at a convenient location between the excitation system and the brush rigging. After the leads are open, only cables of almost exactly equal length are left connected to the brush rigging. The effect introduced by these cables is generally negligible. Figure 11.42 illustrates the overall connection setup for the spinning RSO test.

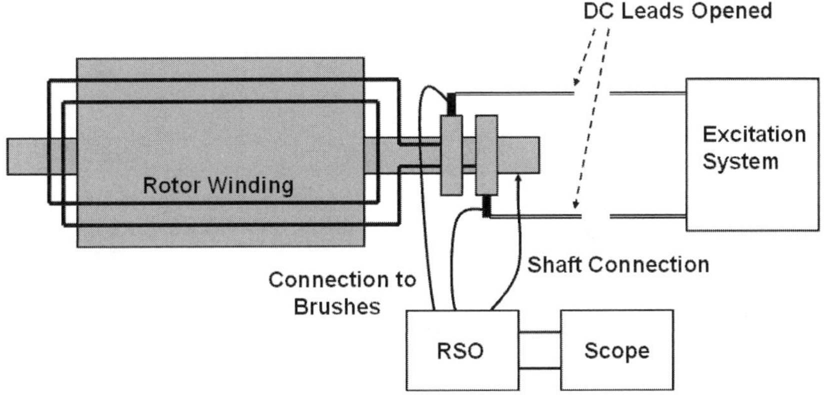

Fig. 11.42 Spinning RSO test connection setup.

Figures 11.43 to 11.49 depict samples of RSO test readings taken on a two-pole turbo-generator rotor.

11.7.8 Shorted Turns Detection by Open-Circuit Test

Producing an open-circuit saturation curve and comparing it to the design open-circuit saturation curve may identify a shorted turn condition of significant magnitude. If the field current required to produce the rated terminal voltage has increased from the original design curve, then a short would likely be present (Fig. 11.50).

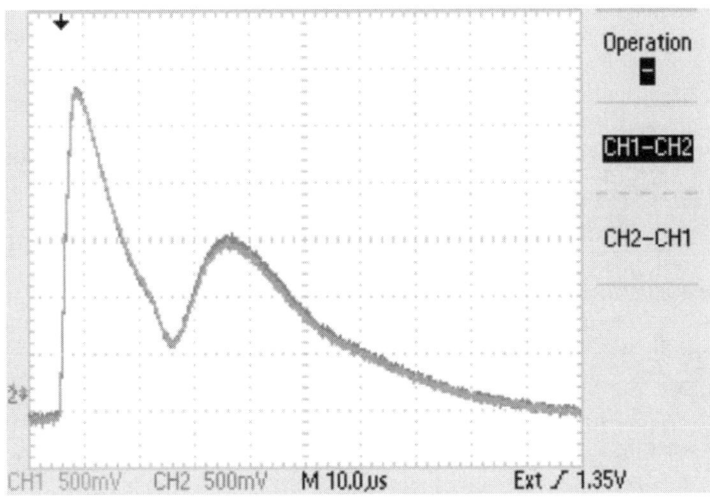

Fig. 11.43 No shorted turns, dual traces superimposed.

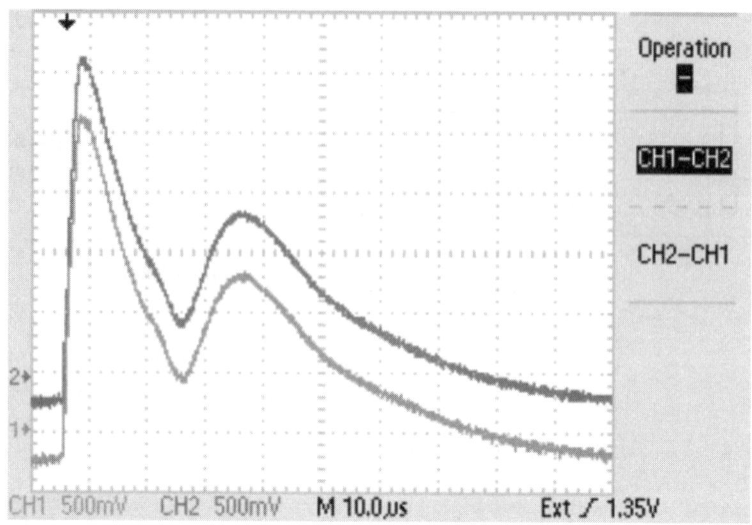

Fig. 11.44 No shorted turns, dual traces separated.

The number of shorted turns may be determined by the ratio of the new field-current value over the design field-current value. However, due to the high number of turns in a typical rotor winding, the changes in open-circuit voltage due to a single shorted turn in the field winding may go unnoticed since the measurement is too small for a positive identification.

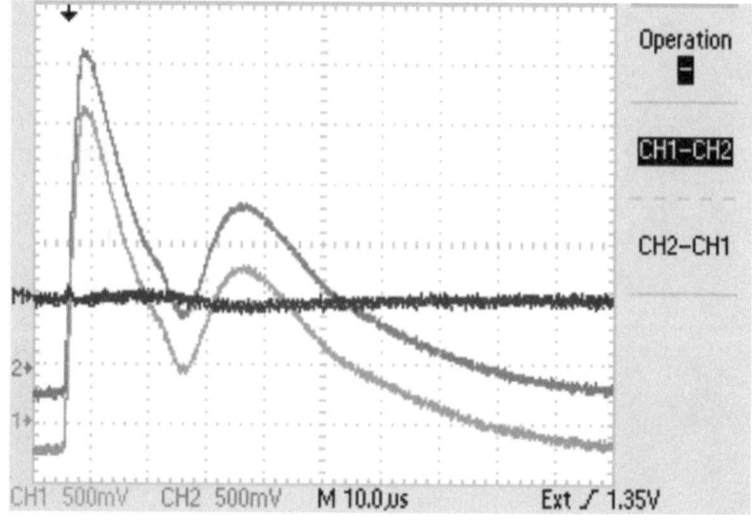

Fig. 11.45 No shorted turns, difference trace.

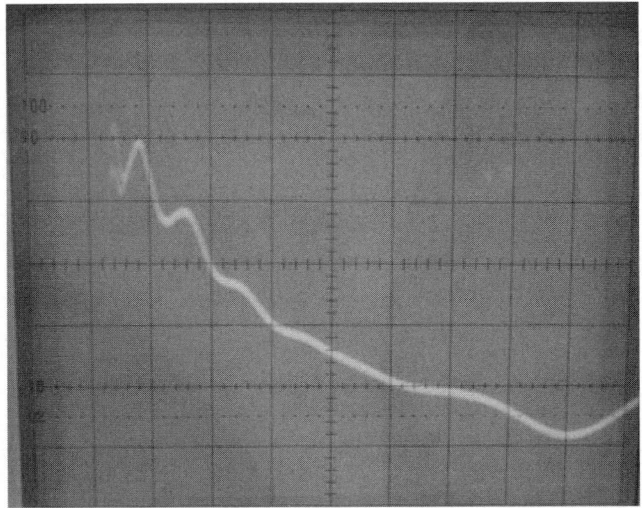

Fig. 11.46 One shorted turn, coil #1; dual traces superimposed.

The open-circuit stator voltage versus field-current characteristics can be measured in all synchronous machines. This curve, taken with the machine spinning at synchronous speed, is unique for each machine. In principle, the OC test allows detection of shorted turns in brushless machines, for which RSO techniques are too difficult to perform, and always entails partial disconnection of the rotor leads.

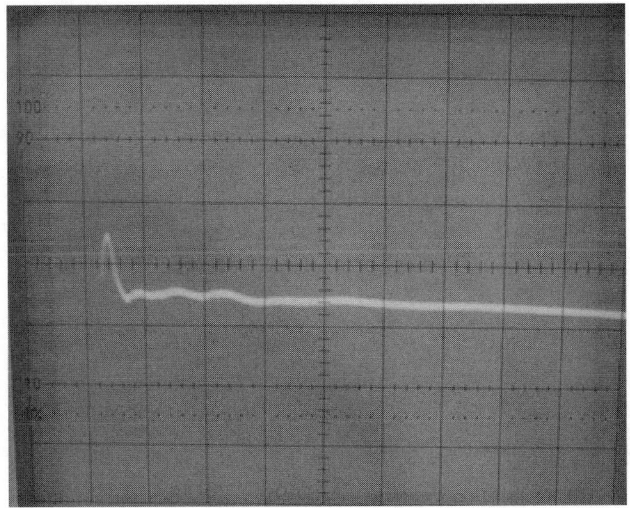

Fig. 11.47 One shorted turn, coil #1; difference trace.

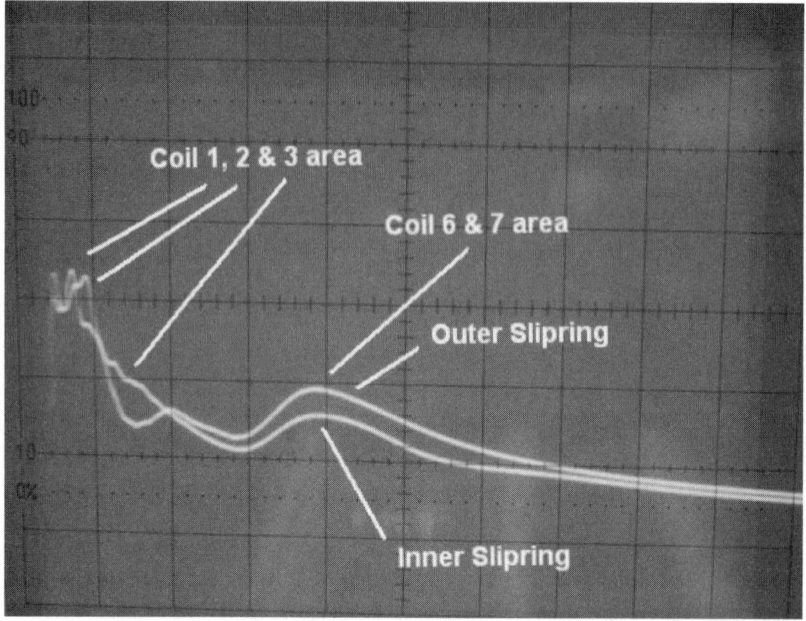

Fig. 11.48 Multiple shorted turns, dual traces superimposed.

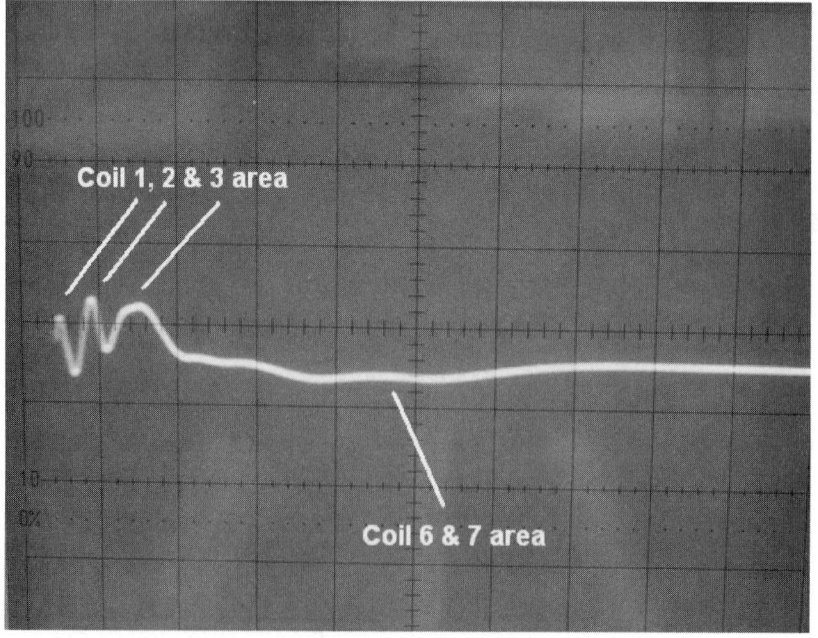

Fig. 11.49 Multiple shorted turns, difference trace.

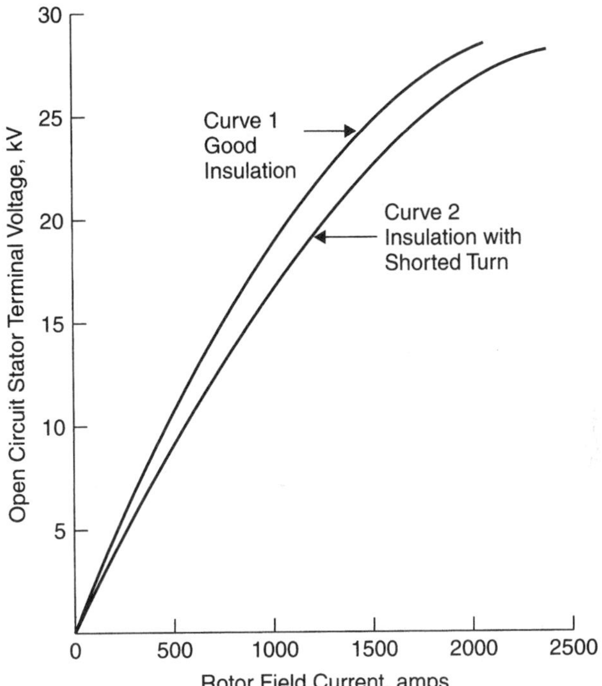

Fig. 11.50 Shorted turns detection by open-circuit saturation curve. (Courtesy of EPRI.)

11.7.9 Shorted Turns Detection by Winding Impedance

Impedance measurements while the machine is decelerating or accelerating can also be used to detect a speed-dependent shorted turn. Any sudden change in the readings may indicate a shorted turn being activated at that speed. A gradual change of impedance of more than 10% may also indicate a solid short (Fig. 11.51).

The method of measurement of winding impedance is a combination of dc and ac. The first measurement is the actual dc resistance of the winding from the positive to the negative collector ring. This can be done with a *Kelvin bridge,* a *doctor,* or a *micro-ohm meter*. When using the ductor or micro-ohm meter, at least 10 amperes of current should be applied to the winding for best results. The next measurement is that of the ac resistance by using a *Variac* or *Powerstat* to apply approximately 120 volts ac at the power frequency. By measuring the current delivered as the speed is varied from zero to rated, one can make successive readings of voltage and current and use those in the following formula to produce the curve of Figure 11.51:

$$\text{Winding Impedance} = \frac{\text{Applied ac voltage}}{\text{Measured ac current}}$$

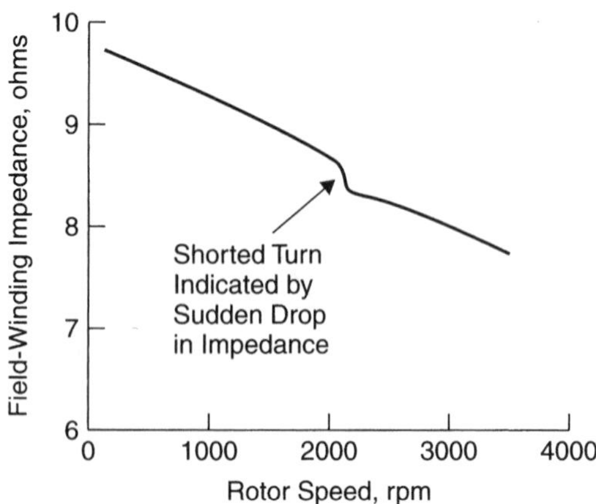

Fig. 11.51 Shorted turns detection by impedance testing. (Courtesy of EPRI.)

If there are speed-dependent shorted turns, the resistance of the winding will decrease and the ac current should increase suddenly when the short becomes active. The overall winding impedance will drop as per the above formula.

11.7.10 Shorted Turns Detection by Low-Voltage dc or Volt Drop

This test is designed to determine the existence of shorted turns in the rotor winding. The test is entirely different when performed on salient-pole rotors than on cylindrical (round) rotors.

In salient-pole machines, a "pole-drop" test is done. In this test, the resistance across each individual pole is measured by the V/I method, namely, by applying a voltage of around 100 to 120 volts, 60 Hz, to the entire winding, and then measuring the voltage drop across each pole. A pole with lower voltage drop will indicate a shorted turn or a number of shorted turns.

In either salient-pole or round-rotor machines, the shorted turns are often speed dependent (i.e., they might disappear at standstill). To partially offset this phenomenon, it is recommended to repeat the pole-drop test a few times with the rotor at several angles. The gravity forces exerted on the vertically located poles may activate some short circuits between turns, which might not show up when in, or close to, the horizontal position.

In round rotors, the individual windings are generally not accessible, unless the retaining rings are removed. Therefore, detection of shorted turns in not always possible by this method.

In cases in which the retaining rings are removed, a more in-depth volt-drop method other than a simple pole-drop test may be applied. The following technique will allow determination of the number of shorts and their location. Also, the rotor can be rotated for successive 90° angles to change the loading on the rotor winding and insulation and detect flaws that may be pressure sensitive.

From past experimentation, there is a clear difference between the use of ac and dc for volt-drop testing for best resolution on measurements, to locate the position of any shorts. As it turns out, both ac and dc are required at different stages. ac testing provides the best resolution for measuring initial pole balance and for coil-to-coil volt drops. However, when testing for turn-to-turn shorts on an individual coil, it then becomes clear that dc testing provides a better resolution and makes the estimation of the location of shorts along the winding clearer. This is because the dc measurements do not have any inductive effects (as the ac does) from the adjacent turns and the volt drops are more significant as purely resistive. That is to say, the lowest volt drop at the point of the fault is easier to see in the measurements.

To allow the tester to identify exact fault locations, the initial readings should be done as six points using dc testing. All of these are on the end winding due to ease of access. Three measurements are done at each end of the rotor. In this way, one can discriminate between faults that are in the slot and end winding, and determine which side of the rotor they are (Fig. 11.52). If further discrimination is required because the short is deemed to be in the slot portion of the rotor, insulated probes may be used to go down a center radial vent of the rotor winding (or any axial location where there are vents) and divide the slot portion into segments as well.

Figure 11.52 shows a sample sheet that was developed to visualize the measurements and to help locate the fault locations more easily. It is generally more efficient to record all the measurements in this way.

Figure 11.53 is a plot of actual trial tests with numerous variations of shorts applied to a test rotor at locations in the slot, turbine-end, and collector-end end windings. These help to characterize the combinations of shorts that may be found in a rotor.

The examples in Figure 11.53 are merely rough samples. Basically, a single short is very easily detected as are two shorts, separated or not. When there are more than two shorts on a single coil, it becomes more difficult to determine short locations that lie in between the two outer shorts. It is also better to use spiral graphs to locate the short locations, but the simple straight-line chart illustrates the points more clearly.

With regard to the test levels and equipment, the requirements will be different for different rotors and vary from two- to four-pole rotors. However, the range for ac testing is up to 10 amperes and 50 volts. The ac equipment is small and easily carried by hand. The dc equipment to energize the winding is larger and more difficult to obtain. The requirements are approximately 150 amperes at less than 20 volts dc.

11.7.11 Shorted Turns Detection by Low-Voltage ac or "C" Core Test

A C-shaped, wound core is required to carry out this test, together with a voltmeter, wattmeter, and single-phase power supply (Fig. 11.54). Shorted turns are detected by sharp changes in the direction of wattmeter readings (Fig. 11.55).

Rotors with damper windings or with the wedges short-circuited at the ends to form a damper winding have to be disconnected at the ends. This operation requires removal of the retaining rings.

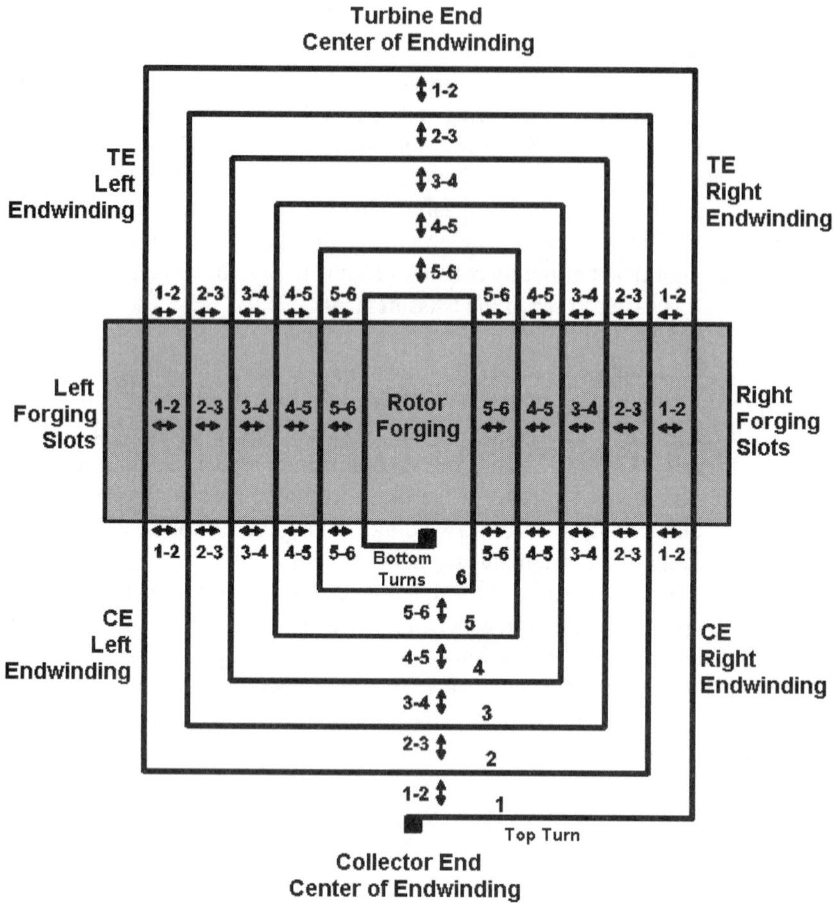

Fig. 11.52 Turn-to-turn volt-drop measurement sheet.

11.7.12 Shorted Turns Detection by Shorted Turns Detector (Flux Probe)

The flux probe is actually a search coil mounted on the stator core by various methods, but located strategically in the airgap. The search coil looks at the variation in the magnetic field produced in the airgap by the rotor as it spins. The energized rotor winding and the slotted effect of the winding arc cause a sinusoidal signal to be produced in the winding face of the rotor. The pole face, on the other hand, has no winding and the signal is more flat since the variation in magnetic field is minimal.

The magnitude of the sinusoidal peaks in the winding face is dependent on the ampere-turns produced by the winding in the various slots. If there is a short in a slot, then the peak of the signal for that affected slot will be reduced. The reduction will be dependent on the magnitude of the short. Therefore, besides learning which

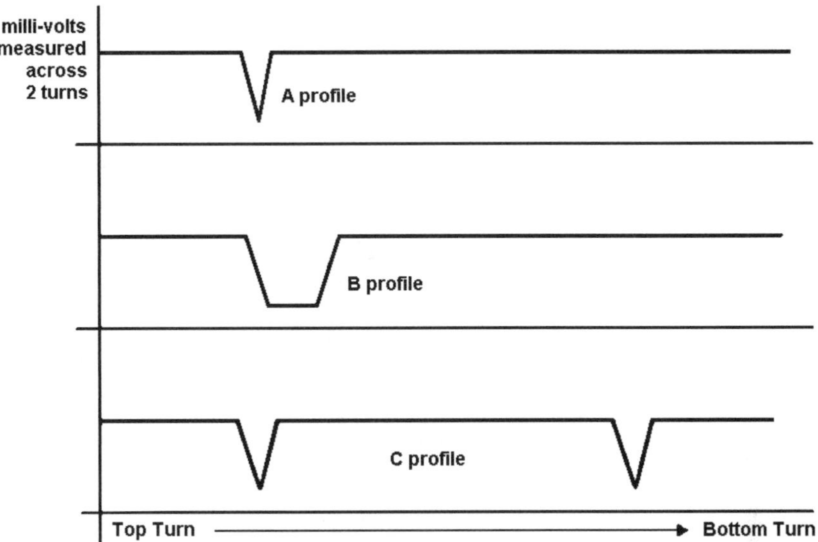

Fig. 11.53 Short locations versus position in the winding from start of the top turn to the end of the bottom turn. Profile A: One short in the endwinding or one short in the slot. Profile B: Two shorts in the end winding at one end, in successive turns; or two shorts in the end winding, one at either end of the rotor, on the same turn; or three shorts, two in the end winding on one end in successive turns and one in the slot on the same turn as one of the end-winding shorts. Profile C: Two shorts in the end winding at one end, one in an upper turn and one lower down; or two shorts in the end winding, one at either end of the rotor, on different turns.

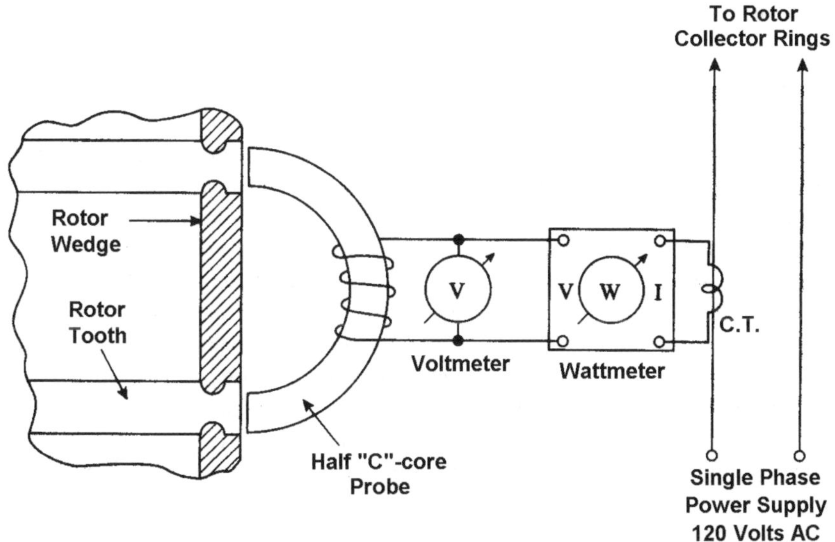

Fig. 11.54 Shorted turns detection by "C" core test. (From Kerszenbaum, *Inspection of Large Synchronous Machines,* IEEE Press, 1996, reprinted with permission.)

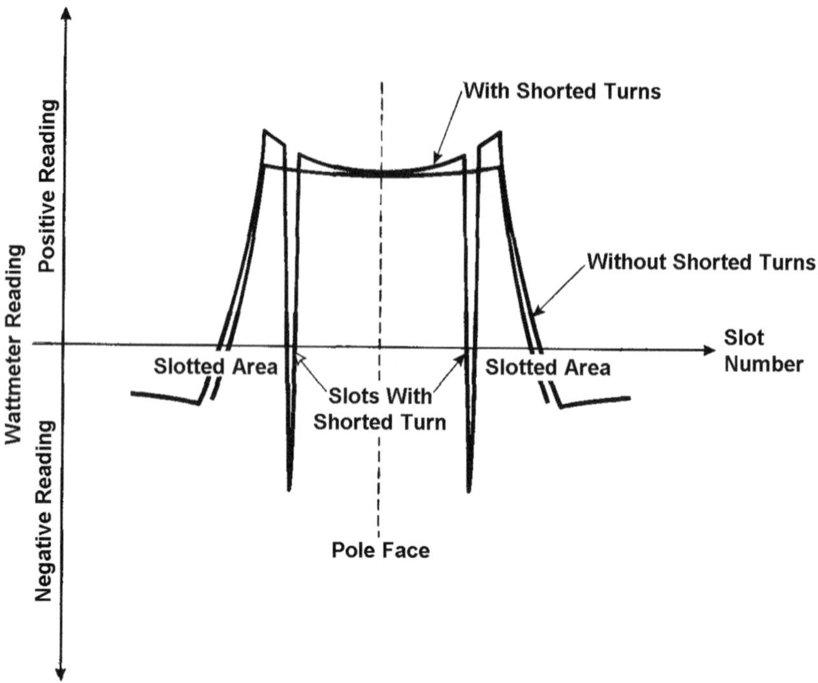

Fig. 11.55 Shorted turns detection by "C" core test. (From Kerszenbaum, *Inspection of Large Synchronous Machines,* 1996, IEEE Press, reprinted with permission.)

slot the short is in, an estimate of the number of shorted turns can be made fairly accurately.

Problems due to saturation effects at full load can occur in analyzing the data and most OEMs now have a dedicated monitor connected to the flux probe to automate the analysis process. This allows the flux probe and monitor to act as a stand-alone sensor to alarm when a short turn is detected and notify the operator for investigation.

Probably the most effective method for the detection of shorted turns in solid rotors is the flux probe method. This maps the flux of the machine as it rotates, indicating possible shorts as changes in the measured waveform. Its main advantage is that it works with the rotor online, capturing the speed-dependent shorts. Its main disadvantages are the expertise required to analyze the recorded waveforms and the fact that the machine has to be deenergized and degassed for the installation of both core-mounted and wedge-mounted types of probes. New commercially available units intended for online continuous operation include software that analyzes the waveform and alerts to a possible shorted-turn condition.

Although the flux probe is probably the best online method of shorted turns detection, there is still the issue of main magnetic field interference with the rotor slot leakage field that is used to capture the shorted turns data via the flux probe. This issue was discussed in Section 5.5.4. Because of the interference of the main magnetic field with the measurement of the rotor cross slot or leakage flux for shorted turns detection, it is required to adjust the rotor load angle to the point of minimum interference; this is called the *zero crossing flux density* or ZCFD (refer to Figs. 5.42 and 5.43). This is the center of the stator and rotor main field interaction, where the flux goes from positive to negative and vice versa and there is the least amount of interference with the leakage field of the rotor slots.

As mentioned in Section 5.5.4, it is not usual to set up specific testing on a regular basis to get a full spread of rotor angles, so this is usually only done when there is a known problem and a specific shorted turns test by flux probe is scheduled. System demands on the unit do not usually allow significant changes to test all ZCFD points for all slots and, therefore, a preplanned test is usually required to accomplish this.

To set up a planned flux probe test such that all rotor slots are checked for shorted turns at the ZCFD point, it is required to calculate the load points needed to "see" each slot when there is minimum main-fieled interference. The parameters required for the calculation are shown in Figure 11.56 and used in the following formula to obtain the actual load points:

$$\text{Load angle, } \delta = \tan^{-1}\left\{\frac{\text{MW}}{\text{Mvar} + [(V_t/V_{t\,\text{rated}}) \times \text{MVA}_{\text{rated}} \times \text{SCR}]}\right\}$$

where

MW = operating power output

Mvar = operating reactive power

V_t = operating terminal voltage

$V_{t\,\text{rated}}$ = rated terminal voltage

MVA = rated apparent power

SCR = short-circuit ratio

But before this can be done, it is also required to know what the slot angles are for the particular rotor being tested, and this is usually available from a rotor cross-section drawing such as the example in Figure 11.57. If the cross-section drawing is not available, then one can make the assumption that the pole face accounts for one-third of the 90° arc of a two-pole rotor and the winding face the other two-thirds. For a four-pole rotor this, is worked out over a 45° arc and so all values need to be divided by two to get the correct load points on the capability curve.

It quickly becomes obvious when one starts to work out the required load points that to obtain the ZCFD for each slot, the entire MW load range must be traversed, as well as a significant portion of the MVar load range. For the slots where the rotor angle is largest, at rated load and rated power factor, MW should be held constant for the calculation and the MVars calculated to find the load point. For the slots where the load angle is smaller, below rated load, the MVars should be held constant and MW calculated to find the load point.

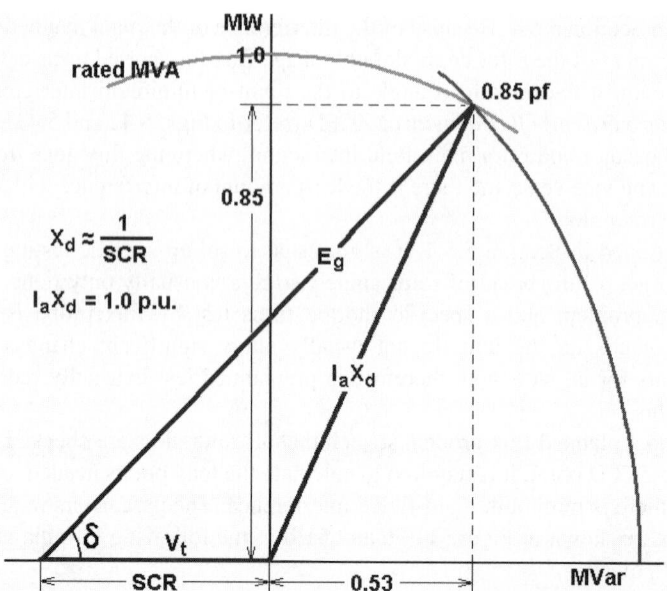

Fig. 11.56 Capability chart representation of the calculation for rotor load angle to determine the zero crossing flux density angle for a rated load of a two-pole machine with a 0.85 pf, given a specific short-circuit ratio and rated load.

Since the operating terminal voltage is not known in this application of the formula above, one can assume rated terminal voltage for a close approximation, and the formula then reduces to the following:

$$\text{Load angle, } \delta = \tan^{-1}\left[\frac{\text{MW}}{\text{Mvar} + (\text{MVA}_{\text{rated}} \times \text{SCR})}\right]$$

The load angle will be known for each slot from the rotor drawing and the SCR from the machine data sheets, as well as the MVA rating of the generator. All one needs to do is hold either MW or MVar constant and vary the other to get a set of ZCFD curves on a capability curve so that the flux probe test can be carried out (Figure 11.58).

Example of Using the Flux Probe and Low dc Voltage Tests to Determine Location of Multiple Shorted Turns. The following example presents a well-documented case of a successful major troubleshooting effort geared to remove a number of shorted turns in the field winding of a large turbo-generator rotor. This case is presented to demonstrate that even a difficult problem can be resolved in a positive manner when it is addressed methodologically technically and managerially. Although in this example the technical side of the effort will be presented, the project management and coordination between the station and the OEM was critical in achieving the

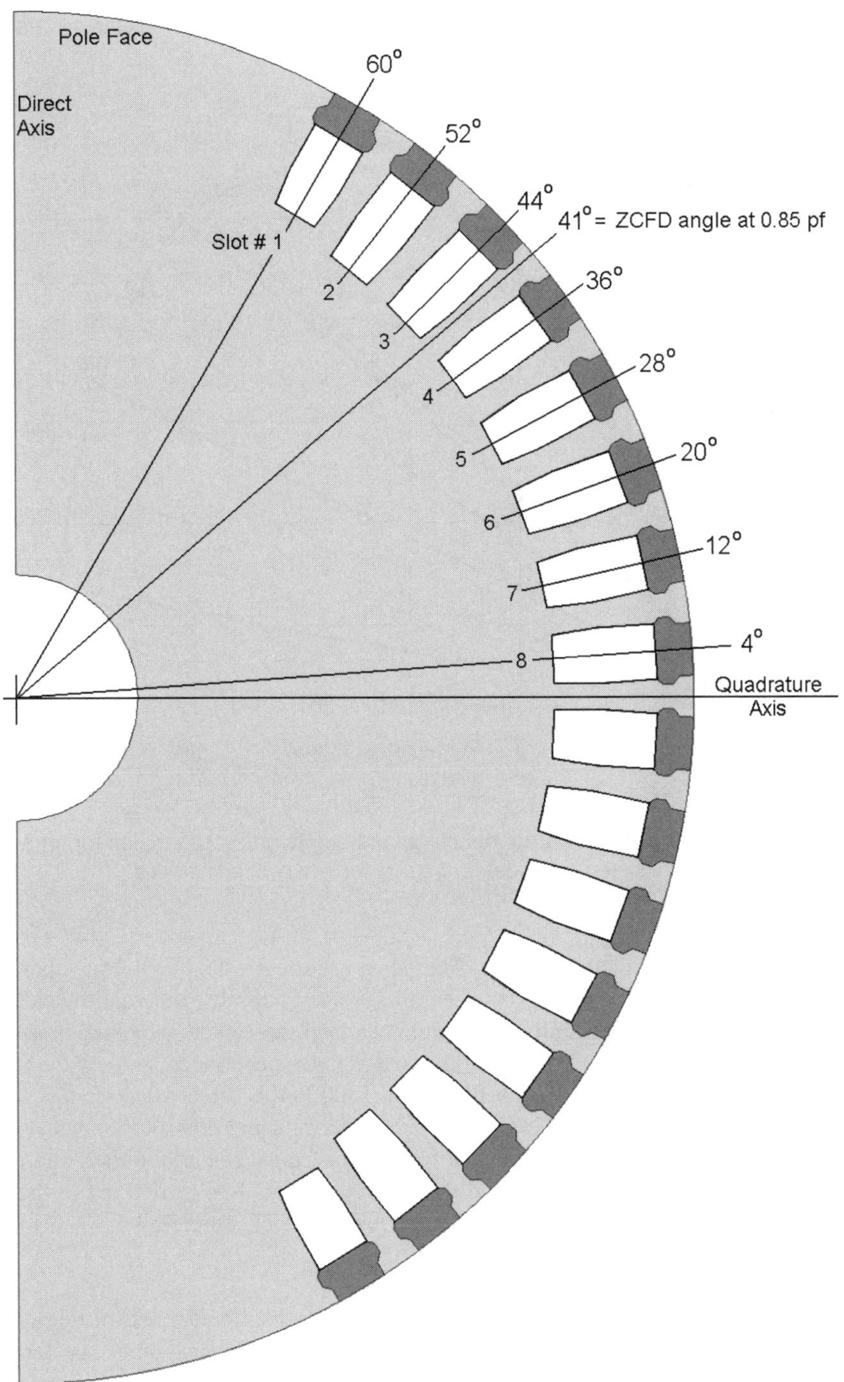

Fig. 11.57 Schematic drawing of a two-pole rotor with eight slots, showing the approximate rotor load angles required to obtain the zero crossing flux density for each slot, for clear shorted turns detection.

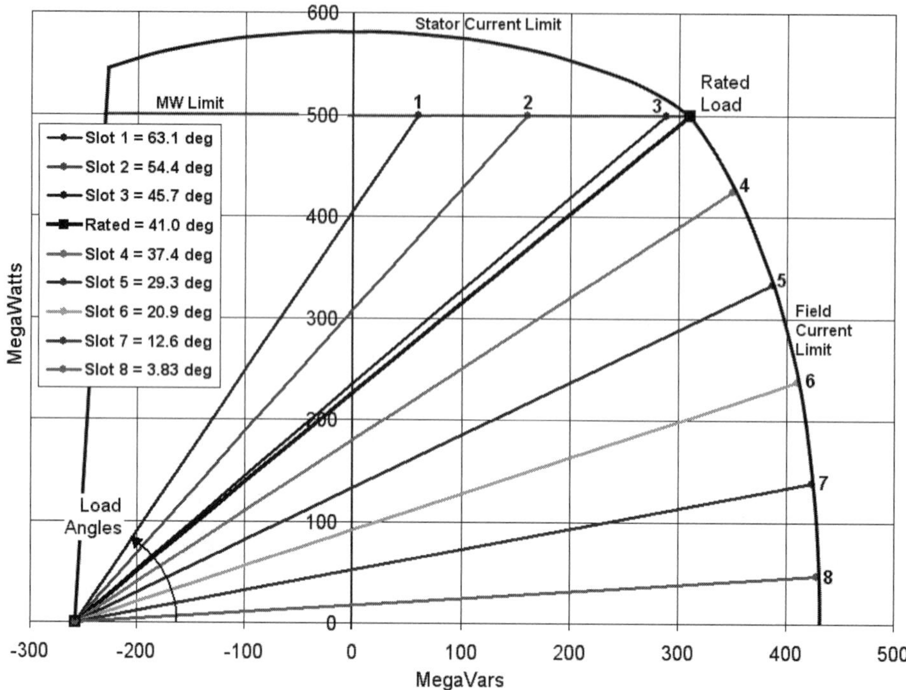

Fig. 11.58 This figure shows the resulting load points and rotor angles for an actual test carried out for shorted turns determination for a two-pole machine.

set goals within schedule and budget. There is no shortcut when it comes to spending the time in coordinating the job between all involved parties, inside and outside the station, as well as ordering all required equipment and spare parts, including those that may be required for certain low-probability but high-cost contingencies.

The machine in question has an 1800 rpm, 1300 MVA, direct-water-cooled stator and hydrogen-cooled rotor. The weight of the rotor is about 200 metric tons. Transportation to an OEM or suitable third-party shop was considered as too risky and time-consuming. Therefore, it was decided that the repair would be carried out in the station, on the turbine deck. Being an open-air turbine deck, suitable large tents were temporarily installed on the deck (Fig. 11.59).

Development of the Shorted Turns. This unit had suffered a loss-of-lubrication event that weakened its windings because of the large vibration and quick deceleration accompanying the incident. This happens because due to the partial loss of support of the coils by the retaining-ring at low speeds, the end windings of a turbo-generator are prone to damage during incidents that produce large tangential forces during fast deceleration. In the months following this event, two step increases in the vibration levels of both bearings of the generator were recorded (Figs. 11.60 to 11.64).

Fig. 11.59 Rotor in tent placed on the turbine deck. Strict FME controls were established, making the area no less "clean" than similar FME areas at the OEM's shop. A second tent, directly connected to the rotor tent and of similar dimensions, was set up to abut the rotor tent, so that entire coil bars and the full-length wedges could be moved to and from the rotor into the other tent, where the turn insulation and other repair work was performed.

Explanation of the Two Vibration Steps. Following the two step increases in vibration levels, an effort was initiated to try to elucidate the source of the problem. At the beginning, many possible sources for the increased vibration were listed (see Section 11.10.1 for such a list). Because the unit in question is in a nuclear plant and carrying out large generator load changes for testing purposes is not a particularly inviting proposition, it was decided to do a "mini" thermal sensitivity test by changing VAR output while keeping the MW output constant. Figure 11.65 shows the results of the test, which point to a VAR-sensitive rotor.

The most direct measurement of shorted turns is the *rotor flux probe* (also called *airgap flux probe,* or just *flux probe*). A flux probe with capability for reading tangential and radial flux was installed on this machine. Radial flux is the flux linking the rotor with the stator, whereas tangential flux is the rotor-slot leakage flux (Fig. 11.66). Some probes available in the industry are sold with only radial capability. However, it has been shown that depending on load conditions, one or the other may be more sensitive. Thus, a probe that offers the capability to read both radial and tangential flux may be more advantageous. Figure 11.67 shows the differences recorded from the tan-

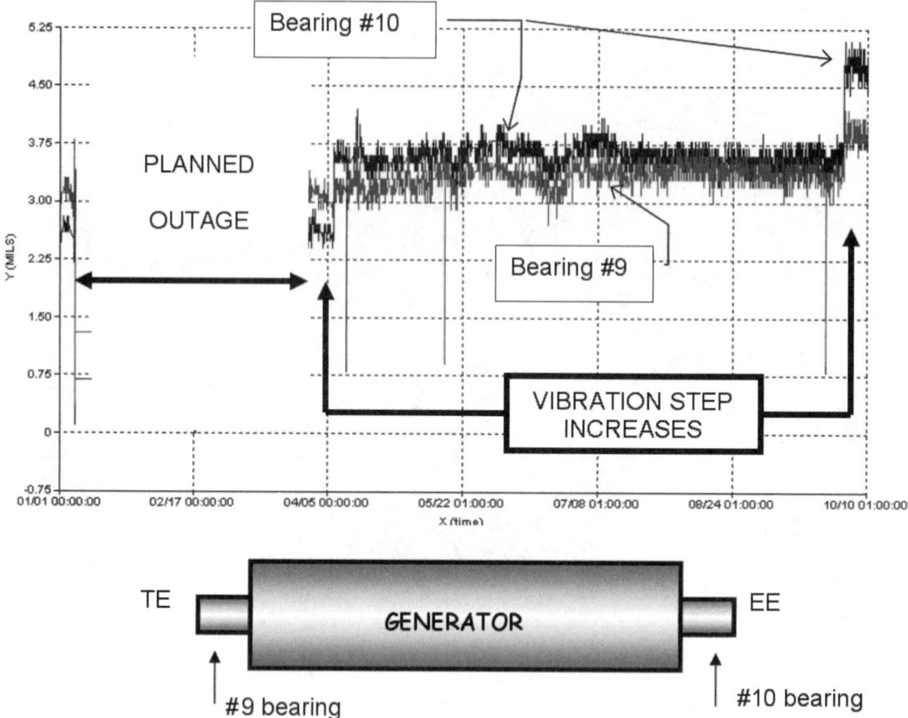

Fig. 11.60 The graph shows the behavior of both generator bearings (turbine end: bearing #9; exciter end: bearing #10). Both bearing vibrations increase in two steps about six months apart. The step changes occurred at exactly the same that time the grid experienced an abnormal condition. In both cases, large swings in power and frequency were experienced at the station switchyard, followed by about 10 to 13 minutes of increased negative-sequence currents. Although the negative-sequence currents were well within the capability of the unit as provided by the relevant IEEE standards, they introduced small torsional oscillations in the generator.

gential and the radial coils of a flux probe. Figure 11.68 shows the obtained proof of the existence of shorted turns using the flux probe.

The vibration of a turbo-generator rotor due to shorted turns in its winding is the result of two forces, both acting on the rotor in the same direction and "pushing" it away from its center of gyration. One force is created by the thermally induced bowing of the forging due to uneven heating of the rotor caused by the shorted turns. The other imbalance is magnetic in nature. It is the result of the asymmetric distribution of the rotor's magnetic field created by the shorts. Figure 11.69 schematically shows how this happens. It so happens that two-pole and four-pole generators react somewhat differently to shorted turns in their field windings. Two-pole machines are more sensitive to thermally induced bowing of the rotor, whereas four-pole machines are more sensi-

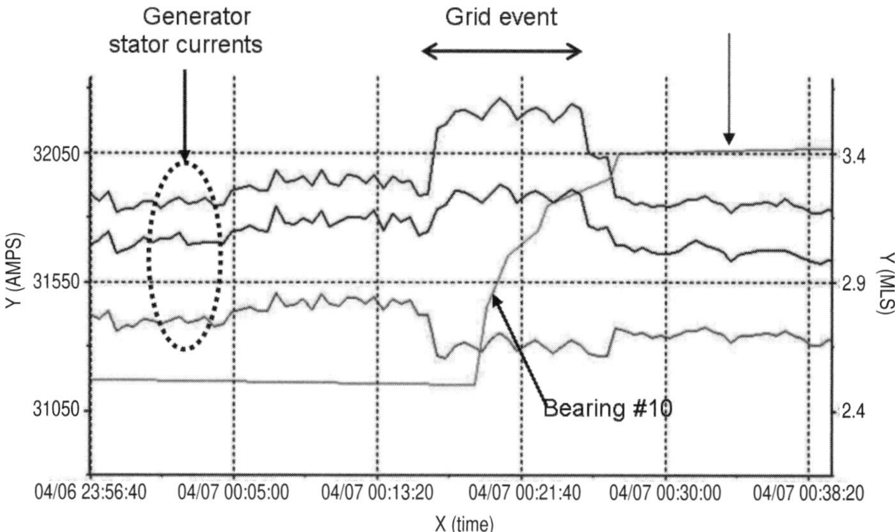

Fig. 11.61 The graph shows the generator terminal currents (in amperes, left ordinate) and the vibration of bearing #10 (generator's exciter end). It can clearly be seen that during the grid disturbance, lasting about 10 minutes, the vibration step occurred about 2 minutes and 45 seconds after the beginning of the disturbance. This was the first instance when the generator bearings went through a step increase in vibrations. The second instance is remarkably similar.

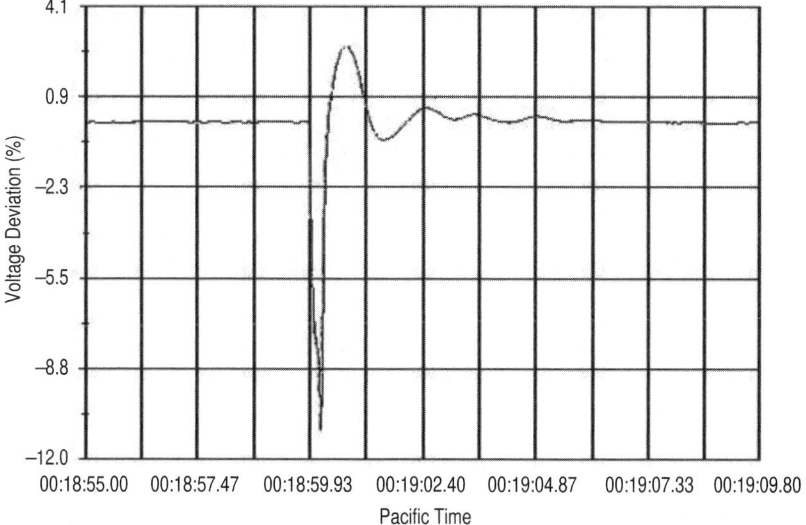

Fig. 11.62 Voltage deviation at the terminals of the generator discussed in this section. It is clearly shown on the graph that the voltage dropped momentarily by up to 10%.

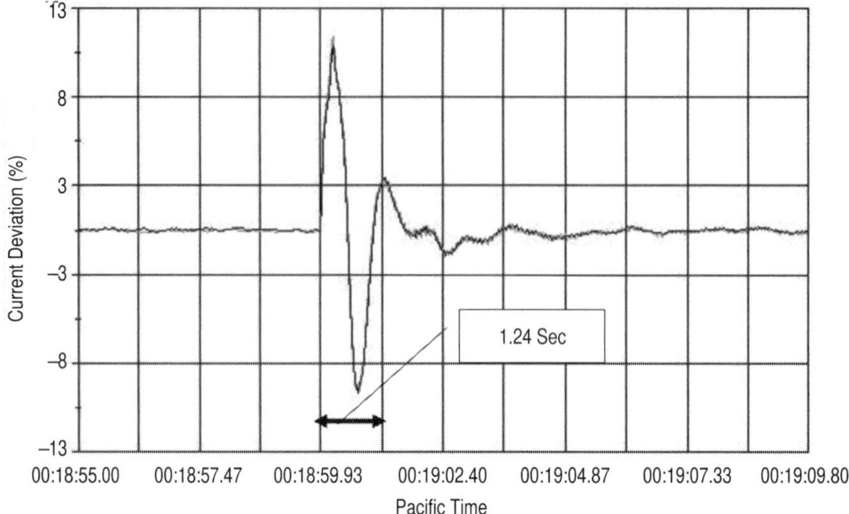

Fig. 11.63 Same grid incident as in previous figure. This graph shows the current changes at the machine terminals during the grid disturbance.

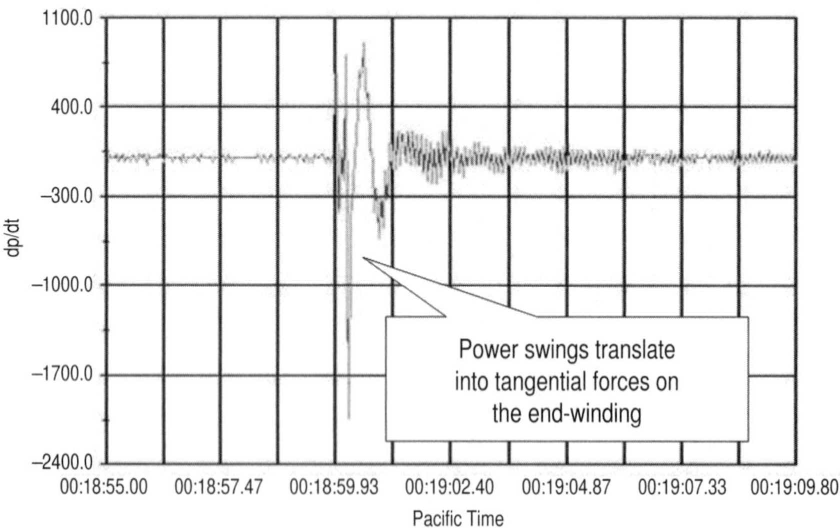

Fig. 11.64 Same grid disturbance as in previous figure. The graph shows generator power swings of up to 2000 MW. This is almost twice the rated output of the machine. These quick and large power changes translated into severe transient forces on the end windings of both the stator and, indirectly, the rotor, due to strong torsional oscillations.

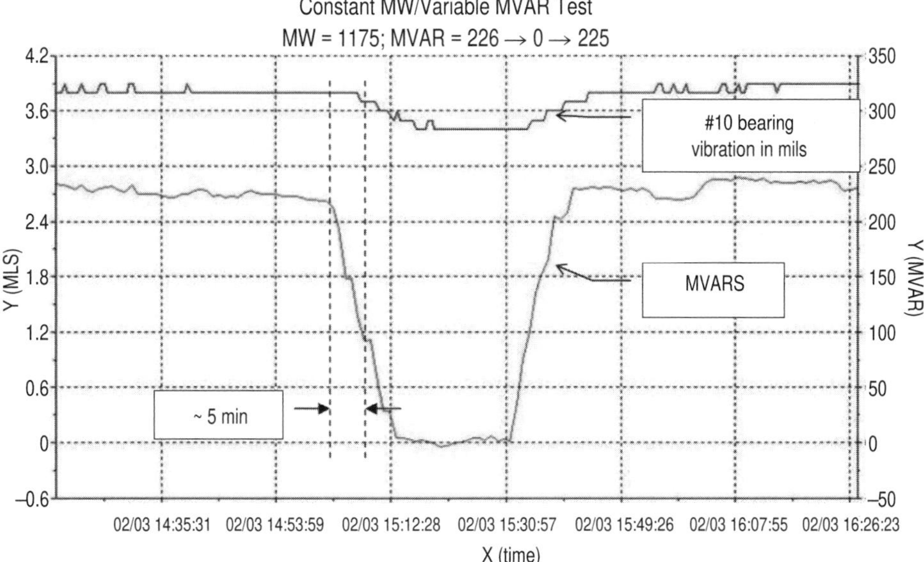

Fig. 11.65 Graphs of generator bearing vibration (measured in displacement mils) versus MVAR loading. A 5 minute delay of the vibration response can be attributed to the thermal component of the rotor imbalance.

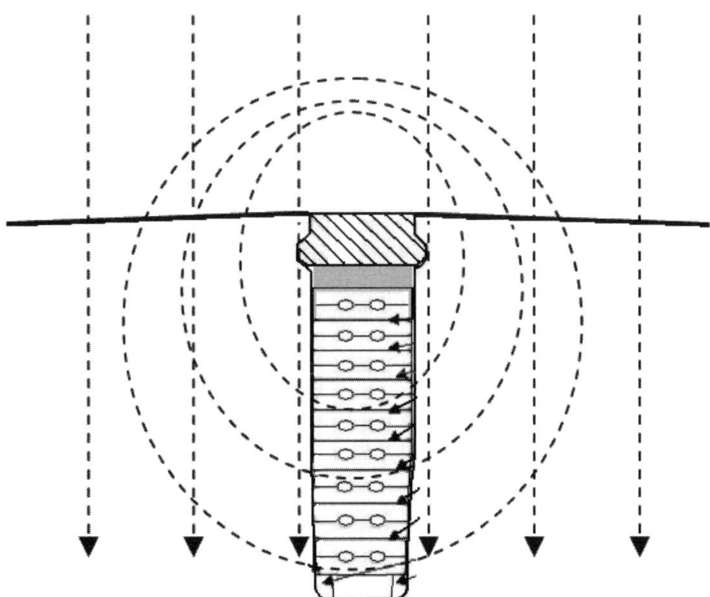

Fig. 11.66 The figure shows a rotor slot with the tangential flux wrapping it and the radial flux parallel to it (with the arrows pointing down).

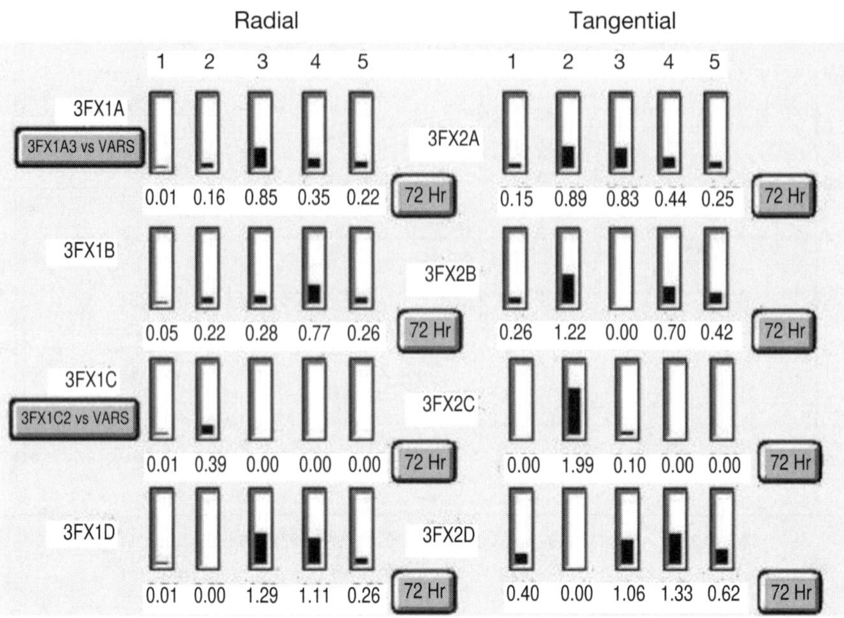

Fig. 11.67 Online readings taken from the rotor flux probe installed in the 1300 MVA, 1800 rpm generator. There is good correlation between radial and tangential readings for most slots. However, in some slots, in particular slot #2, the differential is substantial. Poles are shown in rows and slots in columns.

tive to the unbalanced magnetic pull created by the shorted turns. The explanation is rather simple. In the case of thermal bowing, all things being equal, a rotor at higher rotational speed will cause more intense vibrations with the same displacement (recall that centrifugal acceleration is proportional to the speed to the power of two). On the other hand, the magnetic field of a four-pole machine has one more degree of freedom than a two-pole unit, allowing it to create a much more uneven field distribution around the circumference. The result is a larger magnetic imbalance, all else being equal (Fig. 11.70). Therefore, four-pole generators with shorted turns in their rotor windings may also exhibit an increase in the vibration of the stator (due to the magnetic unbalanced forces established between the rotor and the stator across the airgap). In four-pole rotors with multiple shorted turns in more than one pole (as was the case with the machine studied in this example), the unbalanced magnetic pull can be more striking, resulting in intense vibrations in the stator core and frame (Fig. 11.71).

At this stage, plant personnel recognized that the generator had developed a number of shorted turns in the rotor winding. The next step was to try to identify the location of the shorts so that proper repair planning could be initiated.

Identification of Number of Shorted Turns and Their Location. The ability to retrieving data from the rotor flux probe coupled with some technical ingenuity allowed

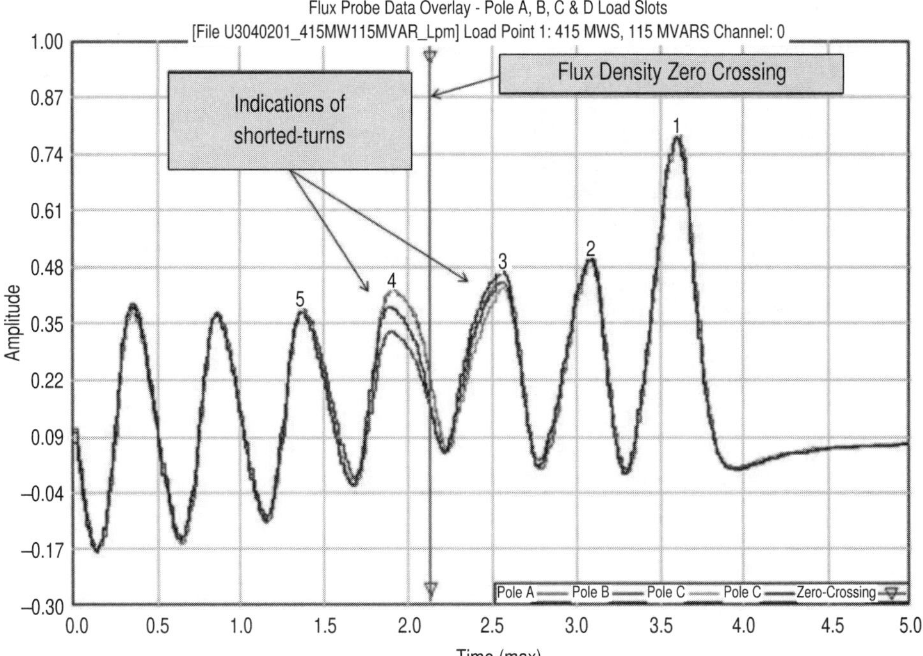

Fig. 11.68 Traces obtained from the rotor flux probe. As explained in Chapter 5, Section 5.5.4, the sensitivity of the flux probe for a particular slot depends on the load point of the machine. In this case, readings were taken with the flux probe as the unit shed load going into an outage, making it possible to obtain sensitive readings for all rotor slots (five per half-pole in this design). The graph in this figure is only one of a number taken. In this graph, in the flux-density-zero-crossing curve between slots 3 and 4, a number of shorted turns are indicated in those slots in a number of poles.

plant personnel to clearly identify the location and number of shorted turns. By analyzing the traces of Figures 11.67 and 11.68, the number of shorted turns where estimated at about 9 or 10. The location was established by adding to the graph of Figure 11.68 a key phasor or reference mark. The key phasor was obtained from the bearing-vibration system. The key phasor is read from a turbine coupling along the unit train. To link the position of the key phasor with a specific pole and slot in the generator's rotor, drawings of the rotor and coupling were used. Figure 11.72 schematically shows how all the dots were connected together to determine the position of each shorted turn.

The result of this effort is shown Table 11.1. The information in Table 11.1 allowed the station and the OEM to prepare the scope of repairs, including required equipment, spare parts, lay-down area, and number of winders and other personnel. This information was key in the successful implementation of the repair plan.

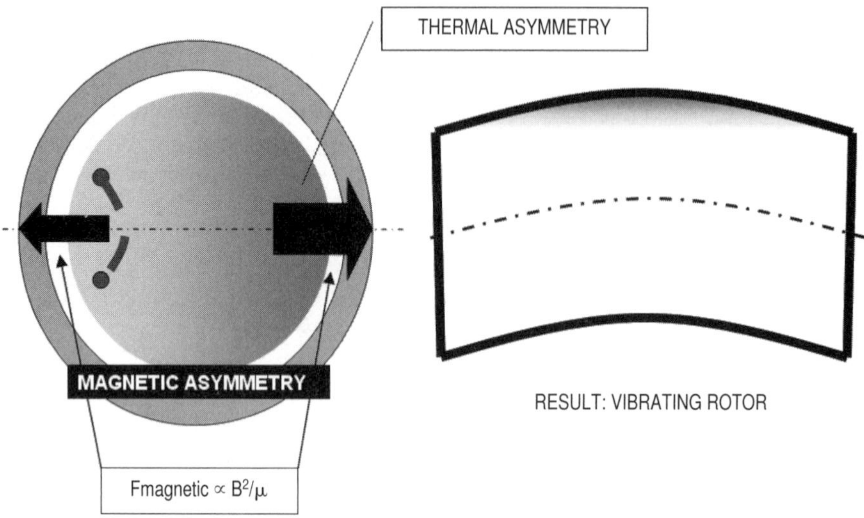

Fig. 11.69 Schematic representation of the two aligned forces causing the vibration of a rotor with shorted turns.

Although the information in Table 11.1 identified in which pole and slot a particular shorted turn was, it neither identified in what side of the pole the short was (each coil rests in two slots, one at each side of a pole) nor which turn was shorted. To obtain this information, the following test was implemented. After both retaining rings were removed the volt drop method was used to take resistance readings between each pair of turns as they leave the slot (Figure 11.73). (See Section 11.7.10 above for details on

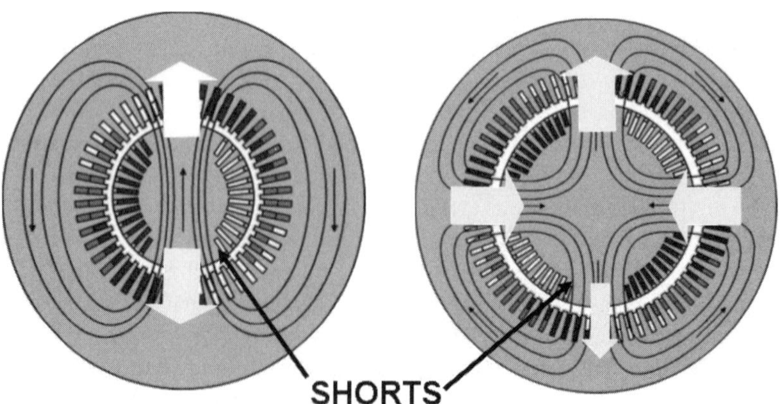

Fig. 11.70 The figure shows how the magnetic field remains quite symmetric in a two-pole rotor with shorted turns, whereas in a four-pole rotor, the asymmetric redistribution is quite obvious.

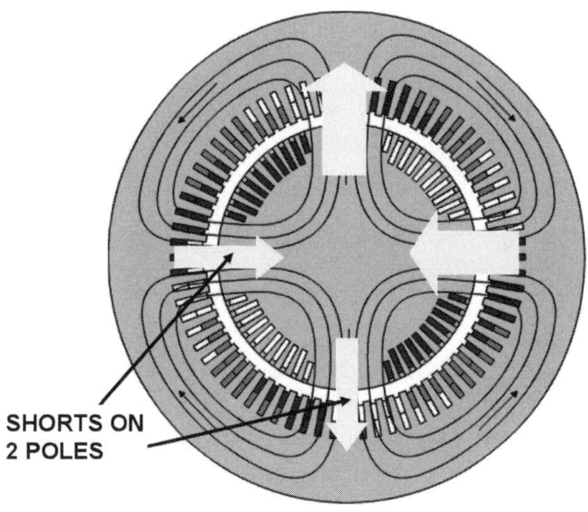

Fig. 11.71 Four-pole rotor with shorted turns in two contiguous poles. The resulting asymmetry in the field distribution is significant.

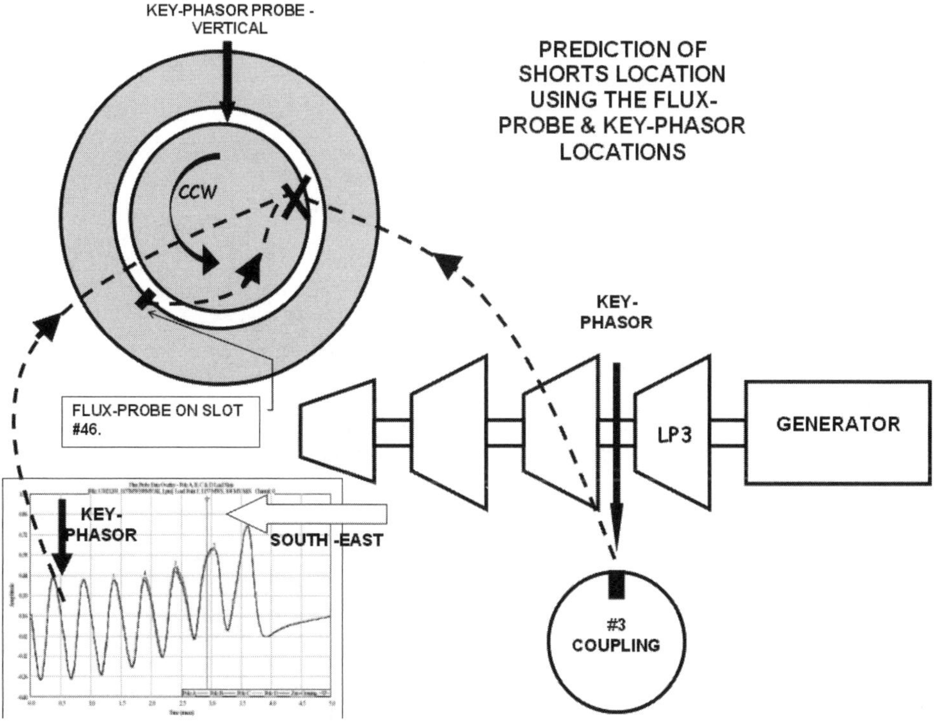

KEY-PHASOR PROBE - VERTICAL

PREDICTION OF SHORTS LOCATION USING THE FLUX-PROBE & KEY-PHASOR LOCATIONS

CCW

KEY-PHASOR

FLUX-PROBE ON SLOT #46.

LP3

GENERATOR

KEY-PHASOR

SOUTH -EAST

#3 COUPLING

Fig. 11.72 The graph schematically shows how drawings, key-phasor readings, and flux-probe readings were combined to identify the location of each shorted turn.

Table 11.1 Predicted location and number of shorts in a 1300 MVA, 1800 rpm rotor. The total number of shorts estimated as nine or ten.

Pole	Coil 1	Coil 2	Coil 3	Coil 4	Coil 5
1	0	0	1	3 or 4	0
2	0	1	0	0	0
3	0	1	0	1	0
4	0	0	2	0	0

the volt drop method to detect shorted turns.) By tabulating and comparing the resistance values found for each slot, the location of each of the shorted turns predicted in Table 11.1 was established. (See Figure 11.53 for typical traces found when shorts are present.) Many shorted turns, in particular in the end winding region, tend to "disappear" once the rotor is brought to standstill. This is because the centrifugal forces that keep the turns in the end winding tightly packed only exist while the rotor rotates at high speed. To compensate for this situation, an RSO can be connected to the terminals of the winding and a rubber hammer used to tap each coil (Fig. 11.74). If a RSO response is observed, it means that the coil being tapped has a shorted turn in that overhang. Steel bands can be applied to the end winding to simulate centrifugal forces and activate any elusive short that may have disappeared when the unit was brought to rest. Also, rotation of the rotor on the pads can elicit response from some shorted turns by shifting the weight on the turns.

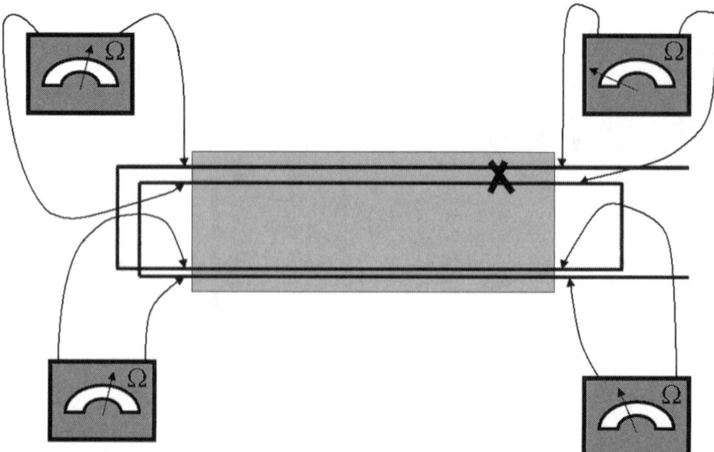

Fig. 11.73 Resistance measurements between each pair of turns as they leave the rotor slot at both sides of the forging allows determination of location of the shorted turn (marked with an X). This technique allows one to identify both the slot where the shorted turn is and the turn shorted in the slot.

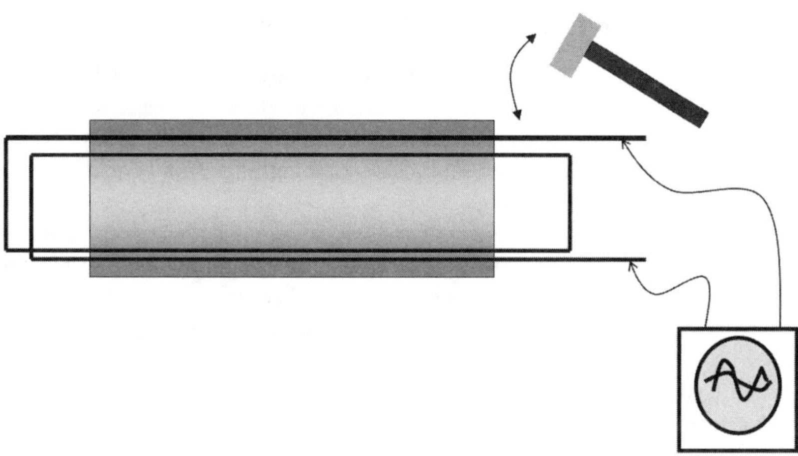

Fig. 11.74 Many shorted turns tend to "disappear" once the centrifugal forces are removed by bringing the rotor to standstill. By taping the overhang with a rubber hammer, it is possible to elicit a response from an RSO connected across the terminals of the field winding. If the RSO responds to the tapping, it indicates that a shorted turn is located in the coil being tapped. This technique helps remove shorted turns from the end windings.

With aid of the tools and techniques mentioned above, all shorted turns were actually found, as tabulated in Table 11.2. The almost perfect correlation with the predicted number and location of the shorts (Table 11.1) is remarkable. This was the result of significant effort in preparation for the repair outage, as well as the foresight of installing airgap flux probes early in the life of the generator. Figures 11.75 and 11.76 show the actual position of the shorts as found, and an example of one of the nine shorts. Once the generator had all its shorted turns repaired and the unit was brought back online, the vibration levels of the generator were back to normal.

Table 11.2 Shorted turns found (in parentheses) versus those predicted. It can clearly be seen that the prediction was extremely accurate.

Pole	Coil 1	Coil 2	Coil 3	Coil 4	Coil 5
1	0	0	1 (1)	3 or 4 (4)	0
2	0	1 (1)	0	0	0
3	0	1 (0)	0	1 (1)	0
4	0	0	2 (2)	0	0

11.7.13 Field-Winding Ground Detection by the Split-Voltage Test

The "split-voltage" test is used locate rotor grounds as a percentage through the field-winding. For this test to be effective, the resistance to ground of the fault must be less than 5% of the balance of the rotor insulation, and the voltmeter must have high input impedance compared to the ground fault. The retaining rings should also be left on in case the ground is to one of the rings.

The test is done by applying up to 150 volts dc, ungrounded, across the slip rings. A measurement of dc voltage is then taken from the rotor coupling at the turbine end of the forging to one of the collector rings. A measurement is next made from the other collector ring and the same location on the rotor coupling at the turbine end. This way, the two voltage measurements can be compared to estimate how far into the winding the ground has occurred. If the two measurements are equal, the rotor ground fault should be found in the middle of the winding. If there is less than 2% difference between the two readings, then the ground could possibly be at the collector rings.

This test is very useful in helping to determine how much dismantling is required to find the ground. Depending on where the ground is located, it can obviously make a big difference in the time expended to find the fault (Fig. 11.77).

11.7.14 Field-Ground Detection by the Current-through-Forging Test

The current-through-forging test is another test used to locate rotor-winding grounds. In this particular application, the test is used to locate the actual "axial" position of the

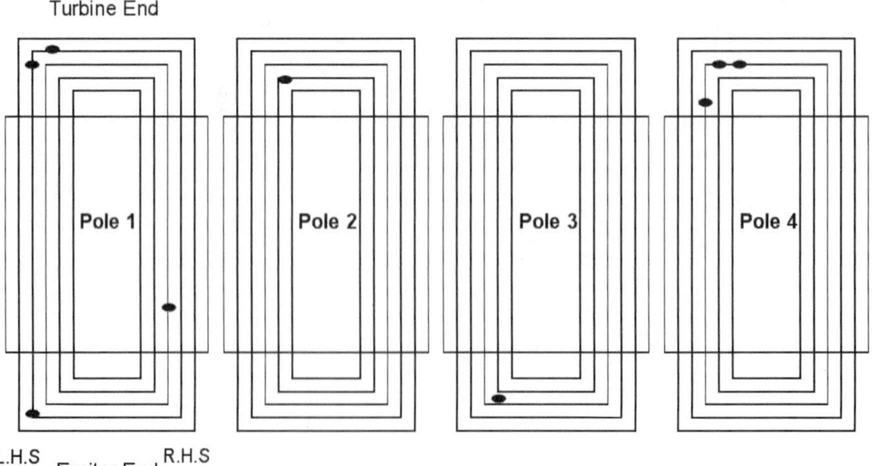

Fig. 11.75 The figure shows the slot location of each shorted turn as found during the repair effort. As expected, almost all shorted turns were found in the end windings. (Courtesy of Alstom Power Inc.)

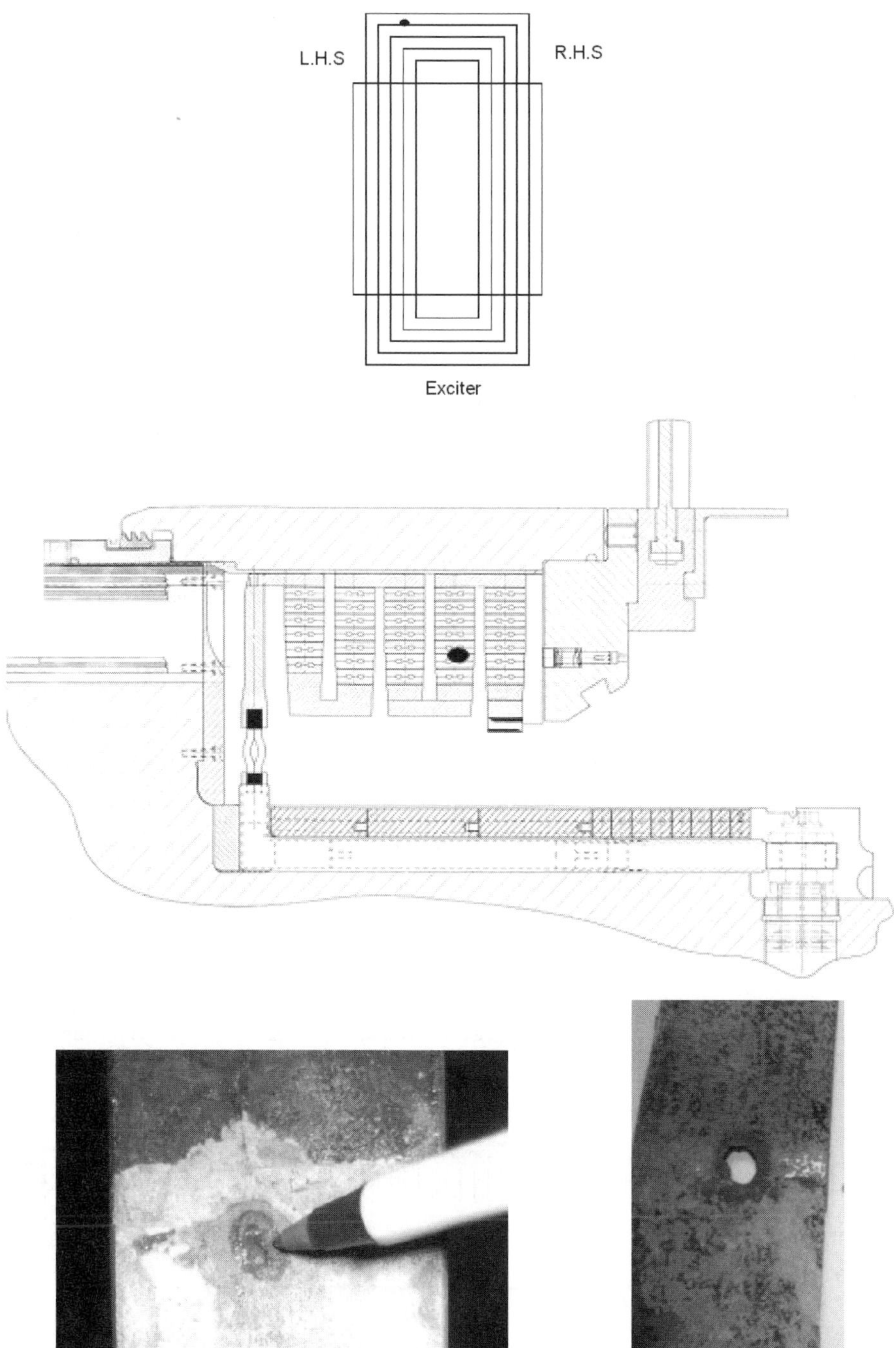

Fig. 11.76 One of the nine shorted turns found during the repair of the 1300 MVA, 1800 rpm unit. The top schematics show the position of the short (Pole 1, Coil 4, Turns 6–7). The bottom left shows the small burn mark on the copper conductor, whereas the bottom right shows the perforated insulation between turns 6 and 7. (Courtesy of Alstom Power Inc.)

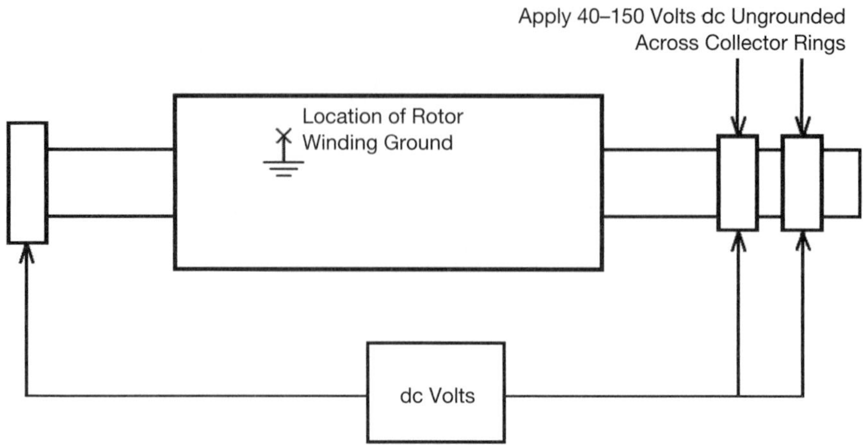

Fig. 11.77 Rotor-ground detection by split-voltage test.

ground. The retaining rings should be left on the rotor in case the ground is at one of the rings.

For this test, a dc current of about 500 amps is put through the forging from the tip of the forging at the slip-ring end to the coupling at the other end. A dc ammeter is used to look for the ground position. This is done by attaching one lead of the ammeter to the most outboard slip ring, and then using the other lead to probe along the axial length of the rotor forging. At the point where the ground is, the current should be zero, or if the current is not zero but only very low, then there will be a polarity change at the ground location (Fig. 11.78).

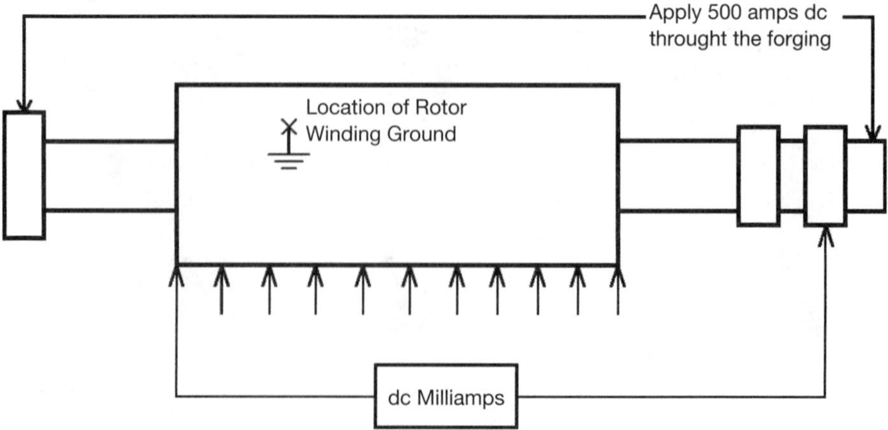

Fig. 11.78 Rotor-ground detection by the current-through-forging test.

11.7.15 Shaft Voltage and Grounding

During operation, voltage may rise on the generator rotor shaft unless the shaft is grounded. The sources of shaft voltage are well established and identified as voltage from the excitation system due to unbalanced capacitive coupling, electrostatic voltage from the turbine due to charged water droplets impacting the blades, asymmetric voltage from unsymmetrical stator core stacking, and homopolar voltage from shaft magnetization. If the electrostatically induced voltages are not drained to ground, they will rise and break down the various oil films at the bearings, hydrogen seals, turning gear, thrust bearing, and so on. The result will be current discharges and electrical pitting of the critical running surfaces of these components. Mechanical failure may then follow.

It should be also be noted that shaft grounding is required to ensure that the shaft is not "floating" and that the magnetically induced voltages, which cannot be eliminated by shaft grounding are insulated against at the bearings and hydrogen seals. It is important that the insulation systems at the bearings and housings be maintained in good condition to prevent large currents from flowing through the bearings and/or hydrogen seals.

Inadequate grounding of the rotor will also allow voltage to build up on the generator rotor shaft. Inadequate grounding may be due to a problem with the shaft grounding brushes caused by wear (requiring replacement of the brushes) or a problem with the associated shaft-grounding circuitry if a monitoring circuit is provided.

High shaft voltages can also be caused by severe local core faults of large magnitude, which impress voltages back on the shaft from long shorts across the core. Protection against shaft voltage buildup and current discharges is provided in the form of a shaft-grounding device, generally located on the turbine end of the generator rotor shaft. The most common grounding device consists of a carbon brush or copper braid, with one end riding on the rotor shaft and the other connected to ground.

Shaft voltage and current monitoring schemes are also provided in many cases to detect the actual shaft voltage level and current flow through the shaft-grounding brushes. This has the advantage of providing warning when the shaft grounding system is no longer functioning properly and requires maintenance. There are numerous monitoring schemes available, and each OEM generally has its own system provided with the turbo-generator set when purchased. For older machines with only grounding and no monitoring, a monitoring system can usually be retrofitted to the existing ground brushes. The OEM should be consulted when upgrading the shaft monitoring. (Refer to Chapter 6 for additional discussion on this topic).

11.8 HYDROGEN SEALS

11.8.1 NDE

The main tests done on the hydrogen seals are liquid penetrant inspection (LPI) for cracks and other surface damage, and ultrasonic testing (UT) for babbit bonding to the seal-ring components.

11.8.2 Insulation Resistance

Megger checks of the seal insulation to ground are done to ensure the rotor shaft is not grounded through the hydrogen seals. The hydrogen-seal ring needs to be insulated from the shaft, to carry out the megger test. This is usually done at only 500 V dc.

11.9 BEARINGS

11.9.1 NDE

The main tests done on the generator rotor bearings are liquid penetrant inspection (LPI) for cracks and other surface damage (Fig. 11.79), and ultrasonic testing (UT) for babbitt bonding to the bearing shell.

11.9.2 Insulation Resistance

Megger checks of the bearing insulation to ground are done to ensure that the rotor shaft is not grounded through the bearings. This is usually done at only 500 V dc. (Refer to Chapter 9 for additional information about bearings.)

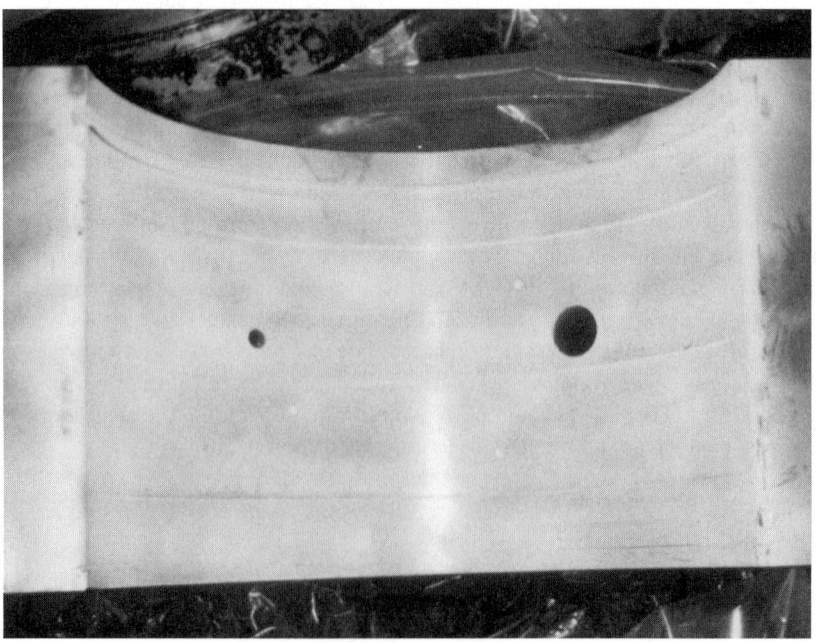

Fig. 11.79 Rotor bearing after liquid penetrant has been removed and the white developer applied. No cracks or abnormalities were found on this half-bearing.

11.10 THERMAL SENSITIVITY TESTING AND ANALYSIS

11.10.1 Background

The thermal stability analysis is performed when vibrations of unknown origin afflict the generator. The purpose of the analysis is to narrow the search for the origin of the vibrations by determining if the vibrations are due to changes in the magnitude of the field current (I_f) in the field winding or something else.

Generator-rotor thermal sensitivity is a phenomenon that occurs in the rotor when its vibration changes as the field current is changed. This has occurred in generators of all manufacturers at one time or another. The thermal sensitivity can be caused by (1) uneven temperature distribution circumferentially around the rotor, (2) winding forces that are not distributed uniformly around the rotor's circumference, or (3) asymmetrical radial gap forces. The primary driver of the second cause is the large difference in coefficients of thermal expansion between the copper coils and the steel alloy rotor forgings and components. If the rotor winding is not balanced both electrically and mechanically around the rotor, the generator rotor will be unevenly loaded, which can cause the rotor to bow and cause vibrations to change.

From experience, it is known that rotor thermal sensitivity rarely affects operation when the unit operates in the region of the rated power factor, because the rotor is below its rated temperature in this region of operation.

Vibrations due to a thermally sensitive rotor are produced mostly at running speed frequency. The vibrations' origin can be further categorized as being "reversible or repetitive" and "irreversible." Reversible vibrations are those in which the vibration vector, when plotted on a polar graph, will not shift (Fig. 11.80). Irreversible vibrations show a shifting vector (Fig. 11.81). These last ones are the most onerous, as they cannot be balanced over the long run. Almost invariably, they result in a winding removal and rewind. To capture the presence of these vibrations, the constant MW test is done in both directions, namely, with the field current changed both ways. Figures 11.80 and 11.81 show what to expect in both cases. In the figures, the test is started in point A and ends at point G.

Field-related vibrations might indicate the presence of one or more of the following:

- Shorted turns in the field winding (the vibration vector is reversible versus I_f for a given number of shorted turns)
- A ratcheting effect (i.e., coils shift to one side of the rotor and the vibration vector becomes irreversible)
- "Sticky coil" (vibration vectors tend to be reversible, but vibration changes might be "jumpy")
- Crease under the retaining ring (vibration vectors may be reversible in the short run but might be of a ratcheting nature in the long run)
- Partial blockage of gas paths inside the rotor (the vibration vector is reversible in the short and medium turns, and irreversible if insulation keeps moving; reversible during the test)

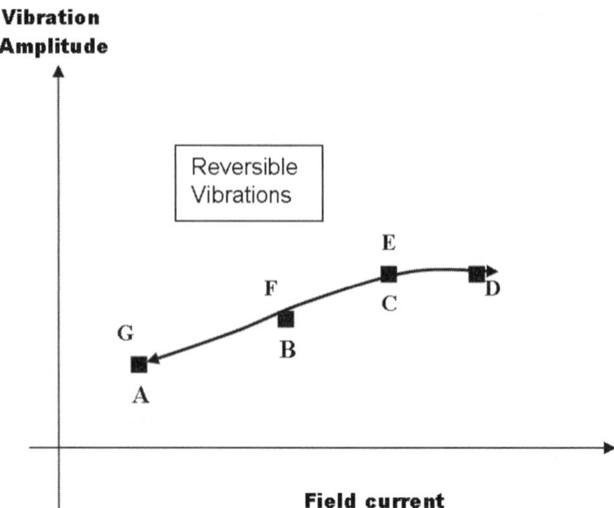

Fig. 11.80 Reversible vibrations in a thermally sensitive rotor. Note that the vibrations have a given value at a given load point. Changing loads will change the vibration, but it will always be the same for a given load. In the rotor of the graph, the field current was changed from A to D back to G(= A). As the field current is decreased, the vibrations came back to their original value for each load point.

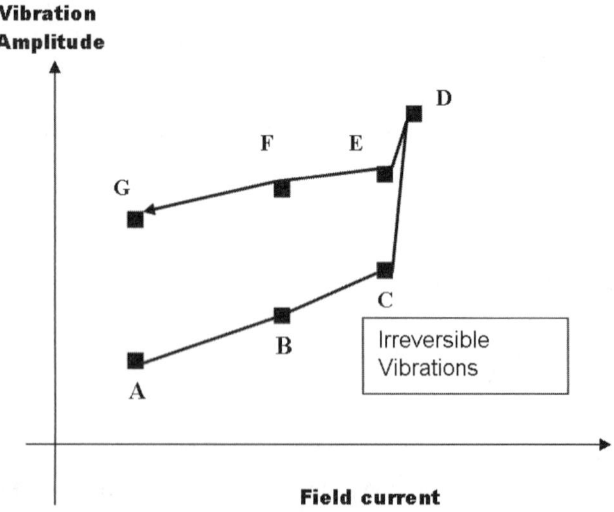

Fig. 11.81 Unlike the situation shown in the previous figure, in the present case the vibrations are not reversible. With every cycle of load they "creep up." This situation has the potential of requiring a major repair of the rotor in short order.

- Rotor stiffness dissymmetry (reversible vibrations)
- Wedges that were partially replaced during overhaul, and are not of identical dimensions, resulting in uneven fit along the circumference (vibrations tend to be reversible)
- Insulation that broke and shifted in the slots (reversible or irreversible during test, depending on severity of insulation damage)
- Nonhomogenous forging (vibration vector is reversible; vibration pattern will show up when unit is new or after repair of the forging, e.g., welding and machining)

If the unit does not indicate a field-current dependency, then most probably vibrations are of a mechanical nature.

Possible causes of rotor vibration are:

- Rotor mechanical imbalance due to a mass shift
  - Retaining-ring movement
  - Balancing weights shift
  - Wedges shift
  - Fan shift
  - Shift of blocking under the retaining rings
  - Any other mass shift
- Crack in the rotor forging or in one of its components
- Loose components

Possible causes of stator/frame vibration are:

- Uneven heating of the stator coils, core, and frame
- Bearing misalignment
- H_2 seals rub
- Coupling misalignment
- Loose footing
- Lose components
- Frame twisted

11.10.2 Typical Thermal Sensitivity Test

In order to segregate between the various sources, a thermal sensitivity test is performed. The following variables are recorded during the test (others may be added to the list). All manufacturers of large generators perform this type of test with some variances, under one name or another. The following list contains the most common approaches. Consult with the OEM of your unit for the test on your machine.

- MW
- MVAR
- Terminal volts
- Terminal current
- Field voltage
- Field current
- Vibration of all bearings of interest
- Temperatures of interest
- Time (hours, minutes) to stamp and correlate all readings

It is important to verify that the unit remains within its capability curve for the entire duration of the test. For that purpose, it is convenient to plan the load points ahead of the test, though it is not critical the actual load points selected during the test closely match those in the plan.

The test has three main parts.

1. *Constant MW and MVAR.* The aim of this test is to capture a rotor that is thermally sensitive due to forging asymmetries (due to the forging process or to machining, welding, etc.) or to stator-frame issues. During this part of the test, the unit's active and reactive power are held constant (AVR manual). The temperature of the generator's hydrogen is changed in steps by controlling flow of cooling water to the hydrogen heat exchangers. Stator water temperature is not changed. It is important to ascertain that the unit remains within its temperature operating limits. Readings are taken when the temperatures are stable, and the time is noted (hours, minutes). This test is done at a reduced load point, to "make room" for temperature increases during the test.
2. *Constant MW.* During this part of the test, the generator's gross MW output must remain constant. With MW constant, the field current (I_f) is changed (thus, the VARs also change), and readings are taken of the variable parameters. The test is repeated for a number of field-current values. After a new I_f is set, the machine is allowed run for a given time to allow temperatures to stabilize, and then the readings are noted and the time is recorded. The field current is changed in both directions to capture "ratcheting" problems.
3. *Constant Field Current (I_f).* During this part of the test, the I_f must remain constant and the AVR must be set to manual. Under this condition, the MW of the generator is changed in steps. After every change, the unit is allowed to stabilize, and readings of the variables of interest are taken and the time is noted.

By analyzing the test results, it is possible to reduce the number of possible causes to one or a very small number. This is very important, as it helps minimize the cost of any required outage and repair.

11.11 HEAT-RUN TESTING

A heat run is a method used to determine the temperature rise of the various parts of the generator during operation. This is usually done for new equipment to ensure that it meets the specified requirements for operation within its design temperature limits. Often, a heat run test is also carried out when a generator is exhibiting problems and a comparison to previous heat run data is required. This type of comparison allows checking for progressive deterioration and the rate of deterioration.

11.11.1 Test Procedure

Three different load points are required as a minimum to obtain information about the generator performance at full load. These should preferably be done at the full MW output of the generator for the minimum leading power factor load point, the unity power factor load point, and the maximum lagging power factor (rated) load point.

Full-load heat run tests should be performed with the generator synchronized to the system and operating normally at maximum rated MW load and rated H_2 gas pressure. At each test load, generator operation should be held constant for at least 30 minutes to allow all generator internal temperatures of the machine to stabilize. Readings should be taken of all generator and auxiliary parameters as follows.

Electrical Parameters:

- Power Output, MW
- Reactive Power, MVar
- Terminal Voltage
- Average Stator Current
- Individual Phase Currents
- Field Voltage
- Field Current
- Rotor Winding Temperature
- Shaft Voltage
- Exciter Power

Hydrogen System:

- H_2 Pressure
- H_2 Purity
- H_2 Dewpoint
- All Cold H_2 Gas Temperatures
- All Hot H_2 Gas Temperatures

Hydrogen Coolers:

- Auxiliary Cooling Water Inlet Temperature to Coolers
- Auxiliary Cooling Water Outlet Temperature, All Coolers

Exciter:

Cooling Air Inlet Temperature

Cooling Air Outlet Temperature

Bearings:

Lube-Oil Inlet Temperature

Lube-Oil Outlet Temperatures

Bearing-Metal Temperatures

Bearing Vibration

Hydrogen Seals:

H_2 Seal Inlet Oil Temperature

H_2 Seal Outlet Oil Temperatures

H_2 Seal Metal Temperatures

Seal-Oil System:

Seal-Oil Supply Pressure

Seal-Oil Inlet Pressures

Seal Oil/H_2 Differential Pressure

Generator Seal-Oil Inlet and Outlet Temperatures

Stator Cooling Water System:

Generator SCW Inlet Temperature

Generator SCW Outlet Temperatures

SCW Conductivity

Generator SCW Inlet and Outlet Pressures

SCW Coolers, Inlet and Outlet Temperatures

Raw Cooling Water to SCW Coolers, Inlet and Outlet Temperatures

Stator Winding Temperatures:

All Hose Outlet Thermocouples or RTDs

All Stator Slot Thermocouples or RTDs

Stator Core Temperatures:

All Stator Core Thermocouples or RTDs

All Core End Plate Thermocouples

All Core End-Plate Flux-Shield Thermocouples

11.11.2 Acceptance Parameters

During the heat-run testing at the specified load points, the following parameters must be met to ensure that the equipment has met the design operational limits.

1. Power, MW. The rated power in MW must me met over the full power-factor range from 0.95 leading to 0.90 (or sometimes 0.85) lagging, without exceed-

ing any Class B temperature limits of 130°C or vibration limits specified on any of the stator, rotor, or excitation components.

2. Reactive, Mvar. The stator must be capable of exporting the design level of MVars at full MW output and importing full design MVars at full MW output, without exceeding any Class B temperature limits of 130°C on any of the stator, rotor, or excitation components, or vibration limits on the stator core and frame and stator windings.
3. Terminal Voltage. The stator should be capable of operating over the full terminal voltage range, +5% and –5%, at full MW output, and MVar and power-factor range, without exceeding any Class B temperature limits of 130°C on any of the stator components. In some cases, a wider terminal voltage range may be applicable, depending on the original machine capability.
4. Average Stator Current. The stator should be capable of operating up to full rated stator current without exceeding 80°C on any measured stator conductor bar water-outlet temperature sensor or on any of the temperature sensors embedded between stator bars in the slot on the stator winding. The measured differential temperature between top and bottom stator bars should not be greater than 10°C.
5. Stator Vibration. The stator end-winding vibrations should not exceed the design limits (usually 4 mils or 100 micrometers peak to peak) for any load point tested over the full power-factor range of the generator.
6. All Cold H_2 Gas Temperatures. The cold hydrogen gas temperature should be less than 40°C during all heat-run testing for acceptance.
7. Generator SCW Inlet Temperature. The stator cooling water inlet temperature should generally be less than 50°C during all heat-run testing for acceptance.
8. All hose outlet thermocouples or RTDs should maintain temperatures less than 90°C over the full MW and MVar range of the generator.
9. All stator-slot thermocouples or RTDs should maintain temperatures less than 90°C over the full MW and MVar range of the generator.
10. All stator core thermocouples or RTDs should maintain temperatures less than 130°C over the full MW and MVar range of the generator.
11. All core-end-plate thermocouples or RTDs should maintain temperatures less than 130°C over the full MW and MVar range of the generator.
12. All core-end flux shunt or flux screen thermocouples or RTDs should maintain temperatures less than 130°C over the full MW and MVar range of the generator.
13. There should be no shorted turns in existence on the installed rotor as measured by a flux probe, if one is installed.

11.12 HYDROGEN LEAK DETECTION

Hydrogen leaks or potential leak areas are discussed in a number of sections of this book. One of the areas where hydrogen may leak is the casing. The casing in hydrogen-cooled machines serves two important functions: It is part of the stator supporting structure, together with the frame, and it is a pressure vessel designed to safely contain the hydrogen inside the machine. There are a number of mechanisms by which hydrogen may escape, such as cracks, manholes, instrument penetrations, pipe connections, leaky gaskets, hydrogen shaft-oil seals, and the like. Following are some of the issues related to casing leaks. Figure 11.82 shows such a leak.

In addition to the loss of hydrogen from the casing in various locations, additional areas where hydrogen may leak from the generator are into the stator conductor bars and, hence, into the stator cooling water and its associated system (see Section 11.4 above for details); and into the seal oil and into the hydrogen coolers and the raw service water. All of these represent a loss of hydrogen from the machine and will cause the casing pressure to decrease. Finding and quantifying the leaks is extremely important as each leak location can have subsequent detrimental effects on machine operation as well as become a risk to the various internal components and a safety hazard to personnel.

11.12.1 Pressure Drop

A pressure test is a global test to ascertain that the casing is holding or not holding hydrogen at operating pressures. There is always some loss of hydrogen pressure in the

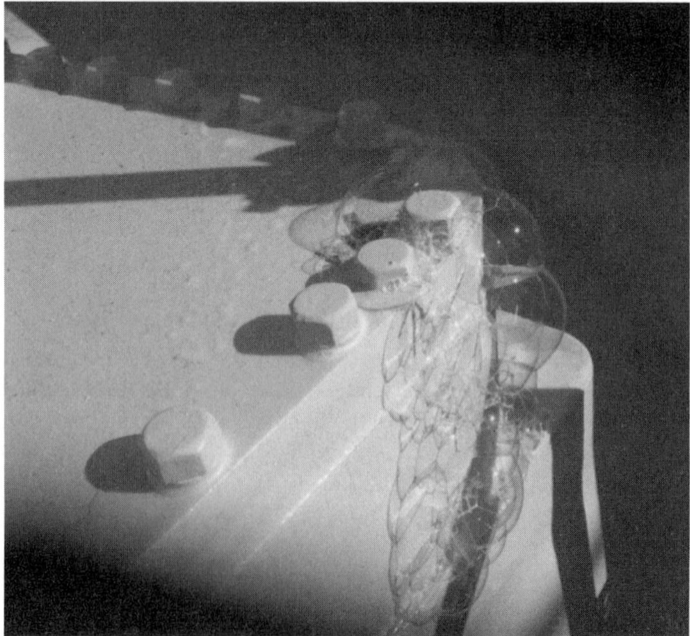

Fig. 11.82 Hydrogen leak through a bolt of one of the hydrogen coolers of a large generator. The bubbles are from a soapy substance used to find the leak.

generator and it is quantified as the amount of leakage of hydrogen from the generator in cubic feet per day. All generators should be capable of operating with less than 500 cubic feet per day loss. When hydrogen does leak out of the generator, whether it be from the casing, into stator cooling water, or into the seal oil, the limit allowable is generally up to 1,500 cubic feet per day from all hydrogen loss sources. Upon reaching this value, investigations for the location of the leakage should begin.

Determination of Casing H_2 Leakage Rate. For leakage from the generator casing, the amount of leakage can be estimated while on the turning gear or at rest as follows:

$$\text{Hydrogen leak rate total. } L_C = 235 \times \frac{V}{H}\left(\frac{P_1 + B_1}{273 + T_1} - \frac{P_2 + B_2}{273 + T_2}\right), \quad \text{cubic feet per day}$$

where

L_C = total H_2 gas loss from the casing, cubic feet per day
V = volume of gas in the generator, cubic feet
H = test duration in hours
P_1 = initial generator gas pressure, inches of Hg
P_2 = final generator gas pressure, inches of Hg
B_1 = initial barometric pressure, inches of Hg
B_2 = final barometric pressure, inches of Hg
T_1 = initial generator gas temperature, °C
T_2 = final generator gas temperature, °C

If the test was done in air, then to find the equivalent leakage of hydrogen at the tested pressure, when the hydrogen is 98% pure, hydrogen leakage at this condition = $3.38 \times L_C$.

It should be noted that the above method is only applicable when the temperature variation is 5°C or less and the pressure variation is 2 inches of Hg or less.

If it is required to estimate the equivalent leak of hydrogen at operating pressure, the following relationship can be used to estimate this.

$$\text{Hydrogen operational leak rate, } L_{op} = (3.38 \times L_C) \times \left(\frac{29.4 + P_o}{29.4 + P_t}\right) \times \left(\frac{P_o}{P_t}\right), \quad \text{cubic feet per day}$$

where

L_{op} = equivalent H_2 gas leakage at operating pressure, cubic feet per day
L_C = total H_2 gas loss from test, cubic feet per day
P_o = operating generator gas pressure, psig
P_t = Average test pressure in psig = $[(P_1 + P_2)/2]/2.036$

Determination of Loss of H_2 into Stator Cooling Water. When the leakage is into the stator cooling water, the amount of leakage is usually estimated by a leak detection system whose output is given in cubic feet per day. The limit for the loss of hydrogen into stator cooling water is generally 50 cubic feet per day. At this value, investigations should begin due to the concern for potential hydrogen gas locking of the stator cooling water flow in the stator winding.

Determination of loss of H_2 into Seal Oil. For leakage into the seal oil, the amount of leakage can be estimated by the following method:

$$\text{Loss of gas into seal oil, } L_{go} = \frac{(S_h \times 60 \text{ minutes} \times 24 \text{ hours})}{7.5} \times P \times Q \times S_f,$$

cubic feet per day

where
L_{go} = total gas loss into seal oil, cubic feet per day
S_h = solubility of H_2 in oil = 7% (0.07)
solubility of air in oil = 10% (0.10)
P = generator gas pressure, atmospheres (5 Atm = 435 kPag = 60 psig)
Q = seal-oil flow into generator in US-GPM (may be up to 7 US-GPM)
S_f = saturation factor = 1.0 for H_2 in oil
saturation factor = 0.5 for air in oil

The leakage should be less than the 500 cubic feet per day normal leakage expected.

11.12.2 SF_6 Gas

Using SF_6 gas is a technique that has been considered in the last few years to find small (and big) H_2 leaks. It actually has been implemented in a number of stations.

The principle of SF_6 leak detection is the fact that a few parts per million of that gas mixed with air or hydrogen are enough to detect a leak using a kind of infrared camera specifically tuned to the wavelength of light absorbed by SF_6. If injected while the unit is running, leaks are very easily seen with the special camera. However, the technique is not widely implemented because of the existing concerns about the noxious components of SF_6 decomposition under electrical discharges. Inside a large generator, there is partial discharge activity (PDA) between the slots and coils, and corona in the end windings. PDA can be a source for decomposition of SF_6 inside a generator. Some of the resulting halides may have detrimental effects on critical components such as retaining rings (halides have been documented as possible causes for stress corrosion cracking of retaining rings). Therefore, most if not all OEMs do not offer open support for using SF_6 as a leak detector in an operating machine. More research is required to evaluate the safety of this technique. The application of SF_6 as a leak detector must also be checked against applicable environmental regulations for each location.

Leak detection by SF_6 can be deemed safe in those instances in which the application is carried out where there are no electrical discharges (e.g., if the machine is deenergized and if SF_6 is purged from the machine before energization).

Fig. 11.83 Hand-held ultrasonic leak detector.

11.12.3 Helium

Using helium for finding leaks in hydrogen-cooled generators goes back a long time. Special sniffers can detect helium with ease. Being also a very light inert gas, it is useful in finding very small leaks. Helium, though, is an invisible gas, and, thus, finding a leak can be a very tedious exercise. Helium is well suited to find leaks in the water boxes or water connections at the end of the stator bars.

11.12.4 Snooping

Applying a soapy mixture to a suspected area will cause bubbles to appear if and where a leak exists. Figure 11.82 shows such a case. It is a cumbersome method, but effective.

11.12.5 Ultrasonic

Special sonic snoopers (Figure 11.83) can pick up the very high-frequency noise created by a hydrogen leak. It is not effective for very small leaks, though.

REFERENCES

1. IEEE/ANSI C50.13-1989, "Requirements for Cylindrical-Rotor Synchronous Generators."
2. IEEE Std 1-2000, "IEEE Recommended Practice—General Principles for Temperature Limits in the Rating of Electrical Equipment and for the Evaluation of Electrical Insulation."
3. IEEE Std 4-1995, "Standard Techniques for High-Voltage Testing."
4. IEEE Std 43-2000, "IEEE Recommended Practice for Testing Insulation Resistance of Rotating Machinery."

5. IEEE Std 95-1977 (R1991), "IEEE Recommended Practice for Insulation Testing of Large AC Rotating Machinery with High Direct Voltage."
6. IEEE Std 100-2000, "The Authoritative Dictionary of IEEE Standards Terms" (7th Edition).
7. IEEE Std 115-1995, "IEEE Guide: Test Procedures for Synchronous Machines. Part 1: Acceptance and Performance Testing, Part II—Test Procedures and Parameter Determination for Dynamic Analysis."
8. IEEE Std 275-1992 (R1998), "IEEE Recommended Practice for Thermal Evaluation of Insulation Systems for Alternating-Current Electric Machinery Employing Form-Wound Preinsulated Stator Coils for Machine Rated 6900V and Below."
9. IEEE Std 421.1-1986 (R1996), "IEEE Standard Definitions for Excitation Systems for Synchronous Machines."
10. IEEE Std 432-1974 (R1991), "IEEE Recommended Practice for Insulation Testing of Large AC Rotating Machinery with High Voltage at Very Low Frequency."
11. IEEE Std 433-1974 (R1991), "IEEE Recommended Practice for Insulation Testing of Large AC Rotating Machinery with High Voltage at Very Low Frequency."
12. IEEE Std 434-1973 (R1991), "IEEE Guide for Functional Evaluation of Insulation Systems for Large High Voltage Machines."
13. IEEE Std 522-1992 (R1998), "IEEE Guide for Testing Turn-to-Turn Insulation on Form-Wound Stator Coils for Alternating-Current Rotating Electrical Machine."
14. IEEE Std 37.102-1995, "IEEE Guide for AC Generator Protection."
15. ISO 3746 (1995), "Acoustics—Determination of Sound Power Level of Noise Sources Using Sound Pressure Survey Method Using An Enveloping Measurement Surface Over a Reflecting Plane" (Second Edition).
16. ISO 7919-2 (1996), "Mechanical Vibration of Non-reciprocating Machines—Measurements on Rotating Shafts and Evaluation Criteria. Part 2: Large Land-Based Steam Turbine Generator Sets" (First Edition).
17. ISO 7919-4 (1996), "Mechanical Vibration of Non-reciprocating Machines—Measurements on Rotating Shafts and Evaluation Criteria. Part 4: Gas Turbine Sets" (First Edition).
18. EPRI Report EL/EM-5117-SR, "Guidelines for Evaluation of Generator Retaining-Rings—Development of Ultrasonic Examination Method for Detection of Stress Corrosion Cracking in Generator Retaining-Rings," April 1987.
19. "Handbook to Assess Rotating Machine Insulation Condition," *EPRI Power Plant Series,* Vol. 16, November 1988.
20. "Turbine Generator Retaining Rings—Supplementary Information on the Inspection of Rings Manufactured in 18 Mn–18 Cr Steel," *Electra Magazine,* October 1998.
21. IEEE C50.13-2005, "IEEE Standard for Cylindrical-Rotor 50 Hz and 60 Hz Synchronous Generators Rated 10 MVA and Above."
22. "Phased Array UT Applications Development at the EPRI NDE Center", Greg Selby, EPRI, NDT.net, October 1999, Vol. 4 No. 10.
23. "Universal Phased Array UT Probe for Nondestructive Examinations Composite Crystal Technology", J. Ritter, Siemens KWU, Erlangen, July 2001.

12

MAINTENANCE

12.1 GENERAL MAINTENANCE PHILOSOPHIES

Over the many years during which the authors of this book have been active participants in the electric power generation business, they have witnessed a rising awareness of maintenance management issues. Improvements in the management of operations and more efficient and cost-effective maintenance practices have gone hand in hand with other significant developments: safety of employees and the public at large, reliability of electricity delivery, and conservation of the environment, for example (Fig. 12.1).

Over these years a number of maintenance-management studies and guidelines were published. References [1 and 2] are just two such important sources, with reference [2] being the latest arrival on the subject. The nuclear industry has taken a big step forward by implementing a program for establishing system reliability based on the development of stochastic trees, in which the discrete probability of failure represents single components. Discovering critical components helps determine those requiring special maintenance attention. Although mostly applied to nuclear plant safety systems, on occasion it extends to certain major plant equipment, such as main transformers and generators.

In general, the maintenance philosophies adopted by the electric power industry can be categorized as belonging to one or another of the following:

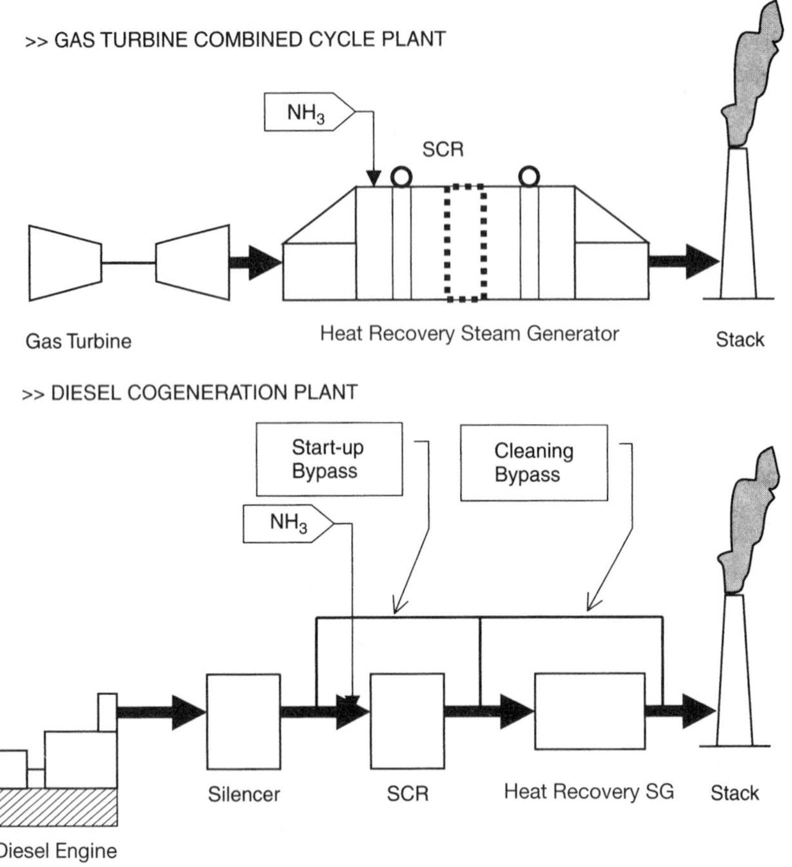

Fig. 12.1 Environmental friendliness. Schematic view of a selective catalytic NOx removal (SCR) system for NO_x reduction installed on both a combustion turbine and a diesel engine. Both are prime movers for utility and industrial generators.

1. Breakdown maintenance
2. Planned maintenance
3. Predictive maintenance
4. Condition-based maintenance (CBM)

An overview of these different approaches to maintenance is discussed below.

12.1.1 Breakdown Maintenance

This type of reactive approach (fix it as required) is usually employed for short-term economic gains with little regard to the future of the specific piece of equipment. For a large and central piece of equipment, such as a generator, it is rarely used in the developed world. It is commonly applied to smaller components when repair is more costly than replacement and loss of the particular component during operation does not disrupt the generation of electric power. Interestingly, with the advent of deregulation of the electric power industry, and with it the requirement for some utilities to divest their generating assets, some units have been operating with little attention to their required maintenance, to the extent they are being run "to breakdown."

12.1.2 Planned Maintenance

Planned maintenance has been in the past and perhaps still is the predominant maintenance philosophy in maintaining critical equipment in power plants.

Planned maintenance entails predictive maintenance and condition-based maintenance. These are based on experience concerning the reliability of the equipment acquired during many years of operation, as well as on load demands, weather, personnel availability, coordination with other plants of the same utility, and so on. For instance, in nuclear power plants, refueling cycles are a major determinant of when maintenance of major equipment, including the generator, is performed. Given this strong constraint, even if the plant, in general, follows predictive or condition-based maintenance, the predicted/required maintenance timing and scope will be adjusted to accommodate the timing of the refueling outage. Figure 12.2 depicts a typical state of affairs found in many stations following CBM practices constrained by planned maintenance schedules.

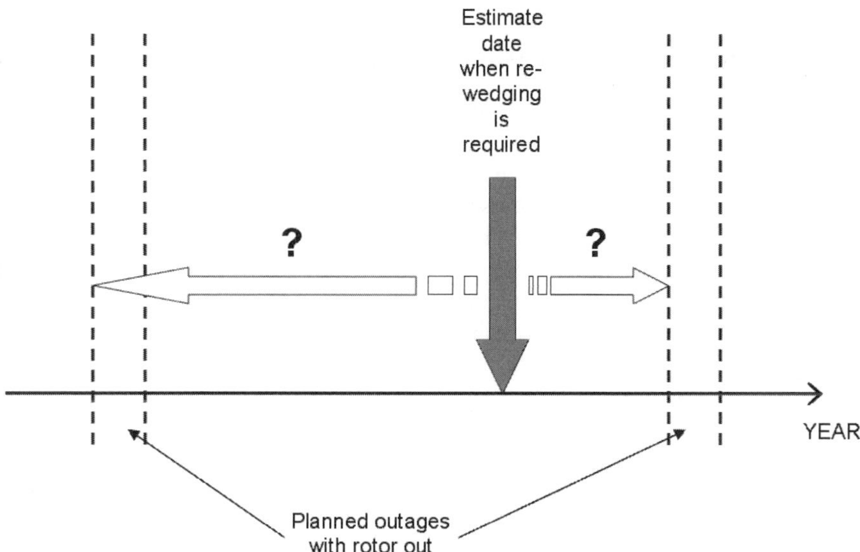

Fig. 12.2 Compromise between condition-based maintenance and planned maintenance requirements. In this example, based on CBM trending, the rewedge activity falls between two outages with the rotor out. The most probable choices are moving the rewedge forward or back; each has its own technical and economic merits and demerits.

12.1.3 Predictive Maintenance

In predictive maintenance, the schedule is based mainly on statistical calculations. These calculations take into account parameters such as mean time to failure (MTTF) of critical components, age of the insulation components, type of insulation (insulation systems), load cycles, and abnormal operation events (e.g., short circuits close to the machine, motoring due to loss of turbine power, and asynchronous operation due to loss of excitation).

Probabilistic risk analysis (PRA) is the science that tries to establish reliability indexes based on stochastic analysis. As stated previously in this chapter, this type of analysis is common to nuclear power stations. It determines probable modes of failure for critical equipment. Some aspects of this analysis spill over into maintenance procedures. For a comprehensive discussion of PRA techniques applied to nuclear power plants, see References [3–6].

It is important to recognize that predictive maintenance, together with planned maintenance, cannot determine in most cases the optimal time to inspect, maintain, and refurbish a specific piece of equipment, in particular something as complex as a large turbo-generator. Planned and/or predictive maintenance has proved to be adequate over many years of operation. However, recent trends of deregulating the electric power industry are pushing utilities and independent power producers (IPPs) to do more with less. In the realm of maintenance, this translates into longer periods between turbine and generator inspections. The result may be lower reliability of operation and more catastrophic and expensive outages.

12.1.4 Condition-Based Maintenance (CBM)

CBM is the most recent approach to guide station personnel in determining when to inspect, maintain, and refurbish a generator and other plant equipment. As the name adequately indicates, condition-based maintenance operations follow a concrete need by a component or apparatus to be refurbished.

Condition-based maintenance can only be applied when equipment is monitored by a number of online, real-time sensors, as well as offline periodic testing routines. For example, the following parameters are candidates for continuous monitoring in a large generator: current, voltage, vibration, partial discharges, stator winding vibration, air-/gasgap flux density, gas humidity, decomposition elements of insulation in the cooling gas, water purity, temperature of core/windings gas, and so on. Additionally, other periodic offline test inputs to CBM activities are current spectrum (for squirrel-cage large induction motors, not generators), electric tests of insulation components, visual inspection of commutators and dc field collectors, and so forth.

Although requiring an initial higher capital investment in instrumentation, CBM is perceived as providing, in the long run, a more reliable and less expensive operation. In power plants with generators of 1000 MVA ratings and higher, the costs of forced outages are such that the expense incurred by the installation of the additional instrumentation is well warranted.

12.2 OPERATIONAL AND MAINTENANCE HISTORY

When planing the inspection and maintenance of a generator (or any other major electrical apparatus), it is critical to refer to the machine's previous operational and maintenance history. Following is a list of parameters that will influence the condition of the generator:

Operating Statistics

- Operating hours
- Number of starts/stops
  - Cold/warm/hot
- Stress events
  - Fast MW swings
  - Fast MVAR swings
  - Trips
  - Sudden forced outages
  - Load rejections
  - Generator rejections
  - Overspeeds
- Descriptive summary of operating problems

Outage and Maintenance Statistics

- Forced outages
- Unit deratings
- Equipment incapability
- Problem components (repairs and/or replacements)
- Internal data
- External data (NERC, CEA, EPRI, etc.)

Identified OEM Generic Equipment Problems

12.3 MAINTENANCE INTERVALS/FREQUENCY

The issue of inspection and maintenance intervals or frequency is one that has come to the forefront in recent years. The advent of deregulation, among other things, created an incentive to extend the period between inspection and maintenance. This important topic is covered in this book in a number of places.

Below we list the most important elements that are determinants of the maintenance intervals, based on the discussion of Section 12.1, above.

Maintenance Planning

- OEM recommendations
  - Generally based on time or operating hours and known generic problems

- Preventive philosophy
  Regular and/or scheduled maintenance
- Breakdown philosophy
  Maintain when forced
- Extending inspection intervals
  Condition-based maintenance facilitates achieving this goal without placing the equipment and operation at risk

Frequency Criteria for Planned Outages

- By number of years of operation
  Generally not practiced nowadays
- Operating hours (e.g., schedule a major outage after approximately 30,000 operating hours)
  Adjust frequency based on operating duty (i.e., increase the frequency on two shifting units)
- Governed by system schedules (i.e., major outage when unit not required by the system)
- Condition-based maintenance
  Applied in a growing number of power plants
  Based on the actual equipment condition
  Allows increased availability by maintenance interval extension and reduces station maintenance costs if a machine is in good condition
  Can also result in generators in poor condition, leading to the need to be overhauled more often out of necessity

12.4 TYPE OF MAINTENANCE

12.4.1 Extent of Maintenance

Following is a brief description of those elements included in a minor outage scope and a large outage scope of work. Obviously, in different conditions the same scope can take vastly different lengths of time to be completed. Thus, more than the duration of the outage, the scope of disassembly of the generator is the true indicator of the extent of the work.

Minor Outage

- No major dismantling of the machine (i.e., rotor not pulled from stator bore)
- Minor consumable component replacement (gaskets, O-rings, etc.)
- Visual inspection
- Usually only a few weeks duration

Major Outage

- Rotor removed
- Hydrogen coolers usually removed (on hydrogen-cooled generators)

- Nondestructive examination (NDE) performed on critical components
- Extensive inspection and testing
- Usually not less than two months long, but getting shorter
- Major repairs and replacements carried out
- Mostly carried out when similar major overhaul is being performed on the prime mover

Fig. 12.3 Exciter rotor. (a) Before and (b) after refurbishment.

12.4.2 Repair or Replacement

On certain occasions, there is no doubt that a faulty component must be replaced because of extensive damage (e.g., a retaining ring with a significant crack). However, most often the choice between replacement and repair is not so obvious. For instance, leaking hydrogen coolers can almost always be fixed or they can be replaced.

The existence of spare parts is a significant advantage by presenting an excellent option for a quick turnaround. The trade-off is unused capital sitting on a shelf. Obviously, it is a cost–benefit issue unless some other issues are involved, such as criticality of availability of that unit to the electric power system.

The following list enumerates those drivers that may come into play when making decisions regarding when to repair or replace, or what spares to carry.

Repairs

- When components are capable of being fixed in a timely fashion
- More economic than replacement
- When reliability of operation is not compromised to any significant extent

Replacement

- When components are not capable of being fixed
- When replacement is more cost-effective
- When reliability of operation could be compromised by repairing one critical component

Component Exchanges

- Carried out for both minor and major components
- From minor and major replacement spares
- Reduction in downtime
- Removed component can usually be repaired and put back in stock as a spare

Decision-Making Factors

- Cost of repair compared with replacement
- Scheduling (time to repair vs. replacement)
- Revenue loss (cost of downtime)
- Replacement energy (usually from more expensive source)
- Contracts in place for the delivery of power that must be honored
- Reliability (operating duty required, e.g., double-shifting)
- Risk to the system (How critical is the equipment to the interconnected network?)
- QA (checks and balances to ensure that the work is done right)
- Contractor selection (who gets the job) based on:
  - Can they do the job to the required standard?
  - Can they meet the required schedule?
  - Is the price right?

12.4.3 Rehabilitation/Upgrading/Uprating

On occasion, the opportunity arises to increase the rating of a turbo-generator unit. Normally, the mechanical components (boiler, reactor, turbines, etc.) are the main drivers for an uprate. If those are considered viable for an output increase, then the next layer of equipment to be investigated is the one comprising the main and auxiliary transformers and generator (see Section 12.8).

Uprating a generator to any significant degree always requires performing calculations to extrapolate operation out to the higher load, based on heat-run testing of the machine at its present maximum load. This basically consists of running the same computer program OEMs use for new machines with the existing and uprated data. Once the redesign is performed, the next decisions to be made are what components must be replaced, which ones must be modified, and what are the costs and time schedule. Given the scope of redesign, it is customary for the operator to go back to the original OEM for the redesign effort, or to the vendor that has inherited the original data following the many mergers and acquisitions that the industry undergone in recent years. The larger the unit, the more critical it is to make sure that the original design and construction data are in possession of those redesigning the generator.

Uprating is not too common an occurrence for turbine-driven units. However, upgrading the generator is certainly a very common event. Upgrading is indicated by a need or desire to increase the availability or reliability of the unit by using components designed and manufactured to newer standards.

Finally, rehabilitation is normally understood to be the refurbishment of the machine, basically with components and techniques identical or similar to the originals, with the purpose of bringing the generator back to its previous condition.

The following list summarizes the aforementioned.

Rehabilitation or Refurbishment

- Generally to the same design
- Some component replacement and some component repair
- Some new materials
- To maintain existing design performance capability
- To extend operating life
- To improve reliability of poor components

Plus Upgrading

- For design improvements
- To install improved materials of the same or new design
- To increase reliability
- To extend operating life

Plus Uprating

- To increase output capability
- Possible if improvements in design changes or materials allow it

Generator Uprating Example Figure 12.4 shows the increased capability of a generator in the reactive-power region of its capability characteristic, after the original field with B-class insulation was replaced by one with F-class insulation. This change permits the unit to output more MVARs for a particular MW load.

In Figure 12.5, the same machine has its capability increased in the MW region by installing a new stator winding with a higher-class insulation (for instance, moving from class B insulation to class F, or from F to H). In this case, note that the rated power factor line is less steep; namely, the point where the field and stator limiting curves meet results in a somewhat higher power factor.

If both the stator and field windings are upgraded from, for example, class B insulation to class F, the result is a generator with a larger capability for delivering active and lagging reactive power (Fig. 12.6). Note that the leading region of the capability curve does not change by changing winding insulation. This region is basically limited by the end-core phenomenon, which does not directly depend on winding ratings.

It is important to recognize, as stated earlier in this chapter, that increasing the rating of a generator requires more design work than simply improving the insulation.

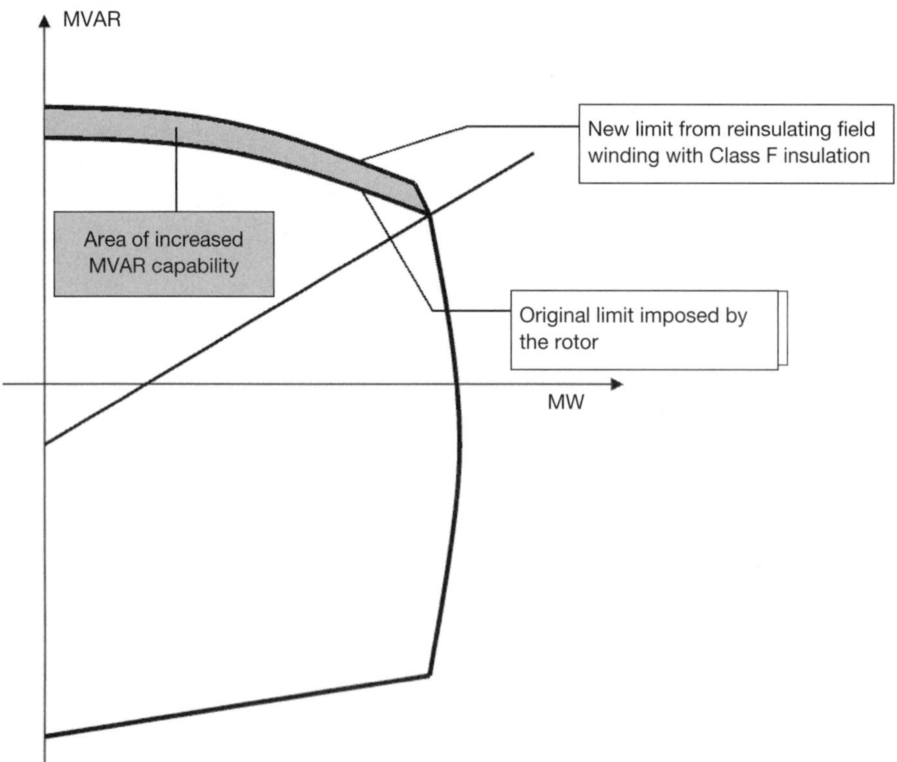

Fig. 12.4 Uprating the capability of a generator in the reactive region by rewinding its field with a higher class insulation, in this case, from class B to class F.

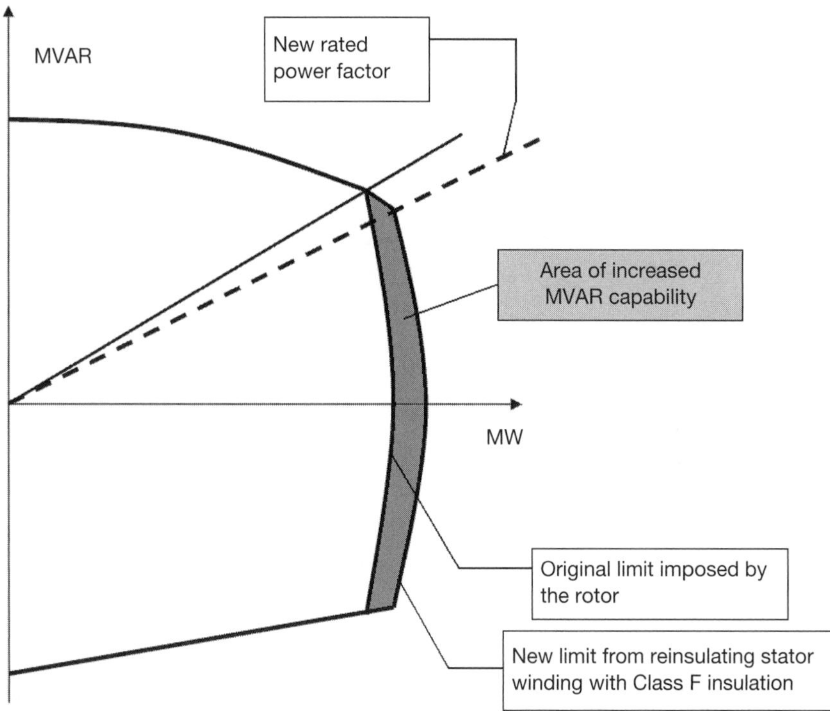

Fig. 12.5 Uprating the capability of a generator in the active region by rewinding its stator with a higher class insulation, in this case, from class B to class F. The new rated power factor is somewhat higher.

One must ascertain that the extra power delivered to the shaft, frame, frame support, and other mechanical components are within the capability of these components. Therefore, it is always advisable to engage the OEM or other capable organization when attempting such an uprate. See Reference [7] for an elaborated treatment of the subject of uprates and other refurbishment options.

12.4.4 Obsolescence

Obsolescence is a major factor for consideration when preparing long-term life-cycle plans (see Section 12.10).

Obsolescence affects every component of the generator and auxiliary systems. However, industry experience clearly shows that the areas more impacted are the excitations systems and protection devices. It is not uncommon for a generator to have part or all of its excitation systems replaced 15 to 20 years after commissioning of the unit. In particular, the automatic voltage regulator (AVR) is most impacted by obsolescence problems. The key reason is the speed in which electronic and power electronic technology is changing, compared with the established and slow-moving art of designing

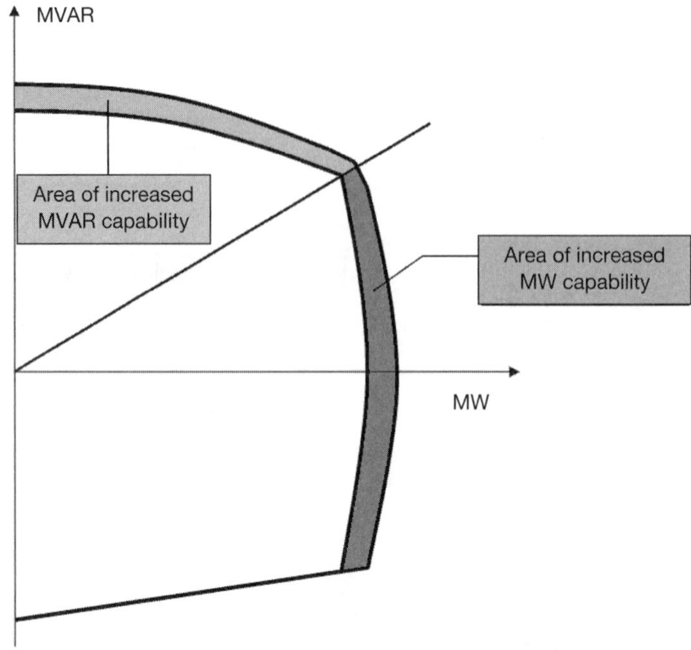

Fig. 12.6 Uprating the capability of a generator in both the reactive and active region by rewinding its field and stator with a higher class insulation, in this case, from class B to class F.

and manufacturing generators and auxiliaries. It is not uncommon for an AVR to become impossible to maintain properly because components are not made anymore, or a laptop running on DOS with the AVR software is broken and no modern laptop will do the job without errors, or the OEM simply does not support this particular design anymore. In fact, otherwise very good AVR systems are often replaced because of lack of OEM support or a dearth of available spare parts.

Obsolescence also affects large equipment of the excitation system, such as main exciters, permanent magnet generators (PMGs), or high-frequency generators (HFGs). Many a station has replaced old rotary exciters with static excitation systems. Many things must be considered before taking this route.

Protection is also affected by obsolescence. Old discrete relays do not have adequate spare parts. The alternative is then changing for new, mostly solid-state instruments, or, as done quite often, changing the entire protection system made up from discrete relays with one or two multifunction relays (see Chapter 6 for examples of discrete and multifunction relays). There other many components and subcomponents that may be affected by obsolescence. Some may require major capital investments and will only be addressed if they fail or are about to fail, and better alternatives are available. For example, an old thermoplastic (asphaltic) stator winding will not be replaced in-kind, because, if nothing else, that type of insulation is not supported by any

OEM or third-party vendor. The same would apply to 18–5 retaining rings. There is no logical reason for anybody to buy a new replacement made of 18–5 material, but even if one wanted to purchase such a ring, there is no reputable manufacturer to make one. Other components affected by obsolescence are generator condition (core) monitors (GCM), hydrogen dryers, stator cooling water systems, rectifying diodes, end-winding support systems, rotor winding insulation and blocking material, and so on and so forth. Obsolescence may also be driven by environmental issues. For instance, old insulation and blocking containing asbestos cannot be replaced by same material.

12.5 WORK SITE LOCATION

Where to perform certain repairs or refurbishments is always an issue for consideration. For example, consider a full rotor rewind. It can be performed at the site, if proper conditions and space are available, or in the vendor's factory. If done at the site, the risk and cost of shipping the rotor can be avoided. This can be a substantial argument if the rotor must be shipped thousand of miles. However, an in-plant rewind will not be full-speed balanced. The result could be a rotor running at a somewhat higher level of vibration than it would if it were rewound at the shop, or having to use trim balancing to achieve acceptable operation. Trim balancing efforts can sometimes be time-consuming and costly to the utility. Other topics to be taken into consideration are QA in situ versus at the shop, cost of labor in situ versus at the shop, expertise availability on a daily basis, and so on. To summarize:

On Site

- Minor and major outage work
- Both generally done on site if possible

Off Site

- Transport equipment to external facility
- Usually for major work such as rotor rewinds
- Usually means the job cannot be done on site
- Or is more cost effective to do off site

12.5.1 Transportation

Transportation is a ubiquitous activity in power plant experience, specifically during construction and outages. It should be something that is not difficult to do in an orderly and safe manner. Nevertheless, it is amazing that so often we hear about a transformer that hit a bridge, or a generator that was excessively bumped during transportation by train or truck. In many cases, exhaustive care is taken by the generator and turbine personnel to minimize the risks during major work on major components, only to later see the equipment being damaged during a transportation incident. Figure 12.7 shows such an unlucky occurrence.

Fig. 12.7 Large turbo-generator rotor that fell off a truck during transportation from a repair facility.

The corollary is that management and technical personnel engaged in shipping large components such as winding bars, cores, and rotors should make sure that a lot of effort and money (and time!) will not go down the drain because of lack of oversight during the preparations for transportation and the moving itself.

G-recorders (or "bump" instruments) can be attached at one or more points of a load to be shipped by truck or train. Nowadays, such devices can have GPS and transmitting systems built-in, allowing oversight of location and transportation conditions, over the Internet, on a continuous basis (see Fig. 12.8).

12.6 WORKFORCE

The following list summarizes the various personnel that may be involved in maintenance activities and related issues. Note that consulting a third-party expert (if such expertise is not available in-house) can result in big savings to the plant. OEMs are by nature of their business very conservative (there is little or nothing to gain for an OEM by "taking risks," i.e., for proposing technical solutions other than the most reliable ones, which tend to be also the most expensive). A third-party expert can help to balance the picture, so that the scope of repairs and/or refurbishment is the most econom-

Points on map of a rotor being transported by rail, automatically generated by the bump instrument and dedicated software. Additional information received indicates if any event over the acceleration threshold occurs.

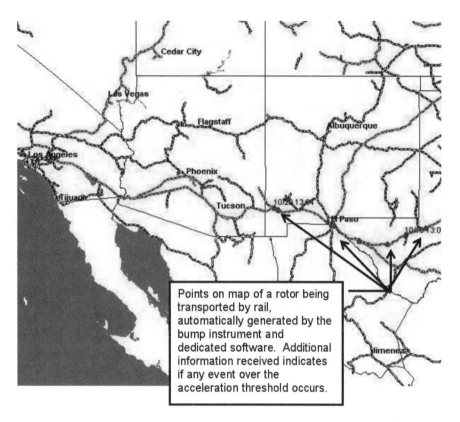

Fig. 12.8 Automatic indication of transportation location by a bump device attached to the cradle of a large rotor being moved by train.

Fig. 12.9 Complete generator loaded onto a Schnabel rail car and ready for transport.

Fig. 12.10 1,300 MVA, four-pole rotor transported by rail in cradle.

ic to the station—one that balances reliability with cost. The goal is to keep costs down and maintain the station's control of the work and outside workers.

In-house

- Station staff
- Other internal resources (e.g., head of office engineering, NDE specialists, scheduler, and outage manager)

OEM

- Major manufacturer
- Advantage of having specific knowledge about the equipment
- Specialized people not available within the operator's own organization
- Specialized tools
- Generally more expensive than in-house resources
- Less risk than using internal and/or non-OEM personnel
- Usually the highest work quality

Specialist Contractors

- Competitor (to OEM) repair facility
- Function-specific contractor (e.g., one that specializes in retaining-ring removal)

- Usually have high-level knowledge about the equipment (oftentimes have previously worked in OEM companies)
- Specialized people (experts)
- Specialized tools (if too expensive to keep at the station for odd jobs that happen once in the lifetime of the unit)
- Generally a lower cost option
- Generally more limited resources than the OEM
- Cannot always cope with the unforeseen (this issue keeps coming back when the scope of work is relatively complex)
- Station must be selective about what work is let out to third-party contractors
- Based on past performance and knowledge of their capability
- May be in conflict with OEM long-term warranties, if and when they still apply

12.7 SPARE PARTS

The list of spare parts a station carries (or should carry) for a particular generator depends on a number of things. For instance, type of unit, number of identical units at the plant, and number of identical units in other stations within the same utility. Other factors may be the size of the unit, the criticality of the unit to the system or the owners, the distance from OEMs or distributor centers, and so forth.

The OEM normally furnishes the basic spare-parts list with the unit. Over the life of the generator, and as the machine ages, the owner may elect to add additional parts to the spare-parts inventory. For instance, with a number of aging identical units, the owner/operator may choose to purchase a new rotor. This new rotor will be used to start a replacement/rewind round, which eventually will end with all rotors in operation having new field windings, plus a spare (with or without a new field winding). Similar activities can be performed regarding hydrogen coolers, terminal bushings, standoff insulators, and so on.

The following is only a summary of what may be included in a spare-parts list:

Minor Spares

- Parts that are needed for regular maintenance
- General consumable materials (e.g., gaskets, O-rings, bolts, and stator rewedge material)
- Parts with high failure or wear rates (e.g., slip ring brushes, ground brushes, and brush holders)
- Parts identified as having the potential to create outage extensions if not available (e.g., insulation tapes, bell-shaped (spring) washers, and PTE hoses)
- Surge capacitors (for those units where used)
- Thermocouples and RTDs (resistance temperature detectors)
- Usually these minor spares are of low cost

Major Spares

- Spare rotor
- Spare stator
- Spare stator bars (including at least two bottom bars and a number equal to those required to lift in the event a bottom bar must be replaced)
- Standoff insulators
- Terminal bushings
- Current transformers (CTs)
- Bearings and/or major bearing components
- Hydrogen seals and/or seal components
- Collector rings and collector-ring sleeve insulation
- Normally these will be of medium to high carrying cost

Auxiliaries

- Excitation components (diodes, rectifiers, etc.)
- Enough spare parts for every single auxiliary. (It would be unfortunate if a station that maintains a long list of major generator spare parts was forced into a long outage, e.g., because a critical lube-oil pump became disabled and spare parts were not available; or a hydrogen dryer was forced out of operation for a long time because a small, inexpensive, but otherwise difficult-to-obtain part was not in stock. The list of these examples can be long.)

12.8 UPRATING

There is considerable interest in today's power industry in trying to uprate existing generators to achieve higher outputs with minimal cost input. In doing so, there is always some trade-off, whether it is in the cost needed to refurbish the existing equipment or suffering advanced aging and reduction in usable expected life and reliability. This section outlines the process involved and problems with carrying out uprating of large turbo-generators.

12.8.1 Drivers for Uprating

The motivation for uprating of any plant equipment is to achieve higher overall power output from the plant. With regard to the generator, this can mean up to as high as 15% additional MVA capability, depending on the existing margins of the generator and its auxiliaries, and the additional output that could be achieved with modifications or upgrading of individual components. In addition, the other consideration is usually to extend the life of the plant and basically achieve a second life for the equipment. Therefore, uprating also includes upgrading of the components to achieve both the increased output and the second life of operation.

Upgrading is a common activity in recent times because of improved technology that allows life extension of generators and other plant equipment by replacing components that limit life, while taking advantage of the inherent ability of other components to operate indefinitely. For example, a rotor forging can often operate for a few lifetimes, but the copper winding and the insulation in the rotor wear out far quicker than the steel forging, which may be capable of operation for decades. Therefore, if one replaces the winding and insulation (a rotor rewind), a second life is easily possible.

The question now arises as to whether there is the possibility of achieving higher output from the equipment, and in this case it is more than just the ability to replace components and keep running. It now means a careful investigation of boilers, reactors, turbines, and the generators and auxiliaries to determine if there is margin for a power increase. If there is, then it may mean significant upgrades to existing components to get a higher rating. For the generator, there are some basic design issues that cause immediate limitations and these need to be studied carefully to determine the extent of uprating possible. The existing stator core and frame and the rotor forging may have specific geometries that cannot be altered unless they are replaced, so any uprating without replacing these components limits what can be done.

When the stator core is reused, it has to be fully tested to ensure that it can operate for another lifetime. Usually, this means another 25 to 30 years. If the stator core interlaminar insulation is in good enough condition for a second lifetime, then a stator rewind is generally required since it is the winding that wears out and largely governs the stator life. The stator frame and casing are not usually a concern but these also need to be checked. Once it is decided to rewind the stator, then a design study must be carried out to determine what changes can be made to the winding design, within the existing geometry, to determine how much additional power output is possible. Of course, there must be coordination with any turbine and boiler upgrading for determination of overall output, but there will be some machines that have more capability than the turbine upgrades will provide, and cases in which the generator will be the limiting factor. Regardless of which component is the limiting factor in any uprating endeavor, it is the most common practice to carry out improvements and modifications to the generator to achieve as much output capability as possible. The premise being that, "you never know what will happened in future . . . so be prepared." Stator and rotor rewinds will generally will not add much incremental cost.

Another factor in deciding to uprate is whether there are so many modifications and upgrades being done that it is simply more cost-effective to replace the generator or one of the major components, rather than reuse and improve the existing components. For instance, if the stator core or the rotor forging do have longevity or long-term reliability issues and have to be replaced in addition to new windings, the incremental cost for that work shown in a "cost/benefit" study may lead to the conclusion that replacement is the best solution for upgrading and uprating.

As well as the above issue, it may be that the existing turbine, boiler, and other equipment can all be uprated, with some modifications and component replacements, but the generator cannot be uprated in its existing geometry to meet the uprating available from the rest of the plant. In this case, the overall economics may dictate generator replacement.

Given that an uprating is possible, the extent of the uprating then depends on the individual machine and what is possible to do to achieve a higher output. This requires a design study to determine what needs to be done. The following sections provide some insight into how this process is carried out and all the considerations that are involved. As a general rule of thumb, the breakdown in Table 12.1 shows the normal range of modifications and technical issues that need to be dealt with for all the generator and auxiliary components in a machine with water-cooled stator windings, for upgrading to as much as 15%.

12.8.2 Uprating Considerations

There are three major concerns with regard to the generator for uprating and a number of additional concerns regarding the auxiliaries and operational parameters. The main concerns for the generator are:

- Temperature increase in the various components
- Increased mechanical stresses due to higher electromagnetic forces
- Increased vibration at higher output

Temperature Increase. The insulation systems are the most important items to be evaluated. Regardless of whether the temperatures exceed the design limits after uprating, higher temperatures will lead to decreased lifetime of the insulation due to heat deterioration. In the cases in which the machine has already operated for many years, it may even be necessary to consider stator and/or rotor rewinding, depending on the future operation that is under consideration. If the machine is near the end of its design life, the probability of this being required becomes higher.

Both rotor current and stator current are increased in any machine uprating. Therefore, the rotor and stator windings generally become the critical components in the

Table 12.1 Table of modifications and changes required for uprating the individual major components a generator with direct water cooled stator windings

<table>
<tr><th>Component/System</th><th>5%</th><th>10%</th><th>15%</th></tr>
<tr><td>Stator winding</td><td>Existing component (some modifications)</td><td colspan="2">Stator rewind (new design winding)</td></tr>
<tr><td>Rotor winding</td><td colspan="2">Existing component (some modifications)</td><td>Rotor rewind (new design winding)</td></tr>
<tr><td>H_2 Coolers</td><td colspan="3">Existing equipment</td></tr>
<tr><td>Stator cooling water system</td><td colspan="2">Existing equipment (control changes)</td><td>Higher capacity system</td></tr>
<tr><td>H_2 cooling system</td><td colspan="2">Existing equipment (control changes)</td><td>Higher capacity system</td></tr>
<tr><td>Excitation system</td><td>Existing equipment</td><td colspan="2">New and larger exciter (power system stabilizer required)</td></tr>
<tr><td>Generator protection</td><td colspan="3">Recoordination of generator protection (new settings)</td></tr>
<tr><td>System stability</td><td colspan="3">Generator and power system stability study required</td></tr>
</table>

consideration of uprating. It then becomes necessary to evaluate the temperature rise of both the rotor and stator windings, respectively. Increased temperature rise in these components during operation will accelerate deterioration of the insulation systems and lead to shortened expected remaining life.

Electromagnetic Forces. With regard to electromagnetic forces, generally there are two concerns and these are mainly for the stator winding. The first is increased radial bar bounce forces in the slot and the second is stress on the stator end winding.

The stator bar bounce forces in the slot are a function of the stator current squared, and the main concern is the ability of the stator winding wedges in the slot to accommodate the higher forces.

With regard to the effect on the stator end winding, it must be considered whether the existing support structure of the stator end winding is strong enough to restrain the higher forces (also proportional to the stator current squared) and vibrations while allowing the winding to thermally expand. Strengthening of the support structure may have to be considered, depending on the degree of uprating. And in addition to strengthening the support structure, the natural frequencies of the end winding as a whole and the high voltage bushings must also be considered. Design changes in the stator winding and support systems may alter the natural frequencies and cause some of them to be too close to the once- and twice-per-revolution frequencies—50 and 100 Hz for a 50 Hz system and 60 and 120 Hz for a 60 Hz system.

Excessive Vibration. For increases in field current, it is necessary to consider the additional vibrations due to thermal elongation of the rotor winding. The increase in temperature in the rotor copper may cause excessive elongation to occur beyond the winding packing clearances; then deformation of the copper would be possible. The ability of the rotor winding to accommodate the additional elongation should be checked.

Additional Concerns. Besides the above considerations about the generator itself, there are auxiliary equipment issues (electrical and mechanical), protective relaying, and stability of the electrical system that have to be considered as follows.

EXCITATION CAPACITY. For an increase in field current, it is necessary to ensure that the excitation system has the current capacity and ceiling voltage capability for the higher output and still remain within design temperature limits.

COOLING SYSTEMS. To suppress the general temperature rise of the hydrogen cooled components, one of the effective methods is to increase the cooling water flow to the hydrogen coolers or increase the hydrogen gas pressure in the machine if allowable.

For the stator conductor bars, increasing stator cooling water flow is also an effective measure to reduce or maintain the temperature rise in the stator bars to some degree.

In both these cases, it is necessary to consider the capacity of the coolers, water supply, and pump.

PROTECTION LIMITS. For uprated operation, it is also necessary to review the protection limits for the stator bar temperatures, overcurrent protection relay, field current, and so on.

STABILITY OF THE ELECTRICAL SYSTEM. In general, system stability is decreased by an uprating operation. It is necessary to consider the generator constants at the uprated condition. The short-circuit ratio may change with significant output increase and this will have an effect on how the generator responds during system events. System studies are often necessary to evaluate any changes required to accommodate proper interaction between the uprated generator and the system.

12.8.3 Component Evaluations

In the uprating process, the following components of the generator are required to be evaluated in detail, to determine the effect of an increase in power output of the machine.

- Temperature rise of the stator winding
- Temperature rise of the field winding
- Electromagnetic forces on the stator bars in the slot and the end winding
- Stator terminal capability
- Excitation capacity
- Stator water cooler capacity
- Stator water cooling pump capacity
- Hydrogen cooler capacity
- Short-circuit ratio and generator constants (for system stability evaluation)
- Core-end cooling
- Isolated phase bus capability
- Main transformer rating
- Main circuit breaker rating

Temperature Rise of the Stator Winding. To evaluate the stator winding temperature rise, all temperature data should be converted to the rated load condition.

Armature current (I_a) is evaluated from the actual data of MW, Mvars, and voltage:

$$\text{Stator current, } I_a = \frac{\text{MVA}}{\sqrt{3}V_t}, \qquad \text{amperes}$$

$$\text{Apparent power, MVA} = \sqrt{(\text{MW}^2 + \text{Mvar}^2)}$$

where
MVA = rated apparent power
MW = real power output
Mvar = reactive power output
I_a = stator current
V_t = terminal voltage

Converting to the new uprated condition increases some parameters, but not others. Stator current and apparent power increase, but terminal voltage remains the same because the winding configuration is not changed. To maintain the same temperature rise of the stator winding, the cooling water flow will need to be increased as shown in the table below.

Parameter	Original	Uprated
MVA	A	$A + x\,\%$
V_t	B	B
I_a	C	$C + x\,\%$
SCW flow F (L/min)	D	$D + y\,\%$
SCW inlet temp (°C)	E	E

To obtain an estimate of the increase in stator winding temperature (T) at the new output, use the following equation:

$$\text{SCW } T_{\text{out uprated}} = \left[(\text{SCW } T_{\text{out rated}} - \text{SCW } T_{\text{in rated}}) \times \left(\frac{I_{\text{a uprated}}}{I_{\text{a rated}}}\right)^2 \times \left(\frac{\text{Flow}_{\text{rated}}}{\text{Flow}_{\text{uprated}}}\right)\right] + \text{SCW } T_{\text{in}}$$

The results of the above require the following to be considered:

- Reducing the stator cooling water inlet temperature
- Increasing the stator cooling water flow
- Checking the stator cooler capability
- Improving the stator end-winding support

Temperature Rise of the Rotor Winding. The original rated condition of the rotor winding is

Original rotor winding temperature rise at rated output = X °C (at rated MVA)

Evaluation for uprating provides the following:

$$\text{Temperature of uprated rotor winding} = \left[\text{rated temperature rise} \times \left(\frac{I_{\text{f uprated}}}{I_{\text{f rated}}}\right)^2\right] + \text{cold } H_2 \text{ gas temperature}$$

The allowable temperature rise by IEEE C50.13 is 65°C for B class insulation.

Uprating is generally allowable if the hot-spot temperature is below the 65°C rise; however, the deterioration of rotor winding insulation will increase at higher operating temperatures.

To reduce the effect, reduce the hydrogen cold gas temperature if possible. Generally, it is usually possible to decrease it by about 5°C without significant detrimental effects from the increased thermal differential between various components.

The other possibility is to increase the hydrogen gas pressure. However, this will depend on the ability of the casing, end doors, fittings, flanges, and, most importantly, the hydrogen seals to handle any increase. This is likely to be only a marginal allowance and would generally be considered along with the temperature as follows:

$$\text{Temperature of uprated rotor winding} = \left[\text{rated temperature rise} \times \left(\frac{I_{\text{f uprated}}}{I_{\text{f rated}}}\right)^2 \times \left(\frac{\text{H}_{2\ \text{pressure rated}}}{\text{H}_{2\ \text{pressure uprated}}}\right)^{0.8}\right] + \text{cold H}_2 \text{ gas temperature}$$

Electromagnetic Forces on the Stator Endwinding. This is usually determined by performing finite-element analysis. From this, the electromagnetic force on the stator bar ends is evaluated.

For uprating from rated condition to an increase in stator current of 5%, the increasing rate of electromagnetic force is generally in proportion to the square of the stator current and would be in the range of about 10% in this case.

Increasing the electromagnetic force affects the stress of the endwinding support structure of the stator end winding and increases it in the same proportion.

It is also necessary to consider natural frequencies as well. Operation at or near resonant frequencies will cause excessive vibration and advanced deterioration of the stator winding insulation and/or the support structure.

Stator Terminal Capability. This should be evaluated in the same manner as the stator winding above, to ensure that the temperature limits are not exceeded and that the terminals can withstand the additional forces during stress events.

Excitation Capacity. The exciter capability must be checked to ensure that it can deliver the current required for an X% increase in output of the machine. It must be capable of doing this without compromising the temperature limits of the excitation equipment or the performance of the overall machine in terms of system response and stability.

Stator Cooling Water Capacity. If the losses to the stator cooling water cooler are 100% at the original output, then the increase in losses due to uprated output is

$$\text{Increase in losses from uprating} = \left(\frac{100\% + X\%\ \text{MVA}}{100\%\ \text{MVA}}\right)^2$$

For example, if the increase in rating is 5%, then the increase in losses to the stator cooling water coolers would be

$$\text{Increase in losses from 5\% uprating} = \left(\frac{105\%}{100\%}\right)^2 - 100\% = 10.25\%$$

Therefore, it is required to study the stator cooling water coolers to see if they can handle the increased losses within the existing stator cooling water flows. If not, then it is required to determine the amount of stator cooling water flow required to maintain design cooler limits. Next, determine whether the stator cooling water pumps can deliver the increased flow.

Stator Cooling Water Pumps. For an uprating, where the loss of stator coil increases 10%, for instance, the following must be considered.

To keep the stator coil temperature the same as the original temperature, it is required to increase the stator water flow to 110%. If the rated flow of the cooling water pump = X gallons/min, the necessary cooling water flow at the original generator output consists of

$$\text{Flow to stator coils} + \text{flow to deionizer} = \text{total flow} = X\ \text{gallons/min}$$

Then, in the case of uprating:

$$\text{Total flow} \times 1.10\ \% = \text{new flow required}$$

The pump capacity for the existing SCW pump would have to be capable of handling this new flow. If it cannot handle the new output, it should be checked to see if the plant cooling water temperature for cooling stator water can be reduced to keep the original water outlet temperature the same.

Hydrogen Coolers. If the loss to the hydrogen coolers is 100% at original output, then for uprating by X% the new loss will be 100% plus an additional amount that includes the rotor copper losses, core losses, and stray losses. The increase in losses will not be directly proportional to the amount of uprating.

The basic requirement is to determine if the coolers have any original margin that would accommodate an increase of X%. It may be that the hydrogen cooling gas temperature needs to be reduced by a certain amount to maintain the original rotor winding temperature rise if no margin exists. This may be accomplished if the plant cooling water temperature for the hydrogen coolers can be reduced by that amount.

Generator Constants for System Stability Evaluation. Various generator constants at rated condition and uprated condition need to be evaluated by using values measured in factory tests before construction. They generally react as follows (from an actual example):

Parameter	Rated	Uprated	% Change
MVA	448	489	+ 9.15
AFNL (no load)	1225	1225	no change
AFSI (SC)	2375	2592	+ 9.14
SCR	0.516	0.473	- 8.33
Xd (sat)	194%	212%	+ 9.28
Xd' (unsat)	35.6%	39%	+ 9.55
Xd'' (unsat)	28.2%	31%	+ 9.93

Stability must be reevaluated because the characteristics change in the table above.

Core-End Cooling. The core-end cooling must be checked to ensure that it can operate under the insulation class limits specified for the specific increase in output of the machine. The manufacturer will likely have to be consultated on this as it is a complicated issue to resolve and must be done as part of the overall core-end design.

Isolated Phase Bus Capability, Main Transformer Rating/Main, and Circuit Breaker Rating. These are outside the realm of the generator calculations but need to be addressed in any uprating endeavor.

12.8.4 Reliability and Effect of Uprating on Generator Life

Stator Winding Insulation. The main deterioration factors affecting machines are absolute temperature and thermal cycling effects, electrical effects, mechanical effects from vibration and high stresses, and environmental effects of the generator atmosphere.

The main deterioration factors that are further affected or accelerated by upgrading are the absolute thermal, thermal cycling, and mechanical effects.

Thermal deterioration increases if the temperature of the stator bar increases. Regarding thermal deterioration, it is said that the 10°C halving rule generally applies (the thermal life of the insulation is halved for every 10°C rise in constant operating temperature; see IEEE Standard 1). It is also known as the Arrhenius rule (Fig. 12.11).

Thermal cycling occurs when the temperature of the stator bars rises and falls due to start/stop operation and load fluctuations. During temperature rise of the stator bars, thermal expansion of the bars occurs. This thermal expansion produces shearing stresses due to the difference in linear expansion coefficients of the conductor and insulation, bending stress at the bar ends, and so on, so upgrading becomes a life-shortening factor.

Electromagnetic forces cause mechanical deterioration due to induced bar vibration. The stator bar ends outside of the core are fixed by supports and ties, and the straight portions of the bars in the core slots are restrained by wedges. By upgrading, the electromagnetic forces will be increased due to higher stator current. When the

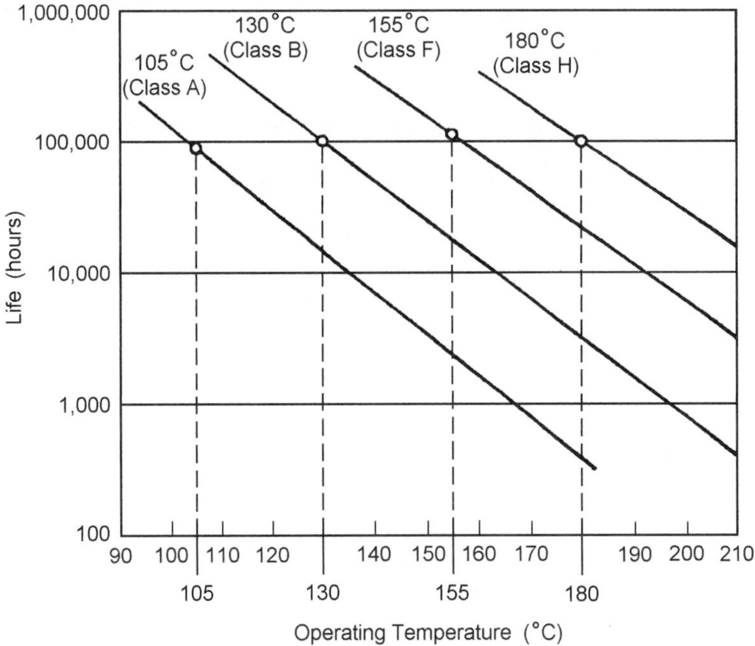

Fig 12.11 The Arrhenius rule graph shows the increase in insulation life when the operating temperature is reduced, for various classes of insulation.

bars restraints in the slot or end winding loosen, the vibration increases due to the electromagnetic forces. Vibration can cause rapid abrasion of the insulation and a high possibility of failure.

The electrical stress to the ground-wall insulating layer does not change after upgrading because the rated voltage does not change.

Environmental deterioration does not become a problem because the stator bars are operated in hydrogen gas, and because moisture is not absorbed.

Field Winding Insulation. The deterioration factors for the rotor insulation are basically the same as for the stator bar insulation. Electrical degradation and mechanical deterioration by electromagnetic force are, however, much less severe. Instead, mechanical deterioration is caused by the centrifugal loading applied on the rotor winding insulation.

The increase in excitation current makes the effects of thermal and heat cycle deterioration prevalent. It is generally seen that insulation life will be shortened more by thermal and heat cycle deterioration than by mechanical effects.

Stator Core Insulation. The stator core is also affected by thermal and heat cycle deterioration and mechanical effects. It is considered that the life of the stator core will be shortened by uprating, but usually this is marginal. In addition, the life of the stator core is usually longer than the stator bars and this means that the life of the stator core does not generally influence the lifetime evaluation of the generator.

12.8.5 Required Inspection and Tests Prior to Uprating

The components most affected by uprating amd upgrading need to be inspected and tested. In addition, opportunity should also be taken to inspect, observe, and test (as necessary) all other machine components that may not be as much affected by any upgrading of the generator. The following are the basic items to consider.

Stator

1. Insulation diagnostic testing is required as follows:

    - Insulation resistance and polarization index
    - Per-phase capacitance
    - Dissipation factor and tan δ tip-up
    - Dielectric absorption and dc ramp-up
    - Partial discharge
    - Corona probing

    The above range of testing comprises both ac and dc testing so that the dielectric (ac) as well as the resistive (dc) characteristics of the insulation are considered in the evaluation. Purely ac or purely dc testing is not sufficient for complete diagnostics.
2. Visual inspection of the stator bars and tapping test of the stator wedges will reveal how tight the stator bars have been held in the slots and if there may have been damaging relative movement. Signs of fretting debris and greasing are important clues. With regard to wedge tightness, specific criteria are applied as follows to determine if stator wedge retightening is required. Rewedging in a particular slot is required if:

    - Less than 75% of the wedges are tight in the slot
    - Three or more adjacent wedges are fully loose

    Retighten any fully loose end wedge. Rigidity evaluation of the stator end winding and support structure are required. This is accomplished by natural frequency measurements for the end winding and support structure.
3. El CID testing of the stator core interlaminar insulation
4. Leakage testing of the stator winding and cooling water system should be done by the following:

- Air pressure decay test with Helium leak detection
- Vacuum decay test

Rotor

1. On-line shorted turns detector test prior to shutdown (flux probe), (if installed)
2. Field winding dc resistance
3. Insulation resistance and polarization index of the field winding insulation
4. RSO (recurrent surge oscillation) test for shorted turns (offline, rings on, and can be done with the rotor rolling as well)
5. Visual inspection with retaining ring removed
6. Nondestructive inspection of the retaining rings by flourescent dye penetrant
7. Nondestructive inspection of the rotor wedges
8. Nondestructive inspection of the rotor shaft

12.8.6 Required Maintenance Prior to Uprating

Due to the fact that most machines being considered for uprating have been operated for many years, it is prudent to ensure that all existing components are in sufficiently good condition for a significant extended life. Therefore, it should be ensured that all the required maintenance and repairs are carried out on the affected components prior to any uprating.

In addition, any new technologies made available by the OEM should also be considered for application at the time of uprating. These may not be required for the uprating but may enhance generator performance and lead to extended life and reliability. The following inspection and repair items may be included:

1. Damaged and aged parts discovered in these inspections should be repaired, including but not limited to the following:

   - Loose stator wedges
   - Loose stator end windings and supports
   - Copper dusting in rotors
   - Shorted turns in rotors
   - Worn collector rings

2. Improvement of reliability by the adoption of new technology (where applicable) should be recommended:

   - Stator wedge system improvements
   - 18 Mn–18 Cr retaining rings

3. Improved generator component monitoring by new technologies available:

- Installation of partial discharge monitoring
- Installation of vibration-monitoring probes for the stators

12.8.7 Heat-Run Testing After Uprating

Heat run tests should be made to determine the temperature rise of the various components of the generator, to ensure that the uprated equipment meets the temperature specifications.

Full-load heat-run tests should be performed with the generator synchronized to the system and operating normally at maximum rated MW load and rated hydrogen gas pressure. At each test load, generator operation should be held constant for at least 30 minutes to allow all generator internal temperatures of the machine to stabilize. Readings should be taken of all generator and auxiliary parameters monitored (as listed below).

Three different load points are generally required as a minimum, to obtain information about the generator performance at full load and to determine that the upgrading has been done successfully, without exceeding the specified temperature rises. These should be the minimum leading power factor load point, the unity power factor load point, and the maximum lagging power-factor (rated) load point.

During all tests conducted, the following parameters should be recorded after 30 minutes stabilization, for each load point.

Electrical Parameters:

- Power, MW
- Reactive power, MX
- Terminal voltage
- Three-phase currents
- Field voltage
- Field current
- Rotor winding temperature
- Shaft voltage
- Exciter power

Hydrogen System:

- H_2 pressure
- H_2 purity
- H_2 dewpoint
- Fan H_2 differential pressure
- All cold H_2 gas temperatures
- All hot H_2 gas temperatures

Hydrogen Coolers:

- Service water inlet temperature to coolers
- Service water outlet temperature from all coolers

Collector Rings and Brush Gear:

- Cooling air inlet temperature
- Cooling air out temperature

Bearings:

- Lube-oil inlet temperature
- Lube-oil outlet temperature
- Bearing-metal temperature
- Bearing vibration

Hydrogen Seals:

- Metal and oil temperatures

Seal-Oil System:

- Seal-oil supply pressure
- Seal-oil inlet pressures
- Seal oil/H_2 differential pressure
- Generator seal-oil inlet and outlet temperatures

Stator Cooling Water System:

- Generator bulk SCW inlet temperature
- Generator SCW outlet temperatures
- SCW conductivity
- Generator SCW inlet and outlet pressures
- SCW coolers, inlet, and outlet temperatures
- Service water to SCW coolers, inlet and outlet temperatures

Stator Winding Temperatures:

- All hose outlet thermocouples or RTDs
- All stator slot thermocouples or RTDs

Stator Core Temperatures:

- All stator core thermocouples or RTDs
- All core-end clamping-plate thermocouples
- All core-end clamping-plate flux screen thermocouples

12.8.8 Maintenance Schedule After Uprating

For reliable operation after uprating and upgrading of the equipment, it is important to inspect for the effect after all the modifications and work are complete. This should be done fairly soon afterward and are somewhat like of warranty-type inspections to determine that all components are working properly. If some problems are found in this inspection, it is necessary to implement corrective action early. This is to reduce the possibility of shortening the life if problem issues remain unattended.

It is generally recommended to carry out an overhaul one year after upgrading and to carry out a major overhaul between 5 (30,000 operating hours) to 8 (50,000 operating hours) years thereafter. For machines with extensive two-shifting, it is recommended to make the inspection closer to the 5 year mark.

For all these inspections, the following is prudent:

- Remove the rotor
- Inspect the inside of the generator visually
- Detailed inspection of the stator and wedge tapping test
- El CID test for the stator core
- Insulation diagnosis test for stator coil
- Nondestructive test for the rotor
- Detailed inspection of the rotor without pulling out the retaining ring

12.9 LONG-TERM STORAGE AND MOTHBALLING

This section discusses the protection of large steam turbo-generators and their auxiliaries when not in use and during storage. It is intended to provide guidelines to adequately protect the generator equipment when mothballed for long periods of time in the unit position or during storage as major spares. Details regarding the requirements for preparation of the equipment for storage, monitoring during storage, and returning the equipment to service are provided.

12.9.1 Reasons for Storage of Generator Equipment

There are two general conditions that may require the storage of large steam turbo-generators and auxiliaries. These are (a) installed in the unit position but incurring delays during construction, or for long-term mothballing, and (b) for storage as major spare components.

Construction Delays and Mothballing. This type of storage involves a generator installed but not put into service due to incompleteness of other essential systems in the station, or when a station is mothballed, requiring long-term storage of the whole machine. Such cases involve storing the machine and auxiliaries in their operational position. The following sections provide the requirements for putting the equipment into storage, monitoring its condition during storage, and for returning the equipment to service. The instructions also deal with the various time spans over which storage will take place.

Major Spare Components. Major spare components that may be stored for an extended and indefinite period of time include stators, rotors, and spare stator windings. These are stored under warehouse-like conditions. A different set of precautions and procedures are required for putting the equipment into storage, monitoring its condition, and removal from storage in preparation for installation. The following sections describe the various methods for both caged and integral machines under these conditions.

12.9.2 General Requirements

General Instructions for Equipment Arriving on Site. The generator components and auxiliaries should all arrive at site with the manufacturer's packing intact. All parts should have been maintained completely dry and clean from the time of dispatch to the point of arrival. A detailed record of the condition of all parts at the time of dispatch and a detailed log of transportation, including incidents that might disturb components or packing, should be available for inspection upon arrival at the site.

Once at the site, a thorough inspection of shipping packings and protective coatings should be made to determine the condition of the equipment. All packings should be

Fig. 12.12 Cradle of a large four-pole rotor being opened for inspection after arrival at shop, to be stored for a year.

opened, inspected, and tested where required, and resealed. This should be done in the presence of the manufacturer's representative for the purposes of the warranty. Any damage should be catalogued and repaired or replaced.

If possible, it is preferable to install and operate all components to ensure that they are in working condition and guarantee commencement of warranty. When this is not possible, as in the case of spares, preparations will be required for storage, and special precautions and procedures will be required. When equipment is installed, it may also have to be stored or mothballed to lie in wait until required for service. This will require different storage methods.

General Storage Environments. For any storage of electrical equipment, there are basic environmental requirements. Electrical equipment in general should be stored indoors. Temperature and humidity should be controlled at typical room temperatures of about 22 ± 5°C. Dust and dirt should be excluded by some method of covering.

All components must be well supported to keep them above ground or floor. It is desirable to use timbers for main support on unmachined or other noncritical surfaces where possible. Where equipment is supported on timbers, waterproof waxed paper should be used to isolate the support surfaces.

Whatever the packing or covering method, provision should be made for access for inspection and testing while ensuring the exclusion of rodents and reptiles.

Components stored as spares should be located in areas with the least amount of traffic as possible and fenced off to keep out unauthorized personnel.

Particular precautions and environments will be dealt with in the text, along with the methods of establishing them.

Protection of Generator Components During Installation. During preparation for installation, the protective coatings on the generator components are removed, leaving them vulnerable to damage. For this reason, the scheduling should be arranged to minimize the time between unpacking the equipment and installation.

Particular attention must be paid to caged cores and generator rotors.

Caged cores should be kept covered by a lightweight tarp up to the point of sliding them into the stator casing, and great care should be taken in protecting the phase connections from contamination or damage. Generator instrumentation should be protected from damage during stator core installation. The casing should be thoroughly cleaned and inspected prior to the stator core installation and the ends covered at all times before the core is inserted, and after that, until time for rotor installation.

Rotors should also be covered with a lightweight tarp after unpacking, until they are slung for insertion. The collector rings and coupling faces should be covered at all times during installation until their involvement in the assembly.

12.9.3 Storage Requirements

Storage of Assembled Generators. The generator, exciter, and auxiliaries will be fully erected on their foundations and electrically and mechanically interconnected with other systems. Storage under this condition may be for short periods, typically up to two years, during construction phases or for long or indefinite time periods during mothballing when not required by the power system.

STATOR. The assembled stator will be a sealed unit ready for operation. For the machine that is newly installed and lying in storage prior to service, this period is usually no more than up to twelve months during normal construction or up to three years due to in-service delays for various reasons. Regardless of the storage period, the machine needs to be protected from the environment.

One method is to place the stator under nitrogen blanketing to provide an inert, nonoxidizing atmosphere for best preservation. A quantity of silica gel should also be placed inside the casing of the machine to continually improve the relative humidity of the atmosphere within. The correct amount of silica gel will depend on the volume of the machine atmosphere and the specifications of the particular brand of silica gel to be used (Section 12.9.6). The nitrogen should be introduced once the silica gel is in place and the machine resealed. It should be kept at a few psi above the atmospheric pressure to ensure that there is no leakage into the machine. The seal-oil system must be kept fully operational to achieve this. The stator windings should be drained and vacuum dried to remove all moisture, as per Section 12.9.6. The stator winding insulation resistance should be measured and a polarization index set up, as per Section 12.9.6.

If the machine is provided with external heaters, they should be used to keep the stator about 5°C higher than the turbine-hall temperature.

For machines that are to be shut down and mothballed for an indefinite time period, the same procedure should be followed.

STATOR CASING. Integral machines are stators whose core frames and outer casings are one welded unit. Caged machines are stators whose core frames are removable from the outer casing. When storing either type as assembled units, no special precautions are required. It is recommended, however, to keep all terminal enclosures sealed, make a visual check of the condition of the instrumentation connected to the casing, and maintain the entire structure as clean as possible.

CORE END-PLATE FLUX-SCREEN COOLING SYSTEM. On machines with core end-plate flux-screen cooling systems, the system should be disconnected for draining and vacuum drying as per Section 12.9.6. This should be done for any length of storage.

The flux screen insulation resistance should be measured and recorded for electrically floating screens.

STATOR WATER COOLING SYSTEM. The stator water cooling system external to the generator, consisting of all pumps, coolers, filters, and pipework, should be disconnected from the generator under any storage condition and completely drained of water. This is in an effort to avoid deterioration of the copper and stainless steel internals of the system components.

All system valves should then be closed and the system reconnected to the generator. All organic-type filters should be removed prior to draining and drying and put away in storage.

Isolate all instruments' air lines.

Check the IR and PI and the pump motors initially, as per Section 12.9.6.

The station service cooling water side of the stator water coolers should be isolated, drained, and dried by blowing dry instrument air through it for at least two hours.

Seal-Oil System. The seal-oil system must be kept operational if a nitrogen blanket is to be used for storage. Oil to the seals is required continually to maintain a nitrogen blanket, and periodically when rolling the generator rotor. The quality of the nitrogen will have to be maintained, as the seal-oil system itself can be a source of moisture if the seal-oil quality is not maintained as well.

The seal-oil temperature is not critical and, therefore, the low-pressure service water side of the seal-oil coolers should be isolated, drained, and dried by blowing dry instrument air through them for at least two hours.

Rotor. The portion of the rotor inside the generator will be protected from atmosphere by the nitrogen blanket to prevent corrosion of critical components such as retaining rings.

Generator heaters and silica gel are used to prevent moisture buildup and condensation on rotor surfaces, and to avoid stress-corrosion attack on sensitive steel components.

The external portion of the rotor is, however, exposed to the turbine hall atmosphere. The slip rings in particular should be covered with a self-amalgamating wrap or tape to completely exclude air and moisture.

An initial IR and PI should be performed, as per Section 12.9.6.

The bearings and hydrogen seals should be fully installed and the oil supply should be operational for periodic rotation of the turbo-generator shaft.

Brush Gear. The brush gear should be removed and stored in a dry, clean area until return to service.

Rotating Exciters. Rotating exciters must be kept warm and dry in a humidity controlled environment. The air supply and discharge system for the exciter should be isolated. A quantity of silica gel (Section 12.9.6.) should be placed in the environment along with thermostatically controlled heaters to maintain the environment at about 5°C above turbine hall ambient temperature.

Initial IR and PI readings should be taken on the stator and rotor windings.

Static Exciters. Excitation cubicles should be kept closed and locked if possible. The internal atmosphere of the cubicles should be maintained warm and dry. This should be done with a thermostatically controlled heater of appropriate size and a quantity of silica gel (Section 12.9.6), consistent with the manufacturer's specification. The main power supply to the exciter should be isolated, as well as the air supply and discharge. The internal atmosphere should be kept at 5°C above ambient temperature.

GENERATOR PEDESTAL. The generator pedestal should be IR tested initially to ensure that the insulation level is high enough and to establish a benchmark for later checks prior to entering service. The pedestal should be clean and clear of any type of storage material. The insulation levels should be as per the manufacturer's equipment manual.

GENERATOR CURRENT TRANSFORMERS AND TERMINALS. The CT enclosures should be maintained warm and dry using thermostatically controlled heaters and silica gel. The internal atmosphere should be kept at 5°C above ambient temperature.

An insulation resistance check on the CTs should be done at 500 V dc using a megger; see Section 12.9.6.

GENERATOR INSTRUMENT PANELS. The generator instrument panels (hydrogen, seal oil, lube oil, and stator water) should be kept closed to exclude contaminants, rodents, and reptiles. Internal atmospheric control is not required.

Storage of Generator Components as Major Spares. Storage of major generator components implies indefinite time periods and will require methods of storage that will ensure that the machine or components will be adequate for service when needed.

This section describes the storage of the stator, rotor, and stator windings as major spares.

STATORS. For storage of a complete stator as a spare, it is required that it be kept indoors in a protected and dry area. The powerhouse is the recommended area for this as it will allow good control over the storage environment and easy access by crane when the machine is required for installation.

1. *Integral Stators.* When storing integral stators, a covering over the entire stator is unnecessary. Due to the nature of its design, only steel end covers are required to be bolted on where the end-door halves would normally be installed. A main access plate should be provided at each end of the machine on each steel plate for easy access for inspection. A fitting should also be provided to allow removal of air and admission of nitrogen. Since the vessel should now be sealed, a slightly positive nitrogen pressure of 2 to 3 psi will better preserve the machine insulations and core. A quantity of silica gel should be maintained inside the vessel to absorb any moisture as per Section 12.9.6. Prior to nitrogen blanketing, the stator windings should be drained and vacuum dried as per Section 12.9.6.
2. *Caged Cores.* When storing caged cores, nitrogen blanketing of the machine is not possible. Once the stator is set down and supported, it is best to build a metal frame house over the entire machine, with polyethylene for walls. A fan and air heater should be installed in one wall to circulate warm, dry air around the stator. This will slightly pressurize the house to exclude dust and

other contaminants. A filter system must be used to prevent debris from outside from entering the enclosure. The air should be kept at about 5°C above outside air temperature to avoid condensation forming. The stator windings should also be drained and vacuum dried as per Section 12.9.6. An initial IR and PI check on the stator windings of all stored stators should be performed and recorded.

ROTORS. A spare rotor should also be stored indoors (e.g., in the powerhouse) to maintain good control over the storage environment. A heavy vinyl bag is an acceptable method for atmospheric control, rather than simply covering the rotor with no environmental control, when being stored indoors (Fig. 12.13). However, the shipping container, if airtight, is also acceptable, especially metal containers.

The utmost care should be taken for any duration of rotor storage since it involves a number of critical components and surfaces. The rotor itself should be prepared as follows before being entirely contained:

- The slip rings must be covered, preferably with some type of self-amalgamating wrap or tape, to totally exclude air and moisture.
- The journals and hydrogen seal collars should be appropriately greased or waxed to inhibit rust and deterioration.
- Most rotor retaining rings are painted to prevent aqueous stress corrosion and rusting. It should be ensured that painted end rings are completely covered. Re-

Fig. 12.13 The rotor shown is stored in a sealed vinyl bag with silica gel placed inside and electrical test leads attached for periodic meggering. This arrangement is suitable only for indoor storage of rotors as it is not well temperature controlled. The rotor behind the bagged rotor is simply covered and being readied for installation in a unit.

taining rings that are not painted should be given a protective coating for storage and it should be ensured that this is in place.

- Cooling vents should be covered to exclude dust and contamination.
- Coupling faces should be greased or waxed to inhibit rust and deterioration.

The rotor should then be maintained in the container for long-term storage. The rotor should always be supported on the noncritical surfaces, specifically the rotor body on rubber-cushioned, formed wooden trestles. Rotors must not be supported at any time on their retaining rings. Silica gel should be used inside the container for humidity control, with humidity indication provided, as rotors are extremely sensitive to rust and stress corrosion. It would be most desirable to keep the rotor temperature at least 5°C above ambient by using a thermostatically controlled heater.

Leads should be brought outside the storage container to IR and PI test the rotor windings so that the container seal need not be disturbed.

Stator Windings. Spare stator windings will normally be shipped in a robust hardwood container with each stator bar well protected and supported on trestles within.

Each conductor should be individually checked for IR and PI, consistent with Section 12.9.6, prior to storage, for confirmation of insulation condition and future benchmarks. The container should be resealed with silica gel placed inside, as per Section

Fig. 12.14 The rotor from Figure 12.12 in its cradle, shown stored for a year in a temperature-controlled "tent."

12.9.6, for humidity control. Temperature control of the environment is not essential if the container is stored inside a temperature-controlled building.

12.9.4 Monitoring and Maintenance During Storage

During the storage period, the equipment must be monitored periodically to ensure that it is maintained in good condition and will be ready for in service when required with minimal remedial effort.

The following section describes the practice to be used for assembled machines and major spares storage to maintain them in good condition. Check sheets are provided in Section 12.9.6.

Monitoring and Maintenance Requirements of Assembled Generators

STATORS. The nitrogen pressure should be checked and recorded on a daily basis until the system can confidently maintain a positive nitrogen pressure of 2 or 3 psi above the seal-oil pressure. After this initial period, the checks can be made and recorded weekly.

To ensure that the heaters are performing, the stator winding temperatures should be sampled and recorded weekly to ensure that the stator atmosphere is maintained at 5°C above turbine hall ambient temperature.

The silica gel should be checked at predetermined intervals according to expected working life, as per product specifications. At these checks, the machine will be opened. The silica gel condition should be recorded and then reactivated for further storage. The IR and PI on the stator windings should be measured and recorded at this time. This will indicate if the present storage conditions are maintaining the equipment warm and dry. If a satisfactory PI is obtained, the machine can be resealed for storage. If not, drying out of the stator windings will be required, by circulating 60°C stator water through the windings for approximately 24 hours. Alternatively, if hot water circulation is not possible, then dc current application through the winding and monitoring the winding temperature is another method to dry the windings.

STATOR CASING. The dew point and internal atmospheric temperature of the casing should be checked weekly and recorded for signs of high humidity or lower temperature.

CORE-END PLATE FLUX-SCREEN COOLING SYSTEM. No special requirements for monitoring are needed.

STATOR WATER COOLING SYSTEM. The rotors of the pumps and motors should be manually rotated each month and an IR and PI check done on the rotor windings and recorded.

SEAL-OIL SYSTEM. The seal-oil system will be in operation if the nitrogen blanket method is being used on the machine internals. The pump supply should be changed each month to allow all sources of supply to be used periodically.

The seal-oil supply should be maintained at the minimum level for nitrogen blanketing and periodic rotor turning.

ROTOR. An IR should be performed on the rotor windings monthly and recorded. The rotor should be barred one-half turn each month to avoid bending distortion of the shaft, unless it is supported (cribbed) under its body.

BRUSH GEAR. Not applicable.

ROTATING EXCITERS. The stator and rotor windings IR and PI should be checked monthly as per Section 12.9.6 and recorded. Monthly temperature and humidity checks should be made and recorded to ensure that the heaters and silica gel are working. The silica gel should be checked at regular predetermined intervals and reactivated as per product specifications.

STATIC EXCITERS. Monthly temperature and humidity checks should be made and recorded to ensure that the heaters and silica gel are working. The silica gel should be checked at regular predetermined intervals as per product specifications and reactivated.

GENERATOR PEDESTAL. All pedestal insulation should be checked every six months by insulation resistance measurement. The area should be maintained clean and free of storage materials or oil spills.

GENERATOR CURRENT TRANSFORMERS AND TERMINALS. An insulation resistance check should be performed monthly and recorded on the CT windings as per Section 12.9.6, using a 500 V megger.

The temperature and humidity of the terminal enclosure should be checked monthly to ensure the space is being maintained warm and dry.

The silica gel should be checked at regular predetermined intervals as per product specifications and reactivated.

GENERATOR INSTRUMENT PANEL. The instrument panel should be opened and visually inspected monthly to ensure the exclusion of rodents, reptiles, and debris.

Monitoring and Maintenance Requirements of Generator Components Stored as Major Spares

STATORS. Machines that are stored as a sealed unit under a nitrogen blanket should be checked and recorded daily for nitrogen pressure until the vessel can confidently maintain a positive 2 or 3 psi pressure. After this initial period, these checks can be made weekly. This type of storage also employs silica gel, which should be checked at regular intervals as per product specifications and reactivated.

For this type of storage and for caged cores, the stator windings IR and PI should be checked monthly as per Section 12.9.6. This will be the best indicator that the storage conditions are adequate to maintain the machine insulations in good condition.

The atmosphere of the storage area or container should be checked monthly to ensure that it is maintained at 5°C above ambient temperature.

ROTORS. The rotor windings should be checked for IR and PI, as per Section 12.9.6, monthly and recorded.

Where applicable, the silica gel should be monitored and reactivated when required.

The atmosphere of the storage container should be checked monthly to ensure that it is maintained at 5°C above ambient temperature.

STATOR WINDINGS. The spare stator windings should be checked for IR and PI during silica gel inspections and reactivation periods. These times will depend on the product specifications for the silica gel. It is unnecessary to disturb the container seal at any other time.

12.9.5 Restoration from Storage

After the mothballing or storage period, when the equipment is required for service, the storage conditions must be reversed.

This section will describe the requirements for demothballing and returning the equipment back to operating condition.

Requirements for Restoration from Storage of Assembled Generators

STATOR. The nitrogen blanket should be removed from the stator by simply leaking it out and purging the machine with air. The silica gel should be removed and kept for future use.

A final IR and PI should be done on the stator winding, as per Section 12.9.6, to ensure good insulation condition. A low PI on the stator winding may require drying out but this should be unnecessary if the machine internals have been kept warm and dry. Drying out of the stator winding can be done by circulating 60°C stator water through the winding for approximately 24 hours. Alternatively, if hot water circulation is not possible, then dc current application through the winding and monitoring the winding temperature is another method to dry the windings.

STATOR CASING. A visual inspection of the casing internals and externals should be made to ensure that everything is in order. All instrumentation should be tested to ensure that it is functioning correctly and repaired if not.

CORE-END PLATE FLUX-SCREEN COOLING SYSTEM. The core-end plate flux-screen cooling system should be reconnected and flushed when the entire stator water system is flushed.

Electrically floating flux screens require a final IR check to ensure that they remain properly insulated.

STATOR WATER COOLING SYSTEM. The stator water system should be reconnected and the filters replaced. The pump motors should be checked for IR and PI before running.

A backward and forward stator water flush should be performed according to the manufacturer's equipment manual prior to placing the machine in service.

The system should actually be recommissioned.

Seal-Oil System. The seal-oil system will be operational and requires no recommissioning since it was in service during the entire storage period.

The seal-oil coolers should be valved back into service on the low-pressure service water side.

Rotor. The collector ring surfaces should be uncovered and prepared for service. An inspection of all visible external rotor parts should be made to ensure that they are in good condition.

A final IR and PI on the rotor windings should be done prior to application of the field as per Section 12.9.6.

Brush Gear. The brush gear should be inspected and reinstalled.

Rotating Exciters. The silica gel and heaters should be removed and the air cooling system returned to operating condition.

A final IR and PI on the rotor and stator windings, as per Section 12.9.6, should be performed to ensure that drying out is not required and the insulation is in good condition.

Static Exciters. The silica gel and heaters should be removed and the ventilation system returned to service. The main power supply should be deisolated.

Generator Pedestal. A final IR check should be performed on all pedestal insulations to ensure that they have been maintained at the working level.

Generator Current Transformers and Terminals. The silica gel and heaters should be removed from the terminal enclosures. A final IR check on the CT windings should be performed at 500 V dc using a megger. See Section 12.9.6.

Generator Instrument Panels. All instrument panels should be visually inspected. Any cleaning or repairs should be done. All instrumentation should be checked for operation and recalibrated if required.

Requirements for Restoration from Storage of Generator Components Stored as Major Spares

Stators. Integral stators that are stored with end-sealing plates for nitrogen blanketing should maintain the end plates in place until the stator is installed. Once installed and ready for connections and rotor insertion, the nitrogen blanket may be removed by simply leaking it off and then removing the end plates. The silica gel should be removed and kept for future use. An IR and PI check on the stator windings should be done to ensure that the winding insulation is in operating condition.

Caged cores will have to have all protective housings and atmospheric control systems removed prior to installation. An IR and PI check should also be done prior to installation.

Regardless of the type of stator, it is required to perform a capacity ac hi-potential test on the stator windings in air at 75% of twice the line-to-line rate plus 1000 volts for one minute. Refer to IEEE Standard 4, "Techniques for Dielectric Tests," and IEEE Standard 433, "Recommended Practice for Insulation Testing of Large ac Rotating Machinery with High Voltage at Very Low Frequency."

ROTORS. In preparation for service, the rotor should be removed from its storage container and all protective coatings removed. A complete inspection of all rotor surfaces should be made for any signs of deterioration, rust, or accumulation of dirt.

An IR and PI check on the rotor should also be performed, as per Section 12.9.6, prior to insertion.

It is also required to perform a one-minute dc hi-potential test on the rotor windings at 2000 volts. Refer to IEEE Standard 95, "Recommended Practice for Insulation Testing of Large ac Rotating Machinery with High Direct Voltage."

The rotor should then be prepared for insertion and the slip rings recovered until ready for brush gear installation.

STATOR WINDINGS. The stator windings should be unpacked and the required bars removed to external trestles for inspection and testing.

Each stator conductor should be vacuum/pressure, IR, and PI tested and the slot-armor surface resistance verified. Any visible flaws should be repaired using compatible materials and the bar retested again. The bars should be kept covered to exclude contamination until installation.

12.9.6 Long-Term Storage Maintenance Procedures and Testing

The following sections provide some guidelines for various procedures and tests that have been used in some utilities over the years. They are by no means the only way to carry out these procedures, but they are tried and true methods that have been used successfully.

Draining and Vacuum Drying Procedure. Draining and vacuum drying is a procedure used on water cooled stator windings. Along with environmental temperature and humidity control, it will be sufficient protection for the winding under any storage conditions.

Experience shows that regardless of the precautions taken, the bare copper surface of the copper winding will always maintain an oxidized surface. This surface does not deteriorate further to any degree over the time periods of storage considered for stator windings.

Draining

1. Isolate the entire stator winding, including phase leads and/or flux-screen cooling circuit (if water cooled flux screens are used) with blanks. A valve should be provided at each end of the circuit to be drained.
2. Pressurize the circuit from one end only using dry instrument air at 30 psi and hold for 30 minutes.
3. Open the opposite end valve and allow the air pressure to assist in draining out the water. Repeat this procedure until all visible trapped water is removed.
4. Allow dry instrument air at 30 psi to flow through the circuit for two hours to complete the draining.

Vacuum Drying

1. Connect a vacuum pump of sufficient capacity for the subject machine to the water circuit at one end and close the opposite end valve.
2. Draw a vacuum of 29″ Hg or better for 24 hours to remove the air and moisture.
3. At the end of the 24 hour period, close the vacuum pump valve and disconnect the pump from the circuit to ensure that no oil from the pump is drawn back into the circuit when the pump is shut down.
4. Open all valves to drain any water collected.
5. Close all valves to reconnect the vacuum pump.
6. Draw a vacuum of 29″ of Hg until dew point readings at the pump outlet indicate that the circuit is dry.
7. Close the pump valve and disconnect the vacuum pump.
8. Remove all blanks and simply cap the circuit openings or reconnect to the stator water package unit.

Silica Gel Maintenance. Silica gel is used in the storage of generators and auxiliary equipment for the purpose of moisture removal and humidity control. It has the following properties, which makes it highly desirable for our purposes:

- High absorptive capacity due to its porosity and large surface area.
- Chemically inert material. The absorptive action is purely physical, with no change in size or shape and no chemical compounds formed.
- Long desiccant life.
- Good physical strength and abrasion resistance.
- Easily and economically reactivated.

When using silica gel for our purpose, it should be purchased on the above basis. It is desirable to obtain an indicating type that will change color when it is exhausted and requires reactivation. The quantity used should be determined by the volume of the container and the specifications of the particular brand of silica gel purchased.

When using the silica gel, the required number of packages should simply be placed inside the container on a clean, dry surface exposed to the environment within. When the gel is exhausted as indicated by color change, it should be reactivated as per instructions and specifications obtained at the time of purchase. Reactivation is usually done by heating for a few hours at about 100°C. The silica gel can be reused after it is fully reactivated.

Insulation Testing and Maintenance. The two principle measurements used are the insulation resistance test and the polarization index test. Both these checks are described in detail in IEEE Standard 43, "Recommended Practice for Testing Insulation Resistance of Rotating Machinery," and should be referred to there.

Monitoring and Maintenance Check Sheet

INSTALLED GENERATORS

Stators:	(a) nitrogen pressure	________
	(b) heaters @ 5°C above ambient temperature	________
	(c) silica gel condition	________
	(d) IR stator windings	________
	(e) PI stator windings	________
Casing:	(a) dewpoint	________
	(b) temperature of internal atmosphere	________
Stator Water System:	(a) rotate pumps and motors	________
	(b) IR motor windings	________
	(c) PI motor windings	________
Seal-Oil System:	(a) change pump supply	________
	(b) oil pressure	________
Rotor:	(a) IR windings	________
	(b) PI windings	________
	(c) turn rotor shaft	________
Rotating Exciter:	(a) IR windings	________
	(b) PI windings	________
	(c) ambient temperature	________
	(d) humidity	________
	(e) silica gel condition	________
Static Exciter:	(a) humidity	________
	(b) ambient temperature	________
	(c) silica gel condition	________
Pedestal:	(a) IR pedestal to ground	________
CT's and Terminals:	(a) IR to ground	________
	(b) ambient temperature	________
	(c) humidity	________
	(d) silica gel condition	________
Hydrogen Panel:	(a) visual inspection	________

SPARE COMPONENTS

Stator:	(a) nitrogen pressure	____________
	(b) heaters @ 5°C above ambient temperature	____________
	(c) silica gel condition	____________
	(d) IR stator windings	____________
	(e) PI stator windings	____________
Rotor:	(a) IR windings	____________
	(b) PI windings	____________
	(c) silica gel condition	____________
	(d) temperature of storage atmosphere	____________
Stator Windings:	(a) IR each conductor	____________
	(b) PI each conductor	____________
	(c) silica gel condition	____________

12.10 LIFE CYCLE MANAGEMENT (LCM)

A life cycle management document (LCM) comprises a long-term strategy for major station equipment. The LCM, in principle, does not deal with everyday problems (e.g., troubleshooting a protection relay, fixing pipe leaks in the cooling water supply, changing gaskets, reclaiming a bad RTD unless they are part of a long-term issue. LCMs are generated to provide management with a long-term strategy, based on in-house and other plants' experience, specific technical information, and an economic cost–benefit analysis. LCM preparation, if done correctly, requires significant input from a number of plant personnel and, if needed, outside individuals. It is an ongoing project. Therefore, LCMs are normally only prepared for critical station components. LCMs are found mostly in nuclear power plants, where manpower is more available than in other types of plants with relatively lean staffing. LCMs are predicated on needs such as regulatory concerns, major risk to safety and/or production, major economic impact, and major replacement cost. Intrinsic-to-equipment drivers include obsolescence, degradation, and chronic and expensive maintenance.

EPRI has published guidelines for the preparation of LCMs [8]. However, LCMs are most effective when written by taking into consideration the specific circumstances of a given station. In general, the LCM will do a "remaining life" analysis on the generator, focusing on major components such as stator, rotor, excitation, AVR, and subcomponents such as stator core, retaining rings, wedges, and so on. Once the "remaining life" issues are described, potential failure modes analyzed, and obsolescence problems uncovered, the LCM prepares a number of life-cycle options. Such options may include alternatives such as refurbishment, replacement, increasing the spare-parts list (including even major items like a spare rotor), or, most probably, a combination of all these.

It is important at the onset of the LCM to delineate what are the systems and components to be included. For instance, auxiliary systems (water, oil, and gas treatment skids) ought to remain outside the scope of the LCM, unless there is an overriding need to include them. Also, the LCM ought to take into account forward-looking maintenance approaches. For instance, does the station encourage the use of condition-based maintenance? Is the station planning for extended periods without a major inspection? Is there a goal of reducing maintenance man-hours? These and other constraints should be taken into consideration when preparing an LCM.

Perhaps the most difficult part of preparing the LCM is assessing remaining life of certain main components such as core, windings, and excitation systems. One can look at industry published data for determining statistically based failure mechanism probabilities of major components and subcomponents. Nevertheless, the LCM should be mainly based on intense scrutiny of the condition of the generator in question. For instance, a rotor winding may be expected to last about 30 years, but a serious look at one's machine may point to a different expected lifespan. The bottom line is that a LCM is first and foremost about a specific system or component, and not a generic document that would apply to many units. One important requirement for an effective LCM is that it should be maintained on a regular basis (i.e., revised based on ongoing changes in the situation of the equipment and/or other elements).

As explained above, a LCM should be tailored based on the individual idiosyncrasies of a given station. Nevertheless, the typical LCM may have the following content:

- Objective
- Approaches to creating and maintaining the LCM
- Executive Summary (for those managers who do not have the time to read the entire LCM; an effective executive summary out to be cogent while concise)
- Component Description
  - Frame
  - Casing
  - Stator core
  - Stator winding
  - Rotor forging
  - Rotor winding
  - Retaining rings
  - Excitation
- Time-line graph (capturing major changes over the life of the machine and the future outlook)
- Machine parameters (a section containing as many macchine constants and graphs as possible)
- Summary of recommendations
- References (industry-wide, OEM, internal to the station)
- Attachments: any report, internal or external, of importance to the LCM

As stated above, other components may or may not be included, such as auxiliaries.

For identical units, a single LCM is probably the most effective way to go. It should include comparative information such as a common graph of wedge looseness for all the identical units. Other issues for a LCM, such as vibration, can also be compared with each other to reveal significant insights on condition and remaining life. For example, Figure 12.16 shows a graph with wedge looseness information for two identical four-pole, 1300 MVA units. The units not only are identical in design, but have almost same number of operating hours, starts, and load-rejection events. Nonetheless, the traces clearly show that each unit has a very different history of wedge looseness.

Figure 12.16 condenses a number of issues that ought to be taken into account in preparing a LCM of a generator.

12.11 SINGLE-POINT VULNERABILITY (SPV) ANALYSIS

Single-point vulnerability analysis is another program found mainly in the nuclear industry. First implemented for safety-related equipment, it has made the leap to the balance of plant systems.

The purpose of a SVP program is to give priority in maintaining those components and subcomponents that may cause major loss of production or/and result in significant risk to safety of plant and personnel. For example, a SPV analysis of a main generator may consider a single protective instrument as a SPV if by failing it can cause a unit trip or significant loss of production. The program requires that a list of SPVs be written by cognizant engineers, and then maintenance standard instructions or mainte-

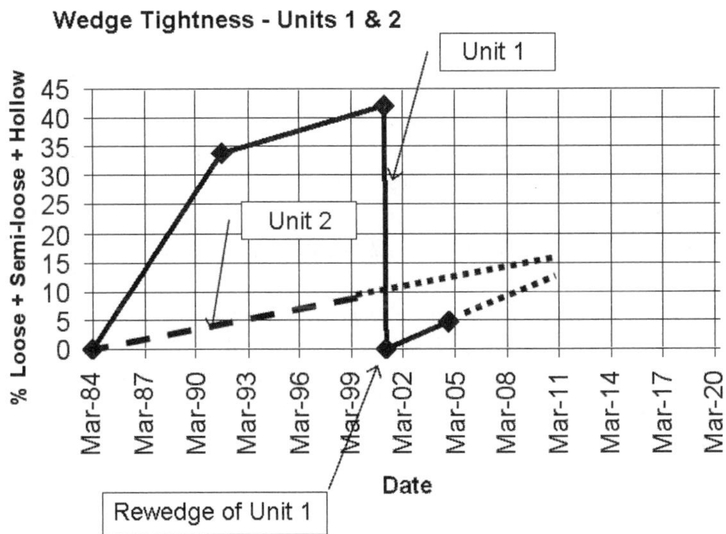

Fig. 12.15 Comparative wedge history of two identical four-pole, 1300 MVA units.

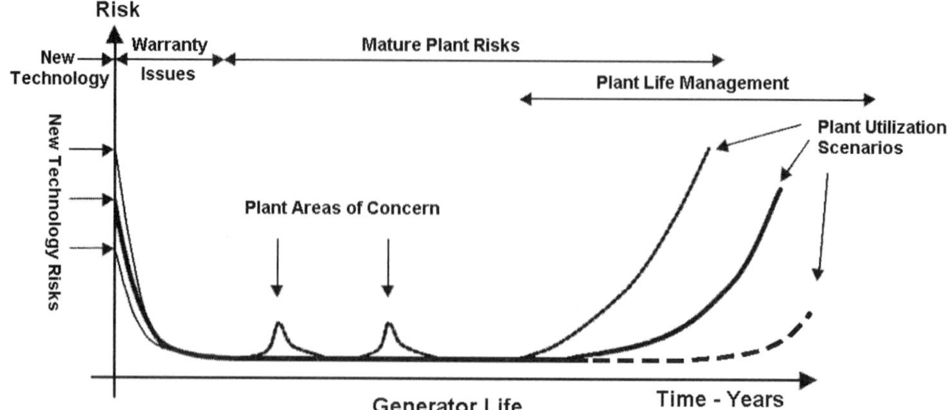

Fig. 12.16 Risk management over the life of the generator components.

nance orders are reviewed to ascertain that these SPV components are maintained in a proper manner. In practical terms for main generators, almost every component has the potential of causing major losses. Thus, as shown in other places in this book, it is highly recommended that they are be given maximum attention.

REFERENCES

1. Organization for Economic Cooperation and Development, *Methods of Projecting Operations and Maintenance Costs for Nuclear Power,* 1995.
2. *Life Cycle Management Sourcebook: Main Generator,* Vol. 5. EPRI, Palo Alto, CA.
3. McCormick, *Reliability and Risk Analysis: Methods and Nuclear Power Applications,* Academic Press, 1981.
4. Fullwood and Hall, *Probabilistic Risk Assessment in the Nuclear Power Industry: Fundamentals and Applications,* Pergamon Press, 1988.
5. Dhillon and Singh, *Engineering Reliability: New Techniques and Applications.* Wiley, 1981.
6. T. Bedford and R. Cooke, *Probabilistic Risk Analysis: Foundations and Methods,* Cambridge University Press, 2001.
7. R. J. Zawoysky and K. C. Tornroos, *GE Generator Rotor Design, Operational Issues, and Refurbishment Options.* GE Power Systems, Publication GER-4212, 2001.
8. EPRI Technical Report #1007423: "Life Cycle Management Planning Sourcebooks," Volume 5: Generator.

INDEX

Note: Information presented in figures and tables is denoted by *f* and *t,* respectively, following the page number.